SR-009
June 2010

Observing Photons in Space

Editors
M.C.E. Huber[1], A. Pauluhn[1], J.L. Culhane[2], J.G. Timothy[3], K. Wilhelm[4] & A. Zehnder[1]

[1] *Paul Scherrer Institut, Villigen PSI, Switzerland*
[2] *UCL/Mullard Space Science Laboratory, Holmbury St. Mary, UK*
[3] *Nightsen, Inc., Tiverton RI, USA*
[4] *Max-Planck-Institut für Sonnensystemforschung, Katlenburg-Lindau, Germany*

INTERNATIONAL
SPACE
SCIENCE
INSTITUTE

Cover: The cover illustration exemplifies the span of the wavelength domains described in the book:

- **γ-rays**: the sensor of *Explorer XI* (top centre, from Pinkau K, History of gamma-ray telescopes and astronomy (2009) Exp Astron 2:157-171, DOI 10.1007/s10686-009-9143-z
- **X-rays**: the *XMM-Newton* observatory (centre)
- **Vacuum ultraviolet**: spectroheliograms observed by *SOHO* (upper left-hand corner)
- **Ultraviolet, visible and infrared**: the *Hubble Space Telescope* (lower right-hand corner)
- **Far-infrared and submillimetre region**: the mirror of *Herschel*, currently the largest astronomical mirror in space (upper right-hand corner)
- **Microwaves**: part of the Cosmic Microwave Background sky as recorded by *WMAP* (lower left-hand corner)

The International Space Science Institute is a Foundation under Swiss law. It is funded by the European Space Agency, the Swiss Confederation, the Swiss National Science Foundation, and the University of Bern. For more information about ISSI see *www.issibern.ch*.

Published for: The International Space Science Institute
Hallerstrasse 6, CH-3012 Bern, Switzerland

By: ESA Communications
Keplerlaan 1, 2200 AG Noordwijk, The Netherlands

Publication Manager: Karen Fletcher

ISSN: 1608-280X
ISBN: 978-92-9221-938-3
Price: €60

Contents

IV DETECTORS

VI GENERAL TECHNIQUES

VII SPACE ENVIRONMENT

VIII EPILOGUE

IX APPENDICES

Foreword

ROGER-MAURICE BONNET AND LODEWIJK WOLTJER

The first revolution in astronomy occurred four centuries ago when Galileo and others turned their telescopes to the sky and discovered a Universe infinitely richer than suspected before. These early discoveries have been celebrated in 2009 during the International Year of Astronomy proclaimed by the United Nations Educational, Scientific and Cultural Organization and the International Astronomical Union. During the last half century or so space astronomy (together with radio astronomy) revealed that radiation from extraterrestrial sources may be observed over a wavelength range of a factor of 10^{20} or more — instead of the factor of 2 to 3 with which early astronomers had to content themselves. This led to the second revolution. At many wavelengths that radiation is absorbed by the Earth's atmosphere. And even in the visible the turbulence and associated refractive effects in the atmosphere and its variable transparency and brightness limited the precision with which observations could be made.

Early space observations were made from rockets (and balloons), but only satellites in orbit could provide the necessary stable platforms for the long observations required for studying faint sources. Other technologies needed for efficiently operating in space were rapidly developed, in part also for reconnaissance satellites. Especially in the detector area the need for remote observations made digital detectors of fundamental importance. At visible wavelengths the CCDs were developed for space observations. Their higher efficiencies and rapid response time were equally beneficial to ground-based astronomy and led to the gradual disappearance of photographic plates for most applications. By now they have also taken over the mass market of cameras for the general public. A similar development has been seen in X-rays. More and more photographic procedures for medical imaging have been replaced by digital techniques. At still higher energies detectors were developed for detecting gamma-rays from nuclear explosions: fortunately none have been detected, but gamma-ray bursts dating from the early days of the Universe have been discovered. It is no exaggeration to say that most of the new discoveries in astronomy have been the result of technological developments. Sometimes the astronomical application came first, at other times astronomers profited from developments made outside astronomy.

While the access from space to the large parts of the electromagnetic spectrum that cannot penetrate the atmosphere has been the most spectacular, even in other wavelength domains the qualitative contribution of space-based instruments has been enormous. The positioning of stars by astrometric methods is gaining in precision by an order of magnitude or more allowing the mapping of our Galaxy in phase space. The radiation from the Sun and its variations, important in climate studies, can only be determined from space with the required accuracy. And the *Hubble* Space Telescope has not only made it possible to obtain the most detailed data for scientific study, but their aesthetic quality has brought the Universe much closer to the general public than before, which also has contributed to a positive funding climate for astronomy in general. So astronomers who a century ago were rather remote from the concerns of humanity are now well integrated into our scientific-technological society.

If the space environment is so superior for astronomical observation, why has one not closed all ground-based observatories? The reason is largely financial. The cost of a large space telescope is such that during the foreseeable future there is no possibility to construct a ten meter space telescope diffraction limited at visible wavelengths. Thus, when a large light collecting power is needed or a high intrinsic angular resolution, we have no choice but to put the corresponding telescope on the ground. In selecting the telescopes which are to operate at a given wavelength, the optimization has to take cost factors into account.

In the first chapter of this book we are reminded of the earlier discussions: should we build ten Palomars or one *Hubble* Space Telescope? This missed the essential point. Ten Palomars would be very useful and, in fact, the world has constructed ten even larger telescopes on the ground during the last sixty years. However, *HST* has a privileged domain of science that only it can do. The conclusion, therefore, is: execute programs of observation with the cheapest instrument that can do so. In addition it is well to remember that the ground-based telescopes have been so successful because of the detectors originally developed for HST and other space missions.

The first eight chapters of the book contain much valuable introductory material about the physical processes which are observed in the different spectral domains. Then follow 27 chapters about techniques for imaging, positioning, spectroscopy, polarimetry and the detectors used for these. Thereafter comes a chapter on calibration, a particularly vital subject in piecing together the data at different wavelengths, and finally some chapters on general aspects of observing in space. According to the editors, the intended level is suitable for undergraduates.

In 2001 was published "The Century of Space Science", a 1846-page elegantly produced book, edited by J.A.M. Bleeker, J. Geiss and M.C.E. Huber (Kluwer Academic Publishers). It described the scientific developments in the field — including the solar system, cosmic-rays and other aspects, but trod lightly on instrumentation. In 2003 ISSI engaged a reflection on the orientation to give to its future activities. One concern was the gradual disappearance of the generation of trained experimentalists who made space research possible and the need for transmitting the know-how accumulated by them to future generations of space scientists. Indeed, too many young researchers obtain degrees by analyzing data without having much of an idea of the instruments with which these have been obtained, sometimes leading to erroneous conclusions. From such considerations came the concept of editing a book containing a series of articles related to experimentation and photon detection in the field of space astronomy, written by the best world experts. The present volume, again including M.C.E. Huber as editor, is particularly welcome for its concentration on instrumental subjects. These two books are very much complementary, with the present one much more affordable for students. The six editors and 40 authors should be congratulated on the result of their efforts.

I. Observing Photons in Space

— 1 —

Observing photons in space

For the truth of the conclusions of physical science, observation is the supreme court of appeals
Sir Arthur Eddington

Martin C.E. Huber[I], Anuschka Pauluhn[I] and J. Gethyn Timothy[II]

Abstract

This first chapter of the book ‘Observing Photons in Space’ serves to illustrate the rewards of observing photons in space, to state our aims, and to introduce the structure and the conventions used. The title of the book reflects the history of space astronomy: it started at the high-energy end of the electromagnetic spectrum, where the photon aspect of the radiation dominates. Nevertheless, both the wave and the photon aspects of this radiation will be considered extensively. In this first chapter we describe the arduous efforts that were needed before observations from pointed, stable platforms, lifted by rocket above the Earth's atmosphere, became the matter of course they seem to be today. This exemplifies the direct link between technical effort — including proper design, construction, testing and calibration — and some of the early fundamental insights gained from space observations. We further report in some detail the pioneering work of the early space astronomers, who started with the study of γ- and X-rays as well as ultraviolet photons. We also show how efforts to observe from space platforms in the visible, infrared, sub-millimetre and microwave domains developed and led to today's emphasis on observations at long wavelengths.

The aims of this book

This book conveys methods and techniques for observing photons[1] in space. ‘Observing’ photons implies not only detecting them, but also determining their direction at arrival, their energy, their rate of arrival, and their polarisation.

Progress in observing photons in space has largely been driven by space astronomy: one rises to an altitude as high as necessary in order to get an unimpeded view of the rest of the Universe from above the atmosphere. The pioneers of space

[I]PSI—Paul Scherrer Institut, Villigen, Switzerland
[II]Nightsen, Inc., Tiverton RI, USA

[1]The name ‘photon’, derived from the Greek $\tau\grave{o}\ \varphi\tilde{\omega}\varsigma$ for ‘the light’, was coined by Lewis (1926), more than 20 years after Einstein's postulate.

astronomy concentrated their efforts on observing at the short wavelengths that are completely absorbed by the atmosphere. These efforts started shortly after World War II. Space observations in the visible, which would avoid the blurring effects by the turbulent atmosphere, were considered less of a priority at the start of the space age, but became possible when the *HST* was launched in 1990 (and repaired in 1993). It is worthwhile noting, however, that in the year before the launch of *HST*, two missions exploiting both the absence of a disturbing atmosphere and the possibility to scan great circles of the celestial sphere were launched, namely *COBE* and *Hipparcos*. Nowadays, infrared and sub-millimetre observations are the prime aim of observational space astronomy; for several reasons, including technology, they took a long time to come to the fore.

Along with space astronomy, and by use of most of its technology, space observations of the Earth have gained importance and momentum over the past decades. Although Earth observations — as well as observations during a fly-by or from orbit around planets and comets — do primarily serve geophysics, they are largely similar to space astronomy. But, as applications, observing strategies and procedures for data handling in this field do differ from those familiar to astronomers, we will summarise these facets in the brief Chapter 39 (Pauluhn 2010).

Given today's efficient support and service to users of observing facilities, the number of astronomers whose knowledge and interest is entirely concentrated on interpreting observations has grown substantially in the past decades. Quite the opposite has happened to the number of scientists who are familiar with, and capable of dealing with, instrumentation. This kind of scientist is, however, needed to innovate in space astronomy by advanced, more sophisticated or otherwise distinctive observing tools. With the long gestation, high productivity and extended life of modern astronomical facilities, the rate of innovation — particularly for space instrumentation — has slowed down. It is now essential that a record of experience and of sound practice be passed on to future generations of instrument builders in order to preserve knowledge and experience gained over half a century.

The authors and editors of this book have accumulated much experience in enabling space astronomy by designing, building, testing and calibrating space-qualified instruments. Some of them have been involved in building instruments early in the space age, others are right now working with the first results of their instruments. The latter are those who have built observing facilities for the more recently accessible infrared and sub-millimetre domains. Some authors cover methods and techniques that will be extensively exploited in space in the future, namely interferometry and polarimetry. Another topic included in this book, is the modus operandi of laser-aligned structures in space; this will make space astronomy even more powerful. Giant laser-aligned structures, moreover, will open a fundamentally new window to the Universe, namely the hitherto untapped information carried by gravitational waves or — to remain in the parlance of this book — by gravitons.[2]

[2] *LISA*, a projected giant laser-aligned structure with a 5 Gm extent, would be able to expand our knowledge of the Universe by sensing the low-frequency gravitational radiation that is continuously emitted by massive objects throughout the Universe. Low-frequency gravitational waves are accessible only in space, where they are not submerged in the geophysical noise (as is the case for ground-based gravitational-wave detectors). Gravitational radiation eventually would let us reach further back in time, beyond the Cosmic Microwave Background (CMB) — the current

One may wonder why this book's title is not simply 'Observing Electromagnetic Radiation in Space', instead of the shorter but perhaps somewhat gauche 'Observing Photons in Space'. As mentioned, our title reflects the history: space astronomy started with the exploration of the X- and γ-ray domains and observations in the ultraviolet wavelength range, thus in these early observations, the photon as an 'energy packet' was the dominating concept. Up to today, dispersionless, i.e., detector-based spectroscopy, or the timing of photon arrival time are methods used for measuring photons not only in the high-energy domain, but also at considerably lower energies, down to the visible range. Nonetheless, both the wave and the particle aspects of electromagnetic (EM) radiation are represented throughout the book. In this sense we interchangeably describe the relevant spectral domains by the energy of the photon, $E_\nu = h\,\nu$, by the wavelength (in vacuo), λ, or by the frequency, ν; where h stands for Planck's constant, and ν and λ are related through the speed of light $c_0 = \nu\,\lambda$. As far as units are concerned, we tried to enforce the rules of the *Système International d'Unités (SI)*, which uses units that are based on terrestrial, i.e., laboratory standards. To further encourage the use of SI we present a short guide to the Système International in the Appendix.

As we intend to preserve the experience of today's generation of scientists who have been involved with instrumentation for future generations, we present the material at a level that is accessible to interested undergraduate students. For them, Chapter 2 (Wilhelm and Fröhlich 2010) on 'Photons — from source to detector' provides an introduction to the physical processes involved in the generation, transmission and imaging of photons, as well as in their spectroscopic and polarimetric analysis.

Throughout the book we emphasise the principles of the methods employed, rather than details of the techniques being used. Overall, the subject is treated from the point of view and to the benefit of the practitioner: potential pitfalls are mentioned and good practice is stressed. We believe that this concept provides a useful guide not only to students, but also to professionals who want to inform themselves about methods being used in a field far from their own.

Why observe photons — and why from space?

To date, observing photons over the entire range of the electromagnetic spectrum is probably the most powerful means to gain an overall understanding of our environment. This applies to the nearest objects as well as to the early phases of the Cosmos that can be observed in the distant Universe. Much of the matter which we see in this way emits or absorbs photons in a gaseous or plasma state. Spectroscopy of the line spectra of matter in these aggregate states can yield temperature, composition and motions of (and in) the objects under investigation. By extending the sensitivity of our instruments into the infrared and sub-millimetre regions, we gain access to the broad-band spectral features that provide information about the composition of solid objects, such as comets, planets, asteroids and

limit to photon observations. Moreover, when direct observation of gravitational radiation from astrophysical sources begins, new tests of general relativity, including strong gravity, will become possible (Will 2009).

dust. And by going into the microwave region, we can reveal the 'structure' of the early Universe, which is contained in the anisotropy and polarisation of the Cosmic Microwave Background (CMB) radiation.

If we remain on the ground, the Earth's atmosphere limits our 'view' to the so-called 'optical' window, i.e., to near-ultraviolet, visible and near-infrared light, whose wavelengths reach from $\lambda \approx 310$ nm to $\lambda \approx 1$ μm.[3] Ground-based observations are also possible in the so-called 'radio' window, i.e., for wavelengths of about 5 mm to 10 m, the latter corresponding to a frequency of $\nu \approx 30$ MHz. At the other end of the spectrum, galactic photons of extremely high energy (> 1 TeV) can be indirectly observed from the ground by studying the secondary events they cause in the atmosphere. In order to directly observe the impinging photons in all spectral domains, however, one has to go to space. The same is true if one wants to overcome the distorting influence of the turbulent atmosphere, the seeing, which blurs our view in the optical window. This deleterious effect, however, can be significantly reduced by the indirect method of interferometry, and in direct imaging by so-called adaptive optics, albeit up to now at infrared wavelengths only.

Another advantage of space astronomy that has already been mentioned, but is pointed out less frequently, is the undisturbed and simultaneous access to opposite parts of the celestial sphere. This was of fundamental importance for *COBE* (Boggess et al 1992) and the astrometry mission *Hipparcos* (Perryman 2009).

The absorption of EM radiation by the terrestrial atmosphere is shown schematically in Figure 1.1. Note that the interstellar medium, in addition, blocks a significant part of the vacuum-ultraviolet (VUV) spectral region, even if the observing instrument is above the atmosphere. This absorption is dominated by the Lyman continuum ($\lambda < 91.2$ nm) of hydrogen, the neutral species of the most abundant element in the Universe. Owing to the proximity of the Sun this absorption is not noticeable in front of our daylight star. In fact, the Sun provides the main example of a stellar spectrum in this domain (cf., Figure 2.1 in Wilhelm and Fröhlich 2010).[4] The onset of atmospheric absorption at the long-wavelength end of the radio window is subject to diurnal variations: it depends on the electron density in the ionosphere, plotted for day- and night-time in Figure 1.2. The figure also shows the various subdivisions of the Earth's atmosphere — troposphere, stratosphere, mesosphere, thermosphere and exosphere — and the general course of temperature and electron density. The atmosphere expands with increasing solar activity through the activity-related increase of the solar irradiance at VUV wavelengths. Thus, the level of the solar VUV irradiance controls the exospheric temperature and hence the extent and density of the upper atmosphere. The change in the thermo- and exospheric temperatures with solar activity is shown schematically in Figure 1.3. Increasing activity causes both an increased irradiance and subsequent absorption of VUV radiation and an increase in the density at a given altitude in the terrestrial atmosphere. Earth-orbiting satellites thus experience an increased drag and a more

[3] The atmosphere also has some windows and stretches of only moderate absorption in the wavelength range between ca. 1 μm and 30 μm (cf., Figures 1.1 and 1.7).

[4] There is, in addition, a sample of nearby stars, where the column density of the interstellar gas is exceptionally low along the lines of sight, and where, consequently, studies in this wavelength region are possible as well (cf., *EUVE*, Bowyer and Malina 1991).

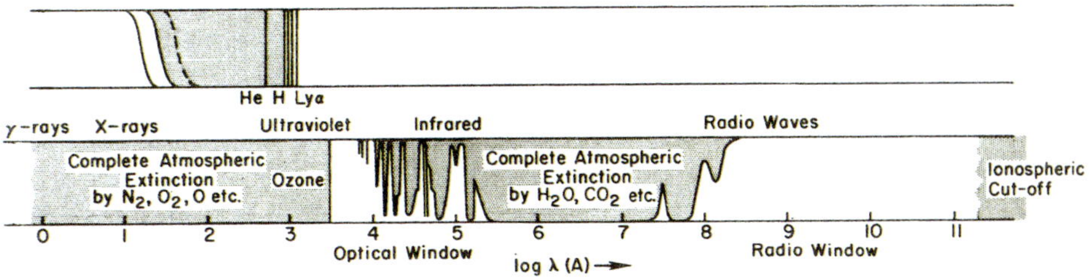

Figure 1.1: Lower panel: Absorption in the terrestrial atmosphere plotted vs. a logarithmic wavelength scale in Ångström. Upper panel: The Lyman series and the ionisation continua of H and He atoms absorb radiation in the interstellar medium, and thus block a part of the spectrum even for instruments carried above the Earth's atmosphere. This latter absorption applies to the radiation of almost all stars, but not to the Sun (from Aller 1963).

Table 1.1: Discriminating spectral domains according to the sub-disciplines of observational space astronomy.

Sub-discipline of observational astronomy	Energies, wavelengths and frequencies
High-energy astrophysics	> 100 keV
X-ray astronomy	from 100 eV to 100 keV
Extra-terrestrial/vacuum-ultraviolet range	from 300 nm to 10 nm
Visible and near-infrared spectral domains	from 30 μm to 0.3 μm
Mid- and far-infrared range	> 30 μm
Cosmic Microwave Background	from 30 GHz to 300 GHz

rapid decay of their orbits. This eventually leads to the destruction of a satellite as it enters the denser lower layers and burns up.[5]

The sub-division of the spectrum into energy, wavelength, or frequency domains varies among research communities.[6] In this book we simply define them through the sub-disciplines of observational space astronomy (cf., Table 1.1).

As we shall see at the beginning of the next section, even balloons provide certain advantages to astronomical observations. More recently still, observing facilities are also placed in orbits around the Sun rather than the Earth.

[5]In the exceptional case of the *HST*, this fate has been postponed several times by re-boosts. As the *HST* was designed for in-orbit servicing, it can be connected to the *Shuttle* and, by use of the *Shuttle* propulsion engines, can be lifted into a higher orbit. This possibility also allowed astronauts, who visited *Hubble* five times, to repair and upgrade *HST*; in the first instance by auxiliary optics, which compensated the accidental spherical aberration of the prime mirror. The *ISS* is regularly boosted, too.

[6]The International Standard Organisation (ISO) has issued a highly-detailed list of definitions of wavelength ranges. This list, however, is accessible only to paying customers.

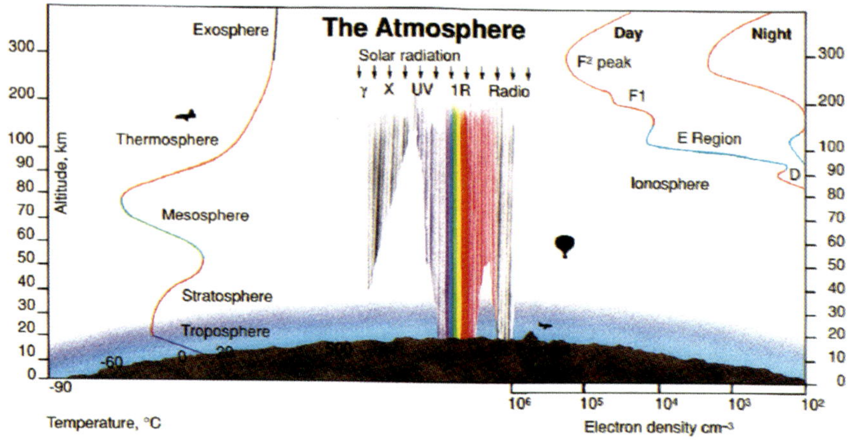

Figure 1.2: The general structure of the terrestrial atmosphere and the diurnal variation of the ionosphere (from the Space Weather Prediction Center www.swpc.noaa.gov).

Enabling technologies

High-altitude balloons floating above 30 km permit access to wavelengths below 300 nm, the atmospheric absorption limit,[7] but measurements at shorter wavelengths require the use of sounding rockets or satellites operating at altitudes above 160 km. However, even at these altitudes residual absorption and emission by the geocorona and by terrestrial airglow emission, respectively, can affect the measurements. The latter effects can be eliminated, if the observing instrument is placed at L1 or L2, i.e., at Lagrange points of the Sun-Earth system, and thus in a heliocentric orbit. L1 and L2 lie on the Sun-Earth line 1.5 Gm away from the Earth,[8] in front of and behind the Earth, respectively. In general the spacecraft are, however, not placed in the Lagrange points themselves, but guided along halo orbits around L1 or L2, where their station keeping is energetically less demanding than if their position were maintained in the Lagrange points themselves.[9]

The first sounding-rocket measurements were made in the mid 1940s with German V2 rockets that had been captured at the end of the Second World War.[10]

[7]High-altitude balloons do not only extend the accessible wavelength range towards shorter wavelengths, but also permit to overcome the seeing caused by turbulence in the lower, denser atmosphere. This was exploited by the 'Stratoscope' project, a telescope (with a diameter $D =$ 30.5 cm), which took photographs of the solar granulation at a wavelength $\lambda \approx 505$ nm at an altitude of 24 km, i.e., above the tropopause (cf., Figure 1.3). With proper optical and thermal design, the theoretical diameter of the diffraction disk, given by the diameter of the primary mirror of the telescope, $1.12\,(\lambda/D) \approx 1.85\times10^{-6}$ rad $= 0.35''$ was actually achieved (Schwarzschild 1959).

[8]1.5 Gm correspond to 1 % of the Earth-Sun distance.

[9]For L1 this also avoids disturbances between the radio downlink of the spacecraft and radio emission from the Sun.

[10]Even before the end of World War II, instrumentation had been prepared in Germany for flights with an A4 rocket. The initiative came from Erich Regener, and was supported by Wernher von Braun. A barrel-shaped container (the 'Regener Tonne') was equipped with barometers and temperature sensors as well as spectrographs to observe the solar spectrum at high altitude.

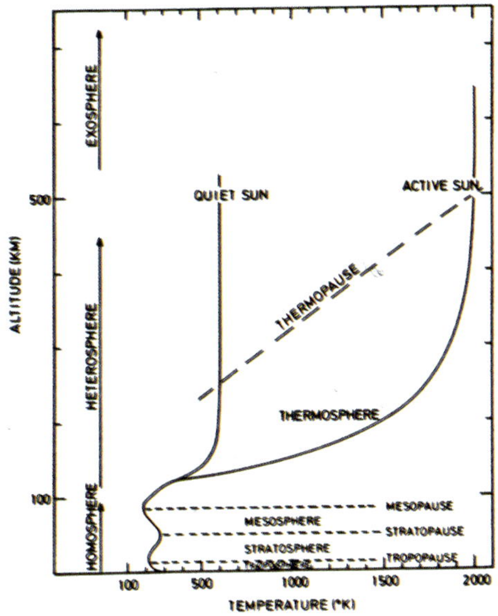

Figure 1.3: Thermospheric temperature variations (Banks and Kockarts 1973), reproduced from Dickinson (1986).

Figure 1.4 shows the first solar spectra obtained in this way. Results from these unstabilized rocket systems were limited and special techniques were required to direct the solar VUV radiation into the spectrographs (de Jager 2001). Tousey et al (1947), who had photographed the spectrum shown in Figure 1.4, reported that "a LiF sphere of 2 mm diameter had been used in place of the [spectrograph entrance] slit." [11] The development in the 1950s and 1960s of sub-orbital and orbital launch vehicles (Russo 2001) brought much progress.[12] Alongside the improvement of the transportation systems three-axis stabilised pointing systems were perfected in another most important effort. This not only permitted recording the VUV spectrum of the Sun with high spectral and spatial resolution (de Jager 2001), but also now offered the opportunity to observe more remote astronomical objects in the VUV (Savage 2001).

Although this will sound curious to today's space researchers, there was no weight limitation: the A4 rocket, which was used by the military as the V2 missile, had been designed to carry a ton of explosives. However, the lack of a pointing system would have allowed exposures to be taken only during the descent on a parachute. Diffusing optics, as described by de Jager (2001), were foreseen to feed light into the spectrometers. To achieve a proper deployment of the parachute, a launch up to an altitude of only ca. 50 km was planned. The payload had been delivered to the launch site, Peenemünde, in early January 1945 and its adjustments and functional tests had been completed on 18 January. At that time, however, the increasing severity of the war prohibited an A4-flight for scientific purposes (cf., DeVorkin 1993; de Jager 2001; Keppler 2003).

[11] Regener and Regener (1934) had earlier recorded the solar spectrum down to 294 nm on a balloon flight that had reached an altitude of 29.3 km.

[12] For a description of the early history of rocket developments see, for example, Bonnet (1992) and Stuhlinger (2001).

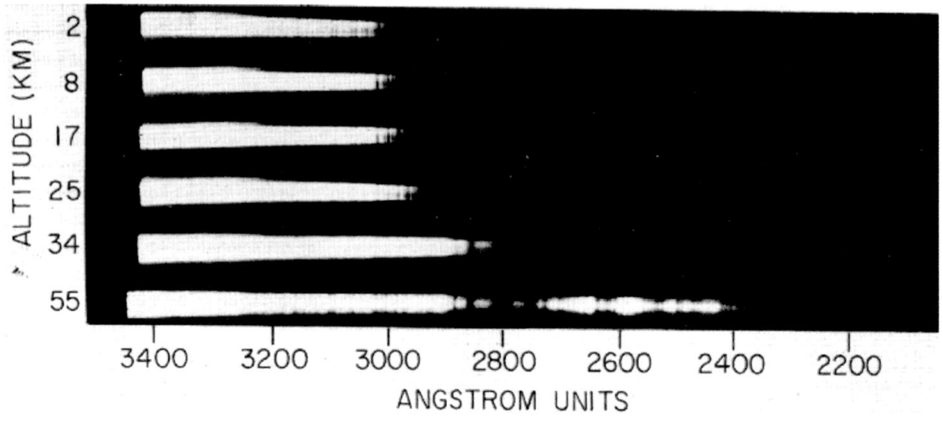

Figure 1.4: The first photograph of the ultraviolet spectrum of the Sun, at an altitude of 55 km. Several exposures were made at different altitudes with a US Naval Research Laboratory spectrograph that was carried in a V2 rocket on 10 October 1946 (from Tousey 1967).

Morton (1966) reported the first VUV spectra of bright hot stars, which were recorded with a dispersion that was high enough for measuring the spectral lines present. Absorption lines in these spectra revealed the existence of stellar winds with velocities ranging from 1400 $\mathrm{km\,s^{-1}}$ to 3000 $\mathrm{km\,s^{-1}}$. At that time the use of objective-grating spectrographs avoided the need for very high pointing accuracy. Pointing stability on the other hand was required to avoid the blurring of the spectrum. This could be achieved with the help of a fine-stabilisation system that operated in conjunction with the attitude control system, and kept the spectrograph platform steady to within ca. $\pm 16''$ with a few excursions to $\pm 50''$ (Morton and Spitzer 1966).[13]

How difficult it was to work without a three-axis attitude control system and with only marginal spatial resolutions is also apparent from the paper describing the first evidence of X-rays from sources outside the solar system (Giacconi et al 1962). This specific observation actually started X-ray astronomy on objects beyond the Sun and was eventually recognised by the first Nobel prize for space astronomy. On the rocket flight in question, the spectral information available from the two functioning Geiger counters with mica windows of two different thicknesses was rather austere as well, so that a thorough discussion rationalising the exclusion of other celestial, near-Earth and solar-system sources was required to corroborate the conclusion.

[13] It is remarkable that seven years later, the same group could report a stability of the *Copernicus (OAO-3)* satellite of better than ± 50 mas over a full orbit of ca. 90 min (Rogerson et al 1973).

The study of photons at γ- and X-ray energies

With nuclear physics being at the centre of contemporary physics, and moreover being based mostly on electronic detection and analysis techniques, it was quite natural that cosmic-ray research — and this meant also γ-ray astronomy — should become part of the early epoch of space research (cf., Chapter 3, Kanbach et al 2010). The 31 first γ-ray photons detected from space (rather than from the terrestrial atmosphere, where they are generated through interaction with cosmic rays) were identified in 1961 by the *Explorer-11* satellite. This was followed in 1968 by *OSO-3*[14] observations of 621 cosmic-ray photons (Kraushaar et al 1972). Figure 1.5 shows the distribution of these photons with galactic latitude. As pointed out by the authors "the peak at the [galactic] equator and the apparently flat plateau over the range $|b^{II}| > 40°$ suggest the existence of two components in the cosmic γ-ray flux[15] — one, which is clearly of galactic origin and another which is isotropic and possibly of extragalactic origin." *OSO-7* then saw the first γ-ray lines generated by solar flares in 1972 and — following the unfortunately abbreviated[16] flight of *SAS-2* — the ESA mission *COS-B* produced from 1975 until 1982 the first sky map, which showed diffuse radiation from the galactic disk caused by the interaction of cosmic rays with the interstellar medium in the Milky Way and point sources such as pulsars, as well as one individual extra-galactic source, namely 3C 273.

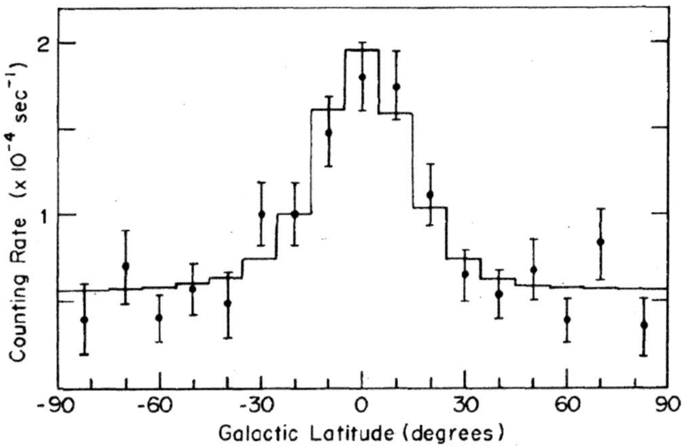

Figure 1.5: Galactic latitude distribution of cosmic γ-ray photons with energies above 50 MeV as observed by *OSO-3* (Kraushaar et al 1972). Photons emerging from the galactic longitude range, $-30° < \ell^{II} < 30°$, had been excluded, as it was surmised that sources in the direction of the galactic centre might be of a special nature.

[14]The *OSO* series of NASA missions was primarily aimed at solar observations. Yet, the *OSO* satellites not only provided a platform for instruments pointed at the Sun but also accommodated a set of instruments, which scanned the sky in the course of six months (see, e.g., Kraushaar 1972).

[15]In modern nomenclature: 'radiance'

[16]Because of a tape-recorder failure

In 1967, the US *Vela* satellites observed the phenomenon of γ-ray bursts; this, however, was kept classified until 1973, because the *Vela* satellites had been built in order to enforce the nuclear test ban treaty, and were meant to detect terrestrial nuclear explosions. The γ-ray bursts remained a puzzle until 1997, when the afterglow of a few such bursts could be followed in X-rays and by optical telescopes. From the redshifts of the γ-ray burst remnants it could be concluded that such events were a very remote phenomenon, which set free extremely high energies: assuming isotropic emission, the energy in a γ-ray burst can be of an order equivalent to a solar mass. It is not surprising then that the most remote object known today was identified through a γ-ray burst — an object with redshift $z = 8.2$. This implies an explosion 13×10^9 light years away, which had occurred at a time when the Universe was about 1/20 of its present age. Indeed, this γ-ray burst was "a glimpse of the end of the dark ages" (Tanvir et al 2009).

A most significant inference of γ-ray astronomy is the absence of a background of γ-rays from nucleon-antinucleon annihilation. This finding empirically excludes a matter-antimatter symmetric Universe (Cohen et al 1998; Pinkau 2009).

The impact of X-ray observations on the entire field of astronomy is summarised in the Nobel lecture — The Dawn of X-ray Astronomy — by Giacconi (2003). A breakthrough occurred, when — after several earlier attempts with threshold-limited sensitivities to detect X-rays from sources other than the Sun — a rocket payload "intended to observe the X-ray fluorescence produced on the lunar surface by X-rays from the Sun and to explore the night-sky for other possible sources" had been launched on 18 June 1962. With this rocket experiment, which was able to detect an X-ray photon irradiance of (0.1 to 1) $\text{cm}^{-2}\,\text{s}^{-1}$, and thus was about 50 to 100 times more sensitive than any experiment flown before, Giacconi et al (1962) discovered the first X-ray source from outside the solar system. The detectors scanned the sky over a band covering the Moon's disk in two energy bands. Figure 1.6 shows the results. The source that gave the large signal was later on named Sco X-1. Since then the sophistication of X-ray observations has increased from scanners

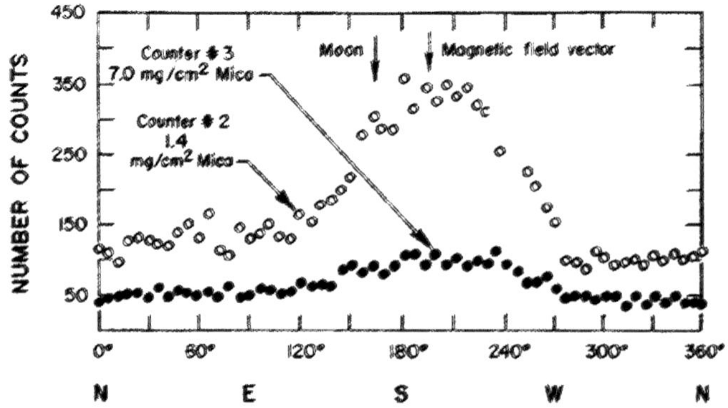

Figure 1.6: The first observation of Sco X-1 and of the X-ray background in the rocket flight of 12 June 1962 (from Giacconi et al 1962). The ordinate shows the number of counts accumulated in 350 s in angular intervals of 6°.

with one or a few Geiger counter detectors with mechanical collimators, i.e., baffles restricting the field of view, to high-resolution telescopes with imaging detectors, and with energy discrimination either by diffraction gratings that disperse the radiation according to wavelength, or by energy-discriminating detectors (cf., Chapter 4, Culhane 2010). From the beginning in 1962 to the present, the sensitivity of X-ray observing techniques has increased by more than nine orders of magnitude, a range comparable to that from the naked eye to the current 10 m ground-based telescopes (Giacconi 2003). High-energy astronomy — i.e., X-ray astronomy and γ-ray astronomy together — has led to the confirmation of fundamental, but originally only surmised, phenomena, such as black holes, or the intergalactic plasma in galaxy clusters.

Although opening the X-ray window to the Universe beyond the Sun was, of course, a momentous discovery, we recall that X-ray observations of the Sun have given us an early insight into the large radiance variations associated with flares and the structure and dynamics of the solar corona, which is consisting of highly inhomogeneous plasma that is contained by magnetic field structures of widely different stability (cf., Blake et al 1963; Friedman 1963; Vaiana and Rosner 1978). And solar γ-rays continue to play an important role in γ-ray astronomy (cf., Chapter 3, Kanbach et al 2010).

Space astronomy in the VUV — an observationally delicate region

The early VUV and soft X-ray spectra and images of astrophysical objects were recorded by photography. Conventional photographic emulsions, however, were not suitable for recording the interesting VUV wavelength range below 200 nm, because the protective gelatine layer absorbed the radiation in question before it reached the silver-halide grains. Schumann (1901) developed emulsions on glass plates with just enough gelatine to hold the silver halide grains. Such emulsions, however, were extremely sensitive to physical contact and, particularly if on a glass substrate, were not suitable for use in space, given the launch environment of space vehicles (cf., Chapter 40, Kent 2010). For the manned *Skylab* flights in 1973 and 1974, which offered the opportunity to have the exposed films returned to Earth by the astronauts, VanHoosier et al (1977) developed suitable containers with special film-based Schumann emulsions.

Photographic emulsions have several drawbacks beyond the paramount obstacle of recovering them from orbit after they have been exposed, namely their limited dynamic range, and the need to develop them before the data can be viewed.[17] This meant, for example, that film-based science operations during the first of the

[17]NASA's early lunar exploration programme was based in part on the use of the unmanned '*Lunar Orbiter*' spacecraft. These were designed to obtain detailed photographs of potential *Apollo* landing sites in the years 1966 to 1967. As no suitable electronic cameras were available at that time, the Moon's surface was photographed from lunar orbit. The exposed film was developed and photoelectrically scanned on board. The images could then be sent back to Earth by telemetry (Nelson and Aback 1971).

three manned observing periods on *Skylab* had to be carried out blind, because results became available only after recovery.

On the other hand, photographic material is inherently two-dimensional, and accordingly permits the simultaneous recording of an image, for example, a field on the sky, or a stigmatic spectrum, i.e., a spectrum which reflects the radiance variations along the entrance slit of the spectrograph. This was, in the first instance, a definitive advantage for the photographic techniques. Nevertheless, for rapidly varying objects, one would need short exposure times and ample photographic material. With the advent of imaging detectors the advantage of photography disappeared; now images could also be downlinked to Earth in near-real time. In addition, photoelectric measurements eventually became the adopted solution for accurately calibrated observations.

Initially, photoelectric detectors were not even one-dimensional, but recorded a single pixel only. For space astronomy they had to be rugged and this required special developments. When used to register the weak solar VUV radiation, they also had to be 'solar blind', i.e., insensitive to the strong visible light stemming from the solar photosphere, which invariably would appear as stray light. In fact, this is also critical for observations of other, relatively cool stars.[18]

Further difficulties arise if one progresses to even shorter wavelengths in the VUV, namely to below ≈ 116 nm, where no rugged window materials are available. Hence the need for purely reflective optical systems and detector systems that can operate without windows.[19]

For the detection of visible and infrared radiation, as well as for the detection of X-rays, the most common detector today is a solid-state device — the charge-coupled device (CCD, see Chapters 23 through 25, Waltham 2010; Holland 2010; Schühle 2010), which however has a relatively slow read-out. As CCDs are not sensitive to ultraviolet radiation, appropriate image intensifiers or fluorescent material have to be added to achieve the desired results for observations of ultraviolet and VUV radiation; and for X-rays thin-film metal filters have to be added to reject radiation of longer wavelength.

Another important imaging detector system, which, on the other hand, is operating in vacuum, is the microchannel-plate (MCP) detector. If operated as a photon counter this detector can also time-tag single photons (cf., Chapter 22, Timothy 2010). The origin of this detector can be traced to the magnetic electron multiplier (MEM), which was based on a continuous-dynode structure. The concept of a continuous-dynode electron multiplier had first been proposed by Farnsworth (1930), but practical devices only became available in the 1960s. The first practical continuous-dynode multiplier was the crossed-field Bendix MEM (Heroux and Hinteregger 1960; Timothy et al 1967, cf. also, Chapter 22, Timothy 2010, where

[18] As there were no 'solar blind' emulsions, alkali films for wavelengths in the VUV or thin metal filters for X-rays had to be used to reject visible light in photographic apparatus, unless the object was a young star, with peak radiation in the ultraviolet. For photographic X-ray observations thin metal films were used not only to reject the visible radiation, but also to selectively transmit a given wavelength region (cf., Vaiana and Rosner 1978).

[19] Finally, the reflectance of normal-incidence optical systems falls dramatically below 50 nm, and use of grazing-incidence telescopes (Brueggemann 1968) and spectrometers (Samson 1967), or synthetic multilayer interference coatings (Spiller 1974) is mandatory (cf., Chapters 9 and 10, Lemaire et al 2010; Lemaire 2010).

the intermediate stage between MEM and micro-channel plates, namely the channeltron, is described as well).

The path towards space telescopes observing in the visible and infrared

Today's most important space telescope, the *HST* has its roots in the first quarter of the 20th century. In his book 'Die Rakete zu den Planetenräumen'[20] Hermann Oberth (1923) had written that "Telescopes of any size could be used in space, since the images of stars do not twinkle..." (cf., Stuhlinger 2001). In reviewing the path of *HST*, then called '*Space Telescope*', Lyman Spitzer (1979) explained: "... the engineering possibilities of launching a complicated device into space and keeping it in operation seemed rather remote at the time of Oberth's suggestion, and few astronomers took these possibilities seriously." However, after World War II, Spitzer (1946) discussed a number of scientific programmes that could be carried out with large reflecting telescopes in space, namely "a) pushing back the frontiers of the Universe and determining galactic distances by measuring very faint stars, b) analysing the structure of galaxies, c) exploring systematically the structure of globular clusters, and d) studying the nature of planets, especially their surfaces and atmospheres" (quoted after Spitzer 1979). We recall here that another Princeton astronomer, namely Spitzer's colleague Martin Schwarzschild (1959) made the first high-resolution observations in the visible by photographing the granulation of the solar atmosphere from a stratospheric balloon in the late 1950s (cf., footnote 7).

Much emphasis today is on the cool parts of the Universe and on the red-shifted Cosmos. As a consequence many, if not most, current and currently-planned major space observatories are aimed at the infrared and sub-millimetre domains: *Spitzer* and *Herschel* are in orbit, *JWST* is scheduled for launch in 2014 and *SPICA* is under advanced study. Moreover, a facility installed on an aeroplane, SOFIA, is now entering service. The road to this state was arduous, as described by Harwit (2001). Certain segments within the infrared domain are accessible, depending on wavelength, from high mountains, aeroplanes or stratospheric balloons (cf., Figure 1.7). This has often led to extensive debates as to whether a given observational problem really required a telescope in space, and has caused serious rifts in the astronomical community. Moreover, infrared detectors and the cryogenics needed to suppress the thermal background of the equipment itself at longer wavelengths, are technically demanding and required long developments. In brief, the long fruition time of infrared space observations can be explained by both observational and technological reasons. This has now been overcome, although operating infrared observing facilities is still a demanding task, especially because of the limited lifetime of the cooling agent, or the need to fly a cooling apparatus (cf., Chapter 37, Rando 2010).

The development of microwave observations in space, in particular the development of *COBE* (cf., Chapter 8, Lamarre 2010) seems to have steadily progressed

[20]Literally: "The Rocket to the Space of the Planets".

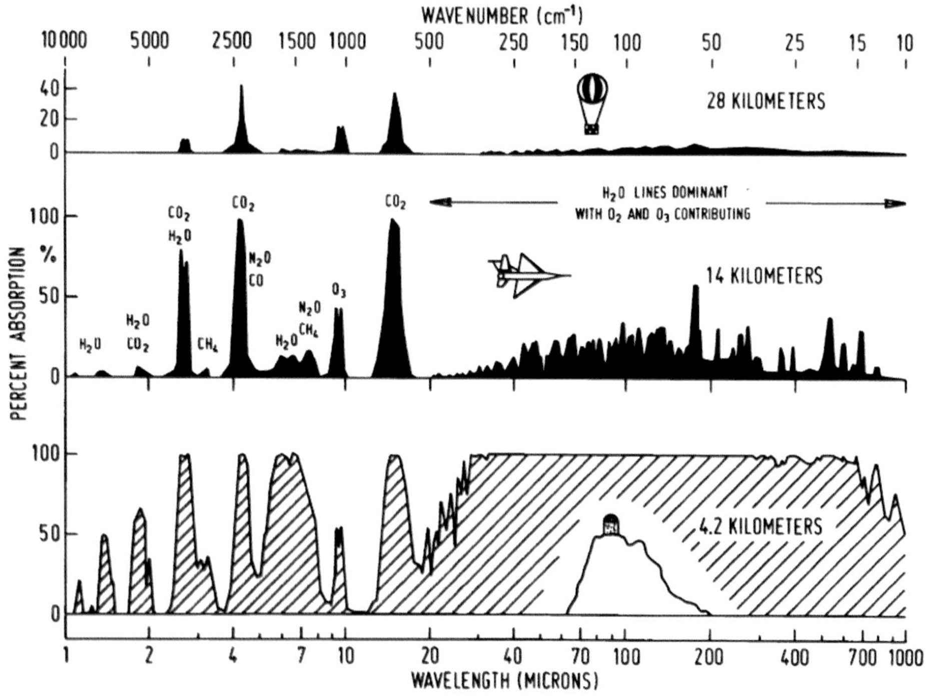

Figure 1.7: Atmospheric absorption in the infrared at a mountain-top site like Mauna Kea and at altitudes of aircraft and balloons (from Harwit 2001).

quite separately from the development of infrared space observatories, and has led to the 2006 Nobel prize being awarded to John C. Mather and George F. Smoot for this work in space astronomy. In the meantime *COBE* was followed by *WMAP*, another NASA mission, and now the ESA mission *Planck* has reached its orbit around L2, after a joint launch in May 2009 together with the infrared/sub-millimetre observatory *Herschel*.

Innovative space technology meets conservative attitude

It might be constructive in the long run to look back to the time when space astronomy was in its infancy, and was considered by many as a somewhat extravagant exercise — the early epoch of space science. Seen from today's astronomy scene, where ground-based and space astronomy are viewed as being complementary, it may be interesting (and, if not disconcerting, now and again amusing) to recall the early attitudes of 'traditional' astronomers towards the opportunities offered by space technology (similarly, by the way, to those of 'traditional' geophysicists and oceanographers). Three quotes from history studies of space astronomy shall illustrate this mercifully forgotten issue.

In her description of starting the space-astronomy programme of NASA, Roman (2001) writes: "Astronomers, practitioners of a very old science, deal with long-lived objects and thus tend to be conservative. Hence, it is not surprising that there were social as well as technical problems to be met in the development of the new NASA astronomy program." Also Stuhlinger (2001) mentions that the value and cost of "ten Palomars vs. one Space Telescope" were often used as arguments against building a space telescope. A special impediment has been the infrared community. Harwit (2001) reports: "Since rocket payloads were relatively expensive, the U.S. ground-based observers feared that the National Aeronautics and Space Administration (NASA) or National Science Foundation (NSF) support for rocket work would drain away funding from their own efforts. Uncertainties about the precise limitations of ground-based techniques often led to acrimonious disputes between ground-based observers and those who wanted to launch sensitive infrared astronomical telescopes above the atmosphere on rockets. Time and again, the more traditional astronomers bluntly recommended that rocket-borne efforts be slashed or pared back to an absolute minimum. Pioneering infrared rocket research was not a happy venture!" Harwit, however, then goes on with a very useful description of how the difficulties were overcome by uniting the infrared space community behind one single project.

There was no great difference between the sceptical attitudes on the two sides of the Atlantic, although the politically motivated support of all space activities in the US led to a great disparity in funding for space science. Space astronomy in Europe started — with the exception of the already mentioned γ-ray mission *COS-B* — relatively late, i.e., long after atmospheric and magnetospheric research. This may be ascribed to similar attitudes. The lack of adequate funding in Europe, however, led advanced practitioners to enter into fruitful collaborations with NASA; this joining of forces did, although it led to brain drain initially, ultimately help both sides. Indeed, observing photons in space — like other space ventures — was an early manifestation of international collaboration beyond continental boundaries. Given the cold war, however, bilateral collaborations across the iron curtain remained restricted to only a few western countries, such as the USA, Austria, France and Sweden, although the Committee on Space Research (COSPAR) held biennial conferences where scientists from East and West could easily meet.

Overall structure of the book and conventions used

The chapter following this introduction presents the fate of photons from source to detector, i.e., the physics of the interaction of photons with matter with an emphasis on the principles of physics being used in detecting and analysing photons. The next six chapters deal with both the physical processes and the characteristics of the radiation that can be observed in the energy-, wavelength- or frequency-domains from γ-rays to radio waves. Examples of the relevant astrophysical objects and the early missions devoted to their study are presented together with resulting key findings. The subsequent eleven chapters are devoted to descriptions of the techniques and systems used for imaging and spectral analysis. An in-depth discussion of detectors used in space astronomy, subdivided into 13 chapters, then precedes three chapters on polarimetry and an additional four chapters on techniques of general importance to space observations, such as calibration, cryogenic systems, laser-aligned structures, and the peculiarities of Earth observations, including radar measurements. Two final chapters, namely a pragmatically-oriented essay on the implications of the space environment that must be considered when building space instrumentation, and a summary and outlook, conclude the book.

In an Appendix we present a brief overview of the most pertinent rules of the Système International. The SI has been used as systematically as possible in the text. Figures which show historical plots, however, were not redrawn, and therefore may show obsolete nomenclature and units. Conversions to SI units are also given in the Appendix. Abbreviations and acronyms of spacecraft are not introduced; they are collected in a list of acronyms at the end of the book. References to internet resources and URLs have been checked to be valid in August 2009. However, the internet remains a volatile resource and its contents and locations are subject to change.

Acknowledgements

On behalf of all editors, we want to gratefully acknowledge the contributions of André Balogh who provided the liaison to ISSI. Thjis de Graauw participated with useful advice at the early stages of the book. Anthony J. Banday, Walter Gear, Hugh Hudson, Anthony Marston and Claudia Wigger acted as outside referees. Many of the book's authors also gave valuable advice as referees for chapters of their colleagues. Karen Fletcher of ESA's Communication and Knowledge Department is thanked for sound professional editorial support. We also wish to thank Brigitte Schutte-Fasler, Andrea Fischer, Saliba F. Saliba, Katja Schüpbach, Silvia Wenger and the other members of the staff at ISSI for actively and cheerfully supporting us whenever we needed to use the infrastructure of ISSI.

Bibliography

Aller LH (1963) The Atmospheres of the Sun and the Stars. New York: Ronald Press Co. p 4

Banks PM, Kockarts G (1973) Aeronomy. New York, NY: Academic Press, 2 vol.

Blake RL, Chubb TA, Friedman H, Unzicker AE (1963) Interpretation of X-ray photographs of the Sun. Astrophys J 137:3–15

Boggess NW, Mather JC, Weiss R (plus 15 authors) (1992) The *COBE* mission— Its design and performance two years after launch. Astrophys J 397:420–429

Bonnet RM (1992) Les Horizons Chimériques, Paris: Dunod

Brueggemann HP (1968) Deep conic cassegrains, in Conic Mirrors. London, New York: The Focal Press, Section VIII:102–111

Cohen AG, de Rújula A, Glashow SL (1998) A matter-antimatter Universe? Astrophys J 495:539–549

Culhane JL (2010) X-ray astronomy: energies from 0.1 keV to 100 keV. ISSI SR-009:73–88 de Jager C (2001) Early solar space research. In: The Century of Space Science (eds JAM Bleeker, J Geiss, MCE Huber) Vol 1, pp 203–223

DeVorkin DH (1993) Science with a vengeance—How the military created the US space sciences after World War II. New York: Springer, 404 pp

Dickinson RE (1986) Effects of solar electromagnetic radiation on the terrestrial environment. In Physics of the Sun, Vol III, Heidelberg: Astronomisches Rechen-Institut, pp 155–191

Farnsworth PT (1930) Electron Multiplier. US Patent number 1 969 399

Friedman H (1963) Ultraviolet and X-rays from the Sun. Ann Rev Astron Astrophys 1:59–96

Giacconi R (2003) The dawn of X-ray astronomy. Rev Mod Phys 75:995–1010

Giacconi R, Gursky H, Paolini FR, Rossi BB (1962) Evidence for X-rays from sources outside the solar system. Phys Rev Lett 9:439–443

Harwit M (2001) The early days of infrared space astronomy. In The Century of Space Science (eds JAM Bleeker, J Geiss, MCE Huber) Vol 1, pp 301–330

Heroux L, Hinteregger HE (1960) Resistance strip magnetic electron multiplier for the extreme ultraviolet. Rev Sci Instrum 31:280–286.

Holland A (2010) X-ray CCDs. ISSI SR-009:409–418

Kanbach G, Schönfelder V, Zehnder A (2010) High-energy astrophysics – energies above 100 keV. ISSI SR-009:57–71

Kent BJ (2010) Implications of the space environment. ISSI SR-009:625–643

Keppler E (2003) Raketenentwicklung und Weltraumforschung—Ein kurzer Abriss der Entwicklung der A4/V2 Rakete und deren erstmalige Anwendung für wissenschaftliche Zwecke durch Erich Regener, Katlenburg-Lindau: Copernicus GmbH

Kraushaar WL, Clark GW, Garmire GP (plus four authors) (1972) High-energy cosmic gamma-ray observations from the *OSO-3* satellite. Astrophys J 177:341–363

Lamarre JM, The Cosmic Microwave Background. ISSI SR-009:149–162

Lewis GN (1926) The conservation of photons. Nature 118:874–875

Malina RF, Bowyer S (1991) In: Extreme Ultraviolet Astronomy, A selection of papers presented at the first Berkeley colloquium on Extreme Ultraviolet Astronomy, University of California, Berkeley, January 19-20, 1989. RF Malina, S Bowyer (eds) New York: Pergamon Press, 397 pp

Morton DC (1966) Far-ultraviolet spectra of six stars in Orion. Astron J 71:172–173

Morton DC, Spitzer L Jr (1966) Line spectra of delta and pi Scorpii in the far-ultraviolet. Astrophys J 144:1–12

Nelson CH, Taback I (1971) Digital lunar orbiter photographic atlas of the Moon. Houston: Lunar and Planetary Institute; see also http://www.lpi.usra.edu/resources/lunar_orbiter/book/contents.shtml

Oberth H (1923) Die Rakete zu den Planetenräumen. München: R. Oldenbourg; and (1960, 1964, 1984) Feucht-Nürnberg, Uni Verlag

Pauluhn A (2010) Earth and planet observations. ISSI SR-009:617–622

Perryman M (2009) Astronomical applications of astrometry—Ten years of *Hipparcos* satellite data, Cambridge: Cambridge University Publishing

Pinkau K (2009) History of gamma-ray telescopes and astronomy. Exp Astron 25:157–171

Piore ER, Harvey GG, Gyorgy EM, Kingston RH (1952) A high vacuum recording spectrograph for the study of radiation from solids in the 100-800 Å range. Rev Sci Instrum 23:8–12

Rando N (2010) Cryogenics in space. ISSI SR-009:593–606

Regener E, Regener VH (1934) Aufnahme des ultravioletten Sonnenspektrums in der Stratosphäre und vertikale Ozonverteilung. Phys Zeitschr 35:788–793

Rogerson JB, Spitzer L Jr, Drake JF (plus four authors) (1973) Spectrophotometric results from the *Copernicus* satellite. I. Instrumentation and performance. Astrophys J 181:L97–L102

Roman NG (2001) Exploring the Universe: Space-based astronomy and astrophysics. NASA SP-4407, Washington: NASA History Office, 77 pp; see also http://spacescience.nasa.gov/admin/pubs/history/Chap3-essay.PDF

Russo A (2001) The space age and the origin of space research. In The Century of Space Science (eds JAM Bleeker, J Geiss, MCE Huber) Vol. 1, pp. 25–58

Samson JAR (1967) Techniques of Vacuum Ultraviolet Spectroscopy. New York: Wiley, pp 56–60

Schühle U (2010) Intensified solid state sensor cameras: ICCD and IAPS. ISSI SR-009:419–428

Schumann V (1901) Ueber ein verbessertes Verfahren zur Herstellung ultraviolettempfindlicher Platten. Ann Physik (Leipzig) 310:349–374

Schwarzschild M (1959) Photographs of the solar granulation taken from the stratosphere. Astrophys J 130:345–363

Space Weather Prediction Center www.swpc.noaa.gov

Spiller E (1974) Multilayer interference coatings for the vacuum ultraviolet. In: Space Optics, BJ Thompson, RR Shannon, eds, Washington, DC: National Academy of Sciences, pp 581–597

Spitzer L Jr (1946) Astronomical advantages of an extra-terrestrial observatory. Project RAND Report, Douglas Aircraft Co., September 1.

Spitzer L Jr (1979) History of the space telescope. Q Jl R astr Soc 20:29–36

Stuhlinger E (2001) Enabling technology for space transportation. In The Century of Space Science (JAM Bleeker, J Geiss, MCE Huber) Vol 1, pp 59–114

Tanvir NR, Fox DB, Levan AJ (plus 59 authors) (2009) A glimpse of the end of the dark ages: the gamma-ray burst of 23 April 2009 at redshift 8.3. arXiv:0906.1577

Timothy AF, Timothy JG, Willmore AP (1967) The performance of open structure photomultipliers in the 1100-Å to 250-Å wavelength region. Appl Opt 6:1319–1326

Timothy JG (2010) Microchannel plates for photon detection and imaging in space. ISSI SR-009:365–391

Tousey R, Strain CV, Johnson FS, Oberly JJ (1947) The solar ultraviolet spectrum from a V-2 rocket. Astron J 52:158–159

Tousey R (1967) Some results of twenty years of extreme ultraviolet solar research. Astrophys J 149:239–252

Vaiana GS, Rosner R (1978) Recent advances in coronal physics. Ann Rev Astron Astrophys 16:393–428

VanHoosier ME, Bartoe J-D, Brueckner GE (plus two authors) (1977) Experience with Schumann-type XUV film on Skylab. Appl Opt 16:887–892

Waltham N (2010) CCD and CMOS sensors. ISSI SR-009:391–408

Wilhelm K, Fröhlich C (2010) Photons — from source to detector. ISSI SR-009:23–54

Will CM (2009) The confrontation between general relativity and experiment. Space Sci Rev, DOI 10.1007/s11214-009-9541-6

— 2 —

Photons — from source to detector

KLAUS WILHELM[I] AND CLAUS FRÖHLICH[II]

Abstract

The central theme of the book "Observing Photons in Space" is the detection and characterization of photons with instruments aboard spacecraft. This chapter presents a global overview of the fundamental processes that accompany photons all the way from their origin in the source region to their detection in our instruments. The radiation of the Sun is taken as example in some cases and is treated in more detail.

Introduction

The goal of observing photons clearly is to learn something about the regions that are emitting the radiation, and it is, therefore, necessary to understand the basics of the interaction of photons with matter at the source, during transit, and at the detection site. The last point is the most important one in the context of this book, but some generation and transfer processes have to be mentioned as well. Many of these aspects are discussed in more detail elsewhere in this book and cross-references are provided.

The interaction of photons with matter is mainly governed by two equations derived by Einstein (1905 a, c). They can be written in a modern format as

$$E_\nu = h\,\overline{\nu} = h\,c_0/\overline{\lambda} \quad \text{and} \quad E = m\,c_0^2 \quad , \tag{2.1}$$

where $h = 6.626\,068\,96(33) \times 10^{-34}$ J s is the Planck constant[1], E_ν is the energy quantum of electromagnetic radiation with mean values[2] of the frequency $\overline{\nu}$ and wavelength $\overline{\lambda}$, and $c_0 = \nu\,\lambda = 299\,792\,458$ m s^{-1} (exact) is the speed of light in vacuum[3] in the mass-energy equivalence equation. The mass of a particle is

[I]MPS—Max-Planck-Institut für Sonnensystemforschung, Katlenburg-Lindau, Germany

[II]PMOD / WRC—Physikalisch-Meteorologisches Observatorium Davos / World Radiation Center, Switzerland

[1]This and other constants are taken from CODATA recommended values of the fundamental physical constants: 2006 (Mohr et al 2008). The CODATA-06 set is available from http://physics.nist.gov/constants.

[2]An exactly defined value of ν would require an infinitely extended photon. In the following discussion, the notations ν and λ will be used to simplify the equations, unless emphasis is placed on the fact that mean values are considered in special cases.

[3]Follows from the definition of the SI base unit "metre"(BIPM 2006).

$m = \gamma m_0$, where $\gamma = (1 - \beta^2)^{-1/2} > 1$ is the Lorentz factor, $\beta = v/c_0$ with $|v| < c_0$ the speed of the particle relative to a system, in which its rest mass is m_0. This mass is believed to be zero for photons[4], but their kinetic mass is $m_\nu = h\,\nu/c_0^2$.

It must, however, be emphasized from the outset that the wave theory of electromagnetic radiation as formulated by Young, Huygens, Fresnel, Maxwell and Hertz is a valid description of most of the observations as long as the energy quantization in Equation 2.1 is of no importance (see, e.g., Born and Wolf 1999; Jackson 1999). Consequently, the laws of physical optics, and even of geometrical optics are relevant to a large extent; in particular, those concerning the propagation of radiation. This permits the use of ray-tracing procedures. In the following, the wave or particle aspects of electromagnetic radiation will, therefore, be emphasized as required. The wave theory is also adequately describing the polarization properties of radiation (treated in detail in Chapters 33 to 35, Stenflo 2010; Hajdas and Suarez-Garcia 2010; Carretti and Rosset 2010), which correspond to the spin $s = 1$ of photons. Major modifications are, however, required during photon emission and absorption processes. The quantum electrodynamics developed by Born et al (1926), Dirac (1927), and others include most of these modifications, and renders a suitable description of photons (cf., e.g., Glauber 2007), albeit their "existence" is not generally accepted (Lamb 1995).

Special mention must be made of entangled states of photons as an important quantum physical phenomenon, although no application in astrophysical observations is known at this stage, other than the two-photon emission in the hydrogen spectrum detected by Brugel et al (1982). In response to the *Gedankenexperiment* of Einstein et al (1935) stipulating that entangled particles no longer interact after their separation, an inequality was formulated by Bell (1964) that would be valid under this condition. Since then, many violations of Bell's inequality, in the formulation of Clauser et al (1969), have been observed. One of the most recent tests was performed over a distance of more than 10 km yielding a speed for the quantum influence of at least $1 \times 10^3\, c_0$ (Zbinden et al 2001). The remaining critique of such experiments is that statistical results obtained by observing ensembles of photons are applied to a single photon pair; thus, it is claimed, it would still be possible to accept the physical reality of a photon (cf., Tiwari 2002; Adenier 2008).

In this context, experiments are of interest that study the structure of photons (Nisius 2000), and those demonstrating that two-photon interference cannot always be understood as the effects of two single photons (Strekalov et al 1998). It is worth noting that bunching of two photons in a symmetric state has been observed as a consequence of their bosonic nature by Hong et al (1987). Concerning the latter behaviour of photons see also Bose (1924) and Ehrenfest (1925).

[4] A zero rest mass follows from the special theory of relativity and a speed of light in vacuum constant for all frequencies. Various methods have been used to constrain the photon rest mass to $m_0 < 1.7 \times 10^{-50}$ kg (cf., Amsler et al 2008, Goldhaber and Nieto 1971). The latter authors also discuss the consequences of a finite rest mass for the existence of longitudinal electromagnetic waves.

Some photon generation processes

The most fundamental source of radiation is a so-called "black body", a system that absorbs all radiation impinging on it, i.e., the absorption coefficient is $\alpha_\lambda = 1$ for all λ. Kirchhoff's law then yields the emission coefficient $\epsilon_\lambda = 1$ at all wavelengths. The absorption coefficient $\alpha_\lambda = 1$ obviously means that a black body, or a representation of it, is a good detector for total irradiance measurements. Depending on its temperature, T, such a system emits radiation according to Planck's law[5] (Planck 1900, 1901) derived from a quantization of the radiation field

$$L_\lambda(T) = \frac{2\,h\,c_0^2}{\lambda^5}\left[\exp\left(\frac{h\,c_0}{\lambda\,k_\mathrm{B}T}\right) - 1\right]^{-1} \quad , \tag{2.2}$$

where $k_\mathrm{B} = 1.380\,6504(24) \times 10^{-23}$ J K^{-1} is the Boltzmann constant and $L_\lambda(T)$ the spectral radiance at the wavelength λ. The corresponding formula for $L_\nu(T)$ is

$$L_\nu(T) = \frac{2\,h\,\nu^3}{c_0^2}\left[\exp\left(\frac{h\,\nu}{k_\mathrm{B}T}\right) - 1\right]^{-1} \quad . \tag{2.3}$$

The Cosmic Microwave Background (CMB) at $T_\mathrm{CMB} = 2.726$ K (Penzias and Wilson 1965; Mather et al 1990, 1994) is an example of black-body radiation (cf., Chapter 8, Lamarre 2010). It is thought to have been released from the "Big Bang plasma", when it changed from the optically thick to thin state at the time of atom formation (Dicke et al 1965). The CMB is probably the earliest electromagnetic radiation released after the Big Bang that is still accessible to us. Since the temperature at the time of atom formation was typically 3000 K, a redshift of z between ≈ 1000 and 1500 (Sunyaev and Zel'dovich 1980) has to be assumed.

A black body is the extreme case of an optically thick medium. In the Universe, many plasma systems emit radiation approximately according to Equation 2.2, at least, in certain wavelength ranges, for which a brightness temperature, T_B, can then be defined in local thermal equilibrium. Other regimes, although still optically thick, deviate from such an equilibrium and require radiation transfer calculations. Regions that are optically thin for some or all wavelengths represent the other extreme, and allow the integration of the radiation along the line of sight (LOS).

Many photon emissions are a consequence of electron transitions in atomic species from a higher energy level, E_2, to a lower one, E_1, as postulated by Bohr in 1913:

$$\overline{E_2} - \overline{E_1} = h\,\overline{\nu_{21}} \quad , \tag{2.4}$$

with $\overline{E_2}, \overline{E_1}$ the levels of the unperturbed states; an assumption that has to be modified in the light of Equation 2.5. The recoil of the emitter leads to an additional frequency shift depending on the details of the process.

In general, these emissions are spontaneous, but can also be stimulated by appropriate photon fluxes (Einstein 1917)—a prerequisite of any laser operation. Equation 2.4 also describes photon absorption if $E_2 < E_1$. The latter process can

[5]Einstein and Stern (1913) argued in favour of a zero-point energy $h\,\nu/2$ of a resonator, first introduced by Planck (1909). The zero-point energy of the electromagnetic field causes, for instance, the Casimir forces (cf., Casimir 1949; Mohideen and Roy 1998).

be considered as an excitation mechanism of molecules, atoms or ions, and will be of importance for the detection of photons. Excitations can also be induced by collisions between atomic species and electrons. Whenever inner electrons of an atom were involved in such transitions, the emitted radiation was historically called Röntgenstrahlung or X-rays. If the radiation was generated by transitions of excited nuclei of atoms, it was called γ-rays. However, in modern terms X-ray and γ-ray radiations are defined by their photon energies.

Additional generation processes of radiation involve accelerated or decelerated particles, in particular electrons, such as synchrotron radiation and bremsstrahlung. The annihilation of matter and antimatter should also be mentioned. In the case of an electron and a positron, the process produces, in general, two γ-ray photons of 511 keV energy, each. A prominent γ-ray line at 2.223 MeV is caused by neutron capture of hydrogen. Details on the thermal and non-thermal radiation of high-energy particles are discussed in Chapters 3 and 4 (Kanbach et al 2010; Culhane 2010). A special case is the $\lambda = 21.106$ cm line generated in the Universe by an electron spin reversal with respect to the proton spin in hydrogen atoms. Rotational transitions in molecular clouds are another source of low-energy photons.

The generated electromagnetic radiation is, as we have seen, released and absorbed as photons, quantized in energy. Einstein (1917) showed that the generation of a photon as a spherical wave is not possible, because a momentum transfer is invariably involved in the emission process. Given a certain atomic transition, the released energy is generally rather well defined resulting in a nearly mono-energetic photon flux, or, in the equivalent description, in waves with a frequency, ν_0, and wavelength, λ_0. Consequently, such a transition can be identified by the detection of a spectral line at λ_0. Atoms, ions and molecules emit a multitude of spectral lines related to their energy levels. Even a summary of the physics of these transitions would be far beyond the scope of this chapter. It suffices to say that the lines allow us to deduce elemental abundances, ionization stages, densities, electron and ion temperatures, and other parameters of the source regions. Selected spectra of the Sun might serve as example here:

Figure 2.1 depicts the vacuum-ultraviolet (VUV) radiance spectrum of a quiet-Sun region with many emission lines and some continua in the wavelength range from 80 nm to 150 nm. At longer wavelengths, the emission lines fade out and absorption features—the Fraunhofer lines—become prominent as a consequence of the thermal structure of the solar atmosphere. This can be seen in Figure 2.2 showing an irradiance spectrum of the Sun (Thuillier et al 2004). In general, these measurements are performed from platforms moving with respect to the source (the Sun). The necessary corrections are treated on page 37. Such spectra provide important information about the composition and other properties of the Sun or a star (see, e.g., Fröhlich et al 1998). Although the lines in Figures 2.1 and 2.2 appear to be rather narrow, they can be resolved into profiles by modern spectrometers furnishing information on motions in the solar atmosphere via the Doppler effect (cf., page 31). Similar measurements can be obtained from stellar and other cosmic objects. The spectrum of α Cen A, for instance, a G2 V star, closely resembles that of the Sun in the same spectral class (Ayres 2000).

Some emission lines are much wider than most of the other spectral profiles, in particular those from transitions of auto-ionization states, where the very short life

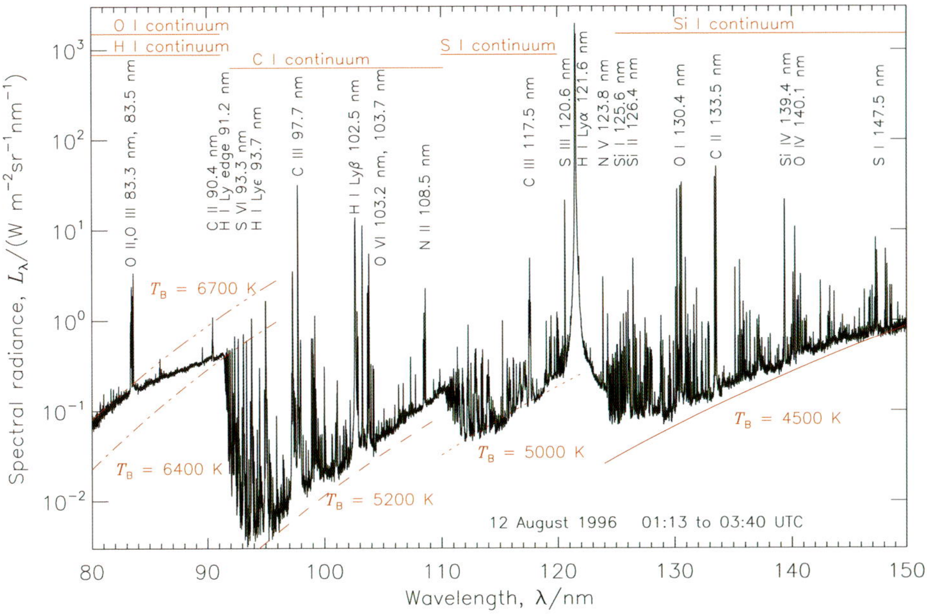

Figure 2.1: Spectral radiance of the quiet Sun in the VUV range from a region near the centre of the disk. Prominent emission lines are marked. The spectral radiances expected for some brightness temperatures, T_{B}, are shown in red as approximations of the continua in the corresponding wavelength ranges (after Wilhelm et al 2002).

times lead to a significant energy spread and a corresponding broadening (Avrett et al 2006). This is related to Heisenberg's uncertainty principle (Heisenberg 1927)

$$\Delta E \, t_{\mathrm{H}} \geq h \quad , \tag{2.5}$$

where t_{H} is the inner time of the decaying system (Aharonov and Bohm 1961), and is consistent with the derivation of Mandelstam and Tamm (1945) who defined t_{H} as the life time of a state with respect to a certain observable (cf., also Hilgevoord 1996, 1998).

The solar spectral irradiance (SSI) in the visible wavelength region can be reliably determined from a high-altitude station. This was, for instance, done in the pioneering work of Labs and Neckel (1962, 1967) from the Jungfraujoch in the Swiss Alps during 1961 and 1965. The final results from these measurements (Neckel 2003) are still an important reference and baseline for the SSI from 330 nm to 1099 nm. Reference spectra can be derived from measurements taken during distinct time periods—similar to the data from Labs and Neckel (1967) and the one compiled by Thuillier et al (2004) for two periods in 1992 and 1994, one of which is shown in Figure 2.2. Such spectra are used for many applications, e.g., in weather and climate models. In the VUV range, that is below about 300 nm, such measurements can only be performed from above the atmosphere of the Earth with space probes, satellites, or rockets. Early measurements were obtained by Hinteregger and Hall (1969). For a review see the corresponding articles in White (1977).

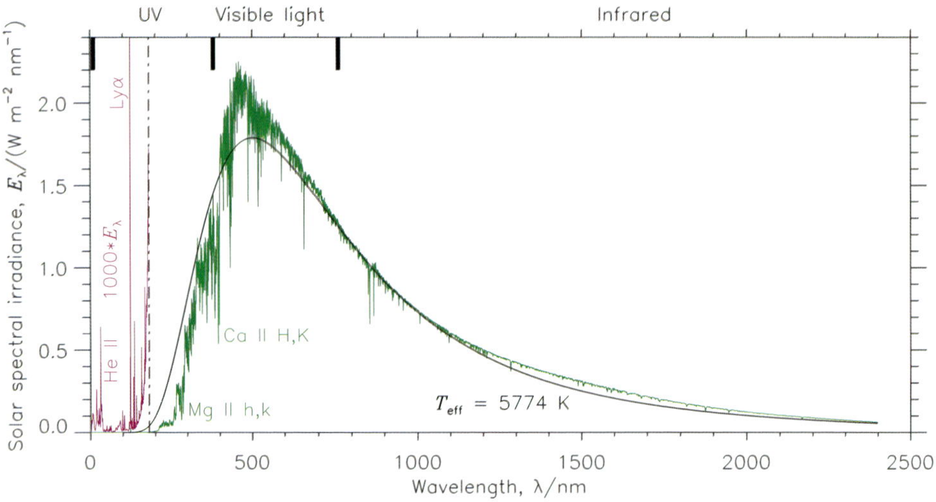

Figure 2.2: SSI, E_λ, of the quiet Sun at 1 ua (Data: Courtesy of G. Thuillier). For wavelengths below $\lambda = 180$ nm, 1000 E_λ is plotted in violet to show the low irradiance values in this range. The H I Lyα line is off the scale (cf., Figure 2.1). The ranges of the UV, visible and infrared radiations are marked on the top, and some prominent emission and absorption lines are indicated (cf., Chapters 5 to 7). The black curve shows the radiation of the Sun as a "black body" with $T_{\text{eff}} = 5774$ K and a radius $R_\odot = 696.35$ Mm. The Sun also emits soft and hard X-rays as well as γ-rays and radio waves, but the contributions to the total solar irradiance (TSI) are very small and highly variable (cf., Figure 2.3).

Continuous measurements in the ultraviolet (UV) became available with the help of satellite instrumentation in late 1970. Although the radiometric uncertainty was still quite large, the relative variability of the solar VUV irradiance could be determined, mainly because it increases substantially with decreasing wavelength. In the visible regime, however, truly continuous measurements started only with the filter radiometers in 1996 on VIRGO/*SOHO* (Fröhlich et al 1997), and with the spectroradiometer SIM in 2003 (Harder et al 2000; Rottman et al 2005)—otherwise one has to rely on measurements at distinct periods of time, such as those mentioned before. The study of the irradiance variations during the 11-year cycle, on the other hand, requires well sampled time series. An important question for an understanding of the underlying physical mechanisms is how the variation of the TSI is spectrally distributed. This is related to the question of how much of the observed irradiance variations—total and spectral—are changes in luminosity of the Sun or only an angular redistribution of radiance into the direction towards the Earth.

Figure 2.3 demonstrates with the help of a few time series the solar irradiance variability during solar cycles 21 to 23 in wavelength ranges from the visible to X-rays. The data were obtained with space experiments using filter radiometers or spectrometers for the wavelength selection. For comparison, the daily averages of the TSI are plotted in Figure 2.3a as a PMOD composite[6] updated from

[6]For more details see Chapter 32 (Fröhlich 2010).

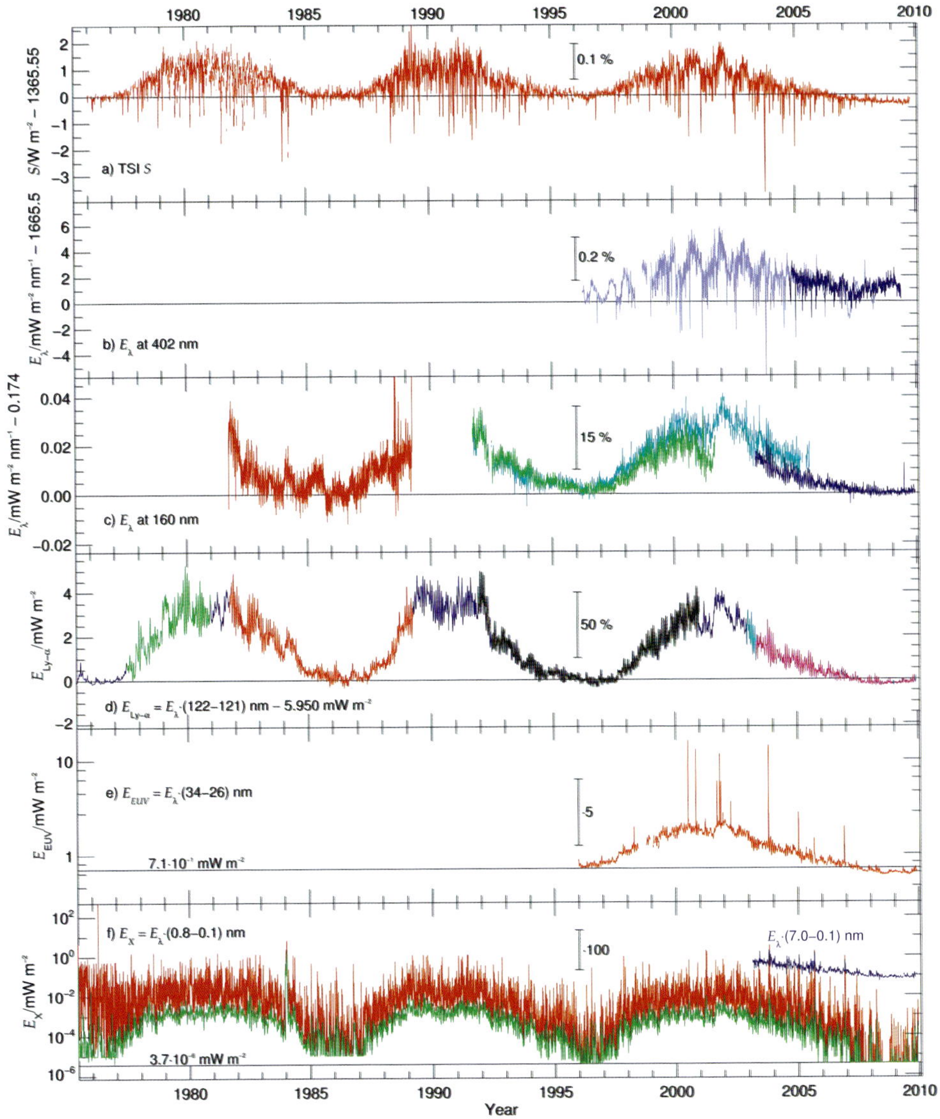

Figure 2.3: **a** TSI composite. The bars in this panel and the following ones indicate relative variations based on the mean values at solar minimum. **b** SSI results from SPM/VIRGO and SIM/*SORCE* in visible light. **c** Time series of SSI in VUV from *SME*, SOLSTICE/*UARS*, SUSIM/*UARS* and SOLSTICE/*SORCE*. **d** Irradiance composite in the range from 121 nm to 122 nm. **e** Extreme-ultraviolet (EUV) irradiance data from SEM/*SOHO* are available for solar cycle 23. The prominent increases correspond to strong flare activity on the Sun. The largest enhancement, for instance, coincides with an X17 flare on 28 October 2003. **f** The XRS/*GOES* X-ray irradiance, and XPS/*SORCE* measurements integrated over a wider range.

Fröhlich (2006). Figure 2.3b shows (in light blue) results from the 402 nm VIRGO channel (Fröhlich and Wehrli 2006) and from SIM (Rottman et al 2005; private communication J. Harder) with a similar spectral resolution (in dark blue). In Figure 2.3c time series are shown of measurements at 160 nm. The values have been adjusted so that they agree at the solar minima. The red curve is from *SME* (Rottman 1988), the green one from SOLSTICE/*UARS* (Rottman 2000), the light blue one from SUSIM (Floyd et al 1998), and the dark blue one from SOLSTICE/*SORCE* (Rottman et al 2006). Fig. 2.3d: Composite derived for the VUV range near 121 nm as compiled by Woods et al (2000) updated until recently: dark blue is from a model (before 1985 from the radio flux at 10.7 cm, later from the Mg II index), green from *AE-E* (Hinteregger et al 1981), red from *SME*, black from SOLSTICE/*UARS*, light blue from SEE (Woods and Eparvier 2006), and magenta from SOLSTICE. This range contains the very bright H I Lyα line at 121.57 nm and reflects its variations. Fig. 2.3e: Data from SEM (Judge et al 2002) integrated over the 26 nm to 34 nm wavelength range encompassing the bright He II line at 30.4 nm. Fig. 2.3f: The XRS/*GOES* X-ray irradiance from 0.1 nm to 0.8 nm (Kahler and Kreplin 1991; Garcia 1994) (green is the daily median and red the daily maxima and minima). For comparison the XPS/*SORCE* measurements are shown in blue (Woods and Rottman 2005).

The variability of the TSI (Figure 2.3a) can now be compared with the corresponding changes of the SSI in the following panels. The relative variations in the visible and near UV ranges are below 1 % during an 11-year cycle. Since it is very difficult to correct the long-term changes of the instrumentation at these wavelengths reliably, the results must still be regarded as preliminary, although the values of the two time series in panel b agree within a few percent. Note in Figure 2.3c, however, the difference between the behaviour of SUSIM (in light blue) and SOLSTICE (in green) near the year 2000, which indicates a deviation of the long-term calibrations of the instruments.

It is interesting that only the TSI is showing a minimum at the beginning of 2008 with a relative decrease of almost 0.025 % with respect to the other minima. This is about a quarter of the solar cycle amplitude. All the spectral data, however, show almost constant minimum values over the last three solar cycles with relative changes of a few percent of their cycle amplitudes. This may indicate that the solar cycle variations of TSI and SSI are mainly due to changes in the solar atmosphere related to the prevailing amplitude of the photospheric magnetic fields expanding up to the corona, and that the long-term variations of the TSI are more likely depending on changes of the structure and possibly the temperature of the Sun. Such changes do not have a strong signature in the spectral data. A change of the photospheric temperature of 1 K produces a relative TSI variation of 0.07 %, nearly comparable to the solar cycle effect, whereas it is less than 0.05 % at 1000 nm and about 0.4 % for the Lyα line seen against a relative solar cycle amplitude of more than 50 % at these wavelengths.

Photons in transit

Most of the astronomical objects of interest are far away from the Earth, and the emitted photons have to travel long distances before their detection on a spacecraft. Magnetic or electric fields, unless they are very strong, do not affect photons in vacuum, but gravitational fields influence their propagation. A ray from a star passing close to the Sun will be deflected by $\approx 2''$ (Mikhailov 1959), and gravitational lensing by galaxies or clusters of galaxies is well known for very distant objects. These effects are treated in the general relativity theory (Einstein 1916) that yields a straight propagation path of light in Euclidean space only if there are no masses. In this context, it should be mentioned that the travel time of a photon passing near the Sun increases by ≈ 100 µs (Shapiro 1964).

Strong gravitational potentials can cause significant redshifts. For instance, the relative shift of a spectral line emitted in the photosphere of the Sun observed from the Earth is

$$\frac{\Delta\lambda}{\lambda_0} = \frac{1}{c_0^2}\frac{G\,M_\odot}{R_\odot} \approx 2 \times 10^{-6} \quad , \tag{2.6}$$

where $G = 6.674\,28(67) \times 10^{-11}\ \mathrm{m^3\,kg^{-1}\,s^{-2}}$ is the Newtonian constant of gravitation, and the mass of the Sun $M_\odot = 1.989 \times 10^{30}$ kg (Einstein 1911; Schiff 1960; Hentschel 2005). The gravitational shift was confirmed in laboratory experiments through the Mössbauer effect by Pound and Rebka (1959) as well as by Cranshaw et al (1960). It was subsequently also observed in the solar spectrum by Blamont and Roddier (1961), Brault (1962, 1963) Snider (1972, 1974) and during a rocket flight to an altitude of 10 000 km (Vessot et al 1980). An accelerated system should, according to the equivalence principle (Einstein 1916), cause a shift as well. This was actually measured, also with the help of the Mössbauer effect, by Hay et al (1960).

A photon travelling in an inertial system without any interaction with matter will neither change its energy, E_ν, nor its momentum

$$m_\nu\, c_0 = p_\nu = \frac{E_\nu}{c_0} = \frac{h\,\nu}{c_0} \tag{2.7}$$

under the assumption of the validity of energy-mass and momentum conservation laws. Such changes can only occur if photons are interacting with matter in that system or when they are observed in other reference systems. The Compton effect is an example (see page 35). Another one is the Doppler effect discussed in the next section. There are other effects that affect photons in the presence of matter, e.g., the Faraday rotation of the plane of polarization caused by a magnetic field along the LOS in a plasma (cf., Hutchinson 2002), the Hanle effect discussed in Chapter 33 (Stenflo 2010), and absorption as well as scattering by dust, gas, or plasma regions. A few scattering processes will be described below, but any details are beyond the scope of this chapter.

Doppler effect and aberration

The Doppler effect was originally discovered and formulated for double-star systems (Doppler 1843; Bolzano 1843), and was later verified for sound waves

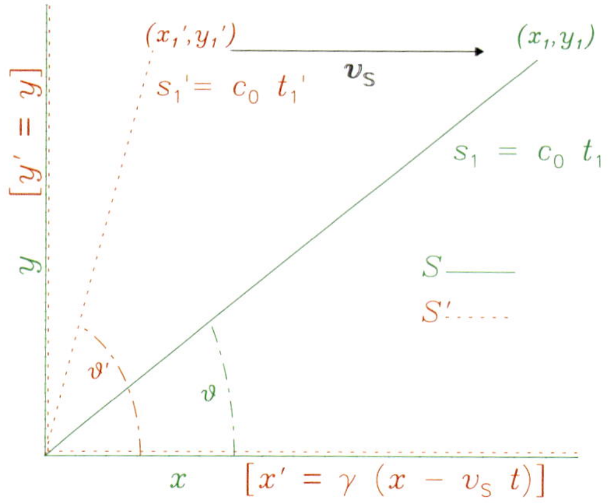

Figure 2.4: The inertial systems S (plotted in green) and S' (red) have parallel coordinate axes. S' is moving with a speed v_S along the positive x-axis of S. Both systems have a common origin at a certain time, $t_0 = 0$, when a photon is emitted there at an angle ϑ in S along a path s_1. The path s_1' in S' will then form an angle ϑ' with the x'-axis. The diagram shows the configuration for the values $\beta = 0.5$ and $\vartheta = 30°$.

propagating in a medium. In the acoustic case, it was necessary to consider the motions of the transmitter and receiver in the medium separately, because their influences are not symmetric.

For velocities v_t and v_r at angles of ϑ_t and ϑ_r with respect to the direction from the transmitter to the receiver, an emitted frequency, ν, will be received as

$$\nu' = \nu \frac{1 - (v_r/c_s) \cos\vartheta_r}{1 - (v_t/c_s) \cos\vartheta_t} \quad , \tag{2.8}$$

where c_s is the speed of sound in the medium.

For electromagnetic waves in vacuum, the special theory of relativity (STR; Einstein 1905b) ascertains that only the relative motion between the transmitter and receiver is of importance. Consider two inertial systems S and S' moving relative to each other with the velocity vector[7], $\boldsymbol{v}_S$. An emitter of photons with $E = h\,\nu$ is at rest at the origin of system S. An observer at rest in S' sees these photons with $E' = h\,\nu'$ arriving from a certain direction. The Doppler effect and the aberration (also following from the STR) answer the questions related to the frequency and direction changes of electromagnetic radiation with a speed of $c_0 = \nu\,\lambda = \nu'\,\lambda'$ in vacuum. The Doppler formula can be written for a monochromatic electromagnetic wave in its most general form (Einstein 1905b):

$$\nu' = \nu\,\gamma\,(1 - \beta\,\cos\vartheta) \quad , \tag{2.9}$$

where ϑ is the angle between the velocity vector and the radiation propagation path. (For $\beta = v_S/c_0$ and $\gamma = (1 - \beta^2)^{-1/2}$, cf., page 23.)

The aberration of light was first noticed by Bradley in 1728. The derivation of the relativistic aberration formula

$$\cos\vartheta' = \frac{\cos\vartheta - \beta}{1 - \beta\cos\vartheta} \quad , \tag{2.10}$$

[7]In equations related to Doppler shifts, there is a danger of confusing the frequency symbol, ν, with the relative speed between the inertial systems, v. Therefore, a subscript is used for v in all these cases. Note that $|v_S| < c_0$, cf., page 23.

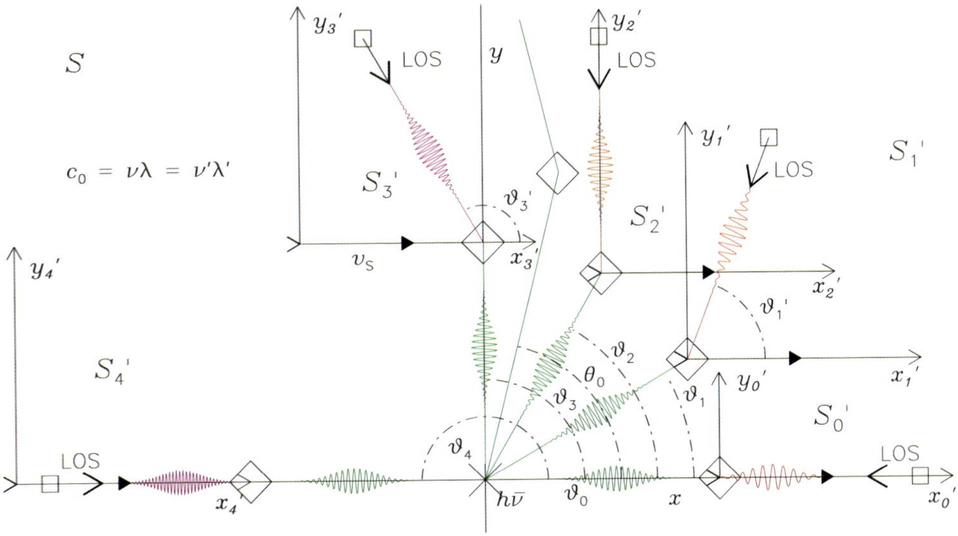

Figure 2.5: Photons are emitted from a source at rest in the inertial system S. The propagation directions from the origin are given by the angles ϑ_0 to ϑ_4. Detectors (indicated by squares) are fixed in the systems S'_k, $k = 0, \ldots, 4$, all moving with a constant velocity, v_S, relative to S parallel to the x axes. The LOS directions of the detectors are indicated by $\vartheta'_k + 180°$ ($\vartheta'_0 = 180°$; $\vartheta'_4 = 0°$). The emitted photons are sketched as green wave packets, and the Doppler-shifted ones in red and violet, respectively. The relative wavelengths and the aberrations are to scale for $\beta = 0.5$. The diamond signs represent some photon-matter interactions in the moving systems. No shift occurs for the angle θ_0 (green line).

follows from the two Lorentz transformations

$$x' = \gamma\,(x - v_S\,t) \quad \text{and} \quad t' = \gamma\,(t - \beta\,x/c_0) \quad , \tag{2.11}$$

with $\cos\vartheta = x/s_0$ and $\cos\vartheta' = x'/s_0'$ in Figure 2.4.

Doppler and aberration effects are conceptually shown for identical wave packets in Figure 2.5 for various emission angles, ϑ_n, from the origin of S, and $\beta = 0.5$ between S and S'_k. The Doppler shifts resulting from Equation 2.9 are to scale, although it must be kept in mind that they only manifest themselves during the interaction of the photons with matter in the systems S'_k.

There is no aberration for $\vartheta_0 = 0°$ and $\vartheta_4 = 180°$, but, according to Equation 2.9, a longitudinal Doppler effect of

$$\nu' = \nu\,\gamma\,(1 \mp \beta) = \nu\sqrt{\frac{1 \mp \beta}{1 \pm \beta}} \tag{2.12}$$

will be observed. On the other hand, there is no Doppler effect for an emission angle θ_0 defined by

$$\cos\theta_0 = (\gamma - 1)/(\gamma\,\beta) \quad . \tag{2.13}$$

However, θ_0 changes to $\theta'_0 = 180° - \theta_0$ in the moving system (cf., green line in Figure 2.5). Hovsepyan (1998) demonstrated these relations in a three-dimensional treatment. According to the STR, a corresponding diagram with stationary S_k' (relative to some laboratory system, for instance) and a moving S (with $-v_S$) would give the same results[8].

Combining Equations 2.9 and 2.10 gives the inverse relativistic Doppler formula

$$\nu = \nu' \gamma \, (1 + \beta \cos \vartheta') \quad . \tag{2.14}$$

With $\vartheta' = 90°$, we obtain the transverse Doppler effect

$$\nu' = \nu / \gamma \tag{2.15}$$

in line with the time dilatation, $t = t'/\gamma$; i.e., a clock in S moving with respect to the observer in S' appears to be delayed. Note that from $\cos \vartheta' = 0$ in Equation 2.10 it follows that $\cos \vartheta = \beta$, and Equation 2.9 reduces to Equation 2.15. The name "transverse Doppler effect" is a little misleading, because the relativistic effect does not depend on the angles ϑ and ϑ', and was, in fact, first observed with canal rays moving in both directions along the LOS (Ives and Stilwell 1938, 1941), and later with two-photon spectroscopy in a similar geometry (Kaivola et al 1985).

A detailed discussion of Doppler measurements from binary stars is given by Kopeikin and Ozernoy (1999). Since Einstein derived the relativistic aberration formula in 1905, an endless discussion on its validity with respect to observations of double stars pursued (e.g., Ives 1950; Eisner 1967); recently summarized and emphasized by Puccini and Selleri (2002). To the confusion might have contributed that Einstein (1905b) defined the angle ϑ' (ϕ' in the original) wrongly, although this is obvious from the discussion that follows in Einstein's paper. Many authors have concluded since that the double-star observations are in conflict with the theory of special relativity. These conclusions are, however, based on a misunderstanding of the physical situation.

Let us consider, without restricting the generality, the simplest configuration: a binary system with equal masses of the components in an orbit coplanar to a circular Earth's orbit. The Sun and the centre of gravity (CG) of the binary system are assumed to be stationary with respect to an inertial system, S'. For observations with largest aberration effects, we then have the requirement that the relevant photons must travel perpendicular to the velocity vectors of the stars and the Earth along the direction from the CG to the Sun, i.e., $\cos \vartheta' = 0$. Yet this does not imply that the light rays are emitted and received at 90° relative to the velocity vectors; a fact that has, of course, been realized by all authors since Bradley for the Earth, but not by all of them for the stars; and there lies the problem.

From Equation 2.10 it follows for the angle of emission

$$\cos \vartheta = \frac{v_\star}{c_0} \quad , \tag{2.16}$$

where $v_\star$ is the orbital velocity of the star considered. For the angle of reception at the Earth, we get

$$\cos \vartheta'' = -\frac{v_E}{c_0} \quad , \tag{2.17}$$

[8]This situation was considered by Fermi (1932) for the non-relativistic case.

with v_E the orbital velocity of the Earth. Although the angle ϑ changes with the reversal of $v_\star$, this does not affect ϑ'', in accordance with the observations.

One could argue that the separation into two aberration effects just performed is a trick to save the relativistic aberration formula. It has to be noted, however, that Puccini and Selleri (2002) (and others) also used a trick in their argument by assuming that $v_\star$ and v_E are the consequences of pure translations. If they were, the stars would indeed be seen from the Earth under

$$\cos\vartheta' = -\frac{v_S}{c_0} = -\frac{v_E + v_\star}{c_0} \tag{2.18}$$

(cf., S_3' in Figure 2.5). The second trick is wrong, because the transit times of the photons from the stars to the Earth are much longer than the orbital period of the binary system, whereas the first trick treats the emission locally, and the approximation of $v_\star$ and v_E as translations is nearly perfect.

Photon scattering

In many cases, the propagation directions of photons interacting with matter are changed in some chaotic manner. This process is called scattering, and does not include effects such as reflection, refraction, etc. Usually a distinction is made between coherent and incoherent scattering processes.

Thomson and Rayleigh scattering

In these cases, the scattered photons have the same energy as the original ones, although other parameters (polarization, flux, etc.) might be affected. Since no energy is transferred, this process is also called elastic scattering. The most important one is scattering on small particles. It is called Thomson scattering for electrons and Rayleigh scattering for other particles small compared to the wavelength. The above formulation assumes that the particles are at rest in the photon-emitting frame. If they are not, the normal Doppler effect (cf., page 31) will occur, but one would still call the process Rayleigh scattering unless there were some additional energy changes. If scattering on larger particles is considered, it is called Mie scattering and exhibits a preferred forward direction.

Compton effect

Photons can interact with electrons through an effect discovered 1922 by Compton (1923). For electrons in matter this interaction is most efficient at energies E_ν between ≈ 10 keV and 1 MeV, for which many electrons in all materials can be considered to be "free" (cf., Chapter 11, Schönfelder and Kanbach 2010). They take up a portion of the energy and momentum of the impinging photon with wavelength λ_0 and cause an incoherent scattering with a wavelength change according to

$$\Delta\lambda = \lambda_1 - \lambda_0 = \frac{h}{m_e\, c_0}\,(1 - \cos\vartheta) = \frac{2\,h}{m_e\, c_0}\,\sin^2\frac{\vartheta}{2} \geq 0 \quad , \tag{2.19}$$

where $m_e = 9.109\,382\,15(45) \times 10^{-31}$ kg is the rest mass of an electron, λ_1 the new wavelength, and ϑ the deflection angle. The wavelength change, $\Delta\lambda$, does not depend on the photon energy and is of the order of the electron Compton wavelength,

$$\lambda_{\mathrm{C}} = \frac{h}{m_{\mathrm{e}}\, c_0} = 2.426\,310\,2175(33)\ \mathrm{pm} \quad . \tag{2.20}$$

As a note, it should be mentioned that relativistic electrons can scatter on low-energy photons to produce high-energy ones through the inverse Compton effect with $\Delta\lambda < 0$ (cf., Chapters 3 and 4, Kanbach et al 2010; Culhane 2010).

Raman effect

Based on an attempt by Darwin (1923) to explain the optical dispersion phenomenon with quantum physical effects, Smekal (1923) predicted *Translationsquantenübergänge* (translatory quantum transitions) induced by mono-energetic radiation in media with a quasi-periodic atomic structure. The interaction of the radiation with matter would then lead to both energy decreases and increases. These changes were experimentally found by Raman (1928) in many substances. This so-called Raman effect can be observed as radiation scattered with

$$\nu_{\mathrm{R}} = \nu_{\mathrm{o}} \pm \nu_{\mathrm{s}} \quad , \tag{2.21}$$

where ν_{o} is the irradiating frequency and ν_{s} is related to the internal oscillations of the medium. Note that, indeed, there are frequency increases as well as decreases.

Fluorescence

In this context, fluorescence might also be mentioned. Here an atom is excited by a high-energy photon and returns to the ground state in several transitions while emitting lower-energy photons. In a unified view, fluorescent and Raman scattering both represent atomic or molecular scattering, for which the initial and final states are different. If such scattering occurs near a resonance, then we speak of fluorescent scattering, while if it is far from a resonance (in the distant line wings), we speak of Raman scattering.

Physical principles of photon detection

The detection process aims at a quantitative determination of the photon flux as a function of the photon energy, its direction, the time of arrival, as well as the polarization characteristics. As for the generation processes, the interactions of photons with matter play a fundamental rôle for their detection.

For the application of the Doppler effect to specific detector designs, we assume that the mass of the receiving device is very much larger than the mass of a photon[9]. Three extreme cases can then be singled out that are of importance in our context:

[9] The details of the transfer of the momentum from the absorbing or emitting systems—usually atomic particles—to the detector device require quantum-mechanical considerations beyond the scope of this chapter (cf., Dicke 1953; Mössbauer 1958).

1. The absorption of a photon by a detector (in general, moving with respect to the radiation source) can be considered as a completely inelastic collision. The total momentum of the detector-photon system will be conserved as well as its energy. Only a certain fraction of the photon energy can be converted into kinetic energy of the detector system in this collision. The remaining portion will be absorbed in other forms of energy. It can be shown that the frequency change of the ensuing Doppler effect just corresponds to the positive or negative kinetic energy transferred to the detector system (cf., e.g., Sommerfeld 1978).

2. The interaction of a photon with a (perfect) mirror as the first element of an optical system represents an elastic collision. This case has been treated by Einstein (1905b) with the result that either the normal Doppler effect is observed, if the photon is subsequently absorbed in the optical system attached to the first element, which will be the usual situation, or, if the photon is observed in the initial system, the Doppler effect can be considered to be applied twice. A mirror receding with v_S in normal incidence will thus cause a frequency change[10] from ν to ν' according to

$$\nu' = \nu \left(\sqrt{\frac{1-\beta}{1+\beta}} \right)^2 = \nu \, \frac{1-\beta}{1+\beta} \quad . \tag{2.22}$$

3. If the first optical element of a photon analyzer is operating in transmission, the momentum and energy transfer required by the Doppler effect will occur at the first encounter with the device. Internally, the Doppler-shifted photon is travelling in the rest frame of the device. Should the final detector, however, be at rest in the initial system, the first Doppler effect would be cancelled by the second one during the absorption.

Irradiance measurements

Conceptually the easiest observation of radiation in space is the determination of the TSI requiring a measurement of the power absorbed by a representation of a black body with known geometry and a relatively crude alignment. The experimental challenge stems from the extreme accuracy required for this measurement, with a relative uncertainty of ideally less than 0.1 %, and even better as far as irradiance variations are concerned (cf., Figure 2.3 and Chapter 32, Fröhlich 2010).

The Sun with a spectral radiance, L_ν, emits perpendicular to a surface element, ΔF, in the frequency and time intervals, $\mathrm{d}\nu$ and Δt, into the solid angle, $\Delta \Omega$, the energy

$$\mathrm{d}Q_\nu = L_\nu \, \mathrm{d}\nu \, \Delta t \, \Delta \Omega \, \Delta F \quad . \tag{2.23}$$

If the radiative properties of the Sun are measured from a moving spacecraft, corrections have to be applied in order to find the quantities for a stationary detector at 1 ua. The normalization with respect to the variable distance is straightforward, because the irradiance varies with the inverse square of the distance, r_E. The velocity correction, however, is more involved:

[10] A typographical error in Einstein's equation has been corrected in Equation 2.22.

Since the speeds relative to the Sun are lower than $v_S \approx 30\ \mathrm{km\,s^{-1}}$, relativistic effects can be neglected. Thus only the radial velocity component, v_r (positive away from the Sun), and linear terms in $\beta_r = v_r/c_0$ are of importance. The intervals Δt, $\Delta\Omega$, as well as ΔF are not dependent on v_S in this case.

Equation 2.23 can be written in photon units as

$$\mathrm{d}N_\nu = \mathrm{d}Q_\nu/(h\,\nu) = [L_\nu/(h\,\nu)]\,\mathrm{d}\nu\,\Delta t\,\Delta\Omega\,\Delta F \quad , \tag{2.24}$$

where $\mathrm{d}N_\nu$ is the number of photons in the interval $\mathrm{d}\nu\,\Delta t\,\Delta\Omega\,\Delta F$. The corresponding emission rate is $\mathrm{d}N_\nu/\Delta t$. If it is assumed that the detector collects all photons emitted into $\Delta\Omega$, the detection rate in the moving system will be

$$\frac{\mathrm{d}N'_{\nu'}}{\Delta t} = (1-\beta_r)\,\frac{\mathrm{d}N_\nu}{\Delta t} \quad , \tag{2.25}$$

where the factor $1-\beta_r$ reflects the geometric correction. Conversion back to energy units yields

$$\mathrm{d}Q'_{\nu'} = (1-\beta_r)\,\frac{\nu'}{\nu}\,\mathrm{d}Q_\nu \tag{2.26}$$

and, with the (classical) Doppler effect (Equation 2.9; $\gamma = 1$),

$$\mathrm{d}Q_\nu = \frac{\mathrm{d}Q'_{\nu'}}{(1-\beta_r)^2} \approx \mathrm{d}Q'_{\nu'}\,(1+2\,\beta_r) \quad , \tag{2.27}$$

which, after integration over all frequencies, provides the TSI velocity correction used in Chapter 32 (Fröhlich 2010). Similarly, irradiance measurements of stars and other cosmic objects can be performed and corrected.

Spectral solar measurements (radiance and Sun-as-a-star observations; see Figures 2.1 to 2.3) are more complicated, because optical and dispersive systems as well as detectors over a wide spectral range are required with calibrated radiometric responsivity. Moreover, potential variations of the spectral parameters have to be considered, caused, for instance, by relative motions between the emitter and observer, requiring additional corrections. From Equation 2.27, evaluated with the help of Equations 2.9 and 2.23, the correction

$$L_\nu = \frac{L'_{\nu'}}{(1-\beta_r)} \tag{2.28}$$

can be deduced for the spectral radiance and

$$L = \frac{L'}{(1-\beta_r)^2} \tag{2.29}$$

for the radiance. If Equation 2.28 is applied to the Planck law in Equation 2.3, the black-body spectral radiance correction is

$$L_\nu(T) = \frac{L'_{\nu'}(T)}{(1-\beta_r)} \quad . \tag{2.30}$$

A short calculation with the substitution $T = T'/(1-\beta_r)$ and application of the Doppler formula then gives

$$L_\nu(T) = \frac{L'_{\nu'}(T')}{(1-\beta_r)^3} \approx L'_{\nu'}(T')\,(1+3\,\beta_r) \quad . \tag{2.31}$$

Direction finding

In this section, most of the interactions of photons with matter can completely be described in terms of the classical concepts of electromagnetic radiation as long as no absorption occurs. A specific aspect of this treatment is that the photon propagation can be reversed. The relevant laws of geometric and physical optics can be found in textbooks.

Reflection

The reflection law for polished (metal, glass or similar) surfaces is

$$\alpha_\mathrm{i} = \alpha_\mathrm{r} \quad , \tag{2.32}$$

where the angle of incidence with respect to the normal is α_i and α_r is the angle of reflection. Both rays and the normal lie in one plane. The reflectivity depends on the material and the wavelength. The radiation that is not reflected is absorbed in the material or transmitted.

Dispersion and refraction

Dispersion and refraction laws are of general interest as they govern the propagation of photons in (transparent) matter. For waves, and electromagnetic waves in particular, the phase velocity is defined by

$$c = \frac{\omega}{k} \quad , \tag{2.33}$$

with the angular frequency $\omega = 2\,\pi\,\nu$ and the wave number $k = 2\,\pi/\lambda$. The group velocity is

$$v_\mathrm{g} = \frac{\mathrm{d}\omega}{\mathrm{d}k} \quad , \tag{2.34}$$

In transparent media, phase and group velocities are in general functions of the frequency of the waves—a phenomenon called dispersion.

In vacuum, however, there is no dispersion and

$$c = v_\mathrm{g} = c_0 \tag{2.35}$$

which follows from

$$c = \frac{\omega_0}{k_0} = \nu_0\,\lambda_0 = c_0 \quad \text{and} \quad v_\mathrm{g} = \frac{\mathrm{d}\omega_0}{\mathrm{d}k_0} = c_0 \quad , \tag{2.36}$$

where the index zero indicates a quantity in vacuum, and, in particular, c_0 the speed of light in vacuum as a fundamental constant (cf., pages 23 and 31). With the magnetic constant of the vacuum (also called permeability) $\mu_0 = 4\,\pi \times 10^{-7}\ \mathrm{N\,A^{-2}}$ (exact), the exact electric constant (or permittivity), ε_0, follows from

$$c_0 = \frac{1}{\sqrt{\varepsilon_0\,\mu_0}} \tag{2.37}$$

in Maxwell's equations.

The interaction of photons in a (transparent) medium with the phonons (a description of oscillations in solid materials) leads to a dependence of the propagation velocity on the frequency $\nu = \nu_0$ (which is the same outside and inside the medium). This can be formulated in the wave description as a change in wavelength:

$$\lambda = \frac{\lambda_0}{n} \quad , \tag{2.38}$$

where n, the refractive index, may depend on λ_0 (or ν_0). We then get for the phase velocity

$$c(\nu_0) = \nu_0\,\lambda = \frac{\nu_0\,\lambda_0}{n\,(\nu_0)} = \frac{c_0}{n\,(\nu_0)} \quad , \tag{2.39}$$

and the group velocity is

$$v_\mathrm{g} = \frac{\mathrm{d}\omega_0}{\mathrm{d}k} = c_0\left(n + \nu_0\frac{\mathrm{d}n}{\mathrm{d}\nu_0}\right)^{-1} = c_0\left(n - \lambda_0\frac{\mathrm{d}n}{\mathrm{d}\lambda_0}\right)^{-1} = c\left(1 + \frac{\lambda}{n}\frac{\mathrm{d}n}{\mathrm{d}\lambda}\right) \quad . \tag{2.40}$$

Note the effect of the substitution according to Equation 2.38 on the form of the equation. A dependence of v_g on ω_0 (and λ_0) will lead to a spreading of a pulse of radiation in the medium. If $\mathrm{d}n/\mathrm{d}\lambda_0 < 0$, we have normal dispersion and $v_\mathrm{g} < c$. Invoking quantum interference effects, Liu et al (2001) brought a light pulse to a complete stop.

Anomalous dispersion with $\mathrm{d}n/\mathrm{d}\lambda_0 > 0$ and $v_\mathrm{g} > c$ can occur under certain resonance conditions of the electromagnetic waves with the atomic elements of the medium. The expression $n_g = n + \nu_0\,\mathrm{d}n/\mathrm{d}\nu_0$ is sometimes called group velocity index (cf., Wang et al 2000). It can become less than one (or even negative), resulting in superluminal light pulse propagation. Under all conditions, however, the signal velocity is defined as the advancement speed of a step function smaller than c_0 (cf., Brillouin 1931), unless evanescent modes are considered (Nimtz and Stahlhofen 2008), which are not relevant in our context. Astronomical observations of propagating fronts with an apparent $v_\mathrm{g} > c_0$, so called superluminal motions, can be understood as interactions of wavefronts or jets with spatial structures at angles of nearly 90°with a LOS close to the jet direction (Brillouin 1960; Falla and Floyd 2002).

If a medium with refractive index n is moving at a speed v_S with respect to a laboratory frame, parallel or anti-parallel to the light propagation direction the phase velocity of light inside the medium (as seen from the laboratory) is

$$c' = \frac{c_0}{n} \pm (1 - \frac{1}{n^2} - \frac{\lambda_0}{n}\frac{\mathrm{d}n}{\mathrm{d}\lambda_0})\,v_\mathrm{S} \quad , \tag{2.41}$$

known as the Fizeau effect (cf., Einstein 1914).

If electromagnetic radiation travels from one transparent medium to another one, Snell's law

$$n_1 \sin\alpha_1 = n_2 \sin\alpha_2 \tag{2.42}$$

applies, where n_1 and n_2 are the indices of refraction in the media, α_1 is the angle of incidence, and α_2 is the angle of the transmitted ray[11]. In general, both the

[11]Note that certain materials exhibit a double refraction.

reflected and the refracted beams are present and partially polarized as given by Fresnel's formulas[12]

$$\phi_{\rm r}^{\perp} = \frac{\sin^2(\alpha_1 - \alpha_2)}{\sin^2(\alpha_1 + \alpha_2)}\,\phi_0^{\perp} \quad , \tag{2.43}$$

$$\phi_{\rm t}^{\perp} = \frac{\sin 2\,\alpha_1\,\sin 2\,\alpha_2}{\sin^2(\alpha_1 + \alpha_2)}\,\phi_0^{\perp} = \frac{2\,\sin\alpha_1\,\cos\alpha_1\,\sin 2\,\alpha_2}{\sin^2(\alpha_1 + \alpha_2)}\,\phi_0^{\perp} \quad , \tag{2.44}$$

$$\phi_{\rm r}^{||} = \frac{\tan^2(\alpha_1 - \alpha_2)}{\tan^2(\alpha_1 + \alpha_2)}\,\phi_0^{||} \quad , \tag{2.45}$$

and

$$\phi_{\rm t}^{||} = \frac{\sin 2\,\alpha_1\,\sin 2\,\alpha_2}{\sin^2(\alpha_1 + \alpha_2)\cos^2(\alpha_1 - \alpha_2)}\,\phi_0^{||} = \frac{2\,\sin\alpha_1\,\cos\alpha_1\,\sin 2\,\alpha_2}{\sin^2(\alpha_1 + \alpha_2)\cos^2(\alpha_1 - \alpha_2)}\,\phi_0^{||} \quad , \tag{2.46}$$

where $\phi_0^{\perp}$ is the incident flux polarized perpendicular to the plane of incidence and $\phi_0^{||}$ that polarized in the plane. $\phi_{\rm r}$ and $\phi_{\rm t}$ are the reflected and transmitted fluxes. For $\alpha_1 = \alpha_2 = 0$ some caution is required in evaluating the limits. Equation 2.43, for instance, can be reformulated with $\sin\alpha_2 = (n_1/n_2)\sin\alpha_1$; $n_2 > n_1$ as:

$$\phi_{\rm r}^{\perp} = \frac{1 - (n_1/n_2)(\cos\alpha_1/\cos\alpha_2)}{1 + (n_1/n_2)(\cos\alpha_1/\cos\alpha_2)} \quad . \tag{2.47}$$

Equation 2.45 shows that the reflected radiation is completely polarized perpendicular to the plane of incidence if $\alpha_1 + \alpha_2 = 90°$ (Brewster's law). A consequence of Snell's law in Equation 2.42 is the total reflection of radiation in a medium with n_2, if the index n_1 is smaller outside of an interface. Photons impinging under an angle α_2, such that $(n_2 \sin\alpha_2)/n_1 > 1$, cannot leave the denser medium. This effect is the basis of the design of fiber optics.

Attenuation

Radiation is attenuated in any homogeneous medium according to

$$\phi(x) = \phi_0\,\exp(-A\,x) \quad , \tag{2.48}$$

where A is the absorption coefficient, ϕ_0 is an initial flux at the position $x_0 = 0$, and x is a linear coordinate along the path of the radiation. Since we are more interested in the transmitted radiation, $\phi(x)$, we can ignore quantum effects in this case, unless single photon experiments are considered. Often the optical depth, τ, is used in this context, giving the path length along which the radiation is attenuated by a factor of e. For homogeneous media, it is $\tau = A^{-1}$.

[12]Fresnel's formulas using a complex index of refraction are given in Chapter 9 (Lemaire et al 2010).

Sine condition and Fermat's principle

For precise imaging in an optical system with a magnification factor M, Abbe's sine condition $(n_1 \sin \delta_1)/(n_2 \sin \delta_2) = M$ must be fulfilled, where δ_1 and δ_2 are the respective slope angles of a ray from the object point to the image point.

Fermat's principle provides a more general theorem by stating that the optical path length between two points is an extremum. In vacuum light propagates on geodesics; it travels, in fact, on null geodesics as the special relativity theory requires that the eigentime of a photon is zero.

Direction finding concepts

The method of finding the direction of an incoming photon depends very much on the spectral domain. For γ-rays, no optical systems can be designed, because the reflectivities and refractive indices of all substances are too low. Modulation by masks and other specialized methods, including timing, are required, which will be discussed in Chapter 12 (Hurford 2010). The same is basically true for the hard X-ray range, although so-called Laue lenses (based on Bragg diffraction) and reflective surfaces in grazing incidence have been employed as well. The reflectivity of smooth surfaces increases in grazing incidence as a consequence of Fresnel's formulas, which show in Equations 2.44 and 2.46 that the transmitted beam becomes evanescent for $\alpha_1 = 90°$ $(\cos \alpha_1 = 0)$.

Multilayer coating of the optical surface is another technique for enhancing the reflectivity in a certain wavelength band, and at the same time providing a crude spectral selection (cf., Chapter 9, Lemaire et al 2010). It relies on interferometric amplification of specific bands and attenuation of others. In the soft X-ray regime, multilayer-coated optics can also be used in normal incidence, in addition to optical elements used in grazing incidence. Here and in the VUV range the reflectivities are not very high and, therefore, as few as possible optical surfaces should be used. Silicon carbide (SiC), for instance, probably the best material for use in the VUV, has a significant reflectivity (≥ 0.3) between 60 nm and 300 nm with a maximum of 0.52 at 160 nm.

No material is available for transmission optics in this range, although aluminium has a relatively high transmission from 17 nm to $\approx$ 84 nm. The transmission limits of some materials at short wavelength are: 155 nm for quartz glass (SiO_2), 122 nm for calcium fluoride (CaF_2); 110 nm for magnesium fluoride (MgF_2); and $\approx$ 105 nm for lithium fluoride (LiF). Above 120 nm metallic mirrors (Au, Ag, Al) with protective coatings of MgF_2 can be used. Some transmission limits at long wavelength are: 4 μm for SiO_2, 14 μm for NaCl and 23 μm for KCl. Although in the near-UV, visible and near-infrared ranges transmission optics with lenses can be employed, most of the imaging systems are based on reflecting devices, which show less spectral defects and can easier be corrected for astigmatism and other geometric aberrations (cf., Chapter 9, Lemaire et al 2010).

In general, it is much more difficult to build a stigmatic instrument for a wide field of view than for a narrow field close to the optical axis. This axis is, in general, the reference for pointing the optical system in a certain direction. The spacecraft and instrument subsystems must guarantee the pointing accuracy and stability required by the science objectives. A summary of design concepts for VUV instruments and many examples are compiled by Wilhelm et al (2004). Towards longer wavelengths, reflecting telescopes of various design are also the main imaging devices. Of practical importance for the design of a telescope are the various stops: the aperture, defining the opening; the field stop, defining the field of view; and the Lyot stop, internally reducing the diffracted radiation from the aperture.

The angular resolution, δ, achievable in the diffraction-limited case is a function of both the wavelength and the aperture of the instrument, in the sense that it improves with decreasing wavelength and increasing dimensions of the aperture according to $\delta/\text{rad} \approx 0.6\,\lambda/r$, where r is the radius of the aperture. At shorter wavelengths, a higher spatial resolution, i.e., a smaller Airy disk, can thus be obtained with smaller apertures. It is clear that this dependence is related to the wave aspects of the electromagnetic radiation and the corresponding Fraunhofer diffraction effects of apertures on a wave field. Whether the diffraction-limited resolution can be achieved depends critically on the polishing quality of the mirror surface both as far as its figure and micro-roughness are concerned.[13]

Telescopes are not only required for defining the direction of an incoming photon, and providing imagery in a certain field of view, but also as photon collectors, in particular for faint sources. They operate as radiant flux concentrators. It should, however, be noted that the radiance is constant in an optical system. For high-resolution spatial, temporal, and spectral observations even the solar disk might be considered as a faint source, and the corresponding measurements would be photon-limited unless very large apertures are employed, which are not easy to implement on space missions. Since the radiance of the Sun, in contrast to the irradiance, is not dependent on the observing distance, the situation does not improve by flying closer to the Sun.

Energy discrimination

The next task, after having defined the direction of a photon, is the determination of its energy. Again the energy range, in which the observations are being performed, has a significant influence on the experimental methods to achieve this objective. In the high-energy range, the measurement has to be made in the detector systems themselves (cf., Chapter 21, Smith 2010). In the very low-energy range, i.e., in the microwave energy, discrimination will also be accomplished by the detector systems (cf., Chapter 31, Wild 2010). In the soft X-ray, UV, visible and infrared regimes, dedicated spectral devices can be employed, mostly diffraction gratings, but also prisms and filters, to perform the energy discrimination before the radiation reaches the detector (cf., Chapter 10, Lemaire 2010). The decrease of the index of refraction with increasing wavelength (normal dispersion) is utilized for the spectral analysis in prisms.

The interaction of photons with (non-transparent) material structures is governed by diffraction. In practical applications, Fraunhofer diffraction with parallel rays is distinguished from Fresnel diffraction with inclined rays. Diffraction can

[13] Telescopes usually collect radiation in the far-field region of a source, where the diffraction limit is of importance. In the near-field region, however, focusing beyond this limit is possible (cf., e.g., Lerosey et al 2007).

completely be understood in the framework of the electromagnetic wave theory. Of fundamental importance in our context is the grating formula (given here for the simple case of a Fraunhofer geometry in reflection)

$$\kappa\,\lambda = d\,(\sin\alpha + \sin\beta) \tag{2.49}$$

where κ is the order of diffraction, d the spacing of the grating, α the angle of incidence and β the angle of reflection (cf., Chapter 10, Lemaire 2010, for a more general approach). Since the spacing, d, can be varied in very wide ranges, spectral discrimination by gratings, in particular by reflection gratings, is useful in many wavelength regimes. Even natural spacings, d, given by the structure of crystals can be invoked in the Bragg diffraction:

$$\kappa\,\lambda = 2\,d\,\sin\theta \quad , \tag{2.50}$$

where θ is the glancing angle.

In order to avoid many reflecting surfaces in VUV application, gratings are often designed in such a way that they can perform, in addition, the functions of optical elements, such as concave mirrors. Filters and prisms obviously are operating in transmission and thus are restricted in their applications.

In most cases, grating and prism devices are designed as slit spectrographs. They provide a good spectral resolution in a wide range, but use the photons collected by the telescope rather inefficiently. In particular, the imaging of extended areas in spatially-resolved observations requires a scanning motion of the slit perpendicular to its length, which is very time consuming. Slitless spectrographs are more effective in this context, but can only be used for strong spectral lines in an otherwise relatively flat spectral regime, because spatial and spectral regimes are superimposed. Two-dimensional spectroscopy employing interferometers is the choice for fast instruments operating in a restricted wavelength range (cf., Chapters 16 to 19, Lindegren 2010; Quirrenbach 2010; Griffin and Ade 2010; Title 2010). As long as transmission optics are available, such systems are relatively easy to implement, but in the VUV range the design and development of interferometers are challenging tasks.

For all systems depending on interference methods, the coherency of the radiation under study is of critical importance. The standard coherence condition for Young's double slit configuration is $2\,d\,\sin\epsilon \ll \lambda$, where d is the separation of the slits, ϵ is the half-opening angle of the beam and λ is the wavelength of the radiation (Verdet's coherence condition). The same condition is applicable to slit spectrometer using gratings, where d is then equal to the slit width. If, however, the spectrometer provides a stigmatic image of the slit on the detector, the optical performance of the spectrometer has to be considered in detail. It is of interest here that the time of arrival of photons is correlated in coherent radiation (Hanbury Brown and Twiss 1956a). This can be used in an intensity interferometer to measure radii of stars (Hanbury Brown and Twiss 1956b). These observations had a great influence on the development of the quantum theory of optical coherence (Glauber 1963). It is also noteworthy that interference of radiation from two lasers has been observed (Laradour et al 1993).

Photon detectors

Modern detectors mounted on spacecraft systems have to transform, in some way, a photon impact into digital information that can be recorded and telemetered to the ground. Most of them are two-dimensional devices composed of many pixels, i.e., picture elements, of a certain size. Pixels can be geometrically defined, but virtual pixels generated by the readout algorithm are also possible. This pixel size has to be adjusted to the capabilities of the telescope and spectrometer, and then determines the resolution in both dimensions corresponding in general, to the spatial and spectral domains. Resolving powers are usually defined for both regimes by $p_{\mathrm{x}} = x/\delta x$ and $p_{\lambda} = \lambda/\delta\lambda$, respectively. Equating δx and $\delta\lambda$ with the pixel size and considering the sampling theorem would indicate a spatial resolution of $\approx 2\,\delta x$ and a spectral of $\approx 2\,\delta\lambda$. It is of great importance for the analysis of spectra that the shift and broadening of spectral lines can be determined with sub-pixel resolution, provided the line width is wider than several pixels. The details depend on the signal-to-noise ratio, but typically the shift can be measured to one tenth of a pixel (cf., Wilhelm et al 1995). An important distinction between types of detectors is whether they are photon-counting devices or integrating systems.

Quantum efficiency

A critical performance measure of any photon detector is its quantum efficiency (QE) as a function of wavelength, $\eta(\lambda)$. The definition of this quantity in the literature varies; here we use the ratio:

$$\eta = \frac{C}{N} \tag{2.51}$$

where C is the number of counted events and N the number of incoming photons. The efficiency is in general a strong function of the photon energy, but may depend on other parameters as well. In general, the QE has to be determined by applying a radiometrically-calibrated photon flux and comparing it with the output counts (cf., Chapter 36, Huber et al 2010). However, Rarity et al (1987) performed QE measurements without a calibration standard. They used the parametric down conversion of a photon with wavelength λ_0 into two photons with λ_1 and λ_2 in a non-linear crystal. Two detectors properly placed would count the down-converted photons as $N_1 = \eta(\lambda_1)\,P$ and $N_2 = \eta(\lambda_2)\,P$, where P is number of photon pairs created. The coincidence counts would be

$$C = \eta(\lambda_1)\,\eta(\lambda_2)\,P \tag{2.52}$$

and thus $\eta(\lambda_1) = C/N_2$ and $\eta(\lambda_2) = C/N_1$.

Photon absorption

A brief survey of the primary interaction processes is given here, all of which are based on the absorption (or rather the demolition) of the photon to be detected.[14]

[14] For completeness it should be mentioned that non-demolition measurements of single photons were accomplished (Nogues et al 1999), as well as electric field measurements at sub-photon levels (Brune et al 1994; Foster et al 2002).

The absorption is a strong function of the photon energy and, of course, heavily depends on the detector material under consideration. It is clearly related to the attenuation of radiation treated on page 41, but here we are more interested in the effects of the individual photon absorption event. Once a photon interacts with the detector system, various amplification and measuring schemes have to be employed as detailed in the other chapters of this book.

Photo effect The photo effect, or more exactly, the external photo-electric effect, is one of the most important interaction processes of photons with opaque matter in photon detectors. The effect was discovered by Hertz (1887) and Hallwachs (1888), studied in detail by Lenard (1902), explained by Einstein (1905a) and experimentally confirmed by Millikan (1916). A photon with energy $h\,\nu$, impinging on the surface of a material, can release an electron with a maximal energy of

$$W_{\rm max} = h\,\nu - W_{\rm A} \quad , \tag{2.53}$$

where $W_{\rm A}$ is an electron work function characteristic of a given material[15]. Typical values are between $W_{\rm A} \approx 2$ eV and 6 eV.

Pair Production At energies above $E_\nu = 1.02$ MeV, i.e., twice the equivalent rest mass of an electron, $m_{\rm e}$, a photon can produce an electron-positron pair in the electric field of a nucleus. The nucleus is required for momentum conservation. The pair production efficiency increases with the photon energy and the interaction cross-section is proportional to Z^2, with the Z charge number of the nucleus. The electron-positron pairs then have to be recorded by appropriate means (cf., Chapter 11, Schönfelder and Kanbach 2010). Pair production also occurs in the Coulomb field of electrons albeit with much smaller cross-sections.

Solid-state detectors and scintillation counters In solid-state detectors, with donor and acceptor regions forming n-p junctions, the internal photo effect creates electrons and electron holes that can be collected and counted (cf., e.g., Chapters 25 and 26, Schühle 2010; Schühle and Hochedez 2010).

In the X-ray and γ-ray regimes, scintillation detectors are in frequent use (cf., Chapter 21, Smith 2010). The high-energy photons, through the internal photo effect, ionize atoms in certain crystallic materials. The recombination events of the atoms and electrons then produce a large number of optical photons that can be collected by standard photomultipliers. Na J and Cs J crystals have frequently been employed in the past. Modern detectors work with $Gd_2\,Si\,O_5$ (GSO) and $Bi_4\,Ge_3\,O_{12}$ (BGO) crystals, which are not hygroscopic and have different rise and decay times facilitating pulse discrimination (cf., e.g., Takahashi et al 2007).

Photon detectors based on superconductivity Many modern photon detectors employ superconductive material. Although this chapter cannot discuss the physics of superconductivity in any detail, a few important concepts will be mentioned as well as their application in photon detectors.

[15]Note that time-delayed photo effects have been observed for semi-transparent photocathodes that are not in line with Equation 2.53 (Billard and Burns 1983).

Cooper pairs Cooper (1956) suggested that electrons (as fermions) in superconducting metallic materials form pairs—later to be called Cooper pairs—due to an electron-phonon interaction at low temperatures, which could overcome the Coulomb repulsion of the electrons. The resulting bosons would be responsible for the superconductive properties, but might easily split into two electrons by an external excitation. Even photons with an energy of a few millielectronvolts can break up a Cooper pair into the electrons, which then have to be collected by suitable circuitry. This could, for instance, be a superconducting tunnel junction (STJ); see Perryman et al (1993), Peacock et al (1996), and also Chapter 27 (Martin and Verhoeve 2010).

Microcalorimeters Superconducting devices can also be employed to measure the heat input of single photons from the infrared to γ-ray ranges. Typical detectors are designed as transition edge sensors (TES) (cf., Chapters 28 and 29, Porter 2010; Eaton 2010). The absorbed photon energy is thermalized and leads to a large increase in resistance of the detector material that is biased close to the superconducting-normal transition edge. Details on the energy resolution and dead times achieved can be found in Irwin et al (1998) and Romestain et al (2004).

Heterodyne receivers In the sub-millimetre regime and at longer wavelengths, the photon concept is only of theoretical interest. The receivers and detectors are designed for the analysis of long-wavelength electromagnetic radiation employing heterodyne techniques (cf., Chapter 31, Wild 2010). A low noise temperature is of paramount importance in such systems. Superconductor-insulator-superconductor (SIS) tunnel junctions with their low band gaps are useful in this context.

Closing summary

The interaction of photons with matter during both the emission and the absorption processes is considered in this chapter as an introduction to the following chapters. The physics involved in the transfer phase of the photons from the astronomical source region to their detection near the Earth is outlined as well. The radiation from the Sun is discussed in some detail as an example. The physics of entangled photons, on the other hand, is only briefly mentioned, as applications to astrophysical observations are rather limited at this stage.

Bibliography

Adenier G (2008) Quantum entanglement, fair sampling, and reality: Is the moon there when nobody looks? Am J Phys 76:147–152

Aharonov Y, Bohm D (1961) Time in the quantum theory and the uncertainty relation for time and energy. Phys Rev 122:1649–1658

Amsler C, Doser M, Antonelli M (plus 170 authors)(2008) Review of particle physics. Phys Lett B 667:1–1340

Avrett EH, Kurucz RL, Loeser R (2006) Identification of the broad solar emission features near 117 nm. Astron Astrophys 452:651–655
Ayres TR (2000) The *SOHO*-stellar connection. Sol Phys 193:273–297
Bell JS (1964) On the Einstein–Podolsky–Rosen paradox. Physics 1:195–200
Billard TC, Burns G (1983) Time-delayed photoelectric effect. Nature 306:247–248
Blamont JE, Roddier F (1961) Precise observation of the profile of the Fraunhofer strontium resonance line. Evidence for the gravitational redshift on the Sun. Phys Rev Lett 7:437–440
Bolzano B (1843) Ein paar Bemerkungen über eine neue Theorie in Herrn Professor Chr. Doppler's Schrift. Ann Phys (Leipzig) 136:83–88
Born M, Wolf E (1999) Principles of optics. Electromagnetic theory of propagation, interference and diffraction of light (7th edition). Cambridge University Press, Cambridge New York
Born M, Heisenberg W, Jordan P (1926) Zur Quantenmechanik. II. Z Phys 35:557–615
Bose SN (1924) Plancks Gesetz und Lichtquantenhypothese. Z Phys 26:178–181
Brault JW (1962) The gravitational redshift in the solar spectrum. PhD Diss, Princeton University
Brault J (1963) Gravitational redshift of solar lines. Bull Am Phys Soc 8:28
Brillouin L (1931) Die Quantenstatistik. Verlag von Julius Springer, Berlin
Brillouin L (1960) Wave propagation group velocity. Academic Press Inc, New York, London
Brugel EW, Shull JM, Seab CG (1982) The ultraviolet spectrum of Herbig-Haro object 2H. Astrophys J 262:L35–L39
Brune M, Nussenzveig P, Schmidt-Kahler F (plus four authors) (1994) From Lamb shift to light shift: Vacuum and subphoton cavity fields measured by atomic phase sensitive detection. Phys Rev Lett 72:3339–3342
Bureau International des Poids et Mesures (BIPM) (2006) Le Système International d'Unités (SI). 8e édition, Sèvres
Carretti E, Rosset C (2010) Polarization measurements of the Cosmic Microwave Background. ISSI SR-009:573–582
Casimir HBG (1949) Sur les forces van der Waals–London. J Chim Phys 46:407–410
Clauser JF, Horn MA, Shimony A, Holt RA (1969) Proposed experiment to test local hidden-variable theories. Phys Rev Lett 23:880–884
Compton AH (1923) A quantum theory of the scattering of X-rays by light elements. Phys Rev 21:483–502
Cooper LN (1956) Bound electron pairs in degenerate Fermi gas. Phys Rev 104:1189–1190
Cranshaw TE, Schiffer JP, Whitehead AB (1960) Measurement of the gravitational red shift using the Mössbauer effect in Fe^{57}. Phys Rev Lett 4:163–164
Culhane JL (2010) X-ray astronomy: energies from 0.1 keV to 100 keV. ISSI SR-009:73–88
Darwin CG (1923) The wave theory and the quantum theory. Nature 111:771–773
Derishev EV, Aharonian FA, Kocharovsky VV (2007) Off-axis emission from relativistic plasma flows. Astrophys J 655:980–988
Dicke RH (1953) The effect of collisions upon the Doppler width of spectral lines. Phys Rev 89:472–473

Dicke RH, Peebles PJE, Roll PG, Wilkinson DT (1965) Cosmic black-body radiation. Astrophys J 142:414–419

Dirac PAM (1927) The quantum theory of the emission and absorption of radiation. Proc Roy Soc London A 114:243–265

Doppler CA (1843) Über das farbige Licht der Doppelsterne und einiger anderer Gestirne des Himmels. Abh königl böhm Ges Wiss 2:465–482

Eaton HAC (2010) Infrared imaging bolometers. ISSI SR-009:477–483

Ehrenfest P (1925) Energieschwankungen im Strahlungsfeld oder Kristallgitter bei Superposition quantisierter Eigenschwingungen. Z Phys 34:362–373

Einstein A (1905a) Über einen die Erzeugung und Verwandlung des Lichtes betreffenden heuristischen Gesichtspunkt. Ann Phys (Leipzig) 322:132–148

Einstein A (1905b) Zur Elektrodynamik bewegter Körper. Ann Phys (Leipzig) 322:891–921

Einstein A (1905c) Ist die Trägheit eines Körpers von seinem Energieinhalt abhängig? Ann Phys (Leipzig) 323:639–641

Einstein A (1911) Über den Einfluß der Schwerkraft auf die Ausbreitung des Lichtes. Ann Phys (Leipzig) 340:898–908

Einstein A (1914) Bemerkungen zu Paul Harzers Abhandlung "Über die Mitführung des Lichtes im Glas und die Aberration". Astron Nachr 1999:7–10

Einstein A (1916) Die Grundlage der allgemeinen Relativitätstheorie. Ann Phys (Leipzig) 354:769–822

Einstein A (1917) Zur Quantentheorie der Strahlung. Phys Z 18:121–128

Einstein A, Stern O (1913) Einige Argumente für die Annahme einer molekularen Agitation beim absoluten Nullpunkt. Ann Phys (Leipzig) 345:551–560

Einstein A, Podolsky B, Rosen N (1935) Can quantum-mechanical description of physical reality be considered complete? Phys Rev 47:777–780

Eisner E (1967) Aberration of light from binary stars—a paradox? Am J Phys 35:817–819

Falla DF, Floyd MJ (2002) Superluminal motion in astronomy. Eur J Phys 23:69–81

Fermi E (1932) Quantum theory of radiation. Rev Mod Phys 4:87–132

Floyd L, Reiser P, Crane P (plus three authors) (1998) UV measurements from SUSIM on *UARS*. Sol Phys 177:79–86

Foster GT, Smith WP, Reiner JE, Orozco LA (2002) Time-dependent electric field fluctuations at the subphoton level. Phys Rev A 66:033807-1–12

Fröhlich C (2006) Solar irradiance variability since 1978: Revision of the PMOD composite during solar cycle 21. Space Sci Rev 125:53–65

Fröhlich C, Wehrli C (2006) Comparison of the WRC85 solar spectral irradiance with RSSV1 and the SPM of VIRGO/*SOHO*. In: *SORCE* Science Meeting, 19-22 September 2006, San Juan Islands, Washington, USA. Poster available from ftp://ftp.pmodwrc.ch/pub/Claus/SORCE-2006/SSI_poster.pdf

Fröhlich C, Crommelynck D, Wehrli C (plus seven authors) (1997) In-flight performances of VIRGO solar irradiance instruments on *SOHO*. Sol Phys 175:267–286

Fröhlich C, Huber MCE, Solanki S, von Steiger R (eds) (1998) Solar composition and its evolution – From core to corona, Space Sci Ser ISSI, Vol 5. Kluwer Academic Publishers, Dordrecht, and Space Sci Rev 85, Nos. 1-2, 1998

Fröhlich C (2010) Solar radiometry. ISSI SR-009:525–540

Garcia HA (1994) Temperature and emission measure from *GOES* soft X-ray measurements. Sol Phys 154:275–308
Glauber RJ (1963) The quantum theory of optical coherence. Phys Rev 130:2529–2539
Glauber RJ (2007) Quantum theory of optical coherence (selected papers and lectures). Wiley-VCH Verlag, Weinheim
Goldhaber AS, Nieto MM (1971) Terrestrial and extraterrestrial limits on the photon mass. Rev Mod Phys 43:277–296
Griffin MJ, Ade PAR (2010) Narrow-band imaging by use of interferometers. ISSI SR-009:313–326
Hajdas W, Suarez-Garcia E (2010) Polarimetry at high energies. ISSI SR-009:559–571
Hallwachs W (1888) Über den Einfluss des Lichtes auf elektrostatisch geladene Körper. Ann Phys Chem 269:301–312
Hanbury Brown R, Twiss RQ (1956a) Correlation between photons in two coherent beams of light. Nature 177:27–29
Hanbury Brown R, Twiss RQ (1956b) A test of a new type of stellar interferometer on Sirius. Nature 178:1046–1048
Harder J, Lawrence GM, Rottman G, Woods T (2000) Solar spectral irradiance monitor (SIM). Metrologia 37:415–422
Hay HJ, Schiffer JP, Cranshaw TE, Egelstaff PA (1960) Measurement of the red shift in an accelerated system using the Mössbauer effect in Fe^{57}. Phys Rev Lett 4:165–166
Heisenberg W (1927) Über den anschaulichen Inhalt der quantentheoretischen Kinematik und Mechanik. Z Phys 43:172–178
Hentschel K (2005) Testing relativity. In: Physics before and after Einstein (M. Mamone Capria ed.). IOS Press, Amsterdam pp. 163–182
Hertz H (1887) Über einen Einfluss des ultravioletten Lichtes auf die elektrische Entladung. Wied Ann 31:983–1000
Hilgevoord J (1996) The uncertainty principle for energy and time. Am J Phys 64:1451–1456
Hilgevoord J (1998) The uncertainty principle for energy and time. II. Am J Phys 66:396–402
Hinteregger HE, Hall LA (1969) Solar extreme ultraviolet emissions in the range 260 to 1300 Å observed from *OSO-III*. Sol Phys 6:175–182,
Hinteregger HE, Fukui K, Gilson BR (1981) Observational, reference and model data on solar EUV, from measurements on *AE-E*. Geophys Res Lett 8:1147–1150
Hong CK, Ou ZY, Mandel L (1987) Measurement of subpicosecond time intervals between two photons by interference. Phys Rev Lett 59: 2044–2046
Hovsepyan YI (1998) Some notes on the relativistic Doppler effect. Phys Uspekhi 41:941–944
Huber MCE, Pauluhn A, Timothy JGT, Zehnder A (2010) Calibration. ISSI SR-009:585–592
Hurford GJ (2010) X-Ray imaging with collimators, masks and grids. ISSI SR-009:223–234
Hutchinson IH (2002) Principles of plasma diagnostics (2nd edition). Cambridge University Press, Cambridge New York Melbourne Madrid Cape Town

Irwin KD, Hilton GC, Wollman DA, Martinis JM (1998) Thermal-response time of superconducting transition-edge microcalorimeters. J Appl Phys 83:3978–3985
Ives HE (1950) Extrapolation from the Michelson-Morley experiment. J Opt Soc Am 40:185–191
Ives HE, Stilwell GR (1938) An experimental study of the rate of a moving atomic clock. J Opt Soc Am 28:215–226
Ives HE, Stilwell GR (1941) An experimental study of the rate of a moving atomic clock. II. J Opt Soc Am 31:369–374
Jackson JD (1999) Classical electrodynamic (3rd edition). John Wiley & Sons, New York etc.
Judge DL, Ogawa HS, McMullin DR (plus two authors) (2002) The *SOHO* CELIAS/SEM EUV database from SC23 minimum to the present. Adv Space Res 29:1963–1968
Kahler SW, Kreplin RW (1991) The NRL *SOLRAD* X-ray detectors – A summary of the observations and a comparison with the SMS/*GOES* detectors. Sol Phys 133:371–384
Kaivola M, Poulsen O, Riis E, Lee SA (1985) Measurements of the relativistic Doppler shift in neon. Phys Rev Lett 54:255–258
Kanbach G, Schönfelder V, Zehnder A (2010) High-energy astrophysics – energies above 100 keV. ISSI SR-009:57–71
Kopeikin SM, Ozernoy LM (1999) Post-Newtonian theory for precision Doppler measurements of binary star orbits. Astrophys J 523:771–785
Labs D, Neckel H (1962) Die absolute Strahlungsintensität der Sonnenmitte im Spektralbereich $4010 \leq \lambda \leq 6569$ Å. Z Astrophys 55:269–289
Labs D, Neckel H (1967) Die absolute Strahlungsintensität der Mitte der Sonnenscheibe im Spektralbereich $3288 \leq \lambda \leq 12\,480$ Å. Z Astrophys 65:133–185
Lamarre JM, The Cosmic Microwave Background. ISSI SR-009:149–162
Lamb WE, Jr (1995) Anti-photon. Appl Phys B 60:77–84
Lemaire P (2010) Normal- and grazing-incidence gratings and mountings used in space. ISSI SR-009:193–205
Lemaire P, Aschenbach BA, Seely JF (2010) Space telescopes. ISSI SR-009:165–191
Lenard P (1902) Über die lichtelektrische Wirkung. Ann Phys (Leipzig) 313:149–199
Lerosey G, de Rosny J, Tourin A, Fink M (2007) Focusing beyond the diffraction limit with far-field time reversal. Science 315:1120–1122
Lindegren L (2010) High-accuracy positioning: astrometry. ISSI SR-009:279–291
Liu C, Dutton Z, Behroozi CH, Hau LV (2001) Observation of coherent optical information storage in an atomic medium using halted light pulses. Nature 409:490–493
Louradour F, Reynaud F, Colombeau B, Froehly C (1993) Interference fringes between two separate lasers. Am J Phys 61:242–245
Mandelstam LI, Tamm IE (1945) The uncertainty relation between energy and time in non-relativistic quantum mechanics. J Phys (USSR) 9:249–254
Martin DDE, Verhoeve P (2010) Superconducting tunnel junctions. ISSI SR-009:441–457
Mather JC, Cheng ES, Eplee RE, Jr. (plus 18 authors) (1990) A preliminary measurement of the Cosmic Microwave Background spectrum by the *Cosmic Back-*

ground Explorer (COBE) satellite. Astrophys J 354:L37–L40
Mather JC, Cheng ES, Cottingham DA (plus 20 authors) (1994) Measurement of the Cosmic Microwave Background spectrum by the *COBE* FIRAS instrument. Astrophys J 420:439–444
Mikhailov AA (1959) The deflection of light by the gravitational field of the Sun. Mon Not R Astr Soc 119:593–608
Millikan RA (1916) A direct photoelectric determination of Plank's "h". Phys Rev 7:355–390
Mohideen U, Roy A (1998) Precision Measurement of the Casimir force from 0.1 to 0.9 μm. Phys Rev Lett 81:4549–4552
Mohr PJ, Taylor BN, Newell DB (2008) CODATA recommended values of the fundamental physical constants: 2006. arXiv:0801.0028
Mössbauer RL (1958) Kernresonanzfluoreszenz von Gammastrahlung in Ir^{191}. Z Physik 151:124–143
Neckel H (2003) On the Sun's absolute disk-center and mean disk intensities, its limb darkening, and its 'limb temperature' (λλ330 to 1099 nm). Sol Phys 212:239–250
Nimtz G, Stahlhofen AA (2008) Universal tunneling time for all fields. Ann Phys (Berlin) 17:374–379
Nisius R (2000) Photon structure from deep inelastic electron-photon scattering. Phys Rep 332:165–317
Nogues G, Rauschenbeutel A, Osnaghi S (plus three authors) (1999) Seeing a single photon without destroying it. Nature 400:239–242
Peacock A, Verhoeve P, Rando N (plus nine authors) (1996) Single optical photon detection with a superconducting tunnel junction. Nature 381:135–137
Penzias AA, Wilson RW (1965) A measurement of excess antenna temperature at 4080 Mc/s. Astrophys J 142:419–421
Perryman MAC, Foden CL, Peacock A (1993) Optical photon counting using superconducting tunnel junctions. Nucl Instrum Methods Phys Res Sect A 325:319–325
Planck M (1900) Über irreversible Strahlungsvorgänge. Ann Phys (Leipzig) 306:69–122
Planck M (1901) Über das Gesetz der Energieverteilung im Normalspektrum. Ann Phys (Leipzig) 309:553–563
Planck M (1909) Zur Theorie der Wärmestrahlung. Ann Phys (Leipzig) 336:758–768
Porter FS (2010) X-ray calorimeters. ISSI SR-009:459–475
Pound RV, Rebka GA (1959) Gravitational red-shift in nuclear resonance. Phys Rev Lett 3:439–441
Puccini GD, Selleri F (2002) Doppler effect and aberration of light from the point of view of absolute motion. Nuovo Cim 117 B:283–293
Quirrenbach A (2010) Interferometric imaging from space. ISSI SR-009:293–311
Raman CV (1928) A new radiation. Ind J Phys 2:387–398
Rarity JG, Ridley KD, Tapster PR (1987) Absolute measurement of detector quantum efficiency using parametric down-conversion. Appl Opt 26:4616–4619
Richards PL (1994) Bolometers for infrared and millimeter waves. J Appl Phys 76:1–24

Romestain R, Delaet B, Renaud-Goud P (plus four authors) (2004) Fabrication of a superconducting niobium nitride hot electron bolometer for single-photon counting. New J Phys 6:129-1–15

Rottman GJ (1988) Observations of solar UV and EUV variability. Adv Space Res 8:53–66

Rottman G (2000) Variations of solar ultraviolet irradiance observed by the *UARS* SOLSTICE – 1991 to 1999. Space Sci Rev 94:83–91

Rottman G, Harder J, Fontenla J (plus three authors) (2005) The spectral irradiance monitor (SIM): Early observations. Sol Phys 230:205–224

Rottman GJ, Woods TN, McClintock W (2006) *SORCE* solar UV irradiance results. Adv Space Res 37:201–208

Schiff LI (1960) On experimental tests of the general theory of relativity. Am J Phys 28:340–343

Schönfelder V, Kanbach G (2010) Imaging through Compton scattering and pair creation. ISSI SR-009:207–222

Schühle U (2010) Intensified solid state sensor cameras: ICCD and IAPS. ISSI SR-009:419–428

Schühle U, Hochedez J-F (2010) Solar-blind UV detectors based on wide band gap semiconductors. ISSI SR-009:429–439

Shapiro II (1964) Fourth test of general relativity. Phys Rev Lett 26:789–791

Smekal A (1923) Zur Quantentheorie der Dispersion. Naturwissenschaften 43:873–875

Smith DM (2010) Hard X-ray and γ-ray detectors. ISSI SR-009:345–364

Snider JL (1972) New measurement of the solar gravitational red shift. Phys Rev Lett 28:853–856

Snider JL (1974) Comments on two recent measurements of the solar gravitational red-shift. Sol Phys 36:233–234

Sommerfeld A (1978) Optik, Verlag Harri Deutsch, Thun, Frankfurt/Main.

Stenflo JO (2010) Stokes polarimetry of the Zeeman and Hanle effects. ISSI SR-009:543–557

Strekalov DV, Pittman TB, Shih YH (1998) What we can learn about single photons in a two-photon interference experiment. Phys Rev A 57:567–570

Sunyaev RA, Zel'dovich YB (1980) Microwave background radiation as a probe of the contemporary structure and history of the universe. Ann Rev Astron Astrophys 18:537–560

Takahashi T, Abe K, Endo M (plus 67 authors) (2007) Hard X-ray detector (HXD) on board *Suzaku*. Publ Astron Soc Japan 59:35–51

Thuillier G, Floyd L, Woods T (plus four authors) (2004) Solar irradiance reference spectrum. In: Geophysical Monograph 141: Solar variability and its effect on climate, American Geophysical Union, Washington DC, USA, pp. 171–194

Title AM (2010) Michelson interferometers. ISSI SR-009:327–338

Tiwari SC (2002) Relativity, entanglement and the physical reality of the photon. J Opt B Quantum Semiclass Opt 4:S39–S46

Vessot RC, Levine MW, Mattison EM (plus ten authors) (1980) Test of relativistic gravitation with a space-borne hydrogen maser. Phys Rev Lett 45:2081–2084

Wang LJ, Kuzmich A, Dogariu A (2000) Gain-assisted superluminal light propagation. Nature 406:277–279

White OR (1977) The solar output and its variations. Colorado Associated University Press, Boulder, USA

Wild W (2010) Coherent far-infrared / submillimeter detectors. ISSI SR-009:503–523

Wilhelm K, Curdt W, Marsch E (plus 13 authors) (1995) Some design and performance features of SUMER. Proc SPIE 2517:2–11

Wilhelm K, Schühle U, Curdt W (plus four authors) (2002) Solar vacuum-ultraviolet radiometry with SUMER. ISSI SR-002:145–160

Wilhelm K, Dwivedi BN, Marsch E, Feldman U (2004) Observations of the Sun at vacuum-ultraviolet wavelengths from space. Part I. Space Sci Rev 111:415–480

Woods TN, Rottman G (2005) XUV photometer system (XPS): Solar variations during the *SORCE* mission. Sol Phys 230:375–387

Woods TN, Eparvier FG (2006) Solar ultraviolet variability during the *TIMED* mission. Adv Space Res 37:219–224

Woods TN, Tobiska WK, Rottman GJ, Worden JR (2000) Improved solar Lyman-α irradiance modeling from 1947 through 1999 based on *UARS* observations. J Geophys Res 105:27 195–27 216

Zbinden H, Brendel J, Tittel W, Gisin N (2001) Experimental test of relativistic quantum state collapse with moving reference frames. J Phys A Math Gen 34:7103–7109

II. Energy and Wavelength Regions

— 3 —

High-energy astrophysics – energies above 100 keV

GOTTFRIED KANBACH[I], VOLKER SCHÖNFELDER[I] AND ALEX ZEHNDER[II]

Abstract

γ-rays represent the most energetic photons of the electromagnetic spectrum. Astronomy with γ-rays, therefore, allows us to study the most compact, energetic and violent objects in the Universe. These are neutron stars, stellar and massive black holes, supernovae and their remnants. An overview of the history of γ-ray astronomy, the production mechanisms for γ-rays, and an overview of the main results achieved are given.

Introduction

γ-ray astronomy at present ranges from about 100 keV to more than 10 TeV — these are eight orders of magnitude in photon energy. This range is comparable to that from radio to optical astronomy. Thus it is plausible that in γ-ray astronomy we are dealing with a variety of different objects and phenomena. Some of these objects even have their maximum of luminosity at γ-ray energies. It is, therefore impossible to understand the physics of these objects without knowing their γ-ray properties.

γ-rays cannot penetrate the Earth's atmosphere without interaction; the interaction length of rays in air is about 50 $\mathrm{g\,cm^{-2}}$ for photon energies of 10 MeV and shorter at lower and higher energies. Considering the total depth of the atmosphere of 1000 $\mathrm{g\,cm^{-2}}$, it is clear that cosmic γ-rays cannot reach the ground without many interactions in the atmosphere. γ-ray astronomy is separated into two main branches: spaceborne and ground-based. Spaceborne γ-ray astronomy ranges from about 100 keV to typically 50 GeV to 100 GeV. The γ-ray telescopes in this range are carried by balloons or satellites to the top of or outside the Earth's atmosphere. Above 100 GeV the photon fluxes of cosmic sources are so small that reasonable numbers of photons cannot be collected by space telescopes due to their limited detection areas, but cosmic γ-rays can be detected from the ground. At such high energies the primary γ-rays form electromagnetic cascades in the Earth's atmosphere, which propagate through the atmosphere in the direction of the original γ-ray. The secondary electrons and positrons of these cascades

[I]MPE—Max-Planck-Institut für extraterrestrische Physik, Garching, Germany
[II]PSI—Paul Scherrer Institut, Villigen, Switzerland

are observed at ground level by large arrays of large-area optical telescopes via the emitted Čerenkov radiation. Much progress has been achieved during the last decade in this young branch of γ-ray astronomy. However, in this book we restrict the coverage to spaceborne γ-ray astronomy.

Time-line development of γ-ray astronomy from space

The history of γ-ray astronomy from the early 1960s till now is illustrated in Figure 3.1. The first real cosmic γ-ray detections were those made by *Explorer-11* and *OSO-3* in 1961 and 1968 (Kraushaar et al 1965, 1972). Both experiments were designed to measure high-energy γ-rays above 50 MeV. *Explorer-11* detected in total 31, and *OSO-3* 621 cosmic γ-ray photons. The *OSO-3* results already showed clear evidence for γ-ray emission from the Milky Way. The cosmic γ-ray burst phenomenon was discovered at about the same time — namely in 1967 by the network of *Vela* satellites of the US Department of Defense, which were designed to monitor nuclear tests in the atmosphere. This discovery was not made public for several years, and it was in 1973 (Klebesadel et al 1973), that it was first publicized as a new astronomical phenomenon, whose origin remained a puzzle for another 25 years. Also in the 1960s and early 1970s there were several low-energy γ-ray instruments on cis-lunar missions like the *Ranger* (Metzger et al 1964), and especially the *Apollo* missions (Trombka et al 1973). From these measurements a diffuse cosmic background component was discovered, whose energy spectrum showed a bump at photon energies of a few megaelectronvolts, which exceeded the true spectrum by about a factor of ten. Later it was shown that it was actually an artifact caused by instrumental background radiation. This experience emphasizes again the importance of a careful accounting of systematic background sources in the range of γ radiation.

Galactic γ-ray astronomy at high energies took a major step forward with the two satellite missions *SAS-2* and *COS-B* (> 30 MeV and 70 MeV, respectively). Apart from the study of the diffuse galactic γ-ray emission (Mayer-Hasselwander et al 1982), these two missions also provided the first real source detections. The strongest source features were found from the Crab and Vela pulsars and from a source called Geminga, which remained unidentified for another twenty years. Finally in 1992, Geminga was identified as a pulsar as well. These sources can be clearly seen on the *COS-B* and *SAS-2* sky maps. The *COS-B* source catalogue contained in total 25 objects, of which only one (3C 273) was extragalactic (Swanenburg et al 1981).

In the field of cosmic γ-ray line spectroscopy a milestone was achieved by the Germanium spectrometer onboard *HEAO-III* with the first detection of a γ-ray line related to processes of nucleosynthesis, i.e., the creation of new (in this case radioactive) isotopes in astrophysical fusion processes: the 1.809 MeV line from radioactive ^{26}Al which was discovered from the general direction of the galactic centre region (Mahoney et al 1984).

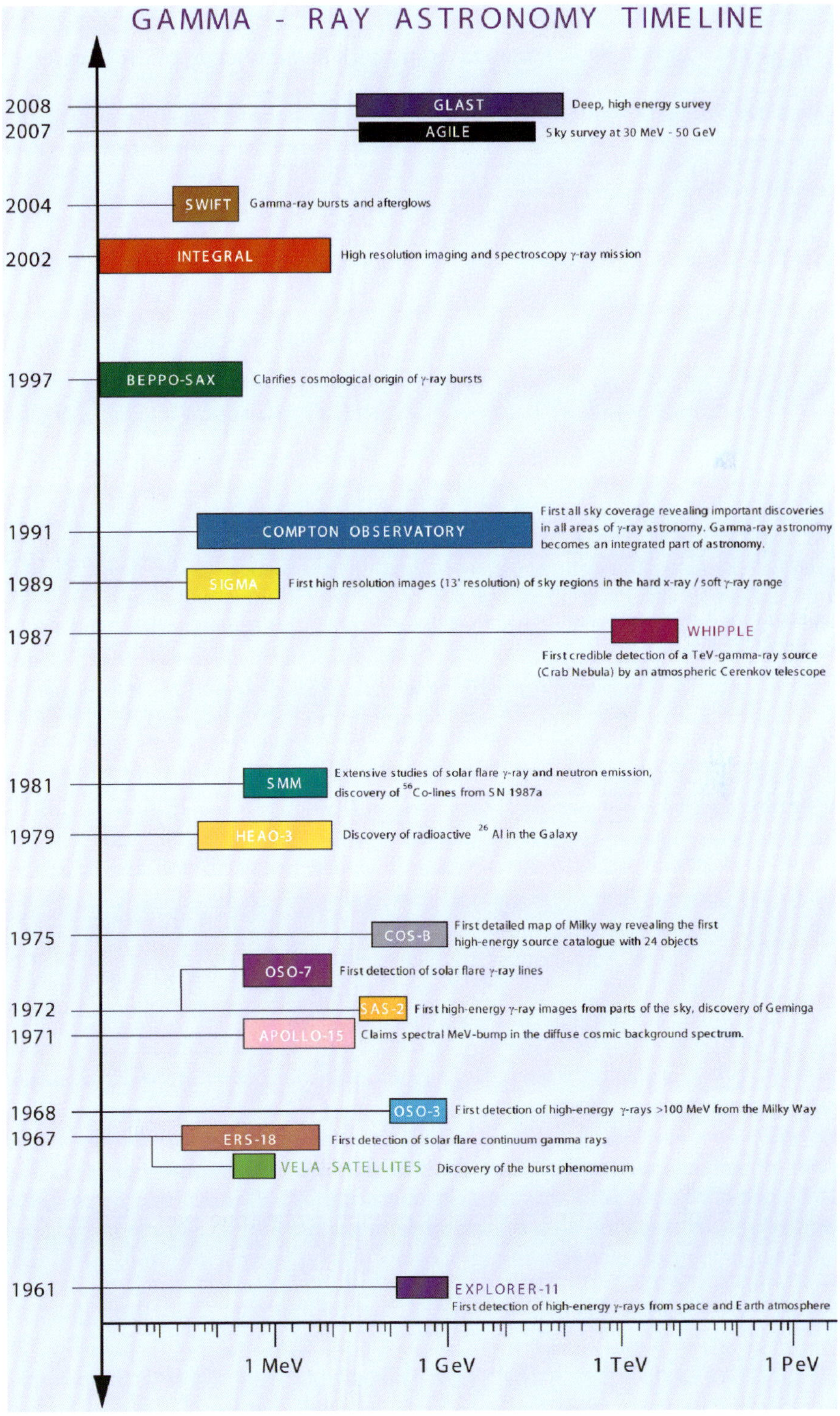

Figure 3.1: The history of observational γ-ray astronomy from the early 1960s until the launch of the *Fermi* Gamma-ray Space Telescope, formerly *GLAST*.

The first γ-ray line measurements from the Sun were made in 1972 by *OSO-7* from the flare of 4 August 1972. Apart from the continuum emission four lines were clearly detected: the annihilation line at 511 keV, the 2.2 MeV neutron capture line, and nuclear interaction lines at 4.4 MeV and 6.1 MeV from excitation of carbon and oxygen nuclei. Pure continuum emission in the γ-ray domain from solar flares up to at least 3.7 MeV had been already detected earlier — in May 1967 — with a scintillation spectrometer on *ERS-18* (Gruber et al 1973). The scintillation γ-ray spectrometer on the *SMM*, which was launched in 1981 and stayed in orbit for nine years, later provided a wealth of information on solar flare γ-ray emission in lines and continuum up to 10 MeV. In addition, however, it also made important studies of cosmic γ-ray lines, e.g., the 511 keV annihilation line and the 1.809 MeV ^{26}Al line from the Galaxy, and, in particular, it succeeded in an impressive detection of nucleosynthetic lines from ^{56}Co decay in SN 1987a in the Large Magellanic Cloud (Leising and Share 1990).

In the 1990s, γ-ray astronomy matured. This "golden epoch" started with the launch of the French telescope SIGMA on board the Russian *GRANAT*. SIGMA allowed imaging in the transition region between X-ray and γ-ray astronomy (mainly around 100 keV) with an unprecedented angular resolution of the order of $10'$. It mainly observed the galactic centre region, where it detected about 30 sources showing a great variety of compact objects, especially systems with stellar black hole candidates (Vargas et al 1997). The wealth of exciting discoveries of the Compton Observatory *CGRO*, which was launched in April 1991 and stayed in orbit for nine years, established the role of γ-ray astronomy as an important branch of astronomy and astrophysics, in general. Thanks to *CGRO*, γ-ray astronomy became an integrated part of astronomy.

The presently known celestial γ-ray sources cover a large variety of different objects, such as the Sun, isolated spin-down pulsars, accreting binaries with stellar neutron stars and black holes, supernovae, supernova remnants, the interstellar medium, normal galaxies, radio galaxies, Seyfert galaxies, quasars and the famous γ-ray burst sources. The question of the origin of the cosmic diffuse background is also of fundamental interest. During the last decade, a few very successful missions with specific scientific objectives followed. The *BeppoSAX* mission — launched in 1997 — succeeded in observing X-ray afterglow emission of a few γ-ray burst sources (Costa et al 1997). The subsequent observations of these objects at optical wavelengths clearly established the extragalactic origin of these sources. The mission *RHESSI*, launched in 2002 and still in operation (Lin et al 2002), is a solar mission, which is mainly devoted to study flares with a spatial resolution of a few seconds of arc and high resolution γ-ray line spectroscopy. The mission *INTEGRAL* (launched also in 2002 and still operating) is devoted to high resolution γ-ray line spectroscopy up to a few megaelectronvolts and to source imaging with an angular resolution of a few minutes of arc in the hard X-ray/soft γ-ray range (around 100 keV).

In the post-*BeppoSAX* era γ-ray burst astronomy has concentrated on the observation of afterglow emission throughout the electromagnetic spectrum. The prime objective of the missions *HETE II* (launch 2000) and *SWIFT* (launch 2004) was to locate the burst positions with high precision within the shortest possible time and to minimize the time-lag between the burst onset and the start of the afterglow observation. Thanks to *SWIFT* the different origins of long and short bursts could be clarified: short bursts come from the coalescence of compact objects like neutron stars and/or black holes, whereas long bursts have their origin in the collapse of very massive stars leading to the formation of a hyper-accreting black hole.

Much progress can be expected in the very near future in the area of high energy γ-ray astronomy (> 50 MeV): in 2007 the *AGILE* mission was launched. It is able to observe sources at a sensitivity level which is comparable to that of EGRET. On 11 June 2008 the *Fermi* Gamma-ray Space Telescope (20 MeV to 100 GeV) was launched. *Fermi* (formerly termed the *GLAST* mission) is about thirty times more sensitive than EGRET and is expected to detect several thousands of new high energy γ-ray sources.

γ-ray production

To produce γ-rays with energies in the 0.1 MeV to few teraelectronvolt regime, sources in thermal equilibrium emitting black body radiation are not sufficient, since unrealistically high temperatures ($> 10^{10}$ K) would be needed. On the other hand, given a relativistic particle population, several well-known mechanisms can give rise to γ-rays that are briefly discussed below. For an extensive review of γ-ray emitting processes see the text books by Hayakawa (1969) or Longair (1992).

Acceleration of charged particles

An electromagnetic field exerts a force on a moving charged particle. The resulting acceleration causes the particle to radiate. The character of the radiation and the name given to it depend on the nature of the accelerating force. If the electron is accelerated in the electrostatic field around a nucleus, the resulting radiation is called bremsstrahlung. It is called cyclotron or synchrotron radiation when the acceleration takes place in a magnetic field for non-relativistic or relativistic electrons. The most common mechanism for the production of relativistic particle populations is Fermi acceleration: in shock fronts particles can gain large momentum by crossing the shock repeatedly. The resulting particle energy spectrum follows a power law and can reach very high energies.

Bremsstrahlung

Bremsstrahlung is emitted during encounters of charged particles, e.g., in ionized plasma. The radiating particle is free, both before and after the deflection. Hence, also the term free-free radiation is commonly used. As the changes in the energy of the charged particle due to the collision are continuous, bremsstrahlung is characterized by a continuous spectrum over a very wide range that extends into γ-rays. Collisions between particles of the same charge-to-mass ratio do not produce bremsstrahlung, at least not in the dipole approximation. Astrophysical plasmas contain electrons with a distribution of velocities. The emission properties of a thermal plasma, with a Maxwellian velocity distribution, give a thermal bremsstrahlung distribution, mainly in the X-ray regime, whereas γ-rays are produced by highly relativistic electrons impinging on ambient protons (electron-proton) or of fast protons with ambient electrons (proton-electron). For a given proton energy, the electron-proton bremsstrahlung decreases for large scattering angles, whereas for proton-electron bremsstrahlung the distribution is almost independent of scattering angle. For the same Lorentz factor $\gamma_{\rm L}$, the maximum photon energy for proton-electron bremsstrahlung is about $2\gamma_{\rm L}$ larger than for the electron-proton bremsstrahlung.

Synchrotron emission

Charged particles moving in a magnetic field experience a constant acceleration, and thus emit electromagnetic radiation. The emitted power goes as $(1/m_0^2)\,\gamma_{\rm L}^2\,B^2$, where m_0 is the particle rest mass and B the magnetic field strength. Because of the mass dependence, synchrotron losses are much more important for electrons than for protons. For relativistic particles the main emission power is beamed into an angle of order $2/\gamma_{\rm L}$. Typical observational signatures of synchrotron emission are smooth spectra over a large range of wavelengths without emission lines. The spectra can be approximated by power laws that usually turn over at low and high frequencies. High-energy electrons will lose their energy first, therefore the electron distribution will turn over at the high-energy end of the spectrum. The emission shows strong linear polarization of up to 70 %. A distribution of electrons whose energies are distributed according to a power law with index p will produce a synchrotron spectrum that is also a power law, but with an index $\frac{p-1}{2}$.

Inverse Compton scattering

In an encounter between a high-energy photon and a low-energy electron, the electron receives energy described by Compton scattering. Conversely, high-energy electrons can also up-scatter a low-energy photon to high energies, the so-called inverse Compton effect. The mean energy gained by the photon in one collision is $\Delta E_\gamma \approx \frac{4}{3}\beta^2\,\gamma_{\rm L}^2\,E_\gamma$. If the electrons have a Lorentz factor of $\gamma_{\rm L} \approx 10^3$, photons can be up-scattered from the Cosmic Microwave Background, or they can be produced by the electrons themselves via synchrotron radiation in a magnetic field.

Nuclear line emission

Nuclear γ-ray line observations also provide unique opportunities to trace accelerated nuclei. Low-energy ($<$ 100 MeV per nucleon) accelerated particles, interacting with ambient matter produce excited nuclei. They decay promptly via characteristic γ-ray emission into their ground state. If the nuclei are radioactive, they will decay with their characteristic lifetimes in their daughter products emitting delayed nuclear γ-ray lines. Important nuclear γ-ray lines are given in Table 3.1.

Annihilation

The annihilation of a positron and an electron can either occur directly (with both free and bound electrons), producing two photons at 511 keV, or it can form

Table 3.1: Prompt and radioactive nuclear γ-ray lines including the main production and decay channels. From Diehl (2001) and Ramaty et al (1979). (Abbreviations: gs: ground state, *: exited state, SN: Supernovae, N: Novae.)

Isotope	Energy E_γ/keV	Production	Decay	Mean life
^{56}Fe	847	^{56}Fe(p, p′)^{56}Fe*	^{56}Fe* $\rightarrow^{56}$ Fe$_{gs}$	9.7 ps
^{24}Mg	1369	^{24}Mg(p, p′)^{24}Mg*	^{24}Mg* $\rightarrow^{24}$ Mg	1.8 ps
^{20}Ne	1634	^{20}Ne(p, p′)^{20}Ne*	^{20}Ne* $\rightarrow^{20}$ Ne$_{gs}$	1.2 ps
^{28}Si	1779	^{28}Si(p, p′)^{28}Si*	^{28}Si* $\rightarrow^{28}$ Si$_{gs}$	0.7 ps
^{12}C	4438	^{12}C(p, p′)^{12}C*	^{12}C* $\rightarrow^{12}$ C$_{gs}$	0.06 ps
^{16}O	6129	^{16}O(p, p′)^{16}O*	^{16}O* $\rightarrow^{16}$ O$_{gs}$	24 ps
^{2}H	2223	n(p, γ)^{2}H	Radiative n-capture	
^{7}Be	478	N	^{7}Be $\rightarrow^{7}$ Li*	77 d
^{56}Ni	1238, 847	SN	^{56}Ni $\rightarrow^{56}$ Co $\rightarrow^{56}$ Fe*	111 d
^{57}Ni	122	SN	^{57}Ni $\rightarrow^{57}$ Co $\rightarrow^{57}$ Fe*	390 d
^{22}Na	1257, 511	N	^{22}Na $\rightarrow^{22}$ Ne* + e$^+$	3.8 a
^{44}Ti	1156, 68,78	SN	^{44}Ti $\rightarrow^{44}$ Sc $\rightarrow^{44}$ Ca*	89 a
^{26}Al	1809, 511	Star, SN, N	^{26}Al $\rightarrow^{26}$ Mg* + e$^+$	1.04 Ma
^{60}Fe	1173, 1332	SN	^{60}Fe $\rightarrow^{60}$ Co*	2.0 Ma

positronium, either the ortho- or para-states. Para-positronium decays into two photons, producing a line at 511 keV, while ortho-positronium decays into a three-photon continuum below 511 keV. If dark matter, which constitutes most of the mass of galaxies, is made of super-symmetric particles, the centres of galaxies should emit very high-energy γ-rays produced by the self-annihilation of super-symmetric neutralinos.

Particle decay

Relativistic protons which interact with ambient matter or photon fields can produce a π_0 meson that decays into two photons in flight. Given a sufficiently high energy of the primary particles, these photons can reach teraelectronvolt energies and beyond. Neutrinos are also produced through this kind of interaction, mainly via the decay of charged pions. The appearance of the high-energy (> 100 MeV) γ-ray sky is dominated by diffuse emission from the galactic disk originating from π_0 decay following the interaction of high-energy (> 1 GeV per nucleon) protons and nuclei with the interstellar gas.

Overview of the γ-ray sky

γ-ray astronomy has opened a new window to astronomy. The fascinating results obtained so far reveal a Universe of extremes, often far removed from the experiences of classical astronomy. Acceleration and interactions of very energetic particles in environments dominated by extreme magnetic fields, temperatures, and strong gravity can be observed directly through secondary radiation. The γ-

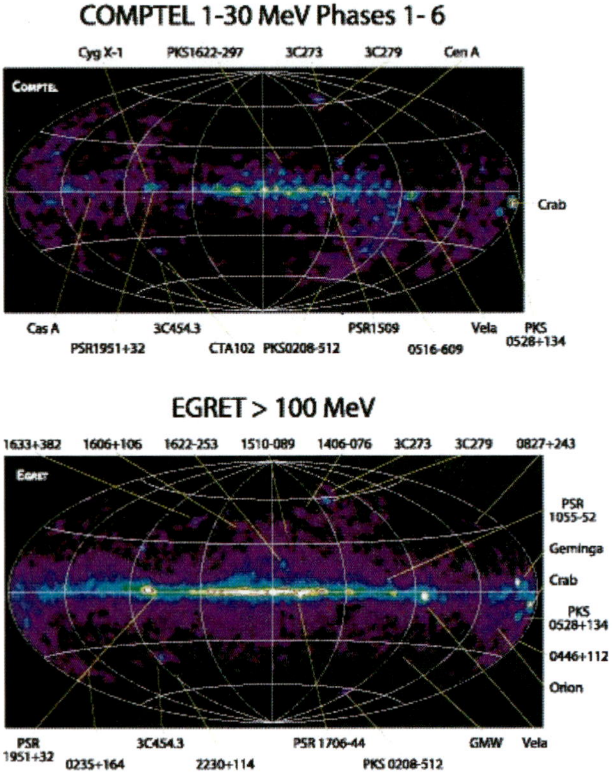

Figure 3.2: COMPTEL and EGRET all-sky maps shown in a galactic coordinate system. The dominant continuum emission along the galactic disk comes from cosmic-ray / gas interactions in interstellar space. Superimposed on the large-scale emission are point-like sources (a few are identified and indicated by their names, but many of the galactic point sources remain unidentified).

ray photons, with their high penetrating power, can leave the production sites more freely than other forms of radiation. The conditions of relativistic physics around black holes and neutron stars, nuclear processes in massive stars exploding as supernovae, and high-energy phenomena on galactic and super-galactic scales can thus be studied and provide insights into phenomena that can not be reproduced in terrestrial laboratories—they can only be observed in nature!

All-sky surveys

Highlights of the *CGRO* mission are the all-sky maps of γ-ray continuum radiation and of spectral line emissions. Figure 3.2 shows the continuum maps of COMPTEL (1 MeV to 30 MeV) and EGRET (above 100 MeV). The dominant galactic continuum emission comes from the interaction of cosmic rays with the gaseous interstellar medium. Superimposed on the large-scale galactic emission are

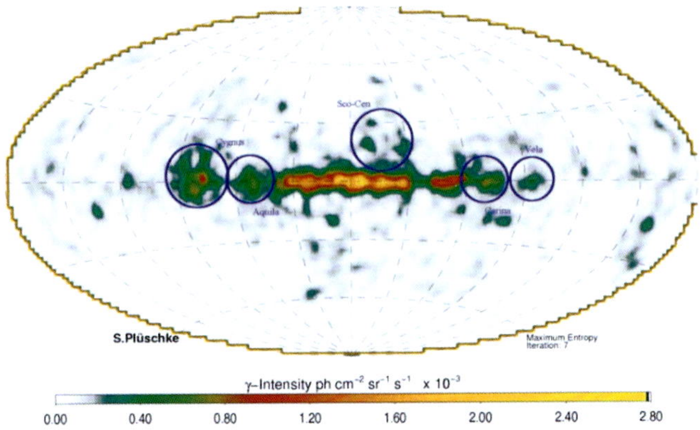

Figure 3.3: Regions of intense 1.809 MeV radiation from radioactive ^{26}Al in this COMPTEL all-sky map indicate the sites of ongoing nucleosynthesis in massive stars and supernova events.

point-like sources, like Crab, Geminga, Vela, and Cygnus. At mid and high latitudes many of the labelled γ-ray sources are identified as active galaxies with nuclei harbouring super massive black holes (AGN). The γ-ray bright AGN are of a type called blazars, which stands for a combination of the AGN type of BL Lac galaxies and to blaze, i.e., shine brilliantly. A different view of the sky is afforded by restricting the energy of the radiation to the narrow band of a γ-ray line. The most prominent line comes from the radioactive isotope ^{26}Al at 1.809 MeV, which is generated in massive stars and in supernova explosions. The ^{26}Al radioactivity is dispersed from the stellar sources into the interstellar medium and, with its half life of $\approx 10^6$ years, maps the production of new nuclei in the Galaxy on this time scale. The COMPTEL ^{26}Al map, displayed in Figure 3.3, shows prominent regions of recent nucleosynthesis in the Galaxy. γ-ray lines from other radioactive isotopes (^{60}Fe, ^{56}Co, ^{57}Co, ^{44}Ti) have also been detected by OSSE and COMPTEL and more recently with the spectrometer on *INTEGRAL*.

Cosmic radioactivity is not only detectable by direct nuclear γ-ray lines but also via positron-electron annihilation, which shows up in a line at 511 keV. The recent sky map from the *INTEGRAL* SPI telescope, shown in Figure 3.4, indicates a strong extended annihilation source from the inner Galaxy (Weidenspointner et al 2008). The one-sided extent of this antimatter emission towards the southern Galaxy (right side of the map) corresponds to a population of binary stars that is similarly off-centre. This observation powerfully suggests that these low mass X-ray binaries are responsible for a major amount of positron production.

γ-ray sources

A functional definition of a γ-ray point source is based on the angular resolution of the telescopes employed: significantly enhanced emission above the surrounding background radiance that conforms spatially to the point spread function of the in-

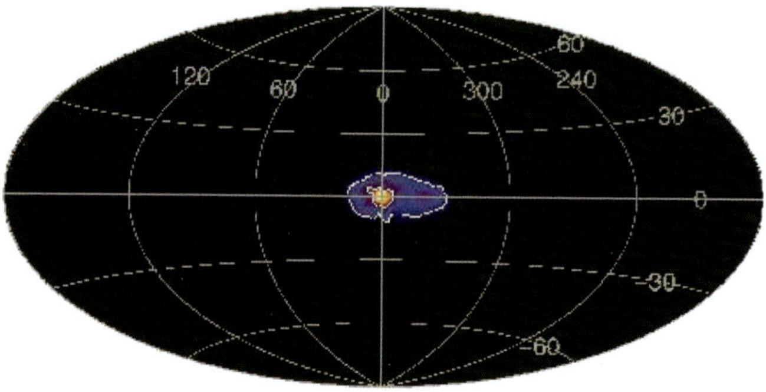

Figure 3.4: A map of the radiance of 511 keV radiation from the annihilation of positrons and electrons in the interstellar medium.

strument can be called an unresolved or point source. The true nature of the source, e.g., as a star-sized object or as a small cloud or supernova remnant, can not yet be discerned clearly with the current γ-ray telescopes. Additional astronomical, spectral, or temporal information has to be introduced to reveal a potential identification. The γ-ray surveys shown in Figure 3.2 were analyzed for point sources by subtracting the diffuse galactic and cosmic background emission and correlating the excess emission with the specific point spread functions. The resulting catalogues of significant sources were published by Schönfelder et al (2000) and Hartman et al (1999) for COMPTEL and EGRET.

A combined all-sky map of the sources detected by COMPTEL and EGRET is shown in Figure 3.5. It contains nearly 300 objects in total. Among the EGRET

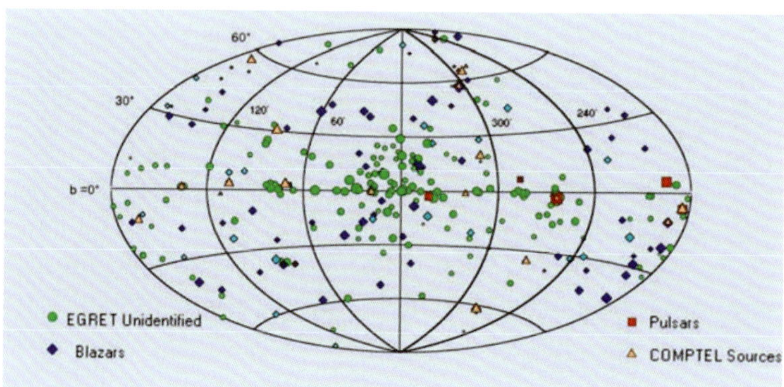

Figure 3.5: A sky map of high-energy point-sources derived from the all-sky surveys of COMPTEL (0.75 MeV to 30 MeV) and EGRET (> 100 MeV). The observations were carried out during the *CGRO* mission from 1991 to 2000.

sources found in the energy range > 100 MeV, seven to ten spin-down pulsars are clearly or tentatively identified by their timing signature as pulsating sources (Kanbach 2002). Nearly one third of all sources ($\approx$ 50 to 70), mostly at high galactic latitudes, are identified as γ-ray blazars (Mukherjee et al 1997). Two thirds of the EGRET sources are still unidentified. Among the COMPTEL sources in the energy range (0.75 to 30) MeV we also find spin-powered pulsars ($\approx$ 4 to 5) and γ-ray blazars ($\approx$ 10), but in addition accretion driven X-ray binaries, a radio galaxy and special γ-ray line sources (supernova remnants and regions of massive star associations).

γ-ray spectroscopy: continuum and nuclear lines

γ-ray spectra carry essential information on the nature of the radiating sources. Aside from wide-band continuum spectra, typically generated in bremsstrahlung, inverse Compton scattering, synchrotron radiation, or π_0 decay, γ-ray lines from nuclear transitions or matter-antimatter annihilation are telltale signatures of the radiating material. Nuclear level transitions can either be excited promptly by reactions of energetic particles or in the course of a radioactive decay chain. To understand the complex line spectra, a de-convolution of the spectra and modelling of the source is often required. Two typical examples are given below: a *RHESSI* solar flare spectrum (see Figure 3.6) and a composite spectrum of the AGN 3C279 (see Figure 3.7). The different processes involved are given in the figure captions.

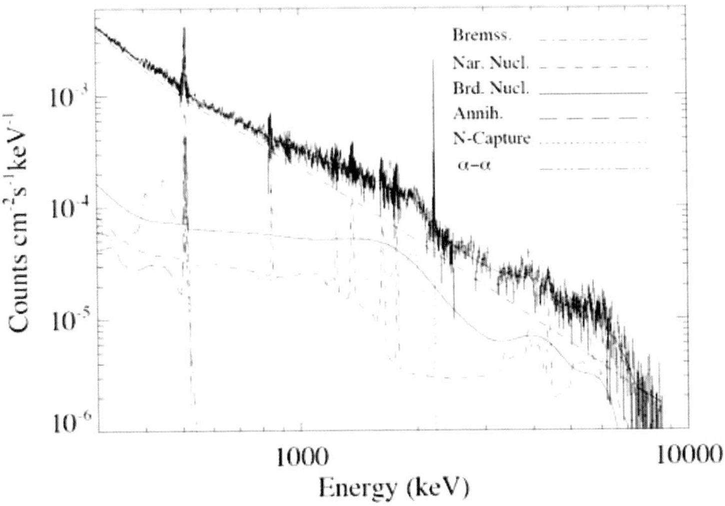

Figure 3.6: Count spectrum of the 23 July 2002 solar flare with fits revealing different components. Brems: non-thermal bremsstrahlung (electron energy); Nar. Nucl: narrow nuclear lines; Brd. Nucl: broad nuclear lines (ion energy and abundance, coronal ambient composition), Annih: annihilation line (positron density), N-Capture: thermal neutron capture (photospheric ^{3}He), $\alpha - \alpha$:^{7}Be and ^{7}Li lines produced in fusion of accelerated α-particles with ambient ^{4}He (from Share et al 2004).

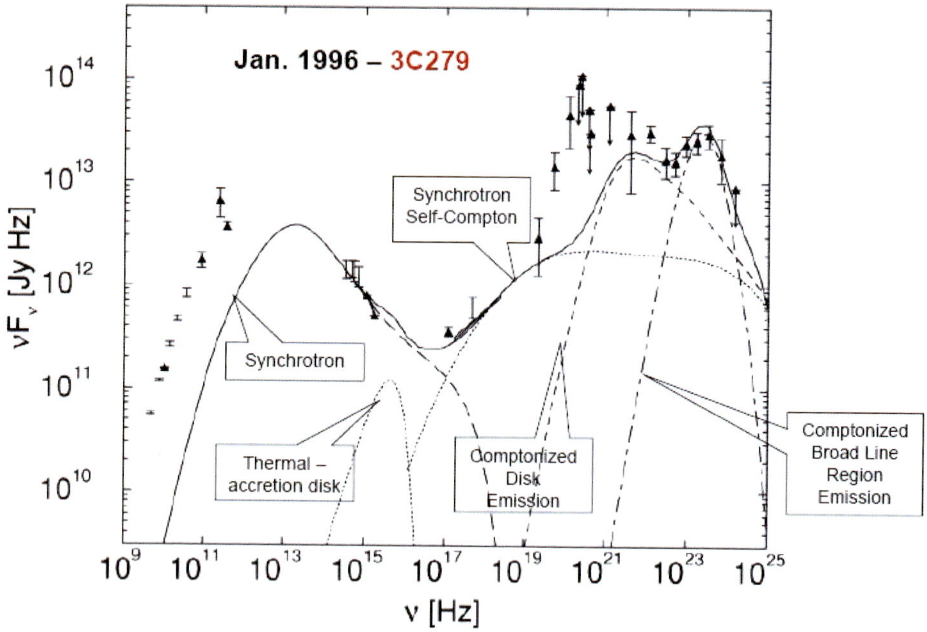

Figure 3.7: Fit to a simultaneously measured broad-band spectrum (from radio to γ-ray frequencies) by electron-induced processes of the quasar 3C279. The total fit (solid line) is the sum of five components. These are from left to right: 1. Synchrotron emission of electrons accelerated along the jet-axis (which is directed towards the observer and which lies perpendicular to the accretion disk around the massive black hole); 2. Thermal emission from the inner part of the accretion disk; 3. Inverse Compton emission produced by relativistic electrons in the jet with their self-generated synchrotron photons (synchrotron-self Compton component); 4. Inverse Compton emission by relativistic electrons in the jet with unscattered disk photons; 5. Inverse Compton emission by relativistic electrons in the jet with disk photons which were scattered into the jet by broad-line clouds.

γ-ray bursts

Since their serendipitous discovery in 1968, γ-ray bursts (GRB) have been intensively researched with space γ- and X-ray telescopes as well as with optical observatories on the ground. The BATSE instrument on *CGRO* provided the first all-sky map of these events. The highly isotropic distribution of GRBs indicated already that they are not of galactic origin. Since the discovery of optical counterparts (prompt emission and afterglows) of GRBs, it is known that most come from cosmological distances. To date the most distant GRB was located at a redshift of $z = 6.29$ (Kawai et al 2006). The currently active GRB space missions (*SWIFT*, *INTEGRAL*, etc.) continue to supply surprising and novel facts about GRB events: we distinguish at least two types of events—long duration (> 2 s) events with soft spectra, and short (< 2 s) hard bursts. The long events are generated in the core collapse of massive stars into a black hole. A fireball directed along a jet with a total energy of $\approx 10^{44}$ J to 10^{45} J is emitted and creates many observed characteristics of these GRBs. The short events could be the result of the merger of two neutron stars in a binary after the orbital momentum has been radiated away in gravitational waves. GRBs are the most powerful, although short lived, explosions known in astronomy. They can be detected across the visible Universe and already serve as important sources for cosmology.

Outlook

The short overview of γ-ray astronomy given in this chapter cannot encompass the richness of high-energy astronomy in all details. However the results testify to the tremendous efforts made in the past; they have stimulated new theories and plans for new γ-ray astronomy missions. In 2002, ESA's *INTEGRAL* mission ($\approx$ 20 keV to several megaelectronvolts) started to observe with several instruments dedicated to imaging and spectroscopy. *INTEGRAL* will probably be operated beyond 2010. In 2007 the small pair-imaging telescope *AGILE* (Italy) was launched and provided the first exciting results of an all-sky survey in 2008, more than a decade after the EGRET survey. The next high-energy mission *Fermi* (15 MeV to 100 GeV), a NASA project with world-wide international participation, was launched in June 2008.

The energy region above the *INTEGRAL* range (> 300 keV) and below the effective onset of *Fermi* (< 100 MeV) is generally considered to be a still under-explored but scientifically very promising spectral range. At these energies detectors based on Compton interactions or low-energy pair-creation telescopes have to be used. This range was investigated by the pioneering COMPTEL instrument on *CGRO* but no follow-up mission with greatly increased sensitivity is currently scheduled. Several proposals for advanced Compton/pair-creation telescopes (0.3 MeV to 50 MeV, Kanbach et al 2005, Greiner et al 2008, Boggs et al 2006) or a focusing kiloelectronvolt telescope based on a crystal Laue lens (Knödlseder et al 2006) are discussed and have been forwarded to the funding agencies. The outlook for a realization of such a mission in the near future is, however, not promising.

For several years, ground based imaging atmospheric Čerenkov telescopes with large collecting mirrors (10 m to 20 m diameter) have covered the γ-ray energy range above $\approx$ 50 GeV with increasing sensitivity, albeit with rather small fields of view. The observatories HESS (Hoffmann et al 2003), MAGIC (Baixeras et al 2003; Holder et al 2006), and Cangaroo (Kubo et al 2004) have performed the first galactic surveys and found, to date, more than 60 sources in this energy range (e.g., Aharonian et al 2006, for an updated catalog see http://tevcat.uchicago.edu/). The sources include supernova remnants as potential sites of cosmic-ray production, pulsar wind nebulae, and BL Lac type galaxies.

Multi-wavelength observations of high-energy sources using space-based telescopes as well as ground based facilities covering a wide range of the electromagnetic spectrum will be essential for the future of high-energy astrophysics. In the γ-ray range we aim especially to study and understand the non-thermal physics of our cosmic origins, the synthesis of the elements, and the evolution and fate of stars and stellar systems.

Bibliography

Aharonian F, Akhperjanian AG, Bazer-Bachi AR (plus 96 authors) (2006) The H.E.S.S. survey of the inner galaxy in very high energy gamma rays. Astrophys J 636:777–797

Baixeras C (2003) The MAGIC telescope. Nucl Phys B Proc Suppl 114:247–252

Boggs SE, Kurfess J, Ryan J (plus 30 authors) for the larger ACT collaboration (2006) The Advanced Compton Telescope Mission. ArXiv Astrophysics e-prints arXiv:astro-ph/0608532

Bothe W, Kolhörster W (1929) Das Wesen der Höhenstrahlung. Z Phys 56:751–777

Chupp EL, Forrest DJ, Higbie PR (plus three authors) (1973) Solar gamma ray lines observed during the solar activity of August 2 to August 11, 1972. Nature 241:333–335

Costa E, Frontera F, Heise J (plus 23 authors) (1997) Discovery of an X-ray afterglow associated with the γ-ray burst of 28 February 1997. Nature 387:783–785

Gehrels N, Fichtel CE, Fishman GJ (plus two authors) (1993) The Compton Gamma Ray Observatory. Sci American 269:38–45

Greiner J (2008) GRIPS-Gamma-Ray Burst Investigation via Polarimetry and Spectroscopy. In: Galassi M, Palmer D, Fenimore E (eds) AIP Conf Ser 1000:279 620–623

Gruber DE, Peterson LE, Vette JI (1973) Observation of MeV gamma-ray events during May 1967 from ERS-18. NASA Special Publication 342:147

Hartman RC, Bertsch DL, Bloom SD (plus 24 authors) (1999) The third EGRET catalog of high-energy gamma-ray sources. Astrophys J Suppl 123:79–202

Hartman RC, Böttcher M, Aldering G (plus 71 authors) (2001) Multiepoch multiwavelength spectra and models for blazar 3C279. Astrophys J 553:683–694

Hayakawa S (1969) Cosmic ray physics. Nuclear and astrophysical aspects. Interscience Monographs and Texts in Physics and Astronomy, New York: Wiley-Interscience, 1969

Hess VF (1913) Über den Ursprung der durchdringenden Strahlung. Physik Zeitschr 14:610–617

Hofmann W, HEGRA Collaboration (2003) Search for TeV gamma-rays from the Andromeda galaxy and for supersymmetric dark matter in the core of M31. In: International Cosmic Ray Conference, International Cosmic Ray Conference, vol 3 pp 1685–1688

Holder J, Atkins RW, Badran HM (plus 75 authors) (2006) The first VERITAS telescope. Astroparticle Physics 25:391–401

Hurley KC (2003) Swift GRB MIDEX mission. Proc SPIE 4851:1173–1179

Kanbach G (2002) Gamma-ray pulsars. In: Becker W, Lesch H, Trümper J (eds) Neutron Stars, Pulsars, and Supernova Remnants, 91–99

Kanbach G, Andritschke R, Zoglauer A (plus nine authors) (2005) Development and calibration of the tracking Compton/Pair telescope MEGA. Nuclear Instruments and Methods in Physics Research A 541:310–322

Kawai N, Kosugi G, Aoki K (plus 25 authors) (2006) An optical spectrum of the afterglow of a γ-ray burst at a redshift of $z = 6.295$. Nature 440:184–186

Klebesadel RW, Strong IB, Olson RA (1973) Observations of gamma-ray bursts of cosmic origin. Astrophys J Lett 182:L85–L88

Knödlseder J (2006) GRI: the gamma-ray imager mission. Proc SPIE 6266:626623, DOI 10.1117/12.671451

Kraushaar W, Clark GW, Garmire G (plus three authors) (1965) Explorer XI experiment on cosmic gamma rays. Astrophys J 141:845–863

Kraushaar WL, Clark GW, Garmire GP (plus four authors) (1972) High-Energy Cosmic Gamma-Ray Observations from the OSO-3 Satellite. Astrophys J 177:341–364

Kubo H, Asahara A, Bicknell GV (plus 48 authors) (2004) Status of the CANGAROO-III project. New Astronomy Review 48:323–329

Lamb DQ, Ricker GR, Atteia JL (plus 37 authors) (2004) Scientific highlights of the HETE-2 mission. New Astronomy Review 48:423–430

Leising MD, Share GH (1990) The gamma-ray light curves of SN 1987A. Astrophys J 357:638–648

Lin RP, Dennis BR, Hurford GJ (plus 63 authors) (2002) The Reuven Ramaty High-Energy Solar Spectroscopic Imager (*RHESSI*). Sol. Phys.210:3–32

Longair MS (1992) High energy astrophysics. Vol.1: Particles, photons and their detection. High Energy Astrophysics, by Malcolm S. Longair, pp. 436. ISBN 0521387736. Cambridge, UK: Cambridge University Press, March 1992.

Mahoney WA, Ling JC, Wheaton WA, Jacobson AS (1984) HEAO 3 discovery of Al-26 in the interstellar medium. Astrophys J 286:578–585

Metzger AE, Anderson EC, van Dilla MA, Arnold JR (1964) Detection of an Interstellar Flux of Gamma-Rays. Nature 204:766–767

Michelson PF (2003) Instrumentation for the Gamma-ray Large Area Space Telescope (GLAST) mission. Proc SPIE 4851:1144–1150

Mukherjee R, Bertsch DL, Bloom SD (plus 19 authors) (1997) EGRET Observations of high-energy gamma-ray emission from blazars: an update. Astrophys J 490:116–135

Ramaty R, Kozlovsky B, Lingenfelter RE (1979) Nuclear gamma-rays from energetic particle interactions. Astrophys J Suppl 40:487–526

Schönfelder V (ed) (2001) The Universe in gamma rays, Springer Verlag, Berlin

Share GH, Murphy RJ (2004) Solar Gamma-Ray Line Spectroscopy - Physics of a Flaring Star. In: Dupree AK, Benz AO (eds) Stars as Suns : Activity, Evolution and Planets, IAU Symposium, vol 219, pp 133–144

Swanenburg BN, Bennett K, Bignami GF (plus 11 authors) (1981) Second *COS-B* catalog of high-energy gamma-ray sources. Astrophys J Lett 243:L69–L73

Tavani M, Barbiellini G, Argan A (plus 55 authors) (2003) The AGILE instrument. Proc SPIE 4851:1151–1162

Trombka JI, Metzger AE, Arnold JR (plus three authors) (1973) The cosmic γ-ray spectrum between 0.3 and 25 MeV measured on *Apollo 15*. Astrophys J 181:737–746

Vargas M, Paul J, Goldwurm A (plus 13 authors) (1997) Sigma Observations of high energy steady sources and transients in the Milky Way. In: Winkler C, Courvoisier TJL, Durouchoux P (eds) The Transparent Universe, ESA Special Publication, vol 382, pp 129–136

Weidenspointner G, Skinner G, Jean P (plus eight authors) (2008) An asymmetric distribution of positrons in the galactic disk revealed by γ-rays. Nature 451:159–162

— 4 —

X-ray astronomy: energies from 0.1 keV to 100 keV

J. Len Culhane[I]

Abstract

Following a short introduction to the cosmic and solar branches of the subject, the thermal and non-thermal radiation processes for the production of both cosmic and solar X-rays are briefly outlined. Methods for X-ray detection are described along with reflecting systems for collecting, focussing and dispersing X-rays. The use of CCDs and cryogenic detectors for non-dispersive spectroscopy is also discussed. The present state of the art in the topic is presented through the description of several recent and current space missions in both the solar and cosmic fields.

Introduction – the background to X-ray astronomy

The Sun was identified as a source of X-rays in 1949 in a rocket flight of X-ray sensitive film behind a thin beryllium filter by Burnight (1949) from the US Naval Research Laboratory, but prospects for a broader discipline based on computing the X-ray flux from the Sun placed at the distance of the nearest star did not appear encouraging. Thus the detection of a bright source in Scorpio along with evidence for a diffuse isotropic X-ray flux by Giacconi et al (1962) transformed the situation and opened the way for what has since become a mainstream branch of astronomy.

Following the detection of the first cosmic X-ray source, the subject proceeded through successive rocket flights of typically 5 min duration. These were the first surveys and by late 1970 about 50 X-ray sources had been detected. The mechanically collimated proportional counters used had poor angular resolution and so specialist observations using other methods, e.g., lunar occultation of sources and modulation collimation (Oda 1965), were carried out for position and source extent measurement. The former method allowed the extent of the Crab Nebula X-ray source to be determined (Bowyer et al 1964) while the latter reduced the position uncertainty for Sco-X1 (Gursky et al 1966) and allowed its identification with a 13th magnitude star with old nova spectral characteristics (Sandage et al 1966). Thus the early development proceeded through a combination of sky surveys to discover new sources and observations of the key properties of individual objects.

[I]UCL/MSSL—University College London/Mullard Space Science Laboratory, UK

This dual approach continued until the deployment of large grazing-incidence X-ray telescopes allowed a measure of unification for the subject.

While the Sun was the first known astronomical X-ray source, its angular size was too small for simple collimated detectors to resolve detail on the surface although measurements of the solar flare plasma temperature were made given the brightness and small size of the flare regions (Culhane et al 1964). However, the Sun soon became the first target for the newly developed X-ray imaging telescopes. An early rocket flight of such a telescope (Giacconi et al 1965) obtained images on photographic film with an angular resolution of $< 1'$, a figure significantly better than that previously achieved with simple pinhole cameras (Blake et al 1963). A major advance followed the launch and operation of the NASA *Skylab* manned mission in May 1973. A grazing-incidence telescope with photographic film registration and a resolution of $2''$ observed the Sun during three manned periods in May, July/August and October/November 1973 (Vaiana et al 1973). The resulting images transformed our understanding of the solar corona. In the following years, a combination of X-ray imaging and high-resolution spectroscopy steadily advanced the solar discipline.

In this chapter, we will outline the processes, both thermal and non-thermal, that give rise to X-ray emission in the Sun and in a wide variety of cosmic X-ray sources. We will then briefly survey the methods used for collimation, detection, focussing and spectral analysis, dispersive and non-dispersive, of X-rays. Finally, we will illustrate the facets of the subject with an account of key solar and cosmic X-ray space missions.

The X-ray spectrum: radiation processes

Thermal radiation processes

The Sun and many other cosmic X-ray sources contain bodies of high-temperature plasma that are optically thin to their own radiation, quasi-isothermal and with their constituent particles — mainly electrons and protons — having equilibrium Maxwellian velocity distributions. Temperatures range from around 10^6 K for the solar corona to $\leq 10^8$ K for clusters of galaxies with supernova remnant plasmas being at intermediate values. Such plasmas give rise to thermal emission through several processes.

Continuous emission spectra arise mainly through bremsstrahlung or free-free and radiative recombination or free-bound radiation. In the free-free process, electrons are accelerated in the Coulomb fields of protons or heavier ions and as a result radiate photons. The spectral shape goes as $T_{\rm e}^{-1/2} \exp[-h\,\nu/(k_{\rm B}\,T_{\rm e})]$, with $T_{\rm e}$ the electron temperature of the plasma, ν the frequency and $h\,\nu$ the photon energy. The emission is proportional to $n_{\rm i}\,n_{\rm e}\,T_{\rm e}^{-1/2}$, where $n_{\rm i}$ and $n_{\rm e}$ are the electron and ion densities.

For the free-bound process, an electron is captured into an ion bound state and excess energy appears as a continuum photon. The overall spectral shape is similar to the free-free continuum but punctuated by discontinuities or absorption edges at the binding energies of the different ions. Thus the contribution of each of the principal ions to the continuum must be calculated separately and will depend on the ion abundances. These in turn will depend on both the original element abundance and on the balance between ionisation and recombination processes in the plasma.

At plasma temperatures $T_e = 20$ MK the spectrum is dominated by line emission. For low densities, $n_e \leq (10^{11}$ to $10^{12})$ cm^{-3} typical of many plasma environments in the Universe, e.g., solar and stellar coronae, ions are in the ground state but electrons can be excited to higher levels by free electron collisions with ions. De-excitation is by the electron returning to the ground state with photon emission, even for the so-called *forbidden transitions* with long radiative decay times that would not be observed in high-density laboratory plasmas. As for continuum processes, radiant power density $L_{\rm line}$ depends on n_i and n_e. It is often written as $0.8\, n_e^2$, and so for a volume V of emitting plasma, $L_{\rm line}$ depends on $\int n_e^2\,{\rm d}V$ — the plasma emission measure. However, the fraction $n_i/n_{\rm element}$ depends on T_e through the ionisation and recombination rates where collisional and autoionisation are balanced by radiative and dielectronic recombination. As T_e increases, more bound electrons are removed by collisions. Thus for the distributed high-temperature gas in galaxy clusters with $T_e \approx 50$ MK, Fe ions are typically in H-like or He-like states. In contrast, for the solar and stellar coronae, a range of temperatures is encountered extending from ≈ 1 MK for quiet parts of the atmosphere to (20 to 30) MK in flare plasma. This broad range in T_e results in rich emission line spectra which for iron will include lines of ionisation states from Fe^{+7} to Fe^{+25}.

X-ray emission line observations provide a powerful tool for the study of plasma properties. An ionisation balance calculation allows estimation of T_e based on the ion states detected in the gas. Consideration of the line excitation physics allows direct T_e and n_e estimates from measurement of line radiance ratios. Observation of Doppler shifts in the line peak provides estimates of plasma velocity. Many different flow phenomena have been detected in the solar corona. For equilibrium isothermal plasma, the measured value of T_e, e.g., from line ratios, should equal T_i, the ion temperature, which can be estimated from the width of the line profile. However, the line width is often larger than that implied by the T_e value. Measurement of the difference allows the identification of plasma turbulence or other non-thermal processes. Finally, element abundances can be estimated from the line radiances. This is particularly important for cosmic X-ray sources such as supernova remnants and clusters of galaxies.

Examples of emission spectra for solar active region (AR) plasma and for the gas in a cluster of galaxies are given in Figures 4.1 and 4.2. The solar spectrum was obtained by the EUV Imaging Spectrometer (EIS; Culhane et al 2007) on the Japanese *Hinode* spacecraft while the cluster observation used the Reflection Grating Spectrometer (RGS; den Herder et al 2001) on the ESA *XMM-Newton* spacecraft. For solar spectra, plasma properties are usually determined by direct comparison of line radiances with calculated values from the CHIANTI database (Dere et al 1997; Landi et al 2006). With multi-thermal plasma, the database can be used to generate a best-fitting differential emission measure, ${\rm d}(\int n_e^2\,{\rm d}V)/{\rm d}T$. For cosmic X-ray sources, plasma parameters are derived from the measured spectra by fitting with a range of thermal and power-law model spectra calculated using the SPEX package (Kaastra et al 1996).

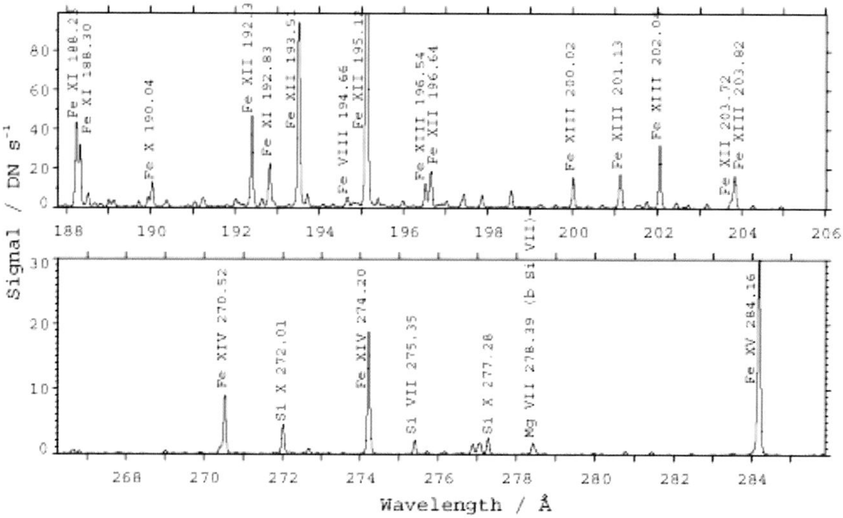

Figure 4.1: *Hinode* EIS EUV spectra of a solar active region (Young et al 2007). Ion species present indicate T_e from ≈ 2 MK to 3 MK.

Non-thermal radiation processes

Non-thermal radiation processes arise in situations where electrons that have been accelerated to high energies, e.g., ≥ 20 keV, do not exhibit a Maxwellian velocity distribution and are more usually described by a power-law energy spectrum of the form $\mathrm{d}N/\mathrm{d}E = k\,E^{-\alpha}$, where k is a proportionality constant. A beam of such electrons when injected into a dense plasma will produce a continuous X-ray photon spectrum by the non-thermal bremsstrahlung process. Such spectra are frequently studied in solar flare observations where the photon spectra are inverted in an effort to deduce the original electron spectra. Recent examples of this technique have been provided by observations with the *RHESSI* instrument (Lin et al 2002).

If electrons are in circular motion around a magnetic field, photons are emitted because the electrons are continuously accelerated. For non-relativistic electrons, this process is called cyclotron radiation. Electrons emit at their gyrofrequencies given by $\nu_g = e\,B/(2\,\pi\,m_e)$, where B is magnetic field strength and m_e the electron rest mass. Polarization is circular when viewed along the field and linear in the perpendicular direction. For increased electron velocity, $v \geq 0.1\,c_0$, the radiation shifts to higher harmonics and for a population of electrons, a continuum will result. This process is important in solar flares, but at microwave radio frequencies. However, at X-ray energies it has been observed in emission due to Landau quantization at $E \approx 58$ keV in the spectrum of the neutron star binary Her X-1, by Trümper et al (1978) who estimated the magnetic field strength as approximately 5×10^8 T. For even higher electron energies with velocities approaching the speed of light that are propagated in a magnetic field, a continuous X-ray spectrum can arise through the synchrotron process. Emission is tightly beamed in the direction of motion with angular width $\approx 1/\gamma$, where γ, the Lorentz factor, is defined in Chapter 2 (Wilhelm

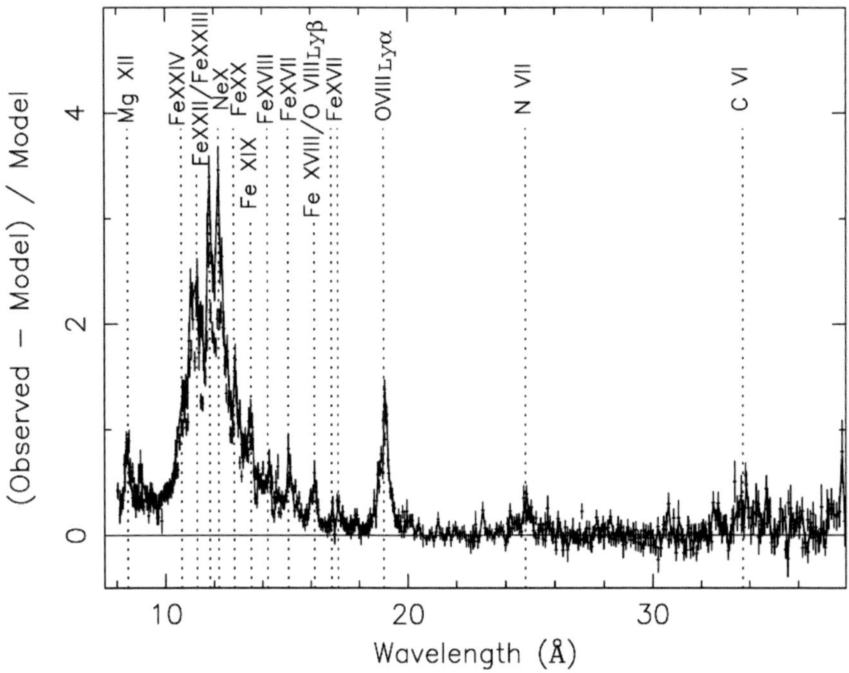

Figure 4.2: *XMM-Newton* RGS X-ray spectrum of the M87 cluster core (Werner et al 2006). Model fitting shows plasma components with T_e values of 10 MK and 20 MK.

and Fröhlich 2010). The radiation is linearly polarised and the spectrum consists mainly of high harmonics. Again an electron population distributed in a strong magnetic field will emit a continuous spectrum. The best known astrophysical example is provided by fast rotating neutron stars or pulsars, e.g., in the Crab nebula. Electrons at $\approx 10^4$ km from the neutron star centre are rotating at essentially the speed of light and so generate synchrotron X-ray pulses. While the X-rays from the hot gas heated by a supernova shock are thermal, there is evidence for a small number of such objects, e.g., SN1006 (Koyama et al 1995), of X-ray synchrotron emission from high-energy accelerated electrons around the edge of the expanding shock front. Finally, synchrotron X-ray emission is probably produced in the jets that are ejected along the axes of massive black holes at the centres of active galactic nuclei (AGN).

Compton processes are important for photon-electron interactions both in absorption and emission. In Compton scattering, wavelength increases and frequency or photon energy decreases. While there is a dependence on the angle between the outgoing photon and its original direction, we have on average $\Delta\nu/\nu \approx h\,\nu/(m_e\,c_0^2)$. This interaction is important in the detection of hard X-ray photons of $E \approx 30$ keV to 100 keV and in the scattering or Comptonisation of these photons in optically thick astrophysical environments. The reverse or inverse Compton process operates with a low-energy photon gaining energy by interacting with a high-energy electron. Here the final photon energy or frequency depends on the electron energy or

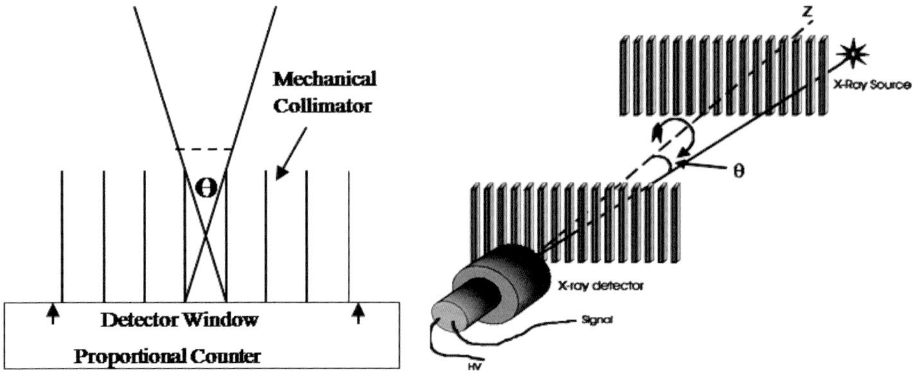

Figure 4.3: Schematic diagram of a simple mechanical collimator (left) and an outline of the principle of the rotation modulation collimator (RMC) (right).

$\gamma^2 \approx \nu_{\rm final}/\nu_{\rm initial}$. An astrophysically important example is given by the Sunyaev-Zel'dovich effect in which cosmic microwave background photons are increased in energy by interacting with the free electrons in the high-temperature plasma in clusters of galaxies.

Outline of X-ray photon handling techniques

Detection and collimation

The photoelectric effect has until recently formed the basis for X-ray photon detection. In *gas-filled detectors*, e.g., proportional counters, the quantum efficiency (QE) is the product of entrance window transmission and gas absorption. Energy is converted to electron-ion pairs with $\omega \leq 30$ eV per ion pair. Early on these detectors were used to provide non-dispersive spectroscopy with $E/\Delta E_{\rm FWHM} \approx 3\sqrt{E/{\rm keV}}$ or ≤ 10 for $E = 10$ keV. Statistical fluctuations in E/ω and in the detector gas amplification factor limited the energy resolution. *Scintillation detectors* are used at energies ≥ 20 keV. Photons are detected in a scintillating crystal. The liberated electrons radiate visible light that interacts in a semi-transparent photocathode and generates electrons within a photomultiplier. An amplified charge pulse results from detection of the original X-ray photon. While this pulse is proportional to photon energy, the fluctuations in the intermediate processes set the energy resolution at about three times worse than that of the proportional counter. *Solid-state semiconductor detectors* have the advantages of greater stopping power and thus higher QE than the gas-filled detectors. For Si, $\omega = 3.6$ eV per electron-hole pair with improved energy resolution due to the reduction in initial charge fluctuations. Non-dispersive energy resolution in semiconductor and cryogenic detectors is further discussed in Chapters 24 and 28 (Holland 2010; Porter 2010).

X-rays with $E \leq 20$ keV are absorbed in quite small thicknesses of material and have refractive indices $n \leq 1$, so conventional optical systems are not appropriate. Thus the early X-ray astronomy instruments used simple mechanical collimators of "egg box" or honeycomb construction (Figure 4.3). The smallest achievable value of Θ was $\approx 1^\circ$, and hence source positions could be at best determined to a fraction of a degree for bright sources. In the rotation modulation collimator (RMC; Oda 1965, Figure 3), the detector is placed behind a matching pair of grids made from high-Z X-ray absorbing material and the system is rotated around the z-axis. Intensity for a single off-axis source is modulated with frequency as a function of the source off-axis angle, θ, while the signal phase depends on the source's azimuthal position. Finally the relative modulation by grid pairs with different angular resolutions allows estimates of source size. These principles have been applied in the *RHESSI* solar instrument (Lin et al 2002) where nine grid pairs and cooled Ge detectors allow estimates of position, radiance and size for solar hard X-ray sources. These systems are further discussed in Chapter 12 (Hurford 2010).

Focussing and dispersion

Although X-rays can not be handled by normal-incidence optics, given a typical metal refractive index $n \approx 0.995 = 1 - \delta$ we obtain the critical angle θ_c for total external or grazing-incidence reflection from $\cos\theta_c \approx \sqrt{2\,\delta}$ and find $\theta_c \approx 0.5^\circ$ to 2° for the 1 keV to 10 keV range. Two reflections are necessary for aberration-free image formation and a confocal paraboloid-hyperboloid configuration can achieve this (Wolter 1952; see also Chapter 9, Lemaire et al 2010). Given the very small wavelength and angle values it is necessary to figure and polish the reflecting surfaces with extreme care and precision.

The constraints of small grazing-incidence angles can be relaxed for $\lambda \geq 5$ nm by using coatings of alternate high- and low-Z material deposited on a substrate. The reflection process is similar to Bragg diffraction. Since the layers are continuous, unlike the isolated atom diffracting centres in a crystal, good reflectance values are possible. Using 40 layer pairs of Si and Mo, Windt (1998) achieved a fractional reflectance > 60 % at a wavelength of 13.5 nm and at near-normal incidence. However such performance is available in a restricted wavelength range. Peak reflectance increases and pass band narrows as the number of layer pairs increases but eventually a limit is reached due to absorption in the multilayer structure. Reflection over a broad range of higher photon energies can be achieved at very small grazing angles by the use of depth-graded multilayers as described further in Chapter 12 (Hurford 2010).

Dispersive elements — diffraction gratings and Bragg crystals — with comparatively low throughput, have worked well for the solar corona but it has been more difficult to achieve adequate sensitivities for cosmic sources. Significant progress has been made following the launch of *Chandra* (July, 1999) and *XMM-Newton* (December, 1999). Both used diffraction gratings which for *Chandra* were transmission structures mounted behind the nest of four concentric grazing-incidence telescope elements. When the gratings are in position, X-rays are diffracted through an angle β to the focal plane detectors according to the relation $\sin\beta = n\,\lambda/d$, where n is the order number and d the grating period. The transmission structures, consisting of small gold bars mounted on thin polyamide membranes, can be removed from the telescope beam to allow reflected radiation to reach the focal plane directly. High and low-energy gratings operate from 0.1 keV to 10 keV with $E/\Delta E$ from 1000 to

60. In contrast *XMM-Newton* has fixed reflection grating arrays behind two of its three telescope modules. The gratings intercept 40 % of the beam and disperse the radiation to a separate CCD detector while the remaining radiation reaches the telescope focal planes. Here a dispersion equation $m\,\lambda = d\,(\cos\beta - \cos\alpha)$ applies where α and β are the angles between the incident and dispersed rays and the gold coated SiC grating surface. Spectral order values are $m = -1, -2, \ldots$. Spectral resolution is in the range $\lambda/\Delta\lambda \approx 60$ to 800 for a wavelength range 0.5 nm to 3.5 nm. With its large effective area at the higher X-ray energies, the *XMM-Newton* telescope was optimised for the spectroscopy of fainter sources (see Figure 4.2).

For solar coronal spectroscopy at $E > 1$ keV, Bragg spectrometers have excellent wavelength resolution ($\lambda/\Delta\lambda \approx 10^4$). Given a low throughput and the difficulty in connecting them to X-ray optical systems, they have been mainly used for solar flare studies where the curved crystal spectrometer, which simultaneously registers a range of wavelengths, has had considerable success. However, use of multilayer coatings on a normal-incidence mirror and grating in the *Hinode* EIS provides an instrument operating at longer wavelengths (15 nm to 30 nm) that combines high efficiency with $2''$ angular resolution and $\lambda/\Delta\lambda \approx 4000$. The available emission lines allow plasma temperatures in the range 0.5 MK to 30 MK to be studied, though raster scanning imposes a cadence penalty. An EIS spectrum of an active region is shown in Figure 4.1.

Non-dispersive spectroscopy: CCDs and cryogenic detectors

In non-dispersive systems used for spectroscopy, an X-ray photon is absorbed in a medium and converted either to charge or to heat where the latter is described as phonons or quantised vibrations of the crystal lattice. These detectors have the advantages of high QE and an ability to register spectra simultaneously over a broad energy range. In Si CCD devices, energy resolution depends on the energy per electron-hole pair, the electronic noise and the Fano factor which describes the effect of uncertainty in the heat transferred to the crystal lattice. $E/\Delta E$ values of 15 to 60 can be achieved for 0.3 keV $< E <$ 10 keV. More detail is given in Chapter 25 (Schühle 2010). These devices are very useful for studies of the Fe lines at 6 keV to 7 keV in the spectra of AGN. CCD cameras have been used on *Chandra* and *XMM-Newton*.

A major advance came from adaptation of the IR bolometer principle to single photon spectroscopy (McCammon et al 1984). With absorption of an X-ray photon at $T \leq 1$ K, thermal phonons are created where the mean phonon energy is $\leq$ 0.1 meV. Thus the fluctuation in phonon number is $\approx 10^3$ times less than for the charge created in a Si CCD with $\omega = 3.6$ eV. In the absence of noise, energy resolution is set by phonon fluctuations in the absorber and the link to the cold sink giving $\Delta E \approx \sqrt{k_B\, T_S^2\, C_v}$, where T_S is the sink temperature and C_v is the absorber heat capacity. A semiconductor thermometer bonded to the absorber generates a temperature impulse for each detected photon. Absorber equilibrium is restored in a time $\tau = C_v/G$, where G is the thermal conductance of the link to the cold sink. A system of this design for the Japanese *Suzaku* mission has shown an energy resolution of 6.7 eV in orbit for 5.9 keV photons (Kelley et al 2007). Thus the figure for $E/\Delta E \approx 900$ is similar to that achieved with dispersive spectrometers. An alternative approach (Irwin 1995) involves the use of a transition edge thermometer with a photon absorber to form a transition edge sensor (TES). Energy resolutions of ≈ 3 eV for $E = 5.9$ keV have been achieved with these devices. Microcalorimeters are discussed further in Chapter 28 (Porter 2010).

The superconducting tunnel junction (STJ) consists of two superconducting layers separated by a thin insulating barrier. It operates on a different principle by which the photon energy is converted to free electrons by breaking Cooper pairs. The number of electrons created by the absorption of a photon of wavelength λ in a tantalum/aluminium device (energy gap $E_{\mathrm{Ta}} = 0.664$ meV) with a critical temperature of 4.5 K is given by $N(\lambda) = 7 \times 10^5\,(\mathrm{nm}/\lambda)\,(E_{\mathrm{Ta}}/\mathrm{meV})$. A 100 nm photon will create 10 000 electrons and allow the extension of non-dispersive spectroscopy to UV and visible wavelengths. So far, resolution figures of ≈ 11 eV have been achieved for $E = 5.9$ keV and so optimum application for STJs may be at very soft X-ray and UV wavelengths. However, the use of materials with lower band gaps, e.g., Mo (0.139 meV), Hf (0.02 meV), may be possible in the future. STJs are discussed more fully in Chapter 27. For all of these non-dispersive devices, the key challenge is the fabrication of arrays for use at telescope focal planes in a manner that preserves the single-device energy resolution figures.

X-ray astronomy missions

In this section we will briefly outline the characteristics of some space missions so as to illustrate the methods of solar and cosmic X-ray astronomy observation.

Solar observations

The *Yohkoh* mission, launched by Japan's ISAS in August 1991 (Ogawara et al 1991), provides a good example of a well-balanced approach to solar X-ray studies. The payload included a grazing-incidence Soft X-ray Telescope (SXT; Tsuneta et al 1991), a multi-grid Hard X-ray Telescope (HXT; Kosugi et al 1991), a Bragg Crystal Spectrometer (BCS; Culhane et al 1991) and a wide-band non-dispersive X-ray spectrometer (WBS; Yoshimori et al 1991), where the latter two instruments had no imaging capability and were used for solar flare spectroscopy. The grazing-incidence SXT used a front-illuminated CCD and interchangeable filters to give X-ray images in the 0.3 nm to 5.0 nm range with $\leq 3''$ spatial resolution. It had a co-aligned white-light lens system on axis so that photospheric images could be registered on the same CCD thus ensuring good X-ray to visible solar image registration. The novel HXT instrument used 64 different grid pairs or subcollimators which, following Fourier deconvolution, gave images in the 15 keV to 100 keV range with $\approx 5''$ spatial resolution. The BCS had four curved crystals that registered the emission lines of S^{+14}, Ca^{+18}, Fe^{+24}, and Fe^{+25} with $\lambda/\Delta\lambda$ in the range 3000 to 6000. Finally, the WBS used a combination of gas proportional counters and two scintillation detectors for flare studies in the 3 keV to 100 MeV range. Following ten years of observation in orbit, *Yohkoh* has had a major impact on the subject. The role of magnetic reconnection and related plasma dynamics in solar flares has been clarified. Some coronal magnetic structures that are important in CMEs have

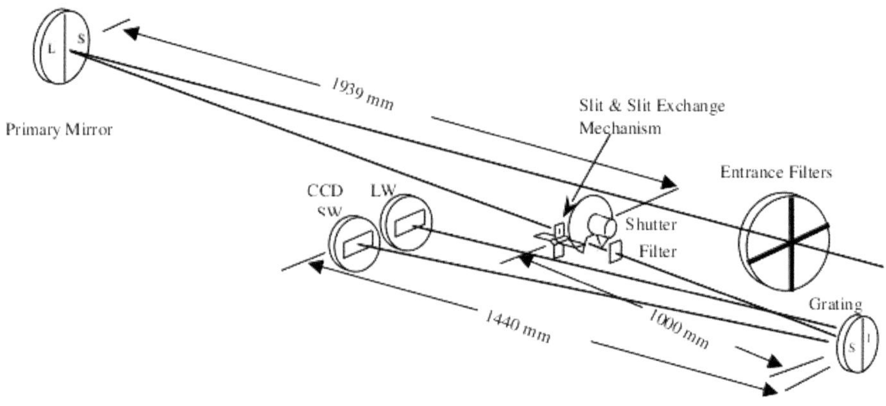

Figure 4.4: Optical layout of the *Hinode*-EIS spectrometer. Components are labelled with their relative distances. L/LW and S/SW refer to long and short wavelength bands.

been identified and the evolution of the solar corona through a complete solar cycle has been observed.

More recently the multilayer coating technique has been used very successfully in the manufacture of normal-incidence EUV imaging telescopes for the *SOHO* and *TRACE* missions that were launched in December 1995 and April 1998. The *SOHO* EUV Imaging Telescope (EIT; Delaboudinière et al 1995) employs Si and Mo layers to register the spectral lines of He II (30.4 nm, 0.08 MK)[1], Fe IX, Fe X (17.1 nm, 1.3 MK), Fe XII (19.5 nm, 1.6 MK), and Fe XV (28.4 nm, 2 MK), thus allowing full-Sun images to be obtained of chromospheric and coronal plasma. When the telescope performance is convolved with the 21 μm pixels of the back-illuminated CCD, the angular resolution is 5″. *TRACE* (Handy et al 1999), which stopped operation in June 2010, also used a normal-incidence multilayer telescope system with somewhat enhanced reflectivity due to the use of Mo_2C and Si layers. Three of its four channels registered the EUV lines of Fe^{+8}, Fe^{+11} and Fe^{+14} while the fourth was optimised for UV lines with $\lambda \geq 120$ nm and did not have a multilayer coating. *TRACE* had 1″ angular resolution with a 0.5″ pixel size but unlike EIT its field of view was restricted to 8.5′ by 8.5′. While *SOHO* is in orbit around the L1 point at 1.5×10^6 km from Earth with an unrestricted view of the Sun, *TRACE* was in a Sun-synchronous near-Earth orbit and so had annual eclipse seasons of ≤ 20 min per orbit for one to two months each year. These two missions have already made a number of important discoveries. Both *Yohkoh* and *SOHO*-EIT have observed dimming regions in the corona following the launching of CMEs (e.g., Hudson et al 1996), while EIT has discovered a new kind of global wave phenomenon whose origins remain controversial (Thompson et al 1999). The superior angular resolution of *TRACE* has allowed detailed studies of coronal active region structures along with the detection of a variety of wave modes propagating in such structures (e.g., Nakariakov et al 1999).

[1] transition wavelength and formation temperature

Another example of multilayer coated EUV optics used for solar spectroscopy is provided by the EIS instrument on *Hinode*. Launched in September 2006 into a Sun-synchronous orbit, the spacecraft included a 0.5 m aperture optical telescope and a grazing-incidence X-ray telescope with $\leq 2''$ spatial resolution. This telescope has a larger effective area than the *Yohkoh* telescope mentioned above and responds to longer wavelength X-rays (Golub et al 2007, Kano et al 2008). An optical layout diagram of the EUV spectrometer is given in Figure 4.4 (Culhane et al 2007). Both primary mirror and grating are coated with matching and separately optimised Si/Mo multilayers that respond in two pass bands; 17 nm to 21 nm and 25 nm to 29 nm with the radiation being detected by a pair of thinned back-illuminated CCDs with a relative QE ≈ 60 % at 26 nm. Although the chosen multilayer properties restrict these wavelength ranges, the large number of detectable emission lines allows plasma with temperature ranging from 0.4 MK to 30 MK to be observed. In addition, a resolution of $\lambda/\Delta\lambda \approx 4000$ at 18.5 nm allows plasma velocity measurements to ± 3 km s^{-1} with ≈ 200 counts in the line profile.

Using EIS, Harra et al (2008) observed outflows from active regions that may indicate an additional source of the slow solar wind while the detection of non-thermal line broadening in active regions (Doschek et al 2007) could provide important clues to the coronal plasma heating mechanism.

X-ray astronomy: surveys

The use of collimated proportional counters in orbit provided the first systematic surveys of the X-ray sky. The *Uhuru* mission launched in December 1970 (Giacconi et al 1971), led to the 4U catalogue of 339 sources. Two collimated detectors with fields of view of $0.52^\circ \times 5.2^\circ$ and $5.2^\circ \times 5.2^\circ$ and responding in the 2 keV to 20 keV range were mounted back-to-back in a spacecraft which rotated once per 720 s. The narrow field of view of one detector was scanned in a direction perpendicular to the spin axis to provide better source position accuracy. In addition to the source catalogue, *Uhuru* discovered X-ray emitting binary systems that contain either a neutron star or a black hole. Here the emission arises from heated gas that has been accreted from the companion star and raised to high temperature following capture in the deep gravitational potential well of the few solar mass degenerate object.

Since *Uhuru*, there have been about ten missions with primary emphasis on X-ray surveys. However with the growing maturity of the discipline, focus has shifted from specialist missions to large general-purpose X-ray observatories which are used both to survey sky fields and to study targets of particular interest and importance. The *ROSAT* or Röntgen-Satellite (Trümper 1982) was a German-led mission with US and UK participation designed to conduct an all-sky survey followed by pointed observations of specified sources. Launched in June 1990, it included a large grazing-incidence telescope with four nested paraboloid-hyperboloid pairs made from Zerodur. With a 2° field of view and a response band 0.1 keV to 2.5 keV, the telescope had a resolution of $2''$ and a location accuracy of $10''$ in the survey mode. The all-sky survey, conducted in the first six months of the mission, resulted in a catalogue of $\approx 150\,000$ X-ray sources with close to 500 EUV sources discovered by the accompanying wide-field EUV telescope. During its almost nine year lifetime positions were established for a further 100 000 sources observed during the pointed phase of the mission. Scientific highlights included structural studies of supernova remnants and clusters of galaxies, detection of X-rays radiated from the cooling surfaces of isolated neutron stars and the discovery of X-ray emission from comets.

X-ray astronomy: high-resolution imaging and high-throughput spectroscopy

In 2009 the current state of the art in X-ray astronomy is represented by two large observatory class missions — *Chandra* and *XMM-Newton* — which have been in orbit since 1999. Their elliptical orbits, with periods of 64 h for *Chandra* and 48 h for *XMM-Newton*, allow both spacecraft to remain above the Earth's trapped radiation regions and to be free of Earth occultation for 65 % to 70 % of the time. They employ grazing-incidence Wolter I telescopes whose characteristics are summarized in Table 4.1. The *Chandra* telescope has four nested reflectors made from low-expansion glass that is highly polished to achieve ≤ 0.3 µm surface roughness which in turn leads to an angular resolution of $0.5''$ FWHM. In contrast, the *XMM-Newton* telescopes are made by a replication method by which thin gold-coated nickel shells are carefully separated from highly-polished mandrels with 58 separate shells being accurately assembled in each single telescope module. Inability to achieve *Chandra*-like surface roughness levels leads to an inferior angular resolution of $5''$. However, the effective reflecting area at 6 keV to 7 keV, where the emission lines of highly ionised Fe are found, is more than 10 times greater than that of *Chandra*. Thus *Chandra* emphasises angular resolution for source positioning while *XMM-Newton* achieves high photon collecting power for spectroscopy at the expense of angular resolution.

Chandra uses CCDs and a microchannel plate (MCP) array (Chapter 22, Timothy 2010) to register reflected X-rays. When used for imaging, the fields of view are $17' \times 17'$ (CCD) and $31' \times 31'$ (MCP). There are also detector arrays that are used to register dispersed spectra when the transmission gratings are placed behind the telescope mirrors. The gratings and their spectroscopic performance have already been described previously. Low- and high-energy grating arrays respond in the en-

Table 4.1: Features of the *Chandra* and *XMM-Newton* telescopes.

	Chandra	*XMM-Newton*
Mirrors	4 nested	58 nested
Aperture dia., D/m	1.2	0.7
Focal length, f/m	10	7.5
Geom. area, a/cm^2	1100; 1 module	6000; 3 modules
Grazing angles	$27'$ to $51'$	$18'$ to $40'$
Coating	iridium	gold
Resolution (FWHM)	$0.5''$	$5''$
E_{max}/keV	10	10

ergy ranges 0.08 keV to 2 keV and 0.8 keV to 8 keV. A review of the first two years of observation together with a description of the mission instrumentation has been presented by Weisskopf et al (2002). *Chandra* is a wide-ranging observatory facility whose impact on astronomy is now on a par with that of the *Hubble* telescope.

XMM-Newton includes three of the mirror modules described in Table 4.1. The field of view is $\approx 14'$ in radius. For two of the modules, ≤ 60 % of the reflected radiation is imaged by front illuminated MOS CCDs while the remaining 40 % is dispersed by the reflection grating arrays. For the third module, all of the reflected radiation is registered on a pn CCD camera which, because of its greater depletion depth, has good quantum efficiency up to 10 keV. The reflection grating stacks that are placed behind the other two modules have been discussed above. In addition to the X-ray telescope systems, the payload also includes a 0.3 m aperture optical telescope which covers the central $17'$ square region of the X-ray telescope field of view. Operating in the 170 nm to 650 nm range, it can detect a $B = 23.5$ magnitude star in ≈ 1000 s — a sensitivity equivalent to that of a 4 m optical telescope on Earth. The mission and its operations are outlined by Jansen et al (2001).[2]

Conclusions

In the 45 years since the discovery of Sco-X1, X-ray astronomy has become a mainstream branch of the subject. Relevant to high-energy processes, X-ray observations are important on scales ranging from the solar system to that of the Universe itself. Thanks to the use of grazing-incidence telescopes, sensitivity has been increased by more than a factor 10^9. The development of cryogenic non-dispersive spectrometers will allow detailed studies of the physics of even the faintest sources. Continued progress will require significant increase in effective photon collecting area for $E \leq 15$ keV. The ESA *XEUS* concept, involving mirror and detector spacecraft separation of ≈ 50 m and relative navigation with ≈ 0.1 mm precision in orbit at L2 (see Chapter 38), offers one solution for achieving a gain in sensitivity of $\approx 10^2$ over *Chandra*. The use of graded thickness multilayer coatings will extend the capability for high-sensitivity observations to $E > 100$ keV. Cryogenic spectrometers with $E/\Delta E = 1000$ will largely replace dispersive systems. X-ray astronomy is crucial for our understanding of the Universe and will continue to be vigorously pursued.

Bibliography

Blake RL, Chubb TA, Friedman H, Unzicker AE (1963) Interpretation of X-ray photograph of the Sun. Astrophys J 137:3–15

Bowyer S, Byram ET, Chubb TA, Friedman, H (1964) Lunar occultation of X-ray emission from the Crab Nebula. Science 146:912–917

Burnight TR (1949) Soft X-radiation in the upper atmosphere. Phys Rev 76:165

[2]This paper forms part of a special issue of Astronomy and Astrophysics which also includes descriptions of the instrumentation and covers a broad range of the early scientific output.

Culhane JL, Pounds KA, Willmore AP, Sanford PW (1964) The solar X-ray spectrum below 14 Å. Space Research 4:741–758

Culhane JL, Hiei E, Doschek GA (plus 24 authors) (1991) The Bragg crystal spectrometer for *Solar-A*. Sol Phys 136:89–104

Culhane JL, Harra LK, James AM (plus 36 authors) (2007) The EUV Imaging Spectrometer for *Hinode*. Sol Phys 243:19–61

Delaboudinière J-P, Artzner GE, Brunaud J (plus 25 authors) (1995) EIT: Extreme-ultraviolet Imaging Telescope for the *SOHO* mission. Sol Phys 162:291–312

den Herder JW, Brinkman AC, Kahn SM (plus 35 authors) (2001) The Reflection Grating Spectrometer on board *XMM-Newton*. Astron Astrophys 365:L7–L17

Dere KP, Landi E, Mason HE (plus two authors) (1997) CHIANTI – An atomic database for emission lines. Astron Astrophys Suppl Ser 125:149–173

Doschek GA, Mariska JT, Warren HP (plus six authors) (2007) Nonthermal velocities in solar active regions observed with the Extreme-Ultraviolet Imaging Spectrometer on *Hinode*. Astrophys J 667:L109–L112

Giacconi R, Gursky H, Paolini FR, Rossi BB (1962) Evidence for X rays from sources outside the Solar System. Phys Rev Lett 9:439–443

Giacconi R, Reidy WP, Zehnpfennig T (plus two authors) (1965) Solar X-ray image obtained using grazing-incidence optics. Astrophys J 142, 1274–1278

Giacconi R, Kellogg E, Gorenstein P (plus two authors) (1971) An X-Ray scan of the galactic plane from *Uhuru*. Astrophys J 165:L27–L35

Gursky H, Giacconi R, Gorenstein P (plus five authors) (1966) A measurement of the location of the X-ray source SCO X-1. Astrophys J 146:310–316

Golub L, Deluca E, Austin G (plus 26 authors) (2007) The X-Ray Telescope (XRT) for the *Hinode* mission. Sol Phys 243:63–86

Handy BN, Acton LW, Kankelborg CC (plus 45 authors) (1999) The Transition Region and Coronal Explorer. Sol Phys 187:229–260

Harra LK, Sakao T, Mandrini C (plus five authors) (2008) Outflows at the edges of active regions: Contribution to solar wind formation? Astrophys J 676:L147–L150; Attn: Erratum. Astrophys J 677:L159

Holland A (2010) X-ray CCDs. ISSI SR-009:409–418

Hudson HS, Acton LW, Freeland SL (1996) A long-duration solar flare with mass ejection and global consequences. Astrophys J 470:629–635

Hurford GJ (2010) X-ray imaging with collimators, masks and grids. ISSI SR-009:223–234

Irwin KD (1995) An application of electrothermal feedback for high resolution cryogenic particle detection. Appl Phys Lett, 66, 1998–2000

Jansen F, Lumb D, Altieri B (plus 12 authors) (2001) *XMM-Newton* Observatory. I. The spacecraft and operations. Astron Astrophys 365:L1–L6

Kaastra JS, Mewe R, Nieuwenhuijzen H (1996) SPEX: a new code for spectral analysis of X & UV spectra. in UV and X-ray Spectroscopy of Astrophysical Plasmas, eds K Yamashita and T Watanabe, Tokyo Univ Press, p. 411–414

Kano R, Sakao T, Hara H (plus 15 authors) (2008) The *Hinode* X-Ray Telescope (XRT): Camera design, performance and operations. Sol Phys 249:263–279

Kelley RL, Mitsuda K, Allen CA (plus 40 authors) (2007) The *Suzaku* high resolution X-ray spectrometer. PASJ 59:77–112

Kosugi T, Masuda S, Makishima K, (plus seven authors) (1991) The Hard X-ray Telescope (HXT) for the *Solar-A* Mission. Sol Phys 136:17–36

Koyama K, Petre R, Gotthelf EV (plus four authors) (1995) Evidence for shock acceleration of high-energy electrons in the supernova remnant SN1006. Nature 378:255–258

Landi E, Del Zanna G, Young PR (plus three authors) (2006) CHIANTI – An atomic database for emission lines. VII. New data for Xrays and other improvements. Astrophys J Suppl Ser 162:261–280

Lemaire P, Aschenbach BA, Seely JF (2010) Space telescopes. ISSI SR-009:165–191

Lin RP, Dennis BR, Hurford GJ and 63 other authors (2002) The Reuven Ramaty High-Energy Solar Spectroscopic Imager (*RHESSI*). Sol Phys 210:3–32

McCammon D, Moseley SH, Mather JC, Mushotzky RF (1984) Experimental tests of a single-photon calorimeter for X-ray spectroscopy. J Appl Phys 56:1263–1266

Nakariakov VM, Ofman L, Deluca EE (plus two authors) (1999) *TRACE* observation of damped coronalloop oscillations: Implications for coronal heating. Science 285:862–864

Oda M (1965) High-resolution X-ray collimator with broad field of view for astronomical use. Appl Opt 4:143

Ogawara Y, Takano T, Kato T (plus five authors) (1991) The *Solar-A* Mission — an overview. Sol Phys 136:1–16

Porter FS (2010) X-ray calorimeters. ISSI SR-009:459–475

Sandage A, Osmer P, Giacconi R (plus nine authors) (1966) On the optical identification of SCO X-1. Astrophys J 146:316–321

Schühle U (2010) Intensified solid state sensor cameras: ICCD and IAPS. ISSI SR-009:419–428

Thompson BJ, Gurman JB, Neupert WM (plus seven authors) (1999) *SOHO*/EIT observations of the 1997 April 7 coronal transient: Possible evidence of coronal Moreton waves. Astrophys J 517:L151–L154

Timothy JG (2010) Microchannel plates for photon detection and imaging in space. ISSI SR-009:365–391

Trümper J, Pietsch W, Reppin C (plus three authors) (1978) Evidence for strong cyclotron line emission in the hard X-ray spectrum of Hercules X-1 Astrophys J 219:L105–L110

Trümper J (1982) The *ROSAT* Mission. Adv Space Res 2:241–249

Tsuneta S, Acton L, Bruner M (plus seven authors) (1991) The Soft X-ray Telescope for the *Solar-A* Mission. Sol Phys 136:37–67

Vaiana GS, Davis JM, Giacconi R (plus four authors) (1973) X-ray observations of the solar corona: Preliminary *Skylab* results Astrophys J 185:L47–L51

Weisskopf MC, Brinkman B, Canizares C, (plus four authors) (2002) Five years of observations with the *Chandra* X-ray observatory PASP 114:1–24

Werner N, Böhringer H, Kaastra JS (plus three authors) (2006) *XMM-Newton* high-resolution spectroscopy reveals the chemical evolution of M 87. Astron Astrophys 459:353–360

Wilhelm K, Fröhlich C (2010) Photons—from source to detector. ISSI SR-009:23–54

Windt DL (1998) Multilayer films for figured X-ray optics. Proc SPIE 3448:371–382

Wolter H (1952) Spiegelsysteme streifenden Einfalls als abbildende Optiken für Röntgenstrahlen. Ann Phys 10:94–114

Yoshimori M, Okudaira K, Hirasima Y (plus seven authors) (1991) The Wide Band Spectrometer on the *Solar-A*. Sol Phys 136:69–88

Young PR, Del Zanna G, Mason HE (plus nine authors) (2007) EUV emission lines and diagnostics observed with *Hinode*/EIS. Publ Astron Soc Japan 59:S857–S864

— 5 —

The extra-terrestrial vacuum-ultraviolet wavelength range

J. Gethyn Timothy[I], Klaus Wilhelm[II], and Lidong Xia[III]

Abstract

Electromagnetic radiation in the vacuum-ultraviolet (VUV) and extra-terrestrial range at wavelengths from 10 nm to 300 nm is absorbed in the upper atmosphere by ozone, molecular and atomic oxygen, and molecular nitrogen. Observations at wavelengths down to ≈ 200 nm can be carried out from stratospheric balloons, and observations below 200 nm require space platforms operating at altitudes above 250 km. The VUV spectral region contains emission lines and continua arising from plasma at formation temperatures ranging from about 10^4 K to more than 10^7 K. This chapter describes the wide range of plasma diagnostic techniques available at VUV wavelengths, and the development of instrumentation for studies of the high-temperature solar outer atmosphere and astrophysical plasmas. Finally, the prospects for future studies are briefly discussed.

The early days

It has been known for many years that the outer atmosphere of the Sun is significantly hotter than the 5800 K temperature of the photosphere, particularly from observations of forbidden coronal lines at 637.5 nm (Fe X—red line) and 530.3 nm (Fe XIV—green line) by Lyot (1937), Grotrian (1939), and Edlén (1943). The Sun must emit strong ultraviolet (UV) radiation from these regions at wavelengths shorter than those of the visible spectrum ($\lambda \approx 400$ nm to ≈ 800 nm). However, ground-based efforts to explore this wavelength regime have met with only limited success. This is because of the onset of very strong absorption at wavelengths below ≈ 290 nm by ozone at altitudes between about 15 km and 35 km (Massey and Boyd 1960).

At shorter wavelengths below ≈ 200 nm molecular oxygen in both the Herzberg dissociation continuum and the Schumann-Runge bands causes complete absorption at higher altitudes, and atomic oxygen and molecular nitrogen at wavelengths below ≈ 90 nm cause absorption at altitudes above 160 km. The altitude at which

[I]Nightsen, Inc., Tiverton RI, USA
[II]MPS—Max-Planck-Institut für Sonnensystemforschung, Katlenburg-Lindau, Germany
[III]School of Space Science and Physics, Shandong University, China

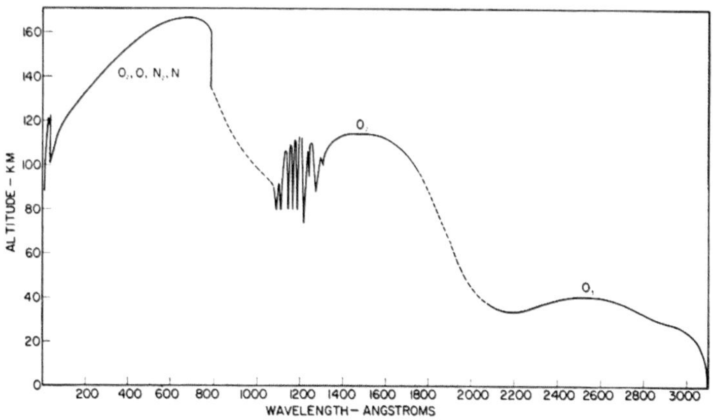

Figure 5.1: The altitude at which the fraction of the UV radiation is reduced by a factor of 1/e (from Friedman 1959; reproduced from Savage 2001).

the fraction of the vacuum-ultraviolet (VUV) radiation is reduced by a factor of 1/e is shown in Figure 5.1.

Atmospheric effects

Currently there is considerable concern about the increasing depletion of the ozone layer, caused by photo-dissociation of chlorofluorocarbon compounds (CFCs), which could expose the biosphere to increasing levels of UV-A (315 nm to 380 nm) and UV-B (280 nm to 315 nm). This has led to a phase-out and ban on the production and use of CFCs by the US Environmental Protection Agency in 1994 and 1996 under extensions to the 1987 Montreal Protocol.[1] A critical example of ozone depletion is the annual formation of an ozone "hole" over Antarctica. UV-B radiation is known to cause skin cancer, and significant increases in skin cancer and melanoma have been recorded in Chile as the direct result of the expansion of the ozone "hole" (Abarca and Casiccia 2002). UV-A radiation may also play a role in causing melanomas, although the effects are mediated by reactive oxygen species (de Gruijl 2002).

VUV radiation at wavelengths below ≈ 120 nm is responsible for the heating of the thermosphere and the formation of the E and F regions of the ionosphere (Timothy et al 1972). However, the very strong hydrogen Ly α emission line at 121.6 nm coincides with a window in the O_2 absorption cross-section, which allows the radiation to penetrate to below 90 km, where together with VUV radiation at wavelengths longer than ≈ 180 nm and soft X-rays, it forms the D region of the ionosphere (Nicolet and Aikin 1960).

[1] For details see http://www.epa.gov/ozone.

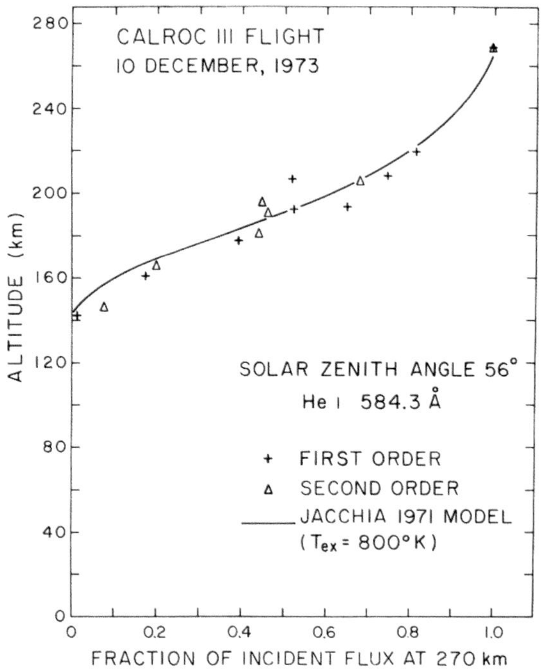

Figure 5.2: Comparison of a measured He I 58.4 nm absorption profile with a calculated profile from a Jacchia model atmosphere (Timothy et al 1975).

An example of the fitting of a Jacchia model profile (Jacchia 1971, 1977) to a measured sounding rocket profile of the strong He I 58.4 nm resonance line is shown in Figure 5.2.

Observing platforms and enabling technologies

Whereas measurements at wavelengths down to $\approx$ 200 nm are possible from high-altitude balloons floating at altitudes above 40 km, measurements at shorter wavelengths require the use of sounding rockets, satellites orbiting at altitudes well above 160 km, or space probes. Even at altitudes above 200 km residual absorption by the geocorona and terrestrial airglow emission can affect the measurements. The initial measurements were directed at studies of the solar VUV emission because of the relatively high irradiances. Studies of fainter astrophysical objects followed at a later date as the technology improved.

The earliest sounding rocket measurements were made in the US in the late 1940s using, first, captured German V2 rockets, and then US-developed Aerobee rockets. The data obtained with these unstabilized rocket systems were limited and special techniques were required to direct the solar VUV radiation into the spectrometers (de Jager 2001; Wilhelm 2003). Limited measurements of the solar

VUV spectrum were obtained down to ≈ 200 nm. The development in the 1950s and 1960s of more powerful sub-orbital and orbital launch vehicles (Russo 2001) was the key to future progress. The development of two-axis and three-axis pointing systems, first for sounding rockets and, later, for orbiting satellites, allowed not only the exploration of the solar VUV, but also permitted the first observations of VUV emission from astrophysical objects.

The early spectra and images at VUV wavelengths were recorded using photographic emulsions. Conventional emulsions have extremely low sensitivities at VUV wavelengths because of absorption by the gelatin layer containing the silver halide grains. Special emulsions for VUV studies, where the silver halide grains were located on or very near the surface of the gelatin were developed by Schumann (1901). The early Schumann emulsions were fabricated on glass plates, which are not suitable for use in the launch environment for space vehicles, and are extremely sensitive to contamination and are destroyed by physical contact. In the years leading up to the *Skylab* flight in 1973, a series of special film-based Schumann emulsions and their holders were successfully developed (VanHoosier et al 1977). Since the solar VUV radiation arises in the hot outer atmosphere of the Sun and is many magnitudes fainter than the intense visible light radiation from the photosphere (cf., Figure 2.2 in Chapter 2, Wilhelm and Fröhlich 2010), the development of "solar blind" detector systems insensitive to visible light was critical to the fields of VUV solar and astrophysical studies (see Chapters 22, 25, and 26, Timothy 2010; Schühle 2010; Schühle and Hochedez 2010). Further, at wavelengths below ≈ 116 nm no rugged window materials are available. Hence there is a need for reflective optical systems, and for detector systems that can operate without windows at shorter wavelengths. Finally, the reflectance of normal-incidence optical systems falls dramatically below ≈ 50 nm, and the use of grazing-incidence telescopes (Brueggemann 1968) and spectrometers (Samson 1980), or synthetic multilayer interference coatings (Spiller 1974), is mandatory.

Physical processes

The solar VUV spectrum at wavelengths longer than about 200 nm is fundamentally similar to the solar visible light spectrum, namely, a radiation continuum with superimposed Fraunhofer absorption lines. Below about 200 nm the spectrum is densely populated with strong emission lines and continua formed at temperatures from about 4×10^3 K to greater than 10^7 K. The corresponding energetic photons are responsible for critical processes in the atmosphere of the Earth, such as photodissociation of key atmospheric molecules, the formation of the different layers of the ionosphere, and the heating of the thermosphere (Timothy et al 1972). In addition, the very wide range of excitation and ionization temperatures makes this spectral region ideal for studies of the hot outer regions of the atmospheres of the Sun and other stars, and for the investigation of other high-temperature and dynamic phenomena in the Universe.

In order to use the VUV emission lines for plasma diagnostics, the processes leading to the formation of the emission lines and continua must be fully understood. Wilhelm et al (2004) have presented a detailed review (with many references) of the processes of spectral line formation and plasma diagnostic techniques, specifically for studies of the solar chromosphere, transition region, and corona. We review briefly here some of the key issues.

Spectral line formation

Some of the radiation in the VUV range is emitted by atoms, notably by neutral hydrogen and helium, but many of the VUV spectral lines are generated by ions of heavier elements. One of the key processes for producing the ions is electron impact ionization:

$$X^{+Z} + e^- \longrightarrow X^{+(Z+1)} + 2e^- \quad , \tag{5.1}$$

where X^{+Z} is a Z-fold charged ion of element X and e^- on the left-hand side is an electron with sufficient kinetic energy, W, to provide the required ionization energy, W_0. This process has to compete with recombination to lower charge states and further ionization to higher charge states.

VUV spectroscopic plasma diagnostics

The wide range of ionization energies at VUV wavelengths renders this radiation extremely useful for a number of plasma diagnostic techniques. These include:

Emission measures

Many plasmas in the Universe are optically thin, and therefore the radiation emitted is seen by a detector as integrated along the line of sight (LOS) of length z_0. In general, excited atoms and ions spontaneously emit photons as the main depopulation process. The probability for a radiative transition from state j to state i is given by the Einstein coefficient, A_{ji}. The radiant power density[2] in a spectral line at λ_{ji} is

$$\varphi(\lambda_{ji}) = \Delta\varepsilon_{ij}\, A_{ji}\, n_j \quad , \tag{5.2}$$

where $\Delta\varepsilon_{ij}$ is the energy difference between the states i and j and n_j is the number density of the radiating species in state j. The quantity $\varphi(\lambda_{ji})$ can be obtained from the spectral radiant power density, φ_λ, by integrating over the line profile and deduction of the background:

$$\varphi(\lambda_{ji}) = \int\limits_{\lambda_{ji}-\delta\lambda/2}^{\lambda_{ji}+\delta\lambda/2} (\varphi_\lambda - b_\lambda)\, \mathrm{d}\lambda \quad . \tag{5.3}$$

The background, b_λ, can be determined from the shape of the spectrum near the spectral line.

An emission measure (EM) (Pottasch 1963) can then be defined as

$$\int\limits_{z_0} n_\mathrm{e}^2\, \mathrm{d}z = \frac{\langle\varphi(\lambda_{jg})\rangle\, z_0}{\gamma\, \Delta\epsilon_{gj}\, G(T_\mathrm{e})}\, \frac{n_\mathrm{H}}{n_\mathrm{X}}\, \frac{n_\mathrm{e}}{n_\mathrm{H}} = \frac{4\pi\, L(\lambda_{jg})}{\gamma\, \Delta\epsilon_{gj}\, G(T_\mathrm{e})}\, \frac{n_\mathrm{H}}{n_\mathrm{X}}\, \frac{n_\mathrm{e}}{n_\mathrm{H}} \quad , \tag{5.4}$$

[2]Often called in the literature "emissivity", which is, however, the ratio of the emission of a surface to the corresponding black-body radiation.

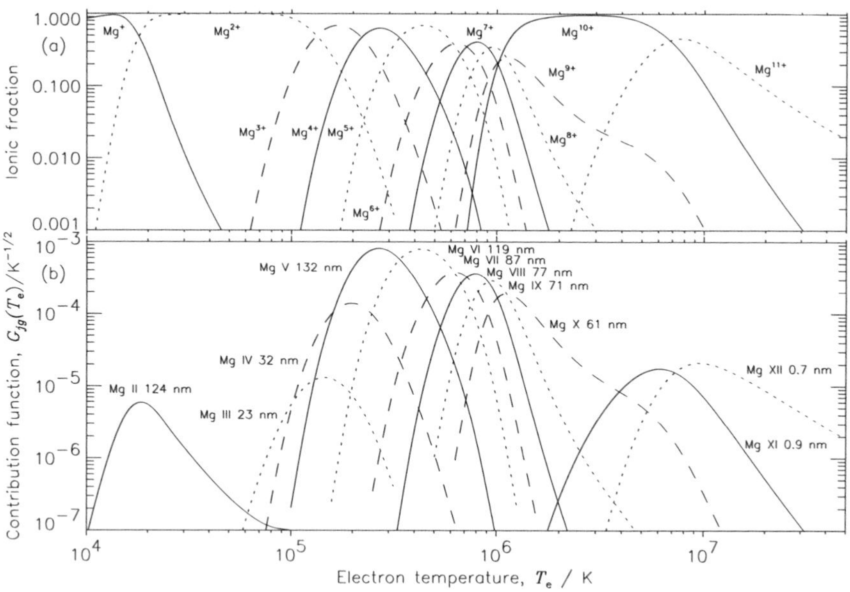

Figure 5.3: (a) Ionic fractions of magnesium ions as a function of the electron temperature of a plasma in ionization equilibrium. (b) The corresponding contribution functions for emission lines of magnesium ions in the VUV and soft X-ray wavelength ranges.

with n_e the electron density, $\langle\varphi(\lambda_{jg})\rangle$ the average power density along the LOS, and $L(\lambda_{jg}) = \langle\varphi(\lambda_{jg})\rangle\, z_0/(4\,\pi)$ the observed radiance of the emission line at λ_{jg}. The element abundance of element X with respect to hydrogen is n_X/n_H and the number density of hydrogen relative to the electron density is n_H/n_e (cf., Wilhelm et al 2004). The factor γ can be obtained from atomic physics calculations. The contribution function, $G(T_e)$, is defined by (see, e.g., Mariska 1992):

$$G(T_e) = \frac{n_g}{n_X} \frac{1}{\sqrt{T_e}} \exp\left(\frac{-\Delta\varepsilon_{gj}}{k_B T_e}\right) \quad , \tag{5.5}$$

where k_B is the Boltzmann constant and n_g/n_X the ionic fraction (Mazzotta et al 1998). The contribution function has its maximum at the formation temperature of the line. As an example we show in Figure 5.3a the ionic fractions of the element magnesium as a function of the electron temperature, T_e, taken from Mazzotta et al (1998), and in Figure 5.3b the corresponding contribution functions calculated with the help of Equation 5.5. The assumptions made above include that the plasma is iso-thermal along the LOS at a certain temperature, T_e. If there are temperature gradients along the LOS, a differential emission measure (DEM), $\Phi(T_e)$, must be used (Athay 1966). The integration is then performed over a certain temperature interval, $T \pm \Delta T/2$, in which the spectral line is formed:

$$\int_{z_0} G(T_e)\, n_e^2\, dz = \int_{\Delta T} G(T_e)\, \Phi(T_e)\, dT_e \quad , \tag{5.6}$$

with the DEM defined by

$$\Phi(T_\mathrm{e}) = n_\mathrm{e}^2 \frac{\mathrm{d}z}{\mathrm{d}T_\mathrm{e}} \quad . \tag{5.7}$$

If the plasma is optically thick, direct information can only be obtained about the "surface" by analyzing the spectral radiance. In intermediate cases, radiative transfer processes within the plasma have to be considered as well.

Density-sensitive line ratios

Collisionally-excited states may be depopulated by non-radiative processes, in particular, if they are metastable. The radiant power density is thus reduced for lines from such states as a function of n_e. The ratio

$$R_{12} = \frac{\varphi(\lambda_1)}{\varphi(\lambda_2)} \tag{5.8}$$

of an allowed emission line at λ_1 and a metastable line at λ_2 is consequently changing with n_e, and can be calculated from atomic physics principles. From Equation 5.4, we may conclude that this can be accomplished by measuring the ratio of the line radiances

$$R_{12} = \frac{L(\lambda_1)}{L(\lambda_2)} \quad . \tag{5.9}$$

This method is especially useful if both lines are emitted by particles of the same species and ionization stage, because abundance and ionization variations will have no effect (cf., e.g., Doschek et al 1997).

The electron density so determined can, under certain assumptions, be compared with the mean density obtained from the emission measure analysis and provide information on the so-called "filling factor" (cf., Dere et al 1987).

Temperature-sensitive line ratios

If two excited states, j_1 and j_2, from which emission lines originate, have very different energy levels, their relative collisional excitation becomes a function of the electron temperature, T_e, and the ratio can again be obtained from atomic physics calculations. A comparison with the measured ratio of the line radiances then provides information on the electron temperature (cf., e.g., David et al 1998).

Element abundances

Two lines at λ_1 and λ_2 from different species, but with very similar contribution functions, stemming from the same location, and thus generated at the same electron density, emit radiant power densities, $\varphi(\lambda_1)$ and $\varphi(\lambda_2)$, or, from Equation 5.4, the corresponding line radiances (adjusted for any variations of $\Delta\varepsilon_{ij}$ and γ). Their ratio can be used to obtain the element abundance ratio $n_\mathrm{X}/n_\mathrm{H}$. Variations of abundances are caused both in the corona and in the solar wind, in particular, by the first-ionization potential (FIP) effect (Feldman and Laming 2000).

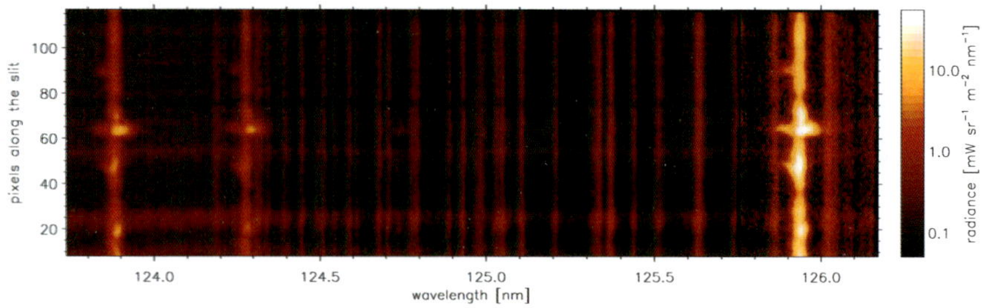

Figure 5.4: Doppler shifts in transition region lines caused by upward and downward flows in an explosive event with speeds up to 100 $\mathrm{km\,s^{-1}}$ (observation from the SUMER instrument on *SOHO*, described in Table 5.1).

Ion motions and temperatures

The non-relativistic Doppler formula

$$v = c_0 \frac{\lambda - \lambda_0}{\lambda_0} = c_0 \frac{\Delta\lambda}{\lambda_0} \quad , \tag{5.10}$$

where c_0 is the speed of electromagnetic waves in vacuo and λ_0 is the rest wavelength of an emission line, is adequate for most of the speeds encountered in solar and stellar atmospheres. Equation 5.10 allows a conversion of measured line shifts or spectral profiles[3] into plasma motions along the LOS, either as average bulk velocity, or as the corresponding component of unresolved small-scale turbulence.

With m_i the mass of the ion and $\Delta\lambda_\mathrm{D}$ the Doppler width[4], thermal and non-thermal contributions, e.g., turbulence, affect the profile according to

$$\Delta\lambda_\mathrm{D} = \frac{\lambda_0}{c_0}\sqrt{\frac{2\,k\,T_\mathrm{i}}{m_\mathrm{i}} + \xi^2} \quad , \tag{5.12}$$

where T_i is the ion temperature and ξ the most probable non-thermal LOS velocity (see, e.g., Mariska 1992). Examples of emission lines affected by Doppler shift are shown in Figure 5.4.

Atomic databases

As the quality of the observations improved, the development of atomic databases for spectroscopic diagnostics in the VUV began (see, e.g., Mason and Monsignori Fossi 1994). A highly detailed computer program called CHIANTI, using atomic data and transition rates, has been developed to allow the computation of synthetic spectra and the application of plasma diagnostic techniques. The first version was

[3]Any instrumental line broadening is assumed to have been taken out before.

[4]For a Gaussian profile of an emission line at λ_0, the Doppler width, $\Delta\lambda_\mathrm{D}$, is related to the standard deviation, σ, and the full-width at half maximum (FWHM), $\Delta\lambda_\mathrm{FWHM}$, through the equation:

$$\Delta\lambda_\mathrm{D} = \sigma\sqrt{2} = \Delta\lambda_\mathrm{FWHM}/(2\sqrt{\ln 2}) \quad . \tag{5.11}$$

Table 5.1: Key solar physics orbital and deep space missions from 1980 through 2008.

Year	Mission	Instrument	Parameters	Reference
1980	*SMM*	Ultraviolet Spectrometer Polarimeter (UVSP)	115 nm ... 360 nm $\Delta\lambda$: 2 pm	Woodgate et al (1980)
1985	*Spacelab 2*	High Resolution Telescope and Spectrograph (HRTS)	Stigmatic spectra 120 nm ... 170 nm	Brueckner et al (1986)
		Coronal Helium Abundance Experiment (CHASE)	17.1 nm ... 133.6 nm	Lang et al (1990)
		Solar Ultraviolet Irradiance Monitor (SUSIM)	120 nm ... 400 nm $\Delta\lambda$: 0.15 nm and 0.5 nm	VanHoosier et al (1988)
1991	*UARS*	Solar Ultraviolet Irradiance Monitor (SUSIM)	120 nm ... 400 nm $\Delta\lambda$: 1.1 nm ... 5 nm	Lean et al (1992) Brueckner et al (1993)
		Solar Stellar Irradiance Comparison Experiment (SOLSTICE)	120 nm ... 420 nm	Rottman and Woods (1995)
1993	*Spartan 201*	Ultraviolet Coronal Spectrograph (UCS)	103.2 nm, 103.7 nm 121.6 nm	Kohl et al (1994)
1995	*SOHO*	Coronal Diagnostic Spectrometer (CDS)	15 nm ... 80 nm	Harrison et al (1995)
		Extreme-ultraviolet Imaging Telescope (EIT)	17.1 nm ... 30.4 nm	Delaboudinière et al (1995)
		Solar EUV Monitor (SEM)	30.4 nm, 17 nm ... 70 nm	Hovestadt et al (1995)
		Solar Ultraviolet Measurements of Emitted Radiation (SUMER)	46.5 nm ... 161 nm	Wilhelm et al (1995)
		Ultraviolet Coronagraph Spectrometer (UVCS)	103.2 nm, 103.7 nm 121.6 nm	Kohl et al (1995)
1998	*TRACE*	Cassegrain Telescope	17.1 nm ... 160 nm	Handy et al (1999)
2001	*TIMED*	Solar EUV Experiment (SEE)	0.1 nm ... 195 nm	Woods et al (1998)
2006	*Hinode*	EUV Imaging Spectrometer (EIS)	17 nm ... 21 nm 25 nm ... 29 nm	Culhane et al (2007)
2007	*STEREO*	SECCHI EUV Imager (EUVI)	17.1 nm, 19.5 nm, 28.4 nm, 30.4 nm	Wülser et al (2004)

published in 1996 (Dere et al 1997), and an updated version was published in 2006 (Landi et al 2006). CHIANTI has not only been used widely for solar studies, but also for studies of coronae of cool stars, a wide range of other stars, and non-stellar objects such as supernova remnants. A comparison of an observed active region spectrum and a synthetic spectrum produced by use of the CHIANTI code is shown in Figure 5.5 (Mason et al 1997).

Solar physics observations

The Sun is the only star where detailed VUV observations of the surface structure have been possible to date. A strong programme of solar observations at VUV wavelengths has continued from 1946 to this day (de Jager 2001; Wilhelm et al 2007). Fueled by the awareness of the unique properties of the outer solar atmosphere as an astrophysical laboratory and by increasing evidence for solar effects on the terrestrial atmosphere and possible effects on the terrestrial climate, a clear bifurcation in the types of observation has evolved: first, spectral-imaging studies of the solar atmosphere with increasing spatial and spectral resolutions, and, second, measurements of the solar VUV irradiance with increasing radiometric accuracy. Wilhelm et al (2004) have presented a list of some of the important solar observa-

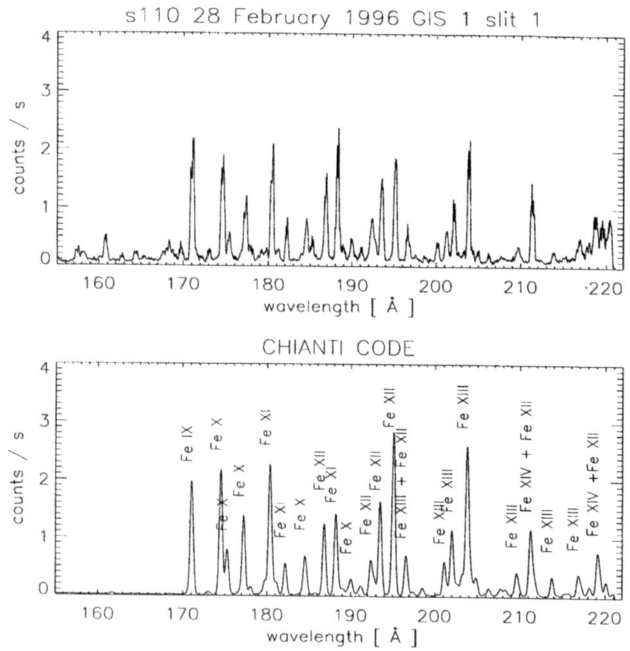

Figure 5.5: Comparison of observed (top) and synthetic (bottom) active region spectra (from Mason et al 1997).

tions during the period from 1946 to 1973, together with descriptions of the key instruments.

There were a number of notable advances during this period. Hinteregger et al (1964) and Hall et al (1965) recorded the first measurements of the VUV spectrum below ≈ 30 nm, including the discovery of the strong emission lines of Fe IX through Fe XIII at wavelengths between 17 nm and 22 nm (see Figure 5.5). The sounding rocket spectrometers employed grazing-incidence optics and open-structure photoelectric detector systems. Reeves and Parkinson (1970) produced the first detailed atlas of spectroheliograms at VUV wavelengths between 30 nm and 140 nm using the Harvard College Observatory (HCO) instrument on the *OSO 4* mission. The spectroheliometer employed a normal-incidence platinum-coated telescope mirror and a gold-coated normal-incidence diffraction grating. The detector was a solar-blind open-structure photon-counting magnetic electron multiplier (see Chapter 22). Also with the help of *OSO 4* Timothy and Timothy (1969) recorded the first long-term irradiance variations in the strong He II Ly α line at 30.3 nm.

The most comprehensive solar-physics mission of the 1970s was the *Skylab* Apollo Telescope Mount (ATM). A detailed overview of this mission has been given by Tousey (1977), see also Bartoe and Brueckner (1975). This mission, coming at the end of the Apollo programme, was unique in a number of ways. First, the solar instruments were the largest ever flown. Second, it was the first mission where on-board astronauts controlled the different solar instruments using both

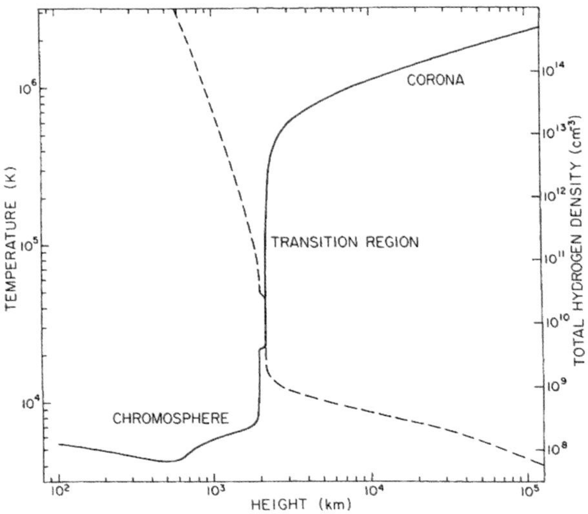

Figure 5.6: Temperature and density variations as functions of height in the solar chromosphere, transition layer, and corona in a quiet region of the atmosphere (from Withbroe and Noyes 1977).

photographic and photoelectric detector systems in a series of coordinated observing programmes. Third, it was one of the first missions to use sounding rocket under-flights to successfully recalibrate some of the instruments during the mission (see, e.g., Reeves et al 1977).

The many results from ATM included the first direct observations of the chromospheric network (Reeves et al 1974), the expanded transition region over polar coronal holes (Huber et al 1974), high angular-resolution[5] ($\approx 2''$) whole disk spectroheliograms (Tousey et al 1977), measurements of VUV emission line profiles (Bartoe et al 1977), and the identification of coronal holes as the source of the high-speed streams in the solar wind (Krieger et al 1973).

The results of all of these early studies led to a general understanding of solar atmospheric conditions. Mean height profiles of the temperature and hydrogen density of the solar atmosphere in a quiet region are shown in Figure 5.6. Nevertheless, the details of the mechanism(s) for heating the coronae of the Sun and other stars remain an active field of research to this day.

It was also learned that the solar atmosphere is both highly structured and highly dynamic. The general structure of the magnetic field configuration in a quiet region of the transition region and lower corona is shown in Figure 5.7. Studies have been presented both of the heating of the solar corona (Kuperus 1969), and of the structure of the transition region (Mariska 1986). However, what became obvious from these early results was that the solar atmosphere is extremely dynamic, not only in small and large scale magnetic reconnection events such as bright points

[5] An angular resolution element of $1''$ at 1 ua corresponds to spatial scale of ≈ 725 km on the solar disk.

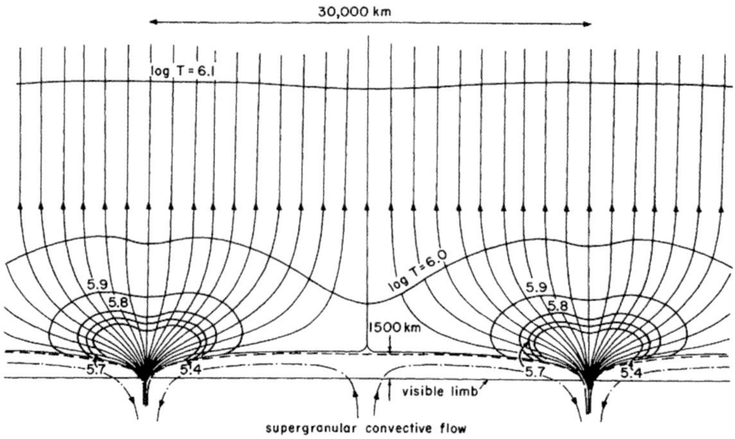

Figure 5.7: Magnetic field configuration associated with the chromospheric network and temperature contours in a quiet region (after Gabriel 1976; reproduced from Withbroe and Noyes 1977).

and solar flares, but also in the "quiet" atmosphere. Observations on timescales of seconds or less are thus required to fully address the physics.

Further, the basic phenomena controlling the outer atmosphere and heating the solar corona take place at the smallest spatial scales (< 100 km), requiring observations from 1 ua with angular resolutions of 0.1″ or smaller.

During the late 1970s, there was a dearth of orbiting solar VUV missions. However, during this period there was a series of sub-orbital sounding rocket flights, particularly those measuring the solar irradiance (see, e.g., Mount et al 1980). A key development was the invention of the Tandem–Wadsworth spectrograph by Bartoe and Brueckner (1975). This led to a series of flights from 1975 through 1992, resulting in more than 100 publications (see, for example, Dere et al 1987). Also, during this period Kohl et al (1978) produced an atlas of the solar spectrum in high resolution at wavelengths between 225.2 nm and 319.6 nm. A number of solar irradiance instruments were also flown on sounding rockets (Mount et al 1980).

This period ended in 1980 with the launch of the *SMM*. The key solar VUV missions from 1980 to 2008 are listed in Table 5.1.

All of the early VUV solar-physics instruments were on platforms in low Earth orbit. This led to about 60 min of solar observations, followed by 30 min in eclipse behind the Earth. The maxim in those early days was that all interesting solar phenomena occurred while the spacecraft was occulted by the Earth!

This all changed dramatically in 1995 when the ESA/NASA *SOHO* spacecraft was placed in a halo orbit around the L1 Lagrange point, where the solar and terrestrial gravitational fields were in balance with the centrifugal force. This allowed *SOHO* to observe the Sun continuously, and also monitor the solar wind upstream of the Earth's magnetosphere. The data from *SOHO* were spectacular and observations continue to this day.[6] Early results can be found in the book "The First Results from *SOHO*" (Fleck and Svestka 1997).

In 1998, *TRACE* was launched into a Sun-synchronous polar orbit (Handy et al 1999). *TRACE* employed a single Cassegrain telescope with multilayer-coated optics and a lumogen-coated charge-coupled device (CCD) detector to obtain an angular resolution of $1''$ ($0.5''$ pixel size) at selected VUV wavelengths. The *TRACE* image quality is superb. The mission obtained its last science image in June 2010.[7]

The *TIMED* mission was launched in December 2001 and has been obtaining solar VUV irradiance data from $\approx$ 0.1 nm to $\approx$ 195 nm since early in 2002 with the Solar EUV Experiment (SEE) (Woods et al 1998).[8]

In September 2006 the Japanese *Solar-B* mission, renamed *Hinode* ("Sunrise"), was launched into a Sun-synchronous Earth orbit at an altitude of 680 km. The primary instrument is a 0.5 m diameter visible-light telescope, the largest solar optical instrument yet operating in space. Complementing the primary instrument is the EUV Imaging Spectrometer (EIS) covering the wavelength ranges 17 nm to 21 nm, and 25 nm to 29 nm, with an angular resolution of $1''$ (Culhane et al 2007).[9]

Finally, a deep space mission, launched on October 2006, is *STEREO*, consisting of two identical spacecraft, one flying forward of the Earth, and the second flying behind the Earth. At separation angles between about 10° and about 15°, the two spacecraft produced high quality stereoscopic images of the solar atmosphere. As the separation angles increased, the spacecraft were able to observe progressively more of the solar atmosphere not visible from Earth, thereby giving advance warning of the development of active regions. Finally, at angles near 90° with respect to the Earth they can directly view coronal mass ejections leaving the Sun and headed towards Earth.[10] VUV observations at wavelengths of 17.1 nm, 19.5 nm, 28.4 nm, and 30.4 nm are being carried out with the Extreme Ultraviolet Imager (EUVI), which has a field of view (FOV) out to 1.7 $R_\odot$ (Wülser et al 2004).

Astrophysical observations

VUV astrophysics studies from space developed at a much slower rate than solar observations, primarily because of the much lower irradiance levels and the need for critical stabilization of the observing platform. Savage (2001) has given an excellent review of the technological challenges and vicissitudes of the early observing programmes. The astrophysics VUV missions from 1966 to 2008 are listed in Table 5.2. Morton and Spitzer (1966) made the first spectral observations from three-axis stabilized sounding rockets, using a VUV spectrometer with its own separate fine stabilization system. The recording medium was photographic emulsion, and the early observations concentrated on bright hot stars (see, e.g., Morton 1967).

[6] Daily images can be seen at http://sohowww.nascom.nasa.gov.

[7] Full information and results are available at http://trace.lmsal.com. (*TRACE* and many other space programmes have an open data policy.)

[8] Information on the *TIMED* mission can be found at http://timed.jhuapl.edu.

[9] Details can be found at http://solarb.msfc.nas.gov.

[10] See http://stereo.gsfc.nasa.gov.

Table 5.2: Key sub-orbital and orbital astrophysics missions from 1966 to 2008.

Year	Mission	Instrument	Parameters	Reference
1966	Three-axis stabilized sounding rockets	Spectrometer	126 nm ... 172 nm $\Delta\lambda \approx 0.1$ nm	Morton and Spitzer (1966)
1972	*TD 1-A Spacecraft*	UV sky survey telescope	135 nm ... 255 nm	Boksenberg et al (1973)
		Stellar Ultraviolet Spectrometer	Three bands of 10 nm at 211 nm, 254.5 nm, 282.5 nm; $\Delta\lambda$: 0.17 nm	de Jager et al (1974)
1972	OAO 3	Telescope and Spectrometer	95 nm ... 145 nm, $\Delta\lambda$: 5 pm; 165 nm ... 300 nm, 2.5 pm	Rogerson et al (1973)
1976	BUSS (upgraded)	Telescope, echelle spectrometer	200 nm ... 340 nm $\Delta\lambda$: 10 pm	Hoekstra et al (1978)
1978	*IUE*	Telescope, echelle spectrometers	115 nm ... 195 nm $\Delta\lambda$: 13 pm 190 nm ... 320 nm $\Delta\lambda$: 20 pm	Boggess et al (1978a)
1990	*HST*	Goddard High-Resolution Spectrograph (GHRS)	115 nm ... 320 nm $\lambda/\Delta\lambda : 2 \times 10^3 \ldots 2 \times 10^5$	Brandt et al (1994)
		Faint Object Camera (FOC)	Filters: 123 nm ... 215 nm	Greenfield et al (1991)
1993	First HST Servicing Mission *(HSM 1)*	Corrective optics package (COSTAR) installed		Hartig et al (1993)
		Wide Field Planetary Camera WFPC 1 replaced by WFPC 2	0.12 µm ... 1.0 µm	Trauger et al (1994)
		GHRS performance significantly improved		Robinson et al (1998)
		Faint Object camera (FOC) performance significantly improved		Jedrezejewski et al (1994)
1997	*HSM 2*	GHRS replaced with Space Telescope Imaging Spectrograph (STIS)	115 nm ... 1 µm $\lambda/\Delta\lambda$: upto 1×10^5	Woodgate et al (1998)
1992	*EUVE*	Grazing Incidence Telescopes and Spectrometers	6 nm ... 75 nm	Sirk et al (1997) Bowyer (1997)
1999	*FUSE*	Normal incidence telescopes and spectrometers	90.5 nm ... 118.7 nm $\lambda/\Delta\lambda$: $2 \times 10^4 \ldots 2.5 \times 10^4$	Sahnow et al (2000)
2002	*HSM 3B*	FOC replaced with Advanced Camera for Surveys (ACS)	115 nm ... 1.1 µm	Ford et al (2003)
2003	*GALEX*	Normal incidence Telescope and grisms	135 nm ... 175 nm, 175 nm ... 280 nm $\lambda/\Delta\lambda$: 100 ... 300	Martin et al (2003)

It soon became apparent that the short observing time (≈ 300 s) available from sounding rockets was a severe limiting factor. The development of orbiting satellites and high-altitude balloons for astrophysical observations then began in the US and in Europe. In the US the *OAO* series of spacecraft (Rogerson 1963) began with the successful launch of *OAO 2* (the attitude control system of *OAO 1* having failed after achieving orbit) in December 1968. *OAO 2* contained instruments from Wisconsin and the Smithsonian Astrophysical Observatory (SAO) (Code 1969). The Wisconsin instrument included two scanning spectrometers as well as photometers (Code et al 1970), while the SAO instrument was equipped for a VUV sky survey.

The major advance came with the launch of *OAO 3* in August 1972, which was renamed *Copernicus*. The Princeton instrument employed an 80 cm diameter Cassegrain telescope with a scanning VUV spectrometer. The wavelength range from 165 nm to 300 nm was covered with a resolution of 2.5 pm. Open-structure photomultiplier tubes with opaque KBr photocathodes were also used for the first time allowing the wavelength range from 95 nm to 145 nm to be covered with a resolution of 5 pm (see Rogerson et al 1973).

The European Space Research Organization (ESRO) launched the *TD-1A* satellite in March 1972. *TD-1A* contained a UV sky survey telescope covering the wavelength range from 135 nm to 255 nm (Boksenberg et al 1973), and a stellar UV spectrometer covering wavelength bands around 211 nm, 254.5 nm, and 282.5 nm (de Jager et al 1974).

The Balloon-borne Ultraviolet Stellar Spectrograph (BUSS) evolved from earlier work in the US and the UK, and was a joint development by the US and The Netherlands. BUSS was designed to cover the wavelength range from 200 nm to 340 nm using an echelle spectrograph with a wavelength resolution of 10 pm (Hoekstra et al 1978). The upgraded instrument was first flown in 1975, and BUSS flights continued into the 1980s when the development of the *HST* rendered it obsolete.

In March 1978, *IUE* was launched into a geosynchronous orbit. *IUE* was a joint development by NASA, the UK Science Research Council (SRC), and the European Space Agency (ESA). The instrument employed a 45 cm diameter Ritchey–Chrétien telescope and two echelle spectrometers. The detectors were UV to visible light image tube convertors coupled to SEC Vidicon image tubes. The wavelength range from 115 nm to 195 nm was covered with a wavelength resolution of 13 pm, and the range from 190 nm to 320 nm was covered with a resolution of 20 pm (Boggess et al 1978a). The in-flight performance has been described by Boggess et al (1978b). Although suffering a number of attitude control system failures, *IUE* continued operating until September 1996, producing over 104 000 UV spectra.[11]

The launch of *HST* in April 1990 marked a seminal moment in optical and VUV astrophysics. Not only does *HST* employ a 2.4 m aperture Ritchey–Chrétien telescope, it can accommodate one radial and four axial science instruments and is configured for on-orbit servicing and instrument replacement by astronauts (see Edelman 1990). Further, the telescope optics are coated with Al protected by a thin MgF_2 overcoat for good UV reflectances, and the pointing accuracy is better than $0.01''$. However, in addition to suffering from the vagaries of the *Space Shuttle (STS)* programme, *HST* has not been without its own problems. The most serious of these was that, shortly after starting science observations, it was discovered that the primary mirror (although its surface was excellently polished) had nevertheless an incorrect shape leading to spherical aberration in the image.

Included in the primary instrument package were the Goddard High Resolution Spectrograph (GHRS) (Brandt et al 1994), and the Faint Object Camera (FOC) (Greenfield et al 1991). GHRS covered the spectral range from 115 nm to 320 nm with spectral resolutions ranging from 2×10^3 to 1×10^5. For the first time GHRS used multi-element detectors to improve the observing efficiency, namely two 1×512 pixel pulse-counting Digicon arrays (Ebbets and Garner 1986). The FOC employed an electron-bombarded silicon (EBS) television tube and several VUV filters between 123 nm and 215 nm.

The solutions to the spherical aberration problem were to replace the radial instrument, the Wide Field and Planetary Camera (WFPC 1), with WFPC 2 which had a corrective optical system installed, and to replace the High Speed Photometer with the Corrective Optics Space Telescope Axial Replacement (COSTAR) which added corrective optics to the axial instruments (Hartig et al 1993). This

[11] Full details of the mission can be found at http://archive.stsci.edu/iue.

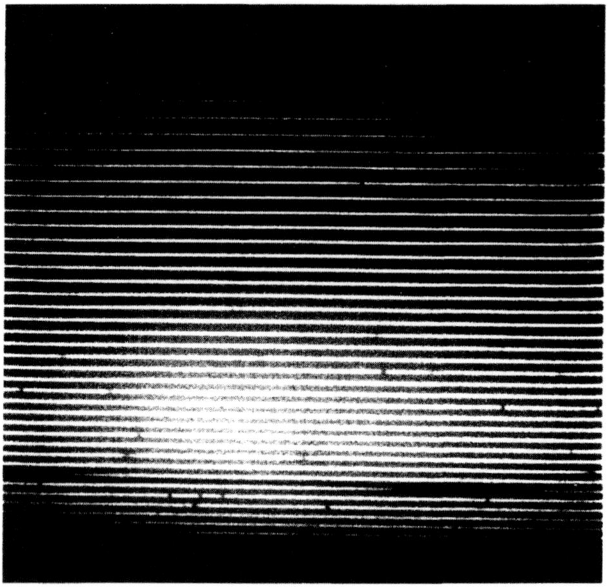

Figure 5.8: Echelle spectrum of BD+28°4211 recorded with the STIS FUV MAMA detector. The spectrum covers the range from 115 nm to 170 nm with a nominal resolving power of 4.6×10^4. Broad Ly α absorption is seen in two echelle orders near the bottom of the format, along with numerous narrow stellar absorption lines (from Kimble et al 1998).

was achieved during the first *HST* servicing mission *(HSM 1)* in December 1993. The WFPC 2 is designed for diffraction-limited imaging at wavelengths between $\approx$ 120 nm and 1.0 µm over a wide FOV (Rodgers and Vaughan, 1993; Trauger et al 1994). The detectors used are four contiguous 800 × 800 pixel CCDs with 15 µm × 15 µm pixels. COSTAR improved the imaging qualities of the GHRS and the FOC (Robinson et al 1998; Jedrzejewski et al 1994).

The GHRS was replaced by the Space Telescope Imaging Spectrograph (STIS) during *HSM 2* in February 1997. STIS is a complex multi-mode imaging spectrometer covering the wavelength range from $\approx$ 115 nm to 1.0 µm (Woodgate et al 1998). Kimble et al (1998) have described the on-orbit performance of STIS. It employs a 1024 × 1024 pixel CCD to cover the wavelength ranges from 305 nm to 555 nm and 550 nm to 1000 nm, a 2048 × 2048 pixel multi-anode microchannel array (MAMA) (see Chapter 22) pulse-counting detector with a Cs Te photocathode to cover the wavelength range from 165 nm to 315 nm (near ultraviolet; NUV), and a 2048 × 2048 pixel MAMA with an opaque Cs I photocathode to cover the wavelength range from 115 nm to 170 nm (far ultraviolet; FUV). These large-format imaging detector systems, together with the direct-imaging, long-slit spectroscopy, and short-slit echelle spectroscopy modes, give STIS a tremendous spectral and spatial "grasp", as shown in Figures 5.8 and 5.9.

Figure 5.9: Composite UV image of Saturn recorded with the FUV MAMA detector. Molecular hydrogen is color-coded blue, while Ly α emission, bright in the auroral zones, is color-coded red. Taken from http://opposite.stsci.edu.

The FOC was replaced with the Advanced Camera for Surveys (ACS) during *HSM 3b* during March 2002. ACS has three camera systems, the Wide Field Camera (WFC) and the High Resolution Camera (HRC) covering the wavelength range from 200 nm to 1100 nm, and the Solar Blind Channel (SBC) covering the wavelength range from 115 nm to 170 nm (Clampin et al 2000; Ford et al 2003). ACS employs a mosaic of two 2048×4096 pixel CCDs to give a FOV of $202'' \times 202''$.

Information on all of the *HST* instruments and observations can be found at the Space Telescope Science Institute (STScI) web site.

It was widely believed for many years that the opacity of interstellar hydrogen would make VUV observations impossible at wavelengths below 91 nm. However, the opacity decreases at shorter wavelengths and the distribution of hydrogen was found to be highly uneven. The *EUVE* observatory was launched in June 1992 to study the wavelength range from 7 nm to 700 nm. *EUVE* employed a set of grazing-incidence telescopes (Sirk et al 1997), and variable line spacing grazing-incidence grating spectrometers (Bowyer 1997). During its lifetime, which ended in 2001, *EUVE* observed over 700 sources of extreme ultraviolet (EUV) radiation.[12]

All of the VUV astrophysics missions, with the exception of *Copernicus*, had lower wavelength limits around 115 nm due to the use of windows or filters. The *FUSE* observatory was launched in June 1999 in order to make high-resolution ($\lambda/\Delta\lambda \approx 2 \times 10^4$) measurements at wavelengths between 90.5 nm and 118.7 nm

[12]Details of the mission can be found at http://archive.stsci.edu/euve.

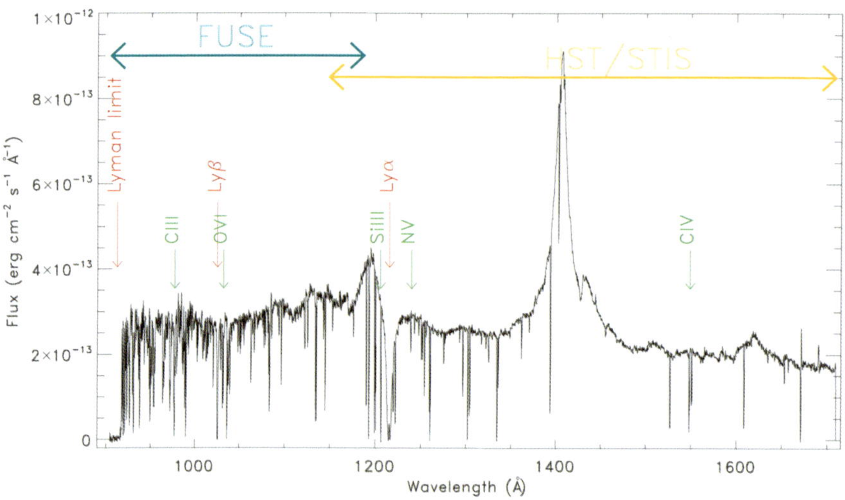

Figure 5.10: The VUV spectrum from the quasar PG 1211+143 observed by *FUSE* and *HST*/STIS (courtesy C. Danforth, data from Danforth and Shull, 2008).

(Sahnow et al 2000). *FUSE* employed normal-incidence SiC optics and open-structure double delay line (DDL) microchannel plate (MCP) detectors (see Chapter 22, Timothy 2010). *FUSE* had a series of specific science objectives and also made measurements that were complementary to those of *HST* (see, e.g., Figure 5.10).[13]

Finally, *GALEX* was launched in April 2003. *GALEX* employs a 50 cm astigmatism-corrected Ritchey-Chrétien telescope and grisms to record imaging sky surveys in two wavelength bands in the range from 135 nm to 280 nm, and spectroscopic sky surveys in this range with a spectral resolution of ≈ 100 (Martin et al 2003).[14]

The future

The future for solar VUV observations seems very solid at this time; *SOHO* and *TIMED* continue their observing programmes. The *STEREO* spacecraft pair has an adequate separation to continue producing unique stereo VUV images. *Hinode* is performing its observing programme, although there are some problems with the high data-rate telemetry downlink. Further, *SDO* was successfully launched by NASA on 11 February 2010 and the instruments are now undergoing on-orbit checkout in preparation for a five-year mission.[15] In addition, the *Solar Orbiter* payload has been defined, including an EUV imager and a spectrometer, which will resolve Doppler shifts down to 2 km s^{-1} and, by approaching the Sun down to 0.22 ua, will resolve solar spatial features to about 150 km.

[13]Details of the *FUSE* mission can be found at http://fuse.pha.jhu.edu.

[14]Full information on the *GALEX* mission can be found at http://galex.caltech.edu.

[15]See http://sdo.gsfc.nasa.gov.

In contrast, the future for astrophysics VUV missions, in the near term was, until recently, highly uncertain, and in the later future, frankly, bleak. *GALEX* continues its observing programme after five years in orbit. However, the *FUSE* mission has been terminated by the total failure of the attitude control system, STIS is currently disabled by the failure of the backup low-voltage power supply, which has closed the shutter, and the ACS has had a catastrophic failure which only permits the SBC to continue operating. Further, *HST* is currently operating on the redundant Side B of the Science Instruments Command and Data Handling system (SICDH), following the failure, after 18 years of operation, of Side A.

The near-term future has now improved dramatically following the success of *HSM4* in May 2009. The STIS and ACS instruments were repaired successfully. Also, two new instruments, namely COS (see Green et al 2003) and WFC3 (see Kimble et al 2008) were installed. Further major upgrades were made to the telescope. Six new gyros, new batteries, a new SICDH, and a new fine guidance system were installed. Finally, the thermal insulation was repaired. NASA has now started to release the full results of *HSM4* (see http://hubblesite.org). Hopefully, we can now look forward to a further five to eight years of VUV astrophysics observations with *HST*. Beyond *HST*, the future for VUV astrophysics missions remains highly uncertain.

The next generation space telescope, namely the *JWST*, is designed for observations at infrared wavelengths, and beyond that, informal proposals for future VUV astrophysics missions are being developed in the US and Europe. However, given the priorities and budget constraints of NASA and ESA, it seems unlikely that another major VUV astrophysics mission can be started before about 2015 at the earliest. It may well be that the challenge is picked up by other space-faring nations.

Acknowledgements: The research on this chapter has benefited from extensive use of the data in NASA's Astrophysics Data System (ADS), administered by the Smithsonian Astrophysical Observatory.

Bibliography

Abarca JF, Casiccia CC (2002) Skin cancer and ultraviolet-B radiation under the Antarctic ozone hole: Southern Chile, 1987–2000. Photodermatology Photoimmunology Photomedicine 18:294–302

Athay RG (1966) Theoretical line intensities. V. Solar UV emission lines of heavy elements. Astrophys J 145:784–795

Bartoe J-DF, Brueckner GE (1975) New stigmatic, coma-free, concave grating spectrograph. J Opt Soc Am 65:13–21

Bartoe J-DF, Brueckner GE, Purcell JD, Tousey R (1977) Extreme ultraviolet spectrograph ATM experiment S082B. Appl Opt 16:879–886

Boggess A, Carr FA, Evans DC (plus 30 authors) (1978a) The *IUE* spacecraft and instrumentation. Nature 275:372–377

Boggess A, Bohlin RC, Evans DC (plus 31 authors) (1978b) In-flight performance of the *IUE*. Nature 275:377–385

Boksenberg A, Evans RG, Fowler RG (plus 11 authors) (1973) The ultraviolet sky-survey telescope in the *TD-1A* satellite. Mon Not R Astron Soc 163:291–322

Bowyer S (1997) The unique variable line space spectrometers on the *EUVE* satellite: In-flight performance and selected scientific results. Proc SPIE 2517:97–106

Brandt JC, Heap SR, Beaver EA (plus 22 authors) (1994) The Goddard high resolution spectrograph: Instrument, goals, and science results. PASP 106:890–908

Brueckner GE, Bartoe J-DF, Cook JW (plus two authors) (1986) HRTS results from *Spacelab 2*. Adv Space Res 6:263–272

Brueckner GE, Edlow KL, Floyd LE (plus two authors) (1993) The solar spectral irradiance monitor (SUSIM) experiment on board the Upper Atmosphere Research Satellite (*UARS*). J Geophys Res 98:10695–10712

Brueggemann HP (1968) Deep conic cassegrains. In: Conic Mirrors. The Focal Press, London, New York, Section VIII:102–111

Clampin M, Ford H, Bartko F (plus 22 authors) (2000) Advanced camera for surveys. Proc SPIE 4013:344–351

Code AD (1969) Photoelectric photometry from a space vehicle. PASP 81:475–487

Code AD, Houck TE, McNall JF (plus two authors) (1970) Ultraviolet photometry from the Orbital Astronomical Observatory. I. Instrumentation and operation. Astrophys J 161:377–388

Culhane JL, Harra LK, James AM (plus 36 authors) (2007) The EUV Imaging Spectrometer for *Hinode*. Sol Phys 243:19–61

Danforth CW, Shull JM (2008) The low-z intergalactic medium. III H I and metal absorbers at $z < 0.4$. Astrophys J 679:194–219

David C, Gabriel AH, Bely-Dubau F (plus three authors) (1998) Measurement of the electron temperature gradient in a solar coronal hole. Astron Astrophys 336:L90–94

de Gruijl FR (2002) Photocarcinogenesis: UVA vs UVB radiation. Skin Pharmacology and Applied Skin Physiology 15:316–320

de Jager C, Hoekstra R, van der Hucht KA (plus five authors) (1974) The orbiting stellar ultraviolet spectrophotometer S 59 in ESRO's *TD-1A* satellite. Astrophys Space Sci 26:207–262

de Jager C (2001) Early space research. In: The Century of Space Science Vol 1 (eds JAM Bleeker, J Geiss, and MCE Huber) Kluwer Academic Publishers, Dordrecht, 203–223

Delaboudinière J-P, Artzner GE, Brunaud J (plus 24 authors) (1995) EIT: Extreme-ultraviolet Imaging Telescope for the *SOHO* mission. Sol Phys 162:291–312

Dere KP, Bartoe J-DF, Brueckner GE (plus two authors) (1987) Discrete subresolution structures in the solar transition zone. Sol Phys 114: 223–237

Dere KP, Landi E, Mason HE (plus two authors) (1997) CHIANTI – an atomic data base for emission lines. Astron Astrophys Suppl Ser 125:149–173

Doschek GA, Warren HP, Laming JM (plus five authors) (1997) Electron densities in the solar polar coronal holes from density sensitive line ratios of Si VIII and Si X. Astrophys J 482:L109–L112

Ebbets DC, Garner HW (1986) Dead-time effects in pulse-counting Digicon detectors. Proc SPIE 627:638–644

Edelman LL (1990) The *Hubble* Space Telescope mission, history, and systems. Proc SPIE 1358:422–441

Edlén B (1943) Die Deutung der Emissionslinien im Spektrum der Sonnenkorona. Z Astrophys 22:30–64

Feldman U, Laming JM (2000) Element abundances in the upper atmospheres of the Sun and stars: Update of observational results. Physica Scripta 61:222–252

Fleck B, Švestka Z (1997) The first results from SOHO. Kluwer Academic Press

Ford H, Clampin M, Hartig G (plus 34 authors) (2003) Overview of the Advanced Camera for Surveys on-orbit performance. Proc SPIE 4854:81–94

Friedman H (1959) Rocket spectroscopy. J Geophys Res 64:1751–754

Gabriel AH (1976) A magnetic model of the solar transition region. Phil Trans R Soc London Ser A 281:339–352

Green JC, Wilkerson E, Morse JA (2003) Status and performance of the cosmic origins spectrograph. Proc SPIE 5164:17–23

Greenfield P, Paresce F, Baxter D (plus nine authors) (1991) In-flight performance of the Faint Object Camera of the Hubble Space Telescope. Proc SPIE 1494:16–39

Grotrian W (1939) Zur Frage der Deutung der Linien im Spektrum der Sonnenkorona. Naturwissenschaften 27:214–214

Hall LA, Schweizer W, Heroux L, Hinteregger HE (1965) Solar XUV spectrum of March, 1964. Astrophys J 142:13–15

Handy BN, Acton LW, Kankelborg CC (plus 45 authors) (1999) The Transition Region and Coronal Explorer. Sol Phys 187:229–260

Harrison RA, Sawyer EC, Carter MK (plus 36 authors) (1995) The Coronal Diagnostic Spectrometer for the Solar and Heliospheric Observatory. Sol Phys 162:233–290

Hartig GF, Crocker JH, Ford HC (1993) Status and optical performance of the Corrective Optics Space Telescope Axial Replacement. Proc SPIE 1945:17–24

Hinteregger HE, Hall LA, Schweizer W (1964) Solar XUV spectrum from 310 Å to 55 Å. Astrophys J 140:319–327

Hoekstra R, Kamperman TM, Wells CW, Werner W (1978) Balloon-borne ultraviolet stellar echelle spectrograph. Appl Opt 17:604–613

Hovestadt D, Hilchenbach M, Bürgi A (plus 30 authors) (1995) CELIAS—Charge Element and Isotope Analysis System for *SOHO*. Sol Phys 162:441–481

Huber MCE, Foukal PV, Noyes RW (plus five authors) (1974) Extreme–ultraviolet observations of coronal holes: Initial results from *Skylab*. Astrophys J 194:L115–L118

Jacchia LG (1971) Revised static models of the thermosphere and exosphere with empirical temperature profiles. Smithsonian Astrophysical Observatory SR 332

Jacchia JG (1977) Themospheric temperature, density and composition: New models. Smithsonian Astrophysical Observatory SR 375

Jedrzejewski RI, Hartig G, Jakobsen P (plus two authors) (1994) In-orbit performance of the COSTAR-corrected Faint Object Camera. Astrophys J 435:L7–L10

Kimble RA, Woodgate BE, Bowers CW (plus 49 authors) (1998) The on-orbit performance of the Space Telescope Imaging Spectrograph. Proc SPIE 3356:188–202

Kimble RA, McKenty JW, O'Connell RW, Townsend JA (2008) Wide Field Camera 3: A powerful new imager for the Hubble Space Telescope. Proc SPIE 7010:1–12
Kohl JL, Parkinson WH, Kurucz RL (1978) Center and limb solar spectrum in high spectral resolution 225.2 nm to 319.6 nm. Harvard-Smithsonian Center for Astrophysics, Cambridge
Kohl JL, Gardner LD, Strachan L, Hassler DM (1994) Ultraviolet spectroscopy of the extended solar corona during the Spartan 201 mission. Space Sci Rev 70:253–261
Kohl JL, Esser R, Gardner LD (plus 37 authors) (1995) The Ultraviolet Coronagraph Spectrometer for the Solar and Heliospheric Observatory. Sol Phys 162:313–356
Krieger AS, Timothy AF, Roelof EC (1973) A coronal hole and its identification as a source of a high velocity solar wind stream. Sol Phys 29:505–525
Kuperus M (1969) The heating of the solar corona. Space Sci Rev 9:713–739
Landi E, Del Zanna G, Young PR (plus three authors) (2006) CHIANTI – An atomic database for emission lines. VII: New data for X-rays and other improvements. Astrophys J Suppl Ser 162:261–280
Lang J, McWhirter RWP, Mason HE (1990) The interpretation of the spectral line intensities from the chase spectrometer on *Spacelab 2*. Sol Phys 129: 31-81
Lean J, VanHoosier M, Brueckner G (plus three authors) (1992) SUSIM/*UARS* observations of the 120 to 300 nm flux variations during the maximum of the solar cycle—Inferences for the 11-year cycle. Geophys Res Lett 19:2203–2206
Lyot B (1937) Quelques observations de la couronne solaire et des protubérances en 1935. L'Astronomie 51:203–218
Mariska JT (1986) The quiet solar transition region. Ann Rev Astron Astrophys 24:23–48
Mariska JT (1992) The Solar Transition Region. Cambridge University Press, Cambridge
Martin C, Barlow T, Barnhart W (plus 48 authors) (2003) The Galaxy Evolution Explorer. Proc SPIE 4854:336–350
Mason HE, Monsignori Fossi BC (1994) Spectroscopic diagnostics in the VUV for solar and stellar plasmas. Astron Astrophys Rev 6:123–179
Mason HE, Young PR, Pike CD (plus four authors) (1997) Application of spectroscopic diagnostics to early observations with the *SOHO* Coronal Diagnostic Spectrometer. Sol Phys 170:143–161
Massey HSW, Boyd RLF (1960) The ozone layer. In: The Upper Atmosphere. Hutchinson of London, 129–138
Mazzotta P, Mazzitelli G, Colafrancesco S, Vittoria N (1998) Ionization balance for optically thin plasmas: Rate coefficients for all atoms and ions of the elements H to NI. Astron Astrophys Suppl Ser 133:403–409
Morton DC (1967) The far-ultraviolet spectra of six stars in Orion. Astrophys J 147:1017–1024
Morton DC, Spitzer L Jr (1966) Line spectra of delta and pi Scorpii in the far-ultraviolet. Astrophys J 144:1–12
Mount GH, Rottman GJ, Timothy JG (1980) The solar spectral irradiance 1200–2550 Å at solar maximum. J Geophys Res 85:4271–4274

Nicolet M, Aikin AC (1960) The formation of the D region of the ionosphere. J Geophys Res 65:1469–1483

Pottasch SR (1963) The lower solar corona: Interpretation of the ultraviolet spectrum. Astrophys J 137:945–966

Reeves EM, Parkinson WH (1970) An atlas of extreme-ultraviolet spectroheliograms from *OSO-IV*. Astrophys J Suppl Ser 21:1–409

Reeves EM, Foukal PV, Huber MCE (plus five authors) (1974) Observations of the chromospheric network: Initial results from ATM. Astrophys J 188:L27–L29

Reeves EM, Huber MCE, Timothy JG, Withbroe GL (1977) Photometric calibration of the EUV spectroheliometer on ATM. Appl Opt 16:849–857

Robinson RD, Ake TB, Lindler DJ (plus 16 authors) (1998) The Goddard High Resolution Spectrograph: Post–COSTAR characteristics. PASP 110:68–78

Rodgers DH, Vaughan AH (1993) Development of the second generation Wide Field Planetary Camera for Hubble Space Telescope. Proc SPIE 1920:288–296

Rogerson JB (1963) The Orbiting Astronomical Observatories. Space Sci Rev 2:621–652

Rogerson JB, Spitzer L, Drake JF (plus four authors) (1973) Spectophotometric results from the *Copernicus* satellite. I Instrumentation and performance. Astrophys J 181:L97–L102

Rottman G, Woods T (1995) SOLSTICE technique for measuring long-term solar variability. Proc SPIE 2583:347–355

Russo V (2001) The space age and the origin of space research. In: The Century of Space Science Vol 1 (eds JAM Bleeker, J Geiss, MCE Huber) Kluwer Academic Publishers, Dordrecht, 25–58

Sahnow DJ, Warren Moos H, Ake TB (plus 32 authors) (2000) On-orbit performance of the Far Ultraviolet Spectroscopic Explorer (*FUSE*). Proc SPIE 4013:334–343

Samson JAR (1980)—second printing from 1967— in Techniques of Vacuum Ultraviolet Spectroscopy. Pied Publications, Lincoln, 56–60

Savage BD (2001) Early ultraviolet spectroscopy from space. In: The Century of Space Science Vol 1 Kluwer Academic Publishers, Dordrecht 287–300

Schühle U (2010) Intensified solid state sensor cameras: ICCD and IAPS. ISSI SR-009:419–428

Schühle U, Hochedez J-F (2010) Solar-blind UV detectors based on wide band gap semiconductors. ISSI SR-009:429–439

Sirk MM, Vallerga JV, Finley DS (plus two authors) (1997) Performance of the extreme ultraviolet explorer imaging telescopes. Astrophys J Suppl Ser 110:347–356

Schumann V (1901) Ueber ein verbessertes Verfahren zur Herstellung ultraviolettempfindlicher Platten. Ann Physik 310:349–374

Spiller E (1974) Multilayer interference coatings for the vacuum ultraviolet. In: Space Optics (eds BJ Thompson, RR Shannon), National Academy of Sciences, Washington, DC, 581–597

Suri AN, Chu EL, Forrest DJ, Rein C (1975) Observations of solar gamma ray continuum between 360 keV and 7 MeV on August 4, 1972. Sol Phys 43:415–429

Timothy AF, Timothy JG (1969) Long term intensity variations in the solar helium II Lyman alpha line. J Geophys Res 75:6950–6958

Timothy AF, Timothy JG, Willmore AP, Wager JH (1972) The ionchemistry and thermal balance of the E- and F-regions of the daytime ionosphere: An experimental study. J Atmos Terr Phys 34:969–1035

Timothy JG, Chambers RM, D'Entremont AM (plus two authors) (1975) A sounding rocket spectroheliometer for photometric studies at ultraviolet wavelengths. Space Sci Instrum 1:23–49

Timothy JG (2010) Microchannel plates for photon detection and imaging in space. ISSI SR-009:365–391

Tousey R (1977) Apollo Telescope Mount of *Skylab* — an overview. Appl Opt 16:825–836

Tousey R, Bartoe J-DF, Brueckner GE, Purcell JD (1977) Extreme ultraviolet spectroheliograph ATM experiment S082A. Appl Opt 16:870–876

Trauger JT, Ballester GE, Burrows CJ (plus 15 authors) (1994) The on-orbit performance of WFPC 2. Astrophys J 435:L3–L6

VanHoosier ME, Bartoe J-DF, Brueckner GE, (plus two authors) (1977) Experience with Schumann-type XUV film on *Skylab*. Appl Opt 16:887–892

VanHoosier ME, Bartoe J-DF, Brueckner GE, Prinz DK (1988) Absolute spectral irradiance 120 nm – 400 nm. Astrophys Lett Comm 27, 163-168

Wilhelm K, Curdt W, Marsch E (plus 13 authors) (1995) SUMER—Solar Ultraviolet Measurements of Emitted Radiation. Sol Phys 162:189–231

Wilhelm K (2002) Spectroradiometry of spatially-resolved solar plasma sructures. ISSI SR-002:37–50

Wilhelm K (2003) Past and recent observations of the solar upper atmosphere at vacuum-ultraviolet wavelengths. J Atmos Sol Terr Phys 65:167–189

Wilhelm K, Dwivedi BN, Marsch E, Feldman U (2004) Observations of the Sun at vacuum ultraviolet wavelengths from space. Part I: Concepts and instrumentation. Space Sci Rev 111:415–480

Wilhelm K, Marsch E, Dwivedi BN, Feldman U (2007) Observations of the Sun at vacuum ultraviolet wavelengths from space. Part II: Results and interpretations. Space Sci Rev 133:103–179

Wilhelm K, Fröhlich C (2010) Photons — from source to detector. ISSI SR-009:23–54

Withbroe GL, Noyes RW (1977) Mass and energy flow in the solar chromosphere and corona. Ann Rev Astron Astrophys 15:363–387

Woodgate BE, Tandberg-Hanssen EA, Bruner EC (plus 11 authors) (1980) The Ultraviolet Spectrometer and Polarimeter on the Solar Maximum Mission. Sol Phys 65:73–90

Woodgate BE, Kimble RA, Bowers CW (plus 41 authors) (1998) The Space Telescope Imaging Spectrograph design. PASP 110:1183–1204

Woods TN, Bailey S, Eparvier F (plus nine authors) (1998) *TIMED* Solar EUV Experiment. Proc SPIE 3442:180–191

Wülser JP, Lemen JR, Tarbell TD (plus 31 authors) (2004) EUVI: The *STEREO*—SECCHI extreme ultraviolet imager. Proc SPIE 5171:111–122

— 6 —

The visible and near-infrared domain

Steven V.W. Beckwith[I]

Abstract

Space observations at visible and near-infrared wavelengths are free from the absorption, high foreground radiation and wavefront distortion caused by the Earth's atmosphere. The absence of these effects in space permits observations of higher sensitivity and stability with better angular resolution and with a larger dynamic range than their counterparts from the ground. Observations from the *HST*, the *ISO* and *Spitzer* satellites and survey satellites such as *IRAS* are still unsurpassed by ground-based telescopes for their sensitivity, radiometric stability, angular resolution over large fields of view, dynamic range and complete spectral coverage. This chapter discusses the inherent limitations of ground-based astronomy that are overcome by space observations and lists some of the spacecraft that have demonstrated these advantages for scientific study.

Introduction

Because the Earth's atmosphere transmits throughout most of the visible and infrared domain between 0.35 µm and 25 µm (generally referred to by astronomers as "optical and infrared" or OIR), astronomy flourished for millennia without access to space. Even in the space age, ground-based telescopes continue to provide us with OIR observations, especially at wavelengths shorter than 2 µm, where it is possible to make sensitive measurements with large telescopes through the Earth's atmosphere. Prior to the development of electronic detectors, space observations of night-sky objects were inefficient and took limited advantage of the lack of atmosphere from spacecraft. But as detection techniques improved, and computers and electronic imaging detectors combined to replace human observers even on ground-based telescopes, space-based observations quickly surpassed anything possible from the Earth's surface. Today, no ground-based telescope approaches the combination of natural sensitivity and resolution allowed by observations in space with equivalent telescope area. The enormous advances made by orbiting observatories such as *HST* show that the disadvantages of the atmosphere make it all but inevitable that many if not all OIR observational techniques will be used in space at some time, once we develop the means to put very large telescopes in orbit.

[I]The University of California, Oakland CA, USA

The Earth's atmosphere imposes two strong limits on OIR terrestrial observations:

1. The combination of thermal radiation, molecular airglow and scattered light from the Moon and artificial lighting create foreground emission (normally called background by astronomers) that provides a fundamental limit to the sensitivity of astronomical observations from the ground. The salient example is the difficulty of seeing stars during the daytime owing to the bright sky.

2. Rapidly changing variations of the index of refraction in turbulent air distort the wavefront of a celestial source, scrambling information about the source at high spatial frequencies and degrading the natural resolution of a telescope by one to two orders of magnitude. The wavefront distortions, called *seeing*, can be corrected in principle over much of the OIR domain using a technique called Adaptive Optics (AO), but AO is applicable to a limited range of observations and does not reliably correct the atmosphere at all wavelengths nor over a very wide field of view.

The practical difficulties associated with AO are sufficiently daunting that they always compromise the information content of terrestrial observations: AO-corrected telescopes limit the field of view, dynamic range, radiometric stability, resolution and astrometric accuracy of an observation from the ground relative to observations in space. These limits are especially severe for observations requiring wide-field imaging, such as surveys of large regions of the sky.

The combination of detection technology and the development of large rockets able to launch ever-larger telescopes into orbit drove the evolution of OIR space telescopes. Increasingly, spectroscopic and interferometric measurements come up against background limits, as improvements in technology produce instruments that detect and analyze nearly all of the incident light with no added noise. In fact, light detectors are currently approaching perfection over most of the OIR domain. As reflecting optics also perform nearly ideally, increased detection sensitivity will in future require larger telescope areas.

Space also affords specialized observations such as polarimetric observations of the Sun with high spatial resolution in the visible. Solar space observations have given insight into the behaviour of magnetic fields in a stellar atmosphere (Chapters 19 and 33, Title 2010; Stenflo 2010).

The following sections illustrate how spacecraft with OIR payloads have allowed modern astronomy to approach the limits of nature.

Background radiation

Background radiation from the atmosphere and zodiacal dust in the vicinity of the Earth limit the sensitivity of all telescopes. Figure 6.1 shows the rate of photons, $\mathrm{d}N_\gamma/\mathrm{d}t$, seen by ground and space-based telescopes at the highest useful angular resolution for unresolved sources. An observation that records N_γ photons has an irreducible noise from the statistical fluctuations in the photon stream: $\sigma_N = \sqrt{N_\gamma}$. The much larger rate of background photons seen by ground-based telescopes means

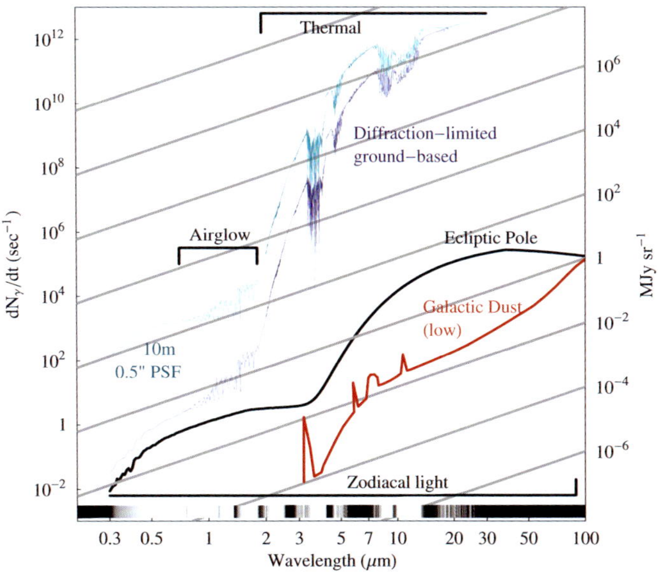

Figure 6.1: Comparison of background photon rates for a 10 m ground-based telescope under ideal conditions, both in the limit of perfect correction for atmospheric phase distortion and in very good seeing, with a diffraction-limited space telescope of the same size, in the direction of the ecliptic pole (black line) and with a low level of galactic dust (red line). At wavelengths longer than $\approx 60\,\mu$m, galactic dust emission becomes an important background source. The scale on the right gives the background radiance in 1 MJy sr^{-1} and the bar at the bottom of the graph is a grayscale representation of the absorption by the terrestrial atmosphere.

that the noise will always be higher than in an equivalent space-based observation in the background limit, higher by orders of magnitude at wavelengths where thermal emission (285 K) is strong. Spatial as well as rapid temporal variations in the airglow and sky radiance create an additional source of noise, introducing systematic uncertainties of order N_γ that are much larger than the statistical fluctuations.

The background radiation for space telescopes located within a few million kilometres from Earth is dominated by the zodiacal dust. Sunlight scattered by small dust particles dominates the background shortward of a few micrometres. At longer wavelengths the background is caused by thermal emission from dust at temperatures close to those on Earth. However, the thin layer of zodiacal dust near the Earth has an opacity of $\approx 10^{-7}$, i.e., orders of magnitude smaller than even the best transmission of the air. Accordingly, the thermal background emission in space is much smaller than that seen from the ground. The zodiacal emission could be avoided if a space telescope were situated well above the ecliptic plane or at orbital distances beyond the asteroid belt, i.e., beyond ≈ 3 ua from the Sun.

Space telescopes operating at wavelengths shorter than a few micrometres take full advantage of the reduced background even if the optics are at ambient temperatures. For longer wavelengths, the telescope optics must be cooled to reduce thermal emission from the optical surfaces so that it reaches the zodiacal light limit. Cooling to below 50 K, such as planned for the *JWST*, will allow the telescope to reach the zodiacal light limit up to about 20 μm; observations at longer wavelengths require greater cooling, typically to less than 10 K, as with *ISO* or *Spitzer*.

Wavefront distortion (atmospheric refraction and "seeing")

Atmospheric refraction causes the path of a light ray to deviate from a straight line as it passes through the atmosphere. This is because the air density varies as a function of altitude. Objects observed from the ground appear to be higher in the sky than their actual positions, an effect which worsens as the object approaches the horizon. Atmospheric refraction also disperses the observed radiation for very high angular resolution studies, with blue wavelengths being more affected than red wavelengths. Two counter-rotating prisms, Risley prisms, are used to correct this effect (see, for example, Horch et al 1994).

A more serious problem affecting ground-based observations is the twinkling of starlight, caused by the disturbance of the wavefront when the light passes through the Earth's atmosphere. Turbulence in the air creates rapid variations in the index of refraction, inducing differential phase delays of several micrometres across distances of a few decimetres perpendicular to the direction of the incident light rays. These phase differences create a corrugated wavefront from an initially plane wave and introduce variations in the wave amplitude as well. The corrugated wavefront spreads the image at the focal plane of a telescope, producing a pattern of speckles over an area with a diameter of $\approx 1''$, with the size of each speckle being approximately equal to the diffraction-limited point-spread function (PSF) of the telescope. The speckle pattern changes on a timescale of 10 ms at wavelengths in the visible part of the spectrum (Hardy 1998).

Figure 6.2 shows a short exposure image of the double star, ζ Booetis, revealing the speckle pattern. The inset shows the resolution that would be achieved in space with the same telescope. This image is typical of ground-based images without any correction for atmospheric distortion. The impact of the atmosphere is wavelength dependent. The wavefront distortion comes about from approximately fixed time delays in the arrival of the wave at different points across the wavefront, but the phase delay varies nearly inversely with the wavelength. The correlation length of the phase, r_0, varies with wavelength as $\lambda^{6/5}$. At a wavelength of 10 μm, a small telescope like the one used for the image in Figure 6.2 would be nearly diffraction-limited with single speckle that moved around in the focal plane due to changes in the tilt of the wavefront: the correlation length r_0 ($\lambda = 10\,\mu\mathrm{m}$) is almost five times larger than the diameter of the telescope, $D_{\mathrm{tel}}/r_0 \approx 0.2$, and the wavefront is well-correlated across the pupil.

One of the best ways to illustrate the impact of a distorted wavefront on the image is using the *Strehl ratio*. The Strehl ratio is defined as the maximum intensity

Figure 6.2: The image (black on white) of the double star ζ Booetis taken around $\lambda = 550$ nm on the 2.6 m Nordic Optical Telescope shows identical speckle patterns for the two stars separated by $0.8''$. Each set of speckles is spread out over a region approximately $0.4''$ across ($r_0 \approx 0.4$ m). The inset pictures (white, with a false-colour grey scale, on black) show how the stars would look with the same telescope at its diffraction limit.

in an image divided by the maximum intensity in a diffraction-limited image, and it is normally used as a measure for long-exposure images where the speckle pattern is averaged to produce an approximately Gaussian-shaped PSF with a width equal to the so-called *seeing angle*. The Strehl ratio of the image in Figure 6.2 is about 0.017 ($D_{\rm tel}/r_0 \approx 7$). In the limit $r_0 \ll D_{\rm tel}$, the Strehl ratio is approximately equal to $(r_0/D_{\rm tel})^2$.

Figure 6.3 shows how the Strehl ratio varies with wavelength for three different ground-based telescopes: a 2.4 m and a 10 m telescope under good seeing ($0.5''$), and a 10 m telescope with an adaptive optics system that reduces the RMS wavefront errors to 0.2 µm, the projected state of the art within 10 years. The Strehl ratio for uncompensated telescopes under excellent conditions falls below 0.1 at wavelengths of about 2 µm and 5 µm, respectively, for the first two ground-based telescopes mentioned above. Even the best AO corrected telescope has Strehl ratios substantially smaller than 1 at all visible wavelengths. Since a space telescope will generally have a Strehl ratio equal to 1 at all wavelengths, the advantage of space for imaging is apparent from this figure. By comparing Figure 6.3 with Figure 6.1, one concludes that space telescopes have an advantage even if AO techniques can remove most of the atmospheric distortion: AO correction works best at infrared wavelengths where, however, the atmospheric background is high. Only in a narrow window between about 1 µm and 2 µm can an AO compensated telescope achieve a sensitivity comparable to a space telescope, and then only when the ground-based telescope has a substantially larger aperture. In practice, the use of large ground-based telescopes brings an advantage only to observations that require very high spectral resolution — a situation where the background rates are reduced.

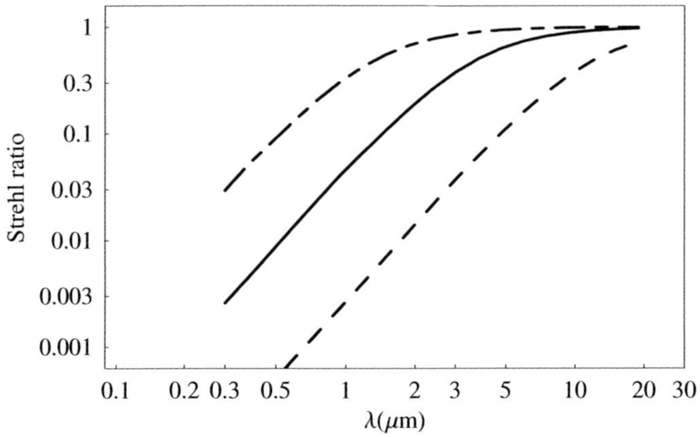

Figure 6.3: The Strehl ratio (peak radiance in the real image of a point source relative to the peak radiance in the diffraction-limited case) variations for 2.4 m (solid) and 10 m (dashed) telescopes in 0.5″ seeing and an AO-corrected ground-based 10 m telescope with residual wavefront errors of 200 nm (short-long dash), i.e., the best projected performance ten years from now. The 2.4 m *HST* achieves a Strehl ratio of 1 throughout this wavelength range.

Sources of radiation

Starlight is the primary source of radiation in the OIR domain. The integrated visible and infrared light from the Milky Way and more distant galaxies produces a global irradiance second only to that of the cosmic microwave radiation in importance. Figure 6.4 shows the relative contributions of different energy sources over the entire spectrum. Stellar spectra are approximately blackbodies with narrow lines from ions, atoms, and molecules superimposed, the relative distribution and line strengths depending on the effective temperature and composition of the stellar atmosphere. The effective temperatures of normal stars range from approximately (2000 to 40 000) K, resulting in maximum flux densities between about 0.1 µm and 2 µm. The Sun has an effective temperature of ≈ 5780 K, its spectral irradiance therefore peaks near 0.5 µm, if expressed as L_λ (i.e., in units of $\mathrm{W\,m^{-2}\,nm^{-1}}$) or near 0.9 µm, if expressed as L_ν (i.e., in units of $\mathrm{W\,m^{-2}\,Hz^{-1}}$) . For regions of the sky where starlight is the dominant energy source, the OIR range contains the majority of the radiation and is almost always critical to studying the physical conditions of the matter.

In addition to thermal radiation from stars, there are two other important sources of continuous radiation, namely synchrotron radiation from ionized gas in the presence of magnetic fields and bremsstrahlung from ionized gas in H II regions, typically referred to as *free-free* radiation. Almost all synchrotron sources that can be observed in the OIR domain are near the low-frequency limit; their flux density is proportional to the −1/3 power of wavelength (Ginzburg and Syrovatskii 1965). Free-free radiation from a plasma at temperature T varies as $\exp[-h\,c_0/(\lambda\,k_\mathrm{B}\,T)]$,

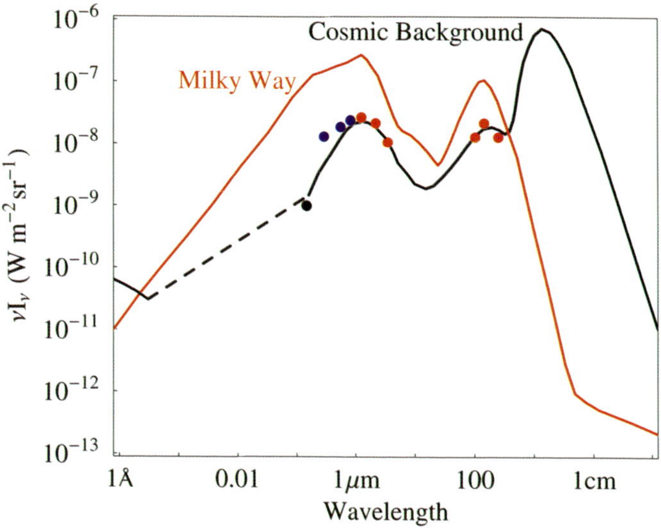

Figure 6.4: The average global spectral radiance (multiplied by the frequency) in the solar neighborhood (from Wright 2008). The cosmic background (black) and Milky Way (red) lines show the separate contributions from extragalactic and galactic sources. Data from Gruber et al (1999), Madau and Pozzetti (2000), Hauser and Dwek (2001), Franceschini et al (1997), Stanev and Franceschini (1998), Bernstein et al (2002) and Wright (1996).

with T approximately 10^4 K for H II regions, but is often several times hotter in planetary nebulae and other ionized regions (Allen and Cox 2000).

Transitions between the electronic energy levels of most atoms and molecules have energies of a few electronvolts, corresponding to wavelengths of a few hundred nanometres, squarely within the OIR domain. Molecular bond-strengths are between a few hundred millielectronvolts and a few electronvolts, placing the vibrational transitions in the infrared part of the OIR region. With the exception of molecular hydrogen, pure rotational transitions of molecules are in the radio spectrum. Spectroscopic observations in the OIR include atomic and molecular lines from stars, from gas in circumstellar regions, and from gas in the interstellar medium (ISM); they occur in emission and absorption. Table 6.1 lists the wavelengths of a small sample of atomic and molecular lines that are frequently used to study the physical conditions in the interstellar medium.

Small solid particles — interstellar dust — can resonantly absorb and emit light over relatively narrow wavelength ranges. These resonant features arise from vibrational excitations of atoms and radicals bound to the surfaces of the particles, and have fractional bandwidths, $\Delta\lambda/\lambda$, of order 0.1, i.e., they are much wider than the lines from discrete energy transitions in single molecules but nevertheless considerably narrower than continuous sources such as thermal radiation. Many of these features are useful to understanding the size and composition of interstellar and circumstellar dust (Spitzer 1978; Sellgren 1984; Léger and Puget 1984). The most

Table 6.1: Selected interstellar lines.

Nebular emission		ISM absorption			
Atom / ion	λ/nm	Ion	λ/nm	Molecule	λ/nm
H	434.0 (H_γ)	Na	330.2; 330.3	CH	313.8
	486.2 (H_β)		589.0; 589.6		314.3; 314.6
	656.2 (H_α)	K	766.5		387.9
	1281.8 (P_β)		769.9		388.6; 389.0
	1875.1 (P_α)	Ca	422.7		430.0
	2165.5 (B_γ	Ca^+	393.4	CN	387.4–7[a]
	2625.2 (B_β)		396.9	CH^+	344.7
	4051.2 (B_α)	Ti^+	307.3		357.5
O^+	372.6; 372.9		322.9		374.5
O^{++}	495.9		324.2		395.8
	500.7		338.4		423.4
N^+	654.8	Fe	372.0	$C^{13}H^+$	423.2
	658.4		386.0		

[a] several lines within the range indicated

salient features at ultraviolet wavelengths are indicative of graphite (Mathis 1990) and several complexes at 3.3 µm and 7.7 µm that arise from Polycyclic Aromatic Hydrocarbons, PAHs (Puget and Léger 1989).

Dust absorbs light along the line of sight, but it can also scatter light. Absorption and scattering are selective in wavelength. Scattering occurs most strongly for short wavelengths owing to the small size of the dust particles (van de Hulst 1981). Scattered light thus becomes bluer and transmitted light becomes redder relative to its source. As such, the colour of radiation gives information about interstellar dust, as well as its original source, usually stars.

In summary, then, space observations have several advantages that bring qualitative changes to traditional astronomical methods: freedom from atmospheric absorption gives access to observations covering large, continuous wavelength regions; the very high signal stability permits the detection of tiny time variations; the high contrast — virtually unattenuated by the intervening medium — lets us study faint light in the immediate vicinity of bright stars and quasars; moreover, a stable diffraction-limited PSF allows the study of individual stars in distant galaxies and enables an extraordinary astrometric accuracy on distant stars and clusters of stars (Chapter 16, Lindegren 2010). And finally, space observations have very high sensitivity at red and infrared wavelengths, where the terrestrial backgrounds are high.

Freedom from atmospheric absorption also means that the entire spectrum of a star, including the Sun (Chapter 32, Fröhlich 2010), is available to measure bolometric luminosity directly. Moreover, all lines are accessible for sources at any redshift. This is an important advantage for deriving samples of galaxies at cosmological distances; ground-based samples normally have gaps in redshift intervals where the prominent lines are blocked by the Earth's atmosphere. Ultraviolet lines of high-redshift objects, such as quasars, may be seen through the Earth's atmosphere in the visible and infrared, but in low-redshift objects these lines can only be seen with space telescopes. Furthermore, lines from molecules like H_2O, CO_2, and OH that are strong in the atmosphere are easily seen from space but unobservable from the ground.

The increased sensitivity of space telescopes at near-infrared wavelengths has also been important to study objects such as supernovae at high redshifts, where the locally emitted blue light is in the infrared (Riess et al 2004). The deepest images of the Universe come from the 2.4 m *HST* (Williams et al 1996; Beckwith et al 2006): the reduced background and small PSF provide deeper images than can be made with even the largest telescopes on the ground.

The stable diffraction-limited PSF of the *HST* also made it possible to study individual Cepheid stars in Virgo-cluster galaxies, providing a measure of the Hubble constant with sufficient accuracy to settle the long running controversy about the age of the Universe (Freedman et al 2001). The same advantage allows observations of solar luminosity stars in M31 to derive the ages of the several populations that make up that galaxy by observing the so-called main-sequence turnoff in the Hertzsprung–Russell diagram (Brown et al 2004). The combination of a small PSF with high spectral resolution provided the first unambiguous rotation curve in the immediate vicinity of the nucleus of a galaxy, confirming that only a Black Hole had sufficient mass density to explain the high velocities of the gas (van der Marel et al 1997). Moreover, the high radiometric stability of space allowed the detection of an atmosphere around the extrasolar planet HD 209458b using differential spectroscopy on the light curve, as the planet transited the face of its host star (Charbonneau et al 2002) — one of the more impressive observations in modern astronomy. Finally, the smallest exoplanet COROT-Exo-7b, with a radius of 1.6 Earth radii, which has been concluded to be a truly Earth-like, rocky planet, was discovered by the *COROT* satellite.

Selected space missions

Optical and infrared missions entered the space age only after the first missions explored wavelengths utterly unobservable from the ground, notably at X-ray and ultraviolet wavelengths. Since most of the OIR region could be observed from the ground — and because infrared detectors were less developed (or information about them was classified) — there was less pressure from the OIR community to take on the challenges and expense of space. But the enormous advantages of continuous transmission, reduced background radiation, and freedom from atmospheric wavefront distortion inevitably drew OIR observers to create space observatories. The resulting missions produced a rapid expansion of our knowledge about the Universe and at the same time increased the popularity of astronomy and space science. Images of the Cosmos taken by the *HST* and images from the surface of Mars and Titan returned by probes from the surfaces of these planets created a public engagement in the space programme rivaled only by the initial Moon landings of the *Apollo* programme.

The National Space Science Data Center (NSSDC) at NASA's Goddard Space Flight Center provides a useful reference for space missions that can be found at http://nssdc.gsfc.nasa.gov/. The following subsections give a brief overview of the missions that covered the visual/near-infrared bands.

IRAS

- Launch: 25 January 1983, Delta from Vandenberg AFB, USA
- Orbit: 900 km LEO
- Optical system: 0.6 m Ritchey–Chrétien telescope, cooled by liquid helium
- Wavelength range: 8 μm to 120 μm
- Instruments: four-band photometry, low-resolution spectrometer
- Mass: 1100 kg
- Lifetime: 10 months

The *IRAS* was the first Explorer-class satellite designed to survey the entire sky at infrared wavelengths between 8 μm and 120 μm. *IRAS* was a joint mission between the United States (NASA), the Netherlands, and the United Kingdom.

IRAS contained a 0.6 m Ritchey–Chrétien telescope cryogenically cooled with liquid helium below 10 K. An array of 62 detectors in the focal plane covered four broad wavelength bands centred at (12, 25, 60, and 100) μm. The survey was carried out by rotating the satellite at a constant angular velocity perpendicular to the satellite-Sun vector, and detecting sources as they transited the fixed array of detectors in the focal plane.

IRAS detected approximately 350 000 sources including many that had well-known positions but had never been seen before in the infrared. Source positions were accurate to about 30″, providing a rich catalogue that is still used as a reference for infrared irradiances of stars and galaxies. Following the sky survey, *IRAS* carried out pointed observations, where a low-resolution spectrometer provided supplementary spectra for many of the more interesting sources, until the cryogen was depleted on 21 November 1983.

Details of the *IRAS* mission and its first scientific results were published by Neugebauer et al (1984), and by further authors in the 1 March 1984 issue of Astrophysical Journal Letters.

Hipparcos

- Launch: 8 August 1989, Ariane 44LP from Kourou, French Guiana
- Orbit: 507 km to 35 888 km elliptical
- Optical system: 0.3 m Schmidt telescope
- Mass: 1025.0 kg
- Lifetime: 3.5 a

Hipparcos took advantage of the wavefront stability above the atmosphere to measure the astrometric positions of 120 000 stars for parallaxes, proper motions and positions with an accuracy of 2 mas. Two fields of view 58° apart were imaged through a single telescope onto a focal plane, which consisted of a regular grid of 2688 transparent parallel slits. The spacecraft spun slowly around an axis at 12 revolutions per day to scan the stars across the focal plane. The grid modulated the intensity of light from the stars which was detected by a photon-counting image tube to detect the phase difference of the modulated light from the two separated fields, thus providing relative positions of stars in the two fields to high accuracy. The large angular separation of the two fields of view reduced the systematic uncertainties that would build up when constructing an astrometric reference system from relative positions of stars within a small field of view on the sky.

In addition to this main detection system, another photomultiplier system detected light from a beam splitter in the optical path to measure the Johnson B- and V-band photometric magnitudes of another 400 000 stars down to 11th magnitude and with a positioning accuracy of 50 mas.

Hipparcos revolutionized the field of astrometry. It vastly improved our knowledge of stellar distances out to about 100 pc and allowed statistical studies of stellar properties with much higher precision than had been possible by any previous ground-based observations.

The results of the past ten years of exploitation of the *Hipparcos* satellite data are comprehensively summarized by Perryman (2009).

COBE

– Launch: 18 November 1989, Delta from Vandenberg AFB, USA
– Orbit: 900 km LEO
– Instruments: Differential Microwave Radiometer (DMR), Far-InfraRed Absolute Spectrophotometer (FIRAS), Diffuse InfraRed Background Experiment (DIRBE)
– Mass: 2200 kg
– Lifetime: 4.1 a

COBE was designed to measure the diffuse radiation from the Cosmos between 1 µm and 1 cm wavelength over the entire sky. Its main mission was to see if the Cosmic Microwave Background closely followed a Planck function at 2.7 K, as expected from the Big Bang cosmological theory. *COBE* then revealed subtle spatial variations in the cosmic background radiation of one part in 10^5 across the sky. In addition, the short wavelength detectors of the DIRBE instrument observed the diffuse infrared background radiation between wavelengths of 1 µm and 300 µm. Thus *COBE* also provided a wide-field sky survey at near-infrared wavelengths.

COBE scanned the sky by rotating 1 min^{-1} about its symmetry axis, oriented at 94° to the Sun-Earth line. It covered the entire sky every six months in this manner, allowing redundant measurements of every direction to reduce uncertainties.

The two principal investigators for *COBE* instruments, John Mather and George Smoot, were awarded the 2006 Nobel Prize for physics for their discoveries with *COBE*.

The instrument and its performance — two years after launch — has been described by Boggess et al (1992) and the results of all four years of observations were reported by Bennett et al (1996).

HST

– Launch: 25 April 1990, Space Shuttle from Cape Canaveral, USA
– Orbit: 690 km circular low Earth orbit (LEO)
– Optical system: 2.4 m $f/24$ Ritchey–Chrétien telescope
– Wavelength range: 0.115 µm to 2.5 µm
– Instruments: cameras, spectrometers and interferometers that are exchanged during visits by astronauts
– Mass: 11600 kg
– Lifetime: ongoing 19 a after launch

The *HST* was one of NASA's four flagship missions called the Great Observatories and was a joint NASA/ESA collaboration both for construction and operation. It was designed as an astronomical observatory to provide diffraction-limited images from ultraviolet to near-infrared wavelengths using a suite of instruments at different spectral resolutions and fields of view.

Launched in 1990, its performance during the initial three years in orbit was hampered owing to a misfigured primary mirror. New instruments installed in 1993 on the first servicing mission corrected this flaw, allowing *Hubble* to become one of the most productive and best known of all space missions. It remains operational through 2009; another servicing mission is planned for 2009 to extend the mission.

Hubble employs a number of novel technologies allowing it to achieve its full resolution and sensitivity. Chief among these are a set of six gyroscopes with unprecedented accuracy and an interferometric Fine Guidance System that uses stars to stabilize the pointing. Together, these subsystems allow the telescope to point anywhere on the sky with an RMS jitter of less than 5 mas. *HST* employs a number of innovative subsystems, including six nickel-hydrogen batteries to power the observatory during the half-hour period in Earth shadow every orbit, and it was the first observatory satellite to use the TDRSS communication system for its telemetry.

Hubble has contributed to nearly every field of astrophysics and many in solar system research as well. To explore the harvest from *HST* one should consult the web site http://www.stsci.edu/hst/.

IRTS

– Launch: 13 March 1995 by an HII rocket from the Tanagashima Space Center, Japan, returned by the Space Shuttle
– Orbit: LEO
– Optical system: 15 cm liquid helium cooled telescope
– Wavelength range: 1 µm to 1000 µm
– Instruments: Near-InfraRed Spectrometer (NIRS; 1.4 µm to 4.0 µm), Mid-Infrared Spectrometer (MIRS; 4.5 µm to 11.7 µm), Far-Infrared Line Mapper (FILM; 145 µm, 155 µm, 158 µm, 160 µm), Far-InfraRed Photometer (FIRP; 150 µm, 250 µm, 400 µm, 700 µm)
– Lifetime: 20 d

IRTS was one of the early satellites launched by the Japanese space agency, JAXA, designed primarily to carry out an infrared spectroscopic survey of 7 % of the sky and provide spectral energy distributions of the objects at far-infrared wavelengths using the FIRP instrument. *IRTS* had a small telescope and short lifetime, thus representing a first step as JAXA entered into the more recent era of large aperture space telescopes for astronomical research (Murakami et al 1996).

ISO

– Launch: 17 November 1995, Ariane 4 from Kourou, French Guiana
– Orbit: 1000 km to 70 500 km elliptical (period 24 h; of which ca. 17 h outside the radiation belts)
– Optical system: 0.6 m telescope cooled with superfluid helium
– Wavelength range: 2.5 μm to 240 μm
– Instruments: two spectrometers (SWS and LWS), a camera (ISOCAM) and an imaging photo-polarimeter (ISOPHOT) jointly covered wavelengths from 2.5 μm to around 240 μm
– Mass: 1800 kg (launch mass, including the liquid helium, 2400 kg)
– Lifetime: 28 months

The European Space Agency's *ISO* was the first true orbiting infrared observatory. *ISO* was a follow-on to the *IRAS* survey and a precursor to the *Spitzer* observatory. *ISO's* 60 cm telescope was cooled with superfluid helium to allow natural background-limited sensitivity at far infrared wavelengths. It had a field of view of $20'$ and stabilized pointing to an accuracy of $5''$.

ISO was an important advance over *IRAS* in space observations of the infrared sky. It observed many sources discovered earlier to ascertain their luminous output, chemical composition, structure, and nature. *ISO* was especially important for its spectroscopy of the interstellar medium in the Milky Way and many distant galaxies, advancing our understanding of infrared luminous galaxies undergoing bursts of star formation. It showed spectra of circumstellar regions around young stars that matched those of solar system objects, establishing the chemical similarity of these regions — and it demonstrated that water is present everywhere in the Universe. The *ISO* science legacy is summarized in a series of papers edited by Cesarsky and Salama (2005).

MSX

– Launch: 24 April 1996, Delta II from Vandenberg AFB, USA
– Orbit: 900 km, polar, near-Sun synchronous
– Wavelength range: 4.2 μm to 26 μm
– Instruments: SPIRIT III, a five-colour, high-spatial resolution scanning radiometer and a six-channel high-spectral resolution Fourier-transform spectrometer; UVISI, five spectrographic imagers and four UV/visible imagers; and Space-Based Visible, a 16 cm visible-band imaging telescope
– Mass: 2700 kg

– Lifetime: 10 months

The *MSX* was a military test project sponsored by the Ballistic Missile Defense Organization (BMDO) to demonstrate the feasibility of identifying and tracking ballistic missiles midway through their trajectories. It employed three instruments covering the range 0.11 µm to 28 µm with multispectral capability producing data cubes that combine spatial and spectral information. *MSX* also carried out some aeronomic and auroral observations for civilian use, in addition to its military mission (cf., Mill and Guilmain 1996).

Spitzer

– Launch: 25 August 2003, Delta 7920H ELV from Cape Canaveral, USA
– Orbit: Earth-trailing, heliocentric
– Optical system: 0.85 m telescope cooled with liquid helium
– Wavelength range: 3 µm to 180 µm
– Instruments: four-channel array camera (IRAC), imaging photometer (MPS), low-resolution spectrometer (IRS)
– Mass: 865 kg
– Lifetime: ≥ 5 a

Spitzer, previously known as *SIRTF*, is a pointed general-purpose infrared telescope similar to *ISO* built by NASA as the fourth of its 'Great Observatories' programme.

The 0.85 m Ritchey–Chrétien telescope is cooled with liquid helium to reduce thermal backgrounds. Its five-year lifetime is made possible by using radiative cooling of its outer shell to reduce the heat load on the cryogen tank, and it uses an Earth-trailing heliocentric orbit to further reduce thermal loads from the Earth and provide a high field of regard. *Spitzer*'s pointing system is based on celestial-inertial three-axis stabilized control.

The three infrared instruments provide a combination of imaging and spectroscopy: IRAC is a four-channel camera at (3.6, 4.5, 5.8, and 8) µm; MPS is a multi-channel photometer with detectors at (24, 70, and 160) µm; IRS is a spectrometer providing continuous coverage between 5 µm and 40 µm. *Spitzer* remains operational at the time of writing.

Spitzer has followed on to the science of *IRAS* and *ISO* by vastly increasing our understanding of distant galaxies, especially those luminous in the infrared, the interstellar medium, circumstellar regions, especially those with the precursors to planetary systems, and extra-solar planetary systems (Werner et al 2004, and all following articles in the same volume). One especially noteworthy result achieved from observations with *Spitzer* was the detection of the thermal radiation from an extra-solar planet for the first time, putting constraints on the planet's temperature and size (Charbonneau et al 2005).

COROT

– Launch: 27 December 2006 by a Soyuz-Fregat launcher from Baikonur, Kazakhstan
– Orbit: 896 km circular polar orbit, allowing continuous observations of two regions in the sky for more than 150 days each
– Optical system: 0.27 m telescope
– Instrument: a wide-field (2.8°× 2.8°) two-part camera operating in the visible — one for each mission objective, namely exoplanet search and asteroseismology
– Mass: 630 kg
– Lifetime: nominally 2.5 a

COROT, a mission led by the French Space Agency (CNES), with contributions from ESA, Austria, Belgium, Germany, Spain and Brazil, uses a wide-field telescope that is designed to detect tiny changes in the brightness of nearby stars. The mission's main objectives are the search for exoplanets by detecting the dimming of the light from a star as a planet passes in front of it, as well as the study of stellar interiors by the method of asteroseismology, with the aim to calculate the star's precise mass, age, and chemical composition (Auvergne et al 2009).

COROT is the first mission capable of detecting rocky planets that are several times larger than Earth, around nearby stars. In spring 2008, *COROT* discovered planet COROT-Exo-7b,whose passage in front of the stellar disk dims the star to 99.65 % of its normal brightness (Léger et al 2009). Based on this observation in space, supplemented by observations from the ground, it was concluded that COROT-Exo-7b, which orbits a K0V star (T = 5300 K) within 20 h, was a truly Earth-like, "rocky" planet. Given its surface temperature of over 1000 °C, however, the surface most likely is covered by lava.

Bibliography

Allen CW, Cox AN (2000) Astrophysical quantities. Berlin: Springer

Auvergne M, Bodin P, Boisnard L, Buey J-T, Chaintreuil S, CoRoT team (2009) The *CoRoT* satellite in flight: description and performance: 2009arXiv0901.2206A, DOI: 10.1051/0004-6361/200810860

Beckwith SVW, Stiavelli M, Koekemoer AM (plus 12 authors) (2006) The *Hubble* ultra deep field. Astrophys J 132:1729–1755

Bennett CL, Banday AJ, Górski KM (plus seven authors) (1996) Four-year *COBE*1 DMR Cosmic Microwave Background observations: Maps and basic results. Astrophys J 464:L1–L4

Bernstein RA, Freedman WL, Madore BF (2002) The first detections of the extragalactic background light at 3000, 5500, and 8000 Å. I. Results. Astrophys J 571:107–128

Boggess N W, Mather J C, Weiss R (plus 15 authors) (1992) The *COBE* mission — Its design and performance two years after launch. Astrophys J 397:420–429

Brown TM, Ferguson HC, Smith E (plus four authors) (2004) Age constraints for an M31 globular cluster from main-sequence photometry. Astrophys J 613:L125–L128

Cesarsky CJ, Salama A, eds (2005) *ISO* science legacy. Space Sci Rev 119:Nos 1–4

Charbonneau D, Allen LE, Megeath ST (plus eight authors) (2005) Detection of thermal emission from an extrasolar planet. Astrophys J 626:523–529

Charbonneau D, Brown TM, Noyes RW, Gilliland RL (2002) Detection of an extrasolar planet atmosphere. Astrophys J 568:377–384

Freedman WL, Madore BF, Gibson BK (plus 12 authors) (2001) Final results from the *Hubble* Space Telescope key project to measure the Hubble constant. Astrophys J 553:47–72

Franceschini A, Aussel H, Bressan A (plus seven authors) (1997) Source-counts and background radiation. ESA SP-401:159–167

Fröhlich C (2010) Solar radiometry. ISSI SR-009:525–540

Ginzburg VL, Syrovatskii SI (1965) Cosmic magnetobremsstrahlung (Synchrotron radiation). Ann Rev Astron Astrophys 3:297–350

Gruber DE, Matteson JL, Peterson LE, Jung GV (1999) The spectrum of diffuse cosmic hard X-rays measured with *HEAO* 1. Astrophys J 520:124–129

Hardy JW (1998) Adaptive optics for astronomical telescopes. Oxford University Press

Horch E, Heanue JF, Morgan JS, Timothy JG (1994) Speckle imaging with the MAMA detector: Preliminary results. Astron Soc Pacific Publ 106:992–1002

Hauser MG, Dwek E (2001) The cosmic infrared background: Measurements and implications. Ann Rev Astron Astrophys 39:249–307

Léger A, Puget JL (1984) Identification of the "unidentified" IR emission features of interstellar dust? Astron Astrophys 137:L5–L8

Léger A, Rouan D, Schneider J (2009) Transiting exoplanets from the *CoRoT* space mission — VIII. CoRoT-7b: The first super-Earth with measured radius. Astron. Astrophys. (submitted for publication in a special issue with ca. 80 CoRoT papers)

Lindegren L (2010) High-accuracy positioning: astrometry. ISSI SR-009:279–291

Madau P, Pozzetti L (2000) Deep galaxy counts, extragalactic background light and the stellar baryon budget. Mon Not R astr Soc 312:L9–L15

Mathis JS (1990) Interstellar dust and extinction. Ann Rev Astron Astrophys 28:37–70

Mill JD, Guilmain BD (1996) The *MSX* mission objectives. Johns Hopkins Appl Phys Lab Tech Dig 17:4–10

Murakami H, Freund MM, Ganga K (plus 25 authors) (1996) The *IRTS* (Infrared Telescope in Space) mission. Publ Astron Soc Japan 48:L41-L46

Neugebauer G, Soifer BT, Beichman CA (plus seven authors) (1984) Science 224:14–21

Perryman MAC (2009) Astronomical applications of astrometry: A review based on ten years of exploitation of the *Hipparcos* satellite data. Cambridge: Cambridge University Press

Puget JL, Léger A (1989) A new component of the interstellar matter — Small grains and large aromatic molecules. Ann Rev Astron Astrophys 27:161–198

Riess AG, Strolger L-G, Tonry J (plus 16 authors) (2004) Type Ia supernova discoveries at z > 1 from the Hubble Space Telescope: Evidence for past deceleration and constraints on dark energy evolution. Astrophys J 607:665–687

Sellgren K (1984) The near-infrared continuum emission of visual reflection nebulae. Astrophys J 277:623–633
Spitzer L (1978) Physical processes in the interstellar medium. New York: Wiley-Interscience
Stenflo JO (2010) Stokes polarimetry of the Zeeman and Hanle effects. ISSI SR-009:543–557
Stanev T, Franceschini A (1998) Constraints on the extragalactic infrared background from gamma-ray observations of MRK 501. Astrophys J 494:L159–L162
Title AM (2010) Michelson interferometers. ISSI SR-009:327–338
van de Hulst HC (1981) Light scattering by small particles. New York: Dover
van der Marel RP, de Zeeuw PT, Rix H-W, Quinlan GD (1997) A massive black hole at the centre of the quiescent galaxy M32. Nature 385:610–612
Williams RE, Blacker B, Dickinson M (plus 14 authors) (1996) The *Hubble* deep field: Observations, data reduction, and galaxy photometry. Astron J 112:1335–1384
Werner MW, Roellig TL, Low FJ (plus 25 authors) (2004) The *Spitzer* Space Telescope mission. Astrophys J Suppl 154:1-9
Wright EL (1996) On the density of primordial black holes in the galactic halo. Astrophys J 459:487–490
Wright EL (2008) personal communication

— 7 —

The mid- and far-infrared range: radiation emission processes from interstellar dust and gas

A.G.G.M. Tielens[I]

Abstract

The thermal-infrared wavelength region contains the spectral signatures of solid-state compounds as well as a variety of lines emitted by atoms and molecules in the gaseous state. Together, these signatures provide unique diagnostic probes of the physical and chemical conditions of the cool and dusty Universe. This article summarizes the contribution of space-based observations in the thermal infrared to our understanding of the Universe. In particular the formation of stars and planets, the characteristics of interstellar dust and of polycyclic aromatic hydrocarbons as well as the interstellar media of galaxies are discussed.

Introduction

All cool and dusty objects in the Universe emit infrared (IR) radiation. Atoms and neutral molecules and dust attain temperatures in the range of about 5 K to 1000 K and therefore emit most of their energy at mid- and far-IR wavelengths. Generally, the near-IR range is defined as extending from 0.75 µm to 2.5 µm and the mid-IR range runs from 2.5 µm to about 25 µm. The far-IR region covers wavelengths from 25 µm to 100 µm, and is then going over into the sub-millimetre domain. The spectra of cool interstellar and circumstellar regions show absorption or emission, characteristic for dust compounds. In addition, broad emission bands are often present due to fluorescence of large molecules, as well as narrow emission lines that can be assigned to individual atomic and molecular transitions. These features provide 'fingerprints' of the absorbing or emitting compounds. Indeed, infrared spectroscopy is an excellent astronomical tool for studying the chemical composition and the physical characteristics of dust and gas in space. Based upon extensive laboratory studies, observed spectral features of dust can be identified with definite minerals and carbonaceous compounds and the abundances of these species can be derived. For a gas, the presence of multiple lines allows determination

[I]Space Sciences Division, NASA ARC, Moffett Field, USA
current address: Leiden Observatory, The Netherlands

both of the abundance of the emitting species and of the excitation conditions of the gas (such as temperature and density).

Over the last three decades our understanding of interstellar dust and gas has increased dramatically. To a large extent this has been driven by rapid developments in IR detector technology. Sensitive IR spectrometers operating in all the IR windows are now standard at all major ground-based infrared observatories. The opening of the full infrared window by space missions further increased our knowledge of interstellar dust. The Low Resolution Spectrometer (LRS) on *IRAS* — a slitless spectrometer sensitive from 7.5 µm to 23 µm with a resolving power of $R \approx 20$ — has provided a first systematic overview (Olnon et al 1986) of the spectral complexity of interstellar and circumstellar dust.[1] The Short Wavelength Spectrometer, SWS (de Graauw et al 1996), and Long Wavelength Spectrometer, LWS (Clegg et al 1996), on board *ISO*, represented the next big step forward by providing complete 2.5 µm to 200 µm spectra of a multitude of sources — essentially all IR-luminous galactic sources — with resolutions ranging from 100 in the long wavelength range to up to ≈ 2000 at shorter wavelengths. The Infrared Spectrometer (IRS) on board *Spitzer* — operating from 5.2 µm to 38 µm at low spectral resolution (60 to 130) and from 10 µm to 37 µm at moderate resolution ($R \approx 600$) — brought further improvements (Houck et al 2004). With its superior sensitivity, the IRS permits systematic spectroscopic studies of dust in typical sources in nearby galaxies and in bright galactic nuclei out to redshifts of ≈ 2.

In the next section, we will review the observational techniques and the advantages and disadvantages of space-based versus ground-based platforms in the thermal IR. Subsequently, we will briefly review the physical processes leading to emission and absorption in the IR and illustrate the impact of space-based observations in the thermal IR on our view of the Universe. We shall focus on star and planet formation, crystalline silicates, interstellar polycyclic aromatic hydrocarbons (PAH) and on galaxies and ultraluminous IR galaxies (ULIRG).

Thermal infrared and space

Ground-based astronomy is limited by the atmosphere in the IR. Telluric absorption in the rotational and ro-vibration bands of atmospheric gases completely blocks transmission at wavelengths between ≈ 30 µm and 300 µm and allows observations only in narrow windows shortwards and longwards of this (Figure 7.1). In the mid-IR, windows are the M band from ≈ 4.5 µm to 5.2 µm, the N band from ≈ 8 µm to 13 µm, and the Q band from (18 to 23) µm, but these windows are still riddled with narrow atmospheric features. In the sub-millimetre range, there are several windows: including around (340, 410, 650, 690 and 800) GHz. Going to space eliminates the telluric absorption and allows full spectral coverage over a wide wavelength range. The SWS and LWS on *ISO* have used this to survey the 2.4 µm to 200 µm range, the IRS on *Spitzer* has sampled the 5 µm to 40 µm range, while *Herschel* will cover the full far-IR and sub-millimetre range (60 µm to 600 µm).

[1] available through http://irsa.ipac.caltech.edu/IRASdocs/surveys/lrs.html

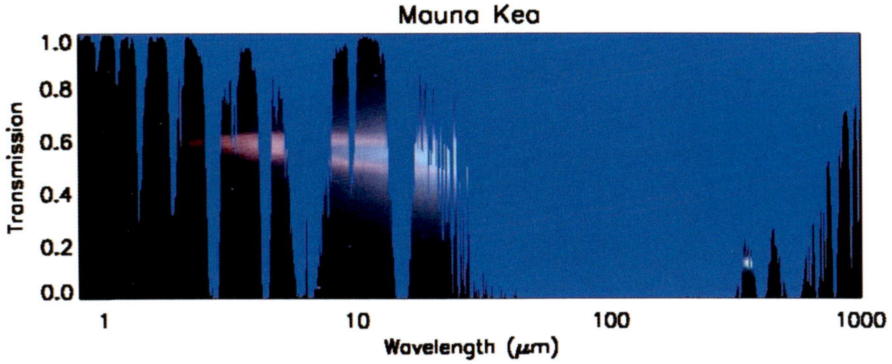

Figure 7.1: Atmospheric transmission in the thermal IR from an excellent ground-based site (Mauna Kea). Because of molecular absorptions, over most of the mid- and far-IR (5 μm $< \lambda <$ 1 mm) the atmosphere is completely opaque, except for a few narrow windows in the mid-IR and sub-millimetre ranges. Background image: the Sombrero galaxy at IR wavelengths obtained by the IRAC instrument on *Spitzer*, courtesy NASA/JPL-Caltech.

In addition, ground-based observations are limited by thermal emission from telescope and atmosphere. A bright source may only have a contrast of 10^{-3} relative to this background emission. This thermal background changes rapidly and on small scales. It dominates as background noise over Poisson noise and further hampers detection. As a result, special 'chopping' and 'nodding' techniques are required to extract the source signal as the difference between emission from the object location and nearby patches of sky. Chopping the secondary mirror at a rapid rate (a few hertz) compensates for the sky noise: the signal at the source position (including the sky) is compared with that at a nearby patch of (blank) sky (typically at a distance of some $15''$). Nodding involves moving the source position into the other chop-beam and repeating the chop-sequence at a much slower (nodding) rate (typically once a minute) to correct for small differences in telescope optics between the two chop positions. In space, active cooling by use of a cryostat or passive cooling by use of sunshields cuts down the thermal background considerably and leads to a great gain in sensitivity. The low and stable background from space allows sensitive IR studies throughout the thermal IR. Figure 7.2 compares the sensitivity for line observations with space-based and ground-based IR and sub-millimetre observatories, illustrating the gain in sensitivity. This is typically well over two orders of magnitude. One disadvantage of space-based operations is the severe limitation on telescope aperture: *ISO* and *Spitzer* had mirrors with diameters less than 1 m because they had to fit within a dewar. *Herschel* is and *JWST* will be passively cooled. *Herschel* has a solid 3.5 m primary mirror which could be fit within the Ariane nose cone; *JWST*, on the other hand, will have to deploy its primary mirror of 6.5 m diameter after launch.

The cold environment of space prevails over the larger ground-based apertures. *Spitzer*, for example, is well over an order of magnitude more sensitive

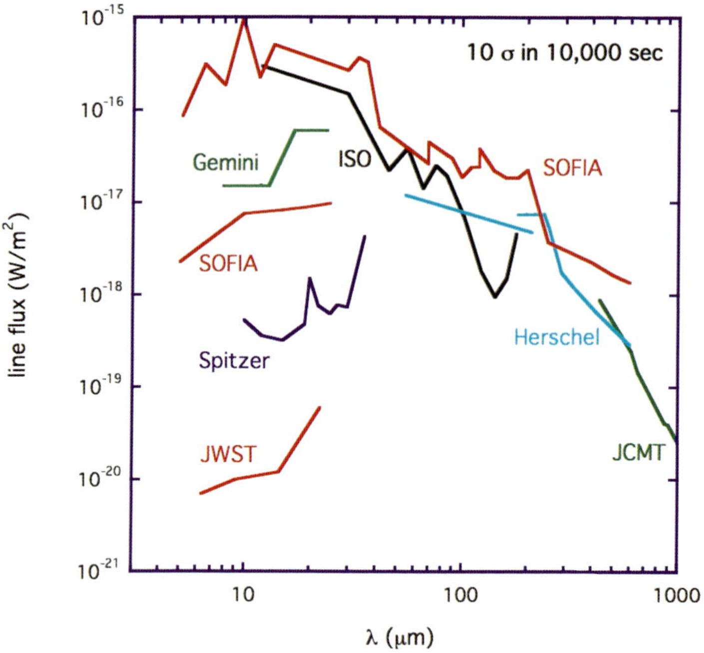

Figure 7.2: A comparison of sensitivities for line detection from point sources by space-based and ground-based observatories throughout the IR and sub-millimetre range. The curves correspond to the line flux (i.e., the irradiance in an isolated spectral line) required to achieve a 10 σ detection in 10^4 s for the different observatories. Detection of wider features depends on the resolution of the instrument. These are for the grating instruments on *JWST* $R \approx 3000$, on *Spitzer* $R \approx 600$, on *Gemini* $R \approx 200$ in N and $R \approx 100$ in Q, on *ISO* $R \approx 1500$ for $\lambda < 45\,\mu m$, $R \approx 200$ for $\lambda > 45\,\mu m$, *Herschel* $R \approx 1500$, and SOFIA $R \approx 300$ and 10^5. For the sub-millimetre heterodyne systems on *Herschel*, SOFIA, and for ground-based observations, a line width of 1 km/s ($R = 3 \times 10^5$) has been assumed.

than an 8 m-class ground-based telescope in the mid-IR (Figure 7.2). In the sub-millimetre region, heterodyne techniques (see Chapter 31, Wild 2010) have reached the quantum-noise limit and high spectral resolution leads to high sensitivity. In this wavelength range, sensitivities are very similar for space-based and ground-based observatories: the advantage of increased aperture size for ground-based telescopes is nullified by the limited atmospheric transmission ($\approx$ 0.2 to 0.6). ALMA, the Atacama Large Millimeter Array, will be some three orders of magnitude more sensitive at long wavelengths than single-dish telescopes because of sheer collecting area and moreover will provide the high spatial resolution of an interferometer. Nevertheless, even with ALMA fully operational, space offers the distinct advantage of contiguous coverage of the full far-IR and sub-millimetre wavelength regions and permits, for example, unbiased spectral surveys as well as the observation of lines from key species such as H_2O, the light hydrides, and the (modestly redshifted) line at 157.4 μm in the C II spectrum.

We note that, for space-based missions, even if a telescope is passively cooled, the instruments will need active cooling to much lower temperatures. Cryogen then becomes a limitation, and lifetimes of IR space missions are typically two to five years. Likewise, because of the long design and construction phases as well as the lack of access during the mission, science instrumentation generally lags behind current technology. In the far-IR, besides the increase in detector sensitivity still possible, an increase in the size of the telescope area and the use of heterodyne arrays provide other clear avenues for increasing observing "speed", in the latter case, particularly when mapping large areas on the sky. Thus, for extended objects, the 5×5 element spectrometer array PACS on *Herschel* gains substantially over the LWS on *ISO*. Of course, mapping on a space-based observatory always comes at a hefty penalty in terms of pointing overheads. A nimble airborne telescope such as SOFIA holds a substantial advantage for (spectroscopic) mapping of (bright) very extended emission regions, actually throughout the thermal IR. Airborne observatories such as SOFIA can also react more rapidly to changing technology (e.g., Moore's law, in principle, would provide a factor of 10 in observing "speed" every 5 to 10 years). SOFIA thus provides an important testbed for future instrumentation and a driver for technology development in the thermal IR. For science objectives which require high spectral resolution, SOFIA is already very competitive in the sub-millimetre range. Future heterodyne instruments throughout the thermal IR as well as heterodyne arrays will reinforce SOFIA's distinct and unique scientific niche in high-resolution spectroscopy. SOFIA also holds an advantage in that weight and cryogen restrictions are largely non-existent on a Boeing 747 aircraft, which accomodates instruments of larger volume and mass than those designed for use in space. Nevertheless, as Figure 7.2 demonstrates, the future of moderate-resolution spectroscopy in the mid- and far-IR resides in space-based observatories such as *JWST*, and missions such as *SPICA* and *SAFIRE*.

Physical processes

Interstellar gas is heated by ultraviolet (UV) photons ($h\nu > 6$ eV) through ionization and dissociation of atoms and molecules. Neutral and ionic atomic gases emit primarily through emission in fine-structure lines of the dominant elements when collisional excitation is followed by radiative decay. For molecular gas, pure rotational transitions provide efficient cooling, and molecular gas therefore is generally relatively cold. The energy-level separation will have to match the thermal energy of the gas for efficient collisional excitation of the upper level. In the two-level approximation, the critical density of a transition $i \rightarrow j$ is given by $n_{\rm cr} = A_{ij}/\gamma_{ij}$ with A_{ij} the Einstein coefficient and γ_{ij} the collisional de-excitation coefficient of the transition. For a multilevel system, the radiative and collisional rates have to be summed over all possible downward transitions (cf., Tielens 2005). The critical electron density is the density at which downward radiative and collisional transitions are equal. Below the critical density every upward collision is followed by photon emission, this leads to cooling. At higher densities collisional de-excitation takes over and cooling is suppressed.

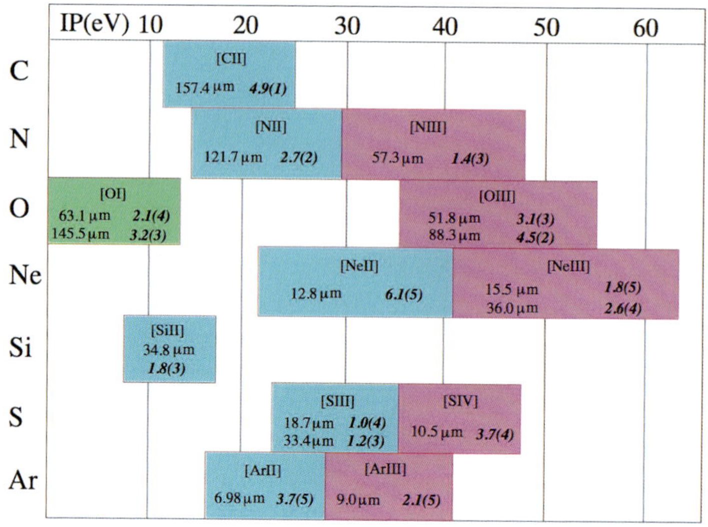

Figure 7.3: IR atomic fine-structure lines for different ionization stages of some relevant elements. The horizontal axis shows the energy range over which ionization can occur. Underneath each atomic or ionic species, the wavelength and critical density of the transition(s) is indicated. The critical density for each transition is expressed as $n_e/\mathrm{cm}^{-3} = a \times 10^b$, where a and b are given by $a(b)$ in italics. It has been evaluated at a temperature of 7000 K appropriate for an H II region. (Figure reproduced with permission from Martín-Hernández et al (2002).)

As a rule of thumb, a line will give information on gas with temperatures and densities matching its energy-level separation and critical density. The abundance of atoms in a given ionization stage, however, depends on density and temperature as well. Figure 7.3 summarizes the characteristics of atomic and ionic fine-structure transitions. In H II regions, photo-ionization and electron recombination set the ionization balance and the abundance of an ionic species. This is controlled by the energy required to ionize a given species. Because of their low excitation energy, the IR fine-structure lines are not sensitive to the temperature of ionized gas. The ratio of IR fine-structure lines, where two lines from the same ionization stage of given element are available, nevertheless provides a good handle on the density of the emitting gas in the range spanning the two critical densities involved. Thus, the [O III] lines are used to study gas with densities in the range $\approx(10^2$ to $10^4)$ cm^{-3}, while the [Ne III] lines are sensitive to somewhat higher densities $\approx(10^4$ to $10^5)$ cm^{-3}. The ratios of lines from adjacent ionized stages are very sensitive to the ionizing radiation field. The lines from Ne, Ar and S include the dominant ionization stages of these elements and hence, when combined with H I recombination lines, provide a direct handle on elemental abundances. *ISO* SWS and LWS have provided a coherent data set for galactic H II regions which have been analyzed along these lines (Martín-Hernández et al 2002). Although the *Spitzer* IRS has more limited spectral coverage, its superior sensitivity has permitted a systematic

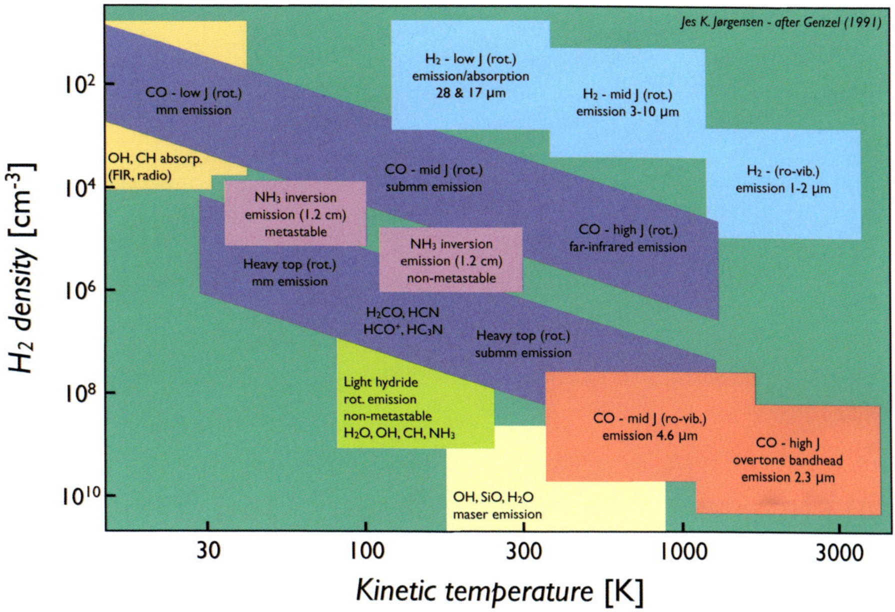

Figure 7.4: An overview of the molecular lines and the range of physical conditions in molecular clouds for which they are effective probes. Figure from J.K. Jørgensen based upon a figure from Genzel (1991).

study of elemental abundance gradients in galaxies such as M83 and M33 (Rubin et al 2007).

Figure 7.4 illustrates the range in temperature and density where specific IR transitions will be important. These reflect the energy-level separation and critical density of the transitions involved. As an example, mid-J rotational transitions (with J being the rotational quantum number of the molecule, cf., Herzberg 1959) of CO occur at shorter wavelengths than low-J transitions and hence they probe warmer and denser gas. Likewise, as a hydride, H_2 transitions occur at much shorter wavelengths than those of the heavier molecule CO. However, as a homonuclear molecule, H_2 has only quadrupole-allowed transitions with much lower Einstein coefficients than dipole-allowed transitions of heteronuclear molecules. Hence, the pure rotational H_2 transitions originate from warmer gas than low-J CO transitions, but the densities are quite similar. Molecular gas can always provide information on the physical conditions in the emitting gas at any temperature or density; if several transitions of a species can be measured, the physical conditions of the emitting gas can be determined more reliably. In addition, atomic or molecular abundances can be measured, and more specifically, the organic inventory of space may be determined.

Molecular species can also radiate in molecular ro-vibrational transitions due to fluorescence, if pumped through UV absorption via electronic transitions. Of particular importance are the CH and CC stretching and bending modes in large

PAH molecules. In the interstellar medium (ISM), a PAH molecule typically absorbs a UV photon every year. Rapid internal conversion after the absorption of the UV photon leaves the molecule in a high vibrational excitation, from where it cools through IR emission on a timescale of ≈ 1 s.

Interstellar dust grains are in radiative equilibrium with the local UV and visible radiation field at temperature ranges from approximately 15 K in the diffuse ISM, to about 75 K at 0.1 pc from an O star, to up to the sublimation temperature of the material (typically 1000 K) near, i.e., 0.1 ua to 1 ua away from a protostar (cf., Tielens 2005). The peak of the emission from interstellar dust therefore shifts from the far-IR to the mid-IR, depending on location. Dielectric interstellar dust grains such as silicates and oxides show strong resonances due to various stretching and bending modes. These bands are characteristic for a material and thus can be used for identification purposes. In addition, if the temperature of the emitting grains is known, the column of dust can be calculated as well.

The Universe at mid- and far-infrared wavelengths

Star and planet formation

In a seminal paper, Adams et al (1987) explained the classes of spectral energy distributions uncovered by *IRAS* and from ground-based photometric studies of sources embedded in molecular clouds as an evolutionary progression in the birth of low-mass protostars. In this way, five classes were recognized: starting with a quiescent prestellar core, followed by a collapse phase with a central object and surrounding circumstellar disk, a clearing of the collapsing envelope by accretion and a strong stellar wind, a "naked" protostar with surrounding planetary gas and dust disk, and finally a star surrounded by a planet/planetesimal disk. Subsequent studies have confirmed this general scheme for low-mass star and planet formation, have filled in many of the details, and raised it to paradigm status.

ISO has revealed the important influence of flaring and shadowing on the spectral energy distribution of intermediate-mass protostars, the Herbig AeBe stars (Meeus et al 2001). In addition, *ISO* was unique in opening up moderate-resolution spectroscopy over a wide wavelength range and this has revealed the ubiquitous presence of a wide variety of emission features including crystalline silicates and PAH molecules in regions of star and planet formation. *Spitzer* has extended these observations to disks around low-mass protostars as far down as brown dwarfs! Much progress has been made in our understanding of star and planet formation since then and this field is too rich to be reviewed here. The interested reader is referred to van Dishoeck (2004) and Werner et al (2006).

Here, we do want to stress the importance of spectroscopy over a wide wavelength range. The 2.4 µm to 45 µm spectrum of one of the strong shock peaks in the Orion KL region (Figure 7.5; Rosenthal et al 2000) illustrates the richness and diagnostic value of spectroscopy in the mid-IR. Several different emission complexes contribute to the observed spectrum from this region of massive star formation. First, there is the emission from the warm molecular gas heated by a shock driven by the powerful outflow from the massive protostar, which is deeply embedded in

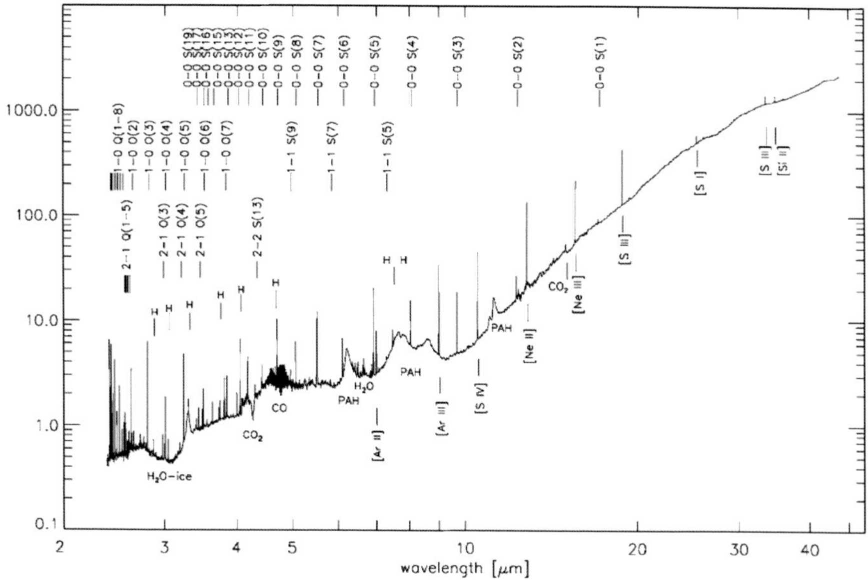

Figure 7.5: The SWS 2.4 μm to 45 μm spectrum of Peak 1 in Orion (Rosenthal et al 2000) reveals a wealth of spectral detail including a multitude of rotational and ro-vibrational lines from H_2 and other molecules, atomic fine-structure lines, HI recombination lines, emission from PAHs and absorption by molecular ices and silicates. Shown on the ordinate axis is the flux density in 10^{-9} Jy/sr.

the KL nebula. This gives rise to the H_2 as well as the CO and H_2O emission lines. The cold molecular cloud material in which this shock is propagating is probed by the H_2O and CO_2 molecular ice and silicate absorption features. In the foreground, the ionized gas associated with the Orion H II region (M42) powered by Θ^1C — a newly formed star which has already disrupted its natal cloud — gives rise to the atomic fine-structure lines (notably [Ar II], [Ar III], [Ne II], [Ne III], [S III], [S IV]) and the H I recombination lines. The atomic gas in the photodissociation regions (PDR) separating the ionized gas from the molecular cloud contributes to the prominent PAH emission features, the [Si II] fine-structure line, and to the ro-vibrational emission from the higher vibrational states of H_2. Together with their different excitation energies and critical densities, this spectrum has been used to determine the physical conditions in the various emission zones in this complex region and to unravel the energetic interaction of newly formed massive stars with their environment (Rosenthal et al 2000).

Crystalline silicates

One of the greatest surprises of the *ISO* mission was the incredible richness of the circumstellar silicate spectra at long wavelengths. *Spitzer* has extended measurements to much fainter objects that *ISO* could not probe, including disks around

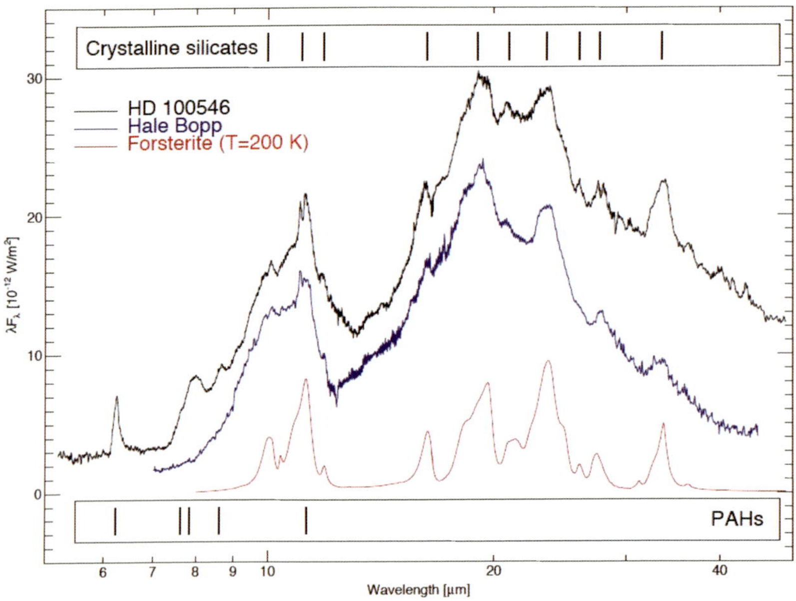

Figure 7.6: Comparison between the *ISO*-SWS spectrum from comet C/1995 O1 (Hale-Bopp) (Crovisier et al 1997) with the spectrum of the young star, HD 100546 (Malfait et al 1998). The bottom trace shows the IR spectrum of forsterite (Mg_2SiO_4) measured in the laboratory (Koike et al 1998). As this comparison shows, most of the observed bands are due to small forsterite grains. Prominent crystalline bands are indicated by tick marks at the top. Note that the spectrum of HD 100546 also shows weak spectral signatures of PAH bands in this wavelength range. (The ordinate shows the product of wavelength and spectral irradiance.)

T Tauri stars and recently also brown dwarfs, late-type objects in external galaxies, ULIRG nuclei, and active galactic nuclei (AGN) toroids (Bouy et al 2008; Bouwman et al 2008; Markwick-Kemper et al 2007; Sloan et al 2006; Armus et al 2007; Spoon et al 2007). Invariably, the 10 μm to 45 μm spectra of sources with circumstellar dust show a great number of features due to crystalline olivine and pyroxene (cf., Figure 7.6; Waters et al 1996; Waters 2000; Waters et al 2000). These features are ubiquitous and, besides O-rich AGB stars and their descendants, they have now been observed in such diverse objects as luminous blue variables (e.g., η Car), Herbig AeBe stars, nominally C-rich objects such as the Red Rectangle and BD +30 3639, and comet Hale-Bopp (Crovisier et al 1997; Malfait et al 1998; Molster et al 2002a; Waters et al 1998a,b). While some objects are dominated by forsterite (Mg_2SiO_4) (e.g., Hale-Bopp and HD 100546; Figure 7.6), other objects (e.g., AGB stars) show strong bands due to enstatite ($MgSiO_3$) as well.

These narrow features clearly imply a crystalline carrier. The study of the IR characteristics of crystalline (olivine and pyroxene) silicates by Koike et al (1993, 1998) has been instrumental in the analysis of these long-wavelength features. Extensive laboratory studies have also been performed by Jaeger et al (1994, 1998). Based on these studies, the prominent bands at 23.6 μm and 33.6 μm can be attributed to olivine while the bands at (26.2, 32.9, 35.8, 40.6 and 43.1) μm are assigned to pyroxene (Waters et al 1996; Jaeger et al 1994, 1998; Molster et al 2002b).

Herschel's contributions to this field will be limited in view of its (long) wavelength coverage. However, the PACS instrument on *Herschel* is well suited to study the 69 μm band, which is revealing in terms of the composition (particularly regarding the Mg/Fe ratio) and temperature of emitting olivine grains (Molster et al 2002b; Bowey et al 2002). Due to the interaction of the cation and anion, the exact peak position of all of these modes is sensitive to the particular metal present with shifts from 0.3 μm in the 10 μm region to 1 μm to 2 μm at longer wavelengths (Farmer 1974). In addition, the peak position — as well as the width of the bands — is sensitive to the temperature of the emitting grains. The 69 μm band is spectrally isolated and provides an ideal probe of these effects. Finally, *JWST* will have the sensitivity and spectral coverage to probe the characteristics of interstellar silicate grains for galaxies out to redshifts of ≈ 2.5; *Spitzer* has shown that this is a promising field of research (Armus et al 2007; Markwick-Kemper et al 2007; Spoon et al 2007).

Interstellar PAHs

IRAS has discovered widespread mid-IR emission in the Galaxy even far from illuminating stars where no warm dust was expected: the so-called IR cirrus (Low et al 1984). At that time, ground-based and airborne spectroscopy had already revealed the presence of broad emission features at (3.3, 6.2, 7.7, and 11.3) μm in the spectra of some objects illuminated by strong UV sources such as HII regions, planetary nebulae and reflection nebulae. This cirrus was generally thought to trace these same emission features. Subsequently, *ISO* unambigously demonstrated that these IR emission features dominate the mid-IR spectra of C-rich post-AGB objects and planetary nebulae, the PDRs associated with HII regions powered by O stars and those associated with reflection nebulae illuminated by late B stars, planetary disks associated with young stars (such as Herbig AeBe stars and T Tauri stars) in the somewhat later stages of evolution, and starburst regions associated with galactic nuclei (Boulanger et al 1998, 2000; Peeters et al 2002; Verstraete et al 1996, 2001; Sloan et al 2007; Acke and van den Ancker 2004; Geers et al 2006; Habart et al 2006). In addition, while stretched to the limit of its capabilities, *ISO* showed that the mid-IR spectrum of the cirrus is dominated by these IR emission features (Mattila et al 1996). *Spitzer* has shown that this is a general characteristic of the ISM in all spiral galaxies (Flagey et al 2006; Engelbracht et al 2005; Brandl et al 2006; Sloan et al 2007). Likewise, the spectra of ULIRGs are often bright in these IR emission features (Genzel et al 1998; Armus et al 2007). *Spitzer* has now detected these features in such sources out to redshifts of ≈ 3 (Yan et al 2005; Lutz et al 2007; Pope et al 2007; Rigby et al 2007).

Together these observations reveal the incredible richness of the mid-IR emission spectrum of the ISM of galaxies (Figure 7.7). Besides the well-known IR emission features at (3.3, 6.2, 7.7, 8.6, and 11.3) μm, the observed interstellar spectra show

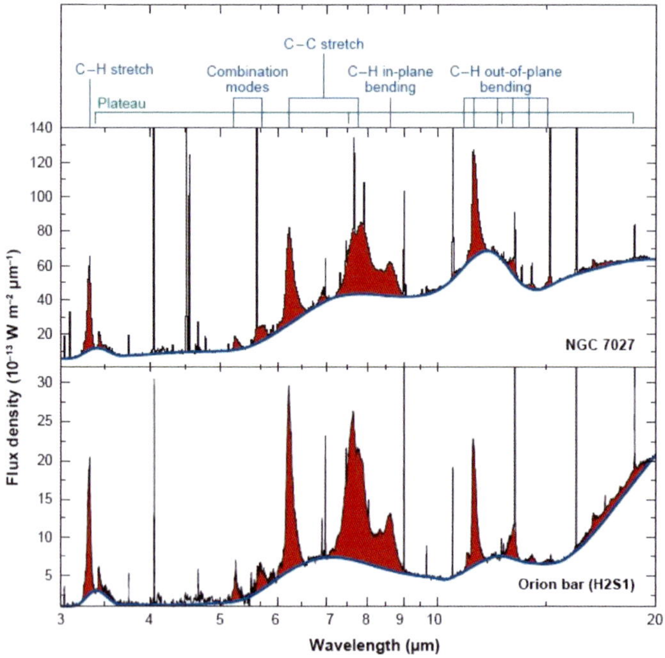

Figure 7.7: The mid-IR spectra (spectral irradiance vs. wavelength) of the PDR in the Orion Bar and in the Planetary Nebulae NGC 7027 are dominated by a rich set of emission features. Assignments of these features with vibrational modes of PAH molecules are labelled at the top. Figure adapted from Peeters et al (2002).

a wealth of weaker features, including bands at (3.4, 3.5, 5.25, 5.65, 6.0, 6.9, 10.5, 11.0, 12.7, 13.5, 14.2, and 16.4) µm. Moreover, many of the well-known features shift in peak position, vary in width, and/or show substructure (Peeters et al 2002; van Diedenhoven et al 2004). These IR emission features are due to IR fluorescence from PAH molecules containing some 50 C-atoms. Assignments for the various modes involved are shown in Figure 7.7. The observed variations of these IR emission features imply a sensitivity to the local physical conditions, and this property is now beginning to be employed as a diagnostic tool for astronomy (Galliano et al 2008). Overall, driven by these observational developments as well as by laboratory and theoretical studies, it has become clear over the past decade that PAH molecules are an abundant and important component of the interstellar medium throughout the Universe (Tielens 2008).

Galaxies and ULIRGs

IRAS opened the realm of external galaxies for studies in the 10 µm to 100 µm range and discovered emission from tens of thousands of normal and active galaxies. *ISO* followed up with detailed imaging, spectroscopy, and spectrophotometry of

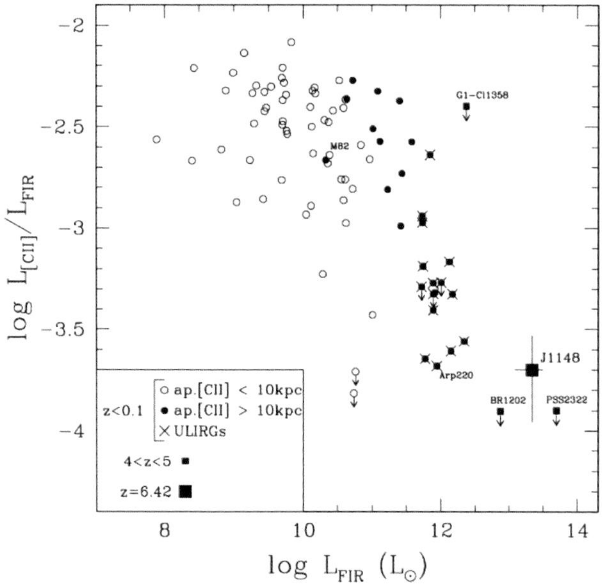

Figure 7.8: The ratio of the [C II] 158 μm line to the far-IR continuum as a function of the IR luminosity. Open (closed) circles are galaxies for which the aperture encompasses less (more) than 10 kpc. Crosses mark ULIRGs. The filled squares are high-redshift objects. The large filled square represents the most distant known quasar, SDSS J114816.64+525150.3 at a redshift of $z = 6.42$ (Maiolino et al 2005). Figure adapted from Malhotra et al (2001).

many galaxies detected by *IRAS*, as well as deep surveys in the mid- and far-IR. *Spitzer* with its superior sensitivity has extended this by dedicated surveys of the emission characteristics of nearby normal and starburst galaxies and by probing even farther back in the history of the Universe. *COBE* has shown that the [C II] 158 μm line is the dominant emission line of the Milky Way with a luminosity of $5 \times 10^7\ L_\odot$; e.g., about 0.003 of the total IR luminosity of the Milky Way is emitted in this single line (Bennett et al 1994). This is a very general feature of galaxies: in a sample of 60 normal, star forming galaxies, the [C II] line is in general the dominant IR cooling line. The [O I] 63 μm line is a close second and in a handful of galaxies even takes over (Malhotra et al 2001). The origin of the [C II] line is controversial (Hollenbach and Tielens 1999). Theoretically, because of their density and temperature, it is expected that (Spitzer-type) H I clouds (e.g., the cold neutral medium, CNM) will radiate most of their energy through the [C II] line (cf., the section above on physical processes or Dalgarno and McCray 1972). This is supported by measurements of the [C II] emission from high-latitude clouds by use of sounding rockets (Bock et al 1993). However, *COBE* demonstrated that the irradiance of the [N II] line correlates with the irradiance of the [C II] line (to the 1.5 power), suggesting that the low-density ionized gas (the warm ionized medium, WIM) contributes a portion of the observed [C II] emission (Heiles 1994). In addition, given the high observed irradiance of [O I] (where ions in the upper level have a critical density of 2×10^5 cm^{-3}) in the sample of star-forming galaxies studied by *ISO*-LWS, a substantial fraction of the observed [C II] emission likely originates from dense, bright PDRs associated with molecular cloud surfaces near regions of massive star formation (Hollenbach and Tielens 1999). Theoretical models have been developed based upon these three premises (CNM, WIM, and PDR origin) and all are in reasonable agreement with the *COBE* observations of the Milky Way.

Herschel (and later SOFIA) permits the study of Doppler shifts of the [C II] and [O I] emission from the Milky Way, nearby normal galaxies, starburst galaxies and ULIRGs. This will provide a better understanding of the origin of these lines. Understanding of the origin of the [C II] line on a galactic scale has recently received additional impetus with the detection of this line in the spectrum of the most-distant quasar, J1148, at a redshift of 6.42 with the 30 m IRAM telescope (Maiolino et al 2005). The observed flux of this line — in conjunction with other PDR tracers (e.g., CO $J = 7$ to $J = 6$; Bertoldi et al 2003) — has been interpreted as evidence for vigorous star formation (3000 $M_\odot$/a) in the host galaxy. Because of the high luminosity in this single spectral line, the [C II] line has often been considered as a key tracer of star formation in the early Universe, particularly for heavily obscured galaxies. Thus, one of the three key scientific goals of the ALMA project is to use the [C II] line to probe star formation in the high-redshift Universe. However, the irradiance of the [C II] line in extreme star-formation environments such as ULIRGs is not well understood (Malhotra et al 2001; Luhman et al 2003). Typically, the strength of this line is a factor of ten too small relative to the far-IR dust continuum in these environments if compared with normal and starburst galaxies (see Figure 7.8). While that might indicate that the gas in ULIRG evironments is denser than in normal or starburst galaxies, other tracers (e.g., CO $J = 1$ to $J = 0$, [C I] $J = 1$ to $J = 0$, and PAH emission) do not seem to be compatible with this solution (Luhman et al 2003). Possibly much of the (ionizing and non-ionizing) ultraviolet flux is absorbed by dust in the H II region and reradiated as far-IR dust continuum rather than as ionic or neutral atomic fine-structure lines or by PAHs (Luhman et al 2003). In this and many other ways, the star-formation environment of ULIRGs may resemble that of hypercompact or ultracompact H II regions (Martín-Hernández et al 2002; Peeters et al 2004; Lahuis et al 2007).

Achievements and outlook

The past IR space missions — *IRAS, ISO, Spitzer, Akari* — have provided us with a wealth of data and great new insights into the detailed physics and chemistry of gas and dust in space and thereby on the origin and evolution of the interstellar medium of galaxies. As detailed in this chapter, these missions have, however, also raised numerous new questions. In the near future, SOFIA will start regular operations and bring IR spectroscopy over a wide wavelength range and at a wide range of spectral resolutions. *Herschel*, launched in May 2009, has now opened the far-IR and sub-millimetre range for systematic studies of the Universe. The more sensitive *JWST*, slated to be launched in 2014, will be able to probe even deeper. With this prospect, and with potential new missions such as *SPICA* and *SAFIRE* appearing on the horizon, the future for infrared astronomy looks bright, indeed.

Bibliography

Acke B, van den Ancker ME (2004) ISO spectroscopy of disks around Herbig Ae/Be stars. Astronom Astrophys 426:151–170

Adams FC, Lada CJ, Shu FH (1987) Spectral evolution of young stellar objects. Astrophys J 312:788–806

Armus L, Charmandaris V, Bernard-Salas J (plus 13 authors) (2007) Observations of ultraluminous infrared galaxies with the infrared spectrograph on the *Spitzer* Space Telescope. II. The IRAS Bright Galaxy Sample. Astrophys J 656:148–167

Bennett CL, Fixsen DJ, Hinshaw G (plus nine authors) (1994) Morphology of the interstellar cooling lines detected by *COBE*. Astrophys J 434:587–598

Bertoldi F, Cox P, Neri R (plus eight authors) (2003) High-excitation CO in a quasar host galaxy at $z = 6.42$. Astronom Astrophys 409:L47–L50

Bock JJ, Hristov VV, Kawada M (plus seven authors) (1993) Observation of forbidden C II 158 micron emission from the diffuse interstellar medium at high Galactic latitude. Astrophys J Lett 410:L115–L118

Boulanger F, Abergel A, Bernard JP (plus 12 authors) (1998) The nature of small interstellar dust particles. Astron Soc of the Pacific Conf Ser 132:15–23

Boulanger F, Abergel A, Cesarsky D (plus four authors) (2000) Small dust particles as seen by *ISO*. ESA Spec Publ 455:91–98

Bouwman J, Henning T, Hillenbrand LA (plus eight authors) (2008) The formation and evolution of planetary systems: Grain growth and chemical processing of dust in T Tauri systems. ArXiv e-prints 802, 0802.3033

Bowey JE, Barlow MJ, Molster FJ (plus six authors) (2002) The 69-μm forsterite band as a dust temperature indicator. Mon Notes Royal Astr Soc 331:L1–L6

Bouy H, Huelamo, N, Pinte C, plus 15 authors (2008), Structural and compositional properties of brown dwarf disks: the case of 2MASS J04442713+2512164. ArXiv e-prints, 803, 0803.2051

Brandl BR, Bernard-Salas J, Spoon HWW (plus 12 authors) (2006) The Mid-Infrared Properties of Starburst Galaxies from Spitzer-IRS Spectroscopy. Astrophys J 653:1129–1144

Clegg PE, Ade PAR, Armand C (plus 60 authors) (1996) The *ISO* Long-Wavelength Spectrometer. Astronom Astrophys 315:L38–L42

Crovisier J, Leech K, Bockelee-Morvan D (plus five authors) (1997) The spectrum of Comet Hale-Bopp (C/1995 01) observed with the Infrared Space Observatory at 2.9 AU from the Sun. Science 275:1904–1907

Dalgarno A, McCray RA (1972) Heating and Ionization of HI Regions. Ann Rev Astronom Astrophys 10:375

de Graauw T, Haser LN, Beintema DA (plus 57 authors) (1996) Observing with the *ISO* Short-Wavelength Spectrometer. Astronom Astrophys 315:L49–L54

Engelbracht CW, Gordon KD, Rieke GH, Werner MW (2005) Metallicity effects on mid-infrared colors and the 8 μm PAH emission in galaxies. Astrophys J Lett 628:L29–L32

Farmer VC (1974) (ed.) The IR spectra of silicates. London: Mineralogical society

Flagey N, Boulanger F, Verstraete L (plus three authors) (2006) Spitzer/IRAC and ISOCAM/CVF insights on the origin of the near to mid-IR Galactic diffuse emission. Astronom Astrophys 453:969–978

Galliano F, Madden SC, Tielens AGGM (plus two authors) (2008) Variations of the mid-IR aromatic features inside and among galaxies. Astrophys J 679:310–345

Geers VC, Augereau JC, Pontoppidan KM (plus ten authors) (2006) C2D *Spitzer*-IRS spectra of disks around T Tauri stars. II. PAH emission features. Astronom Astrophys 459:545–556

Genzel R (1991) Physical conditions and heating/cooling processes in high mass star formation regions. NATO ASIC Proc 342:155–220

Genzel R, Lutz D, Sturm E (plus nine authors) (1998) What powers ultraluminous IRAS galaxies? Astrophys J 498:579–605

Habart E, Natta A, Testi L, Carbillet M (2006) Spatially resolved PAH emission in the inner disks of Herbig Ae/Be stars. Astronom Astrophys 449:1067–1075

Heiles C (1994) On the origin of the diffuse C(+) 158 micron line emission. Astrophys J 436:720–727

Herzberg G (1959) Molecular Spectra and Molecular Structure, I. Spectra of Diatomic Molecules. Princeton NJ: D van Nostrand, p.68

Hollenbach DJ, Tielens AGGM (1999) Photodissociation regions in the interstellar medium of galaxies, Rev Mod Phys 71:173–230

Houck JR, Roellig TL, van Cleve J (plus 32 authors) (2004) The Infrared Spectrograph (IRS) on the *Spitzer* Space Telescope. Astrophys J Suppl 154:18–24

Jaeger C, Mutschke H, Begemann B, Dorschner J, Henning T (1994) Steps toward interstellar silicate mineralogy. 1: Laboratory results of a silicate glass of mean cosmic composition. Astronom Astrophys 292:641–655

Jaeger C, Molster FJ, Dorschner J (plus three authors) (1998) Steps toward interstellar silicate mineralogy. IV. The crystalline revolution. Astronom Astrophys 339:904–916

Koike C, Shibai H, Tuchiyama A (1993) Extinction of olivine and pyroxene in the mid infrared and far infrared. Mon Notes Royal Astr Soc 264:654–658

Koike C, Suto H, Tuchiyama A, Shibai H, Tanabe T (1998) The spectra of pyroxenes in mid and far infrared. IAU Symposium 191:305

Lahuis F, van Dishoeck EF, Blake GA (plus three authors) (2007) c2d *Spitzer* IRS spectra of disks around T Tauri stars. III. [Ne II], [Fe I], and H_2 gas-phase lines. Astrophys J 665:492–511

Low FJ, Young E, Beintema DA (plus seven authors) (1984) Infrared cirrus - New components of the extended infrared emission. Astrophys J Lett 278:L19–L22

Luhman ML, Satyapal S, Fischer J (plus five authors) (2003) The [C II] 158 micron line deficit in ultraluminous infrared galaxies revisited. Astrophys J 594:758–775

Lutz D, Sturm E, Tacconi LJ (plus seven authors) (2007) PAH emission and star formation in the host of the $z \approx 2.56$ Cloverleaf QSO Astrophys J Lett 661:L25–L28

Maiolino R, Cox P, Caselli P (plus ten authors) (2005) First detection of [C II]158 µm at high redshift: vigorous star formation in the early Universe. Astronom Astrophys 440:L51–L54

Malfait K, Waelkens C, Waters LBFM (plus three authors) (1998) The spectrum of the young star HD 100546 observed with the Infrared Space Observatory. Astronom Astrophys 332:L25–L28

Malhotra S, Kaufman MJ, Hollenbach D (plus ten authors) (2001) Far-infrared spectroscopy of normal galaxies: physical conditions in the interstellar medium.

Astrophys J 561:766–786

Markwick-Kemper F, Gallagher SC, Hines DC, Bouwman J (2007) Dust in the wind: Crystalline silicates, corundum, and periclase in PG 2112+059. Astrophys J Lett 668:L107–L110

Martín-Hernández NL, Peeters E, Morisset C (plus nine authors) (2002) ISO spectroscopy of compact H II regions in the Galaxy. II. Ionization and elemental abundances. Astronom Astrophys 381:606–627

Mattila K, Lemke D, Haikala LK (plus five authors) (1996) Spectrophotometry of UIR bands in the diffuse emission of the galactic disk. Astronom Astrophys 315:L353–L356

Meeus G, Waters LBFM, Bouwman J (plus three authors) (2001) *ISO* spectroscopy of circumstellar dust in 14 Herbig Ae/Be systems: Towards an understanding of dust processing. Astronom Astrophys 365:476–490

Molster FJ, Waters LBFM, Tielens AGGM, Barlow MJ (2002a) Crystalline silicate dust around evolved stars. I. The sample stars. Astronom Astrophys 382:184–221

Molster FJ, Waters LBFM, Tielens AGGM, Koike C, Chihara H (2002b) Crystalline silicate dust around evolved stars. III. A correlations study of crystalline silicate features. Astronom Astrophys 382:241–255

Olnon FM, Raimond E, Neugebauer G (plus 26 authors) (1986) *IRAS* catalogues and atlases - Atlas of low-resolution spectra. Astronom Astrophys s 65:607–1065

Peeters E, Hony S, Van Kerckhoven C (plus four authors) (2002) The rich 6 to 9 µm spectrum of interstellar PAHs. Astronom Astrophys 390:1089–1113

Peeters E, Spoon HWW, Tielens AGGM (2004) Polycyclic aromatic hydrocarbons as a tracer of star formation? Astrophys J 613:986–1003

Pope A, Chary RR, Alexander DM (plus six authors) (2007) Mid-infrared spectral diagnosis of submillimeter galaxies ArXiv e-prints, 711, arXiv:0711.1553

Rigby JR, Marcillac D, Egami E (plus ten authors) (2007), Mid-infrared spectroscopy of lensed galaxies at 1<z<3: The nature of sources near the MIPS confusion limit ArXiv e-prints, 711, arXiv:0711.1902

Rosenthal D, Bertoldi F, Drapatz S (2000) *ISO*-SWS observations of OMC-1: H_2 and fine structure lines. Astronom Astrophys 356:705–723

Rubin RH, Simpson JP, Colgan SWJ (plus six authors) (2007) *Spitzer* observations of M83 and the hot star, H II region connection. Mon Notes Royal Astr Soc 377:1407–1418

Sloan GC, Devost D, Bernard-Salas J (plus two authors) (2006) The unusual silicate dust around HV 2310, an evolved star in the Large Magellanic Cloud. Astrophys J 638:472–477

Sloan GC, Jura M, Duley WW (plus nine authors) (2007) The unusual hydrocarbon emission from the early carbon star HD 100764: The connection between aromatics and aliphatics. Astrophys J 664:1144–1153

Spoon HWW, Marshall JA, Houck JR (plus five authors) (2007) Mid-infrared galaxy classification based on silicate obscuration and PAH equivalent width. Astrophys J Lett 654:L49–L52

Tielens AGGM (2005) The Physics and Chemistry of the Interstellar Medium. The Physics and Chemistry of the Interstellar Medium, by A. G. G. M. Tielens, ISBN 0521826349. Cambridge, UK: Cambridge University Press

Tielens AGGM (2008) Interstellar polycyclic aromatic hydrocarbon molecules. Ann Rev Astronom Astrophys 46:289–337

van Diedenhoven B, Peeters E, Van Kerckhoven C (plus four authors) (2004) The profiles of the 3–12 micron polycyclic aromatic hydrocarbon features. Astrophys J 611:928–939

van Dishoeck EF (2004) *ISO* spectroscopy of gas and dust: From molecular clouds to protoplanetary disks. Ann Rev Astronom Astrophys 42:119–167

Verstraete L, Puget JL, Falgarone E (plus three authors) (1996) SWS spectroscopy of small grain features across the M17-Southwest photodissociation front. Astronom Astrophys 315:L337–L340

Verstraete L, Pech C, Moutou C (plus six authors) (2001) The aromatic infrared bands as seen by *ISO*-SWS: Probing the PAH model. Astronom Astrophys 372:981–997

Waters LBFM (2000) The life cycle of dust: an *ISO* view. ESA SP 456:39–44

Waters LBFM, Molster FJ, de Jong T (plus 34 authors) (1996) Mineralogy of oxygen-rich dust shells. Astronom Astrophys 315:L361–L364

Waters LBFM, Beintema DA, Zijlstra AA (plus six authors) (1998a) Crystalline silicates in planetary nebulae with [WC] central stars. Astronom Astrophys 331:L61–L64

Waters LBFM, Cami J, de Jong T (plus 12 authors) (1998b) An oxygen-rich dust disk surrounding an evolved star in the Red Rectangle. Nature 391:868

Waters LBFM, Molster FJ, Hony S (plus five authors) (2000) ISO Spectroscopy of Circumstellar Dust. Thermal emission spectroscopy and analysis of dust, disks, and regoliths 196:3–14

Werner M, Fazio G, Rieke G, Roellig TL, Watson DM (2006) First fruits of the *Spitzer* Space Telescope: Galactic and solar system studies. Ann Rev Astronom Astrophys 44:269–321

Wild W (2010) Coherent far-infrared/submillimeter detectors. ISSI SR-009:503–523

Yan L, Chary R, Armus L (plus six authors) (2005) *Spitzer* detection of polycyclic aromatic hydrocarbon and silicate dust features in the mid-infrared spectra of $z \approx 2$ ultraluminous infrared galaxies. Astrophys J 628:604–610

— 8 —

The Cosmic Microwave Background

Jean-Michel Lamarre[I]

Abstract

The Cosmic Microwave Background (CMB) is one of the pillars on which the Big Bang theory relies. High-quality maps of this radiation strongly constrain the history of the Universe and its content. Such maps require accurate and sensitive measurements of tiny random features on a strong uniform background. Stray radiation must be rejected extremely efficiently and the response and noise of the instrument should be known to better than a percent, while full sky maps are needed for optimal data reduction. This is better achieved outside the atmosphere in the conditions of space. In a field where fundamental physics and astrophysics are closely related, the major advances came from the *COBE* and *WMAP* spacecraft, and further progress is now expected from the *Planck* mission and also from recently developed ground-based experiments with dedicated goals achievable on fractions of the sky.

From ground-based discovery to space exploration

The Cosmic Microwave Background was generated a few hundred thousand years after the Big Bang, when ionized hydrogen was cooling down enough to recombine and become transparent. Photons were scattered a last time by the "last scattering surface" and then travelled for about 14×10^9 years until the present. The expansion of the Universe has redshifted the CMB radiation from the infrared to the millimetre wave range, in which we can detect them in our epoch.

Mapping the CMB provides a picture of the past Universe, or most exactly of the part that is visible to us now. One can expect from its observation to gain an impressive wealth of information and a large number of ways to question our understanding of physics and astronomy. It should unveil features that may verify or falsify the most currently accepted principles. Its simple existence is additional evidence in favour of the Big Bang theory. The light emitted by the last scattering surface informs us about the physical conditions (density and temperature) prevailing only 3×10^5 a to 4×10^5 a after the Big Bang, but also about what happened earlier. This can be compared with the fact that we can study accurately the internal structure of the Sun by observing the movements of its surface. The

[I]LERMA, Observatoire de Paris and CNRS, Paris, France

difference is that this concerns the Universe as a whole, from its earliest moments up to now. Promises of major discoveries were more than kept by two space missions, *COBE* and *WMAP*, which have produced high-quality maps of the sky at millimetre wavelengths. A third one, the *Planck* mission, has a strong potential for new discoveries. Although the CMB is the most important source of photons in the sky outside the solar system, its first detection came rather late, probably because the signal from ground-based radio receivers results from the co-addition of receiver noise, signal picked-up from the antenna environment, thermal emission from the antenna and the atmosphere, and radiation from the sky. Understanding and separating signal components with a high-quality antenna was a full-time job at which Penzias and Wilson (1965) were occupied when they discovered an "excess antenna temperature" that was not identified with any of the first three components. They published this result, for which they won the 1978 Nobel Prize for physics. After the discovery of the CMB, it became clear that observing from above the atmosphere was an efficient way to get rid of the atmosphere itself and of the thermal emission from the immediate environment of the antenna. Dicke differential radiometers aboard a high altitude U2 airplane measured a small dipole in the CMB radiation (Smoot et al 1977). But even at the altitude of stratospheric balloons, i.e., 30 m to 40 km, the residual atmospheric emission proved to remain a strong source of uncertainty in large parts of the spectrum (Woody and Richards 1981). These experiments have confirmed the richness of the domain and proved the efficiency of new instrumental concepts that found their full potential only when brought to space on the *COBE* spacecraft.

The Cosmic Background Explorer (*COBE*)

COBE, launched in November 1989 into a Sun-synchronous orbit, included two microwave experiments (Boggess et al 1992). Both instruments have unveiled, with an angular resolution of 7°, an image of the microwave sky nearly uncontaminated by the radiation from Earth or from other sources in the sky. The Far Infrared Absolute Spectrophotometer (FIRAS) has measured the sky radiation between 30 GHz and 600 GHz with a spectral resolution of 30 GHz (Mather et al 1990). The measured CMB spectrum (Figure 8.1, right) is near to that of a black body at (2.725 ± 0.001) K (Fixsen and Mather 2002). Deviations from a perfect black body are less than 50 parts per million of its peak emission. This very pure spectrum ruled out a number of alternative theories developed to explain the CMB without a Big Bang. The differential maps (Figure 8.1, left) produced by the Differential Microwave Radiometer (DMR) show an incredibly featureless sky (Smoot et al 1992). *COBE* has revealed that the observable Universe is uniform at very large scales, which has strong theoretical implications. To produce such a homogeneous image, a link must exist between regions of the last scattering surface that are too far from each other to have interacted since the Big Bang. This could have happened only in a very early period, before a phase of extremely rapid inflation that separated these regions. It is only by increasing the contrast thousands of times (Figure 8.1, middle left) that one can see a dipole component. The spectrum of the dipole has been measured by FIRAS and its shape is consistent with the

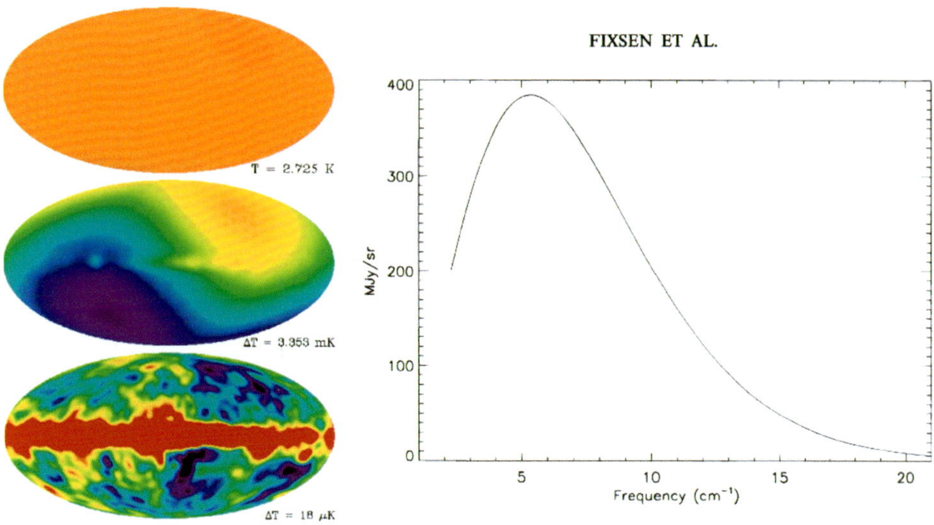

Figure 8.1: Results from the *COBE* spacecraft show that microwave sky is occupied by a nearly featureless source with a spectrum near to perfect black body at 2.7 K, the Cosmic Microwave Background radiation. The FIRAS results (right, Fixsen et al 1996) show the measured spectrum and the theoretical one from a black body. Maps reconstructed from the FIRAS and the DMR data (left, from http://lambda.gsfc.nasa.gov/product/cobe/) show a very uniform sky (upper left). It is only by increasing the contrast thousands of times that one can see the 3 mK dipole component (middle left). Further contrast enhancement reveals that there are, in addition to galactic emission, very small but significant deviations from uniformity (DMR data, lower left).

assumption that it is produced by a Doppler shift of the monopole. Its value of (3.372 ± 0.007) mK implies that the Sun's peculiar speed with respect to a co-moving frame is (371 ± 1) km s^{-1}. Further contrast enhancement reveals a very small but significant random "anisotropy", i.e., deviations from uniformity, with RMS amplitude of (29 ± 1) K at the angular resolution of $10°$ (Bennett 1996). Such anisotropies were keenly expected as a major sign that we understood the nature and the origin of the Cosmic Microwave Background radiation (CMBR). They were expected to be the remnants of quantum fluctuations during the first moments of the Big Bang. And without them, it would have been very difficult to understand how such an extremely uniform source could have produced the highly structured Universe that we can observe now in our neighbourhood. J. Mather and G. Smoot received the 2006 Nobel Prize for physics for the results from the *COBE* spacecraft. Although its angular resolution was only $7°$ and the signal to noise only 2 per beam on anisotropies, *COBE* has produced a wealth of major discoveries.

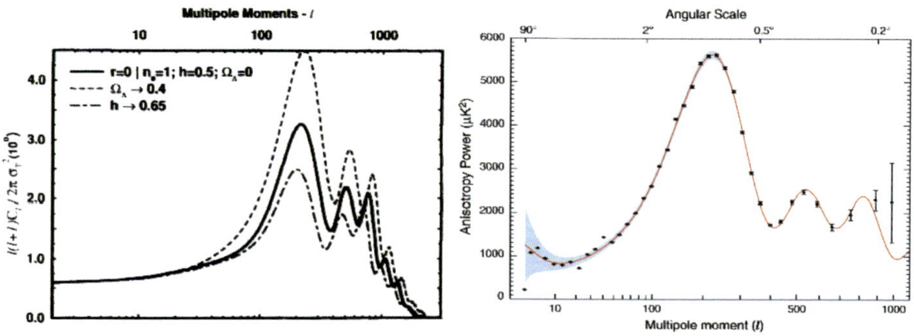

Figure 8.2: Left (Bond et al 1994): Power spectra as a function of "l" for scale-invariant models, with various values given to the cosmological constant ($\Omega_\Lambda = 0\,;\, 0.4\,;\, 0.5$), Hubble constant ($H = 0.5$ and 0.65), and baryonic content ($\Omega_{\mathrm{B}} = 0.5$ and 0.3). Right (Hinshaw et al 2009): $C(l)$, measured by *WMAP* based on the five year data including the best fit model (see Table 8.3).

Entering the era of precision cosmology with *WMAP* and *Planck*

Imprints of the early Universe history and seeds of future large structures were expected to be observable in the CMB anisotropy (Silk 1968; Sunyaev and Zeldovich 1970). On angular scales $\geq 1°$, the CMB probes fluctuations in the gravitational potential (Sachs–Wolfe effect, Sachs 1967) while smaller scales probe the sound waves prior to recombination. Even before the discovery of the CMB anisotropies by *COBE*, it was understood that high-sensitivity measurements could be and had to be made by new types of instruments. It was shown by Bond et al (1994) and Jungman et al (1996) that accurate, high resolution measurements of the CMB could be used to determine many of the cosmological parameters of our Universe not fixed by the various models of cosmology, for example its content (mass, energy), its geometry (curvature), and its dynamics (past and future of the expansion), as shown in Figure 8.2. These constraints can be used alone or in conjunction with other tracers of the large scale properties of the Universe. They provide strong indications on what happened during the first moments of the Big Bang, well before CMB photons are emitted. This is possible in particular because the physics that takes place at and behind the last scattering surface is amazingly simple and well understood (Sunyaev and Chluba 2007). In a way similar to solar physicists who study the internal structure of the Sun by observing its visible surface, cosmologists analyse the imprint left by physical phenomena that happened long before and are not directly observable. The sensitive data is the $C(l)$, i.e., the power spectrum of the anisotropy map, that must be measured at moments l of the spherical harmonics ranging from zero (mean brightness of the sky) to about 3000 (features smaller than $7'$) and a sensitivity of a few microkelvins. A first view of the $C(l)$ was obtained through a number of remarkable ground-based and balloon-borne experiments. This requires a huge improvement with respect to *COBE*. The Boomerang (de Bernardis et al 2000), Maxima (Hanany et al 2000) and Archeops (Benoît et al 2003) balloon-borne experiments have shown images of the CMB on small fractions of the sky with good signal-to-noise ratio. Ground-based experiments, such as CBI (Pearson et al 2003; Readhead et al 2004), ACBAR (Arcminute Cosmology Bolometer Array Receiver, Runyan et al 2003; Kuo et al 2004), VSA (Scott et al 2003), and DASI (Halverson et al 2002) have unveiled the high-resolution part of the CMB spectrum.

The CMB anisotropy consists of faint features that have to be distinguished from (a) the foregrounds, i.e., mainly our Galaxy and external galaxies, (b) the thermal emission of the Earth and other solar-system objects that are huge warm sources, (c) the self-emission of the spacecraft and the instrument themselves, especially their variation with time and (d) any additional signal of any origin, such as cosmic rays, the local magnetic field, electromagnetic interferences, etc. The stationary part of the noise from the instrument can be evaluated by powerful statistical tools, but its non-Gaussian and non-stationary parts and the other sources of systematics are much more difficult to evaluate and to remove from the data. They have to be reduced as much as possible. This difficulty has been considered as the major design driver of CMB experiments. In addition, an optimal interpretation of the $C(l)$ will rely on full sky maps with a pixel size of less than $0.1°$. All these requirements can be met properly only with a space mission with adapted design and mission profile, although very significant observations can be achieved from the ground. This has been undertaken by NASA and ESA with the spacecraft *WMAP* (Bennett et al 2003) and *Planck* (The *Planck* consortia 2005). Both experiments have benefitted from a formal positive decision in spring 1997, a similar date in very different programmatic contexts, NASA favouring immediate start and early launch, while ESA was looking for synergy between *Planck* and *Herschel* and therefore a later launch. Consistently, they illustrate very different strategies. *WMAP* intended to take advantage of the available and proven technology to get major new results as soon as possible, while *Planck* was designed to perform at the limits set by physics, which required more risky and longer developments. We detail hereafter some of the requirements for a "precision cosmology" mission and the solutions used in each case, which is a good illustration of how two missions can be very similar, because the constraints are the same, yet have different ambitions and schedules.

Common requirements and different ambitions

Orbit and general configuration

Figure 8.3 shows the *WMAP* and *Planck* spacecraft at about the same scale and Table 8.1 gathers the main parameters of the two missions. The Lagrange point L2 of the Sun-Earth system was found to offer major advantages over a low Earth orbit to meet the requirements of a CMB mission. The Earth and the Moon are far enough ($\approx 1.5 \times 10^6$ km) and their thermal radiation does not contaminate the faint signal from the anisotropies. In addition, they are approximately aligned with the Sun, and a rather natural configuration is obvious: the solar panels are used as a screen protecting the spacecraft from the Sun, the Earth and

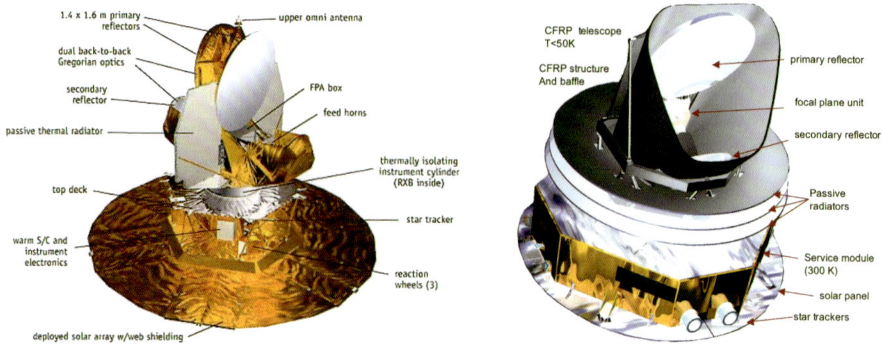

Figure 8.3: *WMAP* (left, from http://map.gsfc.nasa.gov/mission/observatory.html) and *Planck* (right) were designed for the same orbit and rather similar attitude with respect to the Sun.

the Moon. The telescopes' optical axes are nearly perpendicular to the Sun. The rotation of the spacecraft around an anti-solar axis ensures a natural coverage of the whole sky in six months, while the spacecraft follows the general movement of the L2 point around the Sun with the Earth. This configuration is extremely favourable for efficient passive cooling: 90 K for *WMAP* and less than 50 K for *Planck*, which reduces the thermal emission of the telescopes on the receivers and provides an excellent environment for cooling the *WMAP* receivers and for the cryogenic coolers of *Planck*. Finally, this orbit is thermally very stable. All these features minimize the parasitic signals that are the bane of instruments dedicated to measuring statistical properties of the sky. For both spacecraft, full maps of the sky have to be reconstructed from the signal produced by at least six months of data. The noise of the receivers contaminates these maps. Especially, the very slow post-detection noise (the so-called $1/f$ noise) makes stripes that must be removed by software with an efficiency that depends on the details of the scanning strategy. Map-making software relies on redundant observations of the same part of the sky with different scans, which allows the removal of drifts in the signal of instrumental origin. The *WMAP* strategy takes advantage of the three-axis attitude control and of the precession of the rotation axis to cover a wide stripe in the sky in one hour, while *Planck* scans successively the same circles on the sky at one revolution per minute during 45 min. This second option is more demanding in terms of signal stability (see the section on thermal design) but offers powerful and efficient solution to the removal of systematics by comparing 45 successive measurements of the same circle of the sky.

Choice of the frequency range and receiver technology

In principle, the CMB can be observed from radio frequencies (in the gigahertz range) up to about one terahertz, the frequency at which the 2.7 K black-body emission spectrum vanishes. When increasing the sensitivity, the foregrounds will be detected and start to contaminate the maps. Fortunately, the peak of the CMB

Table 8.1: Main features of *WMAP* and *Planck* spacecraft.

	WMAP	*Planck*
Sky coverage	Full sky	Full sky
Optical system	Back-to-back Gregorian, 1.4 m × 1.6 m primaries	Gregorian, 1.5 m × 1.8 m primary
Scan angle	±70° from rotation axis	85° from rotation axis
Detection	HEMT amplifiers	HEMT amplifiers + 100 mK bolometers
Polarization sensitive	I, Q, U Stokes parameters	I, Q, U Stokes parameters (not at 545 GHz and 857 GHz)
HEMT radiometric system	Sky/sky pseudo-correlation differential	Sky/4 K load differential
Spin modulation	≈ 7.57 mHz spacecraft spin	≈ 16.6 mHz spacecraft spin
Precession modulation	≈ 0.3 mHz spacecraft precession	No precession
In-flight calibration	Amplitude from dipole modulation	Amplitude from dipole modulation
Beam knowledge	Calibration on Jupiter	Calibration on planets
Required temperature	About 90 K	Telescope 50 K ; HEMT 20 K ; bolometers 0.1 K
Cooling system	Passive cooling	Passive + H_2 J-T + He J-T + 0.1 K Dilution
Attitude control	3-axis controlled, 3 wheels, gyros, star trackers, Sun sensors	Spin controlled, thrusters, star trackers, gyros
Power	419 W	2000 W
Mass	840 kg	1450 kg
Launch	Delta II 7425-10 on 30 June 2001 at 3:46:46.183 EDT	Ariane 5 dual launch (with HSO) – 14 May 2009
Orbit	1° to 10° Lissajous orbit about second Lagrange point, L2	Lissajous orbit about second Lagrange point, L2
Trajectory	3 Earth-Moon phasing loops, lunar gravity assist to L2	Direct way to L2 orbit
Design lifetime	27 months = 3 month trajectory + 2 a at L2	17 months = 3 month trajectory + 14 months at L2

and the minimum of foreground emissions coincide in the 30 GHz to 300 GHz range, with a rather clean window in the 60 GHz to 100 GHz range (Figure 8.4). The choice of the frequency coverage is a major driver of the design. With the sensitivity of *Planck*, the foregrounds are detectable and constitute a significant source of error if not removed from the data. The *Planck* choice was to associate in a single mission a High Frequency Instrument (HFI) (Lamarre et al 2003), based on the use of bolometers cooled to 100 mK, and a Low Frequency Instrument (LFI) (Valenziano et al 2007) using HEMT (High Electron Mobility Transistor) receivers at 20 K. The nine channels of *Planck* (3 LFI and 6 HFI) covering the 30 GHz to 1 THz range make up a self-consistent instrument, providing the measures needed to identify all astrophysical components and remove them from the map to recover the original CMB signal. At frequencies lower than 100 GHz, the best detectors are amplifiers using HEMT that can work at room temperature, although their sensitivity improves at low temperatures. This is the *WMAP* (Jarosik et al 2003)

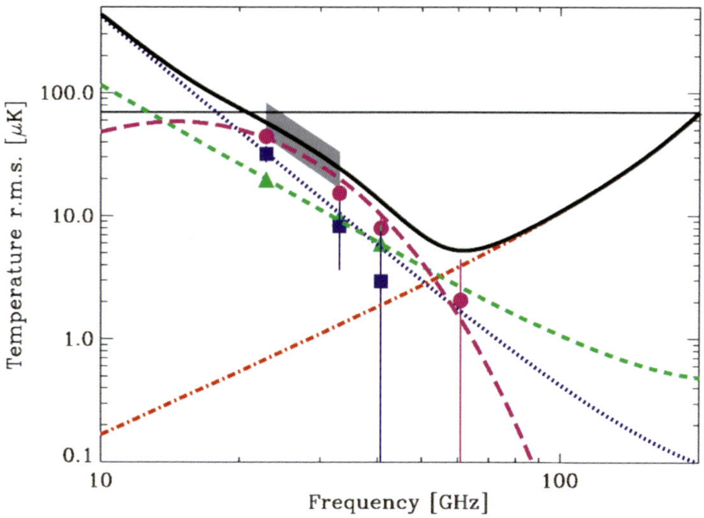

Figure 8.4: RMS fluctuation spectrum (thermodynamic temperature units) of foreground components at 1° resolution (Davies et al 2006). Synchrotron (blue squares), free-free (green triangles) and anomalous dust-correlated emission (magenta circles) add to the dust (red dot-dashed line). CMB is shown at 70 μK RMS. Total foregrounds (black solid curve) show a clear minimum around 70 GHz. HEMTs are competitive up to 70 GHz, while bolometers are more sensitive at higher frequencies.

and *Planck*-LFI option (Cuttaia et al 2004) with receivers cooled at 90 K and 20 K, respectively. At higher frequencies, bolometers (Bock et al 1995; Jones et al 2003) offer a sensitivity up to 30 times better, which allows them both to measure more pixels with better angular resolution and to improve the signal to noise for each pixel. Higher frequencies consistently provide higher angular resolutions for a given telescope diameter. Very sensitive bolometers have to be cooled to 100 mK, which had never been done in orbit before *Planck* and is a major constraint on the system. In most of the bolometer channels of *Planck*, the sensitivity is limited by the fundamental limits set by the photon statistics of the observed source.

Spatial frequencies of interest and optical systems

The CMB anisotropies contain information at all spherical harmonics up to the moment $l = 3000$, which is a great incentive to measure accurately on all angular scales, from 360° to less than 0.1°. High-moment harmonics gather significantly less energy and are fully measurable with the sensitivity of *Planck*. In addition, the beam width filters out the highest spatial frequencies, which consistently prevents *WMAP* from measuring harmonic moments larger than about 1000. The main characteristics of all the frequency channels of the two missions are gathered in Table 8.2. The measurement of low-moment harmonics requires the coverage of the whole sky and an efficient rejection of parasitic sources that could be picked up

Table 8.2: Channels of *WMAP* and *Planck*.

	WMAP					*Planck*								
LFI/HFI						LFI			HFI					
HEMT amps.	x	x	x	x	x	x	x	x						
Bolometers									x	x	x	x	x	x
Centre freq. (GHz)	23	33	41	61	94	30	44	70	100	143	217	353	545	857
Bandwidth (GHz)	5.5	7.0	8.3	14	20	6	8.8	14	33	47	72	116	180	283
No. of unpol. dets.	0	0	0	0	0	0	0	0	0	4	4	4	4	4
No. of polarized detectors	2	2	4	4	8	4	6	12	8	8	8	8	0	0
Angular resolution (FWHM, arcmin)	53	40	31	21	13	33	29	14	9.5	7.2	5	5	5	5
Average $\Delta T/T^a$ per pixel (μK/K)						2.5	3.0	5.2	2.5	2.2	4.8	15	NA	NA
System temp. (K)	29	39	59	92	145	11	16	29						
RJ sensitivity[b] per freq. band (μK $s^{\frac{1}{2}}$)	800	800	500	600	400	110	120	130	14	12	10	8	10	7

[a]for a 14 month mission; [b]Rayleigh-Jeans sensitivity

by diffraction or in any other way. For these reasons, going to space was a major step towards accurate CMB measurements.

The control of far side lobes is an important issue that has to be achieved by an appropriate design of the RF system. At millimetre wavelengths, there is no "black paint" similar to what is used in the optical and infrared ranges to absorb stray radiation and one has to extrapolate the techniques of radio engineering. *WMAP* and *Planck* have rather similar telescopes (Fargant et al 2000; Page et al 2003), i.e., off-axis Gregorian-like arrangements were chosen, avoiding any sharp-edged diffracting object in the beam and therefore induced scattering of power in the far side lobes. This design allows for a sufficient focal plane area with a compact configuration that helps fit the fairing envelopes. The cost of compactness is that the beams on the sky are far from circular, which adds a serious difficulty to the data reduction. Both telescopes are coupled to the receivers by corrugated horns producing nearly Gaussian beams. The edge taper, i.e., the response to sources at the edge of the mirrors with respect to that at the centre is exceptionally low for radio systems (−20 dB to −35 dB), which provides a low level of spill-over, i.e., the fraction of the beam sensitive to flux from outside the main mirror is low. This is critical for avoiding contamination of the low-moment harmonics by pick-up in the side lobes of the large structures of the sky, such as our Galaxy.

WMAP receivers are cooled down to about 90 K by a rather classical passive cooling, taking advantage of the favourable conditions and exceptional stability offered by the L2 orbit, intending to reduce systematic effects of thermal origin to a minimum. Temperature fluctuations are induced only by the annual variation of the distance to the Sun and by bursts of solar activity. In order to cool its bolometers to 0.1 K, *Planck* needs a complex chain of coolers. The passive cooling, based on a three-stage V-groove radiator and by a blackened baffle and structure around the telescope (Riti et al 2003), provides a cooling power of 1.5 W at about 60 K. The LFI HEMT amplifiers are cooled to 20 K by Joule-Thomson (J-T) expansion of hydrogen activated by a sorption pump (Prina and Bhandari 2001). The cooling

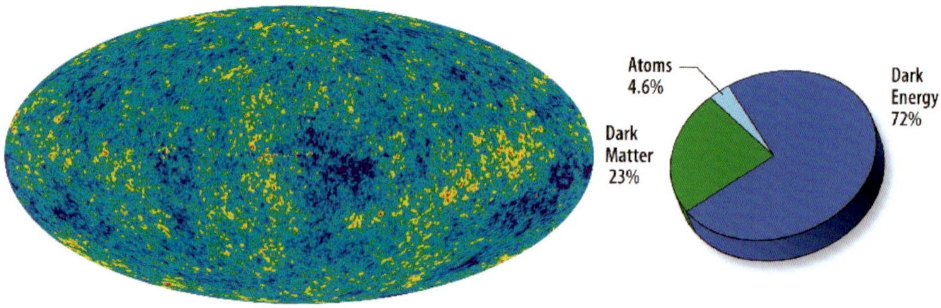

Figure 8.5: *WMAP* data reveal tiny (sub-millikelvin) features that will develop later to become the large structures in the Universe (e.g., clusters of galaxies). Interpreting this image tells a lot about the nature and the history of the Universe and deeply challenges our understanding of physics, as illustrated by the pie chart of the content of the Universe (mass + energy), where "current" matter (atoms) is only 4.6 % of the total (figures from *WMAP* web site, http://map.gsfc.nasa.gov/news/).

Table 8.3: Predictions from the five year data of *WMAP* best-fit Lambda Cold Dark Matter (ΛCDM) model (Hinshaw et al 2009).

Age of the Universe	(13.69 ± 0.13) Ga
Hubble constant H_0	$(71.9^{+2.6}_{-2.7})$ km s^{-1}/Mpc
Mean redshift of CMB photons	1090.51 ± 0.95
Mean time of radiation-matter decoupling	(380.08 ± 5.84) ka
Vacuum energy density Ω_Λ	0.742 ± 0.030
Amount of baryonic matter	(4.41 ± 0.30) %
Amount of dark matter	$(21.4 \pm 0.2.7)$ %

to 4.5 K of the HFI focal plane unit external box is obtained by J-T expansion of helium driven by mechanical pumps, while 1.6 K and 0.1 K stages are based on a complex ^{3}He/^{4}He dilution cooler (Triqueneaux et al 2006). Temperature fluctuations induced by the coolers are a potential source of systematic effects. Extremely stable temperatures are required for the 20 K stage for LFI (100 mK p-p), the 4 K stage of HFI (10 μK Hz$^{-1/2}$) and for the plate of the bolometers (20 nK Hz$^{-1/2}$), achieved with sophisticated active temperature control (Leroy et al 2006). These features contribute to the complexity of the *Planck* spacecraft whose development was undertaken by a large international collaboration involving numerous institutes and space agencies (e.g., ESA, NASA, CNES, ASI) while *WMAP* could be developed within the simpler frame of a NASA MIDEX mission.

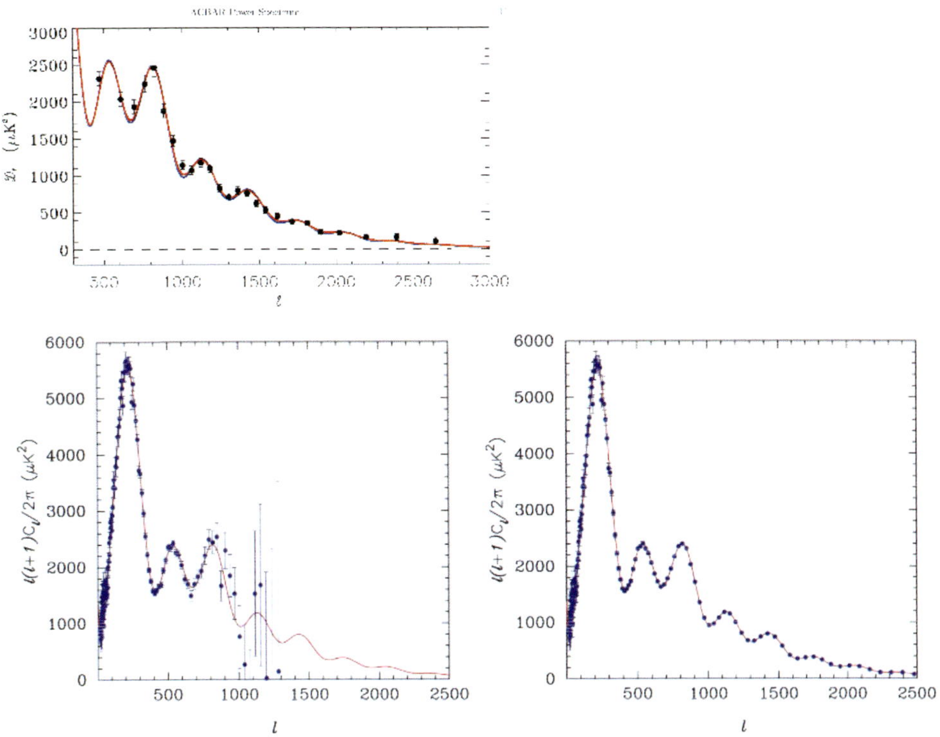

Figure 8.6: Accurately measuring the $C(l)$ curve should be achieved in the first decade of the 21st century, as shown by the 3-year *WMAP* results (Spergel et al 2007) (lower left), the ACBAR measurements (Reichardt et al 2008) from the South Pole (upper left) and a simulation of the *Planck* data (lower right). Lower figures are from Balbi (2007).

A universe to be deciphered

The picture of the Universe given by *COBE* and *WMAP* (Spergel et al 2007) is consistent with the Big Bang theory and the Lambda Cold Dark Matter model (ΛCDM, assuming a cosmological constant different from zero), established in the frame of general relativity and the standard model of particle physics. It gives a good knowledge of the main parameters needed to describe the large scale geometry and the history of the Universe inside this frame (see Table 8.3). These results by themselves, consistently with those obtained by large sky surveys and supernovae surveys, converge to describe a reaccelerating Universe with a large cosmological constant and dominated by dark matter and dark energy (Figure 8.5). Dark matter is detected only by its gravitational interaction with visible matter. Dark energy is the energy needed to reaccelerate the Universe although we expect gravitation to slow down its expansion. We do not know the nature of this matter and we ignore everything on dark energy, the way it interacts, its history if any, etc. This opens large and exciting fields of physics that can be tackled by more accurate measurements of the CMB, as planned with the spacecraft *Planck* and also ground-based experiments (Reichardt et al 2008 and Figure 8.6). Accuracies expected from *Planck* will open new windows to check the validity of the most accepted or unconventional models. Accurate measurements of polarization, expected both from *Planck* and ground-based experiments, should give independent confirmation of what temperature maps tell us, and also provide entirely new tests on the physics of inflation, by the detection of the signature of gravitational waves on the last scattering surface.

Bibliography

Balbi A (2007) Cosmology from *Planck*. New Astron Rev 51:281–286

Bennett CL, Banday AJ, Gorski KM (plus three authors) (1996) Four-year *COBE* DMR Cosmic Microwave Background observations: Maps and basic results. Astrophys J 464:L1–L4

Bennett CL, Bay M, Halpern M (plus 12 authors) (2003) The Microwave Anisotropy Probe mission. Astrophys J 583:1–23

Benoît A, Ade P, Amblard A (plus 65 authors) (2003) The cosmic microwave background anisotropy power spectrum measured by Archeops. Astron and Astrophys 399:L19–L23

Bock JJ, Chen D, Mauskopf PD, Lange AE (1995) A novel bolometer for infrared and millimeter-wave astrophysic. Space Science Rev 74:229–235

Boggess NW, Mather JC, Weiss R (plus 15 authors) (1992) The *COBE* mission – Its design and performance two years after launch. Astrophys J 397:420–429

Bond JR, Crittenden R, Davis RL (plus two authors) (1994) Measuring cosmological parameters with cosmic microwave background experiments. Phys Rev Lett 72:13–16

Cuttaia F, Battaglia P, Terenzi L (plus 13 authors) (2004) Analysis of the pseudocorrelation radiometers for the low frequency instrument onboard the *Planck* satellite. Proc SPIE 5498:756–767

Davies RD, Dickinson C, Banday AJ (plus three authors) (2006) A determination of the spectra of Galactic components observed by the Wilkinson Microwave Anisotropy Probe. Mon Not Roy Astr Soc 370:1125–1139

de Bernardis P, Ade PAR, Bock JJ (plus 33 authors) (2000) A flat Universe from high-resolution maps of the cosmic microwave background radiation. Nature 404:955–959

Fargant G, Dubruel D, Cornut M (plus five authors) (2000) Very wide band telescope for *Planck* using optical and radio frequency techniques. Proc SPIE 4013:69–79

Fixsen DJ, Cheng ES, Gales JM (plus three authors) (1996) The Cosmic Microwave Background spectrum from the full *COBE* FIRAS data set. Astrophys J 473:576

Halverson NW, Leitch EM, Pryke C (plus ten authors) (2002) Degree Angular Scale Interferometer first results: A measurement of the Cosmic Microwave Background angular power spectrum. Astrophys J 568:38–45

Hanany S, Ade P, BalbiA (plus 19 authors) (2000) MAXIMA-1: A measurement of the Cosmic Microwave Background anisotropy on angular scales of $10'$–$5°$.

Astrophys J 545:L5–L9

Hinshaw G, Weiland JL, Hill RS (plus 18 authors) (2009) Five-year Wilkinson Microwave Anisotropy Probe observations: Data processing, sky maps, and basic results. Astrophys J 180:225–245

Jarosik N, Bennett CL, Halpern M (plus 12 authors) (2003) Design, implementation, and testing of the microwave anisotropy probe radiometers. Astrophys J Suppl 145:413–436

Jones WC, Bhatia R, Bock JJ (plus two authors) A polarization sensitive bolometric receiver for observations of the Cosmic Microwave Background. Proc SPIE 4855:227–238

Jungman G, Kamionkowski M, Kosowsky A, Spergel DN (1996) Cosmological-parameter determination with microwave background maps. Phys Rev D 54:1332–1344

Kuo CL, Ade PAR, Bock JJ (plus eleven authors) High-resolution observations of the Cosmic Microwave Background power spectrum with ACBAR. Astrophys J 600:32–51

Lamarre JM, Puget JL, Bouchet F (plus 23 authors) (2003) The *Planck* High Frequency Instrument, a third generation CMB experiment, and a full sky submillimeter survey. New Astron Rev 47:1017–1024

Leroy C, Arondel A, Bernard JPh (plus ten authors) (2006) Performances of the *Planck*-HFI cryogenic thermal control system. Proc SPIE 6265:13(pp. 62650H)

Mather JC, Cheng ES, Eplee RE (plus 18 authors) (1990) A preliminary measurement of the cosmic microwave background spectrum by the Cosmic Background Explorer (*COBE*) satellite. Astrophys J 354:L37–L40

Page L, Jackson C, Barnes C (plus twelve authors) (2003) The optical design and characterization of the Microwave Anisotropy Probe. Astrophys J 585:566–586

Pearson TJ, Mason BS, Readhead ACS (plus 23 authors) (2003) The anisotropy of the microwave background to $l = 3500$: Mosaic observations with the Cosmic Background Imager. Astrophys J 591:556–574

Penzias AA, Wilson RW (1965) A measurement of excess antenna temperature at 4080 Mc/s. Astrophys J 142:419–421

Prina M, Bhandari P, Bowman RC (plus three authors) (2002) Performance prediction of the *Planck* sorption cooler and initial validation. Proc Cryogenic Engineering Conf AIPC 47:1201–1208

Readhead ACS, Mason BS, Contaldi CR (plus 18 authors) (2004) Extended mosaic observations with the Cosmic Background Imager. Astrophys J 609:498–512

Reichardt C, Ade PAR, Bock JJ (plus 15 authors) (2008) High resolution CMB power spectrum from the complete ACBAR data set. arXiv:0801.1491 to appear in Astrophys J

Riti JB, Dubruel D, Nadarassin M (plus five authors) (2003) *Planck* payload module design and performance. Proc SPIE 4850:749–763

Runyan MC, Ade PAR, Bhatia RS (plus 18 authors) (2003) ACBAR: The Arcminute Cosmology Bolometer Array Receiver. Astrophys J Supp 149:265–287

Sachs RK, Wolfe AM (1967) Perturbations of a cosmological model and angular variations of the microwave background. Astrophys J 147:73

Scott PF, Carreira P, Cleary K (plus 23 authors) (2003) First results from the Very Small Array - III. The cosmic microwave background power spectrum. Mon Not

Roy Astr Soc 341:1076–1083

Silk J (1968) Cosmic black-body radiation and galaxy formation. Astrophys J 151:459

Smoot GF, Gorenstein MV, Muller RA (1977) Detection of anisotropy in the cosmic black-body radiation. Phys Rev Lett 39:898–901

Smoot GF, Bennett CL, Kogut A (plus 25 authors) (1992) Structure in the *COBE* differential microwave radiometer first-year maps. Astrophys J 396:L1–L5

Spergel DN, Bean R, Doré O (plus 19 authors) (2007) Three-year Wilkinson Microwave Anisotropy Probe (*WMAP*) observations: Implications for cosmology. Astrophys J Suppl 170:377–408

Sunyaev RA, Zeldovich YB (1970) Small scale fluctuations of relic radiation. Astrophys and Space Sciences 7:3–19

Sunyaev RA, Chluba J (2007) The Richness and the beauty of the physics of cosmological recombination. arXiv:0710.2879

The *Planck* consortia (2005) *Planck*, the science programme. ESA-SCI 1, astro-ph/0604069

Triqueneaux S, Sentis L, Camus Ph (plus two authors) (2006) Design and performance of the dilution cooler system for the *Planck* mission. Cryogenics 46:288–297

Valenziano L, Sandri M, Morgante G (plus 19 authors) (2007) The low frequency instrument on-board the *Planck* satellite: Characteristics and performance. New Astron Rev 51:287–297

Woody DP, Richards PL (1981) Near-millimeter spectrum of the microwave background. Astrophys J 248:18–37

III. Techniques and Systems

— 9 —

Space telescopes

PHILIPPE LEMAIRE[I], BERND ASCHENBACH[II] AND JOHN F. SEELY[III]

Abstract

A brief survey of normal and grazing incidence space telescope types is given. The optimization of telescope efficiency either by dedicated single, bi-layer or multi-layer coatings is described. An outlook of solar and stellar coronagraphs is included.

Introduction

Telescopes for astronomical observations have been in use for four centuries by now. Even for telescopes used in space which covers a much shorter time period it is not possible to give a detailed and complete description in a few pages, although the optical characteristics of the normal-incidence space (NI) telescopes are derived from ground-based telescopes. In this article, the most important parameters are recalled but for more details and complementary information see books on optics, such as Astronomical Optics (Schroeder 2000).

The choice and specification of the optically relevant parameters of each individual telescope are related to the scientific objectives and the constraints imposed by the celestial object under study and/or the capacity of the spacecraft in terms of mass, volume and electrical power which eventually can be supplied.

To fully design a telescope appropriate to the scientific objectives the following topics should be considered and weighed against each other:
– stellar observations (point-like sources), imagery or/and spectroscopy,
– interstellar medium and/or galaxy observations in wide-field mode (extended sources),
– solar observations, imagery or/and spectroscopy,
– spectral range or wavelength domain and efficiency,
– angular resolution and field of view (FOV),
– spacecraft parameters (size and mass allocations, pointing system performance, choice of spacecraft orbit,...),
– acceptable level of scattered light and baffling requirements.

[I]IAS—Institut d'Astrophysique Spatiale, Paris, France
[II]MPE—Max-Planck-Institut für extraterrestrische Physik, Garching, Germany
[III]SSD/NRL—Space Science Division, Naval Research Laboratory,Washington DC, USA

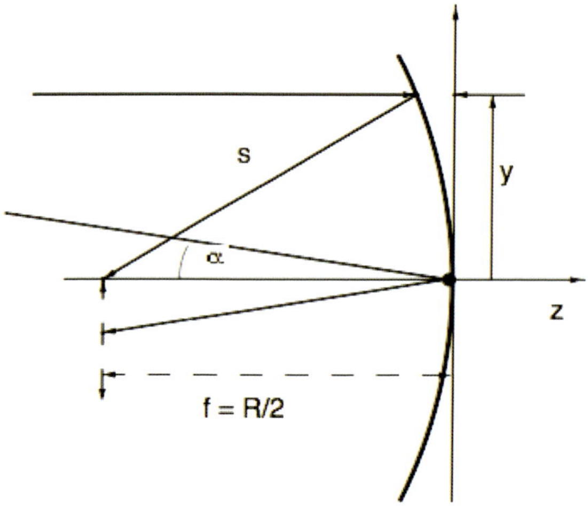

Figure 9.1: Paraboloid telescope.

In the following sections, NI reflective telescopes are considered for use only if the mirror coating reflectivity (single or multilayer) permits efficient observations. At shorter wavelengths, i.e., shorter than about 70 nm (EUV), grazing-incidence (GI) telescopes could be considered as an alternative. At even shorter wavelengths, i.e., shorter than about 5 nm up to hard X-rays with photon energies up to about 100 keV, GI telescopes are the only choice. To increase the mirror reflectivity both for NI and for GI mirrors, multilayer coatings have successfully been developed. A special case of telescopes are the coronagraphs, which have been in use for more than fifty years of observing the solar corona and which are now being developed to observe very faint stars or planets nearby a brighter star.

Normal-incidence telescopes

One-mirror telescope

The one-mirror telescope (mostly an off-axis paraboloid; Figure 9.1) has been developed to feed spectrometers in the VUV wavelength range where the use of only one reflecting surface provides better efficiency. It also offers a high-resolution instantaneous small field of view (FOV). The image is being scanned while maintaining the paraboloid's image quality, which can be achieved by either a rotation of the mirror around its vertex in case of a small field, or by rotation around its focus for a large field, or by rotation of the whole instrument (see SUMER/*SOHO*, Wilhelm et al (1995) and EIS/*Hinode*, Culhane et al (2007)). The paraboloid field curvature (Petzval surface) is $k_\mathrm{p} = 2/R$, where R is the paraboloid radius.

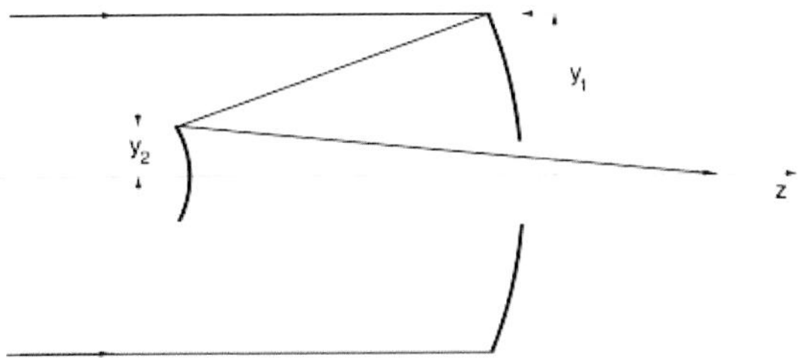

Figure 9.2: Cassegrain telescope.

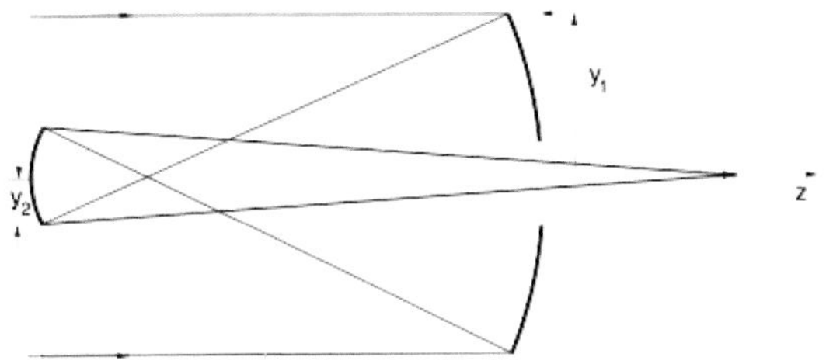

Figure 9.3: Gregorian telescope.

Two-mirror telescopes

Telescopes employing two mirrors are used to feed spectrometers or to make imagery (through wavelength filters) with small or/and wide FOV from the near-infrared to the extreme-ultraviolet wavelength range. A high-quality image over a small field is provided by a Cassegrain-type telescope — a compact telescope, Figure 9.2, e.g., *Copernicus* (Rogerson et al 1973), LPSP (Artzner et al 1977) and LASP (Bruner 1977) on *OSO-8* — or by a Gregorian-type telescope with intermediate primary focus that can be used as field stop, Figure 9.3, e.g., HRTS/*Spacelab 2* (Brueckner et al 1986), SOT/*Hinode* (Shimizu 2004). A high-quality image over a large field (up to 1°) can be obtained by a Ritchey–Chrétien-type telescope, e.g., *IUE* (Boggess et al 1977), *HST* (Jenkner 1990) and EIT/*SOHO* (Delaboudinière et al 1995). Small-field pointing can be controlled by actuating the secondary mirror while large-offset pointing is done with the full-telescope pointing. The Petzval curvature is given by $k_\mathrm{p} = 2\,((1-\rho)/\rho)/R_1$ where $\rho = R_2/R_1$. For a Gregorian telescope ρ is less than zero and the Petzval surface is convex as seen from the secondary mirror, while for a Cassegrain telescope the Petzval surface is concave.

Telescope parameters

The characteristic parameters of the telescopes for the most frequently used combinations of two mirrors are described in Table 9.1, where m_s is the secondary

Table 9.1: Telescope characteristics.

m_s	k_h	ρ	Type	Secondary
> 1	> 0	> 0	Cassegrain	convex
$= 1$	> 0	∞	Cassegrain	flat
0 to 1	> 0	< 0	Cassegrain	concave
< 0	< 0	< 0	Gregorian	concave
< 0	> 1	> 0	inverse Cassegrain	concave

magnification, k_h is the ratio of heights at mirror margins and ρ is the ratio of mirror radius of curvature. The relations between the parameters are:

$$m_s = \frac{\rho}{(\rho - k_h)} \ , \ \rho = \frac{m_s\, k_h}{(m_s - 1)} \ , \ k_h = \frac{(1+\beta)}{(m_s+1)} \quad , \tag{9.1}$$

where β is the back focal distance in units of f_1(primary mirror focal length). If K is the conic constant, the classical type telescope (Cassegrain or Gregorian) has

$$K_1 = -1 \ , \ K_2 = -\left[\frac{(m_s+1)}{(m_s-1)}\right]^2 \quad , \tag{9.2}$$

whereas the aplanatic type telescope (Ritchey–Chrétien) has

$$K_1 = -1 - \frac{2\,(1+\beta)}{m_s^2\,(m_s-\beta)} \ , \ K_2 = -\left(\frac{m_s+1}{m_s-1}\right)^2 - \frac{2\,m_s(m_s+1)}{(m_s-\beta)(m_s-1)^3} \quad . \tag{9.3}$$

Point-spread function and relative encircled energy

The image quality of a perfect telescope with circular aperture is given by the point-spread function (PSF) which provides the image extension in one direction. The normalized PSF is expressed by the following equation:

$$\frac{I_\nu}{I_0} = \frac{1}{(1-\epsilon^2)^2}\left[\frac{2\,J_1(\nu)}{\nu} - \epsilon^2\frac{2\,J_1(\epsilon\nu)}{\epsilon\nu}\right]^2 \quad . \tag{9.4}$$

Another way to qualify the image is the relative encircled energy (EE) which gives the energy distribution within a point image. The relative EE is obtained through the equation:

$$\begin{aligned} \mathrm{E_E}(\nu_0) \quad &= \quad \frac{1}{1-\epsilon^2}\left[1 - J_0^2(\nu_0) - J_1^2\,(\nu_0) + \epsilon^2(1 - J_0^2(\epsilon\nu_0) - J_1^2(\epsilon\nu_0)\right] \\ &\quad - \frac{2\,\epsilon}{1-\epsilon^2}\left[\int_0^{\nu_0} J_1(\epsilon\,\nu)\frac{2\,J_1(\nu)}{\nu}\mathrm{d}\nu\right] \quad , \end{aligned} \tag{9.5}$$

where ϵ is the obscuration ratio (ratio of secondary to primary mirror diameters) and $\nu/\pi = (2/\lambda)(a/R)r$, with a the radius of the main mirror, R the radius of

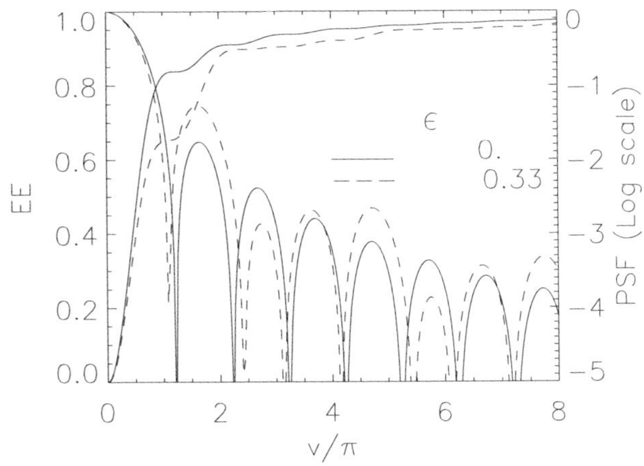

Figure 9.4: PSF (lines undulating along abscissa) and EE (lines increasing monotonically along abscissa) for a perfect image with obscuration factor ϵ of 0 and 0.33.

curvature of the main mirror, r the distance from the focal point in the image plane. J_0 and J_1 are the Bessel functions of zero and first orders. Figure 9.4 gives the variations of the PSF and EE with $\epsilon = 0$ (off-axis single or two-mirror telescope) and $\epsilon = 0.33$ (the obscuration of the *HST*).

Classical aberrations for a centred pupil

The classical and other aberrations are treated in detail by Schroeder (2000). A summary is given in the following subsections.

Paraboloid telescope

The transverse sagittal coma (C_{ST}), the transverse astigmatism (A_{ST}), the angular astigmatism (A_{SA}) and the angular sagittal coma (C_{SA}) are given by the relations:

$$C_{\mathrm{ST}} = -\frac{y^2}{R^2}\,\Theta\,s\;,\;\; A_{\mathrm{ST}} = -\frac{2\,y}{R}\,\Theta^2\,s\;,\;\; C_{\mathrm{SA}} = \frac{\Theta}{16\,F^2}\;,\;\; A_{\mathrm{SA}} = \frac{\Theta^2}{2\,F}\;, \qquad (9.6)$$

where F is the system focal ratio (f/D, telescope focal length over its aperture diameter), R is the system radius ($2\,f$), y is the ray height at mirror margin, Θ is the angular field and $s = f$ on the telescope axis (focal distance, otherwise it is the distance between the mirror vertex and the focus).

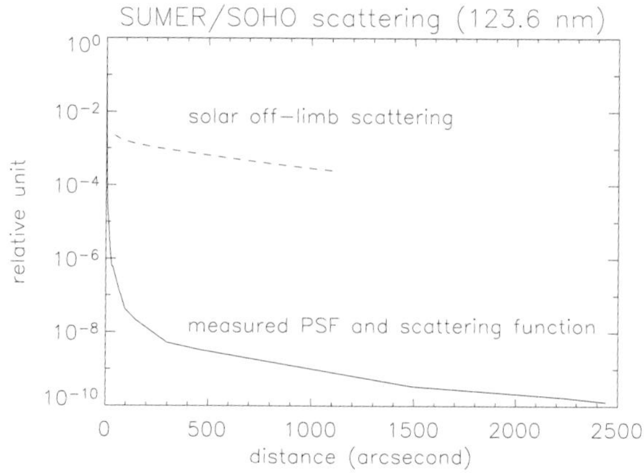

Figure 9.5: SUMER telescope scattering. The solid line gives the measured extension of the PSF combined with the scattering function (Saha and Leviton 1993) and the dotted line the computed and measured solar (32′ diameter) off-limb scattering.

Two-mirror telescope (Cassegrain or Gregorian)

For the classical two-mirror telescope the C_{SA} and A_{SA} aberrations and the angular distortion (D_{IA}) are defined as:

$$C_{\mathrm{SA}} = \frac{\Theta}{16\,F^2}\,, \quad A_{\mathrm{SA}} = \frac{\Theta^2}{2\,F}\frac{m_{\mathrm{s}}^2+\beta}{m_{\mathrm{s}}\,(1+\beta)}\,,$$
$$D_{\mathrm{IA}} = \frac{\Theta^3(m_{\mathrm{s}}-\beta)(m_{\mathrm{s}}^2-1)(m_{\mathrm{s}}+3\beta)}{4\,m_{\mathrm{s}}^2(1+\beta)^2}\,, \tag{9.7}$$

where m_{s} and β have been defined in previous sections.

Aplanetic two-mirror telescope (Ritchey–Chrétien)

In the case of the aplanetic (no spherical aberration and no coma) two-mirror telescope the modified A_{SA} and D_{IA} aberrations are:

$$A_{\mathrm{SA}} = \frac{\Theta^2}{2\,F}\frac{m_{\mathrm{s}}(2\,m_{\mathrm{s}}+1)+\beta}{2\,m_{\mathrm{s}}(1+\beta)}\,,$$
$$D_{\mathrm{IA}} = \frac{\Theta^3(m_{\mathrm{s}}-\beta)[m_{\mathrm{s}}(m_{\mathrm{s}}^2-2)+\beta(3\,m_{\mathrm{s}}^2-2)]}{4\,m_{\mathrm{s}}^2(1+\beta)^2}\,. \tag{9.8}$$

Mirror scattering

Optical surface mirror irregularities with spatial mid-frequency errors (roughness) and spatial high-frequency errors (micro-roughness) give rise to wavelength

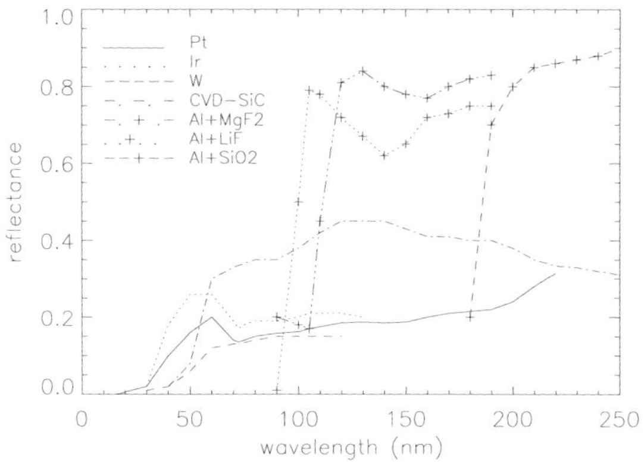

Figure 9.6: Reflectance of several coatings in the VUV at near NI.

dependent angularly scattered light. The dimmed specular reflection I_r is given by the relation:

$$I_r/I_0 = \exp[-(4\,\pi\,\sigma\,\cos\,\alpha/\lambda)^2] \quad , \tag{9.9}$$

where I_0 is the reflection of the mirror with perfect surface, σ is the root mean square (RMS) of the residual surface errors, α is the incidence angle and λ the wavelength. One aspect of the mirror scattering is the redistribution of light far from the specular image in the focal plane, cf., the one-mirror telescope scattering of SUMER in Figure 9.5 (Saha and Leviton 1993). For a point source such as a star, a bright star can mask a weak star or a planet; for extended sources, nearby features (the corona in case of the Sun or weak emitters for nebulae and galaxies) can be left undetectable.

Loss of image quality

Multiple factors can modify the image quality:
– mirror surface deformations by mechanical and/or thermal constraints (the choice of the mirror blank material and the mounting are critical for space application),
– alignment errors (secondary misalignment: tilt or decentre, despace (Schroeder 2000),
– pointing errors (Schroeder 2000),
– degradation of mirror surfaces and coatings by particulates and/or molecular deposition.
Cleanliness control of the optics and minimizing surface contamination is absolutely essential and critical for VUV and shorter wavelength telescopes (especially under the strong solar UV flux (Lemaire 1991) and/or in the case of coronagraphs).

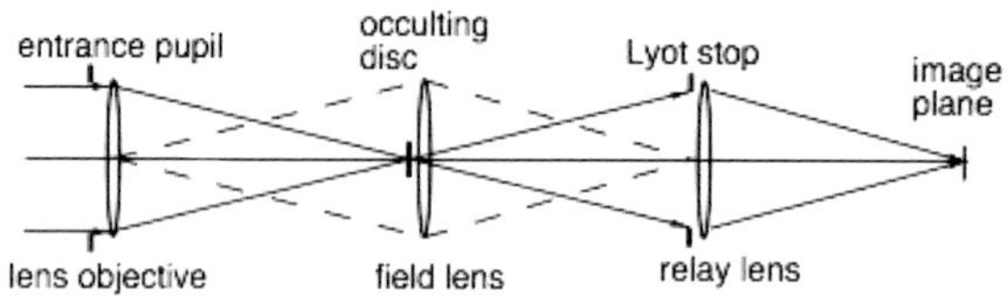

Figure 9.7: The Lyot solar coronagraph scheme.

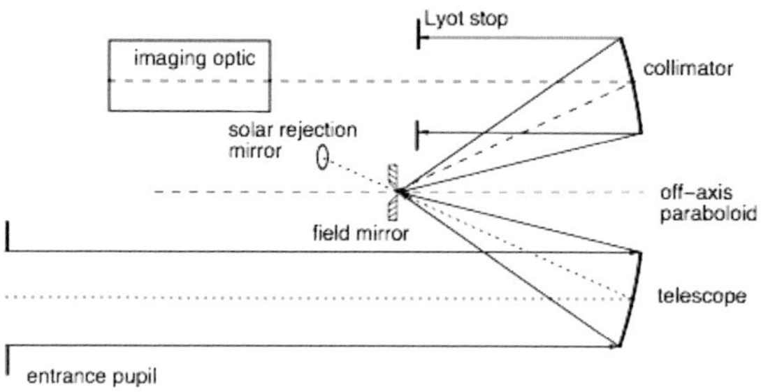

Figure 9.8: Solar reflecting coronagraph C1/LASCO.

Classical mirror coatings

For most applications, working in several wavelength ranges requires wide-band mirror coatings with high efficiency. In the VUV wavelength range, the coating efficiency strongly varies with wavelength (e.g., Figure 9.6) and the number of reflections must be minimized. The most frequently used coatings for the visible wavelength range are Ag and Al+SiO_2, but Al+MgF_2 is used to cover the FUV-visible wavelength range. Au is used in the near infrared.

Coronagraphs

Historically the coronagraph has been developed for solar corona observations. The Lyot coronagraph (Lyot 1932) basic design is shown in Figure 9.7 with a well defined pupil, an occulting disk to mask the image and a Lyot stop (at the pupil image location) to stop the high-intensity diffraction pattern given by the pupil edge. Using the Lyot idea new solar and stellar coronagraphs have been designed.

Solar coronagraphs

Reflecting solar coronagraph

An all-reflecting coronagraph has been designed and operated on *SOHO* (Figure 9.8, Brueckner et al 1995). The telescope part uses an unobstructed off-axis paraboloid.

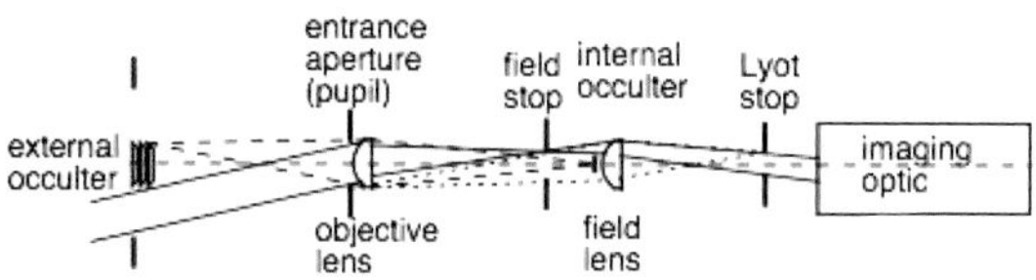

Figure 9.9: Solar refracting coronagraph with external occulter, C2/LASCO.

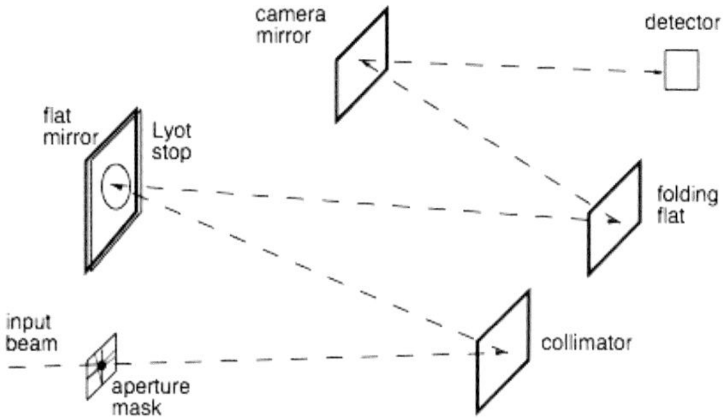

Figure 9.10: Optical scheme of the STIS-*HST* coronagraph (Heap et al 2000).

Solar coronagraph with external occulter

To observe the corona far away from the solar disk with a radiance ratio of $L/L_{\rm disk} \ll 10^{-8}$ an external occulter is used (Figure 9.9) for the C2 and C3 LASCO coronagraphs on *SOHO* (Brueckner et al 1995).

Stellar coronagraphs

To observe faint objects (stars or planets) nearby a bright object several techniques have been studied and are (or will be) implemented in space instrumentation.

Classical Lyot scheme

This technique, derived from the Lyot scheme, has been implemented in the STIS (Space Telescope Imaging Spectrograph) spectrometer (Figure 9.10, Heap et al 2000). An aperture mask stops the specular image of the bright star, and a Lyot stop eliminates the diffraction pattern of the entrance pupil.

Other coronagraph designs

The search, detection and analysis of the atmosphere of Earth-like planets in orbit around a bright star require coronagraphs with very low scatter (about 10^{-10} in the visible or 10^{-7} in the near mid-infrared). These numbers are for a separation of $0.1''$ of the two-component star/planet system located at a distance from the

Earth of 10 pc. A lot of studies are running to search for the optimized design for future missions. Most of them try to combine the advantages of a Lyot design and an interferometric combination of beams, one being π-phase shifted and flipped to obtain a nulling interferometer as suggested by Bracewell (1978). An evaluation of the expected performances of several proposed coronagraphs has been made by Guyon et al (2006).

Grazing-incidence telescopes

The physics of GI reflection

One way to focus and image sources of light is by using curved reflecting surfaces. The interaction of light with matter can be described by the complex index of refraction which describes the change of the properties of the incident electromagnetic wave when crossing the boundary between the two materials involved. The index n reads:

$$n = 1 - \delta - \mathrm{i}\,\beta \quad . \tag{9.10}$$

δ describes the phase change and β accounts for the absorption. The reflection coefficients for p and s polarization are given by the Fresnel equations:

$$r_\mathrm{p} = \left(\frac{E_\mathrm{r}}{E_\mathrm{i}}\right)_\mathrm{p} = \frac{n^2 \sin\theta - \sqrt{(n^2 - \cos^2\theta)}}{n^2 \sin\theta + \sqrt{(n^2 - \cos^2\theta)}} \quad , \tag{9.11}$$

$$r_\mathrm{s} = \left(\frac{E_\mathrm{r}}{E_\mathrm{i}}\right)_\mathrm{s} = \frac{\sin\theta - \sqrt{(n^2 - \cos^2\theta}}{\sin\theta + \sqrt{(n^2 - \cos^2\theta)}} \quad . \tag{9.12}$$

$E_\mathrm{r}/E_\mathrm{i}$ denotes the ratio of the amplitudes of the reflected and incident electric fields and θ is the grazing angle of incidence as measured from the interface plane. For NI, which is the standard in optical telescopes, $\theta \approx 90°$. This approach is correct as long as the assumptions for applying the Fresnel equations are fulfilled. The reflected intensity or reflectivity is then $R_\mathrm{p} = r_\mathrm{p} \times r_\mathrm{p}^*$ and $R_\mathrm{s} = r_\mathrm{s} \times r_\mathrm{s}^*$, where the asterisk denotes the conjugate complex value.

The components of the index of refraction for a vacuum/matter transition are often called the optical constants of the material. In the optical wavelength range, for instance, the real part of the index of refraction is greater than one, but with decreasing wavelength it becomes less than one, which changes the interaction of light with matter dramatically. The reflectivity of the surface at NI decreases rapidly and the mirrors lose efficiency starting in the UV wavelength band. However, applying Snell's law to the incident and refracted light, it turns out that the refraction angle measured from the surface normal is greater than 90° for $n_\mathrm{r} = 1 - \delta < 1$, or that total external reflection occurs for GI angles $\theta \leq \theta_\mathrm{t}$:

$$\cos\theta_\mathrm{t} = 1 - \delta \quad , \tag{9.13}$$

or for $\delta \ll 1$:

$$\theta_\mathrm{t} = \sqrt{2\,\delta} \quad . \tag{9.14}$$

For actual applications a trade-off is to be made in terms of the effective collecting area between the designs of a NI telescope or alternatively a GI telescope. The effective collecting area is the product of the wavelength-dependent reflectivity times the geometric area of the primary mirror projected on the front aperture. Depending on the number of the reflecting optical elements involved, GI telescopes tend to be more efficient for wavelengths shorter than about 30 nm. Furthermore the reflectivity at NI drops so rapidly with decreasing wavelength that for observations at wavelengths shorter than about 15 nm GI is the only choice. This limit can be extended somewhat to even shorter wavelengths by the use of multilayer coatings on the mirror but only over a fairly restricted wavelength band. Multilayer coatings of several hundreds of bi-layers, each a couple of 0.1 nm thick, can also be applied to GI mirrors, thereby extending the photon energy range out to about 100 keV.

The index of refraction or the optical constants can be computed from anomalous dispersion theory. For wavelengths λ or photon energies sufficiently offset from any electron binding energy, a coarse estimate of δ can be made:

$$\delta = \frac{r_{\rm e}}{2\,\pi}\,\frac{N_{\rm A}\,\rho}{A}\,Z\,\lambda^2 \quad , \tag{9.15}$$

where $N_{\rm A}$ is Avogadro's number, $r_{\rm e}$ is the classical electron radius, Z and A are the atomic number and weight, respectively, and ρ is the mass density. For heavy elements for which $Z/A \approx 0.5$, the incidence angle of total reflection for $\delta \ll 1$ can be estimated to:

$$\theta_{\rm t} = 5.6'\,\frac{\lambda}{0.1\ {\rm nm}}\,\sqrt{\frac{\rho}{{\rm g\,cm^{-3}}}} \quad . \tag{9.16}$$

For X-rays, with λ of a few 0.1 nm, $\theta_{\rm t}$ is about 1°. Equation 9.16 suggests the most dense materials are most suitable for reflective coatings, such as gold, platinum or iridium, which have all been used for X-ray space telescope mirrors. However, these materials show a pronounced reduction of reflectivity at energies between 2 keV and 4 keV, because of the presence of M-shell absorption, so that nickel, for instance, despite its lower density has sometimes been preferred, in particular for observations below 4 keV.

The optical constants are related to the atomic scattering factors, the most up-to-date tables of which have been compiled by the Center for X-ray Optics (Henke et al 1993).[1] These tables cover the energy range from 50 eV to 30 keV for the elements with $Z = 1$ to 92, and are a very useful data basis for designing GI optics.

GI telescope configurations

At GI imaging of an extended source or imaging over some extended field requires at least two reflections, i.e., two reflecting surfaces. Single mirrors like GI parabolas suffer from strong coma preventing true imaging. However, such a mirror can still focus and parabolas have been used as 'light buckets'.

There are three different configurations of two-mirror systems which are the Wolter-type systems, the Kirkpatrick–Baez type systems and the focussing collimator or 'lobster-eye' systems.

[1] http://henke.lbl.gov/optical_constants/

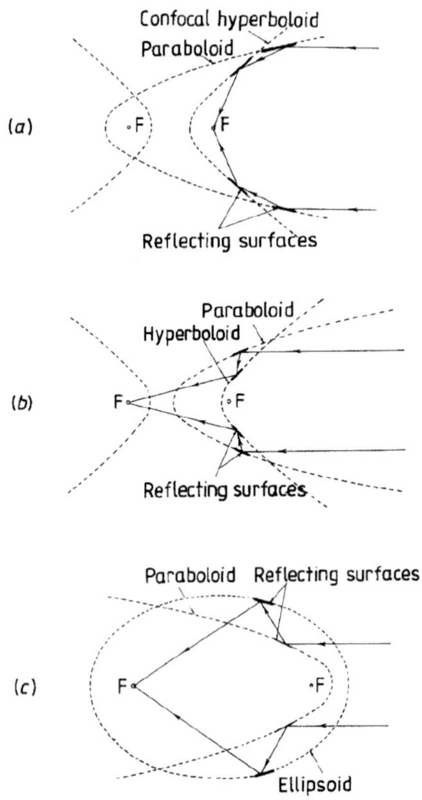

Figure 9.11: Schematic of Wolter telescopes (a) Type I, (b) Type II and (c) Type III.

Wolter telescopes

In 1952, Hans Wolter suggested three different types of imaging telescopes for GI (cf., Figure 9.11), which have become known as Wolter telescopes of Type I, Type II and Type III (Wolter 1952a). The surfaces used encompass a paraboloid, a hyperboloid and an ellipsoid. Type I and Type II make use of a paraboloid and a hyperboloid, type III combines a paraboloid and an ellipsoid. In each case the two mirrors involved are arranged in a coaxial and confocal manner. The main difference between the three types is the ratio of focal length to total system length, i.e., the minimum physical length of the telescope. The focal length of a Type I system is practically given by the distance from the paraboloid/hyperboloid intersection plane to the system focus. Therefore the physical telescope length always exceeds the focal length by the length of the paraboloid. This system has been mostly used in space observations because of its compactness, simple configuration as far as the interface to the mounting structure is concerned, and because it provides free space to easily add further telescopes inside and outside. These telescopes with multiple components are called nested systems. They increase the collecting area substantially.

Single Type I systems have been used for solar X-ray observations, whereas for astronomical EUV and X-ray observations, for which collecting area is of utmost importance, nested systems have been used, cf., the *Einstein* observatory (Van Speybroeck 1979; Giacconi et al 1979) *EXOSAT* (de Korte et al 1981), *ROSAT* (Aschenbach 1988), *ASCA* (Serlemitsos et al 1995) and *Suzaku* (Serlemitsos et al 2007), *Chandra* (Van Speybroeck 1988; Van Speybroeck et al 1997), and *XMM-Newton* (Aschenbach et al 2000), as well as the JET-X telescope of the *Swift* mission (Citterio et al 1996; Burrows et al 2005). For instance, each of the three X-ray telescopes on board of *XMM-Newton* (Aschenbach et al 2000) accommodates 58 nested paraboloid-hyperboloid Wolter Type I mirror shell pairs.

The Wolter Type II system is a true telescopic system, for which the focal length can be much longer than the physical length of the telescope. These systems are useful for feeding spectrometers which require large dispersion.

Wolter-type systems are free of spherical aberration, but still suffer from coma aberration, astigmatism and field curvature. In a second paper Wolter presented the equations for grazing telescopes which exactly obey the Abbe sine condition, which eliminates coma. This is achieved by very small corrections (sub-micrometre to 1 μm) of the axial mirror profile from its nominal second-order shape. The exact surface shape has been derived by Wolter by extending the solutions to GI that Schwarzschild had already obtained for NI in 1905 (Schwarzschild 1905). Therefore, these systems are named Wolter–Schwarzschild telescopes (Wolter 1952b). They supersede the Wolter systems in off-axis imaging performance if used at longer wavelengths, i.e., in the EUV and the soft X-ray band. Wolter–Schwarzschild Type I telescopes were flown on the *EUV-Explorer* (Bowyer and Green 1988; Bowyer and Malina 1991) and the *ROSAT*-WFC (Willingale 1988). A Wolter–Schwarzschild Type II system fed the spectroscopic telescope of the *EUV-Explorer*, and the CDS telescope on board of *SOHO* (Harrison et al 1995) is of the Wolter–Schwarzschild Type II.

The maximum degree of nesting, and therefore the highest throughput relative to the entrance aperture area, is achieved with mirrors as thin as possible. Hundreds of thin foils or sheets representing the mirrors make up the telescopes used in the *ASCA* and *Suzaku* missions. The parabolic/hyperbolic shape of the Wolter Type I mirrors is approximated by straight cones. The perfect image of an on-axis point source is lost but the imaging capability is preserved. Cone approximation of the Wolter Type I configuration has also been used for the *BeppoSAX* X-ray telescopes (Citterio et al 1986).

Kirkpatrick–Baez telescopes

The first two-dimensional X-ray image ever obtained with GI reflection was taken in the laboratory by Kirkpatrick and Baez (1948). The telescope configuration is shown in Figure 9.12. The incident rays are focused to a line image by a parabolic mirror. On their path to the line focus the rays are reflected by a second parabolic mirror to the point-like focus for rays parallel to the centre lines of the parabolas. The surface planes of the two mirrors are oriented at 90° to each other. In order to

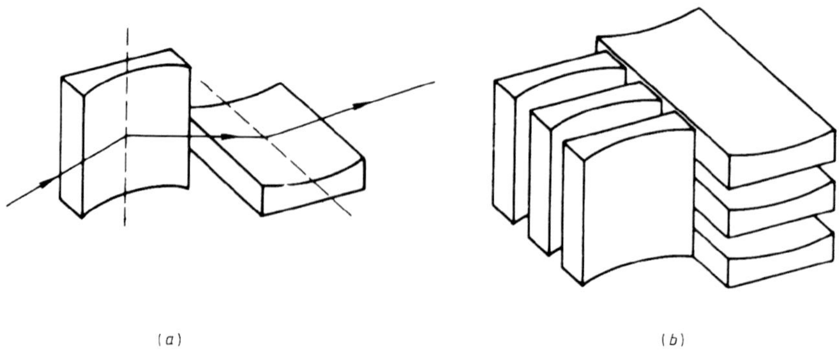

Figure 9.12: Schematic of the Kirkpatrick–Baez telescope. (a) shows a two-mirror combination, (b) displays a stack of several mirrors.

increase the collecting area (the frontal area) a stack of parabolas of translation can be constructed. However, in contrast to the single double-plate system the image of a point-like source starts to become extended, increasing in size with the number of plates involved. Wolter Type I telescopes bend the incident ray direction two times in the same plane, whereas the two bendings in Kirkpatrick–Baez systems occur in two orthogonal planes, which for the same incidence angle on the primary mirror requires a longer telescope.

Focussing collimator or 'lobster-eye' telescopes

The Wolter and the Kirkpatrick–Baez systems have in common a relatively narrow field of view which is practically limited to the grazing angle employed on the individual mirrors. Imaging systems of substantially larger field of view but at systematically reduced on-axis angular resolution have been proposed by Schmidt (1975) and by Angel (1979). Such systems would be ideal for a wide-field imaging monitor.

The principal layout of Schmidt's concept is shown in Figure 9.13. Two stacks of plane mirrors form the telescope. The upper as well as the lower stack consist of a series of plane mirrors, reflecting on both sides, in an orthogonal configuration. The mirrors within each stack are arranged in such a way that their centre lines outline a cylinder. The two cylinders are at right angles to each other, and the crossing of their centre lines is at the origin of the coordinate system. A focus is formed half way between the mirrors and the origin of the coordinate system. The focussing is not perfect because of the finite height of the mirror blades. With such a device a full hemisphere of the sky could be observed simultaneously.

A variation of this design, which provides two-dimensional imaging, has been presented by Angel (1979) (see also Priedhorsky et al 1996, and references therein). The device is composed of many small square-sided tubes with reflecting surfaces. The tubes are based and distributed over the surface of a sphere. The axis of each tube follows a radius vector of the sphere. After a ray has been reflected twice

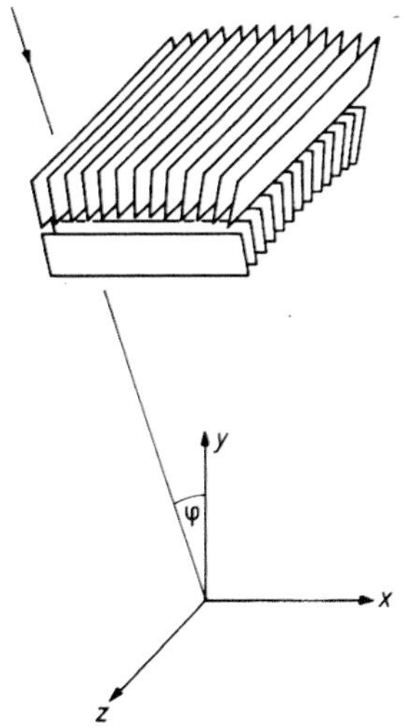

Figure 9.13: Flat-mirror two-dimensional focussing collimator or detached lobster-eye configuration (Schmidt 1975).

within one tube but from adjacent walls a two-dimensional image is formed. The focal surface is a sphere with a radius which is half that of the sphere carrying the tubes. This type of GI optic is actually realized in the reflective eyes of lobsters and shrimps, giving the name to this particular type of X-ray telescope. That the optical principle is very similar to Schmidt's focussing collimator can be seen if the upper and lower mirror stacks of this device are shifted and merged into one section forming square-sided tubes.

Both in the Schmidt and the Angel design there are rays passing through the optics with only one reflection or none at all. They appear as diffuse or line-shaped backgrounds of non-negligible brightness. Per se, the imaging is not perfect, but, ultimately, the angular resolution of such a device is limited by the width of a single tube as seen by the detector, and at second-of-arc resolution the diffraction imposed by the tube width has to be observed, favouring such a system for observing hard X-rays. Such a telescope would have great potential for continuous X-ray monitoring of large fields of the sky.

Design considerations

Parameters

The choice of a particular type of GI telescope depends on the scientific objectives. First of all, the highest photon energy has to be considered, which according to Equation 9.16 defines the optimum grazing angle. Such energies are found, for instance, in the astrophysically important Fe K-lines between 6.4 keV and 7 keV.

The *XMM-Newton* telescope was designed to optimally cover this region and to provide an acceptable area at 2 keV.

Surveying telescopes should have a low f-number, which is equivalent to maximizing the product of the effective collecting area times the field of view. This requires a grazing angle as large as possible but compatible with the average photon energy within the band to be observed. The *ROSAT* telescope was designed along these lines. Furthermore, such a survey telescope should have the best possible angular resolution over a field of view as wide as possible. Wolter Type I telescopes show strong field curvature, which means that the angular resolution increases rapidly with growing field angle (off-axis angle, Van Speybroeck and Chase 1972). The so-called *polynomial telescopes* tend to flatten the field (Burrows et al 1992; Conconi and Campana 2001). They maintain the Wolter configuration but the second-order surface shape is replaced by higher-order polynomials at the expense of degrading the on-axis resolution. Alternatively, for nested systems, the primary mirrors of a Wolter telescope should be kept as short as possible without compromising the free entrance aperture.

Effective collecting area

These trade-offs are usually done by detailed ray-tracing taking into account the reflectivity of the coating, which usually changes substantially with energy. Figure 9.14 shows the effective collecting area of one of the three *XMM-Newton* telescopes. The widest of the 58 nested Wolter Type I mirror pairs has a diameter of 70 cm. The geometric area of the entrance aperture is about 3100 cm^2 and the effective area at the lowest energies (cf., Figure 9.14) is about 1850 cm^2, equivalent to a throughput of less than 60 %. This can be increased by the use of thinner mirrors like foils or thin sheets with which a throughput of more than 80 % has been achieved. The on-axis angular resolution of such foil telescopes, however, is so far limited to about $1'$.

Angular resolution, PSF, EE function

Even for a perfect telescope the light from a point-like object located on-axis is not concentrated in an infinitely small focal spot, but is distributed over an extended image, the surface brightness of which is described by the PSF. The EE function corresponds to the radial integration of the PSF and describes the relative fraction of effective area within some radius. At large field angles the image extent and structure are dominated by geometric aberrations inherent to the telescope design. In contrast, the on-axis PSF is determined by differences between the real and perfect shape of the reflecting surfaces, both in circumferential and in axial direction, as well as by alignment and mounting errors. For high-resolution telescopes, most demanding is the control of the surface shape by appropriate metrology, through which a highly precise feedback to computer controlled grinding and polishing of the relevant surfaces down to residuals of a few 0.1 nm on either the mirrors or mandrels has become possible. Because of GI the tightest requirements have to be observed for surface errors along the mirror meridional or axial profiles. They need to be as low as possible over all spatial frequency scales which

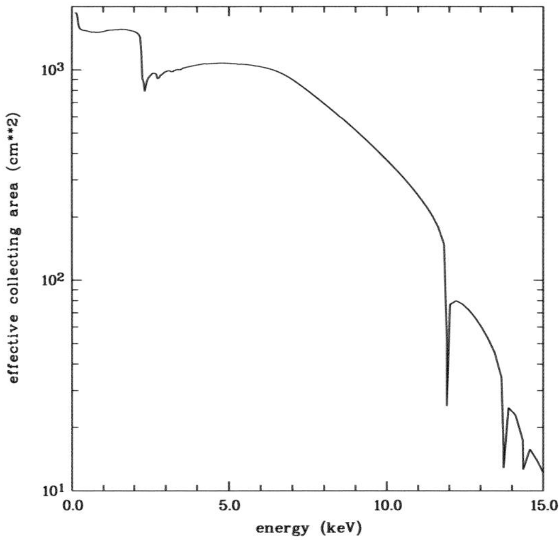

Figure 9.14: On-axis effective collecting area of one of the three *XMM-Newton* X-ray telescopes, as designed. All 58 mirror pairs have been coated with gold, which causes the energy dependence of the area. The jumps and wiggles are due to the binding energies of the N-, M- and L-shell electrons.

range from the full mirror length up to the micrometre range. The same holds for circumferential errors. Low-frequency errors can be considered as geometric slope errors and they can be treated by geometric optics. At high frequencies the errors and their power spectral density distribution can be understood as scattering from randomly rough surfaces (Aschenbach 2005). The fraction I_s/I_0 of scattered X-rays is given by:

$$I_s/I_0 = 1 - \exp\left[-(4\,\pi\,\sigma\,\ \sin\theta/\lambda)^2\right] \quad . \tag{9.17}$$

The micro-roughness σ is the RMS value of the surface height deviations. Micro-roughness reduces the image contrast, produces a loss of flux out of the image core and raises extended wings in the PSF. In the early days of GI mirrors scattering was a major problem because of the low values for σ to be achieved for a reasonable high-contrast image. For instance, for $I_s/I_0 = 0.1$ at X-ray wavelengths ($\lambda = 1$ nm) and GI ($\theta = 1°$) a micro-roughness of $\sigma =$ 0.9 nm is required. Values like 0.3 nm have become possible on highly aspherical GI mirrors and have been realised on the *ROSAT* telescope mirrors.

The shape of the PSF core is generated by geometric errors, the total of which accumulates to less than 0.1″ for the *Chandra* telescope. Figure 9.15 shows the on-axis EEs for the *Chandra* telescope at various photon energies. Energy-dependent scattering, as expected from Equation 9.17, appears to be present.

Baffling

GI telescopes have to be equipped with field stops to baffle the focal plane against stray-light. Otherwise rays from outside the nominal field of view can pass through the telescope by a single reflection from either the primary or the secondary.

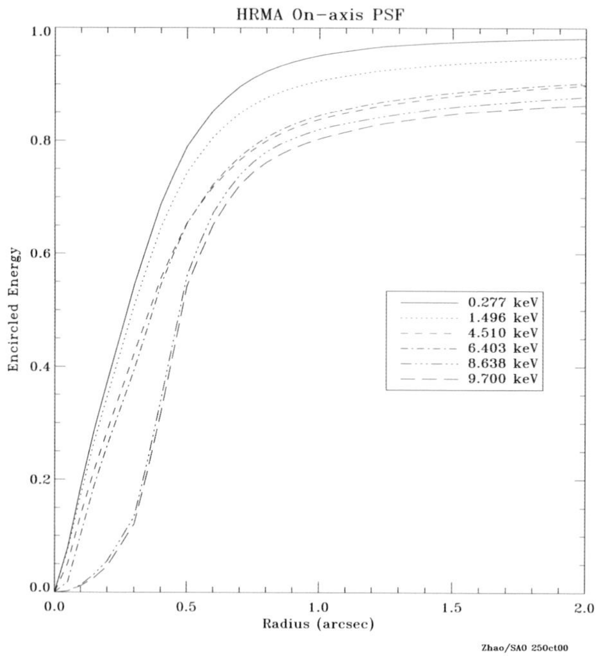

Figure 9.15: EE function of the *Chandra* telescope for different photon energies (Chandra 2003).

There may be even portions of the sky which can be viewed directly from the focal plane detector. Such stray-light increases the background light and produces ghost images. A complete suppression of stray-light requires field stops in several different planes, which are best in front of the telescope aperture and down in the telescope's mirror section. For tightly nested systems with little if any available space between adjacent mirrors a single baffle system in front of the telescope can be used, which, however, reduces but not completely eliminates stray-light.

Multilayer coatings

For wavelengths less than approximately 50 nm, the NI reflectance of a coating consisting of a single layer diminishes with decreasing wavelength as illustrated in Figure 9.6. While the reflectance can be large for grazing angles smaller than the critical angle of total internal reflection, $\theta_t \approx \sqrt{2\delta}$, where δ is a parameter related to the real part of the complex index of refraction $n = 1 - \delta - i\beta$, the values of δ and θ_t become small in the X-ray region resulting in small instrument collection angle and low throughput. Multilayer interference coatings can be utilized to greatly increase the EUV NI reflectance and the X-ray reflectance at grazing angles larger than the critical angle. When applied to solar and astrophysical telescope mirrors, high-reflectance multilayer coatings enable high-throughput and high-cadence observations of transient and faint radiation sources. In addition, the multilayer coating's reflectance profile can be optimized for narrow-waveband imaging and for broad-waveband spectroscopy applications.

The design and fabrication of multilayer interference coatings for the EUV and soft X-ray regions was pioneered by Eberhard Spiller and Troy Barbee. Spiller (1994) wrote an excellent introduction. In this section, we emphasize the utility of multilayers for solar and astrophysical telescopes and for spectrometers and discuss recent developments in the field.

A multilayer interference coating typically consists of alternating layers of two materials with high contrast in the optical constants δ and β, where $1 - \delta$ is the real part of the index of refraction and β is the absorption coefficient. One material has low absorption coefficient and functions as the spacer material, and the other material has relatively high reflectivity and functions as the scattering material. The multilayer coating performs similar to a quarter wave stack, where the electromagnetic waves reflected from the scattering layer interfaces are in phase and constructively interfere when the bilayer thickness d satisfies the Bragg relation $2\,d\,\sin\theta = m\,\lambda$ (neglecting small index of refraction corrections). Thus for NI, $\sin\theta = 1$ and in the first interference order ($m = 1$) the bilayer thickness, the sum of the spacer and scattering layer thicknesses, is equal to half the operating wavelength λ. To a first approximation, the peak of the reflectance profile occurs at the wavelength λ satisfying the Bragg relation, and the width of the reflectance profile is λ/N where N is the number of bilayers that effectively participate in the coating's reflectance and is strongly dependent on the absorption coefficients β of the spacer and scattering layers.

The multilayer coating's reflectance can be calculated using the Fresnel reflectivity coefficient at each layer interface and accounting for absorption in each layer. The effects of micro-roughness and mixing by interdiffusion of the two materials at the interfaces, which reduce the reflectance, can be modelled by a Debye-Waller factor. In order to achieve high reflectance, it is necessary that the micro-roughness and interdiffusion scale lengths be small compared to the wavelength, and a thin barrier layer can be sometimes used to reduce these scale lengths when the spacer and scattering materials do not form discrete, smooth, and stable interfaces. The accuracy of the calculated reflectance primarily depends on the accuracy of the modelling of the layer interfaces and on the accuracy of the optical constants of the layer materials. The optical constants are tabulated in Center (2007), and the IMD code (Windt 1998) is widely used to calculate multilayer reflectances in the EUV and X-ray regions.

NI EUV multilayer coatings

For NI multilayers operating in the EUV region, the spacer material is selected to have an absorption edge at a wavelength below the operating wavelength. Then the spacer material is relatively transmissive at the operating wavelength, and a large number of bilayers (N) can contribute to the reflectance. The scattering material is selected to have good material compatibility with the spacer material, resulting in discrete and stable interfaces, and to have high contrast in the optical constants, resulting in high reflectivity at the interfaces. One of the most widely used material combinations for NI EUV multilayers is Si and Mo (Barbee et al

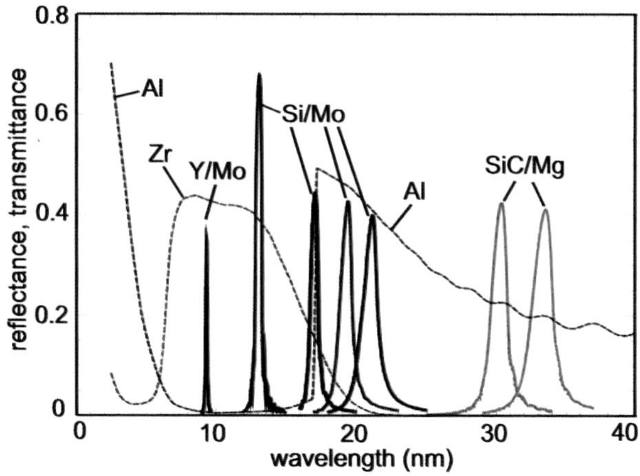

Figure 9.16: Reflectances of the seven AIA multilayer coatings and transmittances of the Al and Zr filters (Soufli et al 2005).

1985). Si is transmissive at wavelengths longer than the Si L edge at 12.42 nm, and Mo has relatively high reflectivity at those wavelengths and has good material compatibility with Si. NI telescopes with highly reflective Si/Mo coatings, or variations, such as Si/Mo_2C, were utilized on numerous rocket missions and subsequent satellite missions for recording solar images with high spatial resolution such as EIT (Delaboudinière et al 1995) and *TRACE* (Handy et al 1999). The peak of each reflectance profile was tuned to the wavelength of a strong emission line in the solar spectrum by varying the bilayer thickness d. The value of the peak reflectance and the width of the reflectance profile were optimized by varying the relative layer thicknesses of the spacer and scattering layers.

The next generation of solar imaging telescopes will implement multilayer coatings other than Si/Mo and related coatings for the purposes of narrowing the reflectance profile, increasing the peak reflectance, improving long-term stability, and imaging spectral lines outside the Si/Mo multilayer range (Windt et al 2004). Shown in Figure 9.16 are the reflectances of the seven multilayers planned for the Atmospheric Imaging Assembly (AIA) on *SDO*. AIA will implement a Y/Mo coating at 9.39 nm as well as Si/Mo coatings for four additional wavelengths in the 13.1 nm to 21.1 nm range (Soufli et al 2005). SiC/Mg coatings were planned for 30.4 nm and 33.5 nm and were replaced by SiC/Si for improved stability.

In addition to greatly increasing the sensitivity of EUV NI imaging telescopes, multilayer coatings have also been used to enhance the performance of EUV spectrometers operating at NI. The multilayer coating is applied to the diffraction grating and to the collection mirror (if utilized in the instrument). While a narrow reflectance profile with high peak reflectance is desired for imaging telescopes, providing isolation of a spectral line with a specific emission temperature, a broader reflectance profile giving wider wavelength coverage of a number of spectral lines is usually desired for spectroscopy. The first such orbital spectrometer, the Extreme-Ultraviolet Imaging Spectrometer (EIS) on the *Hinode* spacecraft, used Si/Mo mul-

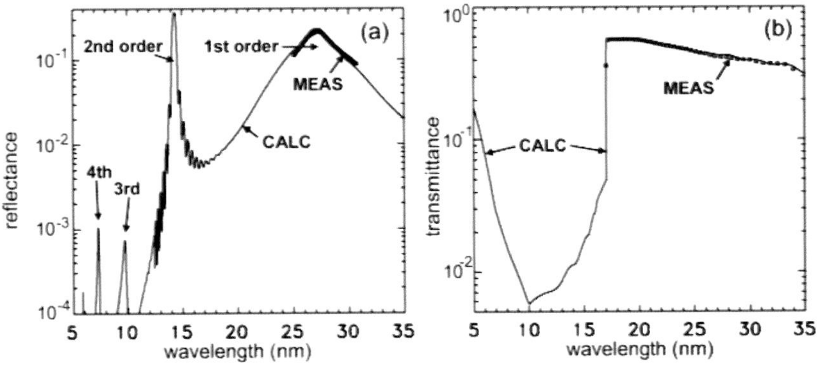

Figure 9.17: (a) Reflectance of the EIS long-band multilayer coatings and (b) transmittance of the EIS aluminum filter.

tilayers on the two halves of a parabolic mirror and a toroidal grating to cover the 17 nm to 21 nm and the 25 nm to 29 nm wavebands (Seely et al 2004).

Multilayer interference coatings can have high reflectance at wavelengths other than the wavelength of the first Bragg order, the usual operating wavelength of EUV imaging telescopes and spectrometers. The reflectance can be large in the higher Bragg orders ($m > 1$) at shorter wavelengths, depending on the optical constants of the multilayer materials. In addition, when the data are recorded by a detector with panchromatic responsivity such as a CCD, the broadband reflectance of UV and visible light from the top layer of the multilayer stack can overwhelm the weaker EUV signal. The out-of-band light is usually blocked by a thin metal filter that transmits the desired EUV radiation and attenuates the shorter wavelengths and the UV and visible light. This is illustrated in Figure 9.17 which shows the reflectance of the EIS long-band (25 nm to 29 nm) multilayer and the transmittance of the 150 nm thick aluminum filter used to block out-of-band radiation. Because of the multilayer design used to optimize the reflectance and bandpass of the coating in the long-band, the Bragg second order at 14.2 nm has higher peak reflectance than the first order at 27 nm; the second order peak is therefore blocked by the filter which has low transmittance below the Al L edge at 17 nm. The higher orders ($m > 2$) have low reflectance owing to absorption in the Si layers at wavelengths shorter than the Si L edge at 12.42 nm.

For multilayer telescopes operating at wavelengths shorter than 17 nm, such as the AIA 9.39 nm and 13.1 nm channels, a zirconium blocking filter can be used as illustrated in Figure 9.16. In general, a multilayer telescope channel operates at a wavelength within a factor of two of the filter absorption edge; otherwise radiation reflected by the multilayer's Bragg second order at half the operating wavelength will be transmitted by the filter and can contaminate the desired Bragg first-order image.

The multilayer reflectances shown in Figure 9.16 indicate that the widths of the reflectance profiles increase with wavelength. This is because fewer bilayers participate in the reflectance owing to increasing absorption with wavelength. The wider reflectance profile can result in contamination of the desired spectral line signal by neighbouring lines and effectively broadens the temperature of the spectral image. It is possible to reduce the reflectance of neighbouring spectral lines by using an anti-reflection coating on the surface of the multilayer, but this has limited effectiveness and can reduce the reflectance at the desired wavelength at the peak of the reflectance profile. Thus the commonly used multilayers are unsuitable for narrow-waveband imaging for $\lambda > 35$ nm, and this leads to the search for new multilayer materials. In particular, it is desirable to develop multilayers for narrow-band imaging of the solar O V line at 63.0 nm, which is emitted from solar regions with 2×10^5 K temperature, and thereby fills the gap between the low-temperature (8×10^4 K) He II 30.4 nm and the hotter ($> 10^6$ K) Fe multilayer telescope observations.

Si/Sc multilayers were developed for the 44 nm to 50 nm wavelength range, operating where Sc is relatively transmissive at wavelengths longer than the Sc M edge at 44 nm, and it was found that the stability was improved by adding thin (< 1 nm) W barrier layers between the Si and Sc layers (Vinogradov et al 2001). The development of the Sc based multilayers suggested that other elements with open N and O shells might have transmission windows at wavelengths > 50 nm. However, the lack of accurate optical constants hindered the effort to identify candidate materials (the values of δ in the database (Center 2007) extend only to 42 nm). A project was initiated to systematically measure the optical constants of the rare earth (lanthanide) elements that have 4f or 5d open shells (Seely et al 2006). It was found that Tb (Kjornrattanawanich et al 2006) and Gd are relatively transmissive at 63 nm and can be used for narrow-band multilayers.

GI X-ray multilayer coatings

While NI multilayer coatings have been developed for wavelengths < 5 nm, particularly for the water window range of interest for biological imaging between the O K (2.3 nm) and the C K (4.4 nm) absorption edges, the peak reflectances are typically less than 10 % and are thus far too low for solar and astrophysical instruments. This is primarily the result of interface micro-roughness and interdiffusion with scale lengths that are significant fractions of the operating wavelength and of the layer thicknesses. Mirrors with metal coatings operating at grazing angles smaller than the critical angle have long been used for imaging X-ray wavelengths < 1 nm (photon energies > 1.2 keV). However, the grazing angles decrease with wavelength and result in a small telescope acceptance angle, relatively low throughput, difficulty of alignment, and severe requirements for smooth mirror substrates and coatings.

A periodic (constant d) multilayer coating can be designed to have relatively high reflectance at grazing angles larger than the critical angle. However, the reflectance profile is quite narrow in wavelength and in angle owing to the large number of bilayers (typically hundreds or more) participating in the reflectance at short wavelengths where absorption is low. This means that different wavelengths will be reflected along the often long and curved GI mirror, reducing the effective collection area for imaging a specific wavelength. The width of the mirror's reflectance profile can in effect be broadened in wavelength and angle by varying the multilayer period d along the mirror surface. Broadening can also be achieved

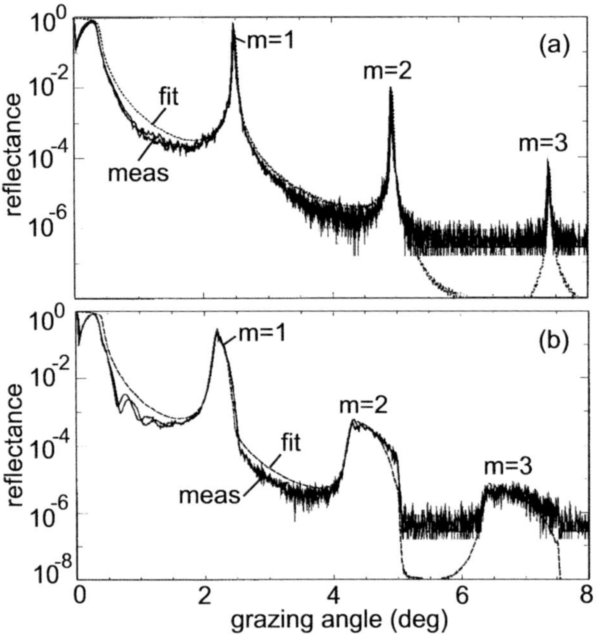

Figure 9.18: Reflectance of (a) periodic W/SiC and (b) depth-graded W/SiC multilayers at 0.154 nm wavelength (Windt et al 2003).

by systematically varying the period with depth into the multilayer, made possible by the large number of bilayers that participate in the reflectance at X-ray wavelengths. In effect, a narrow waveband is reflected by the group of bilayers with thicknesses satisfying the Bragg relation, and a broader range of wavelengths is reflected by properly varying the period laterally or in depth.

Shown in Figure 9.18a is the reflectance of a periodic (constant d) W/SiC multilayer as a function of grazing angle measured at 0.154 nm wavelength (8.05 keV Cu K radiation, Windt et al 2003). The multilayer has 300 bilayers with period 1.80 nm. Also indicated is the fit to the measured data using the IMD code and assuming a 0.23 nm interface parameter. This illustrates the narrow Bragg peaks characteristic of a periodic multilayer. For comparison, shown in Figure 9.18b is the GI reflectance of a depth-graded W/SiC multilayer composed of 1100 bilayers, with periods varying from 1.77 nm to 2.10 nm, and the fit to the data indicates an interface parameter equal to 0.26 nm. Compared to the periodic multilayer, the depth-graded multilayer has much broader reflectance orders with smaller peak reflectances.

Shown in Figure 9.19 is the reflectance as a function of photon energy of the depth-graded W/SiC multilayer measured using synchrotron radiation in the 120 keV to 180 keV energy range and at a grazing angle of 2.09 mrad (Windt et al 2003). This illustrates a broad reflectance profile achieved at a relatively large grazing angle; by comparison the critical angle of a single W layer is 0.46 mrad at 160 keV. Also indicated in Figure 9.19 is the fit to the data points using W

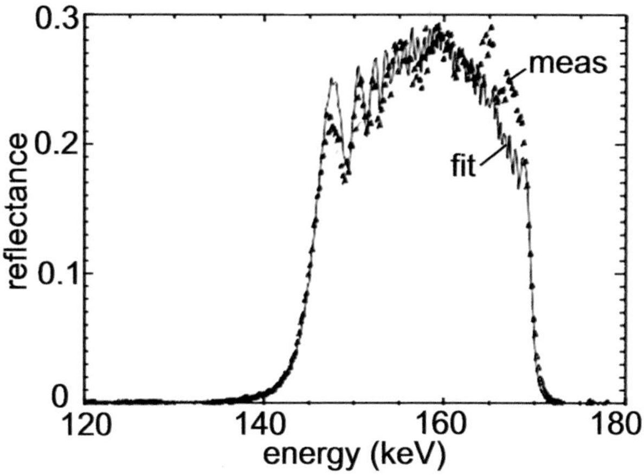

Figure 9.19: Reflectance of a depth-graded W/SiC multilayer measured at 2.09 mrad grazing angle (Windt et al 2003).

and SiC optical constants measured from the reflectance of single W and SiC layers during the multilayer development project. This work demonstrated that it is possible to reliably design and fabricate multilayers, operating at least to 160 keV with a relatively large grazing angle, that are suitable for hard X-ray astrophysical studies.

Conclusion

Space telescopes have fully taken advantage of ground-based telescope developments. Space extends the observational wavelength range to shortest and longest wavelengths which are not available to ground observations. Dedicated telescopes have been built to take advantage of this extension and a few examples have been reported in this paper. In the future, the progress of space astronomy (as in ground-based astronomy) is related to the ability to observe the finest details and the weakest objects.

Bibliography

Angel JRP (1979) Lobster eyes as X-ray telescopes. Astrophys J 233:364–373

Artzner G, Bonnet RM, Lemaire P (plus five authors) (1977) The LPSP instrument on *OSO-8*. I. Instrumentation, description of operations, laboratory calibrations, and pre-launch performance. Space Sci Instrum 3:131–161

Aschenbach B (1988) Design, construction, and performance of the *ROSAT* high-resolution X-ray mirror assembly. Appl Opt 27:1404–1413

Aschenbach B, Briel UG, Haberl F (plus five authors) (2000) Imaging performance of the *XMM-Newton* X-ray telescopes. Proc SPIE 4012:731–739

Aschenbach B (2005) Boundary between geometric and wave optical treatment of X-ray mirrors. Proc SPIE 5900:59000D-1–59000D-7

Barbee TW, Mrowka S, Hettrick MC (1985) Molybdenum-silicon multilayer mirrors for the extreme ultraviolet. Appl Opt 24:883-886

Boggess A, Carr FA, Evans DC (plus 31 authors)(1977) The *IUE* spacecraft and instrumentation. Nature 275:372–377

Bowyer S, Green J (1988) Fabrication, evaluation, and performance of machined metal grazing incidence telescopes. Appl Opt 27:1414–1422

Bowyer S, Malina RF (1991) The extreme ultraviolet explorer mission. Adv Space Res 11:205–215

Bracewell RN (1978) Detecting nonsolar planets by spinning infrared interferometer. Nature 274:780–781

Brueckner GE, Bartoe J-DF, Cook JW (plus two authors) (1986) HRTS results from *Spacelab 2*. Adv Space Res 6:263–272

Brueckner GE, Howard RA, Koomen MJ (plus 12 authors) (1995) The Large Angle Spectroscopic Coronagraph (LASCO). Sol Phys 162:357–402

Bruner EC Jr (1977) The University of Colorado *OSO-8* spectrometer experiment. I: Introduction and optical design considerations. Space Sci Instrum 3:369–387

Burrows CJ, Burg R, Giacconi R. (1992) Optimal grazing incidence optics and its application to wide-field X-ray imaging. Astrophys J 392:760–765

Burrows DN, Hill JE, Nousek JA. (plus 21 authors) (2005) The *Swift* X-ray telescope. Space Sci Rev 120:165–195

Center for X-ray optics. (http://www-cxro.lbl.gov/)

Citterio O, Conti G, Mattaini E (plus two authors) (1986) Optics for X-ray concentrators on board of the astronomy satellite *SAX*. Proc SPIE 597:102–110

Citterio O, Campano S, Conconi, P (plus 11 authors) (1996) Characteristics of the flight model optics for the JET-X telescope on board the *Spectrum-X-Gamma* satellite. Proc SPIE 2805:56–65

Conconi P, Campana S (2001) Optimization of grazing incidence mirrors and its application to surveying X-ray telescopes. Astron Astrophys 372:1088–1094

Culhane JL, Harra LK, James AM (plus 36 authors) (2007) The EUV imaging spectrometer for *Hinode*. Sol Phys 243:19-61

Delaboudinière J-P, Artzner GE, Brunaud J (plus 25 authors) (1995) EIT: Extreme-ultraviolet Imaging Telescope for the *SOHO* mission. Sol Phys 162:291–312

Gardner JP, Mather JC, Clampin M (plus 20 authors) (2006) The James Webb Space Telescope. Space Sci Rev 123:485–606

Giacconi R, Branduardi G, Briel U (plus 28 authors) (1979) The *Einstein/HEAO 2*/X-ray Observatory. Astrophys J 230:540-550

Guyon O, Pluzhnik EA, Kuchner MJ (plus two authors) (2006) Theoretical limits on extrasolar terrestrial planet detection with coronagraphs. Astrophys J Suppl Ser 167:81–99

Handy BN, Acton LW, Kankelborg CC (plus 45 authors) (1999) The transition region and coronal explorer. Sol Phys 187:229-260

Harrison RA, Sawyer EC, Carter MK (plus 36 authors) (1995) The Coronal Diagnostic Spectrometer for the Solar and Heliospheric Observatory. Sol Phys 162:233–290

Heap SR, Lindler DJ, Lanz TM (plus five authors) (2000) Space Telescope Imaging Spectrograph coronagraphic observations of β Pictoris. Astrophys J 539:435–444

Henke BL, Gullikson EM, Davis JC (1993) X-Ray interactions: Photoabsorption, scattering, transmission, and reflection at E = 50 – 30 000 eV, Z = 1 – 92. Atomic Data and Nuclear Data Tables 54(2):181–342

Jenkner H (1990) The Hubble Space Telescope before launch: A personal perspective. Rev Mod Astron 3:297–312

Kirkpatrick P, Baez AV (1948) Formation of optical images by X-rays. J Opt Soc Am 38:766–774

Kjornrattanawanich B, Windt DL, Seely JF, Uspenskii YuA (2006) SiC/Tb and Si/Tb multilayer coatings for extreme ultraviolet solar imaging. Appl Opt 45:1765–1772

de Korte PAJ, Giralt R, Coste JN (plus four authors) (1981) *EXOSAT* X-ray imaging optics. Appl Opt 20:1080–1088

Lemaire P (1991) Sensitivity changes in the CNRS ultraviolet spectrometer aboard *OSO-8*. ESA Journal 15:237–242

Lyot B (1932) Etude de la couronne solaire en dehors des éclipses. Z Astrophys 5:73–95

Priedhorsky WC, Peele AG, Nugent KA (1996) An X-ray all-sky monitor with extraordinary sensitivity. Mon Not R Astron Soc 279:733–750

Chandra Proposers' Observatory Guide, Rev.6.0, TD 403.00.006 pp. 36–38 (2003)

Rogerson JB, Spitzer L, Drake JF (plus four authors) (1973) Spectrophotometric results from the *Copernicus* satellite. I. Instrumentation and performance. Astrophys J 181:L97–L102

Saha TT, Leviton DB (1993) Theoretical and measured encircled energy and wide angle scatter of SUMER demonstration telescope mirror in FUV. Proc SPIE 1945:398–409

Schmidt WKH (1975) A proposed X-ray focusing device with wide field of view for use in X-ray astronomy. Nucl Instr Meth 127:285–292

Schroeder DJ (2000) Astronomical optics. Academic Press, San Diego, London

Schwarzschild K (1905) Untersuchungen zur geometrischen Optik II, Abh Wiss Göttingen Bd IV Nr 2

Seely JF, Brown CM, Windt DL (plus two authors) (2004) Normal-incidence efficiencies of multilayer-coated laminar gratings for the extreme-ultraviolet imaging spectrometer on the *Solar-B* mission. Appl Opt 43:1463-1471

Seely JF, Uspenskii YuA, Kjornrattanawanich B, Windt DL (2006) Coated photodiode technique for the determination of the optical constants for reactive elements: La and Tb. Proc SPIE Proc 6317:OT1–OT9

Serlemitsos PJ, Jahota L, Soong Y (plus 14 authors) (1995) The X-ray telescope on board *ASCA*. Pub Astron Soc Jap 47:105–114

Serlemitsos PJ, Soong Y, Chan K-W (plus 31 authors) (2007) The X-ray telescope on board Suzaku. Pub Astron Soc Jap 59:9–21

Shimizu T (2004) *Solar-B* solar optical telescope (SOT). ASP Conf Ser 325:3–14

Soufli R, Windt DL, Robinson JC (plus eight authors) (2005) Development and testing of EUV multilayer coatings for the atmospheric imaging assembly instrument aboard the Solar Dynamic Observatory. Proc SPIE 5901:173–183

Spiller E (1994) Soft X-ray optics. SPIE Press Book, Vol PM15

Van Speybroeck LP, Chase RC (1972) Design parameters of paraboloid-hyperboloid telescopes for X-ray astronomy. Appl Opt 11:440–445

Van Speybroeck LP (1979) *Einstein* Observatory/*HEAO-B* mirror design and performance. Proc SPIE 184:2–10

Van Speybroeck LP (1988) Grazing incidence optics for the U.S. high-resolution X-ray astronomy program. Appl Opt 27:1398–1403

Van Speybroeck LP, Jerius D, Edgar RJ (plus three authors) (1997) Performance expectation versus reality. Proc SPIE 3113:89–104

Vinogradov AV, Pershin YuP, Zubarev EN (plus six authors) (2001) Structure, thermal stability and reflectivity of Sc/Si and Sc/W/Si/W multilayer X-ray mirrors. Proc SPIE 4505:230–235

Wilhelm K, Curdt W, Marsch E (plus 13 authors) (1995) SUMER – Solar Ultraviolet Measurements of Emitted Radiation. Sol Phys 162:189–231

Willingale R (1988) *ROSAT* wide field camera mirrors. Appl Opt 27:1423–1429

Windt DL (1998) IMD–software for modeling the optical properties of multilayer films. Comput Phys 12:360–370

Windt DL, Donguy S, Hailey CJ (plus seven authors) (2003) W/SiC X-ray multilayers optimized for use above 100 keV. Appl Opt 42:2415-2421

Windt DL, Donguy S, Seely J, Kjornrattanawanich B (2004) Experimental comparison of extreme-ultraviolet multilayers for solar physics. Appl Opt 43:1835-1848

Wolter H (1952a) Spiegelsysteme streifenden Einfalls als abbildende Optiken für Röntgenstrahlen. Ann Phys (Leipzig) 445:94–114

Wolter H (1952b) Verallgemeinerte Schwarzschildsche Spiegelsysteme streifender Reflexion als Optiken für Röntgenstrahlen. Ann Phys (Leipzig) 445:286–295

— 10 —

Normal- and grazing-incidence gratings and mountings used in space

Philippe Lemaire[I]

Abstract

In most spectrometers, the grating is the component chosen to give the necessary spectral resolution either to filter the wavelength band or to obtain the line profiles. An overview of the grating properties and mountings used in space instrumentation is given.

Introduction

Optical gratings have been in use since the 19th century (Fraunhofer 1823). Most of the space spectrometers for the extreme-ultraviolet (EUV) to near-infrared (NIR) wavelength ranges use a grating as the main dispersive element. An overview of grating use is given in the following sections. To obtain more details the reader is referred to specialized optical books, e.g., Schroeder (2000) and Loewen and Popov (1997).

Most grating properties can be described by wave theory. The grating formulae and principles provide the basis from which to build a spectrometer. Several grating types can be built and the selection must be made to fulfill the objectives of the spectrometer (angular and spectral resolution, wavelength range, etc.). Several mountings using normal and grazing incidence gratings are described and some of the inputs needed to optimize the spectrometer designs are reviewed.

Grating formulae and principles

A grating is an interference system with multiple reflective strips or facets (reflection grating, Figure 10.1) or multiple slits (transmission grating). The general equation is:

$$m\lambda = \sigma \ \cos\gamma \, (\sin\beta \pm \sin\alpha) \quad , \tag{10.1}$$

where α is the incidence angle, β is the diffracted angle (in the reflection grating $\beta \geq 0$ if the diffracted and incidence rays are on the same side, reverse for the

[I]IAS—Institut d'Astrophysique Spatiale, Paris, France

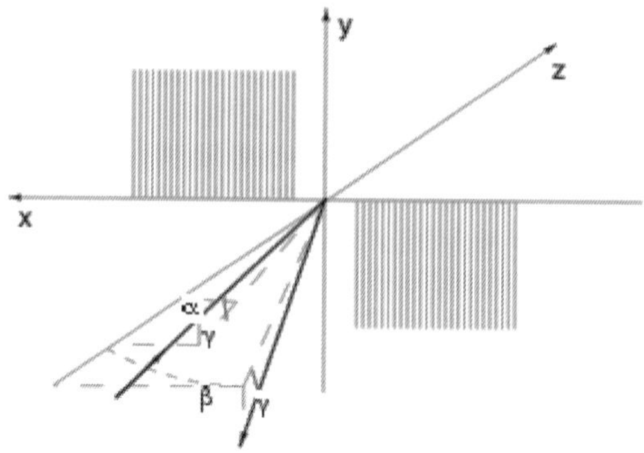

Figure 10.1: Reflective grating. α and β are the incidence and diffraction angles in the xz-plane and γ is the off-plane angle.

transmission grating), the sign is + for reflection and − for transmission, γ is the off-plane angle, σ is the distance between two grooves (or lines), m is the diffraction order and λ the wavelength. The angular dispersion A and spectral line curvature and slope are derived from Equation 10.1:

$$
\begin{aligned}
A &= \frac{\mathrm{d}\beta}{\mathrm{d}\lambda} = \frac{m}{(\sigma\cos\beta\cos\gamma)} \\
\frac{\mathrm{d}\beta}{\mathrm{d}\gamma} &= \frac{\sin\gamma\,(\sin\beta \pm \sin\alpha)}{(\cos\gamma\cos\beta)} = \lambda\, A \tan\gamma \quad .
\end{aligned}
\tag{10.2}
$$

The spectral resolving power R, with W and N the width and the total number of grooves covered by the coherent beam, is given by:

$$
R = \frac{\lambda}{\mathrm{d}\lambda} = \frac{mW}{\sigma} = mN \quad . \tag{10.3}
$$

The normalized intensity distribution i of the reflection grating (Figure 10.2) is the product of the interference function IF and the blaze function BF:

$$
i(\alpha,\beta) = IF \cdot BF = \left(\frac{\sin(N\nu')}{N\sin\nu'}\right)^2 \left(\frac{\sin\nu}{\nu}\right)^2 \tag{10.4}
$$

with $\nu' = \pi\,\sigma\,(\sin\beta + \sin\alpha)/\lambda$ and $\nu = \pi\, b\,(\sin\beta + \sin\alpha)/\lambda$, where b is the width of a groove or a facet (which can be smaller than the distance between two grooves). In a blazed grating each facet is tilted over the grating plane by an angle δ (blaze angle) to optimize the efficiency at the wavelengths given by $\lambda = 2\,\sigma\sin\delta/m$.

Grating types

There are several technical methods to create a grating by reflection or transmission on a blank (Loewen and Popov 1997). The *mechanical ruling with interferometric feedback* is used to obtain triangular or trapezoidal groove profiles either

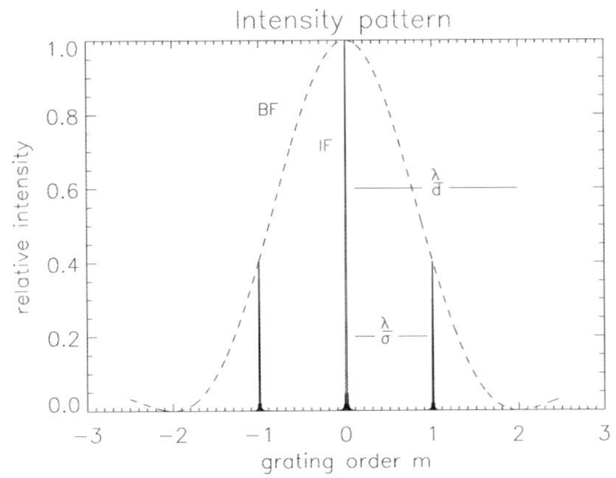

Figure 10.2: Diffracted intensity distribution of a single wavelength.

to produce low blaze angles (echelette working in orders from 1st to 4th or 5th) or high blaze angle (echelle working in orders up to several hundreds). The *holographic grating* working in low orders has near sinusoidal groove profiles with very low scattering. Generated into a photosensitive layer they can be transferred by etching into the underlying material (glass, metal, etc.) and a small blaze angle can be generated during the process. The *lithographic grating* can be generated with a mask in the same way as electronic lithography. The *electron-beam ruling* is a powerful way to draw specific patterns. Variable line spacing used to correct aberrations can be obtained by all these processes.

The *Grism* is a special type of grating operating by transmission or reflection and stuck onto a prism face. A grism can be used to minimize some aberrations in a dispersing converging beam. *Micro-Electro-Mechanical Systems* (MEMS), with two opposite electrostatic combs, interlaced and actuated to modified the distance between the comb teeth, can be used to create a grating with adjustable line spacing in a small range (Tormen et al 2006). This technique (and similar MEMS use) has so far been limited to the IR wavelengths. *Volume Phase Gratings* (VPG), which have been developed for holographic recording, act as transmission Bragg diffraction gratings at NIR wavelengths (Arns et al 1999).

Grating scattered-light performance

For optimal design of a grating the blank and the coating must be selected to provide good scattered-light performance. The grating blank should be made of a rigid and stable material that is easy to polish to $\lambda/10$ or better. BK7 glass, fused silica (such as ultra-low expansion ULE^{TM}), Zerodur^{TM}, SiC-CVD, etc. are the materials most frequently used for blanks. The coating choice must be adapted to the desired wavelength range (as for mirrors). To optimize the reflectivity in the working wavelength range and to limit the pass-band (filtering), multi-layer reflection coatings are the best choice (cf., Chapter 9, Lemaire et al 2010). The scattered radiation of a reflective grating arises from a combination of two effects:

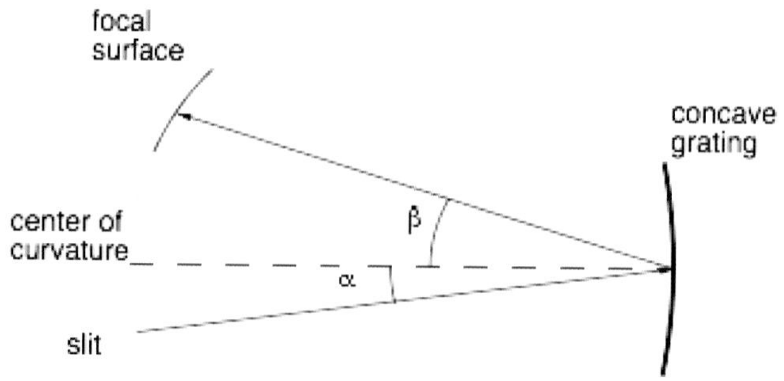

Figure 10.3: The Rowland mounting.

reflection and diffraction. The ripples and micro-ripples (roughness) of the mirror (which can be detected in the zero order) are the first source of scattered radiation. The second is due to defects in the groove shape (along the dispersion) or to irregularities along the grooves (perpendicular to dispersion).

Concave grating mountings

Several concave grating mountings can be used. Descriptions of the Rowland type with spherical or toric gratings, Wadsworth with spherical gratings, tandem-Wadsworth gratings in additive or subtractive dispersion and some other types are given in the following subsections.

Rowland type with spherical grating

In this mounting the slit and the focal surface are on the circle tangential to the grating with a diameter half the radius of the concave grating surface (Rowland circle, $D = R/2$, Figure 10.3). The main aberrations are the tangential astigmatism

$$A_{\mathrm{ST}} = y\,(\sin^2\beta + \sin^2\alpha\cos\beta/\cos\alpha) \quad , \tag{10.5}$$

and the sagittal coma

$$A_{\mathrm{CS}} = y^2(\sin\beta\tan^2\beta + \sin\alpha\tan^2\alpha)/(2\,\mathrm{R}^2) \quad , \tag{10.6}$$

where y is the groove height. The tangential image curvature is $\rho_{\mathrm{TI}} = -2/R$. The Rowland mounting can be used at very large incidence angles to increase the efficiency at XUV-EUV wavelengths (GIS on CDS/*SOHO*, Harrison et al 1995).

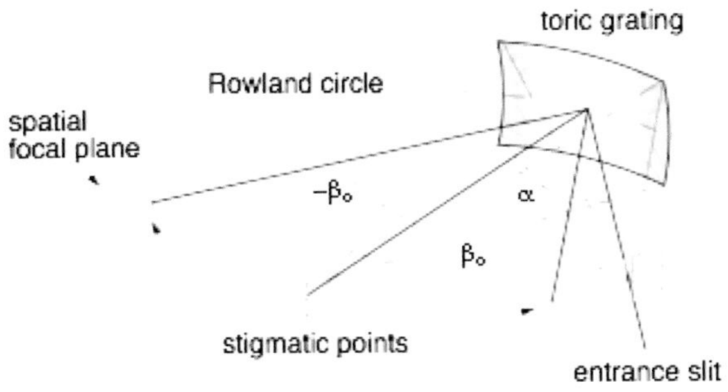

Figure 10.4: The toric grating mounting.

Rowland type with toric grating

The disadvantage of the Rowland mounting with spherical grating is the large astigmatism that limits the angular resolution along the slit length. To overcome this difficulty the use of a toric grating provides two stigmatic points on the Rowland circle for $\pm\beta_0$ diffraction angles (Figure 10.4). The relation between the two radii of curvature R_h (horizontal dispersion) and R_t (perpendicular to dispersion) is $R_\mathrm{h}/R_\mathrm{t} = \cos\alpha\cos\beta_0$. The line between the two stigmatic points is the best angular resolution while the Rowland circle provides the best spectral resolution, cf., UVCS/*SOHO* (Kohl et al 1995) and EIS/*Hinode* (Culhane et al 2007).

Wadsworth mounting

In this mounting the parallel beam strikes the concave grating at an angle α to focus along the grating normal (Figure 10.5). The focal plane distance is given by

$$d_\mathrm{F} = R\cos^2\beta/(\cos\beta + \cos\alpha) \quad . \tag{10.7}$$

The main aberrations are the tangential astigmatism $A_\mathrm{ST} = y\sin^2\beta$ and the tangential coma

$$A_\mathrm{CT} = 3x^2\sin\beta\,[\cos\alpha\,(\cos\beta + \cos\alpha)/\cos^2\beta]/(2\,\mathrm{R}^2) \quad , \tag{10.8}$$

where y and x are the height and the width of illuminated grooves, respectively. The tangential image surface curvature is given by

$$\rho_\mathrm{TI} = -(2/R)(1 + \frac{3}{2}\ \cos\alpha) \quad . \tag{10.9}$$

The Wadsworth grating used as the telescope mirror is an objective grating providing spectral images of the Sun or stars (S082A/*Skylab*, Tousey et al 1977).
The classical Wadsworth mounting uses a parabolic collimator (with a deflecting

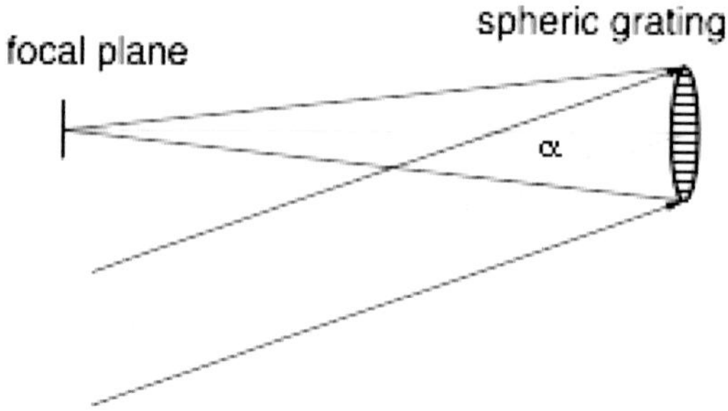

Figure 10.5: The Wadsworth mounting.

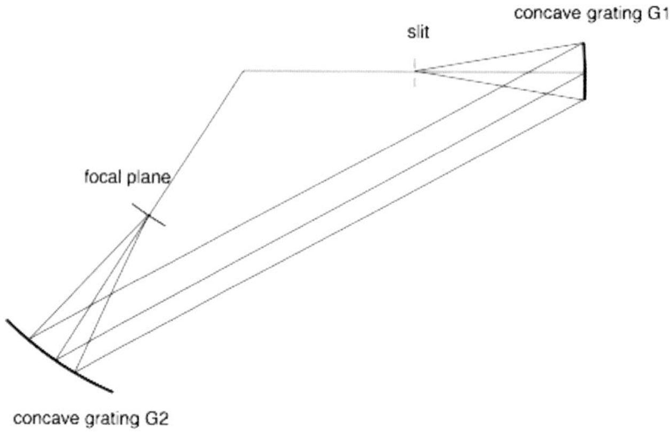

Figure 10.6: The tandem-Wadsworth mounting.

plane mirror able to modify the input angle and so to scan several wavelengths) to feed the grating (SUMER/*SOHO*, Wilhelm et al 1995). The inverse Wadsworth mounting can also be used in spectrometers.

Multi-concave grating mounting

The tandem-Wadsworth in additive dispersion with the same ruling density and focal length (Figure 10.6, HRTS/*Spacelab 2*, Brueckner et al 1986) provides a wide wavelength and angular range with near-zero astigmatism. The tandem-Wadsworth in subtractive dispersion acts as a filter to provide an image in a defined wavelength band (slit imager in HRTS/*SPACELAB2*, Figure 10.7).

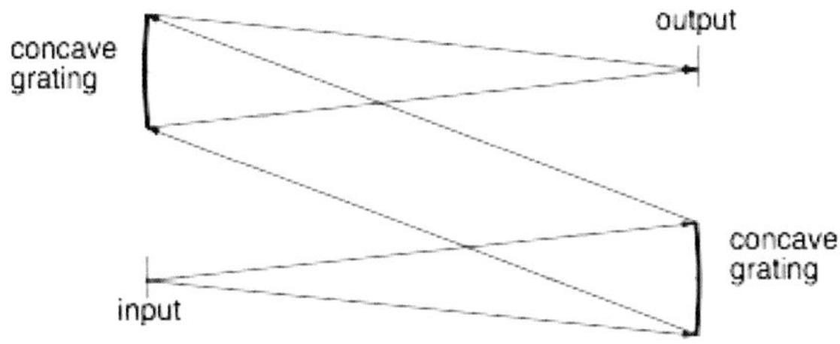

Figure 10.7: The tandem-Wadsworth mounting in subtractive dispersion.

Other concave grating mountings

To reduce the astigmatism and aberrations in the Rowland mounting, the combination of two wavelengths of an aberrated beam in the hologram registration can optimize the holographic mounting (*FUSE* spectrometers, Moos et al 2000). Another way to improve the Rowland mounting is to use the grating with variable line spacing (VLS, *Orpheus*-SPAS spectrometer, Hurwitz et al 1998).

Plane grating mountings

The mountings using a plane grating in a collimated beam require a collimator and a camera mirror to provide a spectral image of the slit in the focal plane.

Czerny–Turner and Ebert–Fastie mountings

As shown in Figure 10.8, the Czerny–Turner mounting uses either an off-axis spherical or paraboloid mirror to collimate the beam intercepted and diffracted by the plane grating. A camera (off-axis spherical or paraboloid) mirror forms the spectral images of the slit in the focal plane. A rotation of the plane grating provides the selection of the wavelength range. When the two mirrors are spherical, residual aberrations (spherical and astigmatism) limit the spectral and angular resolution. In the case of the two off-axis paraboloids the on-axis astigmatism is reduced but the coma increases (LPSP/*OSO-8* polychromator, Artzner et al 1977). To simplify the alignments and to reduce aberrations, the Ebert–Fastie mounting only uses a single mirror divided in two parts, one for the collimator and the other one for the camera mirror (UVS/LASP on *OSO-8*, Bruner 1977). The Czerny–Turner and Ebert–Fastie mountings have a poor angular resolution along the slit and are mainly used in mono- or polychromator modes.

Cross-dispersion echelle mounting

When the plane grating is an echelle with the possibility of multiple grating order superposition, a predisperser can be used to limit the wavelength range. With the observation of near point objects and the use of a two-dimensional detector

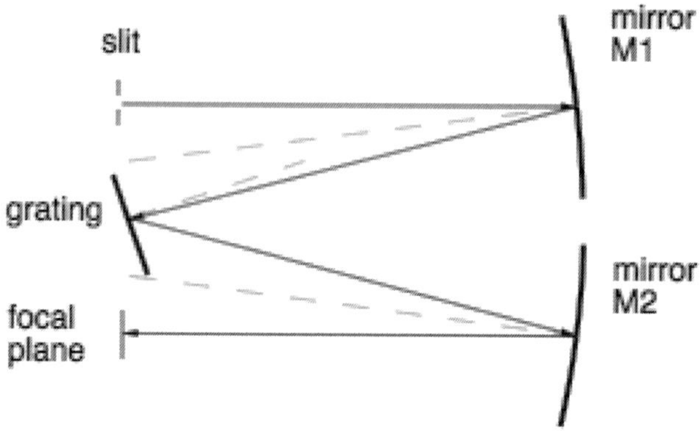

Figure 10.8: The Czerny–Turner mounting.

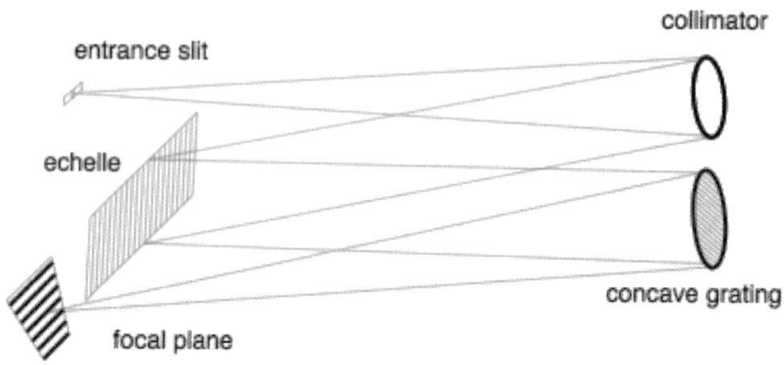

Figure 10.9: The cross-dispersion echelle mounting.

the predisperser can be included in the main spectrometer as a concave grating in Wadsworth mounting cross dispersion either in place of the camera mirror (Figure 10.9, GHRS/*HST*, Brandt et al 1994) or in place of the collimator (inverse Wadsworth, *IUE* spectrometer, Boggess et al 1977). These mountings combine the high spectral resolution and dispersion of the echelle with a large simultaneous wavelength range.

Objective plane grating

An objective plane grating located in the parallel entrance beam of a telescope diffracts the parallel beam which is collected by a concave (spherical or paraboloid) mirror to form dispersed monochromatic images of the observed objects. This scheme was successfully used with a large image anamorphic magnification in RES-C/SPIRIT/*CORONAS*, (Figure 10.10, Zhitnik et al 1992).

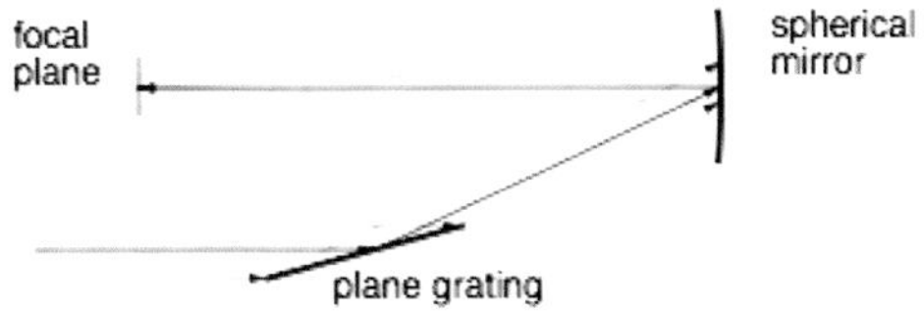

Figure 10.10: The objective plane grating mounting.

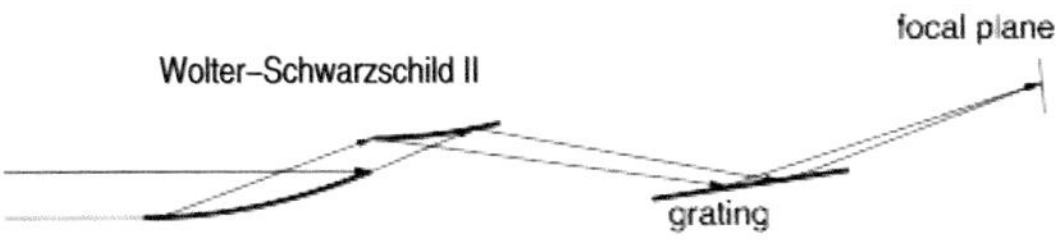

Figure 10.11: The plane grating mounting in a converging beam.

Plane grating in a converging beam

Reflective plane grating in a converging beam

To reduce the number of reflective surfaces (e.g., in EUV) a one-surface plane grating can be used in the converging beam of the telescope. A smooth variation in groove spacing which reduces the aberrations has been implemented in the grazing incidence *EUVE* telescope-spectrometer (Figure 10.11, Bowyer et al 1992).

Transmission plane grating in a converging beam

To obtain a very compact instrument configuration the transmission plane grating can be located in the converging beam before the focus of the telescope to provide multi-wavelength images of a source (e.g., *Shuttle ASTRO1-2*/IUT, Figure 10.12, Stecher et al 1997).

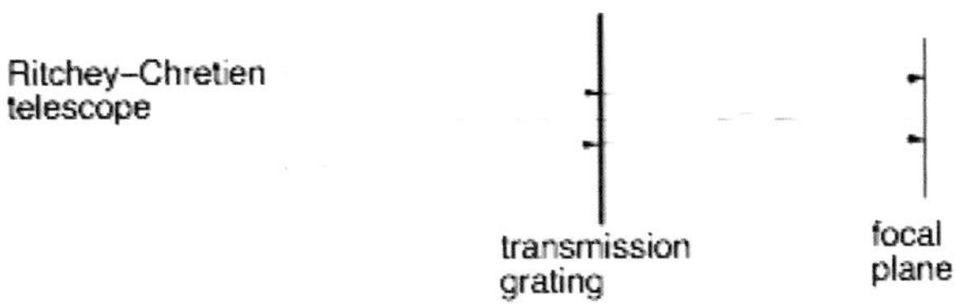

Figure 10.12: The transmission plane grating mounting in a converging beam.

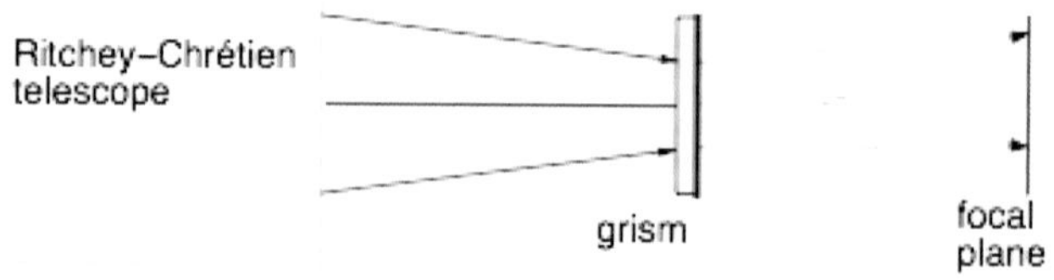

Figure 10.13: The transmission grism mounting in a converging beam.

Grism

The grism is a reflective or transmission plane grating stuck on the face of a prism (grating-prism). In a converging beam the grism improves the aberrated diffracted beam given by a simple plane grating in reducing the coma (Figure 10.13, *GALEX*, Martin et al 1999).

Other grating applications

In a Fourier transform spectrometer the grating can be used to split the normal incidence beam into +1 and −1 grating orders. By inversion the two diffracted beams can be recombined and an interferogram can be built with a variation of the optical path difference. Without any moving parts an all-reflection spatial heterodyne spectrometer, where the interferogram is registered on one dimension of the 2-D detector, provides a wide field to efficiently observe diffuse objects (Harlander et al 1992).

Bragg crystal diffraction

At very short wavelengths, between 0.1 nm and 2.5 nm, a crystal (e.g., KAP — potassium acid phtalate, beryllium, ADP — ammonium dihydrogeno phosphate, quartz, or germanium) can be used as the diffractive element. A collimated X-ray beam is diffracted following the Bragg condition

$$m\lambda = 2\,d\sin\theta \quad , \tag{10.10}$$

where λ is the wavelength, d is the crystal lattice spacing (few 0.1 nm), m is the diffraction order and θ is the glancing angle of incidence with respect to the crystal planes.
In a spectrometer the flat crystal must be rotated to scan the wavelengths, and the bent crystal spectrometer (BCS) is preferred to avoid the rotation (Figure 10.14, Rapley et al 1977). The angular resolution is provided by the collimator resolution while the resolving power can reach several thousands (BCS on *Yohkoh*, Culhane et al 1991). Since the wavelength range of each BCS is small (few 0.1 nm), several

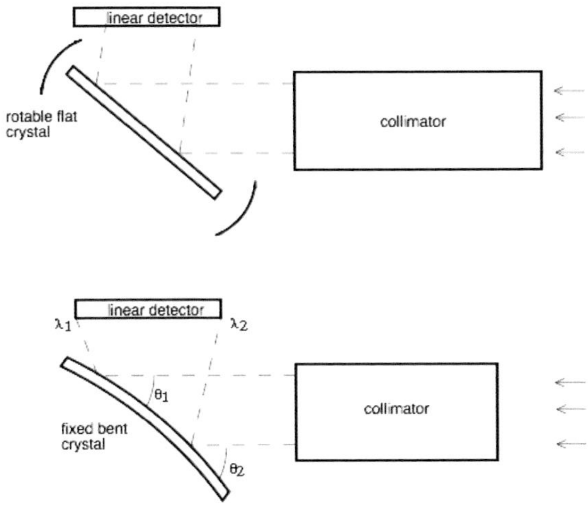

Figure 10.14: Schematic diagram of a flat scanning crystal spectrometer and a fixed bent crystal spectrometer.

BCS are mounted together and the entrance filter is chosen in order to minimize the contribution of long wavelengths.

Comments on grating mountings

The following comments recall a few aspects of the grating mountings which cannot be neglected. To provide spectral and spatial resolution the system has to minimize astigmatism and aberrations in order to be compatible with the detector pixels. The effective spectral resolution of a slit spectrometer is limited by the image of the slit (factor $d\beta/d\alpha$, where $d\alpha$ is the angular size of the slit width seen from the grating). In the best case the effective resolving power is lower than the theoretical resolving power (90 % is a good choice). The baffling inside the spectrometer is critical to avoid a high level of scattered light. With a grating the zero order (grating as a mirror) can be bright and contains the full spectrum reflected by the grating coating and a zero-order light trap is required. In some case, unused grating orders need adapted baffles.

Conclusion

The scientific objectives must provide the guidelines for the design of a grating spectrometer. A non-exhaustive list of questions to be answered and parameters needed to optimize the design is given below:

- purpose of the spectrometer (type of source to be studied, scientific objectives, constraints given by the platform)

- wavelength range to be used
- dispersion (detector pixel size) and resolution needed
- efficiency of the spectrometer
- requirements of a wavelength scan (detector size and number of spectral elements to be observed)
- imaging aberrations tolerated along the image of the slit (slit length) and along the dispersion
- respective location of the slit and the detector (constraints given by the instrument's size, the mechanical stability or the thermal control)
- mechanical dimensions and stability
- adjustment to the telescope (aperture ratio, slit size, etc.)
- choice of the design (with a plane and/or a concave grating) to fill the requirements
- stray light and spectral purity requirements (which guide the choice of the grating type and the design of the baffling system)

Thus, the overall improvement of grating systems in respect of resolution, efficiency, wavelength range and field of view is still in progress.

Bibliography

Arns JA, Colburn WS, Bardan S (1999) Volume phase gratings for spectroscopy, ultrafast laser compressors and wavelength multiplexing. Proc SPIE 3779:313–323

Artzner G, Bonnet RM, Lemaire P (plus five authors) (1977) The LPSP instrument on *OSO-8*. I. Instrumentation, description of operations, laboratory calibrations, and pre-launch performance. Space Sci Instrum 3:131–161

Boggess A, Carr FA, Evans DC (plus 30 authors) (1977) The *IUE* spacecraft and instrumentation. Nature 275:372–377

Bowyer S, Jelinsky P, Christian C, Malina RF (1992) The Extreme Ultraviolet Explorer all-sky survey. ASP Conf Ser 26:613–621

Brandt JC, Heap SR, Beaver EA (plus 22 authors) (1994) The Goddard high resolution spectrograph: Instrument, goals, and science results. PASP 106:890–908

Brueckner GE, Bartoe J-DF, Cook JW (plus two authors) (1986) HRTS results from *Spacelab 2*. Adv Space Res 6(8):263–272

Bruner EC Jr (1977) The University of Colorado *OSO-8* spectrometer experiment. I: Introduction and optical design considerations. Space Sci Instrum 3:369–387

Culhane JL, Hiei E, Doschek GA (plus 24 authors) (1991) The Bragg Crystal Spectrometer for *Solar-A*. Sol Phys 136:89–104

Culhane JL, Harra LK, James AM (plus 36 authors) (2007) The EUV imaging spectrometer for *Hinode*. Sol Phys 243:19–61

Fraunhofer J (1823) Kurzer Bericht von den Resultaten neuerer Versuche über die Gesetze des Lichtes, und die Theorie derselben, Ann Phys (Leipzig) 74:337–378

Harrison RA, Sawyer EC, Carter MK (plus 36 authors) (1995) The Coronal Diagnostic Spectrometer for the Solar and Heliospheric Observatory. Sol Phys 162:233–290

Harlander J, Reynolds RJ, Roesler FL (1992) Spatial heterodyne spectroscopy for the exploration of diffuse interstellar emission lines at far-ultraviolet wavelengths. Astrophys J 396:730–740

Hurwitz M, Bowyer S, Bristol R (plus six authors) (1998) Far-ultraviolet performance of the Berkeley spectrograph during the *ORFEUS-SPAS II* mission. Astrophys J 500:L1–L7

Kohl JL, Esser R, Gardner LD (plus 37 authors) (1995) The Ultraviolet Coronagraph Spectrometer for the Solar and Heliospheric Observatory. Sol Phys 162:313–356

Lemaire P, Aschenbach BA, Seely JF (2010) Space telescopes. ISSI SR-009:165–191

Loewen EG, Popov E (1997) Diffraction grating and applications. Marcel Dekker Inc, New-York

Martin CD, Bianchi L, Donas J (plus nine authors) (1999) The Galaxy Evolution Explorer. ASP Conf Ser 164:182–193

Moos HW, Cash WC, Cowie LL (plus 53 authors) (2000) Overview of the Far Ultraviolet Spectroscopic Explorer Mission. Astrophys J 538:L1–L6

Rapley CG, Culhane JL, Acton LW (plus three authors) (1977) Bent crystal spectrometer for solar X-ray spectroscopy. Rev Sci Instrum 48:1123–1130

Schroeder DJ (2000) Astronomical optics. Academic Press, San Diego, London

Stecher TP, Cornett RH, Greason MR (plus 16 authors) (1997) The Ultraviolet Imaging Telescope: Instrument and data characteristics. PASP 109:584–599

Tormen M, Peter Y-A, Niedermann Ph (plus three authors) (2006) Deformable MEMS grating for wide tunability and high-operating speed. Proc SPIE 6114:108–118

Tousey R, Bartoe J-DF, Brueckner GE, Purcell JD (1977) Extreme ultraviolet spectroheliograph ATM experiment S082A. Appl Opt 16:870-878

Wilhelm K, Curdt W, Marsch, E (1995) (plus 14 authors) SUMER – Solar Ultraviolet Measurements of Emitted Radiation. Sol Phys 162:189–231

Zhitnik IA, Ignatiev AP, Korneev VV (plus ten authors) (1998) Instruments for imaging XUV spectroscopy of the sun on board the *CORONAS-I* satellite. Proc SPIE 3406:1–19

— 11 —

Imaging through Compton scattering and pair creation

Volker Schönfelder[I] and Gottfried Kanbach[I]

Abstract

Compton telescopes and pair-creation telescopes are the most successful instruments used in space-based γ-ray astronomy in the energy range from ≈ 0.3 MeV to ≈ 30 GeV. The principles of measurement of both kinds of telescopes are explained and an overview of early Compton and pair telescopes is given. The properties and capabilities of COMPTEL and EGRET aboard NASA's *CGRO* are described. These two instruments have performed the first-ever all-sky survey in γ-ray astronomy above 1 MeV. The other two *CGRO* instruments OSSE and BATSE have complemented these surveys towards lower energies (for this purpose, the omnidirectionally sensitive BATSE instrument used its capability to monitor hard X-ray sources >20 keV by means of Earth occultation). Finally, the outlook for future Compton and pair creation telescopes is given.

Compton telescopes

Principle

In the 1960s and early 1970s the observation of photons in the energy range from about 1 MeV to 30 MeV was considered an impossible task in astronomy. Instruments relying on photo- and pair production are too inefficient at these energies. The only physical process with reasonable interaction probability is the Compton effect. However, it is impossible to derive energy and arrival direction from a single Compton interaction. This is only possible, if at least two Compton interactions occur.

The classical Compton telescope consists of two detector planes, the first scattering plane D1, and the second scattering plane D2 (Figure 11.1, left side). From the locations of both interactions the direction of the scattered γ-ray is determined, and from the energy losses in both detectors, E_1 and E_2, the scattering angle φ can be derived. Hence, the arrival direction of the primary γ-ray is known to lie on a circle defined by a cone with opening angle φ around the direction of the scattered γ-ray. It is best to choose a low-Z-material for D1 in order to have a high

[I]MPE—Max-Planck-Institut für extraterrestrische Physik, Garching, Germany

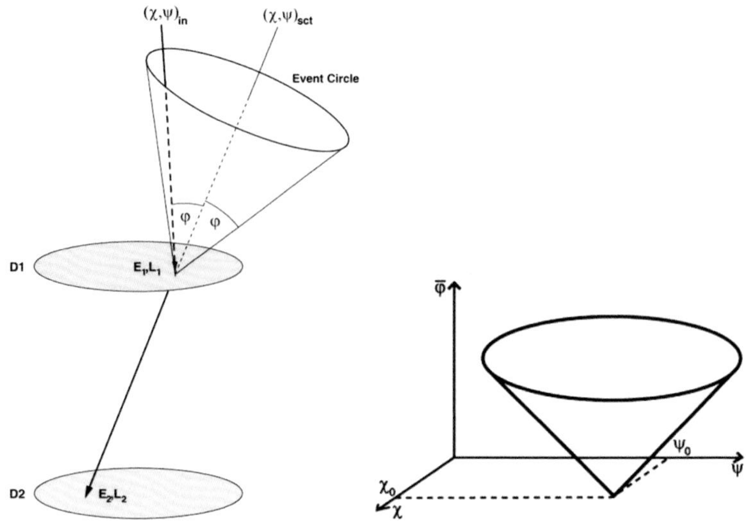

Figure 11.1: Left: principle of Compton telescope. Right: data space of Compton telescope data.

Compton scattering probability; D2 should consist of a high-Z-material to achieve a high absorption probability of the scattered γ-ray.

If the scattered γ-ray energy E_2 is totally absorbed in D2, the energy E_γ of the primary gamma-ray is determined by

$$E_\gamma = E_1 + E_2 \quad , \tag{11.1}$$

and the scattering angle φ by

$$\cos\varphi = 1 - \frac{m_e\, c_0^2}{E_2} + \frac{m_e\, c_0^2}{E_1 + E_2} \quad . \tag{11.2}$$

If the scattered γ-ray is not totally absorbed in D2, then $E_1 + E_2 < E_\gamma$ and the reconstructed scattering angle, called $\bar{\varphi}$, is larger than the true scattering angle φ:

$$\cos\varphi > 1 - \frac{m_e c_0^2}{E_2} + \frac{m_e c_0^2}{E_1 + E_2} = \cos\bar{\varphi} \quad . \tag{11.3}$$

By accepting only doubly scattered events with small scattering angles ($\bar{\varphi} < \bar{\varphi}_{\text{max}}$, where $\bar{\varphi}_{\text{max}}$ can be chosen arbitrarily), the field of view of the telescope can be modified with software.

The energy resolution of a Compton telescope depends on the accuracy of the energy measurements in D1 and D2; the angular resolution depends on two factors: first on the accuracy of measurement of the scattering locations in D1 and D2, and second on the accuracy of the determination of φ (which again depends on the accuracy of the energy measurement). As in case of the pair telescope, where the recoil energy of the nucleus limits the angular resolution, there is also a fundamental limit to the angular resolution of a Compton telescope: it is defined by the orbital energy of bound electrons in the scattering material, which leads to a Doppler broadening of the recoil energy of the Compton scattered electrons (Zoglauer and Kanbach 2003).

The point spread function of a Compton telescope can be described in a powerful way in a 3-dimensional data space defined by two orthogonal celestial coordinates (χ/ψ) describing the direction of the scattered γ-ray, and the scattering angle φ as z-axis (see Figure 11.1, right hand side). Each recorded event is represented by a single point in this data space. In the ideal case, in which the scattered γ-ray is totally absorbed in D2, the pattern of data points from a γ-ray point source with celestial coordinates (χ_0/ψ_0) lies on a cone mantle in the ($\chi/\psi/\varphi$)-space, where the cone apex is at (χ_0/ψ_0), and the cone semi-angle is 45°. The response density along the cone follows the Klein–Nishina cross-section. This idealized cone-mantle response is blurred by measurement inaccuracies in D1 and D2, especially for incompletely absorbed events in D2, which fill the interior of the cone to a certain extent.

The minimum detectable flux for a Compton telescope from a point source depends — apart from the effective detection area and the observation time — mainly on the number of background events (see Equation 11.4). In order to reduce the background, a time-of-flight measurement of the scattered γ-ray from D1 to D2 can be performed. In this way only events are accepted that first scatter in D1 and then in D2. Events that first scatter in D2 and then interact in D1 are rejected. If the material at D1 is chosen properly (e.g., organic liquid scintillator), part of the neutron-induced background can be suppressed by means of pulse shape measurements.

Early Compton telescopes

The first instrument which was described in the literature and which made use of two subsequent Compton interactions was a spectrometer for accurate energy determination of incident γ-rays, but it was not a telescope (Hofstadter and McIntyre 1950). It consisted of two scintillation detectors in backscatter arrangement. Peterson and Howard (1961) described a real double-scattering Compton telescope, which was designed to achieve directional collimation by using the kinematics of two Compton scattering processes. The directionality was achieved by postulating low energy losses in the first of two NaI(Tl) detectors, and high energy losses in the second one. No time-of-flight measurement was performed. The instrument was flown on *OSO-1* on 6 March 1962, but was not successful due to an unexpected very high background (see Greisen 1966).

In 1971, a re-birth of the Compton telescope concept took place in Garching at the Max-Planck-Institute for Extraterrestrial Physics (MPE) (Schönfelder et al 1973). The first Compton telescope at MPE was simple. It consisted of two plastic scintillator arrays separated by a distance of 1.5 m. Downward and upward scattered events could be identified and separated by means of a time-of-flight measurement of the scattered γ-ray. This telescope was operated as a directional collimator only. The first balloon flights with this telescope in 1973 and 1974 were very successful.

Practically at the same time independent work on Compton telescopes started at the University of California, Riverside (White et al 1973). The group there already had a functioning double-scattering neutron balloon experiment which used the time-of-flight technique to measure the energy of the scattered neutrons. In the first balloon flights with this instrument, γ-rays were rejected as background. But in 1972, the interest at Riverside changed from neutrons to γ-rays. From that time onwards their instrument was mainly operated as a Compton telescope.

In the years after 1973 the main activities with Compton telescopes took place at Garching and Riverside. In 1975 the MPE group started building a new balloon telescope which was almost the size of COMPTEL. It was successfully flown three times on balloons. The COMPTEL collaboration started its activities in 1977. It then took nearly 14 years to get COMPTEL into orbit.

COMPTEL

COMPTEL aboard NASA's *CGRO* was in orbit for nine years from 1991 to 2000 (Schönfelder et al 1993). It performed the first all-sky survey in the energy range from 1 MeV to 30 MeV. A photomontage of COMPTEL is shown in Figure 11.2. The upper detector D1 consists of seven modules of liquid scintillator NE 213A. Each module of 28 cm diameter and 8.5 cm thickness is viewed from the sides by eight photomultiplier tubes. The total geometrical area of D1 is 4300 cm^2. The lower detector D2, 1.5 m from D1, consists of 14 modules of NaI(Tl) scintillators. Each module of 28 cm diameter and 7.5 cm thickness is viewed from below by seven photomultipliers. The total geometrical area of D2 is 8600 cm^2. From the relative pulse heights of the photomultiplier tubes in each D1- and D2-module, the locations of the interactions are determined to within about 2 cm. The sum of the photomultiplier signals of each module provides the energy losses E_1 and E_2 (see Equations 11.1 to 11.3). Both D1 and D2 are completely surrounded by veto-domes of 1.5 cm thick plastic scintillator to reject charged particles.

COMPTEL covers the energy range from 0.8 MeV to 30 MeV. Within its large field of view of about 1 sr, its angular resolution is 1.7° to 4.4° FWHM (depending on energy). The relative energy resolution is in the range 5 % to 8 % FWHM, and the effective detection areas are in the range 20 cm^2 to 30 cm^2 (again depending on energy). The sensitivity of COMPTEL is significantly determined by the instrumental background. A substantial suppression is achieved by the combination of the charge-particle veto domes, the time-of-flight measurement technique, pulse-shape discrimination in D1, Earth-horizon angle cuts, and proper event selections in energy- and φ-space.

The analysis of the remaining background rate has shown a direct correlation with the charged particle veto rate, a dependency on the orbital cut-off rigidity and on the orbit altitude, and an increase with mission-time (due to a build-up of radioactive nuclei). Therefore, it has to be concluded that only a small fraction of the background rate is due to diffuse cosmic γ-radiation; most of it is produced in the telescope itself or in the surrounding material next to COMPTEL.

Figure 11.3 illustrates how such processes can simulate double-scatter events in COMPTEL. The first type of instrumental background events are caused by secondary γ-rays which are produced within the COMPTEL field of view either

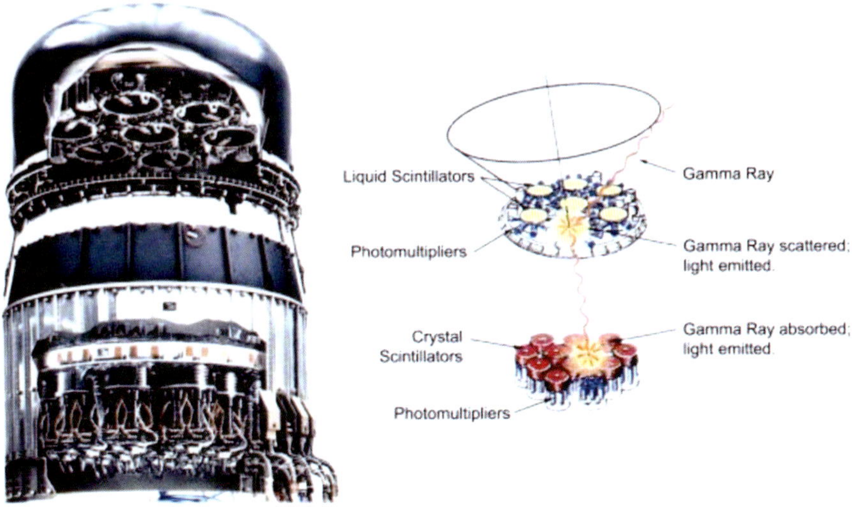

Figure 11.2: Schematic view of COMPTEL.

inside D1 (type A events) or around D1 (type B events). If the production material is inside the veto domes, prompt secondary γ-ray production can only occur via secondary neutrons produced somewhere within the 15 t heavy spacecraft. If the material is outside COMPTEL's veto domes (e.g., in the neighbouring telescopes OSSE or EGRET), then secondary γ-rays can be produced by cosmic-ray interactions or again by secondary neutrons. The second important category of background events are cascade events. These are produced by nuclear interactions which lead to the simultaneous emission of at least two γ-rays: one of them hits D1, and the other one D2 (type C- and type D-events). A significant fraction of these are suppressed by time-of-flight measurements. Finally, cosmic-ray showers in the neighbouring experiments and materials can produce a number of γ-rays which are spatially uncorrelated, but time correlated (type F-events). Again, time-of-flight measurements can suppress significant fractions of these.

The actual total in-flight instrumental background was higher than expected prior to launch by about a factor of four. As a consequence, the sensitivity of COMPTEL was two times lower than expected. Fortunately, however, this loss in sensitivity could be compensated by the four times longer mission lifetime (which originally was planned to be 2.25 years only). The actually achieved point-source sensitivity for a deep observation ($t = 6 \times 10^6$ s) was 6.3×10^{-5} cm^{-2}s^{-1} for continuum emission between 1 MeV and 30 MeV (corresponding to about 3 % of the Crab-flux) and 1.6×10^{-5} cm^{-2}s^{-1} for line emission at 1.157 MeV and 1.809 MeV (^{44}Ti and ^{26}Al lines).

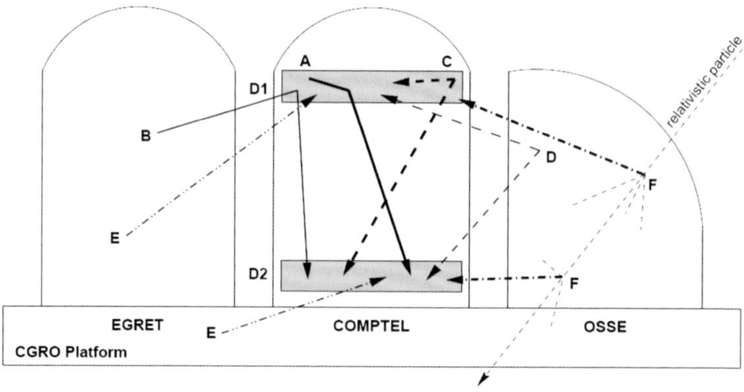

Figure 11.3: Illustration of masking background events in COMPTEL.

Plans for future Compton telescopes

Large efforts are presently being undertaken worldwide to develop the next generation of Compton telescopes. COMPTEL was the first successful Compton telescope put into space. It opened the 1 MeV to 30 MeV range as a new window to astronomy. From COMPTEL we have learned that the sky is rich in phenomena and objects that can be studied around 1 MeV. But it is also true that with COMPTEL we could see only the tip of the iceberg. The sensitivity achieved was still modest. The next telescope would have to be more sensitive by one or two orders of magnitude. Very different concepts of Compton telescopes are presently being studied, investigated and tested. Instead of scintillators other detector materials are also considered and tested, like silicon strip detectors, position-sensitive germanium detectors, Cd-Te detectors, liquid xenon gas detectors, and high-pressure gas detectors. In order to achieve the required improvement in sensitivity, the detection efficiency has to be drastically increased and the background has to be reduced considerably. An increase of the efficiency by a factor of 5 to 10 is possible, if large solid angles (e.g., 2π sr) for the scattered γ-ray are allowed (in the case of COMPTEL the angle was only 0.5 sr). Such telescopes would be necessarily very compact, and it would then no longer be possible to perform a time-of-flight measurement. In these telescopes a rejection of upwards scattered γ-rays could be achieved by the so-called *direction-of-motion* parameter which determines the sequence of interactions from consistency checks of the Compton kinematics.

A very significant background reduction can be achieved by localising the arrival direction of the primary γ-ray on the event circle. This is possible, if the track of the scattered Compton electron is measured as well (like in the proto-type telescope MEGA; Kanbach et al 2003).

To improve the detection sensitivity of point sources, it is most effective to choose detector materials which provide the best possible energy resolution (to minimize the angular resolution element of the telescope). The choice of the satellite orbit, especially its inclination and altitude, has another huge effect on the instrumental background. At least some if not all of these aspects are taken into account in the new investigations and developments mentioned above.

Pair-creation telescopes

The direct conversion of energy into matter was first observed when energetic photons interacted in a bubble chamber and showed, on the basis of magnetic deflection, the creation of a pair of oppositely charged particles: a negative electron and its anti-particle, the positively charged positron. This conversion becomes possible in the field of a nucleus or electron above a photon threshold energy of $E_{\gamma,\min} = 2\, m_{\mathrm{e}}\, c_0^2\, (1 + m_{\mathrm{e}}/M)$, where m_{e} is the electron rest mass, c_0 the speed of light, and M is the target particle's mass, i.e., above $\approx$ 1.022 MeV for conversion on a heavy nucleus and above 2.044 MeV for conversion on an electron. After the conversion the pair electrons essentially carry the remaining photon energy as kinetic energy and their summed momentum vectors correspond to the photon momentum minus the recoil momentum of the conversion target nucleus. This recoil momentum is very hard to determine in practice. Its typical value of 0.5 MeV/c_0 has to be taken as one of the physical limitations to the angular resolution of a pair-creation telescope. The second limit derives from multiple small-angle scattering of the propagating pair electrons, which quickly 'lose' their original direction, especially if they are of low energy and have to pass dense detector materials. The original momenta are of course needed to reconstruct the direction of the primary photon, and it is clear that this will be possible with much better precision for high-energy electrons. A third and more technical aspect is the positional accuracy of the tracking detector. The first two effects are most noticeable at energies below $\approx$ 50 MeV, at which energy the recoil uncertainty would amount already to an angular dispersion of $\approx 2.4°$. The tracker resolution eventually limits the angular accuracy at high energies.

Basic design of pair-creation telescopes

The sequence of detection of high-energy photons in a satellite telescope begins with the conversion of photons into charged particles in a high-Z target material. A high-Z material is preferred since the cross section for pair-creation (the 'stopping power') varies as Z^2. The kinetic energy and momenta of all secondary particles (the (e^+/e^-) pair and a possibly formed electromagnetic shower) have then to be measured with the best available precision. Finally the detector system must be able to reject background radiation like cosmic rays with very high efficiency. It is evident that the first two requirements pose a conflict: a massive conversion detector, which is good for detection efficiency, would lead to severe small-angle scattering of the pair particles and consequently to degraded angular resolution. Amelioration of that conflict leads to a principle of construction where the conversion material is arranged in thin layers, which are interspersed with tracking detectors. The trajectories of the newly created pair particles can then be determined as soon as they emerge from the conversion layer. Energy measurements of the pair particles can be performed in two ways: either the observed small-angle Coulomb scattering

(Molière scattering), which is approximately inversely proportional to the kinetic energy of the particle, can be used at energies up to ≈ 1 GeV; or a massive calorimeter made of scintillators is used to absorb the electromagnetic cascade and measure the energy up to high values. Measures to suppress background events caused by energetic particles are (i) to surround the sensitive detector volume by a plastic scintillator anticoincidence detector and (ii) ensure the sequence of the events by, e.g., measuring the direction of flight of the pair particles by a time-delayed coincidence system or a Čerenkov counter. All the high-energy satellite telescopes reviewed in the following sections are built according to these principles: the core is a layered tracking detector for the conversion of photons and the imaging of secondary electrons. Integrated with the tracking chamber is a trigger telescope that detects the occurrence of a pair-conversion event and initiates the read-out of the detector system. Below the tracker a calorimeter, often made of scintillators and photo-detectors, is used to measure the total energy of particles released in the high-energy interaction. A thin plastic anticoincidence counter surrounds the front and sides of the detector to shield against charged particle background.

The sensitivity for an imaging telescope for point sources (also called 'unresolved sources') is determined by the point-spread function (PSF) and the amount of background that is present under the image of the point source. Let N_s be the number of source photons collected within the PSF of angular size θ. The source with flux F was observed during the time T with a telescope of effective area A_eff so that $N_\mathrm{s} = F\,T\,A_\mathrm{eff}$. The density and rate of background photons on the image is assumed to be n_b so that the total number of background photons within the source image is $N_\mathrm{b} = T\,\pi\,\theta^2\,n_\mathrm{b}$. The statistical significance of an excess of N_s counts above the background can then be written as $n_\sigma = N_\mathrm{s}/\sqrt{N_\mathrm{b}}$. The flux required for an n_σ-significant detection of a source, which is also called the sensitivity limit, is then

$$F_\sigma = \frac{n_\sigma\,\theta}{A_\mathrm{eff}}\sqrt{\frac{\pi\,n_\mathrm{b}}{T}} \quad . \tag{11.4}$$

As we see the sensitivity improves, i.e., a smaller flux F_σ is detectable, when the PSF becomes small, the density of background counts is smaller, the effective area is enlarged, and the observation is performed for a longer time (although only the square-root of time enters). The main characteristics of past and present pair-creation satellite telescopes are listed in Table 11.1.

Early pair-creation telescopes *SAS-2* and *COS-B*

SAS-2 was a NASA mission that carried as a single payload a spark chamber pair-creation telescope (Derdeyn et al 1972). After a series of non-imaging detectors (e.g., *Explorer-11*, Kraushaar et al 1965; and *OSO-3*, 1968; Clark et al 1968) that also detected photons above several 10 MeV via pair creation in stacks of scintillators, the *SAS-2* instrument was the first truly imaging pair-creation telescope. Launched on 15 November 1972 the telescope operated from 19 November 1972 until 8 June 1973, when a technical defect in the power supplies terminated the mission. About 55 % of the sky was surveyed in 27 week-long exposures. The equatorial low-altitude orbit of *SAS-2* ensured a low-background environment and good

Table 11.1: Characteristics of past and present imaging pair-creation telescopes.

Mission	Effective area, $A_{\mathrm{eff}}/\mathrm{cm}^2$ at ≈ 200 MeV	Angular resolution	Field of view	Energy range, E_γ/MeV	Sky coverage
SAS-2 (Nov 72 – Jun 73)	90	$\approx 1°$	0.4 sr	> 35	55 %
COS-B (Aug 75 – Apr 82	45	$\approx 2°$	0.4 sr	> 50	60 %
EGRET/*CGRO* (Apr 91 – Jun 00)	1300	$\approx 1.5°$	0.6 sr	20 to 30 000	100 %
AGILE (launch Apr. 2007)	600	$\approx 1.2°$	≈ 3 sr	30 to 50 000	plan: 100 %
Fermi (GLAST) (launch June 2008)	9000 at 10 GeV	$\approx 3.5°$ at 100 MeV $\approx 0.15°$ at 10 GeV	2.4 sr	20 to 300 000	plan: 100 %

sensitivity although only ≈ 8000 celestial photons were recorded. The structured galactic disk, several point sources including two pulsars, and the extragalactic background were the main results of the *SAS-2* mission (Fichtel et al 1975).

The first ESA scientific mission launched in 1975 was the γ-ray telescope *COS-B* (Bignami et al 1975). Built and operated by a consortium of six European research groups in four countries, *COS-B* was one of the first truly European science missions. Although nearly of the same size as *SAS-2, COS-B* featured an important additional detector in the γ-ray telescope: a calorimeter made of a single CsI(Tl) scintillator. Its depth of 4.7 radiation lengths (r.l., equivalent to a depth of 8.5 cm) allowed the effective measurement of photon energies up to several gigaelectronvolts with a relative energy resolution (FWHM) between 50 % and 100 %. *COS-B* operated until 1982 and deep observations of the galactic disk and of selected extragalactic targets revealed a wealth of high-energy astrophysical results: details of the diffuse galactic γ-ray emission (Mayer-Hasselwander et al 1982), more than 20 point sources and the first extragalactic source, the blazar 3C273 (Swanenburg et al 1981).

EGRET

CGRO carried as its high-energy telescope the 'Energetic Gamma-Ray Experiment Telescope' (EGRET, Kanbach et al 1988). As successor of *SAS-2* and *COS-B*, EGRET was built by a US-German consortium and combined successful features of the small instruments, such as a large stack of digital gas-filled wire spark chambers with interleaved tantalum conversion layers, a fast directional trigger telescope, a NaI(Tl)-PMT (photomultiplier tube) calorimeter of 8 r.l. thickness, and a large anticoincidence detector to veto charged particles. EGRET had a relative energy resolution of ≈ 25 % FWHM and an angular resolution which improved from $\approx 10°$ at 60 MeV to $\approx 0.5°$ at 10 GeV in a large 0.6 sr field of view. EGRET's on-axis effective area of $\approx$ 1300 cm^2 at several hundred megaelectronvolts afforded a sensitivity limit for γ-rays above 100 MeV of $\approx 10^{-7}$ cm^{-2} s^{-1} after a typical two-week-long observation. This was about an order of magnitude better than the sensitivities obtained with *SAS-2* and *COS-B*. *CGRO* was operated until June 2000, although the limited reservoir of spark-chamber gas for EGRET had curtailed its observations to high-priority targets since 1996. In 2000 *CGRO* was deliberately de-orbited because of concerns for a failing attitude system, and crashed into the Pacific Ocean.

All instruments on *CGRO* and especially the 'second generation' telescope EGRET established the role of γ-ray astronomy as an important branch of astronomy and astrophysics. EGRET discovered 271 celestial γ-ray sources covering a variety of objects, ranging from the Sun, to isolated spin-down pulsars, binaries with neutron stars or black holes, supernova remnants, quasars, and γ-ray burst sources. About two thirds of the EGRET sources, however, are still unidentified, which is a fundamental challenge to the next generation of instruments. Figure 11.4 shows scaled schematic views of *SAS-2, COS-B*, and EGRET.

AGILE and *Fermi (GLAST)*

First in the next generation of pair-creation telescopes, which are entirely based on solid-state detector technology, is the small Italian mission *AGILE*, which was launched from India on 23 April 2007 into a nearly equatorial circular orbit of $\approx 2.5°$ inclination and an altitude of 540 km. Taking into account the very low background in the *AGILE* orbit, the expected responsivity to an extragalactic photon source with 5 σ significance in 10^6 s of observation time will be $\approx 10^{-8}$ cm^{-2} s^{-1} (at energies $>$ 100 MeV), which is already a considerable improvement over the EGRET sensitivity.

The *Fermi* mission, formerly known as *GLAST*, was designed to continue the successful observations of the EGRET telescope, albeit with much improved sensitivity. The main instrument on *Fermi* is the Large Area Telescope (LAT). The upper part of *Fermi*-LAT consists of 16 (arranged as 4×4) towers of pair-conversion tracking chambers, each made of 19 stacked silicon-strip detector pairs ($x - y$ resolution) interleaved with lead converter sheets. Each tower measures (33×33) cm^2, giving a total effective detector area of $\approx$ 9000 cm^2 above 1 GeV. The calorimeter, which is located under the tracker, consists of 80 CsI(Tl) scintillator bars per stack viewed by PIN photodiodes. The bars are approximately $(2.3 \times 3 \times 31)$ cm^3 in size and are stacked in layers with alternating perpendicular directions to provide positional information about the shower. LAT is covered over its sensitive field of view with an anticoincidence charged-particle veto shield made of segmented plastic scintillator tiles. The segmentation will reduce the self-veto effect at higher energies resulting from backsplash of shower particles from the calorimeter. Since its launch into a slightly inclined ($\approx$26°) low Earth orbit on 11 June 2008, *Fermi* has been performing an all-sky survey by scanning the celestial sphere continuously with the axis pointed to zenith throughout the orbit.

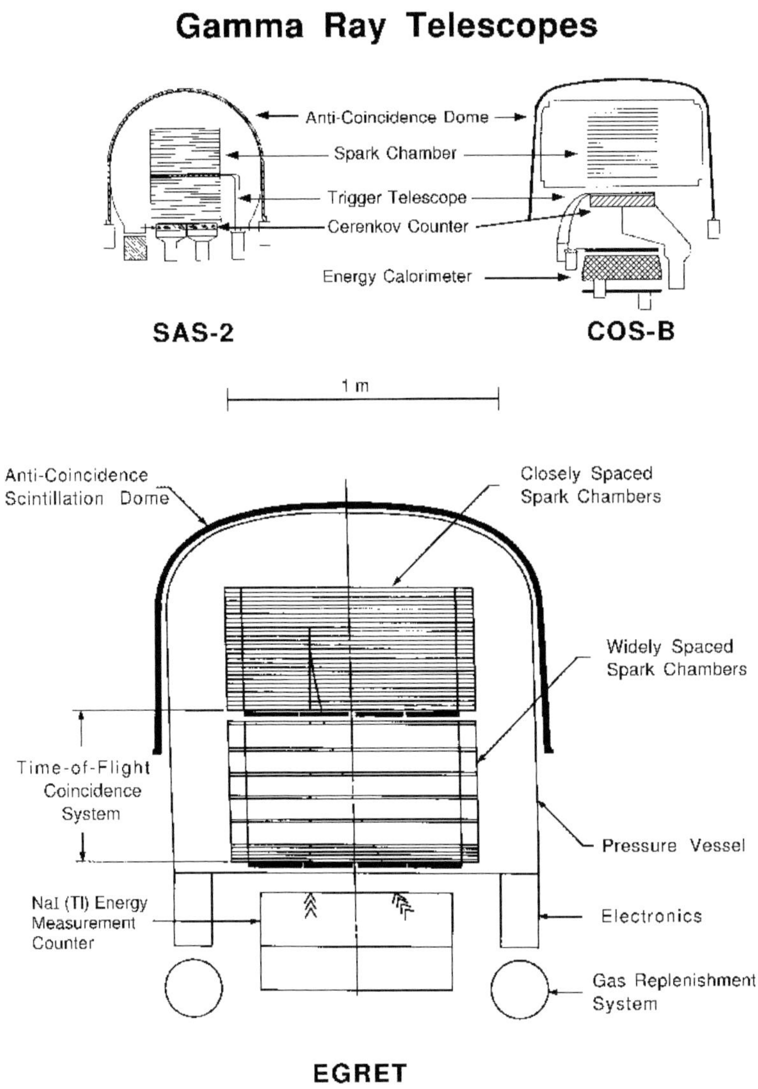

Figure 11.4: Comparison and schematic layout of the pair-creation telescopes *SAS-2*, *COS-B*, and EGRET (to scale).

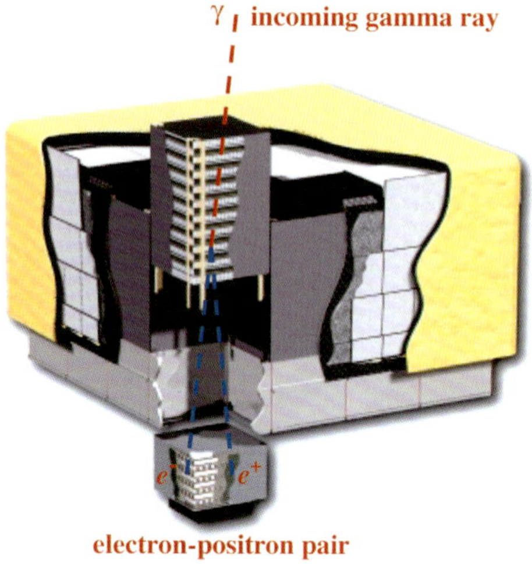

Figure 11.5: Schematic of *Fermi*-LAT. The modular structure of 16 tracker-calorimeter towers is indicated and one tower is highlighted. Photon pair creation occurs in the upper Si-tracker chamber and the energy of the resulting particle shower is measured in the lower calorimeter detector.

Fermi-LAT detects photons from 20 MeV to 300 GeV and can reach point-source sensitivities for photons above 100 MeV of better than 4×10^{-9} cm^{-2} s^{-1} for an observation time of one year in scanning mode.

A schematic view of one of the 16 *Fermi*-LAT towers is displayed in Figure 11.5. *AGILE* may be considered as a mini-version of such a LAT tower.

The Great Observatory *CGRO*

The *CGRO* was one of NASA's four "Great Space Observatories". It was in orbit from April 1991 to June 2000. The 15 t payload was put into an orbit of 450 km altitude by the *space shuttle Atlantis*. Apart from EGRET and COMPTEL the spacecraft platform carried two more instruments, namely OSSE and BATSE. OSSE was a collimated scintillation spectrometer of NaI(Tl)-CsI(Na) phoswich detectors with a field of view of $4° \times 11°$, which mainly covered energies near the transition between X-ray and γ-ray astronomy from 50 keV to about 1 MeV. BATSE was an omni-directionally sensitive Burst And Transient Source Experiment (20 keV to 1.8 MeV), which consisted of eight large-area NaI(Tl)-scintillation detector assemblies, one at each corner of the spacecraft. By comparing the count rates at each of these detectors, the locations of cosmic γ-ray bursters in the sky could be determined to an accuracy of a few degrees.

The four *CGRO*-instruments together covered the energy range 20 keV to 30 GeV, about six orders of magnitude in photon energy. A schematic view of *CGRO* is shown on the left side of Figure 11.6 together with a photograph of the observatory in orbit just prior to release from the shuttle on the right side.

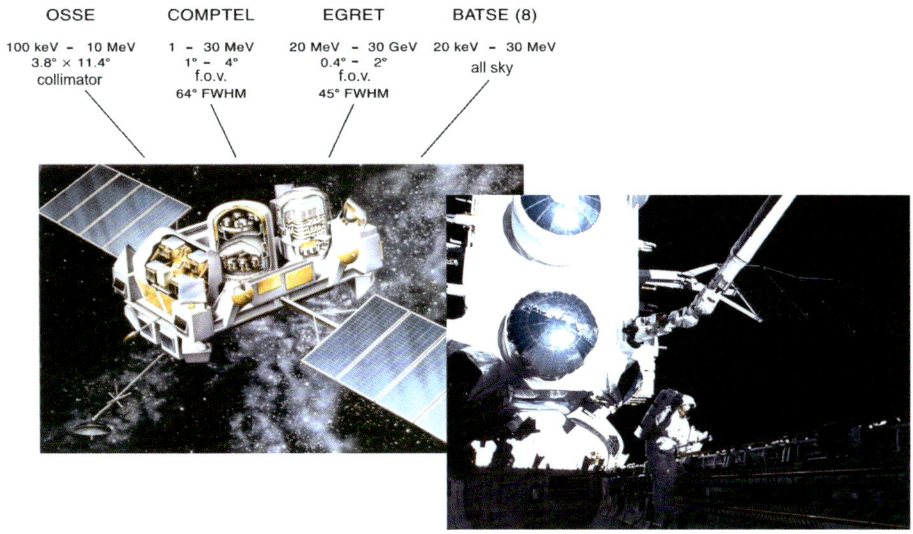

Figure 11.6: Left: schematic view of *CGRO*, right: photograph of *CGRO* in orbit aboard the *shuttle Atlantis*.

The main goal of *CGRO* was to perform the first ever all-sky survey in γ-ray astronomy, and to perform subsequent deep observations of individual sources. Due to the large fields of view of EGRET and COMPTEL (co-aligned in their viewing directions), it took only 1.5 a to perform the full-sky survey. OSSE's field of view was too small for a full-sky survey, but at the end of the mission the instrument was still able to produce an image of the inner radian of the Galaxy by combining all source observations in that region. Within the mission lifetime of *CGRO*, BATSE detected nearly 3000 γ-ray bursts, which were distributed isotropically over the sky. In addition to these burst observations BATSE has been serving the entire high-energy astrophysics community with an unprecedented, nearly continuous, omni-directionally sensitive monitor of celestial sources above 20 keV by means of the Earth-occultation technique. This technique consists of measuring the size of step-like features in the γ-ray count-rate of the BATSE detectors, when a point source in the sky alternatively rises above or sets below the Earth limb (Harmon et al 1992). Measuring the size of the step gives nearly continuous sampling of source intensity and spectrum as a function of time, typically 15 to 30 times a day. The one-day sensitivity of this technique was typically 0.1 of the intensity of the Crab in that energy range.

The most important parameters of any astronomical telescope are its angular resolution and its sensitivity. Figure 11.7 summarizes the angular resolutions of the *CGRO* instruments EGRET and COMPTEL together with those of *AGILE*, *Fermi*, and the *INTEGRAL* instruments. The resolutions are more or less all in the range of degrees. This seems modest, but it is sufficient considering the relatively small numbers of detected sources (nearly 300 in the case of EGRET and 32 in the case of COMPTEL). *Fermi (GLAST)* will probably detect several thousand

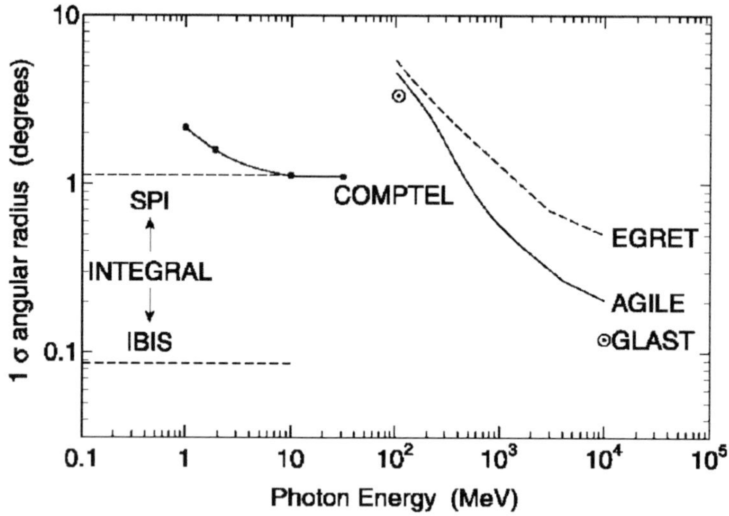

Figure 11.7: Angular resolution of COMPTEL, EGRET, *AGILE* and *Fermi* (formerly *GLAST*) in comparison with other telescopes.

sources. To avoid source confusion, the source identification has to be made above 1 GeV, where the angular resolution is well below 1°.

The continuum sensitivities of COMPTEL, EGRET, *AGILE*, *Fermi* and also those of *COS-B*, *INTEGRAL*, and the ground-based telescopes HESS and MAGIC operating around 1 TeV are shown in Figure 11.8. Obviously the highest sensitivities are achieved at low energies below 100 keV and at high energies above 100 MeV. Two lessons can be learned from this diagram: first, from energies of a few 100 keV to about 50 MeV a sensitivity gap exists (this is essentially the range in which COMPTEL operated). Large efforts are therefore presently being undertaken to bridge that gap, and there seems to be a general agreement that this goal can be achieved by some kind of an advanced Compton telescope. Second, with the launch of *Fermi* in 2008 the energy ranges of space- and ground-based γ-ray astronomy will overlap for the first time.

The implications of progress with respect to angular resolutions and sensitivities for past generations of γ-ray telescopes are illustrated in an impressive way by the all-sky maps shown in this book in Chapter 3 (Kanbach et al 2010) on γ-ray astronomy. There the maps of EGRET above 100 MeV, and those obtained by COMPTEL in the range from 1 MeV to 30 MeV and at 1.809 MeV (^{26}Al line) are presented.

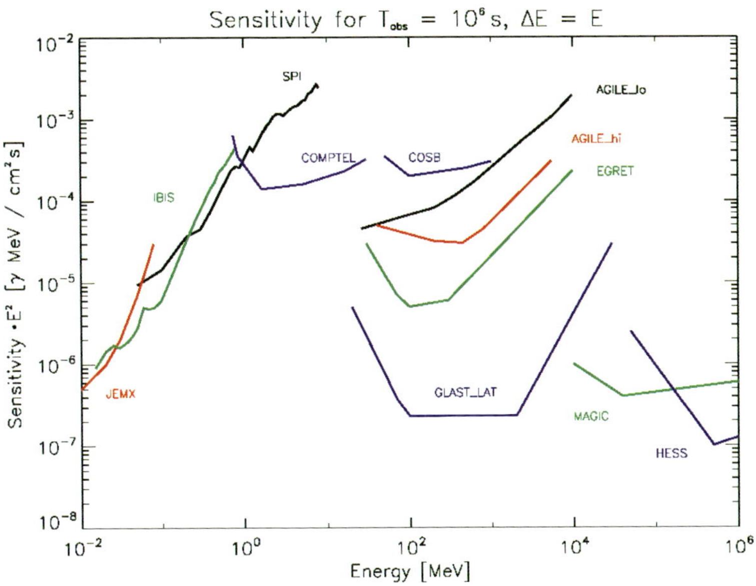

Figure 11.8: Continuum sensitivities in 10^6 s of COMPTEL, EGRET, *AGILE*, and *Fermi* in comparison with other telescopes.

Conclusions and outlook

During the last decades γ-ray astronomy has opened a new window to astronomy. Tremendous difficulties had to be overcome to reach the present status. The results obtained so far have fascinated theorists and observers from all branches of astronomy, and in addition they have stimulated plans for new missions in which the angular resolutions and sensitivities of the telescopes will steadily improve.

Bibliography

Bignami GF, Boella G, Burger, JJ (plus nine authors) (1975) The *COS-B* experiment for gamma-ray astronomy. Space Sci Instrum 1:245–268

Derdeyn SM, Ehrmann CH, Fichtel CE (plus five authors) (1972) *SAS-2* digitized spark chamber gamma ray telescope. Nucl Instrum Meth 98:557–566

Fichtel CE, Hartman RC, Kniffen DA (plus five authors) (1975) High-energy gamma-ray results from the second small astronomy satellite. Astrophys J 198:163–182.

Greisen, K (1966) Experimental gamma-ray astronomy. In Perspectives of Modern Physics (RE Marshak, ed), Wiley, New York, pp.355–382

Harmon BA, Finger MH, Rubin B (plus six authors) (1992) Occultation analysis of BATSE data: Operational aspects. NASA CP-3137:69–75

Hofstadter R, McIntyre JA (1950) Measurement of gamma-ray energies with two crystals in coincidence. Phys Rev 78:619–620

Kanbach G, Bertsch DL, Fichtel CE (plus seven authors) (1988) The project EGRET (Energetic Gamma-Ray Experiment Telescope) on NASA's *Gamma-Ray Observatory (GRO)*. Space Sci Rev 49:69–84

Kanbach G, Andritschke R, Bloser PF (plus three authors) (2003) Concept study for the next generation medium-energy gamma-ray astronomy mission: MEGA. Proc SPIE 4851:1209–1220

Kanbach G, Schönfelder V, Zehnder A (2010) High-energy astrophysics – energies above 100 keV. ISSI SR-009:57–71

Kraushaar W, Clark GW, Garmire G (plus three authors) (1965) *Explorer XI* experiment on cosmic gamma rays. Astrophys J 141:845–863

Kraushaar WL, Clark GW, Garmire G (1968) Preliminary results of gamma-ray observations from *OSO-3*. Can J Phys 46:414–418

Mayer-Hasselwander HA, Kanbach G, Bennett K (plus seven authors) (1982) Large-scale distribution of galactic gamma radiation observed by *COS-B*. Astron Astrophys 105:164–175

Schönfelder V, Hirner A, Schneider K (1973) A telescope for soft gamma ray astronomy. Nucl Instrum Meth 107:385–394

Schönfelder V, Aarts H, Bennett K (plus 27 authors) (1993) Instrument description and performance of the Imaging Gamma-Ray Telescope COMPTEL aboard the Compton Gamma-Ray Observatory. Astrophys J Suppl Ser 86:657–692

Swanenburg BN, Bennett K, Bignami GF (plus 11 authors) (1981) Second *COS B* catalog of high-energy gamma-ray sources. Astrophys J 243:L69–L73

White RS, Koga R, Simnett G (1973) Diffuse galactic and Crab Nebula γ-rays from 1.5 to 10 MeV (Abstract). Proc 13th International Conf on Cosmic Rays 1:7

Zoglauer A, Kanbach G (2003) Doppler broadening as a lower limit to the angular resolution of next-generation Compton telescopes. Proc SPIE 4851:1302–1309

— 12 —

X-ray imaging with collimators, masks and grids

Gordon J. Hurford[I]

Abstract

The basic principles of X-ray and γ-ray imaging for astronomy with non-focusing optics are briefly reviewed. Specific topics include coded masks, and bigrid and multigrid collimators. The advantages and limitations of the various design options are discussed and illustrated with representative examples.

Introduction

At most wavelengths, imaging is based on optics that use reflection or refraction to focus the incident light on a multi-element detector. This provides a direct image of the source. At very short wavelengths (≤ 0.1 µm), however, the physics of the reflection or refraction can render this impractical and other techniques have been employed for hard X-rays and γ-rays (e.g., Ramsey et al 1993). This is particularly true for imaging with high angular resolution ($\approx 4''$ to $6'$) and/or over a large ($\geq 1°$) field of view (FOV).

For imaging beyond the range of focusing optics, it is necessary to resort to collimators, masks and/or grids. By selectively blocking or transmitting incident photons depending on their direction of incidence, they rely on absorption or scattering rather than reflection or refraction to provide an imaging capability. They effectively transform the angular distribution of the incident radiation to either a spatial variation or a temporal variation in the detected photons (spatial or time modulation). Such systems are implemented as a single absorbing layer with apertures (mask), a pair of separated grids (bigrid collimator), or multiple grids (multigrid collimator).

A related approach uses a set of continuous blades oriented parallel to the desired direction of photon travel to restrict the photon transmission to a limited range of angles. Such *slat* or *Soller* collimators can be scanned across the field of view to provide limited angular resolution. They are also useful for coarsely constraining the X-ray FOV (for example to limit background). However, we will not consider them further here.

[I]Space Sciences Laboratory, University of California, Berkeley, USA

The possible configurations of collimator-based imagers are quite diverse and adaptable to a wide range of spacecraft constraints. The choice of collimation approach (single, bigrid or multigrid; temporal or spatial modulation) depends on design drivers such as the available mass and volume, type of attitude control (three-axis or spin stabilized); detector characteristics (particularly its spatial resolution); and the relative scientific importance of angular resolution, FOV and sensitivity. In addition, there are physics-driven constraints on possible combinations of energy range, FOV, angular resolution, and collimator length.

Single-grid systems

Single-grid systems have been used for imaging since long before the advent of modern astronomy. The *camera obscura* consists simply of a small pinhole in an opaque mask. It forms an (inverted) image of a bright object on a distant screen that is otherwise in shadow. The angular resolution of such a device is determined by the combination of single pinhole diffraction and the ratio of the diameter of the pinhole to its separation from the screen. Although originally used for making visible images, such optics can be used at any wavelength provided the pinhole is in a material that is opaque to the photons of interest. As a result, the principle can be applied to X-ray or γ-ray imaging. However, for the weak fluxes characteristic of astronomical sources it is not useful, since the only photons that are detected are those that pass though the pinhole whose size must be small to achieve high angular resolution. As a result, the sensitivity is profoundly limited.

Dicke (1968) overcame this limitation by using a *set* of randomly positioned pinholes in the mask that created a corresponding set of overlapping images (Figure 12.1). For simple sources, the original image can then be recovered by deconvolving the observed pattern. The effective area is now much larger since it is given by the total area of all the apertures that together can be up to ≈ 50 % of the frontal area of the instrument. There are many ways to choose the 'random' pattern of the apertures. One class of such patterns is the uniformly redundant array (URA) illustrated in Figure 12.2. Fenimore's (1980) analysis of a URA response showed the main peak of the angular response is still defined by the geometry of the individual apertures, while the point response function has flat sidelobes. Such URA masks have been extensively used for astronomical applications, most recently on *Swift* (Barthelemey et al 2000) and *INTEGRAL* (Ubertini et al 2003).

As might be expected, the FOV of such systems is given by the ratio of the detector and/or mask size to their separation. The resulting FOV can be quite large (up to ≈ 1 sr) and so enables simultaneous observation of multiple sources. This can be useful for example in survey or burst monitoring applications where the source location is not known a priori. An extensive set of references on coded aperture imaging can be found at http://astrophysics.gsfc.nasa.gov/cai/coded_bibl_short.html.

There are three downsides to imaging with coded masks, however. The first is that the detectors must have spatial resolution that is comparable to or better than the diameter of the mask apertures. This limits the options for detectors by placing a premium on the detector's spatial resolution. In addition to adding complexity and possibly compromising spectral resolution, the detector spatial resolution limits

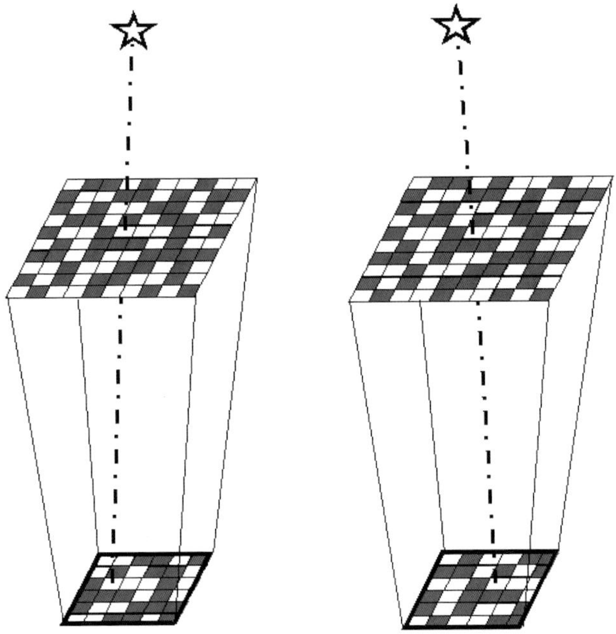

Figure 12.1: Schematic illustration of two identical pinhole arrays (top) casting their shadows on the corresponding detector (bottom). Note how the detected shadow patterns depend on the direction of the incident photon.

the angular resolution achievable by practical sized instrumentation to at best a few minutes of arc. While this may be acceptable for 'point-like' astronomical sources, it is a significant impediment to the use of URAs in solar applications, for example, where angular resolutions of a few seconds of arc are desirable. A second limitation, also relevant to solar applications, is that the response of URAs is degraded for extended sources. Fenimore (1980) provides a clear discussion of the modulation transfer function. The third consideration, in common with most indirect imaging techniques, is that the statistical noise in observations of a given source arises not only from background and its own statistics, but also from all other sources in the FOV. When background does not dominate, this can make it more difficult to detect weak sources in the presence of strong ones.

Bigrid collimators

The modulation collimator (Oda 1965) overcomes the coded mask's angular resolution limitations by replacing the single mask with two or more grids to form a *modulation collimator*. The grids in such systems consist of a large number of parallel X-ray opaque slats separated by X-ray transparent slits. This provides a way to decouple the achievable angular resolution from the spatial resolution of the

Figure 12.2: Uniformly redundant array mask on *INTEGRAL*/IBIS (adapted from http://integral.esac.esa.int).

detector which in turn opens the way to obtaining much higher angular resolution within a given instrument envelope.

Bigrid collimators fall into two broad classes: those that generate a spatial pattern in the transmitted photon flux that depends on their incident direction (spatial modulation) and those that impose a distinctive time-dependence on the transmitted photon flux (time modulation).

In a bigrid collimator using spatial modulation (sometimes called an imaging collimator), the pitch and/or orientation of the front and rear grids differ slightly so that they have slightly different spatial frequencies. As a result, for a given direction of incidence, the transmitted flux forms a large-scale Moiré pattern that has one or a few cycles across the detector (Figure 12.3). The phase of this Moiré pattern (i.e., the location of its maximum) depends sensitively (and periodically) on the incident photon direction, since it goes through a complete cycle with a change of photon direction given by the ratio of the average grid pitch to their separation. Although the Moiré pattern really consists of a large number of narrow stripes (corresponding to the grid pitch), it can be characterized by a detector whose spatial resolution needs to be good enough to see only its large-scale 'envelope'. Therefore the spatial resolution requirement on the detector is determined by the grid diameter, while the angular response of the collimator as a whole is determined by the much smaller grid pitch.

Alternatively, in a time-modulation collimator the top and bottom grids have identical pitch and orientation. If the photon direction with respect to the collimator changes as a function of time, the total transmitted flux also varies. Such time variations can be measured with a detector that need not have any spatial resolution whatsoever, and so it can be chosen or optimized on the basis of other considerations such as spectral resolution or high-energy response. Once again the

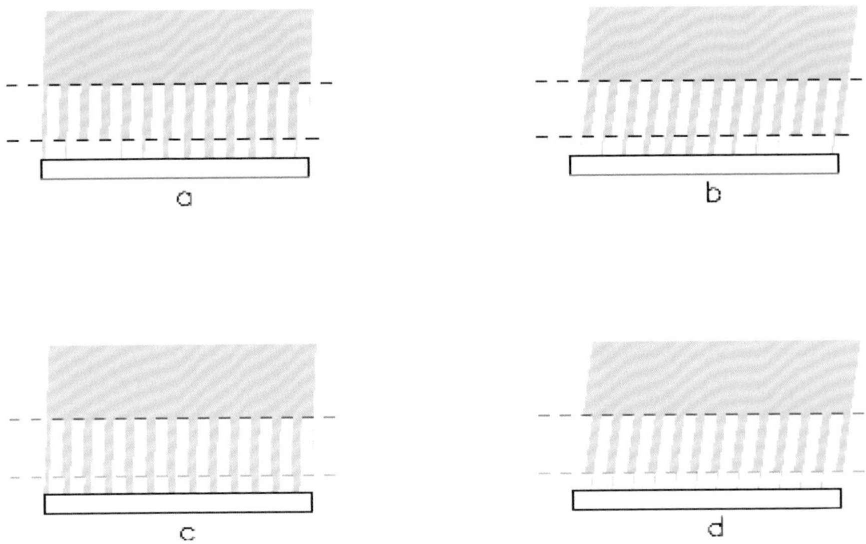

Figure 12.3: Schematic illustration of the transmission of bigrid collimators. The top panels show an imaging collimator for which the front and rear grids have slightly different pitch. For the two incident beam directions shown, the total transmission is about the same, but the location of the smoothed maximum transmission is significantly shifted. Alternatively, for the collimator shown in c and d, the grid pitches are identical. In this case, changing the incident direction affects the total detected flux.

angular resolution (defined as half the period of the response) is given by the ratio of half the grid-pitch to its separation.

Useful insights into the response of a bigrid collimator can be gained by considering the link between its transmission as a function of time (or position for an imaging collimator) and its transmission as a function of incident photon direction. If the slits and slats of the grids are of equal width, then this dependence has a quasi-triangular form. For a point source, the amplitude of this pattern is proportional to its strength, and the timing (or location in the case of spatial modulation) of the transmission maximum depends on source direction. Note that there is no information contained in the period since that is determined by the collimator's design and/or its changing orientation. Considering just the primary sinusoidal component of this pattern, it has been shown (Makishima et al 1977; Prince et al 1988) that the amplitude and phase of this pattern provide a direct measurement of a single Fourier component of the source distribution. Thus the response of a bigrid collimator is a precise mathematical analog to that of a pair of antennas in a radio interferometer, whose correlated signal also measures one such Fourier component. In both cases, images can be reconstructed from a set of Fourier component mea-

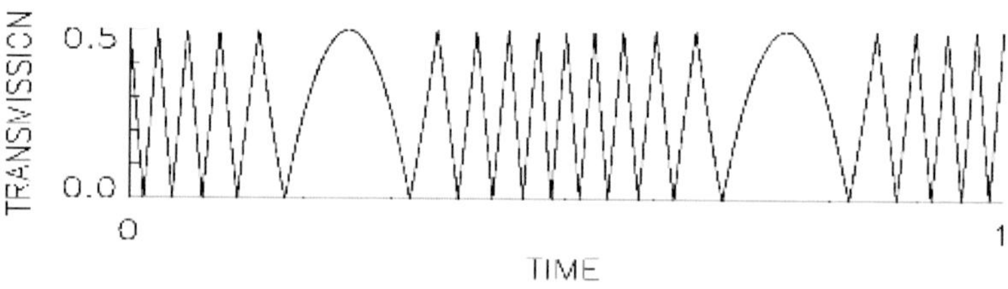

Figure 12.4: Schematic illustration of the time variation of the transmission of a bigrid rotation modulation collimator for a point source. The variations in the period of the modulation arise since, in the plane orthogonal to the grids, the source apparently moves with simple harmonic motion.

surements made at a large number of angular frequencies. This perspective on the response of a bigrid collimator greatly simplifies analyses of its angular response.

Although a time-modulation collimator uses the simpler detector system, some provision must be made to continually move the collimator with respect to the source so as to change the relative direction of the incident photons. This can be done by mechanically rocking the collimator, or by exploiting random motions (as with a badly pointed balloon platform). However, a common approach is to mount the collimator on a rotating spacecraft to form a rotating modulation collimator (RMC, Schnopper et al 1968). As the collimator rotates, its response over a limited range of angles is quasi-periodic (Figure 12.4). Over a half-rotation, however, the amplitude and phase of the modulation measures a set of Fourier components at a complete set of orientations at the spatial frequency determined by the grid pitch and separation. In the parlance of radio interferometry, the system measures spatial frequencies in a circle in the $u\nu$ (spatial frequency) plane in analogy to *Earth rotation synthesis*, except on a much more rapid timescale.

Early implementations of the rotating modulation collimator were used for surveys where the system detected point sources over a relatively wide FOV. Mertz (1968) also suggested using such collimators as imaging devices for extended sources. The most ambitious implementation to date of such an imager is *RHESSI* (Lin et al 2002) which uses a set of nine RMCs to image extended solar flare X-ray and γ-ray sources (Hurford et al 2002), achieving angular resolution as high as $2.3''$. The effectiveness of multiple RMCs in characterizing extended sources is based on the fact that a bigrid collimator cannot modulate X-rays from a source whose angular width is much larger than the collimator resolution. As a result, comparison of the modulation amplitude among collimators with different angular resolution can provide an accurate measurement of source size (e.g., Schmahl and Hurford 2002).

One measure of the performance of an RMC as an imager is the dynamic range of its images, i.e., the ratio of the brightest source in the FOV to the faintest credible feature. The dynamic range can be limited by any of three broad factors. First, a typical set of RMCs measures only a few dozen to a few hundred independent Fourier components, each of which is fully represented by a single complex number. Since the number of observables is so limited, there is an unavoidable limitation on the complexity of the images that can be reconstructed. Second, photon statistics can be a limitation in both astrophysical and solar observations. This translates directly into a statistical uncertainty in each visibility that again limits the achievable dynamic range. Third, as with any instrument, there are inevitable calibration uncertainties that constrain the accuracy of each measurement. In practice any one (or more) of these factors can be the limiting factor, depending on the circumstances.

Although bigrid collimators can provide higher angular resolution than single-mask systems, this comes at some disadvantage in terms of sensitivity. In particular, the nominal average relative transmission of bigrid collimators is ≈ 25 %, compared to ≈ 50 % for single-mask systems. Also, the quasi-triangular modulation shape is not as efficient as that of the (square wave) single-grid systems and so results in some loss of signal to noise. As a result, the choice between single and bigrid systems depends on the relative importance of angular resolution and sensitivity and on the most appropriate detector technology for a given application.

Multigrid collimators

The inclusion of additional grids into a bigrid collimator provides another design option. The additional grids suppress the response at intermediate peaks (Figure 12.5). If enough additional grids are inserted, this can provide a system that has only one response peak across the FOV. A single such collimator can be rastered in time or a fixed array of such collimators can be employed, each with its peak response in a (slightly) different direction. In either case, the result is a direct imaging system, effectively equivalent to a conventional focused telescope in that it has multiple detector elements, each of which is sensitive to a specific area on the source. The angular resolution is still determined by the pitch of the collimator grids. The advantage over the single and bigrid systems is that no image reconstruction is required and so the signal from strong sources does not affect the detection of weak ones. A significant disadvantage, however, is that the available frontal area of the instrument must be shared by many small subcollimators, each with its own look-direction. Alternatively, for scanned systems, a given source is observed with a low duty cycle. In either case, the sensitivity to any given source in the FOV can be very low. The HXIS instrument on *SMM* (van Beek et al 1980) provides a good example of this approach.

Grids

Several technologies have been used to fabricate the grid or mask 'optics' of collimator systems. For coarse grids, mechanical assembly of conventionally machined parts is the typical choice (e.g., Figure 12.2). At intermediate pitches (down to ≈ 1 mm) electron discharge machining is a viable option (e.g., Crannell et al 1991). For fine grids, stacking of photo-etched grid layers (Figure 12.6) has been used to achieve pitches of 35 µm in 1 mm thick material (Lin et al 2002).

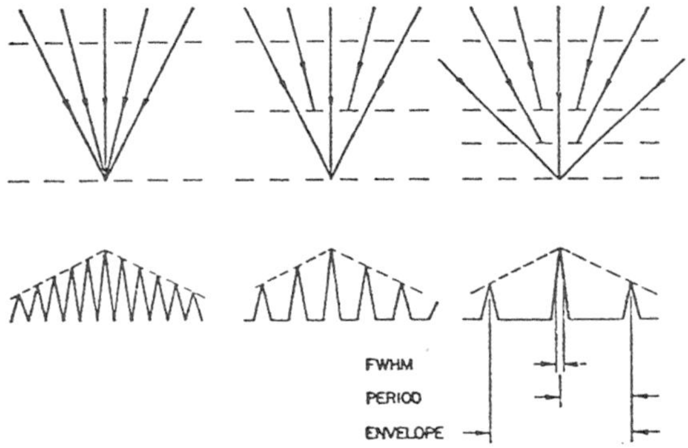

Figure 12.5: Top: Possible photon paths for bigrid and multiple grid collimators, illustrating the periodic maxima in their response and the effect of intermediate grids. Bottom: Corresponding plots of the effective area as a function of incident angle. For a given collimator length, the angular resolution is given by the ratio $\frac{1}{2}$ grid pitch/separation; the response envelope (FOV) is defined by the detector and grid size. The intermediate grids affect the period.

The performance of any collimator or mask system is also subject to two physical limitations. The first limitation is set by a combination of three factors. Specifically, the minimum thickness of the grid is determined by the requirement that the grid be opaque at the maximum energy of interest. This thickness constrains the FOV to an angle given by the ratio of slit width/grid thickness. The slit width in turn is closely tied to the angular resolution ($\frac{1}{2}$ slit pitch / separation). For a collimator of a given length, this combination of factors imposes an unavoidable physics-driven tradeoff among angular resolution, FOV and maximum energy.

The second limitation is diffraction. This sets a lower limit to the energy range since at lower energies the front grid can function as a diffraction grating. A complete analysis of the diffraction response of multigrid collimators (Lindsey 1978) was simplified by Crannell et al (1991) for the case of a bigrid collimator.

Both of these limitations can be relevant in practice. For example, *RHESSI*'s angular resolution above ≈ 1 MeV is limited to $35''$ by a requirement that it maintain a 1° FOV. With a 1.55 m long collimator, *RHESSI* is also prevented from achieving $2.3''$ resolution below 4 keV by diffraction.

Alignment and aspect

The significance of internal alignment and tolerance issues depends on the type of mask or collimator system employed. For mask systems with two-dimensional detectors, the primary requirement is that the relative positions of the mask and detector be known in inertial coordinates to an accuracy small compared to the angular resolution. This must be achieved on timescales that are longer than both the integration time and aspect cadence. The requirement can be met if both the

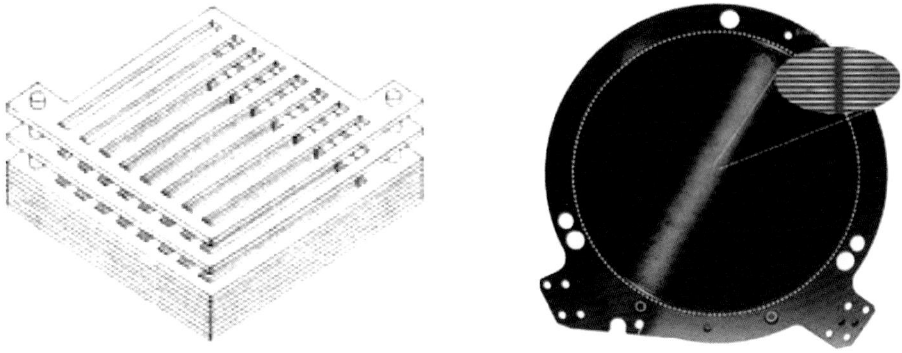

Figure 12.6: Left: Schematic illustration of stacking photo-etched layers to achieve a thick grid with fine slits. Right: *RHESSI* 9 cm diameter, 1 mm thick grid whose slits (inset) have a 35 µm period. The FOV is ≈ 1° (from Lin et al 2002).

metering structure and pointing platform are stable. Alternatively, one can trade mechanical and pointing stability for data rate by using short binning times and a high-cadence aspect system. For high-resolution bigrid collimators this can be done by embedding key elements of the aspect system in the plane of the grids (e.g., Zehnder et al 2003). In this way mechanical flexure of the intermediate metering structure is equivalent to a variation in pointing. With photon tagging or short integration times and a high-cadence aspect system, such variations can then be fully compensated for during analysis.

For bigrid collimators, there is an additional requirement in that the top and bottom grid slits must be parallel (viz., that their relative twist be maintained). However the twist requirement is much coarser than the angular resolution since the former is determined by the ratio of grid pitch to grid diameter (and not to the grid separation). For example, *RHESSI* achieved $2''$ imaging with internal alignment that was controlled to $\approx 1'$ and pointing that was controlled to several minutes of arc. This use of knowledge in place of control of the pointing and internal alignment can greatly simplify the mechanical implementation of a high-resolution collimator.

The alignment requirements for multigrid collimators are much more severe since the intermediate grids must be positioned and maintained to a precision that is small compared to the grid pitch. This requirement was met, however, by the HXIS instrument on *SMM* with grid apertures ≈ 25 µm.

Summary and outlook

Grids and masks have provided the basis for X-ray and γ-ray imaging since the 1960s (Bradt et al 1992). (Also see the list of missions found at http://astrophysics.gsfc.nasa.gov/cai/coded_inss.html). Table 12.1 summarizes the characteristics of a

Table 12.1: Representative instruments.

Mission/ instrument	Year	E_x/keV	Prime imaging objective	Imager type(s)	Area A/cm^2	FWHM resolution	Reference
Ariel-5/ B	1974–1980	0.9 to 18	survey	scanner	≈ 290	0.75°	Villa et al 1976
SMM/ HXIS	1980–1989	3.5 to 30	solar flares	multigrid direct imaging	0.07	8″	van Beek et al 1980
Yohkoh/ HXT	1991–2001	15 to 100	solar flares	64 bi-grid collimators	70	8″	Kosugi et al 1991
HETE-2/ WXM	2000–present	2 to 25	burst location	two 1-D random masks	350	$\approx 10'$	Kawai et al 1999
INTEGRAL/ IBIS	2002–present	15 to 10 000	source identification	URA	2500	12′	Ubertini et al 2003
RHESSI	2002–present	3 to 17 000	solar flares	9 RMCs	90	2.3″	Lin et al 2002
Swift/ BAT	2004–present	15 to 150	burst location	URA	5200	17′	Barthelemy et al 2005

representative set of such instruments. Their capabilities have grown as grid and detector technologies have improved. The techniques do have some significant disadvantages, however. The design requirement that the detector area be comparable to that of the grid or mask makes it much more difficult to reduce background for applications that require high sensitivity. Sensitivity is further affected since the telescope mask or collimator intentionally blocks about half to three quarters of the incident photon flux. In source-limited situations where background is not an issue, the ability to detect weak sources in the presence of strong ones can be limited by the fact that all sources contribute noise to the detection of each source component. Image quality is also significantly constrained for morphologically rich sources.

In the coming years, we can expect masks and collimators to be partially supplanted by technical developments in grazing-incidence optics, especially in low energy applications that require only intermediate resolution and narrow FOVs. In such contexts, focusing optics has a commanding advantage where sensitivity and background rejection are the main drivers or where morphologically complex sources need be imaged.

Nevertheless there will continue to be many applications where mask and grid-based imaging is appropriate. As we have seen, the technique can be adapted to platforms which are three-axis stabilized, rotating or badly pointed (as with balloons); it can be implemented in a wide range of size scales, from compact designs of a few centimeters in scale, to configurations requiring extended booms on scales of meters; it can provide angular resolutions from seconds of arc to degrees over FOVs from $\approx 1^\circ \times 1^\circ$ to ≈ 1 sr; and for a given instrument, the same 'optics' can be used over a wide range of energies, a feature that greatly aids co-location of images and imaging spectroscopy. Therefore, in applications where either compactness, low mass, wide FOV or high-energy response is required, masks, grids and collimators will continue to provide the imaging technique of choice.

Acknowledgement

This work was supported by NASA Contract NAS5-98033.

Bibliography

Barthelemy SD and the *Swift* Instrument Team (2005) Burst alert telescope (BAT) on the *Swift* MIDEX mission. Space Sci Rev 120:143–164

Bradt HVD, Ohashi T, Pounds KA (1992) X-ray astronomy missions. Ann Rev Astron Astrophys 30:391–427

Crannell CJ, Dennis B, Orwig L (plus eight authors) (1991) A balloon-borne payload for imaging hard X-rays and gamma rays from solar flares. AIAA-91-3653:1–11

Dicke RH (1968) Scatter-hole cameras for X-rays and gamma rays. Astrophys J 153:L101–106

Fenimore EE (1980) Coded aperture imaging: the modulation transfer function for uniformly redundant arrays. Appl Opt 19:2465–2471

Hurford GJ, Schmahl EJ, Schwartz RA (plus 11 authors) (2002) The *RHESSI* imaging concept. Sol Phys 210:61–86

Kawai N Matsuoka M, Yoshida A (plus seven authors) (1999) Wide-field X-ray monitor on *HETE-2*. Astron Astrophys Suppl Ser 138:563–564

Kosugi T, Makishima K, Murakami T (plus nine authors) (1991) The Hard X-ray Telescope (HXT) for the *Solar-A* Mission. Solar Phys 136:17–36

Lin RP, Dennis BR, Hurford GJ (plus 63 authors) (2002) The Reuven Ramaty high-energy solar spectroscopic imager (*RHESSI*). Sol Phys 210:3–32

Lindsey CA (1978) Effects of diffraction in multiple-grid telescopes for X-ray astronomy. J Opt Soc Am 68:1708–1715

Makishima K, Miyamoto S, Murakami T (plus four authors) (1977) Modulation collimator as an imaging device, in K A van der Hucht and G Vaiana (eds), New Instrumentation for Space Astronomy, Pergamon Press, New York, pp 277–289

Mertz L (1968) A dilute image transform with application to an X-ray star camera. Proc Symp Mod Opt 17:787–791

Oda M (1965) High-resolution X-ray collimator with broad field of view for astronomical use. Appl Optics 4:143–143

Prince TA, Hurford GJ, Hudson HS, Crannell CJ (1988) Gamma-ray and hard X-ray imaging of solar flares. Sol Phys 118:269–290

Ramsey BD, Austin RA, Decher R (1993) Instrumentation for X-ray astronomy. Space Sci Rev 69:139–204

Schmahl EJ, Hurford GJ (2002) *RHESSI* observations of the size scales of solar hard X-ray sources. Sol Phys 210:273–286

Schnopper HW, Thompson RI, Watt S (1968) Predicted performance of a rotating modulation collimator for locating celestial X-ray sources. Space Sci Rev 8:534–542

Ubertini P, Lebrun F, Di Cocco G (plus 17 authors) (2003) IBIS: The imager on-board *INTEGRAL*. Astronom Astrophys 411:L131–139

van Beek HF, Hoyng P, Lafleur B, Simnett GM (1980) The hard X-ray imaging spectrometer (HXIS). Sol Phys 65:39–52

Villa G, Page CG, Turner MJL (plus three authors) (1976) The Ariel V sky survey instrument and new observations of the Milky Way. Mon Not Roy Astronom Soc 176:609–620

Zehnder A, Bialkowski J, Burri F (plus 12 authors) (2003) *RHESSI* imager and aspect systems. Proc SPIE 4853:41–59

— 13 —

All-sky monitoring of high-energy transients

KEVIN HURLEY[I]

Abstract

The use of triangulation, or arrival-time analysis, to monitor the γ-ray sky, is discussed. A network of spacecraft separated by interplanetary distances can monitor the entire sky continuously for γ-ray bursts and similar astrophysical transients. Single experiments in low Earth orbit can also perform this function, and generally have the advantage of greater sensitivity, at the cost of some loss of spatial and temporal coverage.

Introduction

The γ-ray sky is variable on all timescales with which it has been measured so far, from milliseconds to years and longer. Some sources are recurrent, others exhibit both quiescent emission and bursting behavior, and still others manifest themselves in a single outburst. Simply to discover them requires continuous monitoring of the entire sky, and any attempt to understand them benefits from long periods of uninterrupted observations. In principle, this can only be done far from Earth, since low Earth orbiters are subject to Earth occultation and, in most cases, South Atlantic Anomaly interruptions. In practice, though, achieving the best sensitivity means utilizing the largest possible instruments, and virtually all large, single-experiment monitors are in low Earth orbit; conversely, multi-experiment networks, which must be distributed over interplanetary distances, utilize much smaller experiments. In this chapter, we will consider two of the methods which have been in use over the last several decades to monitor the γ-ray sky at energies above roughly 25 keV and over timescales under about 1000 s. Although the techniques can be adapted to sources within the solar system, the discussion will emphasize objects outside it.

Triangulation

In the early days of X-ray astronomy, Giacconi (1972) suggested the use of triangulation to obtain precise locations for cosmic X-ray sources. Although it was to be many years before this technique was used for this purpose, it did come into

[I]University of California Berkeley Space Sciences Laboratory, Berkeley, USA

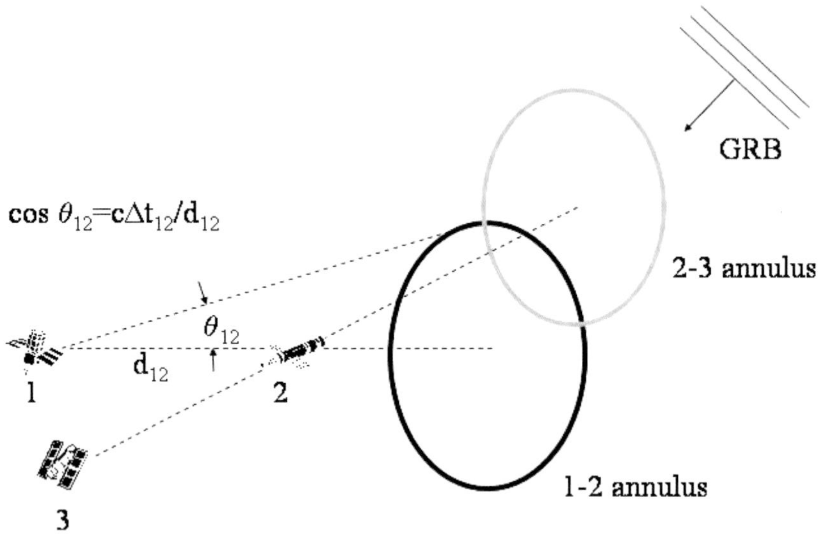

Figure 13.1: The triangulation method. A plane wave resulting from a γ-ray burst is incident on three spacecraft. The difference in arrival times between spacecraft 1 and 2 defines an annulus of possible arrival directions. A third spacecraft gives a second annulus which generally intersects the first one to define two possible arrival directions.

use around the early 1970s for cosmic γ-ray bursts (hereafter GRB), and has been in use almost continually ever since then. Figure 13.1 illustrates the method. A plane wave is incident on three spacecraft. The difference in arrival times between spacecraft 1 and 2, Δt_{12}, and its uncertainty, define an annulus of possible arrival directions by the formula $\cos\theta_{12} = c_0\,\Delta t_{12}/d_{12}$, where c_0 is the speed of light in vacuum and d_{12} is the distance between the spacecraft. (Δt is determined by cross-correlating the light curves recorded on the two spacecraft.) A third spacecraft gives a second annulus which generally intersects the first one to define two possible arrival directions. The third (1–3) annulus intersects the two, but may reduce their areas depending on its associated uncertainty. The ambiguity can be resolved in a number of ways. One is to utilize any inherent directionality in one or more of the individual detectors. Another is by considering whether either of the positions is blocked by a planet, e.g., for an Earth-orbiting or Mars-orbiting spacecraft. A third is to add more spacecraft which are non-co-planar with respect to the first three. The size of the resulting uncertainty box is proportional to the widths of the annuli, which in turn depend mainly on the uncertainty in $\Delta t/(d\sin\theta)$. In practice, the uncertainties in the spacecraft positions are almost always negligible. The plane-wave approximation breaks down for nearby (solar system) objects. When the distances d are in the astronomical unit range, the spacecraft form an interplanetary network, and minute-of-arc accuracy is possible.

Some of the basic principles of the method are the following:

1. The detectors on the various spacecraft do not need to be identical or even similar in area or in type. Good networks have utilized detectors ranging from 20 cm^2 in area to over 100 times this much, and they have utilized materials from inorganic and organic scintillators to germanium. Although dedicated GRB detectors have the advantage of better design, they tend to be more costly; non-dedicated detectors, such as anti-coincidence systems, have the advantage of larger area and higher sensitivity.

2. It helps if the energy ranges of the detectors are similar, but in practice they need only to cover some part of the 25 keV to 100 keV energy range. (The time histories of GRBs are generally different below around 15 keV and above around 100 keV, which makes them difficult to compare.)

3. To achieve accuracies in the minute-of-arc range, the distances between the spacecraft should be in the range of 1 ua and above.

4. The accuracy which can be achieved in cross-correlating two light curves recorded on different spacecraft depends more on the time structure in the light curve than on the number of photons in it. Thus, very short bursts (say 100 ms long) can be localized very accurately, even though the light curves may only contain 50 photons or so.

5. Low Earth orbiting spacecraft will miss at least 30 % of the bursts which they are sensitive to, due to occultation and South Atlantic Anomaly passes. Earth orbiters are an important part of any network, but ideally there should be two or more of them; the current network has four, and a fifth has recently been added.

6. The time resolution of the experiments should be in the 100 ms range or less; the spectral resolution is relatively unimportant.

These constraints place the following requirements on the missions involved.

1. They must have clocks which are accurate to 10 ms or better.

2. The spacecraft coordinates (right ascension, declination, and range) should be known to about $4''$ and 100 km.

3. Data downlinks should be frequent and rapid, since most transient sources benefit from rapid follow-up observations.

The first two requirements rarely pose any major problems, but the third can be difficult to achieve for planetary missions, because light travel times can reach several thousand seconds, and some missions have very little data recovery in their cruise phases.

Calibration is a crucial aspect of any experiment. There are two ways to approach this with an interplanetary network. The first is to send commands to the spacecraft at precisely known times and record their arrival on board. With a knowledge of the light travel time, this can be used to verify the spacecraft clock and range. A more complete, end-to-end calibration is to use the network to triangulate sources whose precise positions are known by other means, such as GRBs with X-ray or optical counterparts, or soft gamma repeaters (SGRs). This calibrates not only the spacecraft clocks, but also their ephemerides, as well as the on-board data handling and ground-based data reduction techniques. Both methods should be used wherever possible.

Results and discoveries

Even before the formation of a network for astrophysical purposes, the *Vela* spacecraft discovered GRBs (Klebesadel et al 1973) and arrival-time analysis was used to get coarse positions for them. Shortly after the formation of the first interplanetary network in 1977, the first giant magnetar flare, from the SGR SGR0525-66 was triangulated to a small uncertainty box, and its source was identified as being contained within the N49 supernova remnant in the Large Magellanic Cloud (Cline et al 1980; Evans et al 1980). The discovery of one of the SGRs (Atteia et al 1983) and the most precise localizations of all of them (Cline et al 1980; Evans et al 1980; Hurley et al 1999a,b,c) have been the results of triangulation. Giant outbursts from Cygnus X-1 were discovered and localized with the help of triangulation (Golenetskii et al 2003), finally fulfilling the suggestion of Giacconi (1972). Until the advent of coded mask imaging detectors, the only way to derive the precise positions of γ-ray burst sources was by triangulation using interplanetary networks.

Single-experiment systems

A single experiment can also monitor a large fraction of the sky. In low Earth orbit instruments can be larger and more sensitive than those which can be placed on interplanetary spacecraft. Compton telescopes, discussed in Chapter 11 (Schönfelder and Kanbach 2010), are one method, and experiments which rely on anisotropic detectors are another. An idealized detector with a flat surface and negligible thickness has a cross-sectional area which is proportional to the cosine of the arrival direction. The ratios of the responses of three or more non-coplanar detectors on a single spacecraft give the arrival direction of a burst. The accuracy with which an uncertainty box can be derived depends on the counting statistics of the detections (rather than the time structure of the light curves) and the systematic uncertainties in the detector calibrations. In addition, backscattering from the spacecraft and, in the case of a low Earth orbiter, the atmosphere, must be taken into account.

The first use of this method was aboard the *Venera 11* and *12* interplanetary spacecraft which were launched in 1977 (the Konus experiment, Mazets and Golenetskii 1981). The detectors were six 50 cm^2 scintillators. In 1991 the method was used aboard the *CGRO* Burst and Transient Source Experiment (BATSE), which utilized eight 2025 cm^2 scintillators. For intense GRBs, the statistical uncertainties were practically negligible, while systematics limited the uncertainty-box sizes to 1.6°, the best accuracy which the method has achieved to date (Briggs et al 1999). Earth-occultation was used to supplement the monitoring capabilities (Ling et al 2000). In 1996, the method was used again for the four 1100 cm^2 scintillators comprising the anticoincidence system of the Phoswich Detection System aboard *BeppoSAX* (the Gamma-Ray Burst Monitor, Guidorzi et al 2004).

The first SGRs were discovered with the Konus experiments (Mazets et al 1979a,b). SGR1627-41 and the bursting pulsar GROJ1744-28 were discovered by BATSE (Woods et al 1999; Kouveliotou et al 1996), and thanks to its great sensitivity, BATSE provided the first unambiguous evidence for the isotropic distribution of GRBs (Meegan et al 1992).

Outlook

There will always be a need for continuous, all-sky monitors, and both single- and multiple-experiment systems will continue to be utilized for the foreseeable future. The Gamma Burst Monitor aboard *Fermi* launched in 2008, is an example of the former, and the 3rd interplanetary network, which began operations in 1990, will continue to operate for at least several more years. The two systems, in fact, work well together; the single-experiment systems have the advantages of good sensitivity and rapid response, while multiple detector systems can refine the uncertainty boxes derived by the single detectors systems (Hurley et al 1999d).

Bibliography

Atteia J-L, Boer M, Hurley K (plus 14 authors) (1987) Localization, time histories, and energy spectra of a new type of recurrent high energy transient source. Astrophys J 320:L105–110

Briggs M, Pendleton GN, Kippen RM (plus four authors) (1999) The error distribution of BATSE gamma-ray burst locations. Astrophys J Suppl 122:503–518

Cline T, Desai UD, Pizzichini G, Teegarden BJ (plus seven authors) (1980) Detection of a fast, intense and unusual gamma ray transient. Astrophys J 237:L1–5

Evans WD, Klebesadel RW, Laros JG (plus seven authors) (1980) Location of the gamma-ray transient event of 1979 March 5. Astrophys J 237:L7–9

Giacconi R (1972) On the use of long-base time-delay measurements in the study of rapidly varying X-ray stars. Astrophys J 173:L79–81

Golenetskii S, Aptekar R, Frederiks D (plus five authors) (2003) Observations of giant outbursts from Cygnus X-1. Astrophys J 596:1113–1120

Guidorzi C, Montanari E, Frontera F (plus four authors) (2004) The *BeppoSAX* GRBM catalog of GRBs. In: The Third Rome Workshop on Gamma-Ray Bursts in the Afterglow Era, Astronom Soc Pac Conf Series 312:39–42

Hurley K, Kouveliotou C, Woods P (plus five authors) (1999a) Reactivation and precise interplanetary network localization of the soft gamma repeater SGR1900+14. Astrophys J 510:L107–109

Hurley K, Kouveliotou C, Woods P (plus five authors) (1999b) Precise interplanetary network localization of a new soft gamma repeater, SGR 1627-41. Astrophys J 519:L143–145

Hurley K, Kouveliotou C, Cline T (plus four authors) (1999c) Where is SGR1806-20? Astrophys J 523: L37–40

Hurley K, Briggs MS, Kippen RM (plus five authors) (1999d), The *Ulysses* supplement to the BATSE 4Br catalog of cosmic gamma-ray bursts. Astrophys J Suppl 122:497–501

Klebesadel RW, Strong IB, Olson RA (1973) Observations of gamma ray bursts of cosmic origin. Astrophys J 182:L85–88

Kouveliotou C, van Paradijs J, Fishman GJ (plus five authors) (1996) A new type of transient high-energy source in the direction of the Galactic Centre, Nature 379:799–801

Ling J, Wheaton WA, Wallyn P (plus seven authors) (2000) A BATSE Earth-occultation catalog of 0.03 – 1.8 MeV gamma-ray source spectra and light curves for phases 1 - 3 (1991 - 1994). Astrophys J Suppl 127:79–124

Mazets E, Golentskii SV, Ilinskii VN (plus two authors) (1979a) Observations of a flaring X-ray pulsar in Dorado. Nature 282:587–589

Mazets E, Golenetskij SV, Guryan YA (1979b) Soft gamma ray bursts from the source B1900+14. Sov Astron Lett 5(6):343–344

Mazets E, Golenetskii S (1981) Recent results from the gamma-ray burst studies in the Konus experiment. Astrophys Space Sci 75:47–81

Meegan C, Fishman GJ, Wilson RB (plus five authors) (1992) Spatial distribution of gamma-ray bursts observed by BATSE. Nature 355:143–145

Schönfelder V, Kanbach G (2010) Imaging through Compton scattering and pair creation. ISSI SR-009:207–222

Woods P, Kouveliotou C, van Paradijs J (plus six authors) (1999) Discovery of a new soft gamma repeater, SGR1627-41. Astrophys J 519:L139–142

— 14 —

Far-infrared imaging and spectroscopic instrumentation

Bruce Swinyard[I] and Wolfgang Wild[II]

Abstract

The subject of the design and implementation of infrared space missions is briefly reviewed and the limitations imposed by the needs of requiring cryogenic instruments and telescopes introduced. We give an introduction to direct detection techniques for imaging and spectroscopy and review the current state of the art in instrumentation and detector technology. We discuss the history of space-borne infrared missions dedicated to photometric surveys, imaging and spectroscopy over the 5 µm to 700 µm waveband starting with the ground-breaking *IRAS* mission launched in 1983 and going up to the latest mission, *Herschel*, launched in May 2009. This mission will also make use of heterodyne techniques for very high spectral resolution. We review the principle of heterodyne detection and briefly describe the HIFI instrument on *Herschel* and how it will take the subject of high-resolution spectroscopy into a new era.

Introduction

The infrared (IR), far-infrared (FIR) and submillimetre wavelength ranges, being adjacent to both the visible and millimetre wavelength ranges, have naturally borrowed instrumental techniques from both sides. Extension into the infrared has not been trivial and scaling with wavelength for optical techniques or with frequency for radio techniques has required dedicated developments. The need to cool detectors, instruments and telescopes has seriously limited progress in IR space instrumentation and capabilities. So far most IR space telescopes have had primary mirrors with diameters smaller than 1 m only[1] and accordingly have achieved angular resolutions at the longer FIR wavelengths that are orders of magnitude below those of the neighbouring wavelength regions (cf., VLT and *HST* or IRAM and ALMA). One can indeed speak of the FIR gap in both spatial resolution and sensitivity. In the coming sections we will present an overview of space instrumentation

[I]Science and Technology Facilities Council, Rutherford Appleton Laboratory, Harwell Innovation Campus, Chilton, UK

[II]SRON Netherlands Institute for Space Research, Groningen, and European Southern Observatory, Garching, Germany

[1]The 3.5 m *Herschel* telescope, launched on 14 May 2009, will reduce the gap, but not reach the resolution achieved by the major ground-based facilities.

for three types of applications: all-sky surveys, imaging and spectroscopy. The selection of technologies for the instrumentation we discuss below was mainly driven by the scientific objectives. These are still heavily compromised by the satellite resources which are, in turn, constrained by available technology and budget. We will present in each of these sections the key IR missions in historical order and summarize the development of space infrared missions and look at how the technology is likely to develop in the coming decades. We start with a discussion of the general problems associated with observing at wavelengths beyond ≈ 15 µm.

Design considerations for space IR missions

Noise due to the self-emission and the natural background

The dominating influence on the design of all IR missions is the need to avoid and control the self-emission from the optics and structural elements of the telescope and instruments themselves. The wavelength at which black-body radiation peaks is given by Wien's displacement law which we can express as $\lambda_{\max} = b/\mathrm{T}$, with $b = 2897$ µm K. As an example the effect of the temperature of the telescope compared to that from the natural zodiacal light is shown for telescopes at various temperatures in Figure 14.1. From this diagram and the displacement law, we see that for an observatory to be operated at 100 µm we need to cool the telescope and optics to well below 30 K to make the contribution from the instrumentation negligible compared to that from the natural background. The actual operating temperature also depends on the emissivity of the optics and the way the detectors view the instrument and structure.

For the remainder of this section we will express signals and noise in electrons per second and electrons to give examples in terms of photo-conducting detectors. The electron emission rate of any surface within the field of view (FOV) of a detector is given in simple terms by

$$S_{\mathrm{back}} = \epsilon\, B(\lambda)\, A\, \Omega\, \eta_{\mathrm{D}}\, \lambda \frac{\Delta\lambda}{h\, c_0} \quad , \tag{14.1}$$

where $B(\lambda)$ is the black-body emission given by Planck's law, ϵ is the emissivity of the surface, $A\Omega$ the étendue of the detector towards the surface, η_{D} is the quantum efficiency of the detector in electrons/photon, λ the wavelength of interest, h is Planck's constant, c_0 the speed of light and $\Delta\lambda$ the bandwidth of the instrument. The electron noise in a photoconductive detector due to background power is then given by

$$N_{\mathrm{e}} = G\sqrt{2 S_{\mathrm{back}} t_{\mathrm{int}}} \quad , \tag{14.2}$$

where G is the photo-conductive gain of the detector (usually $G < 1$), t_{int} is the integration time of the observation and the factor of two is required to account for the recombination noise (see Chapter 30, Raab 2010).

We can now introduce the useful quantity of noise equivalent power (NEP, N_{p}) which — in the context of infrared astronomy — is defined as the equivalent power that will give a signal to noise of one in a one-second integration time (Rieke 2003). NEP is a measure both of the quality of a detector (without the influence of

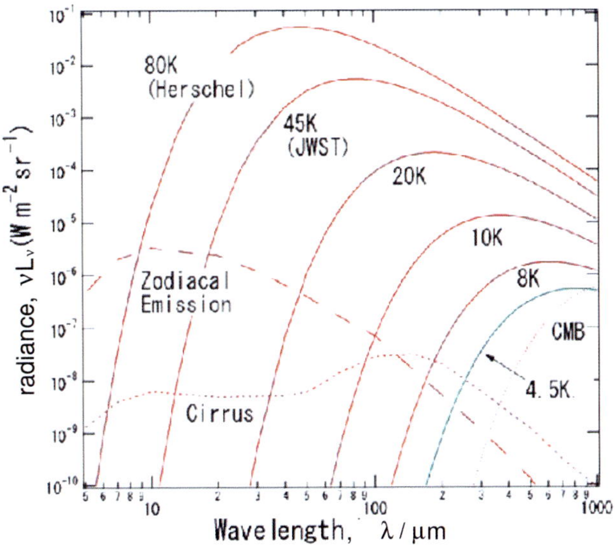

Figure 14.1: Comparison of radiance (scaled with the frequency $\nu = c_0/\lambda$) from telescopes as a function of the wavelength at various temperatures with the natural background from the zodiacal cloud, the galactic cirrus and the Cosmic Microwave Background (Nakagawa 2007).

background power from an external radiant source such as a warm telescope) and of the noise that an external source will generate due to the photon noise. NEP can therefore be used to measure how well-matched a specific detector system is to a particular radiant background. In the case of photo-conducting detectors the detector-specific NEP is defined as:

$$N_\mathrm{p} = \frac{\sqrt{2}\, e\, N_{\mathrm{e},1}}{R_\mathrm{D}} \quad , \tag{14.3}$$

where e is the elementary charge, $N_{\mathrm{e},1}$ is the noise in electrons measured in a one-second integration and R_D is the responsivity, given by

$$R_\mathrm{D} = \frac{e\, \eta_\mathrm{D}\, \lambda}{h\, c_0} \quad . \tag{14.4}$$

The NEP due to the natural background for a typical wideband ($\lambda/\Delta\lambda \approx 3$) Ge:Ga photo-conducting detector operating at 100 μm on a 3.5 m diameter telescope is a few 10^{-18} W Hz$^{-\frac{1}{2}}$. This is of the same order of magnitude as the intrinsic NEP of current state of the art Ge:Ga detectors. In the case of a narrow-band spectrometer (a grating or Fabry–Perot, for instance) the background-limited NEP is much lower, and to take advantage of this the detector performance must be very much improved. Future bolometric detectors operating at temperatures of 0.1 K or lower appear to offer a promising route to obtaining multi-pixel low-noise detectors for the FIR and submillimetre domain (Parkes et al 2004). For a

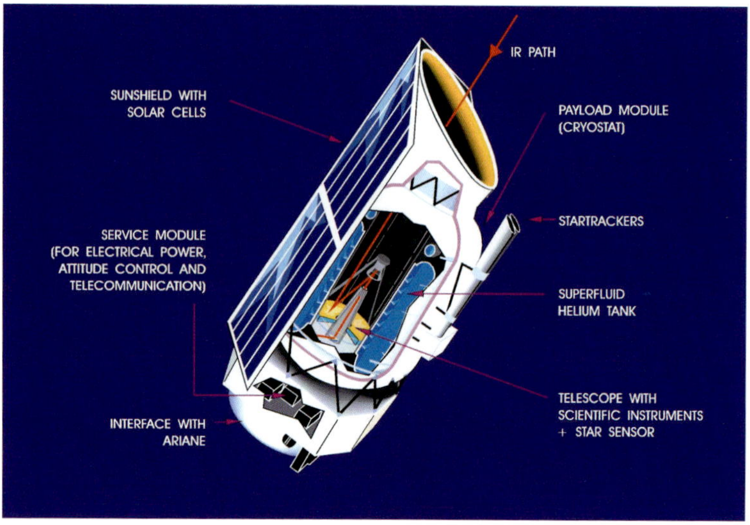

Figure 14.2: ESA's *ISO* which was operational from 1995 to 1998. The telescope and instruments were housed deep within a liquid-helium cryostat.

telescope of a typical relative emissivity of 3 % kept at a temperature of 80 K (such as *Herschel*, see below) the NEP due to the telescope alone is a few 10^{-16} W $\mathrm{Hz}^{-\frac{1}{2}}$ and dominates the noise from both detectors and the natural background. The need to cool both the telescope and the instruments for operation at long wavelengths is therefore obvious.

Cryogenic design

The need to cool telescope and instruments has a large influence on the design of the payload module of an IR satellite and practically determines the orbit into which the satellite must be placed to allow for passive cooling in outer space. In particular cooling of a telescope below 10 K constrains the design as the telescope has to be contained either within a cryostat or within a set of heat shields to provide passive cooling. Passive cooling is only efficient when the shields and radiators on the satellite have a clear and stable view to deep space. This means the best orbits for IR satellites are those far from the Earth, i.e., those with a high apogee, or those that are trailing the Earth or orbiting the L2 Lagrangian point. Nevertheless, low Earth orbits have also been employed. Figure 14.2 shows *ISO*, an ESA mission, launched in November 1995 into a highly eccentric Earth orbit. *ISO* had a telescope with a diameter of only 0.6 m placed within a cylindrical cryostat. This design closely resembles that of the first infrared space observatory, the NASA/Netherlands/UK *IRAS* (Neugebauer et al 1984), and the Japanese *AKARI* mission (Matsumoto 2000). The *Spitzer* mission of NASA employs a modified version of the cryostat design, where the 0.85 m telescope is mounted outside the

Figure 14.3: The ESA/NASA *Herschel* observatory which was launched on 14 May 2009 has a 3.5 m telescope that is passively cooled to between 80 K and 90 K. Only the focal-plane instruments are housed in the liquid-helium cryostat.

liquid helium cryostat, but inside a set of baffles. Both baffles and telescope are launched warm and, once in orbit, passively cooled to 35 K and further cooled to < 5 K by using the helium vapour emanating from the cryostat where the focal plane instruments are housed (Werner et al 2004).

The ESA *Herschel* space observatory, launched into orbit around L2 in May 2009 (Pilbratt 2004), is now the largest space telescope. The 3.5 m telescope (Figure 14.3) is mounted on top of the cryostat with the focal-plane instruments. A single sunshield prevents solar radiation from impinging on the mirrors. The mirrors have cooled to below 90 K in orbit. The lowest temperature achieved was ≈ 82 K and, due to the variation with the distance of the spacecraft from the Sun, it rises to ≈ 90 K at maximum. The instruments, however, are cooled to 1.7 K. They are mounted on an optical bench inside a liquid helium cryostat, whose technology is based on *ISO* heritage.

The use of cryogens for space missions has a number of drawbacks: for safety reasons the cryostat vacuum vessels must be substantial, and therefore heavy; the remote cryostat operation and gas-handling systems are complex and expensive and, most significantly, the cryogens are a finite resource and thus the lifetime of the missions is limited. *IRAS* lasted less than one year; *ISO* two and a half years; the cold phase of *Spitzer* about five years and *AKARI* about two years. *Herschel* has a design lifetime of three and half years. Future missions, such as the proposed Japanese *SPICA* (Nakagawa 2008), will dispense with cryogens, and instead use a combination of efficient baffling and long-lived closed-cycle coolers to bring both the telescope and the focal plane instruments to temperatures < 10 K (Sugita et

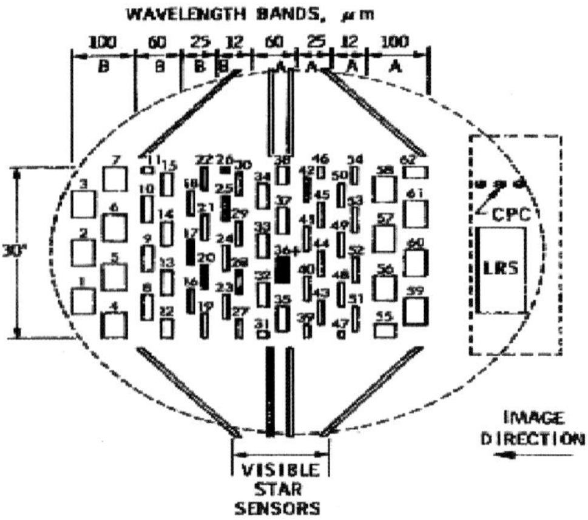

Figure 14.4: The IRAS focal plane layout.

al 2006). This means that, in principle, mission lifetime will only be limited by the common spacecraft constraints of mechanical and electrical failures.

Far-infrared surveys

Two decades of ground-based IR observations, although hindered by narrow and variable atmospheric transmission windows and strong infrared emission from the atmosphere proper, have established the importance of the IR and submillimetre wavelength range for astronomy. It was clear that the first step in space IR astronomy, where the absence of background from the atmosphere affords an increase in sensitivity by several orders of magnitude, would be to make an all-sky survey. A start was made by the US Air Force Cambridge Research Labs (now named AFGL) with a series of nine sub-orbital rocket flights, carrying a 16.5 cm telescope and instrument, cooled by liquid helium. In a total of 30 min of observing time, 90 % of the sky was covered in two bands centred at 11 µm and 20 µm, and smaller fractions at 4 µm (18 %) and at 27 µm (34 %). A first catalogue with 2000 sources was released in 1974. AFGL continued its rocket flights, and another 1000 sources were added. The harvest of less than one hour of sub-orbital observations showed the immense potential of space observations for IR astronomy.

In 1974, an infrared sky survey satellite was proposed to NASA. This proposal was merged with a similar initiative in the Netherlands, and somewhat later the United Kingdom joined in the project that became known as *IRAS*. The main objective was to carry out an all-sky survey in broad wavelength bands centred at (12, 25, 60 and 100) µm. A second objective was to carry out a mid-infrared (MIR) spectral survey of the strongest point sources and more detailed observations by pointed observations. The focal plane assembly, fed by the 0.6 m cooled telescope, contained not only the IR detectors, their masks, spectral filters and field lenses, but also visible detectors and the cold read-out electronics for all detectors. The visible detectors were to provide two-axis spacecraft attitude information from star

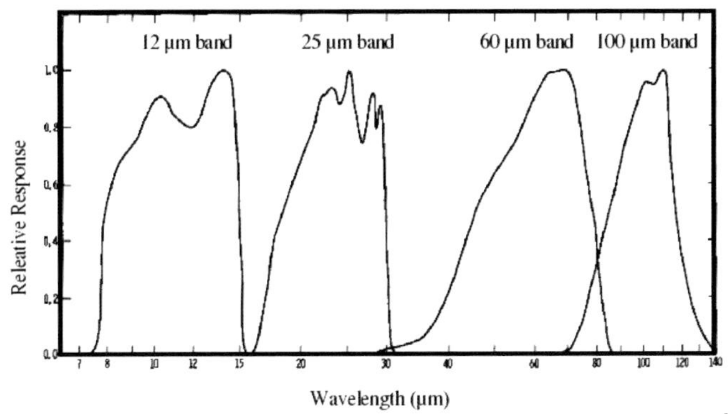

Figure 14.5: The IRAS photometric survey passbands.

crossings and their timing. The infrared detectors were divided over eight modules, two for each wavelength band, and each module contained seven or eight detectors. Figure 14.4 shows the lay-out of the focal plane with the visible and IR detectors and their modules and relative apertures, together with the scan direction. The detector materials used were Si:As (12 µm), Si:Sb (25 µm) and Ge:Ga (60 µm and 100 µm). The transmission bands of the combined filter and detector responses can be found in Figure 14.5.

IRAS was launched into a low Earth, Sun-synchronous polar orbit at 900 km altitude with a period of 103 min. This orbit allowed ground contact twice a day, for 10 min, to transmit the tape-recorded data from the last 10 to 14 hours of observations and to send the commands for the next 10 to 14 hours. The scan strategy was optimised to obtain a sky coverage of 95 % that was 98 % complete and 99.8 % reliable. The scan strategy aimed at redundant coverage with 70 % of the sky surveyed with twelve observation passes. The detector apertures had in-scan dimensions of $0.75'$ to $3'$ and cross-scan $4.5'$ to $5'$. The limiting point-source responsivity was about 0.5 Jy at (12, 25 and 60) µm and 1.5 Jy at 100 µm.

A second mission dedicated to an infrared all-sky survey is *AKARI*, the second infrared astronomy mission built by ISAS (now JAXA), Japan. This mission is similar to *IRAS* but having been launched more than 20 years later, is taking advantage of a new generation of detector technology. The telescope, with a 0.685 m effective primary mirror diameter, feeds the signals to two focal-plane instruments working over the 5 µm to 200 µm range. The Far Infrared Surveyor (FIS) covers the 50 µm to 200 µm range simultaneously in four bands using Ge:Ga photodetector arrays. The four bands cover 50 µm to 70 µm, 50 µm to 110 µm, 150 µm to 200 µm, and 110 µm to 200 µm (see Figure 14.6) with arrays of 2×20 and 3×20 unstressed Ge:Ga detectors for shorter wavelengths, and 2×15 and 3×15 stressed Ge:Ga detectors for the longer wavelengths. Wavelength pass bands are achieved with wire-grid polarisers, dichroic filters and band-pass filters.

Like *IRAS*, *AKARI* is also in a Sun-synchronous orbit. The survey strategy was designed to give sky coverage of 90 % using a cross-scan step between two orbits of $4'$ with an FOV of about $8'$ and pixel sizes of less than $1'$. Each object is

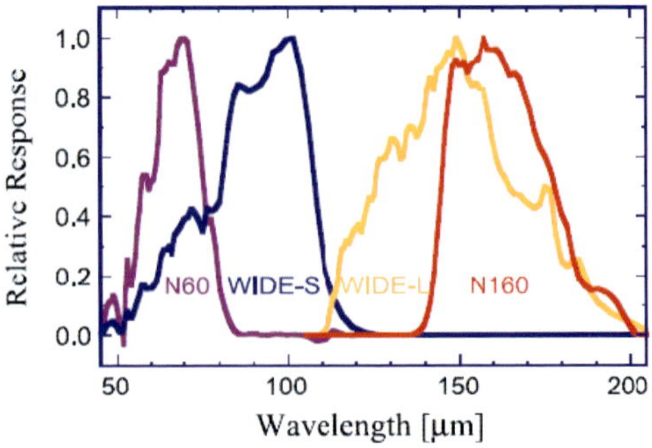

Figure 14.6: *AKARI* wavelength bands.

observed at least twice in a 100 min interval. The point-source flux detection limits at $S/N > 5$ for one scan in the all-sky survey are (2.4, 0.55, 1.4, and 6.3) Jy, one for each of the four bands.

Two other missions are worth mentioning as survey missions in the long-wavelength infrared regime: *COBE* and *Planck* (see also Chapter 8, Lamarre 2010). *COBE* (Boggess et al 1992) was designed to map the spectral distribution of the CMB as well as the spatial temperature fluctuations in the cosmic infrared background. In ten months in 1989/1990 it made an all sky survey at ten wavelengths between 1.25 μm and 240 μm with a large beam size ($42' \times 42'$). In addition to the main survey instrument, the FIRAS (FarInfraRed Absolute Spectrophotometer) instrument made an all-sky spectral survey showing emission from CO and from forbidden transitions in [C I], [C II], [N II]. The observation of the [N II] line at 205.3 μm was the first detection of this important line.

The *Planck* mission, launched together with *Herschel* on 14 May 2009, will carry out an all-sky survey in nine photometric bands ranging from 1 cm to 0.3 mm, using bolometers and radiometers. As a by-product of the CMB subtraction, *Planck* can supply all-sky maps of all the main sources of far-infrared emission with a beam size ranging from $5'$ at the shortest wavelengths to $\approx 30'$ at the longest.

Infrared imaging

Another constraint on the development of FIR observatories is the limited number of pixels available in long-wavelength photo-conducting detector arrays compared to shorter wavelength counterparts. This is because monolithic/thick film array technology has been developed for silicon material and not for the germanium material that provides long-wavelength sensitivity (see Chapter 30, Raab 2010). This has resulted in large-format, high-sensitivity, space-qualified arrays in Si:Ga, Si:As, Si:Sb, covering wavelengths up to about 38 μm. Attempts to develop monolithic Ge:Ga detectors have not been so successful. Therefore photo-conducting detector arrays for wavelengths beyond 50 μm have been constructed by combining single elements. For these reasons, true imaging arrays, i.e., ones that fully spatially sample the point-spread function (PSF) delivered by the telescope, have been difficult to develop for wavelengths longer than ≈ 38 μm. This does not mean, however, that there have not been FIR imaging missions. Indeed both *IRAS* and *AKARI* made images by scanning the spacecraft and reconstructing the sky from the detector signals as a function of time. In this section we will give an overview of the key missions and instrument that have been developed in the 25 years since the launch of *IRAS*, compare progress in the development of arrays in the MIR and FIR and look to the future of imaging in the FIR waveband.

***ISO*:** Two of the instruments on the *ISO* spacecraft had what were termed "cameras". The ISOCAM instrument (Cesarsky et al 1996) operated in the near IR (NIR) to MIR (2.5 μm to 17 μm) and is notable for its use of the first space qualified large format (32 × 32 pixel) imaging arrays. One array used InSb to cover the 2.5 μm to 5 μm band and the other used Si:Ga to cover the 4 μm to 17 μm band. The pixel sizes were relatively large compared to more recent developments (100 μm × 100 μm) and the spatial sampling was therefore limited. Additionally both arrays had various issues associated with slow time response (transients), latent images and, as with all space-borne photoconducting detectors, impacts from charged particles. The ISOPHOT instrument (Lemke et al 1996) operated at longer wavelengths and had two so-called cameras. The C100 was a 3×3 array of unstressed Ge:Ga pixels designed to image the sky over the 50 μm to 100 μm band and the C200, a 2 × 2 array of stressed Ge:Ga pixels operating over the 120 μm to 200 μm band. The modest array sizes meant that the effective pixel size on the sky was $44'' \times 44''$ for C100 and $89'' \times 89''$ for C200. These arrays significantly undersampled the PSF from the telescope and again the spatial sampling was limited necessitating multiple observations to create fully sampled images. The sensitivities of the C100 and C200 cameras were limited by both detector and confusion noise (the latter being the measurement uncertainty caused by other astrophysical sources). The detection limit (i.e., $S/N = 5$) at 90 μm was ≈ 28 mJy and at 160 μm was 140 mJy, both for a 500 s integration. The former is detector-noise limited, the latter confusion limited by the galactic cirrus.

***Spitzer*:** The Multiband Imaging Photometer for *Spitzer* (MIPS) (Rieke et al 2004) was designed to overcome the limitations of previous instruments by using three true imaging arrays to cover photometric bands centred at (24, 70 and 170 μm. The arrays were, respectively, a 128 × 128 pixel Si:As array; a 32×32 pixel unstressed Ge:Ga array and a 2 × 20 pixel-stressed Ge:Ga array. All were sized, together with the design of the optical train, to oversample the PSF of the telescope. In the event only half the 70 μm array was operational due to a cable failure before launch and both the 70 μm and 170 μm arrays suffered from additional detector losses and extraneous noise that reduced their in-flight performance compared to pre-launch predictions. The achieved detector-limited ($S/N = 5$) sensitivity for point sources was 0.11 mJy at 24 μm, 7.2 mJy at 70 μm and 30 mJy at 170 μm for a 500 s integration. The extragalactic confusion limits for the same bands for the *Spitzer* telescope are given by Dole et al (2004) as (0.056, 3.2 and 40) mJy.

***Herschel* and beyond:** The ESA/NASA *Herschel* observatory has as its primary objective to increase the spatial resolution available in the FIR waveband. Given the constraints discussed before it achieves this by using a large (3.5 m) primary mirror but only cooled to ≈ 80 K. The significantly increased spatial resolution of the *Herschel* telescope requires detectors with larger formats in order to correctly sample the PSF. As discussed above, attempts to develop monolithic arrays in Ge:Ga have not been successful, instead the *Herschel* FIR instrument (PACS) instrument uses two bolometric arrays manufactured employing silicon etching techniques, high-resistivity silicon thermistors and fully-integrated readout electronics (Agnese et al 2003). These are cooled to below 300 mK using a dedicated ^{3}He sorption cooler (Duband and Collaudin 1999). Two arrays are used to cover the 60 μm to 130 μm and 130 μm to 200 μm bands. These have 32×64 and 16×32 pixels, respectively, covering an FOV of $1.75' \times 3.5'$. The telescope background limits the intrinsic sensitivity of the instrument to an $S/N = 5$ detection of 8 mJy in a 500 s observation.

However, the confusion-limited responsivity for a 3.5 m telescope is estimated as 0.16 mJy at 70 μm and 10 mJy at 160 μm. A very much colder telescope than *Herschel* is thus required to allow us to see down to these levels and truly explore the unresolved cosmic infrared background. In order to take advantage of such a low-level of background, a new generation of bolometric detectors based on superconducting technology is required (Swinyard and Nakagawa 2009). The recent development of large-format bolometric arrays operating at temperatures below 1 K (Holland et al 2006) for the submillimetre wavelength range shows great promise for further progress and developments to use similar technologies for the FIR are underway for future missions.

Infrared spectrometers

The techniques of optical spectroscopy (prisms, gratings, Fourier transform spectrometers (FTS) and Fabry–Perot interferometers) have also been utilized in IR spectrometers for space missions. For an overview see Table 14.1: we include here instruments operating in the MIR as these are of interest in the development of techniques for the longer wavelengths. The choice of the technique was determined by the resolution required to achieve science objectives, available satellite resources (mass, electrical power, number of connections, physical space, cooling power etc.) and the number of pixels available in the focal plane. There are too many examples to give details of them all and we concentrate on a few to illustrate the design issues associated with each technique.

The *IRAS* Low Resolution Spectrometer (LRS) covered wavelengths between 8 μm and 22 μm with a rather low spectral resolution of $R = \lambda/\Delta\lambda \approx 20$. It operated during the entire survey providing spectra of the brighter point sources. The scanning nature of the mission meant that a passive, slitless prism spectrograph was the best choice of design. The dispersion direction was aligned with the survey scan direction with the field of view determined by a rectangular mask in the focal plane of $6'$ (dispersion) by $15'$ (cross-dispersion). The spectral resolution was determined by a combination of the spectrometer exit slit, the telescope diffraction and the detector electronic filtering. For the short wavelengths three Si:Ga detectors covered each $5'$ of the aperture width and two Si:As detectors covered $7.5'$ each in the long-wavelength section. All detectors used trans-impedance pre-amplifiers with a sampling frequency of 32 Hz, corresponding to a $7.2''$ interval in scan sampling.

Table 14.1: Review of types of spectrometer used in the MIR and FIR.

Type	Resolution-determining element	Achieved resolution, $\lambda/\Delta\lambda$	Wavelength range, $\lambda/\mu m$	Mission-instruments
Prism	prism size and material	≈ 25	8 to 22	*IRAS*-LRS
Grating	grating size	100	2.5 to 12	*ISO*-PHOT-S
		200	42 to 198	*ISO*-LWS
		1500	2 to 43	*ISO*-SWS
		1500 to 2000	57 to 210	*Herschel*-PACS
		4000	5 to 28	*JWST*-MIRI
Fabry–Perot	quality of etalons	9 000	42 to 198	*ISO*-LWS
		20 000 to 30 000	11 to 44	*ISO*-SWS
Fourier transform	mirror travel range	≈ 200 to 1000	100 to 100 000	*COBE*-FIRAS
			200 to 670	*Herschel*-SPIRE
Heterodyne	LO stability / spectrometer	$> 10^6$	530 to 610	*SWAS*, *ODIN*
			157 to 600	*Herschel*-HIFI

More details about the LRS instrument description can be found in Wildeman et al (1983).

Spectroscopy with *ISO*. All four *ISO* instruments had spectroscopy capabilities. Of particular interest in our survey of instrumental techniques are the two spectrometers: *ISO*-LWS and *ISO*-SWS. *ISO*-LWS employed a scanning grating mechanism to provide low-resolution ($\lambda/\Delta\lambda \approx 200$) spectroscopy from 42 μm to 196 μm combined with a pair of Fabry–Perot interferometers to provide high-resolution ($\lambda/\Delta\lambda \approx 9000$) spectroscopy. The Fabry–Perots are discussed in some detail in Chapter 18 by Griffin and Ade (2010) together with a picture of the instrument. The very large wavelength range covered by the instrument was achieved by using a single grating in first order to cover from $\approx$ 90 μm to 197 μm whilst simultaneously using the second order to cover 43 μm to $\approx$ 96 μm. Ten single pixel detectors sensitive to first and second order radiation were interleaved on a circular structure at the focus of a spherical mirror using filters on the detectors to achieve order separation. Four stressed Ge:Ga detectors covered from 100 μm to 197 μm, five unstressed Ge:Ga detectors covered from 50 μm to 100 μm and a single Ge:Be detector covered the 43 μm to 50 μm band. The whole wavelength range was covered by scanning the grating by $\pm 7°$ about its rest position. An entire spectrum was collected in low resolution in a single scan of the grating within $\approx$ 10 min. Integration time was built up by taking repeated scans either over short angular ranges for particular lines or over the entire spectrum. The *ISO*-SWS instrument covered from 2 μm to 43 μm by use of two separate gratings, with the input beam being divided chromatically between the short (2.4 μm to 24 μm) and long (24 μm to 43 μm) wavelength optical trains by a dichroic filter. Similarly to the *ISO*-LWS a set of 15 individual detectors was used, each covering a short section of each

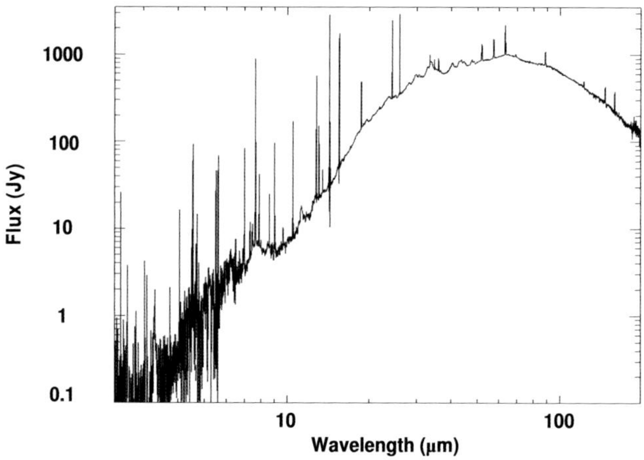

Figure 14.7: Spectrum of NGC 6302 observed by *ISO* (Molster et al 2001). Clearly visible are the numerous emission lines due to fine-structure transitions in the ionized and neutral gas superimposed on the grey body spectrum of the dust surrounding this evolved star.

spectrum necessitating the use of a mechanical scanner to fully sample the spectral resolution. In this case a flat mirror was used to scan the spectrum over the detectors rather than the gratings themselves.

The two *ISO* spectrometers together were the first, and remain the only, infrared facility able to take spectra from the NIR to the FIR, some two orders of magnitude in wavelength, using a single observatory. An example of the remarkable legacy left by the *ISO* spectrometers is shown in Figure 14.7.

***Spitzer*:** The infrared spectrograph on *Spitzer* (Houck et al 2004) covered 5.2 μm to 37.2 μm with a low-resolution mode ($R \approx 80$ to 130) and a medium-resolution mode ($R \approx 600$). The availability of large-format arrays meant that the instrument did not need to contain moving parts using four separate modules to achieve the wavelength coverage and different spectral resolutions. The four 128×128 pixel detector arrays sampled both the spatial (along the slits) and spectral dimensions without the need to scan the gratings. Two Si:As detectors covered from 5 μm to 26 μm and two Si:Sb detectors covered 14 μm to 38 μm in the various modules. Although this instrument had lower spectral resolution than the *ISO*-SWS, it had dramatically better detectors and thus a sensitivity an order of magnitude better than *ISO* allowing it to make discoveries such as the presence of silicate dust in high-redshift galaxies (Weedman et al 2006) and take the first infrared spectrum of a transiting exo-planet (Swain et al 2008).

***Herschel*:** There are two "direct detection" spectrometers on *Herschel*. The SPIRE instrument is an imaging FTS and is described in some detail in Chapter 18. The PACS instrument uses a single grating with Ge:Ga detectors in two 25×16 arrays, one stressed one unstressed, to cover the wavelength range from 52 μm to

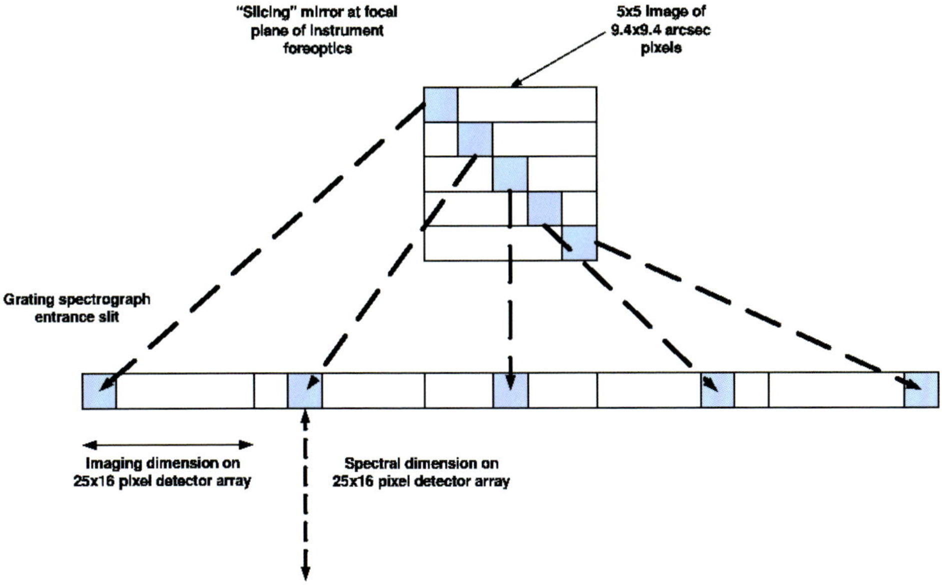

Figure 14.8: Diagram illustrating the operation of the *Herschel* PACS slicing IFU for spectroscopic imaging. The five-element "book stack" slicing mirror and associated optics rearranges the image onto the entrance slit of the (stigmatic) spectrograph. 25 pixels of the detector array in the focal plane of the spectrograph are used for imaging within each slice image and 16 are used to sample the dispersed spectrum from the grating.

210 μm. Three orders from the grating are used to cover the full wavelength range. The spectral resolution varies from $R \approx 1500$ to 3000 and, as with the *ISO*-LWS, the grating needs to be mechanically scanned to provide full spectral sampling as the detectors do not fully sample the spectral resolution element. The major advance with this instrument is the use of a slicing mirror integral field unit (IFU) to provide some spectral imaging capability. A simplified view of the operation of a slicing mirror IFU is shown in Figure 14.8. The sliced image is arranged along the input slit of the grating spectrometer. Single "pixels" of the field are rearranged on different positions along the slit of a (stigmatic) spectrometer, so that the spectrum of the radiation in each one of the pixels is obtained one above the other. This technique is widely used at shorter wavelengths on ground-based instruments and will be used on the MIR spectrometer for the *JWST* (Wright et al 2004). The advantage at shorter wavelengths, compared to the FIR, is that very large-format arrays are available (1024×1024 in the case of *JWST*) which allow a much greater instantaneous wavelength coverage and high spectral sampling in a single grating setting. The small detector arrays available to PACS means that a wavelength interval of only ≈ 0.5 μm is covered by the detector array at each grating position. This means that covering a large wavelength range over a wide field of view will take an unrealistically long time. The imaging FTS systems described in Chapter 18 (Griffin and Ade 2010) have a multiplex advantage in this respect,

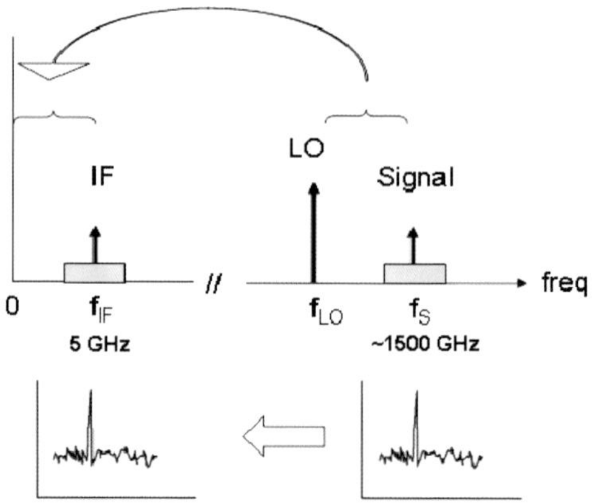

Figure 14.9: Principle of heterodyne detection. The high-frequency signal band is down-converted to the lower IF band without loss of spectral or phase information.

albeit with the drawback of an increased background and reduced sensitivity to an individual line.

Heterodyne receivers

Throughout the radio regime including millimetre, submillimetre, and FIR wavelengths, heterodyne receivers are used to achieve very high to extremely high spectral resolution. The principle of heterodyne detection consists in shifting a high-frequency band (containing the astronomical signal) to a much lower frequency band (the "intermediate" frequency, IF) where the signal can be amplified and spectroscopically analyzed. This frequency down-conversion is done maintaining the spectral and phase information of the original signal (Figure 14.9). In other words, a heterodyne receiver can be viewed as a frequency down-converter. Figure 14.10 shows the layout of a typical heterodyne receiver system. For a more detailed introduction into astronomical heterodyne receivers, see, e.g., Tiuri and Räisänen (1986) and Rieke (2003). The signal at frequency f_S from the telescope is first optically combined by a beam splitter or diplexer with a stable local oscillator (LO) signal at frequency f_{LO} (which is close to f_S) and then coupled to the mixer via an antenna. The mixer produces (among others) the difference frequency $f_{IF} = |f_S - f_{LO}|$ called intermediate frequency (IF) as output which is electronically amplified and analyzed in a spectrometer. The IF output of the receiver is an exact (frequency-shifted) copy of the input spectrum, the spectral resolution of a heterodyne receiver (also called "frontend") is determined by the spectrometer (also called "backend"). The reason for using a mixer as first element in high-frequency (millimetre, submillimetre and FIR) receivers lies in the fact that no suitable electronic components are available at these frequencies. By first down-converting the high-frequency sky signal (in the 0.3 THz to 3 THz range) to the lower IF frequency (typically in the few gigahertz range), amplifiers and other electronic components working in that range can be used. The mixer, local oscillator and first IF amplifier are the

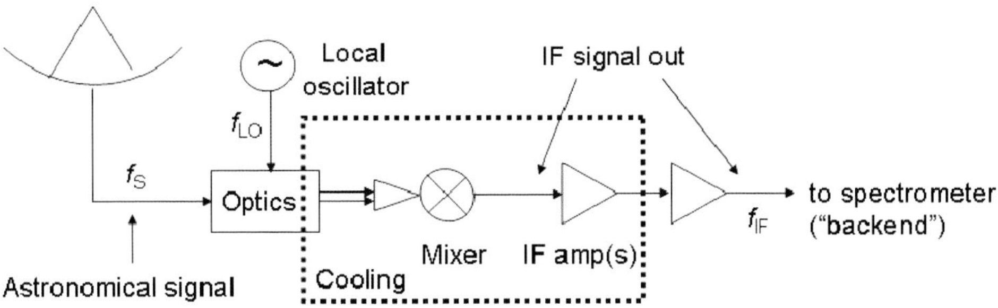

Figure 14.10: Block diagram of a heterodyne receiver (also called "frontend").

most important components of the receiver since they determine to a large part the sensitivity of the whole system. At far-infrared and submillimetre frequencies, three types of mixers are being used: the SIS (superconductor-insulator-superconductor) mixer, the HEB (hot-electron-bolometer) mixer, and the Schottky mixer. The physical principles and properties of these mixers as well as suitable local oscillator techniques are explained in Chapter 31 by Wild (2010). SIS and HEB mixers are superconducting devices and need to be cooled to cryogenic temperatures of 4 K or below for operation. Although Schottky mixers can operate at room temperature they are often cooled to achieve better sensitivity. Cryogenic IF amplifiers are specifically developed by research institutes for use in astronomical heterodyne systems (e.g., López-Fernández et al 2003; Risacher and Belitsky 2003).

Over the past decades, mixers, local oscillators and IF amplifiers — mostly developed by specialized groups in research institutes and universities — have been improved significantly, and today's heterodyne receivers for astronomical use approach the fundamental sensitivity limit of any heterodyne system (the so-called "quantum limit").

We will now very briefly explain the general principle of heterodyne detection which is done by mixing the signal frequency with a local oscillator frequency in a non-linear mixing device to obtain a lower intermediate frequency with the same signal information content. For illustration we consider the mixing of an astronomical signal with amplitude E_{S}, phase ϕ_{S} and angular frequency $\omega_{\mathrm{S}} = 2\,\pi\, f_{\mathrm{S}}$, i.e., $V_{\mathrm{S}} = E_{\mathrm{S}}\cos(\omega_{\mathrm{S}}\, t + \phi_{\mathrm{S}})$, and a local oscillator (LO) signal $V_{\mathrm{LO}} = E_{\mathrm{LO}}\cos(\omega_{\mathrm{LO}}\, t)$, respectively, in a non-linear mixer with a square-law I-V characteristic $I \propto V^2$. In principle any nonlinear device can be used as mixer — here we use a square-law device for the simplicity of illustration. The output current of the mixer is proportional to the square of the input:

$$I \propto (V_{\mathrm{S}} + V_{\mathrm{LO}})^2 = [E_{\mathrm{S}}\cos(\omega_{\mathrm{S}}\, t + \phi_{\mathrm{S}}) + E_{\mathrm{LO}}\cos(\omega_{\mathrm{LO}}\, t)]^2 \quad . \tag{14.5}$$

This expression can be expanded and written as

$$\begin{aligned} I &\propto E_{\mathrm{S}}^2\cos^2(\omega_{\mathrm{S}}\, t + \phi_{\mathrm{S}}) + E_{\mathrm{LO}}^2\cos^2(\omega_{\mathrm{LO}}\, t) + 2E_{\mathrm{S}}E_{\mathrm{LO}}\cos(\omega_{\mathrm{S}}\, t + \phi_{\mathrm{S}})\cos(\omega_{\mathrm{LO}}\, t) \\ &= \frac{1}{2}E_{\mathrm{S}}^2\,[1 + \cos(2\omega_{\mathrm{S}}\, t + 2\phi_{\mathrm{S}})] + \frac{1}{2}E_{\mathrm{LO}}^2\,[1 + \cos(2\omega_{\mathrm{LO}}\, t)] + \\ &\quad E_{\mathrm{S}}E_{\mathrm{LO}}\cos[(\omega_{\mathrm{S}} + \omega_{\mathrm{LO}})\, t + \phi_{\mathrm{S}}] + E_{\mathrm{S}}E_{\mathrm{LO}}\cos[(\omega_{\mathrm{S}} - \omega_{\mathrm{LO}})\, t + \phi_{\mathrm{S}}] \quad . \end{aligned} \tag{14.6}$$

In Equation 14.6 we see that the output of the square-law mixer contains the high-frequency terms $2\,\omega_\mathrm{S}, 2\,\omega_\mathrm{LO}, \omega_\mathrm{S}+\omega_\mathrm{LO}$ and the difference (intermediate) frequency $\omega_\mathrm{S}-\omega_\mathrm{LO}$. The high frequencies are filtered out, and the intermediate frequency term is proportional to the signal amplitude and preserves the phase:

$$I \propto E_\mathrm{S} E_\mathrm{LO} \cos[(\omega_\mathrm{S}-\omega_\mathrm{LO})\,t+\phi_\mathrm{S}] \quad . \tag{14.7}$$

In summary, by combining two frequencies ω_S and ω_LO in a square-law device we have produced the double of each frequency ($2\,\omega_\mathrm{S}$ and $2\,\omega_\mathrm{LO}$), the sum of the two frequencies ($\omega_\mathrm{S}+\omega_\mathrm{LO}$), and the difference of the two frequencies ($\omega_\mathrm{S}-\omega_\mathrm{LO}$). In Equations 14.6 and 14.7 we see that an IF frequency ($\omega_\mathrm{S}+\omega_\mathrm{LO}$) leads to the same result, i.e., a mixer is sensitive to two signal frequency bands separated by $2\,\omega_\mathrm{IF}$, one above the LO frequency and one below the LO frequency. Such a mixer is called a *double-sideband* mixer. It is possible to design a mixer in such a way that only one sideband is down-converted (called a *single-sideband* mixer) or the two sidebands are available separately at two IF outputs (called a *sideband-separating* mixer).

In practice, the LO frequency ω_LO is chosen close to the signal frequency ω_S in order to have a low-IF frequency (for example around 6 GHz). Figure 14.9 is a graphical illustration of the down-conversion of the high-frequency signal to the lower IF frequency. Heterodyne detection has a number of advantages as compared to direct detection. It is the only way to achieve very high to extremely high spectral resolution ($R \geq 10^6$, i.e., a velocity resolution of better than 1 $\mathrm{km\,s^{-1}}$) in the radio to far-infrared range which is crucial for velocity-resolved spectroscopy of line emission of, e.g., interstellar atoms, ions, and molecules (Figure 14.11 shows an example taken with the Swedish *ODIN* satellite). Another important advantage of heterodyne detection is the fact that in shifting the frequency band, the phase of the signal is conserved. A heterodyne receiver can thus detect both amplitude *and* phase, and it is therefore often called a "coherent" receiver, in contrast to direct ("incoherent") detectors where the phase information is lost and only amplitude is detected. Furthermore, the output signal of a heterodyne receiver can be amplified and divided into many copies without adding a significant amount of noise. The properties of phase conservation and signal copies make heterodyne detection very suitable for interferometry. In fact, all existing radio and millimetre wave interferometers use heterodyne receivers. Disadvantages of heterodyne detection include the existence of a fundamental sensitivity limit and a smaller detection bandwidth as compared to direct detection. Figure 14.12 shows the HIFI focal plane unit containing seven dual-polarization receivers operating in double-sideband mode from 480 GHz to 2000 GHz. The fundamental sensitivity limit of heterodyne detection (e.g., Tucker and Feldman 1985; Feldman 1987), the so-called "quantum limit", as imposed by Heisenberg's uncertainty principle, is given by the minimum achievable noise temperature for a single-sideband receiver (e.g., Kerr 1999)

$$T_\mathrm{rec,min} = h\,\nu/k_\mathrm{B} \quad , \tag{14.8}$$

with h the Planck constant, ν the receiver input frequency, and k_B the Boltzmann constant. For a double-sideband receiver, the minimum achievable noise temperature is half the value given in Equation 14.8. State-of-the-art submillimetre and

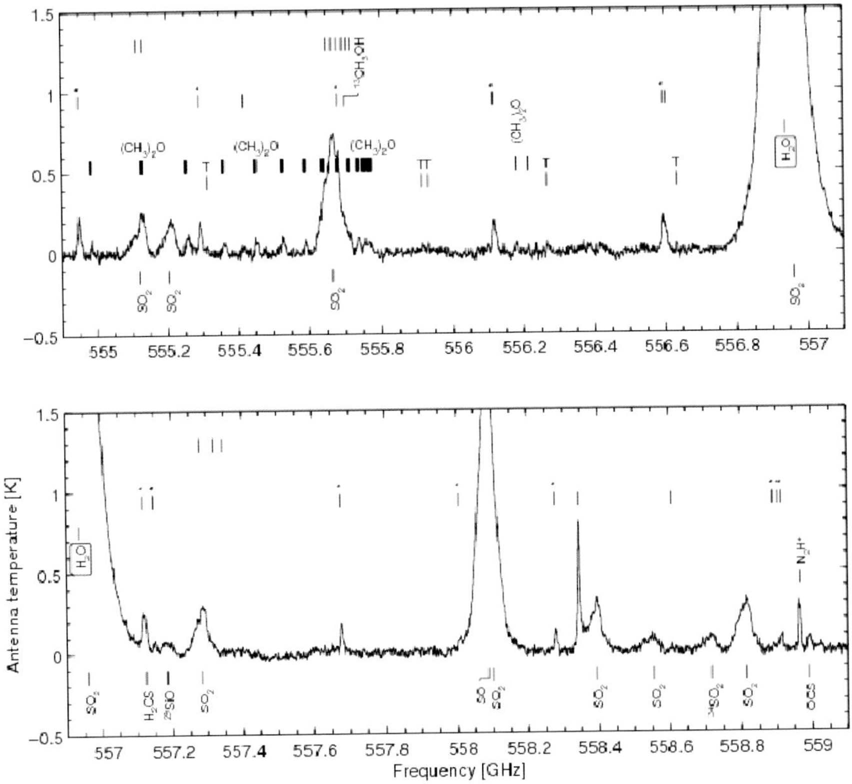

Figure 14.11: Example of a heterodyne spectrum from a wider submillimetre spectral survey by the Swedish *ODIN* satellite, showing the variety of molecules present in the star-forming region in Orion (Olofsson et al 2007; Persson et al 2007).

far-infrared receivers have a sensitivity typically three to ten times $h\,\nu/k_{\rm B}$. For *Herschel*-HIFI SIS and HEB mixers have been developed for use in a space mission. Figure 14.13 shows the achieved flight mixer noise temperatures. They represent a significant improvement in sensitivity (up to a factor of 50) as compared to the previous heterodyne space missions. Submillimetre and far-infrared heterodyne observations are dominated by the noise in the receiving system, i.e., the astronomical signal is much weaker than the noise contribution from the receiver. The detectable signal $\Delta T_{\rm min}$ as function of the system noise $T_{\rm sys}$, the integration time t and detection bandwidth $\Delta\nu$ is given by the radiometer formula

$$T_{\rm min} = \frac{c_1\,T_{\rm sys}}{\sqrt{t\,\Delta\nu}} \quad , \tag{14.9}$$

with c_1 a constant of order unity depending on the observing mode. We see that the required integration time t to detect a signal of a certain intensity depends quadratically on the system noise making the use of low-noise receivers and mixers extremely important. Far-infrared observations are often carried out in an "on-off" mode, i.e., the telescope observes the astronomical source ("on" position) and

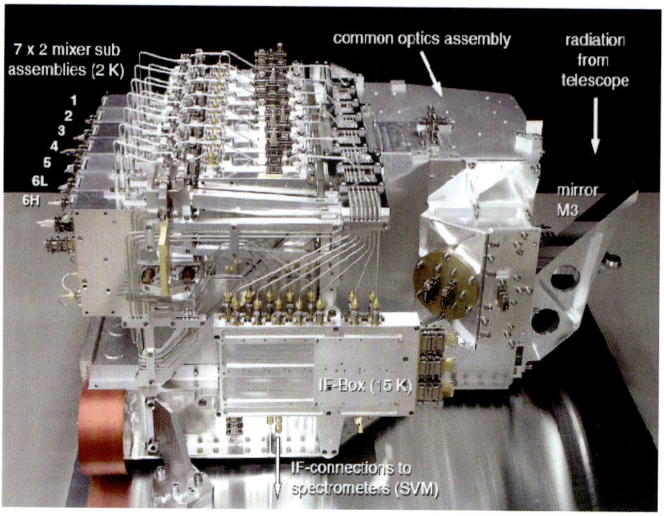

Figure 14.12: The flight-model focal-plane unit of the Heterodyne Instrument for the Far-Infrared (HIFI), an advanced seven-channel heterodyne receiver for observations from 480 GHz to 2000 GHz with the *Herschel* space observatory (Image: SRON).

immediately before and/or after a nearby position on the sky without the source emission ("off" position). The source signal is obtained by subtracting the "off" spectrum from the "on" spectrum. This method allows detecting very small signals on top of a large noise floor. In this way a far-infrared heterodyne system with a noise temperature of, e.g., 100 K to 1000 K (a typical sensitivity) can reliably detect astronomical signals with an intensity in the millikelvin range (i.e., 10^5 to 10^6 times smaller than the system noise).

Table 14.2 gives an overview of astronomical space missions that have used or will use heterodyne receivers (Earth-observation space missions which have used heterodyne receivers are not included). While Schottky receivers have been used on a number of space missions, SIS and HEB mixers have their first space flight with HIFI.

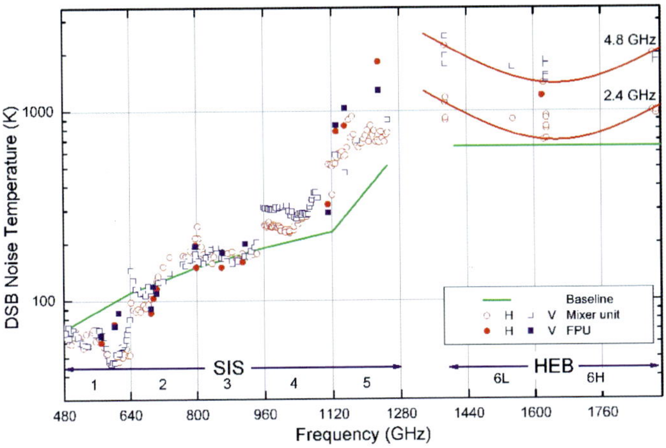

Figure 14.13: HIFI flight mixer performance at mixer level (open symbols) and after integration into the focal-plane unit (from de Graauw et al 2005).

Table 14.2: Astrophysical submillimetre/far-infrared space missions using heterodyne receivers.

Mission	Agency	Launch	Frequency, ν/GHz	Status (2009)
SWAS	NASA	1998	487, 492, 551, 548, 557	completed
ODIN	Sweden	2001	119, 486 to 504, 541 to 580	operational
Rosetta-MIRO	ESA/NASA	2004	190 and 562	on way to comet rendezvous
Herschel-HIFI	ESA	2009	480 to 1900	operational
VESPER	NASA	2012?	455 to 485, 535 to 565	concept
Millimetron	Russia	2017?	$>$ 300, i.e., millimetre, submillimetre	concept

Bibliography

Agnese P, Cigna C, Pornin J-L (plus ten authors) (2003) Filled bolometer arrays for *Herschel*/PACS in millimeter and submillimeter detectors for astronomy. Proc SPIE 4855:108–114

Boggess NW, Mather JC, Weiss R (plus 15 authors) (1992) The *COBE* Mission: Its design and performance two years after the launch. Astrophys J 397:420–429

Cesarsky CJ, Abergel A, Agnese P (plus 63 authors) (1996) ISOCAM in flight Astronom Astrophys 315:L32–L37

De Graauw Th, Caux E, Guesten R (plus nine authors) (2005) The Herschel-Heterodyne Instrument for the Far-Infrared (HIFI). Bull Am Astron Soc 37:1219

Dole H, Rieke GH, Lagache G, Puget J-L (plus 22 authors) (2004) Confusion of extragalactic sources in the mid- and far-infrared: *Spitzer* and beyond. Astrophys J Suppl:154:93–96

Duband L, Collaudin B (1999) Sorption coolers development at CEA-SBT Cryogenics 39:659–663

Feldman MJ (1987) Quantum noise in the quantum theory of mixing. IEEE Trans Magnetics MAG-23(2):1054–1057, DOI 10.1109/20.133940

Griffin MJ, Ade P (2010) Narrow-band imaging by use of interferometers. ISSI SR-009:313–326

Holland W, MacIntosh, M, Fairley (plus 30 authors) (2006) SCUBA-2: a 10 000 pixel submillimeter camera for the James Clerk Maxwell Telescope. Proc SPIE 6275:62751E, DOI: 10.1117/12.671186

Houck JR, Roellig TL, Van Cleve J (plus 32 authors) (2004) The Infrared Spectrograph (IRS) on the *Spitzer* Space Telescope. Astrophys J Suppl 154:18–24

Kerr AR (1999) Suggestions for revised definitions of noise quantities, including quantum effects. IEEE Trans MTT 47(3):325–329, DOI 10.1109/22.750234

Lamarre J-M (2010) The Cosmic Microwave Background. ISSI SR-009:149–162

Lemke D, Klaas U, Abolins J (plus 46 authors) (1996) ISOPHOT – capabilities and performance. Astronom Astrophys 315:L64–L70

López-Fernández I, Gallego-Puyol JD, Diez C (plus two authors) (2003) Wide band ultra low noise cryogenic InP IF amplifiers for the *Herschel* mission radiometers. Proc SPIE 4855:489–500

Matsumoto T (2000), *ASTRO-F* (IRIS) mission. Proc SPIE:4013

Molster FJ, Lim TL, Sylvester RJ (plus six authors) (2001) The complete *ISO* spectrum of NGC 6302. Astronom Astrophys 372:165–172

Nakagawa T (2007) personal communication

Nakagawa T (2008), *SPICA* mission for mid- and far-infrared astronomy. Proc SPIE 7010:70100H-70100H-8

Neugebauer G, Habing HJ, van Duinen R (plus 25 authors) (1984) The Infrared Astronomical Satellite (*IRAS*) mission. Astrophys J 278:L1-L6

Olofsson AOH, Persson CM, Koning N (plus 14 authors) (2007) A spectral line survey of Orion KL in the bands 486-492 and 541-577 GHz with the *Odin* satellite I. The observational data. Astronom Astrophys 476:791-806, DOI: 10.1051/0004-6361:20077229

Parkes W, Gundlach AM, Dunare CC (plus four authors) (2004) Realization of a large-area microbolometer sensor array for submillimeter astronomy applications:

SCUBA-2. Proc SPIE 5498:407–415

Persson CM, Olofsson AOH, Koning N (plus 14 authors) (2007) A spectral line survey of Orion KL in the bands 486–492 and 541–577 GHz with the *Odin* satellite II. Data analysis. Astronom Astrophys 476:807-828, DOI: 10.1051/0004-6361:20077225

Pilbratt G (2004) *Herschel* mission: status and observing opportunities. Proc SPIE 5487:401–412

Raab W (2010) Semiconductors for low energies: incoherent infrared/submillimetre detectors. ISSI SR-009:485–501

Rieke GH (2003) Detection of light: From the ultraviolet to the submillimeter (2nd edition). Cambridge: Cambridge University Press

Rieke GH, Young ET, Engelbracht CW (plus 40 authors) (2004) The Multiband Imaging Photometer for *Spitzer* (MIPS). Astrophys J Suppl 154:25–29

Risacher C, Belitsky V (2003) GaAs HEMT low-noise cryogenic amplifiers from C-Band to X-Band with 0.7 K/GHz noise temperature. IEEE Microwave and Wireless Components Lett 13(3):96-98, DOI 10.1109/LMWC.2003.810116

Sugita H, Nakagawa T, Murakami H (plus five authors), (2006) Cryogenic infrared mission JAXA/SPICA with advanced cryocoolers. Cryogenics 46:149–157

Swain MR, Bouwman J, Akeson RL (plus two authors) (2008) The mid-infrared spectrum of the transiting exoplanet HD 209458b Astrophys J 674 482–497

Swinyard B, Nakagawa T and the SPICA consortium (2009) The space infrared telescope for cosmology and astrophysics: *SPICA* A joint mission between JAXA and ESA. Exp Astron 23:193–219

Tiuri ME, Räisänen AV (1986) Radio-telescope receivers. In: Kraus JD, Radio Astronomy (2nd edition), Cygnus-Quasar Books, Powell, Ohio

Tucker JR, Feldman MJ (1985) Quantum detection at millimeter wavelengths. Rev Mod Phys 57(4):1055–1113, DOI 10.1103/RevModPhys.57.1055

Weedman D, Polletta M, Lonsdale CJ (plus seven authors) (2006) Active Galactic Nucleus and starburst classification from *Spitzer* mid-infrared spectra for high-redshift SWIRE sources. Astrophys J 653:101–111

Werner MW, Roellig TL, Low F (plus 23 authors) (2004) The *Spitzer* space telescope mission. Astrophys J Supp 154:1–9

Wild W (2010) Coherent far-infrared / submillimeter detectors. ISSI SR-009:503–523

Wright GS, Rieke GH, Colina L (plus 15 authors) (2004) The *JWST* MIRI instrument concept. Proc SPIE 5487:653–663

— 15 —

Goniopolarimetric techniques for low-frequency radio astronomy in space

Baptiste Cecconi[I]

Abstract

The principles of space-based radio astronomy are briefly introduced. As the wavelength range considered does not allow the use of focusing systems, goniopolarimetric or direction-finding techniques have been developed. These techniques are presented, and their limitations are discussed. An example from a recent study illustrates the techniques.

Introduction

The frequency band of interest for low-frequency radio astronomy ranges from a few kilohertz to several tens of megahertz (see Figure 15.1a for planetary radio emissions). In this frequency range, the corpuscular description is never used, the electromagnetic wave description is used instead. The sampled wave quantities are then the electric and/or magnetic field components.

Space-based radio instrumentation and interplanetary missions, such as the *Voyager*, *Galileo* or *Cassini* missions, are required due to several constraints: (i) radio emissions below 10 MHz cannot be observed from the surface because of the Earth's ionosphere cutoff; (ii) low-frequency emissions of the outer planets cannot be observed from Earth orbit, because of the solar wind plasma frequency cutoff (see Figure 15.1b); (iii) the space environment of Earth is polluted by man-made radio interferences above 10 MHz. Given this last point, an option not discussed here relies on lunar-based observations from the far side of the Moon.

The angular resolution of a classical telescope, with a "dish" reflector concentrating radio waves onto an array of antennas that will record the image formed at the focal plane of the telescope, is defined by λ/D, where λ is the wavelength of observation and D the aperture size of the telescope. The size required for such a telescope (≈ 1 km to 10^4 km for $\lambda/D \ll 1$) is totally unrealistic for space-based observations. Hence, simple electric dipoles with a length L of a few metres (or tens of meters) are used in general. They have a poor intrinsic angular resolution ($\lambda/L \approx 1$). Goniopolarimetric techniques (allowing retrieval of flux, polarization

[I]LESIA – Observatoire de Paris, Meudon, France

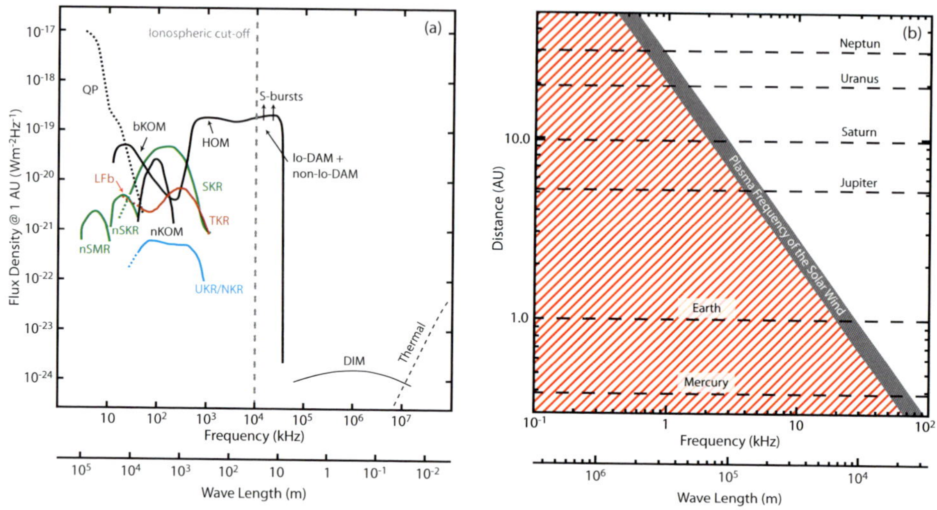

Figure 15.1: **(a)** Typical spectra for planetary radio emissions in the solar system. The radio emissions from Jupiter are in black (nKOM: narrowband kilometric radiation; bKOM: broadband kilometric radiation; QP: quasi-periodic bursts; HOM: hectometric radiation; S-bursts: short (or millisecond) bursts; Io-DAM: Io-controlled decametric radiation; non-Io-DAM: non-Io-controlled decametric radiation; DIM: decimetric radiation; Thermal: black body thermal radiation); from Saturn in green (nSMR: narrowband Saturn myriametric radiation; nSKR: narrowband Saturn kilometric radiation; SKR: Saturn kilometric radiation); from Earth in red (LFb: low-frequency bursts; TKR (or AKR): terrestrial (or auroral) kilometric radiation); and from Uranus and Neptune in blue (UKR/NKR: Uranus/Neptune kilometric radiation). The vertical grey dashed line represents the Earth ionospheric cut-off. Figure adapted from Zarka (1992), updated with Zarka et al (2004); Lamy et al (2008) and Lamy (personal communication 2008). **(b)** Typical range (grey area) of the plasma frequency in the solar wind with respect to the distance from the Sun. The grey area vertical thickness represents the typical interval of the variation of the plasma frequency in the solar wind. As the plasma frequency is a low-frequency cutoff for free-space propagating electromagnetic waves, radio waves cannot propagate in the red hatched area. Figure adapted from Zarka (2000).

state and direction of arrival of an observed electromagnetic wave) have therefore been developed in order to achieve angular resolution from measurements recorded with these antennas. In this chapter, we will first present the antennas and receivers used for space-based radio astronomy, discussing the specific requirements for goniopolarimetric analysis. The different existing goniopolarimetric techniques are presented, as well as their limitations and sources of error. They are illustrated by a recent study: source location of the Saturn Kilometric Radiation with *Cassini* data (Cecconi et al 2009). Another method based on very long baseline interferometry (VLBI) with several spacecraft is also briefly discussed.

Instrumentation

Two key elements of a radio instrument are the antenna and the receiver. The antenna is the sensing device: it transforms the electromagnetic fluctuations of the medium surrounding the spacecraft into voltage fluctuations that can be measured by the receiver. The receiver is the measuring device: it converts the voltage at the antenna feeds into a set of parameters (spectral flux density, amplitude, phase...) that are recorded by a data processing unit (DPU).

The classical space-based instrumentation constraints apply: low mass, low power consumption, small size, reliability, robustness to vibration and radiation. Following recent technological developments, e.g., miniaturization with application specific integrated circuits (ASIC), it is possible to build versatile low-noise radio receivers with very low mass and low power consumption. However, the main limitation still lies in the electrical sensors. They must be light and their length usually implies that they have to be deployed after launch. Hence, their deployment mechanisms have also to be light and reliable.

Depending on the primary objectives of the mission, the spacecraft can be either spinning (plasma, particles and fields science objectives) or 3-axis stabilized (imaging science objectives). The radio science objectives together with the spacecraft attitude will drive the characteristics of the antenna (length, orientation, type, etc.) mounted on the spacecraft and those of the associated radio receiver.

Electrical antennas

Two types of electrical antennas are used for space-based radio observations: "probe" antennas, e.g., *Cluster* (Gustafsson et al 1997) and "wire" antennas, e.g., *Galileo* (Gurnett et al 1992), *Ulysses* (Stone et al 1992), *Wind* (Bougeret et al 1995), *Cassini* (Gurnett et al 2004), *STEREO* (Bougeret et al 2008; Bale et al 2008). The *Geotail* spacecraft (Matsumoto et al 1994) includes a pair of each type. An electrical antenna is characterized by its beaming pattern, defined as the gain of the antenna for each direction of space (Jasik 1961). The emitting and receiving beaming patterns of an antenna are intrinsically identical.

An electrical antenna can be either a monopole or a dipole. In case of a monopole, the instantaneous voltage is probed between the antenna and the spacecraft ground (usually the spacecraft body), while in the case of a dipole, the differential voltage is probed between the two poles of the antenna. Considering the spacecraft as an infinite conductive plane, the monopole and spacecraft body system would be strictly equivalent to a dipole antenna. In reality the spacecraft is smaller than the monopole length and is not planar. The beaming pattern of the effective antenna will thus be altered from that of the physical antenna. The voltage measured at the antenna feeds depends on the integral of the instantaneous currents on the antenna. It thus depends on the antenna length and shape. In Figure 15.2 we show the beaming pattern of a thin boom dipole for various L/λ ratios. When the antenna is short compared to the wavelength[1] ($L < \lambda/10$), the electric field of the

[1] This wavelength range is called the short antenna or quasistatic range.

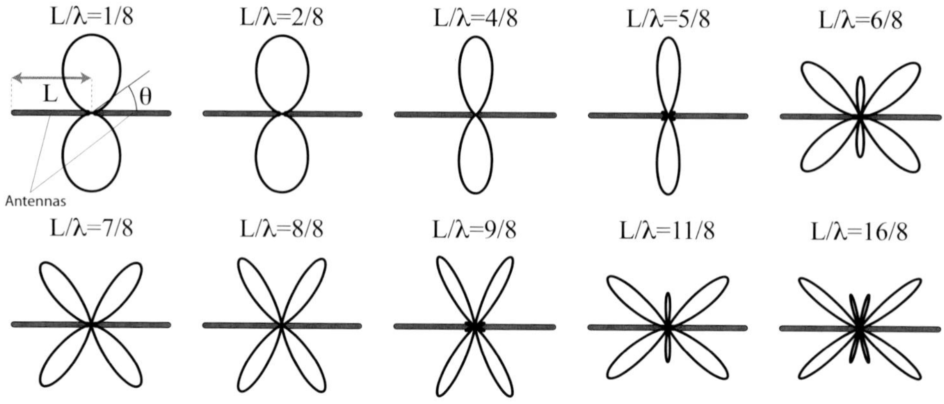

Figure 15.2: Beaming pattern of thin boom antennas for various L/λ values. The definition of angle θ is shown on the upper-left panel. The antenna pattern is single-lobed (with a null in the antenna direction) for short antennas, i.e., up to $L/\lambda \leq 1/2$. When $L/\lambda \lesssim 1/10$ the antenna pattern does not vary with frequency. With long antennas it becomes multi-lobed ($L/\lambda > 1/2$). Figure adapted from Jasik (1961).

wave is almost constant on the length of the antenna. The voltage measured at the antenna feeds is thus:

$$V = \boldsymbol{E}(t) \cdot \boldsymbol{h}_{\text{eff}} \approx E(t) h_{\text{eff}} \sin\theta, \tag{15.1}$$

with θ the angle between the effective antenna vector ($\boldsymbol{h}_{\text{eff}}$) and the wave vector $\boldsymbol{k}$. The same formalism holds when the antenna length is longer ($L \gtrsim \lambda/10$) but the effective antenna vector $\boldsymbol{h}_{\text{eff}}$ becomes complex and its magnitude depends on the wave vector direction.

A probe antenna is usually a dipole, consisting of two electrical probes (usually spherical) attached to the spacecraft by conductive cables with ground shielding. Depending on the antenna design the voltage can be measured at the probe (with preamplifiers placed right at the probe) or at the end of the cables. The actual beaming pattern is changed because of these conductive cables. Probe antennas are used on spinning spacecraft. The centrifugal force maintains them in the spin plane. Their length is usually 10 m to 100 m tip-to-tip (100 m for the *Cluster* spacecraft and the *Geotail* probe antenna dipole).

A monopole wire antenna consists of a single conductive linear probe whose thickness is usually negligible compared to its length. A perfect dipole antenna is a system of two identical wire monopole antennas that are aligned and whose feeds are separated by a short distance compared their length. Wire antennas can be either rigid, self sustained, conductive booms, or centrifugally erected, conductive wires. Boom monopoles are used on 3-axis stabilized spacecraft or as axial antennas on spinning spacecraft. Wire monopoles are used for spin-plane antennas on spinning spacecraft. Boom monopoles (12 m or less for the *Cassini*/RPWS, *STEREO*/Waves, *Ulysses* and *Wind* antennas) are significantly shorter than wire monopoles ($\approx$ 100 m for *Wind* and *Ulysses*), as they are not erected by centrifugal force.

When using an antenna in the quasistatic range, it is possible to find an effective electric dipole that has approximately the same beaming pattern as the combination of the antenna and spacecraft body system. Determining this effective electric dipole is part of the calibration process of the antenna that will be described on Page 268.

Magnetic antennas

Magnetic antennas are usually used below a few 100 kHz. They are useful to determine the propagation characteristics of the measured waves below or close to the local plasma frequency (where the transverse propagation assumption is not true in general). Magnetic antennas consist of coils, whose characteristics (length, diameter, core material, number of loops, etc) depend on the science objectives. However, as the magnetic field component is fainter than the electrical component by a factor of c, with c the speed of light, very efficient preamplifiers are needed. It is thus difficult to achieve the same sensitivity as for electric sensing devices. Furthermore, magnetic coils are more sensitive to electromagnetic interference caused by the spacecraft.

We focus in the following on the radio receivers associated with electrical sensors

Receivers

A radio receiver, consisting of a series of devices that amplify, filter and sample the signals probed by the antennas, provides spectral power series of data. Some radio receivers can record waveform data, but they will not be described here.

Solar or planetary radio emissions are very bursty and show a wide range of flux densities (see Figure 15.1a). Hence, a space-based radio receiver requires low noise and large dynamic range. In modern radio receivers (e.g., *Cassini*/RPWS/HFR (Gurnett et al 2004), *STEREO*/Waves (Bougeret et al 2008)), the amplification of the signal is split into two stages. Preamplifiers are directly connected to the antenna feeds and amplify the signal with a fixed gain before any filtering. Following filtering to the frequency band of analysis, an automatic gain control (AGC) loop adjusts the signal level to adapt it to the sampling analog to digital converter (ADC). This design provides ≈ 90 dB dynamic range for the *Cassini*/RPWS/HFR and *STEREO*/Waves receivers. The instantaneous gain of the AGC loop is recorded, as well as the power measured with the ADC, to allow measurement of the input signal strength. The integration time Δt and frequency bandwidth Δf of the measurements can usually be adjusted. As the relative measurement noise is proportional to $(\Delta t \Delta f)^{-1/2}$, it is not possible to achieve both high temporal and spectral resolution simultaneously.

A radio receiver may have several input channels. A single-channel receiver can only measure the power received by the connected antenna. The first generations of space-based radio receivers were single-channel receivers such as on the *Hawkeye 1* spacecraft (Kurth et al 1975). This type of receiver is also planned for the future *JUNO* mission. Dual-channel receivers can use the signals provided by two antennas simultaneously. Such receivers provide spectral power series for both antennas and, depending on the receiver's design, cross-correlation series of the signals measured on each antenna. The *Cassini*/RPWS/HFR and *STEREO*/Waves receivers include such cross-correlation measurements.

A probe dipole antenna or a perfect wire dipole antenna consists of two identical antenna elements which are aligned in the case of a wire antenna. It is also possible to combine the output of any pair of monopoles. This is called "antenna synthesis". V-shaped wire dipoles are often used, e.g., on *Cassini*/RPWS/HFR or *STEREO*/Waves, and consist of two wire monopoles that are not aligned. The antenna synthesis may also involve a phase shift on one of the input signals. Thus the *Interball*/POLRAD (Hanasz et al 1998a,b) radio receiver has nine channels with a set of three orthogonal antennas (x, y and z). Each channel provides a given combination of the antenna signals: the power densities on each of the three antennas (S_x, S_y, S_z) and the power densities of combined signals in phase and in quadrature (S_{x+y}, S_{y+z}, S_{z+x}, S_{x+iy}, S_{y+iz}, S_{z+ix}).

Synthetic antennas are usually used to simulate antennas with effective directions that would not have met the required constraints, such as: off-axis and off-spin plane antennas on a spinning spacecraft, or antenna directions that would have interfered with the field of view of an imaging instrument on a 3-axis stabilized spacecraft.

The autocorrelation (or cross-correlation) measured on one (or two) antenna(s) is given by:

$$P_{ij} = \langle V_i V_j^* \rangle \tag{15.2}$$

where V_i is the instantaneous voltage at antenna $\boldsymbol{h}_i$, V_j^* is the complex conjugate of the voltage at antenna $\boldsymbol{h}_j$, and $< \cdots >$ is a time averaging operator. If $i = j$, P_{ii} is the power on $\boldsymbol{h}_i$, whereas if $i \neq j$, P_{ij} is the cross-correlation between the signals on $\boldsymbol{h}_i$ and $\boldsymbol{h}_j$. Since the instantaneous voltage is a signal covering a certain spectral bandwidth, the correlations are measured as a spectral power.

Calibration

Antenna calibration consists of determining the antenna beaming pattern. There are three ways to calibrate an antenna response: (i) rheometric analysis, consisting in measuring the electrostatic response of a small model of the spacecraft and its antennas when placed in a tank filled with a dielectric medium (Rucker et al 1996; Oswald et al 2008); (ii) wire-grid electromagnetic modelling (Oswald et al 2008; Fischer et al 2001); (iii) inflight measurement of the antenna response with known radio sources (Vogl et al 2004; Cecconi and Zarka 2005). When using a wire monopole in the quasistatic range, the calibration simply provides the effective length and direction of the equivalent electric dipole.

Another key element of the calibration of an antenna is the determination of its capacitance C_{A}, and that of its mounts and feed, named the base capacitance C_{B}. The electric sensing system, i.e., the antenna itself and its mounts and feeds, is electrically equivalent to a voltage divider. The gain of the antenna is thus proportional to $C_{\mathrm{A}}/(C_{\mathrm{A}} + C_{\mathrm{B}})$.

The receiver calibration provides the parameters required to convert the recorded parameters into the wave parameters probed by the antennas. It thus provides the gain of each of its channels for any operating mode. When the receiver provides cross-correlation measurements, the phase shift between the different channels of the receiver has to be calibrated, too.

Goniopolarimetric techniques

The antennas used in space have a poor intrinsic angular resolution. However, in the quasistatic range, the beaming pattern of a wire antenna has a null in the antenna direction, whereas that of a probe antenna is isotropic. Wire antennas are thus used for goniopolarimetric applications. Goniopolarimetry (GP) is the series of techniques that allows radioastronomers to retrieve the flux, polarization and direction of arrival of an incoming electromagnetic wave. These techniques are also referred to as direction-finding techniques in the literature. however, the name "direction-finding" does not highlight the fact that the direction of arrival and the polarization cannot be retrieved independently. Hence, the term "goniopolarimetry" is more appropriate.

Goniopolarimetric measurements

In the quasi-static frequency range, the voltage induced by an incoming electromagnetic wave is $V_{\mathrm{el}} = \boldsymbol{h} \cdot \boldsymbol{E}$, where $\boldsymbol{h}$ is the effective antenna vector and $\boldsymbol{E} = \boldsymbol{E}_0 \, \mathrm{e}^{i\omega t}$ is the electric field of the wave. A GP radio receiver measures a series of correlation values $P_{ij} =< V_i V_j^* >$. A transverse electromagnetic plane wave (i.e., emitted by a point radio source at infinity) is characterized by its Stokes parameters (Kraus 1966) (S the flux, Q, U, the linear polarization degrees, and V the circular polarization degree) and the wave vector $\boldsymbol{k}$ direction which is defined by its colatitude θ and azimuth ϕ in a given frame. A model for the voltage correlation P_{ij} expression can be found in Lecacheux (1978) or in Cecconi and Zarka (2005) in case of a point source and in Cecconi (2007a) in case of an extended source.

The GP capabilities of a radio receiver depend on the receiver design and especially on the number of independent analysis channels, i.e., the number of simultaneous correlation measurements. For single-channel receivers, only one autocorrelation is measured whereas for dual-channel receivers that are connected to two antennas, up to two autocorrelations (one on each antenna) and a complex cross-correlation may be measured. In order to increase the capabilities of the radio receivers with one or more analysis channels, successive measurements at the same frequency, but on different pairs of antenna, can be performed quasi-instantaneously.

There are two possible ways to achieve GP on space-borne radio receivers: (i) spin demodulation GP on a spinning spacecraft, like *Geotail*, *Wind* and *Ulysses*; and (ii) instantaneous GP on 3-axis stabilized spacecraft, like *Cassini* and *STEREO*. Each of these GP techniques may provide the direction of arrival of the wave, flux, polarization and the angular size of the source. At the time of writing, inversions providing the size of the source exist for spin demodulation GP, while the development of such inversions is in progress in the *STEREO*/Waves team for 3-axis stabilized spacecraft. These inversions will be based on the model-expression of the measurements in case of an extended source (Cecconi 2007a).

Spin demodulation goniopolarimetry

On a spinning spacecraft, the measured correlation is modulated by the spacecraft spin. The minimum signal is obtained when the antenna is pointing closest to the source direction. The result of the measurement's demodulation is the plane containing the spacecraft spin axis and the source. This simple method has been used to perform GP analysis of AKR (terrestrial auroral kilometric radiation) (Kurth et al 1975) using data from the *Hawkeye 1* and *IMP 8* spacecraft, and of solar type III bursts (Gurnett et al 1978) using *Helios 2*, *IMP 8* and *Hawkeye 1* data. The radio receivers used on these spacecraft were single-channel receivers. In order to obtain additional independent measurements, single-channel receivers with antenna switching capabilities were developed for *ISEE-3*, *Wind* and *Ulysses*. These receivers provide series of autocorrelation measurements with different antennas (including synthetic antennas) for a given sampling setup (central frequency, integration time, frequency bandwidth). The number of measurements per spin at a given frequency depends on the receiver setup. For instance, the RAD1 receiver on *Wind* can record up to 16 samples at a given frequency, during one spacecraft rotation (≈ 3 seconds). The more measurements per spin available at a given frequency, the better will be the determination of the wave parameters. The number of measurements must at least equal the number of wave parameters to be determined.

GP inversions involving the fitting (with any minimization process) of the modulated measurement have been proposed (Lecacheux 1978) for (i) one antenna (tilted to the spin axis) and (ii) two antennas (one parallel and one perpendicular to the spin axis). These inversions are adapted to the three spacecraft mentioned above. GP techniques including the size of the source are also available for such receivers (Manning and Fainberg 1980; Fainberg et al 1985). The techniques developed in these two papers were used to derive the source location of solar radio emissions using (i) *ISEE-3* radio data only (Steinberg et al 1984; Dulk et al 1986), (ii) *Ulysses*/URAP data only (Hoang et al 1994; Reiner et al 2006), (iii) *Wind*/Waves data only (Reiner et al 1998b, 2005) and (iv) *Wind*/Waves and *Ulysses*/URAP data performing GP analysis on each data set and then applying triangulation techniques (Hoang et al 1998; Reiner et al 1998a,c). The GP techniques, as described above, have also been applied to the Jovian radio emissions observed by *Ulysses*/URAP during its Jupiter flyby in February 1992 (Reiner et al 1993; Ladreiter et al 1994). The POLRAD experiment on the *Interball* spacecraft (Hanasz et al 1998a,b) is a nine-channel receiver that provides the full set of GP measurements. GP inversions have been built and used on these data to characterize the AKR (Hanasz et al 2003; Panchenko 2004).

Instantaneous goniopolarimetry

For a spinning spacecraft, no meaningful GP result can be obtained on timescales shorter than the spacecraft spin period. In addition the wave parameters must be fixed over the spin period, while natural emissions are often sporadic. Instantaneous measurements require several antennas and multi-channel receivers with cross-correlation capabilities, such as those developed for *Cassini*/RPWS/HFR and *STEREO*/Waves. Such receivers provide four independent measurements (two autocorrelations and real and imaginary parts of the cross-correlation) that can be retrieved instantaneously, and up to nine quasi-instantaneously (Gurnett et al 2004; Cecconi and Zarka 2005; Cecconi 2004; Cecconi et al 2008).

The ideal case of an orthogonal triad of antennas has been studied by Lecacheux (1978) who showed that measurements on three antennas, i.e., seven or nine measurements, are sufficient to obtain the six wave parameters. It was also mentioned that instantaneous measurements on one pair of antennas, i.e., four measurements, are sufficient to derive either the source polarization if the source location is known, or its location if assumptions on the wave polarization are made. Due to both geometrical configuration and electromagnetic interaction between the conducting spacecraft body and the antennas, the effective directions of the antennas mounted on spacecraft are generally not orthogonal one to another. GP inversion taking into account a real antenna triad geometry has been proposed by Ladreiter et al (1995). The proposed GP technique involves a least-square model fitting through singular value decomposition (SVD). A GP inversion using a Powell minimization algorithm was also described in Vogl et al (2004). Although these methods provide useful estimates on the GP result uncertainty, they do not ensure that the solution is unique. Another drawback is the extended computing time needed to invert the data. Hence available computing power is not adequate for long-term or large-scale automatic processing. In order to overcome these limitations, analytical inversions have been developed for use on GP data obtained with three antennas (Cecconi and Zarka 2005).

Sources of uncertainty

Several sources of uncertainty may alter or bias the GP measurements and results. Some of them, such as the signal-to-noise ratio and digitization noise, depend on the receiver setup. The electromagnetic environment also induces noise: galactic background emission (resulting from the free-free interactions of the electrons in the galactic magnetic field) (Dulk et al 2001), local plasma noise and photoelectron noise (Meyer–Vernet and Perche 1989; Zouganelis 2008), and interference lines from the spacecraft. The source variability, its apparent size and the fact that we can observe several sources at the same time, may also induce errors in the results. Complete discussions of these uncertainty sources may be found in Cecconi and Zarka (2005); Lecacheux (1978); Cecconi (2007a); Cecconi et al (2008) and Ladreiter et al (1995).

As the antennas measure the electric field of electromagnetic waves passing in the vicinity of the spacecraft, GP techniques only measure the local direction of the wave vector. When propagation effects occur between the source and the spacecraft, e.g., refraction and diffusion, some modelling of the ray path between the source and the spacecraft is necessary. Particular care is needed when the local plasma frequency along the path of the wave is close to the wave frequency.

Careful data selection is essential in order to obtain accurate goniopolarimetric measurements. In the case of *Cassini*/RPWS/HFR, the selection criteria for a typical accuracy of $\approx 1°$ for directions, 10 % for polarization degrees, and 1 dB for flux, are SNR > 23 dB and $\beta_{ij} > 20°$, where β_{ij} is the elevation of the source

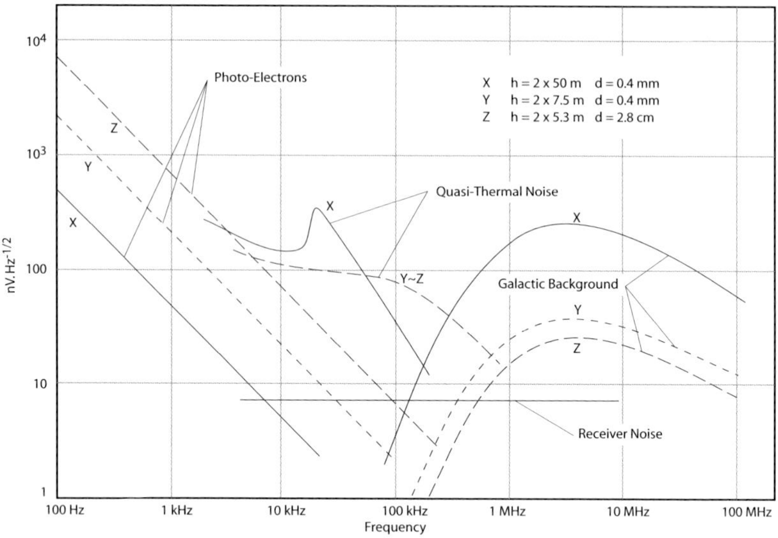

Figure 15.3: Noise sources and noise levels for space-based radio receivers. Interference lines due to other instruments or from power converters can add to the displayed spectra. The noise spectra are traditionally given as a square root of a spectral power, i.e., in units of 1 $\mathrm{nV\,Hz^{-\frac{1}{2}}}$. The noise levels are provided for various antenna characteristics (h is the antenna length and d its diameter). Type X antennas correspond to *Wind* antennas, for instance. Figure adapted from Manning (2000).

direction above the plane defined by the ith and jth pair of antennas (Cecconi and Zarka 2005).

Multi-spacecraft interferometry

Although not properly a GP technique as described in the previous section, it is possible to retrieve wave parameters and source characteristics using waveform receivers on a series of spacecraft with a VLBI technique. This technique does not require GP receivers, but waveform receivers that record the wave electric fluctuations at a high sampling rate (a few tens of kilohertz, typically). The VLBI processing is achieved a posteriori, and provides the apparent location of the source (i.e., not taking into account possible propagation effects between the source and the spacecraft, as with GP techniques) and possibly its apparent size. The flux density and polarization of the wave are only available if each spacecraft can measure them. Such a method was used with the *Cluster* spacecraft, for AKR localization (Mutel et al 2003, 2008).

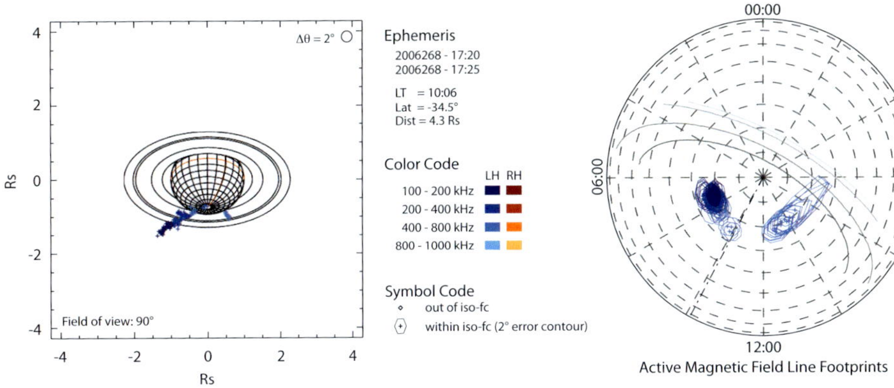

Figure 15.4: SKR source localization with *Cassini*/RPWS/HFR data on day 2006-268, between 17:20 and 17:25, spacecraft event time. The *Cassini* spacecraft was located at 10:06 LT, in the southern hemisphere (-34.5° southern latitude), and at 4.3 Saturn radii from the centre of the planet. Left panel: computed directions of arrival as seen from *Cassini*. Right panel: footprints of the magnetic field lines supporting the radio sources. On that panel, the mixed lines represent the LT of *Cassini*, and the grey curves delimit the visible portion of the $f = f_c$ iso-surface, for each of the four frequency ranges. Left-handed radio emissions are observed (blue) from 100 kHz to 800 kHz. No right-handed (red) emission is observed. This figure is adapted from Cecconi et al (2009).

Example of a goniopolarimetric result: Saturn kilometric radiation source location

Many papers have been published using the GP techniques described in the previous sections. A recent study (Cecconi et al 2009) is highlighted here: the source location of the Saturn kilometric radiation using goniopolarimetric data provided by *Cassini*/RPWS/HFR.

The *Cassini* spacecraft has been orbiting Saturn since July 2004, exploring various orbital distances, latitudes and local times (LT). During the 24 hours of data presented in that study, the radio receiver was operating in two-antenna mode. In order to find the source location, several assumptions were made: (i) the observed radio waves were circularly polarized; (ii) straight propagation between the source and the spacecraft; (iii) an emission process is proposed which implies that the wave is emitted at the local electron cyclotron frequency f_c. With these assumptions, the source is located where the straight line defined by the direction of arrival intersects the $f = f_c$ iso-surface, which is computed using a planetary magnetic field model. Once the source is located, it is possible to compute the footprint location of the active magnetic field line. Figure 15.4 shows an example of GP results and source localization. The observed radio waves are all left-handed (LH) polarized, and come from the southern hemisphere, which is characteristic of an R-X mode emission (Section 3.2 of Zarka 1998). On the left panel of Figure 15.4, there are two series of source locations, on each side of the meridian of *Cassini*. The western part, on the left of the meridian, corresponds to sources that are almost all on the same magnetic field line, whose footprint is at $\approx -68°$ latitude and $\approx$ 07:00 LT, as shown on the right panel. The eastern part, on the right of the meridian of *Cassini*, corresponds to sources at $\approx -70°$ latitude, and between 13:00 and 15:00 LT. The radio source localization process also provides information on the beaming pattern of the source, because the spacecraft has to be in the emission beam of the source to detect the radio wave. The footprint location found in that study is fully consistent with the locus of the ultraviolet aurora.

Conclusion

The GP techniques allow radio astronomers to rebuild the wave parameters from space-based radio instruments, despite the poor angular resolution of electric antennas. From the example presented above, it is clear that the GP techniques provide a very powerful tool that can yield the flux density of the observed radio wave, its polarization state, the direction of the wave vector, and possibly the size of the source. Thanks to careful data selection, it is possible to make radio emission maps (see left panel of Figure 15.4). The main advantage of this imaging technique is that it is possible to build a posteriori any maps with selection on time, frequency, polarization, etc. These imaging capabilities are not possible with classical imaging instruments. The main drawback is that it is usually necessary to make assumptions about the wave (or source) parameters in order to invert GP measurements. However, it is clear from the study of Saturn presented above that GP results can be strongly enhanced when coupled with physical models such as planetary magnetic field and plasma density models along with an understanding of the source emission process.

Acknowledgements

The author wants to thank L. Lamy, R. Prangé and P. Zarka for their helpful discussion during the preparation of the manuscript. He also wants to thank M. C. E. Huber (advised by M. Pick) for soliciting me to write this chapter.

Part of this chapter is based on a presentation "Space-based Radio Measurements: Goniopolarimetry" given at the thirtieth Goutelas Spring school: "Low Frequency Radioastronomy: Instrumentation, Science, Projects", Marcoux, France, June 2007, (Cecconi 2007b).

Bibliography

Bale SD, Ullrich R, Goetz K (plus four authors) (2008) The electric antennas for the *STEREO*/Waves experiment. Space Sci Rev 136:529–547

Bougeret JL, Kaiser ML, Kellogg PJ (plus nine authors) (1995) WAVES: The radio and plasma wave investigation on the *WIND* spacecraft. Space Sci Rev 71:231–263

Bougeret JL, Goetz K, Kaiser ML (plus 40 authors) (2008) S/Waves: The radio and plasma wave investigation on the *STEREO* Mission. Space Sci Rev 136:487–528

Cecconi B (2004) Étude Goniopolarimétrique des émissions radio de Jupiter et Saturne à l'aide du récepteur radio de la sonde *Cassini*. PhD thesis, Univ. Paris 7, Meudon, France

Cecconi B (2007a) Influence of an extended source on goniopolarimetry (or direction finding) with *Cassini* and *STEREO* radio receivers. Radio Sci 42:(RS2003) 1–17

Cecconi B (2007b) Mesures Radio Spatiales: Goniopolarimétrie. In: Zarka P, Tagger M (eds) Comptes rendus de l'Ecole de Goutelas: Radio Astronomie Basse Fréquence, vol 30

Cecconi B, Zarka P (2005) Direction finding and antenna calibration through analytical inversion of radio measurements performed using a system of 2 or 3 electric dipole antennas. Radio Sci 40:(RS3003) 1–20

Cecconi B, Bonnin X, Hoang S (plus eight authors) (2008) *STEREO*/Waves Goniopolarimetry. Space Sci Rev 136:549–563

Cecconi B, Lamy L, Zarka P (plus three authors) (2009) Goniopolarimetric study of the Rev 29 perikrone using the *Cassini*/RPWS/HFR radio receiver. J Geophys Res, DOI 10.1029/2008JA013390

Dulk GA, Steinberg JL, Hoang S, Lecacheux A (1986) Latitude distribution of interplanetary magnetic field lines rooted in active regions. In: Marsden RG (ed) The Sun and the Heliosphere in Three Dimensions, Astrophysics and Space Science Library, vol. 123, pp 229–233

Dulk GA, Erickson WC, Manning R, Bougeret JL (2001) Calibration of low-frequency radio telescopes using the galactic background radiation. Astrophys J 365:294–300

Fainberg J, Hoang S, Manning R (1985) Measurements of distributed polarized radio sources from spinning spacecraft– Effect of a tilted axial antenna– *ISEE-3* Application and Results. Astron Astrophys 153:145–150

Fischer G, Macher W, Rucker HO (plus 11 authors) (2001) Wire–grid modelling of *Cassini* spacecraft for the determination of effective length vectors of the RPWS antennas. In: Rucker HO, Kaiser ML, Leblanc Y (eds) Planetary Radio Emissions V, Austrian Academy of Sciences Press, Vienna, pp 347–356

Gurnett DA, Baumback MM, Rosenbauer J (1978) Stereoscopic direction finding analysis of a type III solar radio burst: Evidence for emission at 2 f_p-. J Geophys Res 83:616

Gurnett DA, Kurth WS, Shaw RR (plus five authors) (1992) The *Galileo* plasma wave investigation. Space Sci Rev 60:341–355

Gurnett DA, Kurth WS, Kirchner DL (plus 27 authors) (2004) The *Cassini* radio and Plasma wave science investigation. Space Sci Rev 114:395–463

Gustafsson G, Boström R, Holback B (plus 24 authors) (1997) The electric field and wave experiment for the *Cluster* mission. Space Sci Rev 79:137–156

Hanasz J, Krawczyk Z, Mogilevsky MM (plus 11 authors) (1998a) Observation of auroral kilometric radiation on the *INTEBALL-2* satellite: The POLRAD experiment. Cosmic Research 36:575

Hanasz J, Schreiber R, de Feraudy H (plus two authors) (1998b) Observations of the upper frequency cutoffs of the auroral kilometric radiation. Ann Geophys 16:1097–1104

Hanasz J, Panchenko M, de Feraudy H (plus two authors) (2003) Occurence distributions of the auroral kilometric radiation ordinary and extraordinary wave mode. J Geophys Res 108(A11):1408 SMP 14:1–14:11

Hoang S, Dulk GA, Leblanc Y (1994) Interplanetary type III radio bursts that approach the plasma frequency: *Ulysses* observations. Astron Astrophys 289:957–971

Hoang S, Maksimovic M, Bougeret JL (plus two authors) (1998) *Wind-Ulysses* source location of radio emissions associated with the January 1997 coronal mass ejection. Geophys Res Lett 25(14):2497–2500

Jasik H (1961) Antenna Engineering Handbook. McGraw-Hill, New York

Kraus JD (1966) Radio Astronomy. McGraw-Hill, New York

Kraus JD, Marhefka RJ (2001) Antennas for all applications, 3rd edn. McGraw-Hill, New York

Kurth WS, Baumback MM, Gurnett DA (1975) Direction-finding measurements of auroral kilometric radiation. J Geophys Res 80:2764–2770

Ladreiter HP, Zarka P, Lecacheux A (1994) Direction finding study of Jovian hectometric and broadband kilometric radio emissions: Evidence for their auroral origin. Planet Space Sci 42:919–931

Ladreiter HP, Zarka P, Lecacheux A (plus five authors) (1995) Analysis of electromagnetic wave direction finding performed by spaceborne antennas using singular-value decomposition techniques. Radio Sci 30:1699–1712

Lamy L, Zarka P, Cecconi B (plus three authors) (2008) Saturn kilometric radiation: average and statistical properties. J Geophys Res 113:A07,201

Lecacheux A (1978) Direction finding of a radiosource of unknown polarization with short electric antennas on a spacecraft. Astron Astrophys 70:701–706

Manning R (2000) Instrumentation for space-based low frequency radio astronomy. Geophysical Monograph, vol 119, AGU, Washington DC, pp 329–337

Manning R, Fainberg J (1980) A new method of measuring radio source parameters of a partially polarized distributed source from spacecraft observations. Space Sci Inst 5:161–181

Matsumoto H, Nagano I, Anderson RR (plus seven authors) (1994) Plasma wave Observations with *GEOTAIL* spacecraft. J Geomagnet and Geoelectricity 46(1):59

Meyer–Vernet N, Perche C (1989) Tool kit for antennae and thermal noise near the plasma frequency. J Geophys Res 94:2405–2415

Mutel RL, Gurnett DA, Christopher IW (plus two authors) (2003) Locations of auroral kilometric radiation bursts inferred from multispacecraft wideband *Cluster* VLBI observations. 1: Description of technique and initial results. J Geophys Res 108(A11):1398 SMP 8–13

Mutel RL, Christopher IW, Pickett JS (2008) *Cluster* multispacecraft determination of AKR angular beaming. Geophys Res Lett 35:L07104 1–6

Oswald TH, Macher W, Rucker HO (plus six authors) (2008) Various methods of calibration of the *STEREO*/Waves antennas. Adv Space Res 43:355–364

Panchenko M (2004) Polarimetry of auroral kilometric radiation with a triaxial nonorthogonal antenna system. Radio Sci 39:RS6010

Reiner MJ, Fainberg J, Stone RG (1993) Source characteristics of Jovian hectometric radio emissions. J Geophys Res 98:18 767–18 777

Reiner MJ, Fainberg J, Kaiser ML, Stone RG (1998a) Type III radio source located by *Ulysses/Wind* triangulation. J Geophys Res 103(A2):1923–1931

Reiner MJ, Kaiser ML, Fainberg J (plus two authors) (1998b) On the origin of radio emissions associated with the January 6-11, 1997, CME. Geophys Res Lett 25(14):2493–2496

Reiner MJ, Kaiser ML, Fainberg J, Stone RG (1998c) A new method for studying remote type II radio emissions from coronal mass ejection-driver shocks. J Geophys Res 103(A12):29 651–29 664

Reiner MJ, Jackson BV, Webb DF (plus three authors) (2005) Coronal mass ejection kinematics deduced from white light (Solar Mass Ejection Imager) and radio (*Wind*/WAVES) observations. J Geophys Res 110:(A09S14) 1–15

Reiner MJ, Kaiser ML, Fainberg J, Bougeret JL (2006) A highly circularly polarized solar radio emission component observed at hectometer wavelengths. Sol Phys 234:301–324

Rucker HO, Macher W, Manning R, Ladreiter HP (1996) Cassini model rheometry. Radio Sci 31:1299–1312

Steinberg JL, Dulk GA, Hoang S (plus two authors) (1984) Type III radio bursts in the interplanetary medium: the role of propagation. Astron Astrophys 140:39–48

Stone RG, Bougeret JL, Caldwell J (plus seven authors) (1992) The unified radio and plasma wave investigation. Astron Astrophys, Suppl Ser 92:291–316

Vogl DF, Cecconi B, Macher W (plus ten authors) (2004) In–flight calibration of the *Cassini*-Radio and Plasma Wave Science (RPWS) antenna system for direction-finding and polarization measurements. J Geophys Res 109:(A09S17) 1–18

Zarka P (1998) Auroral radio emissions at the outer planets: Observations and theories, J Geophys Res, 103, 20159–20194

Zarka P (1992) The auroral radio emissions from planetary magnetospheres: What do we know, what don't we know, what do we learn from them? Adv Space Res 12(8):99–115

Zarka P (2000) Radio emissions from the planets and their moons. Geophysical Monograph, vol 119, AGU, Washington DC, pp 167–178

Zarka P, Cecconi B, Kurth WS (2004) Jupiter's low-frequency radio spectrum from *Cassini*/Radio and Plasma Wave Science (RPWS) absolute flux density measurements. J Geophys Res 109:A09S15

Zouganelis I (2008) Measuring suprathermal electron parameters in space plasmas: Implementation of the quasi-thermal noise spectroscopy with kappa distributions using in situ *Ulysses*/URAP radio measurements in the solar wind. J Geophys Res 113:(A08111) 1–7

— 16 —

High-accuracy positioning: astrometry

Lennart Lindegren[I]

Abstract

The limiting accuracy for measuring the location of an optical image is set by diffraction and photon noise, i.e., by the dual wave–particle nature of light. A theoretical expression for the limiting accuracy is derived under idealized conditions and generalized to take into account more realistic circumstances, such as additive noise and finite pixel size. Its application is discussed in relation to different space-astrometry techniques including grid modulation, direct imaging, and interferometry.

Advantages of space observations for astrometry

Astrometry is the branch of observational astronomy concerned with the accurate determination of the angular coordinates of celestial bodies, and of the temporal changes in the coordinates caused by effects such as proper motion, parallax, orbital motion, and gravitational light deflection. The uses of such data are numerous and wide-ranging; see, e.g., Turon and Robichon (2006). Although modern astrometric techniques embrace the whole electromagnetic spectrum from γ-ray observations (Taff 1988) to radio interferometry (Fomalont 2005), only optical and near-infrared measurements from space are considered here. The launch, in 1989, of the ESA space-astrometry satellite *Hipparcos* (Perryman, ESA 1997) revolutionized optical astrometry and firmly demonstrated the advantages of a space platform for such measurements (Perryman 2009). At about the same time, high-accuracy astrometric observations were started with *HST* (Benedict et al 1994, 2003). The ESA *Gaia* mission (Perryman 2005) is the most ambitious space-astrometry project currently under development; it will survey some 10^9 objects brighter than 20th magnitude at angular accuracies in the 10 µas to 300 µas range. The proposed missions *SIM PlanetQuest* (Nemati 2006; Unwin et al 2008) and *SIM-Lite* (Shao and Nemati 2009) are pointed interferometers aiming at accuracies down to 1 μas or less, albeit for a smaller number of targets ($\simeq 10^4$).

The advantages of space for optical astrometric observations were not universally recognized at the time when *Hipparcos* was under discussion. After all, such measurements had for centuries been made from the ground, with steadily increasing precision and significant potential for improvement (Connes 1979). With the

[I]Lund Observatory, Lund, Sweden

hindsight of the *Hipparcos* experience, the advantages of space can briefly be stated as follows:

- The absence of an atmosphere eliminates the systematic and random effects of refraction and turbulence (seeing).
- Weightlessness eliminates the differential mechanical deformation of the instrument as it is pointed in different directions.
- With proper design of the instrument and spacecraft, space may provide a thermally and mechanically very stable environment.
- At any time, nearly the whole sky is simultaneously accessible from a single observatory, and over a few months the whole sky is accessible.

The last point is perhaps the least obvious advantage, but it is in fact essential for accurate calibration of the instrument and for the determination of absolute parallaxes and an internally consistent global reference frame.

The basic astrometric measurement quantifies the observed direction of a light ray in terms of detector coordinates. These could be the location of the image centre (expressed in pixels and fractions thereof) at a given time, or the phase of a fringe pattern (for an interferometer). Their astrometric interpretation involves a series of transformations, taking into account the geometric calibration of the instrument, its orientation in space (attitude), and the relation between the observed direction and celestial coordinates in some standard reference frame. The first two transformations are highly mission-specific and will not be further considered here, while the last one requires a general-relativistic framework, e.g., as described by Klioner (2003). Only the basic directional observation, in terms of detector coordinates, is presently discussed.

Optical image formation and positional accuracy

The fundamental limit

The limiting positional accuracy (i.e., for an 'ideal' detector) can be derived from elementary considerations of noise and resolution (Falconi 1964; Lindegren 1978). Ultimately, the limit is set by the dual wave–particle nature of light, i.e., by diffraction and photon noise.

With a proper choice of coordinate axes, measurements in orthogonal directions are uncorrelated and it suffices to consider one-dimensional measurements along an axis denoted x (in linear measure) or ξ (in angular measure). Lindegren (2005) showed that a limiting expression for the standard deviation of the angular position ξ can be derived directly from the Heisenberg uncertainty principle, namely

$$\sigma_\xi \geq \frac{\lambda}{4\,\pi\,\Delta x\sqrt{N}} \quad , \tag{16.1}$$

where λ is the wavelength of the radiation, N the number of photons detected, and Δx the root-mean-square (rms) extension of the telescope entrance pupil in the x direction (for example, $\Delta x = D/4$ for a circular pupil of diameter D, $\Delta x = L/\sqrt{12}$

for a rectangular pupil of size L in the x direction, and $\Delta x = B/2$ for a two-element interferometer with projected baseline B in the x direction). Equation 16.1 gives a lower bound which could in principle be approached under ideal circumstances — aberration-free imaging, no background illumination, etc. — and is therefore a useful reference quantity.

Limiting accuracy from diffraction optics

In the Fraunhofer approximation (Born and Wolf 1999) the monochromatic point-spread function (PSF) is

$$P_\lambda(\xi,\eta) = \frac{1}{\lambda^2 A}\left|\iint_{\mathcal{P}} \exp\Big[-\mathrm{i}\,(2\,\pi/\lambda)\big(x\xi + y\eta + W(x,y)\big)\Big]\,\mathrm{d}x\,\mathrm{d}y\right|^2 \qquad (16.2)$$

where (x,y) are linear coordinates in the telescope pupil $\mathcal{P}$, (ξ,η) the corresponding angular coordinates in the image plane, $W(x,y)$ the wavefront uncertainty of the telescope, and A the area of the pupil.

Equation 16.2 is normalized to $\iint_{-\infty}^{+\infty} P_\lambda(\xi,\eta)\mathrm{d}\xi\mathrm{d}\eta = 1$, and therefore gives the probability density function (pdf) for photon detection at $\alpha = \alpha_0 + \xi$, $\beta = \beta_0 + \eta$, given quasi-monochromatic radiation of wavelength λ with the optical centre at (α_0, β_0). Assuming a stationary image with no background illumination and an ideal detector registering the precise coordinates (α_i, β_i) of N photon detection events, the optical centre can be estimated by maximizing the log-likelihood function

$$\ell(\alpha_0,\beta_0) = \sum_{i=1}^{N} \ln P_\lambda(\alpha_i - \alpha_0, \beta_i - \beta_0) \quad . \qquad (16.3)$$

The covariance of the resulting maximum-likelihood estimate $(\hat\alpha_0, \hat\beta_0)$ can be obtained by means of the Cramér–Rao bound (Stuart et al 1998),

$$\mathrm{Cov}\begin{bmatrix}\hat\alpha_0\\ \hat\beta_0\end{bmatrix} \geq \begin{bmatrix} \mathrm{E}\left(\frac{\partial\ell}{\partial\alpha_0}\frac{\partial\ell}{\partial\alpha_0}\right) & \mathrm{E}\left(\frac{\partial\ell}{\partial\alpha_0}\frac{\partial\ell}{\partial\beta_0}\right)\\ \mathrm{E}\left(\frac{\partial\ell}{\partial\beta_0}\frac{\partial\ell}{\partial\alpha_0}\right) & \mathrm{E}\left(\frac{\partial\ell}{\partial\beta_0}\frac{\partial\ell}{\partial\beta_0}\right)\end{bmatrix}^{-1} \quad , \qquad (16.4)$$

where E is the expectation operator. Using Equation 16.3 and introducing $U_\lambda(\xi,\eta) = 2\sqrt{P_\lambda(\xi,\eta)}$, the first matrix element on the right-hand side of Equation 16.4 can be written

$$\mathrm{E}\left(\frac{\partial\ell}{\partial\alpha_0}\right)^2 = N\iint_{-\infty}^{+\infty}\left(\frac{\partial P_\lambda}{\partial\xi}\right)^2 \frac{\mathrm{d}\xi\,\mathrm{d}\eta}{P_\lambda(\xi,\eta)} = N\iint_{-\infty}^{+\infty}\left(\frac{\partial U_\lambda}{\partial\xi}\right)^2 \mathrm{d}\xi\,\mathrm{d}\eta \qquad (16.5)$$

and similarly for the other matrix elements.

The above expressions are valid for arbitrary wavefront aberrations and pupil shapes. Consider now the special case of an aberration-free system, $W(x,y) = 0$, whose pupil is symmetric with respect to both axes. The Fourier double integral in Equation 16.2 is then real and

$$\frac{\partial U_\lambda}{\partial\xi} = \frac{\mathrm{i}\,4\,\pi}{\lambda^2\sqrt{A}}\iint_{\mathcal{P}} \exp\big[-\mathrm{i}\,(2\pi/\lambda)(x\xi + y\eta)\big]\,x\,\mathrm{d}x\,\mathrm{d}y \qquad (16.6)$$

etc. Applying Parseval's identity, and using the symmetry of the pupil, gives

$$\iint_{-\infty}^{+\infty} \left(\frac{\partial U_\lambda}{\partial \xi}\right)^2 \mathrm{d}\xi\, \mathrm{d}\eta = \frac{16\pi^2 \Delta x^2}{\lambda^2} \quad , \tag{16.7}$$

with a corresponding expression in y and η, and zero cross term. The matrix in Equation 16.4 is therefore diagonal, and the Cramér–Rao bound gives the variances

$$\mathrm{Var}(\xi) \geq \frac{\lambda^2}{16\pi^2 \Delta x^2 N}, \qquad \mathrm{Var}(\eta) \geq \frac{\lambda^2}{16\pi^2 \Delta y^2 N} \tag{16.8}$$

in complete agreement with Equation 16.1. For notational convenience, (ξ, η) were here re-introduced for the estimated image centre, instead of $(\hat{\alpha}_0, \hat{\beta}_0)$.

Generalizations

Based on the first equality in Equation 16.5, a number of generalizations can now be introduced. First, it is noted that $P_\lambda(\xi, \eta)$ in Equation 16.5 may include arbitrary wavefront aberrations via Equation 16.2, although the various integrals must then be evaluated numerically. In general, aberrations result in a reduction of the overall PSF slopes in 16.5 and consequently in an increased variance of the estimated centroid coordinate.

Second, for real instruments operating in a finite wavelength band, the monochromatic PSF, $P_\lambda(\xi, \eta)$, must be replaced by the polychromatic PSF

$$P(\xi, \eta) = \frac{\int_0^\infty T_\lambda Q_\lambda P_\lambda(\xi, \eta)\, \mathrm{d}\lambda}{\int_0^\infty T_\lambda Q_\lambda\, \mathrm{d}\lambda} \quad , \tag{16.9}$$

where T_λ is the total optical transmittance and Q_λ the detector quantum efficiency.

Third, further degradation of the image caused by the detector may be included by replacing the optical PSF by an 'effective PSF' (ePSF, Anderson and King 2000)

$$\widetilde{P}(\xi, \eta) = P(\xi, \eta) * P_{\text{pixel}} * P_{\text{diffusion}} * P_{\text{motion}} * \cdots \tag{16.10}$$

obtained through convolution with a series of probability density functions representing the pixel response function, the diffusion of photoelectrons in the CCD silicon substrate, the motion of the image during the exposure, etc. (cf., Page 285).

Finally, let $S(\xi, \eta) = N\widetilde{P}(\xi, \eta)$ be the expected density of stellar photon detections per unit area of the detector and b the expected background density, then

$$\sigma_\xi \geq \left[\iint_{-\infty}^{+\infty} \frac{1}{S(\xi, \eta) + b} \left(\frac{\partial S}{\partial \xi}\right)^2 \mathrm{d}\xi\, \mathrm{d}\eta\right]^{-1/2} \quad . \tag{16.11}$$

Here, b can be taken to represent all constant noise sources, including sky background, detector dark counts, and CCD readout noise (an rms readout noise of r electrons per pixel is equivalent to an increase of the background level by r^2 counts per pixel).

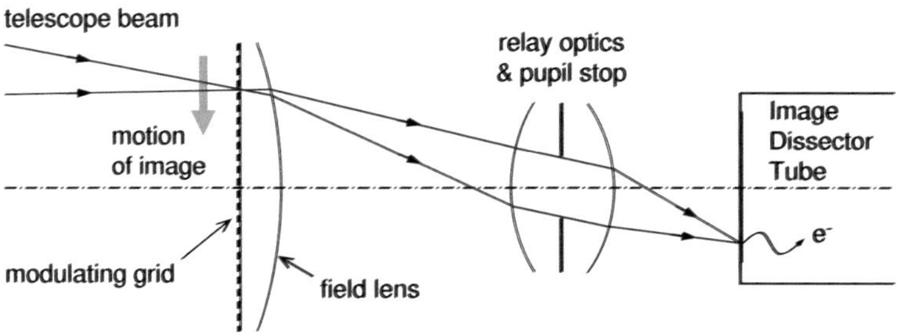

Figure 16.1: Positional encoder for *Hipparcos*: a modulating grid consisting of 2688 transparent slits, covering the central $0.9^\circ \times 0.9^\circ$ of the telescope field of view.

The methodology outlined above allows an estimate of the precision of centroiding on a detected image affected by optical aberrations, sky background, pixelization and other detector imperfections including readout noise. The lower bound in Equation 16.11 can be computed without specifying the centroiding algorithm, and from estimation theory it can be deduced that a good centroiding algorithm (e.g., based on the maximum likelihood method) should come close to this limit, at least when N is not too small. The lower bound is therefore extremely useful for evaluation and optimization purposes, and as an absolute standard for the performance of practical centroiding algorithms.

Example 1: the modulating grid of *Hipparcos*

When the *Hipparcos* satellite design was finalized in the mid-1980s, CCDs were not considered sufficiently stable, or technologically mature, for space-astrometry applications. The solution adopted for *Hipparcos* was to encode the positions of stellar images by means of a highly accurate mask, or modulating grid, occupying the focal surface in front of a photon-counting detector (Figure 16.1). Light modulation was effected through the continuous rotation of the instrument at a nominal speed of $168.75''\ \mathrm{s}^{-1}$. The resulting periodic variation of the transmitted intensity (Figure 16.2) could be accurately modelled by a truncated Fourier series, which for single stars gave the following expression for the photon-count density:

$$S(\xi) = F\big[1 + M_1 \cos(2\pi\xi/s) + M_2 \cos(4\pi\xi/s + \epsilon)\big] \quad . \tag{16.12}$$

Here, ξ is the instantaneous angular coordinate of the image, s the grid period, F the mean stellar count rate (in counts per grid period), and $M_1 \simeq 0.71$, $M_2 \simeq 0.25$ and $\epsilon \simeq 10^\circ$ are instrument 'constants' that have to be accurately calibrated. The grid period $s \simeq 1.2''$ was chosen to give negligible third harmonic in the Fourier series expansion of the detector signal; indeed, the third and higher harmonics are strictly zero for $\lambda > sD/3 \simeq 560$ nm, where $D = 0.29$ m is the diameter of the *Hipparcos* entrance pupil. The slit width $\simeq 0.46''$ was optimized for positional accuracy, taking into account the theoretically predicted variation of F, M_1 and M_2 with the slit width.

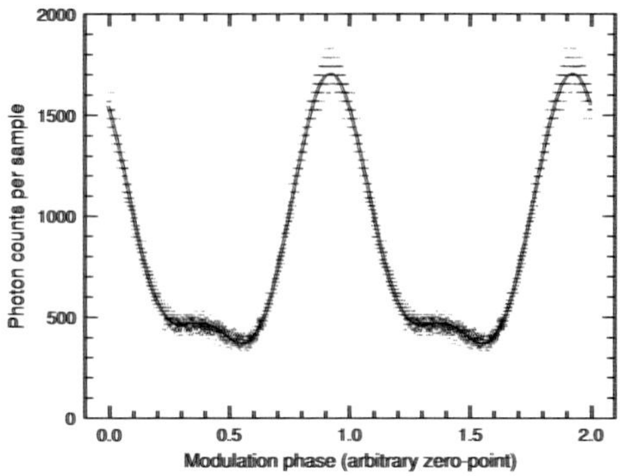

Figure 16.2: Photon counts for a bright (magnitude $\simeq$ 2) star observed with the *Hipparcos* encoder (Fig. 16.1). 2560 consecutive samples are plotted versus the modulation phase (modulo the modulation period). The curve is the fitted Equation 16.12. The discretization prominent for large counts is due to the semi-logarithmic data-compression algorithm coding the counts into 8-bit words.

The performance of the *Hipparcos* positional encoder can be evaluated by inserting Equation 16.12 into Equation 16.11, ignoring the perpendicular coordinate (η). Assuming negligible background ($b \ll F$) and the quoted values for M_1, M_2 and ϵ, we find $(\sigma_\xi)_{\rm Hipp} \simeq 0.330'' N^{-1/2}$, where $N = Ft$ is the total number of photons registered over t modulation periods. This is in fair agreement with values derived from a re-analysis of the *Hipparcos* raw data (van Leeuwen 2007). For comparison, the theoretical limit 16.1 gives $(\sigma_\xi)_{\rm theory} \simeq 0.125'' N^{-1/2}$, assuming $\lambda_{\rm eff} \simeq 550$ nm. Considering in addition that only one third of the photons are transmitted by the grid and relayed to the detector, one can conclude that the grid-modulation technique is not very efficient in absolute terms.

Example 2: CCD imaging for space astrometry

Compared with the *Hipparcos* detection system described above, CCDs have huge advantages in terms of quantum efficiency, broad wavelength coverage, no light loss due to a modulating device, and the possibility to observe many objects simultaneously (Lindegren 2005). The capability of CCDs for high-accuracy differential astrometry from space is well demonstrated using images from the *HST* Wide Field and Planetary Camera 2 (WFPC2) (Anderson and King 2000, 2003) and Advanced Camera for Surveys/Wide Field Channel (ACS/WFC) (Anderson and King 2006), although these cameras were not specifically designed for astrometry. Future space-astrometry missions will use CCDs for direct imaging in drift-scanning mode (*Gaia*) and for fringe tracking in interferometers (e.g., *SIM PlanetQuest*). These instruments represent very different modes of utilizing CCDs for space astrometry.

Direct imaging in pointing mode

High-accuracy astrometry in a field of any appreciable size requires that the image centroids are located to within a small fraction of the pixel size. For example, 1 mas corresponds to about 0.02 pixel both in the WFPC2 and the ACS/WFC. This is in principle not a problem if the image is well-sampled and recorded with sufficient signal-to-noise ratio. The sampling theorem (e.g., Wall and Jenkins 2003) provides a useful rule-of-thumb estimate of the maximum desired pixel size p in terms of the telescope aperture and effective wavelength:

$$p \leq p_{\rm crit} \equiv \frac{\lambda_{\rm eff}}{2\,D} \quad . \tag{16.13}$$

For the *HST* ($D = 2.4$ m) in visual light ($\lambda_{\rm eff} \simeq 550$ nm) this gives $p \leq 24$ mas, or about half the actual pixel size for WFPC2 and ACS/WFC. However, undersampling is not by itself a problem for accurate positioning, provided that the underlying image is simple enough (e.g., consisting of a small number of point sources). A more useful, quantitative criterion can be derived in terms of the limiting positional accuracy as function of p, using Equations 16.10 and 16.11 with the idealized (boxcar) pixel response function

$$P_{\rm pixel}(\xi,\eta) = \begin{cases} 1/p^2 & \text{if } |\xi| < p/2 \text{ and } |\eta| < p/2 \\ 0 & \text{otherwise} \end{cases} \quad . \tag{16.14}$$

Figure 16.3 shows the relative increase in σ_ξ as function of $p/p_{\rm crit}$ for an aberration-free optical system with circular pupil. The effect of the pixelization depends both on the relative bandwidth of the radiation and the level of background noise; three different cases are shown in the diagram. It is seen that the degradation is quite gradual with increasing pixel size, with no specific significance attached to the critical sampling. Thus, some degradation is observed already for well-sampled images ($p/p_{\rm crit} < 1$), while severely undersampled images could still be useful for astrometry. For the WFPC2 and ACS/WFC cameras ($p/p_{\rm crit} \simeq 2$) the pixelization is expected to increase the relative centroiding uncertainty by some 50 %. Optical aberrations and charge diffusion, neglected in this discussion, will increase the uncertainty by some further factor.

To achieve the highly accurate centroiding actually in practice is, however, a different matter. Simplistic centroiding algorithms such as using the moments of the light distribution (Howell 2000) are totally inadequate both in terms of precision and systematic uncertainties. Anderson and King (2000) have demonstrated that the key to successful astrometry with the *HST* cameras is to make an extremely accurate calibration of the ePSF, and subsequently use this for fitting to the observed CCD data. It should be noted that the commonly employed (sub-pixel) dithering techniques (Fruchter and Hook 2002) do not eliminate the information-loss caused by the undersampling, but greatly facilitate the accurate determination of the ePSF.

All CCD detectors used on the *HST* instruments suffer from charge transfer inefficiency (CTI), i.e., from the fact that the transfer of charge from one pixel to the next is not perfect. The measurable effects include loss of flux and systematic centroid shifts depending on (among other things) the position on the chip, the size

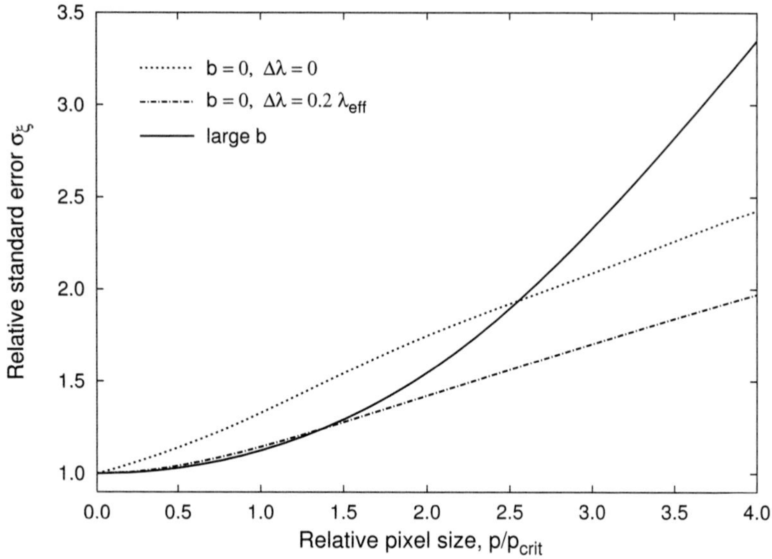

Figure 16.3: Theoretical degradation of the centroiding precision from the pixelization of an aberration-free diffraction image (circular pupil), as function of pixel size. Dashed curves are for no background ($b = 0$) and different bandwidths $\Delta\lambda$; the solid curve is for the large-background case ($b \gg NA/\lambda_{\text{eff}}^2$), where the effect is nearly independent of $\Delta\lambda$.

of the signal, and the background level (Goudfrooij et al 2006; Kozhurina-Platais et al 2007).

Direct imaging in drift-scanning mode

Fast-moving objects can be imaged by operating the CCD in the time, delay and integrate (TDI) mode, also known as drift-scanning. The charges are clocked along the CCD columns at the same (average) speed as the motion of the optical image. The integration time is set by the time it takes the object to move across the CCD. The use of TDI imaging in astronomy is well-known from ground-based programmes such as the Sloan Digital Sky Survey (Gunn et al 1998).

The *Gaia* instrument will use a mosaic of 76 CCDs (SM and AF in Figure 16.4) to detect and record the images of point-like objects traversing the focal plane as a result of the continuous rotation of the satellite at $60''$ s^{-1}. Additional CCDs are used for spectrophotometric and spectroscopic characterization of the objects. All detectors are large-size (45 mm × 59 mm), back-illuminated CCDs operated in TDI mode, providing a maximum integration time of 4.4 s per CCD.

A single CCD detector sweeps a strip of the sky of the same width as the CCD but in principle indefinitely long. By slowly changing the direction of *Gaia*'s spin axis according to a well-defined 'scanning law', the strips from all the CCDs provide multiple coverage of the whole celestial sphere in a period of six months. Two optical fields of view, separated by a 'basic angle' of 106.5° on the sky, are

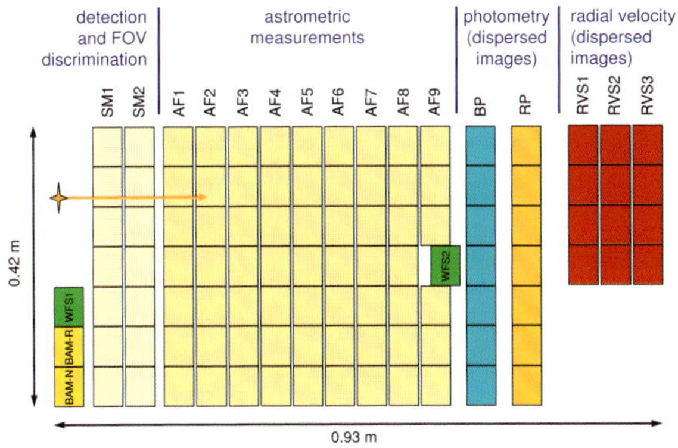

Figure 16.4: Layout of CCDs in the focal plane of *Gaia* (Lindegren et al 2008). Images travel from left to right, crossing in turn the skymappers (SM), astrometric field (AF), blue photometer (BP), red photometer (RP), and the radial-velocity spectrometer (RVS). Also shown are CCDs for the basic-angle monitor (BAM) and wavefront sensors (WFS).

superposed on the single focal plane, although the two skymappers (SM) view the fields separately. This measurement principle means that the centroiding accuracy must be optimized in the along-scan direction (say, ξ), while this requirement is much relaxed in the across-scan direction (Lindegren 2005). The entrance pupils for the *Gaia* telescopes are consequently rectangular (1.45 m $\times$ 0.5 m) with the longer dimension (L) along x, and the pixel size 10 μm $\times$ 30 μm was chosen to match the resulting diffraction image, with the smaller dimension (p) along ξ. Further on-chip binning is employed in the across-scan (η) direction during the readout of the CCDs in the astrometric field.

Since the elementary astrometric observations are thus basically one-dimensional, the relevant response function is the effective line spread function (eLSF), i.e., the marginal density of the ePSF:

$$\widetilde{P}(\xi) = \int_{-\infty}^{+\infty} \widetilde{P}(\xi, \eta)\, \mathrm{d}\eta \quad . \tag{16.15}$$

The theoretical centroiding accuracy can be estimated from the photon count rate and eLSF in complete analogy with the two-dimensional case. The main difference is that the TDI mode introduces some additional smearing of the eLSF due to the relative motion of the optical image with respect to the transported charges during the exposure. This motion has three main causes: the TDI operation itself (the charges move step-wise, while the optical image is in continuous motion); local mismatch in scale due to optical distortions; and real-time attitude-control errors. The TDI stepping can be modelled, for a four-phase CCD, as a convolution with a rectangular function of width $p/4$. The combined smearing from other causes may be modelled as a convolution with a Gaussian function; the estimated standard width is of order 0.13 p. Charge diffusion causes additional, wavelength-dependent smearing (Janesick 2001).

The combination of linear pixel size ($p = 10\ \mu$m), pupil size (L) and focal length ($F = 35$ m) chosen for *Gaia* means that the stellar images are undersampled for $\lambda < 2\,pL/F \simeq 830$ nm. (The total wavelength coverage is about 400 nm to 1000 nm.) This results in some accuracy degradation similar to Figure 16.3. In reality the pupil size and the linear dimensions of the focal plane are determined by practical and budgetary constraints depending on volume, mass, power dissipation, and technical complexity. The chosen focal length is nearly optimal in this respect: a longer focal length would certainly reduce the undersampling factor and hence the astrometric centroiding uncertainty per unit integration time, but it would also decrease the integration time per CCD as well as the mean number of observations per object for a given size of the focal plane. The average accumulated integration time per object scales as F^{-2}, so that the photon-noise factor on the resulting uncertainty is proportional to F. As a result, the optimum F corresponds to slightly undersampled images.

An important consideration for the performance of the *Gaia* mission is the possible accuracy degradation caused by radiation damage to the CCDs (Lindegren et al 2008), i.e., the previously mentioned CTI effects. Extensive laboratory tests on irradiated CCDs show significant charge loss and centroid shifts both along and across the scan. The effects are very reproducible, and therefore possible to calibrate, but their accurate modelling remains a challenging aspect of the *Gaia* data analysis.

Example 3: interferometry

Astrometric measurements made with space interferometers will benefit from the increased angular resolution obtained with baselines B much longer than any single telescope aperture, cf., Equation 16.1 where $\Delta x = B/2$ in this case. For example, the planned *SIM PlanetQuest* has $B \simeq 9$ m, allowing the angles between bright (< 10 mag) target stars to be measured to a precision of $\simeq 1\ \mu$as (individual measurements in the narrow-angle mode). This very high precision is primarily motivated by the search for terrestrial planets (Shao 2006).

The condition for centroiding on an object is that the optical path length from the incident plane wavefront to the detector is the same in the two interferometer arms, i.e., that the optical path difference (OPD) is zero to within a fraction of a wavelength, yielding constructive interference at the beam combiner. The delay in one of the arms is first adjusted until fringes are observed. Then, in order to determine the precise delay corresponding to no OPD, a saw-tooth modulation of the OPD is introduced and the intensity is measured as function of the delay. In the monochromatic approximation the measured intensity at OPD position u can be written (Meier and Folkner 2003)

$$S(u) = F\left[1 + V\cos(k\,u + \phi)\right] \quad , \tag{16.16}$$

where the parameters are the mean flux F (per unit time), fringe visibility (contrast) V, wave number k, and phase ϕ. Least-squares fitting of this fringe model to the CCD pixel values as function of u gives an estimate of all four parameters. The limiting accuracy of the phase determination is readily derived analytically from 16.11, viz.

$$\sigma_\phi \geq \left[\left(1-\sqrt{1-\left(\frac{FV}{F+b}\right)^2}\right)(F+b)\,t\right]^{-1/2} , \qquad (16.17)$$

where t is the integration time and b the background, including CCD readout noise as previously discussed. For projected baseline B the phase uncertainty σ_ϕ translates to the uncertainty in angular position $\sigma_\theta = (\sigma_\phi/2\pi)(\lambda_{\text{eff}}/B)$. With $N = Ft$ denoting the total number of detected photons, and assuming $b \ll F$, it is seen that

$$\sigma_\theta \geq \frac{\lambda_{\text{eff}}}{2\pi B\sqrt{N}}\left(1-\sqrt{1-V^2}\right)^{-1/2} . \qquad (16.18)$$

Compared with Equation 16.1, with $\Delta x = B/2$, this uncertainty is a factor $(1-\sqrt{1-V^2})^{-1/2} \geq 1$ greater, with equality for $V = 1$. Thus, provided the fringe contrast is high and the detector noise small, the performance of the astrometric interferometer in terms of the internal phase uncertainty may approach that of an ideal instrument.

Conclusions

Observations from space offer unique advantages for high-accuracy positioning, or astrometry. In particular, the absence of any atmosphere allows diffraction-limited performance of imaging and interferometric devices. In the photon-counting regime the best possible positional accuracy of such devices can be evaluated from simple and general statistical considerations, as illustrated by the specific examples in this chapter. It can be argued that the actual performance of any real, well-designed experiment should approach this theoretical limit to within a small numerical factor.

Bibliography

Anderson J, King IR (2000) Toward high-precision astrometry with WFPC2. I. Deriving an accurate point-spread function. Publ Astr Soc Pac 112:1360–1382

Anderson J, King IR (2003) An improved distortion solution for the *Hubble* Space Telescope's WFPC2. Publ Astr Soc Pac 115:113–131

Anderson J, King IR (2006) PSFs, photometry, and astrometry for the ACS/WFC. Space Telescope Science Institute, Instrument Science Report ACS 2006-01

Benedict GF, McArthur B, Nelan E (plus 11 authors) (1994) Astrometry with Hubble Space Telescope fine guidance sensor number 3: Position-mode stability and precision. Publ Astr Soc Pac 106:327–336

Benedict GF, McArthur BE, Fredrick LW (plus 17 authors) (2003) Astrometry with the Hubble Space Telescope: A parallax of the central star of the Planetary Nebula NGC 6853. Astron J 126:2549–2556

Born M, Wolf E (1999) Principles of optics: Electromagnetic theory of propagation, interference and diffraction of light (7th ed.). Cambridge: Cambridge University Press

Connes P (1979) Should we go to space for parallaxes? Astron Astrophys 71:L1–L4

Falconi O (1964) Maximum sensitivities of optical direction and twist measuring instruments. J Opt Soc Am 54:1315–1320

Fomalont EB (2005) Radio astrometry: Present status and future. ASP Conf Ser 338:335–342

Fruchter AS, Hook RN (2002) Drizzle: a method for the linear reconstruction of undersampled images. Publ Astr Soc Pac 114:144–152

Goudfrooij P, Bohlin RC, Maíz-Apellániz J, Kimble RA (2006) Empirical corrections for charge transfer inefficiency and associated centroid shifts for STIS CCD observations. Publ Astr Soc Pac 118:1455–1473

Gunn JE, Carr M, Rockosi C (plus 37 authors) (1998) The Sloan digital sky survey photometric camera. Astron J 116:3040–3081

Howell SB (2000) Handbook of CCD astronomy. Cambridge: Cambridge University Press

Janesick JR (2001) Scientific charge-coupled devices. SPIE monograph PM 83. Bellingham, WA: SPIE Optical Engineering Press

Klioner SA (2003) A practical relativistic model for microarcsecond astrometry in space. Astron J 125:1580–1597

Kozhurina-Platais V, Goudfrooij, Puzia TH (2007) ACS/WFC: Differential CTE corrections for photometry and astronomy from non-drizzled images. Space Telescope Science Institute, Instrument Science Report ACS 2007-04

Lindegren L (1978) Photoelectric astrometry — A comparison of methods for precise image location. IAU Coll 48: Modern Astrometry, pp 197–217

Lindegren L (2005) The astrometric instrument of *Gaia*: Principles. ESA SP-576:29–34

Lindegren L, Babusiaux C, Bailer-Jones C (plus ten authors) (2008) The Gaia mission: science, organization and present status. IAU Symp. No. 248, pp 217–223

Meier DL, Folkner WM (2003) SIMsim: an end-to-end simulation of SIM. Proc SPIE 4852:131–142

Nemati B (2006) *SIM PlanetQuest*: status and recent progress. Proc SPIE 6268:62680Q-1–10

Perryman MAC, ESA (1997) The Hipparcos and Tycho Catalogues. ESA SP-1200

Perryman MAC (2005) Overview of the *Gaia* mission. ASP Conf Ser 338:3–14

Perryman MAC (2009) Astronomical applications of astrometry: A review based on ten years of exploitation of the *Hipparcos* satellite data. Cambridge: Cambridge University Press

Shao M (2006) Search for terrestrial planets with *SIM PlanetQuest*. Proc SPIE 6268:62681Z-1–6

Shao M, Nemati B (2009) Sub-microarcsecond astrometry with *SIM-Lite*: A testbed-based performance assessment. Publ Astr Soc Pac 121:41–44

Stuart A, Ord JK, Arnold S (1998) Kendall's Advanced Theory of Statistics, Volume 2A: Classical Inference and the Linear Model (6th ed.). London: Hodder Arnold

Taff LG (1988) Gamma-ray burst astrometry. Astrophys J 326:1032–1035

Turon C, Robichon N (2006) Astrometric surveys. Mem Soc Astron Italiana 77:1073–1080

Unwin SC, Shao M, Tanner AM (plus 33 authors) (2008) Taking the Measure of the Universe: Precision astrometry with *SIM PlanetQuest*. Publ Astr Soc Pac 120:38–88

van Leeuwen F (2007) *Hipparcos*, the new reduction of the raw data. Astrophysics and Space Science Library Vol. 350

Wall JV, Jenkins CR (2003) Practical statistics for astronomers. Princeton Series in Astrophysics. Cambridge: Cambridge University Press

— 17 —

Interferometric imaging from space

ANDREAS QUIRRENBACH[I]

Abstract

Astronomical interferometry, the coherent combination of light from two or more telescopes, can provide images of celestial objects with very high angular resolution. The van Cittert–Zernike theorem forms the theoretical foundation of interferometric measurements and for the reconstruction of images from interferometric data. The closure phase method and phase self-calibration techniques retain useful phase information even if the individual phases are corrupted by instrumental errors. The introduction of a π phase shift in one of the interferometer arms leads to the rejection of on-axis light through destructive interference, enabling observations with extremely high contrast. The most critical technological requirements for space interferometry concern path length control, formation flying, fringe tracking, and specialized optical components. Concepts for space-borne interferometers have been developed covering a wide range of scientific applications and wavelength regimes.

Introduction: Why interferometry, why in space?

The angular resolution of optical systems with size d operated at a wavelength λ is limited by diffraction to $\approx \lambda/d$. High resolution thus implies large optical systems; whenever the desired resolution exceeds the limits of reasonably-sized monolithic telescopes, the entrance aperture has to be split into several smaller collectors, whose separate beams have to be combined coherently. Such systems are commonly called astronomical interferometers; the simplest concept of this kind is shown in Figure 17.1 and discussed in more detail in the following section.

Interferometry at radio wavelengths was introduced in the 1950s and has become the dominant technique in radio astronomy today. Optical/near-infrared interferometry, in contrast, has remained more of a specialty employed to address selected astronomical problems, and to make a limited number of fundamental measurements. This is mainly due to the challenging requirement of measuring and controlling optical paths within the instrument with sub-wavelength precision, and in particular to the difficulties encountered when doing this on the ground through the Earth's atmosphere.

[I]Landessternwarte, Universität Heidelberg, Germany

Space offers a number of advantages, which together enable applications of interferometry that are inconceivable on the ground. The most important of these are:

- *Absence of atmospheric turbulence and dispersion.* Typical path length fluctuations for visible/near-IR light through the atmosphere have amplitudes of tens of micrometres and coherence time scales of milliseconds (the latter scaling with $\lambda^{6/5}$). Fast servo loops are thus necessary to maintain coherence in ground-based interferometers; the need to measure the phase at a high rate limits the sensitivity. Furthermore, dispersion in air introduces a wavelength dependence of the path length, which needs to be compensated carefully. These requirements obviously do not exist in space.
- *Thermally stable and vibration-free environment.* In the absence of atmospheric fluctuations, vibrations and thermal drifts are the dominant sources of path length variations. These are much more easily controllable in space than on the ground.
- *Access to the whole electromagnetic spectrum.* Just like single telescopes, interferometers in space do not suffer from atmospheric extinction blocking access to parts of the electromagnetic spectrum.
- *Low background.* Above the atmosphere, there is little scattered light and no airglow. The whole interferometer can be cooled to reduce the thermal emission of the optics. This is even more important than cooling of monolithic telescopes, since interferometers usually have many more reflecting surfaces in the light path.
- *Free choice of orientation.* In interferometers that are tied to the rotating Earth, the observing geometry changes continuously. This necessitates the inclusion of long delay lines moving at the sidereal rate to compensate the unequal interferometer arm lengths. In space, the geometry can be held fixed for each target, and array layouts are possible in which all interferometer arms have equal lengths, obviating the necessity for long delay lines. On the other hand, measurement and control of the orientation of a telescope array in space is a major challenge, because of the lack of a natural stable platform.

Space interferometry has three principal areas of application: precise astrometry, high-resolution imaging, and detection of faint objects near much brighter ones. While the first of these topics is covered in the preceding chapter (Lindegren 2010; see also Unwin et al 2008), the subject of this chapter is an overview of the second and third. The emphasis is on interferometry in the visible and near-infrared, but most of the principles are equally applicable at other wavelengths.

Principles of interferometry and aperture synthesis

The following sections give a brief introduction to the principles of interferometry and aperture synthesis. For more complete treatments of the subject see the reviews by Shao and Colavita (1992), Quirrenbach (2001), Monnier (2003) and Cunha et al (2007), and the books by Thompson et al (1986) and by Cassen et al (2006).

Visibility measurements

The most important components of an astronomical interferometer are shown schematically in Figure 17.1: two telescopes or siderostats, delay lines that can be moved to equalize the optical path length in the two interferometer arms, a beam combiner, and detectors. When the delay difference $D_{\mathrm{ext}} - D_{\mathrm{int}}$ between the two interferometer arms is exactly zero,[1] the observed signal I_{out} is the sum of signals from the two arms I_1, I_2, plus an interference term:

$$I_{\mathrm{out}} = I_1 + I_2 + 2\sqrt{I_1\, I_2}\, \Re(V(u,v)) \quad . \tag{17.1}$$

Here the visibility $V(u,v)$ is a complex number with modulus $|V| \leq 1$, which depends on the coordinates $(u = x/\lambda\ , v = y/\lambda)$ of the baseline $\boldsymbol{B} = (x,y)$ between the two telescopes, and on the sky brightness distribution, i.e., the structure of the source being observed (see the following section for details). By introducing phase shifts of 0, $\pi/2$, π, and $3\,\pi/2$, one can determine the visibility amplitude $|V|$ and phase $\phi = \arg(V)$. This can either be done simultaneously, by splitting the light into four channels, or sequentially, by varying D_{int} in steps of $\lambda/4$ and counting the photons at each step.[2]

In Michelson arrays with N telescopes the $N(N-1)/2$ visibilities can either be measured by pairwise beam combination, or by bringing the light from all telescopes together on one detector. In the latter "all-on-one" techniques, the fringes from the different baselines have to be encoded either spatially (by using a non-redundant output pupil) or temporally (by using different dither frequencies for the beams from individual telescopes).

It should be noted here that the signal-to-noise ratio (SNR) of all interferometric observables depends not only on the photon count N_{p}, but on $N_{\mathrm{p}}V^2$ for the photon-noise limited regime and on $N_{\mathrm{p}}V$ for the background-limited regime, and that the SNR drops precipitously in the photon-starved regime (see Shao and Colavita 1992 for details). The very drastic loss of sensitivity with source extension may come as a nasty surprise if observations are not planned carefully: even for a modestly resolved star one may find that $V = 0.1$ and thus lose 5 magnitudes of sensitivity compared to a point source.

The van Cittert–Zernike theorem

For a more formal treatment of the interferometer response, it is useful to introduce the *source coherence function* defined as

$$\gamma(\boldsymbol{\xi}_1, \boldsymbol{\xi}_2, \tau) = \langle E(\boldsymbol{\xi}_1, t) E^*(\boldsymbol{\xi}_2, t-\tau) \rangle \quad , \tag{17.2}$$

[1] For simplicity, we will mostly ignore the complications brought about by dispersion for converting between delay, optical path length, and fringe phase.

[2] Note that the there is already a phase shift of π between the two detectors shown in Figure 17.1 due to the additional reflection at the beam combiner for one of them. Thus the sum of the intensities measured by them is independent of V and D_{int}, as required by conservation of energy.

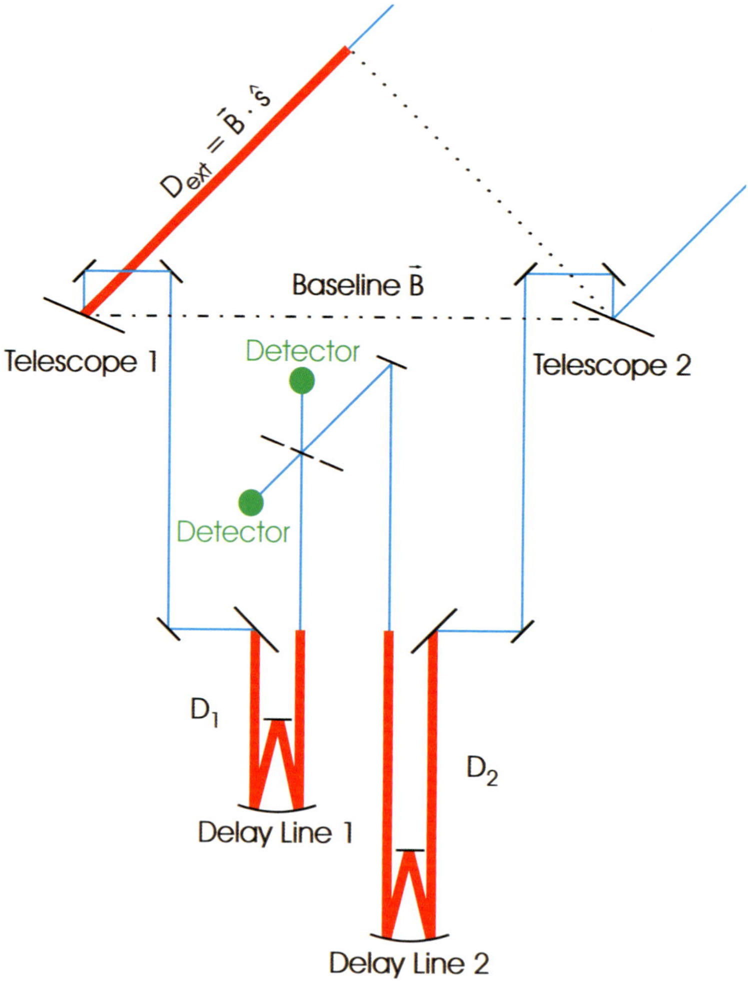

Figure 17.1: Schematic drawing of the light path through a two-element interferometer. The external delay $D_{\mathrm{ext}} = \boldsymbol{B} \cdot \hat{\boldsymbol{s}}$ is compensated by the two delay lines, here $\boldsymbol{B}$ is the baseline vector and $\hat{\boldsymbol{s}}$ the unit vector in the direction of the source. The path lengths D_1, D_2 through the delay lines are monitored with laser interferometers. The zero-order interference maximum occurs when the delay line positions are such that the internal delay $D_{\mathrm{int}} = D_2 - D_1$ is equal to D_{ext}. The beams from the two telescopes are combined with a plate that has a relative transmission of 50 % and a reflectivity of 50 %. The architecture shown is a Michelson interferometer with co-axial beam combination in the pupil plane (cf., Page 301).

where E is the radiation field, $\boldsymbol{\xi}$ the (two-dimensional) direction cosine on the sky, and τ the temporal delay.[3] For $\boldsymbol{\xi}_1 = \boldsymbol{\xi}_2 = \boldsymbol{\xi}$, γ is the time autocorrelation function of the radiation from direction $\boldsymbol{\xi}$; for $\tau = 0$, this is the time-averaged brightness from that direction $B(\boldsymbol{\xi}) = \langle |E(\boldsymbol{\xi})|^2 \rangle$. An extended source is called *spatially incoherent*[4] if $\gamma = 0$ for $\boldsymbol{\xi}_1 \neq \boldsymbol{\xi}_2$; in this case

$$\gamma(\boldsymbol{\xi}_1, \boldsymbol{\xi}_2, \tau) = \gamma(\boldsymbol{\xi}_1, \tau) \cdot \delta(\boldsymbol{\xi}_1 - \boldsymbol{\xi}_2) \quad . \tag{17.3}$$

The electric field of radiation from a source located at position $\boldsymbol{\xi}$ received by a telescope at position $\boldsymbol{u}$ has a phase term $\mathrm{e}^{-2\pi\mathrm{i}\boldsymbol{\xi}\cdot\boldsymbol{u}}$. To compute the interference term between two telescopes located at positions $\boldsymbol{u}_1$, $\boldsymbol{u}_2$, one has to integrate the correlations between all pairs of sky positions $\boldsymbol{\xi}_1$, $\boldsymbol{\xi}_2$ with the appropriate phase terms:

$$\Gamma(\boldsymbol{u}_1, \boldsymbol{u}_2, \tau) = \int_{-\infty}^{\infty}\int_{-\infty}^{\infty} \gamma(\boldsymbol{\xi}_1, \boldsymbol{\xi}_2, \tau)\, \mathrm{e}^{-2\pi\mathrm{i}(\boldsymbol{\xi}_1\cdot\boldsymbol{u}_1 - \boldsymbol{\xi}_2\cdot\boldsymbol{u}_2)}\mathrm{d}\boldsymbol{\xi}_1\mathrm{d}\boldsymbol{\xi}_2 \quad . \tag{17.4}$$

Mathematically, this is the Fourier transform of γ. In the spatially incoherent case, the interferometer response depends only on the *difference vector* $\boldsymbol{u} = \boldsymbol{u}_1 - \boldsymbol{u}_2$ of the telescope positions:

$$\Gamma(\boldsymbol{u}, \tau) = \int_{-\infty}^{\infty} \gamma(\boldsymbol{\xi}, \tau)\, \mathrm{e}^{-2\pi\mathrm{i}\boldsymbol{\xi}\cdot\boldsymbol{u}}\mathrm{d}\boldsymbol{\xi} \quad . \tag{17.5}$$

The interferometer output at zero delay is the *complex visibility* introduced on Page 295; the complex visibility is the Fourier transform of the source brightness distribution:

$$V = \Gamma(\boldsymbol{u}, 0) = \int_{-\infty}^{\infty} \left\langle |E(\boldsymbol{\xi})|^2 \right\rangle \mathrm{e}^{-2\pi\mathrm{i}\boldsymbol{\xi}\cdot\boldsymbol{u}}\mathrm{d}\boldsymbol{\xi} \quad , \tag{17.6}$$

which we can also write as

$$V(u, v) = \int_{-\infty}^{\infty}\int_{-\infty}^{\infty} I(\xi, \eta)\mathrm{e}^{-2\pi\mathrm{i}(u\xi + v\eta)}\mathrm{d}\xi\,\mathrm{d}\eta \quad . \tag{17.7}$$

Here I is the intensity, (ξ, η) are the components of $\boldsymbol{\xi}$, and (u, v) those of $\boldsymbol{u}$. Each observation on one baseline thus measures one Fourier component of the sky brightness distribution. Equation 17.7 is known as the *van Cittert–Zernike theorem*.

Fourier relations, the (u,v) plane, and aperture synthesis

The objective of an imaging observation is the determination of the sky brightness distribution $I(\xi, \eta)$. According to Equation 17.7, this can be achieved by measuring the visibility $V(u, v)$, and performing an inverse Fourier transform:

$$I(\xi, \eta) = \int_{-\infty}^{\infty}\int_{-\infty}^{\infty} V(u, v)\mathrm{e}^{+2\pi\mathrm{i}(u\xi + v\eta)}\mathrm{d}u\,\mathrm{d}v \quad . \tag{17.8}$$

[3] In the absence of dispersion, temporal delay τ and spatial delay $D = c \cdot \tau$ can be used interchangeably.

[4] Most astronomical sources can be considered to be spatially incoherent. Counterexamples are the radiation fields of sources affected by scintillation in the interstellar medium or in the Earth's atmosphere.

This works only, of course, to the extent that the sampling conditions well-known from Fourier theory are fulfilled. First, the resolution in the image plane $(\xi, \eta)_{\min}$ is the inverse of the maximum baselines $(u, v)_{\max}$. Second, and equally important, according to the Nyquist sampling theorem, the detection of large structures up to $(\xi, \eta)_{\max} = 1/(2\Delta)$ requires sampling of the (u, v) plane on a grid with steps of $\Delta = (u, v)_{\min}$. Violating the Nyquist criterion will lead to aliasing; diffuse emission on scales larger than $1/(2\Delta)$ will usually be missed completely by the interferometer.

From these considerations it is clear that images with many independent pixels can only be obtained through a large number of visibility measurements. These do not necessarily have to be carried out simultaneously by a many-baseline array, but can be done sequentially provided that the whole observation does not take longer than the shortest time scale of variability in the source. In space, this *aperture synthesis* or *hyper-synthesis* technique can be implemented conveniently by rotating an array of telescopes around the line of sight, or by linear translation of the telescopes in the aperture plane.

Co-phasing and coherencing

Since the interferometer geometry is almost never known a priori with a precision $\lesssim \lambda$, one has to search the fringes by scanning the delay, and then close a servo loop that tracks their position. Doing this with a precision $\lesssim \lambda/(2\pi)$ is called *co-phasing* or *fringe tracking*; in this case observed phases have a defined meaning. Usually the sensitivity of an interferometer is limited by the fringe-tracking, which has to be done with integration times short compared to the coherence time set by the atmosphere (on the ground) or vibrations and drifts (in space). Once the interferometer is co-phased, the integration time for data taking may be arbitrarily long; the point-source sensitivity is then comparable to the sensitivity of a single telescope with the same aperture.

It is possible, however, to observe fringes over a much larger range of delays; the width of the fringe packet is $\sim \lambda^2/\Delta\lambda$, where $\Delta\lambda$ is the observing bandwidth. Controlling the delay only with this much more relaxed precision is called *coherencing*. In this case phases have no defined meaning (but see the next section), and the integration time has to be kept short enough so that the fringe phase does not vary by more than $\approx 1\,\text{rad}$ during an integration.

Adopting a strategy for fringe tracking (or coherencing, as the case may be) is the crucial first step in the design of any interferometric instrument or observation. Fringe tracking does not necessarily have to be performed at the same wavelength, on the same baseline, or even on the same star as the observation. This opens the following possibilities:

- *On-source fringe tracking.* If the astronomical target is sufficiently bright and provides high visibilities, it can be used directly for fringe tracking. This is frequently the case, for example, in radio interferometry of active galaxies, which have bright unresolved cores.

- *Off-source fringe tracking and phase referencing.* If the object itself is not bright enough or too resolved, a reference source has to be used for fringe tracking. In radio interferometry, it is frequently possible to alternate observations of the target and the reference, and to keep track of the instrumental phase during this process. At shorter wavelengths, the coherence times are usually too short for this approach, and the target and reference have to be observed simultaneously with a *dual-star interferometer*, in which the path length difference between target and reference is compensated with a differential delay line, and monitored with laser metrology.

- *Wavelength bootstrapping.* The target may be faint or extended at the observing wavelength, but much brighter or much more compact at some other wavelength. Then the latter can be used for fringe tracking supporting data taking at the former. In a variant of this technique, a wide bandpass may be used for fringe tracking, and much narrower bands for the scientific data, e.g., for performing interferometric spectroscopy.
- *Baseline bootstrapping.* In arrays with more than two telescopes, it is possible to employ baseline bootstrapping, which uses the signal on the short baselines, where V is high, for the fringe tracking servo, while data are taken on the long baselines, where V and thus the SNR may be much lower. In fact, the array geometry can be optimized to facilitate imaging of extended sources with this technique.

These approaches can obviously be combined with each other, but the propagation of fringe-tracking residuals and optimization of the fringe-tracking SNR has to be considered carefully, especially in multi-telescope arrays (le Poole and Quirrenbach 2002).

Closure phases and self-calibration

If coherencing but not co-phasing is performed, the fringe amplitude can frequently be measured rather reliably, but the fringe phase loses a definite meaning. This means that imaging through Equation 17.8 is no longer possible. Rather frequently it is possible, however, to associate the delay errors with the telescopes in the array, i.e., all baselines that involve a certain telescope are affected by the delay error of that telescope in the same way. Examples of such telescope-based errors are atmospheric path length variations (for ground-based interferometers), or path length variations in the beam train leading from the telescope to the beam combiner. If one considers the baselines formed by three telescopes (i, j, k), the observed phases ϕ'_{ij} are related to the true phases ϕ_{ij} and telescope-based phase errors ψ_i by the following relations:

$$\begin{aligned} \phi'_{ij} &= \phi_{ij} + \psi_i - \psi_j \\ \phi'_{jk} &= \phi_{jk} + \psi_j - \psi_k \\ \phi'_{ki} &= \phi_{ki} + \psi_k - \psi_i \quad . \end{aligned} \tag{17.9}$$

If the delay errors are mostly of this type, one can retain some of the phase information in the form of *closure phases* Φ_{ijk}, which are obtained by summing the phases in the closed triangles of the array:

$$\Phi_{ijk} \equiv \phi'_{ij} + \phi'_{jk} + \phi'_{ki} = \phi_{ij} + \phi_{jk} + \phi_{ki} \quad . \tag{17.10}$$

Table 17.1: Information content in the closure phases as a function of array size.

No. of telescopes	No. of Fourier phases	No. of closure triangles	No. of independent closure phases	Phase information in closure phases
3	3	1	1	33 %
4	6	4	3	50 %
5	10	10	6	60 %
8	28	56	21	75 %
10	45	120	36	80 %
20	190	1 140	171	90 %
50	1 225	19 600	1 176	96 %
N	$\binom{N}{2}$	$\binom{N}{3}$	$\binom{N-1}{2}$	$\frac{N-2}{N}$

Because the phase errors ψ_i cancel each other in this sum, the closure phases Φ_{ijk} are observables that contain phase information on the target, but are unaffected by telescope-based phase errors. The closure phase is the phase of the *bispectrum* or *triple product* $\mathcal{B}$, which therefore is also immune to telescope-based phase errors:

$$\mathcal{B}_{ijk} \equiv V'_{ij} V'_{jk} V'_{ki} = V_{ij} V_{jk} V_{ki} = |V_{ij}|\,|V_{jk}|\,|V_{ki}|\,\mathrm{e}^{\mathrm{i}\,(\phi_{ij}+\phi_{jk}+\phi_{ki})} \quad . \tag{17.11}$$

The number of closed triangles in an array of N telescopes is $\binom{N}{3}$, but only $\binom{N-1}{2}$ of them are algebraically independent of each other.[5] Table 17.1 summarizes the number of closure phases and their information content (i.e., the number of independent closure phases divided by the number of Fourier phases) for a few typical array sizes.

While closure phases retain some of the phase information even when the array cannot be co-phased, their use complicates the data reduction, because the inverse Fourier transform of Equation 17.8 cannot be applied. One usually has to take recourse to iterative image restoration algorithms such as *Högbom CLEAN* (Högbom 1974) or the *Maximum Entropy Method* (e.g., Perley et al 1989). These algorithms use a priori knowledge on the image (e.g., the condition that all pixels must be non-negative, the assumption that the source is contained in a certain area on the sky, or assumptions on the smoothness of the brightness distribution) to derive the most plausible image that is consistent with the measured visibility amplitudes and closure phases. *Phase self-calibration* is an equivalent method that seeks to determine the N unknown parameters ψ_i along with the image from the observed visibility amplitudes $|V_{ij}|$ and phases ϕ'_{ij}.

[5]Note that depending on the beam combination scheme, most or all of the $\binom{N}{3}$ closure phases may be *(photon-)statistically independent* realizations of the $\binom{N-1}{2}$ *algebraically independent* closure phases.

Classification of interferometric concepts

The various beam combination schemes that can be employed in astronomical interferometry may be classified according to several criteria: the beam étendue (single-mode or multi-mode), the beam direction (co-axial or multi-axial), the combination plane (image plane or pupil plane), and the relation between input and output pupils (Michelson or Fizeau configuration).

In a Fizeau interferometer the output pupil is an exact replica of the input pupil, scaled only by a constant factor. This is also known as homothetic mapping between input and output pupil. In contrast, in a Michelson interferometer as depicted in Figure 17.1 there is no homothetic relation between the input and output pupils. This means that Abbé's sine condition is violated, and the object-image relationship can no longer be described as a convolution, because the re-arrangement of the apertures re-arranges the spatial high-frequency part of the object spectrum in the Fourier plane (Tallon and Tallon-Bosc 1992). This has an important consequence for off-axis objects: the image position does not coincide with the white-light fringe position (see Figure 1 in Tallon and Tallon-Bosc 1992). For a non-zero spectral bandwidth this means that the fringe contrast decreases with field angle and the field of view is limited; the maximum size of an image from a Michelson interferometer is $\approx R \equiv \lambda/\Delta\lambda$ resolution elements in diameter. This effect is also known as "bandwidth smearing" in radio astronomy (see, e.g., Thompson et al 1986).

In the focal plane of a properly phased Fizeau interferometer one observes an image of the field, convolved with the point spread function (PSF) of the array. The desired information can thus in principle be obtained rather easily by deconvolution. If the array is very dilute, however, this approach suffers from the fact that the PSF has many rather high side lobes; this spreads the light from a point source over many pixels and reduces the SNR. To avoid this effect, one can magnify the output pupils of the individual telescopes with respect to the arrangement of their centres (Labeyrie 1996). Such a *conformal Michelson array* or *hyper-telescope* still forms a direct image, but only over a limited field; of order N^2 telescopes are needed to obtain a field with N resolution elements across.

In most Michelson array concepts there is no similarity between the input and output pupil at all. These instruments measure the visibilities for all $N(N-1)/2$ baselines in an N-telescope array. The visibilities can either be measured by pair-wise beam combination, or by bringing the light from all telescopes together on one detector. In the latter "all-on-one" techniques the fringes from the different baselines have to be encoded either spatially (by using a non-redundant output pupil) or temporally (by using different delay dither frequencies for the beams from individual telescopes). In any case, the image is computed from the visibilities by a Fourier transform or by the imaging methods described before.

Direct detection and heterodyne interferometry

In the interferometer shown in Figure 17.1, the detectors are placed behind a beam combiner, which mixes the beams from the individual telescopes with appropriate amplitudes and phases. At low frequencies ($\lambda \gtrsim 10$ µm), a different architecture is frequently used: the signal from a local oscillator is mixed with the sky signal at each telescope, and the resulting intermediate frequency (IF) signals correlated with each other electronically (see Thompson et al 1986 for details). Heterodyne interferometers can achieve very high spectral resolution (several 100 000). Further advantages include the ability to amplify the IF signals before correlation, which is especially important in multi-telescope interferometers, and the ability to perform the path length compensation electronically rather than with bulky delay lines. The most important disadvantage of heterodyne systems is their limited bandwidth, restricted by the processing speed of the correlator electronics.

Interferometric spectroscopy

For many astronomical applications, it is very useful to perform *interferometric spectroscopy*, i.e., to obtain visibilities in many wavelength channels. This can be done by coupling the output of the beam combiner into a conventional spectrograph (e.g., Quirrenbach et al 2007). An alternative method is *double-Fourier interferometry* (Mariotti and Ridgway 1988). In a generalization of Equation 17.7, one can write

$$\Gamma(u, v, \tau) = \iiint I(\xi, \eta, \nu) \mathrm{e}^{-2\pi \mathrm{i}(u\xi + v\eta + \tau\nu)} \mathrm{d}\xi \, \mathrm{d}\eta \, \mathrm{d}\nu \quad , \tag{17.12}$$

which shows that the brightness I as a function of position and frequency ν can be obtained from the interferogram, provided that one scans through the fringe packet and takes measurements at many positions of the delay τ. The special case $u = 0$, $v = 0$ corresponds to a classical Fourier transform spectrometer.

Nulling interferometry

The direct detection of extrasolar planets orbiting nearby stars requires imaging with a very high contrast: to enable photometric, spectroscopic, or polarimetric studies of the planet, the light emitted by the host star must be rejected efficiently. At visible wavelengths, this can be done with a coronographic system inside a telescope or with an external occulting disk; in the mid-infrared the required angular resolution is too high for these concepts, and nulling interferometry is the technique of choice.

Basic principles of nulling

In an infrared interferometer the starlight can be suppressed by introducing an achromatic π phase shift in one arm of the interferometer, changing the sign of the interference term in Equation 17.1, so that there is destructive interference for light arriving on-axis (see Figure 17.2); this concept was conceived by Bracewell (1978). The most important characteristic of such a nulling interferometer is the quality of this destructive output, called the null depth d_{N}. In other words, the observed starlight is reduced by a factor d_{N} compared to a constructive interferometer.

The core of a nulling interferometer consists of a beam combiner with a phase shifter, which must introduce the required π phase shift for all wavelengths within

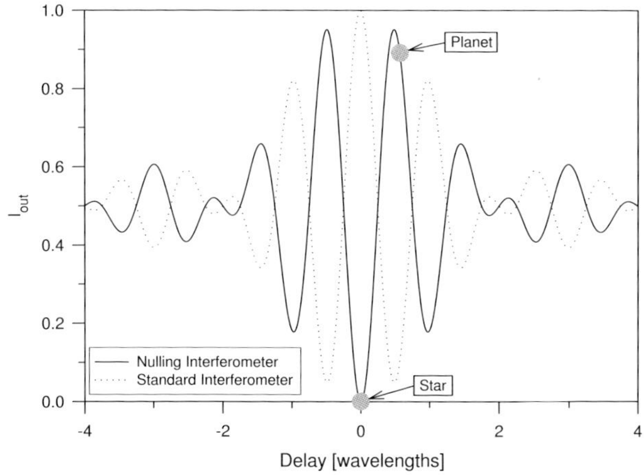

Figure 17.2: Fringe pattern (intensity at the output of the beam combiner) for a standard and a nulling interferometer. The fractional bandwidth $\Delta\lambda/\lambda = 0.5$. The pattern of the nulling interferometer has a central zero; the depths of the first minima of the standard interferometer are 5 %. Fringe tracking ensures that the delay for the star is zero. The resolution of the interferometer is matched to the star-planet separation such that the planet is close to a transmission maximum.

the observing band. This achromatic phase shift can be introduced either by geometric methods (e.g., through-focus field-inversion or field counter-rotation with pairs of periscopes) or with pairs of dispersive glasses. In any case, one has to maintain very precise symmetry in all other parts of the system in order to avoid spurious additional phase shifts.

Null depth

To produce a deep null, the two arms of the interferometer must be made symmetric with very high precision (apart from the π phase shift, of course). Any of the following imperfections can ruin the performance of the nuller: residual phase fluctuations, differences between the two arms in dispersion or polarization properties, a rotation between the two beams, and a mismatch between the two intensities. We denote the residual phase (i.e., the difference between the actual phase and the "best compromise" for all wavelengths within the bandpass and the two polarization states) by $\Delta\phi$, the RMS phase difference due to dispersion mismatch, averaged over the bandpass, by $\Delta\phi_\lambda$, the phase difference between the two polarization states by $\Delta\phi_{\mathrm{s-p}}$, the relative rotation angle between the two beams by α, and the normalized intensity mismatch by $\delta I/I \approx \delta \ln I$. A fairly straightforward analysis shows that the null depth is then given by (Serabyn 2000)

$$d_{\mathrm{N}} = \frac{1}{4}\left[(\Delta\phi)^2 + (\Delta\phi_\lambda)^2 + \frac{1}{4}(\Delta\phi_{\mathrm{s-p}})^2 + \alpha^2 + (\delta \ln I)^2\right] \ . \tag{17.13}$$

From this equation it is evident that very tight control of all instrumental errors is required to achieve a deep null ($d_{\rm N} \approx 10^{-6}$). It has also been tacitly assumed that the wavefronts from each telescope are perfectly flat, so that quantities like $\Delta\phi$ have a well-defined meaning. Because even the best space optics do not deliver sufficiently high wavefront quality, one has to "clean" the wavefront with a modal filter, e.g., with a single-mode fibre. The fundamental mode transmitted by such a fiber corresponds to a perfectly flat wavefront. All other modes are reflected at the fiber entrance, leading to a reduction of I. By using a modal filter one thus needs to control only a relatively small number of degrees of freedom, as indicated by Equation 17.13, rather than the very large number of modes in an aberrated wavefront.

Higher-order nulls

An ideal Bracewell nulling interferometer provides complete rejection of light arriving on-axis; for small off-axis angles $d_{\rm N} \propto \theta^2$. Due to this off-axis leakage, for a star with angular diameter $\theta_{\rm dia}$ the null depth is:

$$d_{\rm N} = \frac{\pi^2}{16}\left(\frac{B\,\theta_{\rm dia}}{\lambda}\right)^2 \quad . \tag{17.14}$$

One therefore has to walk the fine line between making B too small (so that the planet cannot be resolved from its parent star), and making B too large (so that too much light from the stellar disk leaks through the nuller). This problem can be mitigated by creating higher-order nulls (e.g., $d_{\rm N} \propto \theta^4$ or even $d_{\rm N} \propto \theta^6$), which is possible if light from three or more telescopes is combined with appropriate phase shifts. Linear as well as non-linear configurations have been explored for this purpose.

These principles for constructing arrays for higher-order nulling can also be applied to the design of coronographs for high-dynamic range imaging with monolithic telescopes, where light leakage due to the non-zero size of the star is just as important as in nulling interferometry (Kuchner and Traub 2002).

Signal extraction and modulation

The next important question to consider is the way of extracting useful information from a nulling array, which does not produce images, but rather a photon count from a single-pixel detector as a function of time. Temporal modulation of the planetary signal can help to separate it from the stellar leakage. The simplest way of doing this is rotating the interferometer around the line of sight.[6] The information about the position and brightness of the planet is then encoded in the time dependence of the photon count.

The rotational modulation method has a significant difficulty, namely the requirement that the instrumental response has to remain stable over a full rotation

[6] Representing the planet by a point-like sky brightness distribution $B(\xi, \eta) = \delta(\xi_0, \eta_0)$ with $(\xi_0, \eta_0) \neq (0,0)$, one can see from Equations 17.1 and 17.7 that rotating the interferometer (i.e., $u = u_0 \cos\omega\, t, v = u_0 \sin\omega\, t$)) leads to a modulated output signal.

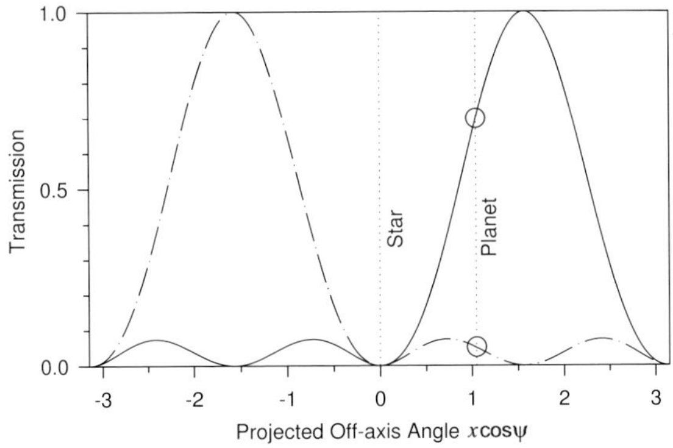

Figure 17.3: Transmission of a linear double-Bracewell interferometer consisting of four telescopes of identical diameter with unit spacing. The horizontal axis is the component of the off-axis angle parallel to the array. The full curve is the transmission for phases (0, $-\pi/2$, π, $\pi/2$), the dash-dotted curve for (0, $\pi/2$, π, $-\pi/2$). Chopping between these states modulates the planetary signal between the two values indicated by circles.

period, which is typically of the order of several hours. Small drifts of the detector sensitivity or background, of the telescope pointing, or other components of the interferometer can mask the presence of a planet or lead to spurious detections. It is therefore highly desirable to modulate the output at a higher rate ($\approx 1\,\text{Hz}$). Combining the light from four identical telescopes with phases (0, π, α, and $\alpha+\pi$) will obviously produce a central null irrespective of the value of α. One can then, for example, quickly alternate between $\alpha = \pi/2$ and $\alpha = -\pi/2$, which generally changes the interferometer response pattern on the sky (see Figure 17.3). This *internal chopping* enables lock-in detection of the planetary signal at the desired high frequency.

Critical technologies for space interferometry

Laser metrology and path length control

Interferometers operating in the visible and infrared normally rely on laser metrology to monitor the internal instrumental path length. Depending on the specific application, the precision requirements range from about $\lambda/20$ (for imaging) down to $\approx 50\,\text{pm}$ (for sub-microarcsecond astrometry). The most challenging task is not building a metrology system that can reach this measurement precision, but rather ensuring that the metrology actually measures the path taken by the starlight. Misalignments, beam walk on actuated optics, non-ideal behaviour of fiducials such as corner cubes and cats' eyes, dispersion between the laser wavelength and the effective observing wavelength, and polarization effects all need to be understood and calibrated.

For applications such as astrometry and high dynamic range imaging, the requirements on *knowing* the actual value of the delay are much more stringent than those on *controlling* it. Nulling, on the other hand, depends not only on measuring the delay, but also on minimizing its deviation from, and fluctuations around, the nominal value. This implies that a real-time servo system is needed with a control bandwidth that exceeds the highest frequency of delay disturbances. Furthermore, even small parasitic polarization or dispersion effects can ruin the null depth (cf., Equation 17.13). This means that these additional degrees of freedom have to be monitored and controlled.

Connected structures and free flyers

Interferometers with modest baselines (up to perhaps 10 m or 20 m) are most easily implemented on a connected structure. This means that the interferometer can be manoeuvered essentially like a single telescope, albeit with high precision requirements on the attitude. The large moment of inertia and consequently high torque needed for turns perpendicular to the long spacecraft axis also needs to be taken into account for the specification of the attitude control system including gyros and thrusters.

When longer baselines are needed, a free-flying architecture has to be adopted, in which each telescope (and, in most concepts, the beam combiner) is carried by its own spacecraft. Formation flying of the array constellation brings about many challenges (see also Chapter 38, Lumb 2010): each spacecraft has to be oriented correctly and manoeuvered to its position in the constellation, and the fringes have to be acquired and tracked. Fortunately the requirements on the formation-flying precision are not as stringent as one might think at first sight, as the fine control of path length and angular alignment can be performed with actuators on each spacecraft, so that the formation-flying tolerances are of order of millimetres or even centimetres even when a nanometre overall precision is required. For aperture synthesis observations (see Page 297), the strategy for filling the (u, v) plane has to be integrated with the formation-flying concept. Here free-flying interferometers offer more flexibility for the telescope arrangement than connected structures, but minimizing the use of propellant requires linear spacecraft trajectories and the capability to perform observations continuously while the telescopes are in motion.

For free-flying interferometers operating in the mid- and far infrared, thermal control is an additional challenge, as radiative coupling between cold parts of a spacecraft with the warm components of its neighbours has to be avoided.

Guiding and fringe tracking

Unlike ground-based interferometers, space-borne arrays are not connected to a stable platform that can be used as a first-order reference for their orientation. In cases where very high absolute accuracy is required (e.g., for astrometry), the guider must itself be an interferometer with precision comparable to the main instrument. In free-flying arrays, the relative positions of the individual spacecraft can be determined with metrology systems, but the overall orientation still needs to be determined as for a structurally connected instrument.

Fringe tracking provides the necessary information on the array orientation for imaging or nulling; the techniques described on Page 298 can all be applied. Off-axis fringe tracking, however, carries a particularly high cost in space. Whereas on the ground long delay lines (of the order of the baseline length B) are needed in any case to compensate for the Earth's rotation, space arrays can be designed with nominally equal arm lengths and thus need only fine actuators to compensate for the tolerances of the spacecraft positioning system. This advantage is partially lost if off-axis fringe tracking is employed, because the differential delay between the target and a reference source located at an off-axis angle θ is of order θB, which typically amounts to many centimetres, re-introducing the need to employ "long-stroke" delay lines, which add a fair number of optical surfaces and mechanisms. To circumvent this difficulty, it has been proposed that the whole interferometer array could be used as a kilometric optical gyro (KOG). As in any optical gyro, two laser beams would be sent clockwise and counter-clockwise around the perimeter of the spacecraft array, so that any rotation of the constellation in inertial space could be deduced from the Sagnac effect. Since the amplitude of the Sagnac effect scales with the area enclosed by the two counter-rotating laser beams, high precision should be achievable in principle; however, detailed studies of the feasibility of KOGs remain to be conducted.

Optical components

Since applications of space interferometry usually demand very high precision, the requirements on the optical quality of all components is extremely high, even by space standards. In addition, beam combiners are very complex systems, which manipulate the light in various ways and have tight alignment tolerances. The functions performed by a beam combiner may include the control of path length, aberrations, and polarization, the introduction of achromatic phase shifts, beam splitting and recombination, injection of metrology beams and more. "Classical" designs of beam combiners therefore tend to lead to crowded and bulky optical assemblies. This makes solutions very attractive that employ fiber optics and/or integrated optics to perform some or most of the functions needed. In contrast to applications in the communications industry, astronomical interferometry needs components that support wide spectral bandwidths and wavelength ranges extending up to ≈ 20 µm. The development of such devices is a high-priority objective for space (and also ground-based) interferometry.

Scientific applications and mission concepts

A wide variety of interferometric missions have been proposed and studied, for wavelengths ranging from the radio range all the way to X-rays. This section provides a brief overview of the most prominent of these concepts, in order to illustrate the potential of space interferometry for a very diverse set of scientific goals.

Darwin / TPF

There is growing interest in rocky planets that orbit their parent stars at a distance where the temperature is such that water can be liquid at the surface, providing favourable conditions for the development of life. ESA and NASA have studied concepts for a mission — known as *Darwin* in Europe and as *TPF* in the US — aimed at performing detailed studies of Earth-like planets in these 'habitable zones' of some 150 solar-type stars (e.g., Quirrenbach 2005). *Darwin/TPF* therefore must have an angular resolution of ≈ 35 mas, corresponding to 0.7 ua at 20 pc. To achieve this in the mid-infrared, an interferometer with baselines of at least 50 m to 100 m is required. The contrast between a terrestrial planet and its parent star at 10 µm is $\lesssim 10^{-6}$, necessitating nulling with excellent depth and stability of the null (see Page 302).

Life on Earth has a profound impact on the composition of the atmosphere, producing abundant oxygen, and methane in extreme thermodynamic disequilibrium. It is plausible to assume that any form of abundant life on an extrasolar planet might transform the planetary atmosphere in a similar way, and thus produce signatures that can be detected spectroscopically (e.g., Des Marais et al 2002). The mid-infrared spectral region ($\approx$ 6 µm to 20 µm) contains a number of important spectral features that can be used to diagnose the presence of H_2O, CO_2, and O_3, which serves as a proxy for O_2 (Angel et al 1986; Léger et al 1996). Methane also has a strong absorption band in this spectral region, but it would be extremely difficult to detect in an Earth-like atmosphere. In the past, however, before the rise of oxygen, biogenic CH_4 may have been more abundant in the Earth's atmosphere by a factor or 100 to 1000; at this level it would also be detectable with *Darwin/TPF* and serve as an indicator of bacterial life.

Stellar Imager

A space-based UV-optical Fizeau interferometer dubbed "*Stellar Imager*" has been studied by NASA (Carpenter et al 2006). 20 to 30 mirror satellites forming baselines of 100 m to 1 km would send the light to a central beam combiner and thus provide 0.1 mas spectral imaging of the surfaces and — via spatially-resolved asteroseismology — interiors of solar-type stars. Additional science goals include the mapping of the interior regions and winds of quasars and other active galactic nuclei, "movies" of nova and supernova explosions, and detailed studies of interacting binary stars.

Far-infrared interferometry

Concepts for space-based far-infrared ($\approx$ 25 µm to 600 µm) interferometers with cryo-cooled ($\approx$ 4 K) telescopes of modest size have been developed in the US and in Europe (e.g., Leisawitz et al 2007). The main science drivers are spectroscopic studies of star and planet formation, with an emphasis on water lines, and surveys of infrared-luminous galaxies across the history of the Universe. A structurally-connected interferometer could provide access to baselines up to tens of metres, with corresponding sub-arcsecond image resolution over a field of view of a few square arc minutes; a far-IR interferometer with components flying in formation could sample baselines up to 1 km. A beam-combining instrument equipped with a scanning optical delay line yielding double-Fourier spectral synthesis data with resolution $R \approx 3000$ is widely viewed as an attractive option. Alternatively, a heterodyne approach would relax the requirements on the formation flying and provide superior spectral resolution (up to $R \approx 300\,000$), but at the cost of a relatively small bandwidth and therefore poorer sensitivity (de Graauw et al 2005).

X-ray interferometry

Laboratory experiments have shown that X-ray fringes can be obtained with grazing-incidence interferometers in a geometry that is in principle suited for astronomical applications (Cash et al 2000). A metre-class interferometer would provide a resolution of ≈ 100 μas at 1 keV, sufficient to image the coronae of nearby stars, but the distance between the optics and the detector would have to be tens of kilometres, necessitating formation flying even though the interferometer itself could be assembled on a single spacecraft. The resolution of an X-ray interferometer with baselines of a few 100 m would be ≈ 100 nas, and could thus resolve the event horizon of the black holes at the centre of our Galaxy, and in active galactic nuclei. The X-ray photon fluxes from these objects are in the range 1 $s^{-1}m^{-2}$ to 1000 $s^{-1}m^{-2}$, making interferometric observations possible in principle. However, because of the very tight tolerances for figuring and aligning the optics, the path from proof of principle to a realistic mission will be very long.

Very low-frequency radio interferometry

The Earth's ionosphere becomes opaque at frequencies below a few tens of megahertz. This frequency range could be explored with an interferometer consisting of 16 identical small satellites equipped with dipole antennas, and distributed in a spherical array 100 km in diameter (Jones et al 2000). With this geometry, the whole sky could be observed continuously with rather good coverage of the (u, v) plane. The science goals of such a mission would range from solar radio astronomy and the Sun-Earth connection to observations of supernova remnants, pulsars, and searches for "fossil" radio galaxies.

Radio very long-baseline interferometry

In radio very long-baseline interferometry (VLBI), telescopes separated by hundreds or even thousands of kilometres are connected "virtually": the signal at each antenna is recorded on tape, the tapes from all antennas are shipped to a correlating center, and the "fringes" are formed in a computer. To overcome the limit on baseline length set by the diameter of the Earth, VLBI arrays have been formed that include one antenna in space. The first such experiment conducted with the 4.9 m radio antenna of the TDRSS communications satellite at 2.3 GHz and 15 GHz showed that a number of extragalactic radio sources are sufficiently compact to be detected on baselines of up to two Earth diameters (Levy et al 1986). The Japanese

HALCA mission (Hirabayashi et al 2000) carrying an 8 m antenna conducted extensive observing programmes at 1.6 GHz and 5 GHz of extragalactic radio sources as well as pulsars and OH masers in conjunction with large ground-based arrays. Unfortunately, the 22 GHz channel of *HALCA*, which might have been scientifically most productive, failed; a successor mission is therefore planned that will extend the frequency coverage up to 43 GHz.

Bibliography

Angel JPR, Cheng AYS, Woolf NJ (1986) A space telescope for infrared spectroscopy of Earth-like planets. Nature 322:341–343

Bracewell RN (1978) Detecting nonsolar planets by spinning infrared interferometer. Nature 274:780–781

Carpenter KG, Schrijver CJ, Karovska M (2006) The Stellar Imager (*SI*) vision mission. Proc SPIE 6268:626821-1–12

Cash W, Shipley A, Osterman S, Joy M (2000) Laboratory detection of X-ray fringes with a grazing-incidence interferometer. Nature 407:160–162

Cassen P, Guillot T, Quirrenbach A (2006) Extrasolar planets. Saas-Fee Advanced Course 31, Springer-Verlag, Berlin, Heidelberg

Cunha MS, Aerts C, Christensen-Dalsgaard J (plus 22 authors) (2007) Asteroseismology and interferometry Astron Astrophys Rev 14:217–360

de Graauw T, Helmich FP, Cernicharo J (plus 15 authors) (2005) Exploratory submm space radio-interferometric telescope. Adv Space Res 36:1109–1113

Des Marais DJ, Harwit MO, Jucks KW (plus seven authors) (2002) Remote sensing of planetary properties and biosignatures on extrasolar terrestrial planets. Astrobiology 2:153–181

Hirabayashi H, Hirosawa H, Kobayashi H (plus 52 authors) (2000) The VLBI space observatory programme and the radio-astronomical satellite *HALCA*. Publ Astron Soc Japan 52:955–965

Högbom JA (1974) Aperture synthesis with a non-regular distribution of interferometer baselines. Astron Astrophys Suppl 15:417–426

Jones DL, Allen RJ, Basart JP (plus 25 authors) (2000) The *ALFA* Medium Explorer Mission. Adv Space Res 26:743–746

Kuchner MJ, Traub WA (2002) A coronagraph with a band-limited mask for finding terrestrial planets. Astrophys J 570:900–908

Labeyrie A (1996) Resolved imaging of extra-solar planets with future 10-100 km optical interferometric arrays. Astron Astrophys Suppl 118:517–524

le Poole RS, Quirrenbach A (2002) Optimized beam-combination schemes for each channel for PRIMA. Proc SPIE 4838:496–502

Léger A, Mariotti JM, Mennesson B (plus four authors) (1996) Could we search for primitive life on extrasolar planets in the near future? Icarus 123:249–255

Leisawitz D, Baker C, Barger A (plus 43 authors) (2007) The Space Infrared Interferometric Telescope (*SPIRIT*): High-resolution imaging and spectroscopy in the far-infrared. Adv Space Res 40:689–703

Levy GS, Linfield RP, Ulvestad JS (plus seven authors) (1986) Very long baseline interferometric observations made with an orbiting radio telescope. Science

234:187–189

Lindegren L (2010) High-accuracy positioning: astrometry. ISSI SR-009:279–291

Lumb DH (2010) Laser-aligned structures in space. ISSI SR-009:607–616

Mariotti JM, Ridgway ST (1988) Double Fourier spatio-spectral interferometry—combining high spectral and high spatial resolution in the near infrared. Astron Astrophys 195:350-363

Monnier JD (2003) Optical interferometry in astronomy. Rep Prog Phys 66:789–857

Perley RA, Schwab FR, Bridle AH (eds) (1989) Synthesis imaging in radio astronomy. Astron Soc Pacific Conf Ser Vol 6, San Francisco CA, 509 pp.

Quirrenbach A (2001) Optical interferometry. Ann Rev Astron Astrophys 39:353–401

Quirrenbach A (2005) Other worlds and life in the Universe. ESA SP-588:59–68

Quirrenbach A, Albrecht S, Vink R (plus three authors) (2007) UVES-I: interferometric high-resolution spectroscopy. ESO Astrophysics Symposia, Berlin: Springer, ISBN 978-3-540-74253-1, pp. 383–394

Serabyn E (2000) Nulling interferometry: symmetry requirements and experimental results. Proc SPIE 4006:328–339

Shao M, Colavita MM (1992) Long-baseline optical and infrared stellar interferometry. Ann Rev Astron Astrophys 30:457–498

Tallon M, Tallon-Bosc I (1992) The object-image relationship in Michelson stellar interferometry. Astron Astrophys 253:641–645

Thompson AR, Moran JM, Swenson GW (1986) Interferometry and synthesis in radio astronomy. Wiley-Interscience, New York

Unwin SC, Shao M, Tanner AM (plus 33 authors) (2008) Taking the measure of the Universe: precision astrometry with *SIM PlanetQuest*. Publ Astron Soc Pacific 120:38–88

— 18 —

Narrow-band imaging by use of interferometers

Matt J. Griffin[I] and Peter A.R. Ade[I]

Abstract

The principles of operation of imaging spectrometers based on the use of Fabry–Perot and Fourier transform interferometers are described. The advantages and disadvantages of the two techniques are discussed, and some examples are given of their application in space-borne instruments for astronomy and Earth observation.

Fabry–Perot interferometers

Principles of Fabry–Perot interferometry

A Fabry–Perot interferometer (FPI) is formed by two plane reflecting surfaces that are parallel to each other with an adjustable spacing. Interference can occur between the multiply reflected beams propagating perpendicularly through the plates.

In the ideal case of two elements with identical reflectivity R, separated by distance d, the fractional transmission of an FPI at wavelength λ is given by the Airy formula (e.g., Born and Wolf 1999),

$$\frac{I_\mathrm{t}}{I_0} = \left[1 + \frac{4\,R}{(1-R)^2}\sin^2(\frac{\delta}{2})\right]^{-1} \quad , \tag{18.1}$$

where I_0 and I_t are the intensities of the incident and transmitted beams, and δ is the phase difference between the interfering waves:

$$\delta = \frac{4\,\pi\,d}{\lambda} \quad . \tag{18.2}$$

The transmission is a maximum, and equal to unity, when $\delta = 2\,n\,\pi$, i.e.,

$$d = \frac{n\,\lambda}{2} = \frac{c_0\,n}{2\,\nu} \quad , \tag{18.3}$$

[I]School of Physics and Astronomy, Cardiff University, UK

where n, a positive integer, is the order of interference, and c_0 is the speed of light. It is remarkable that transmission of $I_t = I_0$ can be achieved through two highly reflecting surfaces. One way of visualising this is that a photon can tunnel through the potential barrier constituted by the device because in the space between the reflectors there is an allowed state with precisely the same energy as that of the photon. For a given thickness, the transmission as a function of wavelength is a series of peaks at wavelengths given by $\lambda = 2\,d/n$. The frequency difference between adjacent orders of interference is a constant, known as the free spectral range, given by

$$\Delta\nu_{\mathrm{FSR}} = \frac{c_0}{2\,d} \quad . \tag{18.4}$$

The FPI transmission as a function of frequency is thus a comb of narrow resonances spaced equally in frequency by $\Delta\nu_{\mathrm{FSR}}$. Assuming that the reflectivity is high so that $(1-R)^2 \ll R$, we can solve Equation 18.1 for the condition that $I_t/I_0 = 0.5$ to find an expression for the FWHM of the transmission peak:

$$\Delta\delta = \delta_1 - \delta_2 = \frac{2(1-R)}{\sqrt{R}} \quad \text{or} \quad \Delta\nu_{\mathrm{FWHM}} = \frac{(1-R)\,c_0}{2\,\pi\,d\,\sqrt{R}} \quad . \tag{18.5}$$

The finesse of the FPI is defined as the ratio of the free spectral range to the FWHM:

$$F = \frac{\Delta\nu_{\mathrm{FSR}}}{\Delta\nu_{\mathrm{FWHM}}} = \frac{2\,\pi}{\Delta\delta} = \frac{\pi\,\sqrt{R}}{(1-R)} \quad , \tag{18.6}$$

and is dictated by the reflectivity of the plates – high reflectivity means high finesse. The spectral resolving power of the FPI operating in order n is given from Equations 18.3 and 18.5 by

$$\frac{\nu}{\Delta\nu_{\mathrm{FWHM}}} = \frac{c_0\,n}{2\,d}\,\frac{2\pi\,d\,\sqrt{R}}{(1-R)\,c_0} = n\,F \quad . \tag{18.7}$$

High spectral resolution is thus achieved with high finesse and with operating in high order. The contrast ratio between maximum and minimum transmission is

$$C = 1 + \frac{4\,R}{(1-R)^2} = \left(\frac{1+R}{1-R}\right)^2 \quad . \tag{18.8}$$

Figure 18.1 shows the relative transmission I_t/I_0 as a function of δ for various values of reflectivity. The transmission varies significantly with thickness or wavelength. As the reflectance of each surface approaches unity (increasing finesse), the widths of the high-transmission regions become very narrow.

The transmission of a practical FPI is less than one, due to finite absorptance of the plates. The shape of the transmission profile is unaffected, but since the device operates on the principle of multiple reflection, the peak transmission is very sensitive to the absorptance. For plates with reflectance R, transmittance T, and absorptance $A = 1 - T - R$, it can be shown (e.g., Hecht 1987) that the peak transmission is

$$\frac{I_{\max}}{I_0} = \left(1 - \frac{A}{1-R}\right)^2 = \left(\frac{T}{1-R}\right)^2 \quad . \tag{18.9}$$

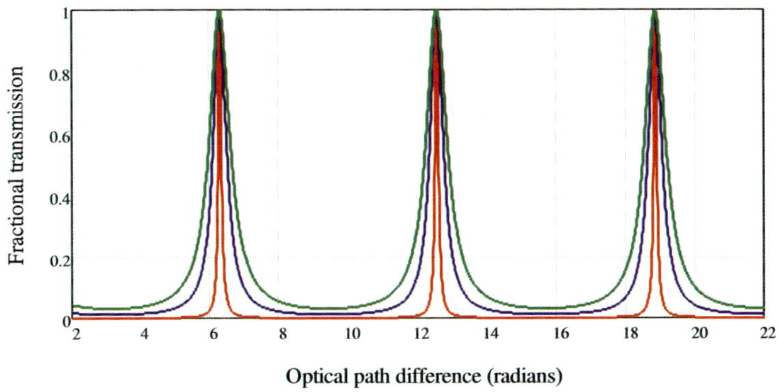

Figure 18.1: Fractional transmission of an ideal FPI as a function of δ/rad for $R = 0.7$, 0.8, and 0.95, with higher R giving narrower peaks.

For example, with $(R, T, A) = (0.95, 0.04, 0.01)$, the peak fractional transmission is 64 %, for (0.95, 0.02, 0.03) it is 16 %, and only 4 % for (0.95, 0.01, 0.04).

To form a tuneable spectrometer, one of the plates is mounted on a translation stage to make the spacing adjustable. It is necessary to suppress all of the resonances except the desired one. This can be achieved using a diffraction grating or a secondary, lower-resolution FPI. The latter method has the advantage that there is no angular dispersion, so that two-dimensional imaging remains possible.

In addition to the reflectivity, the spectral performance of an FPI is sensitive to defects in the flatness or parallelism of the reflectors and to the spread of angles in the incident beam. These effects can be characterised by additional contributions to the overall finesse.

Departures from flatness or parallelism can be modelled by decomposing the FPI into many individual elemental FPIs, all with plane parallel reflectors, whose transmissions are then averaged to give the total transmittance of the FPI (Ulrich et al 1963; Davis et al 1995). The flatness finesse is given by

$$F_{\mathrm{F}} = \frac{\lambda}{2\Delta d} \quad , \tag{18.10}$$

where Δd is the rms surface deviation of the reflectors (due either to non-flatness or departure from parallelism of the plates). The requirements on flatness and parallelism are stringent if high finesse is to be achieved. For instance, a flatness finesse of 100 at 100 μm wavelength requires $\Delta d < 0.5$ μm.

The aperture finesse is determined by the spread of angles propagating through the interferometer, which results in a corresponding spread in phase differences between the interfering beams, as illustrated in Figure 18.2. If the half-angle of the beam in the FPI is θ, corresponding to a solid angle $\Omega = \pi\,\theta^2$, then the additional path difference compared to a perfectly parallel beam is

$$\Delta d = d\left(\frac{1}{\cos\theta} - 1\right) = d\left(\frac{1}{\sqrt{1-\sin^2\theta}} - 1\right) \quad . \tag{18.11}$$

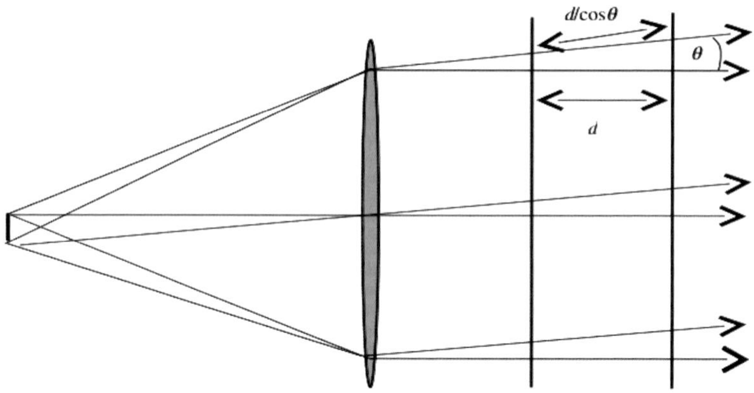

Figure 18.2: Geometry for FPI finesse determined by spread of angles propagating through the FPI.

The solid angle is related to the focal ratio of the collimator, f, by

$$\Omega \approx \frac{\pi}{4\,f^2} \quad . \tag{18.12}$$

For small θ,

$$(1-\sin^2\theta)^{-\frac{1}{2}} \approx (1-\theta^2)^{-\frac{1}{2}} \approx 1+\frac{1}{2}\,\theta^2 \approx 1+\frac{\Omega}{2\pi} \quad , \tag{18.13}$$

giving

$$\Delta d = \frac{d\Omega}{2\pi} \quad . \tag{18.14}$$

The corresponding phase difference is

$$\Delta\delta = \frac{4\pi\,\Delta d}{\lambda} = \frac{2\,d\Omega}{\lambda} = n\,\Omega \quad . \tag{18.15}$$

We can define a corresponding finesse, known as the *aperture finesse*, as

$$F_{\rm A} = \frac{2\pi}{\Delta\delta} = \frac{2\pi}{n\,\Omega} \quad . \tag{18.16}$$

For high finesse, we must therefore operate in high order and minimise the solid angle of the beam by having a collimator with a high f-number.

The overall finesse is given by

$$\frac{1}{F_{\rm tot}^2} = \frac{1}{F^2} + \frac{1}{F_{\rm F}^2} + \frac{1}{F_{\rm A}^2} \quad . \tag{18.17}$$

The performance of the FPI interferometer is critically dependent on the properties of the reflectors. For optical and NIR wavelengths, FPIs are formed from pairs of highly-polished plane-parallel dielectric slabs (e.g., fused silica) with appropriate multi-layer coatings to provide high reflectivity over the desired band. In the far

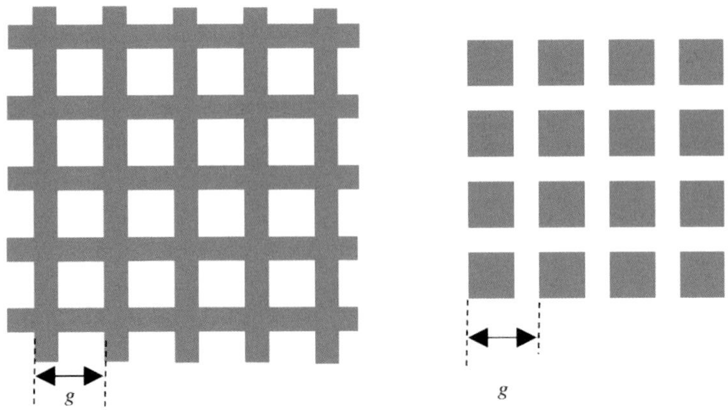

Figure 18.3: Metal mesh grid patterns used for far infrared FPI reflectors. Left: inductive grid; right: the complementary capacitative structure.

infrared, this technique does not work well as the layers become thick and lossy, and problems arise with differential thermal contraction of the layers. The requirements for high reflectivity and low absorption are best met by metallic grids. The grids can be either free-standing or supported on a thin dielectric film (e.g., Ulrich 1967). The most common grid pattern has regularly-spaced square holes with grid spacing g, as shown in Figure 18.3 (left), forming an inductive grid, or the complementary capacitative structure shown in Figure 18.3 (right). An inductive grid can be free standing, but the capacitative grid requires a dielectric substrate on which the metal can be deposited. For efficient operation as an FPI, the wavelength must be $\geq 5\ g$, at shorter wavelengths the grid starts to behave as a diffraction grating. Fixed far-infrared and sub-millimetre filters can be constructed by combining metallic grids with suitable spacings, often in FPI configurations (Ade et al 2006).

An imaging Fabry–Perot spectrometer consists of:

1. a collimator to provide a parallel beam;
2. the FPI with adjustable spacing;
3. an additional element (often another, lower resolution, FPI) to select one order of interference;
4. an additional filter to act as a short-wavelength blocker;
5. a camera to create a focal plane image;
6. a suitable 2-D detector array.

At FIR wavelengths, the interferometer needs to be operated at cryogenic temperatures.

Advantages and disadvantages of Fabry–Perot interferometers

The main advantages of FPIs for the FIR are

1. the simple optical design (fore-optics, collimator and camera), readily suited to an imaging system, and
2. the capability to achieve high spectral resolving power with a compact instrument.

The chief disadvantages are

1. it is difficult to achieve high overall efficiency compared to grating or FTS spectrometers;
2. the broad wings of the instrument response function can present problems in the analysis of rich spectra;
3. the design provides no wavelength multiplexing — mechanical scanning is essential to cover a range of wavelengths;
4. it is difficult to obtain wavelength coverage of more than about an octave with a single device;
5. much attention and care must be paid to the practicalities of order sorting and short-wavelength blocking.

Examples of Fabry–Perot instruments

A number of imaging Fabry–Perot instruments have been built and used on ground-based and airborne telescopes. Most have been based on the tandem FPI concept, in which a low-order FPI is used to order sort a high-resolution FPI. The Far Infrared Imaging Fabry–Perot Interferometer (FIFI, Poglitsch et al 1991) was operated on the KAO and used three cryogenic tuneable FPIs in series giving spectral resolution up to 10^5 in the 40 μm to 200 μm range, with a 5×5 detector array. The Sub-millimetre and Far-InfraRed Experiment, SAFIRE, (Benford et al 2003) is a planned imaging tandem Fabry–Perot instrument for the SOFIA airborne observatory, the successor to the KAO. It will operate between 100 μm and 700 μm with spectral resolution up to 2000 over a field of view of $160'' \times 320''$, covered by a 16×32 array of superconducting bolometers. The South Pole Imaging Fabry–Perot Interferometer (SPIFI, Bradford et al 2002) is a ground-based astronomical spectrometer operating at sub-millimetre wavelengths, using a 5×5 array of bolometric detectors operating at 60 mK. It uses two cryogenic scanning FPIs and has spectral resolving power selectable between 500 and 10 000.

The Long Wavelength Spectrometer (LWS, Clegg et al 1996) on board ESA's *ISO* incorporated two scanning FPIs, one for 45 μm to 90 μm and one for 90 μm to 180 μm, which were order-sorted by a grating and offered resolution ($\lambda/\Delta\lambda$) up to 10 000. The FPIs were located in a three-position interchange wheel, with the third position open so that the instrument could also be operated in grating-only mode for medium-resolution spectroscopy. The optical layout of the LWS is shown in

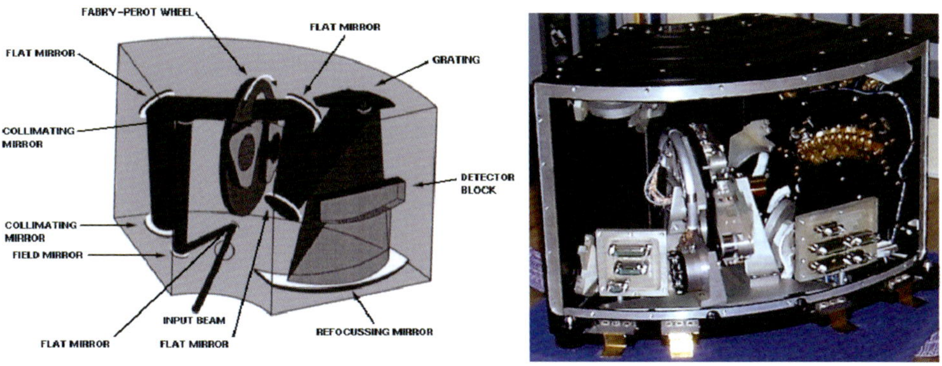

Figure 18.4: Left: optical layout of the *ISO* LWS focal plane instrument; right: photograph of the instrument.

Figure 18.4. The *ISO* Short-Wavelength Spectrometer (SWS, de Graauw et al 1996) used a similar grating-FPI arrangement to provide high-resolution spectroscopy in the 11 µm to 45 µm range.

FPI instruments can also be used to study the chemistry and dynamics of the Earth's atmosphere. An example of such applications is the High Resolution Doppler Imager (HRDI) instrument (Hays et al 1993) on board the *UARS*. This instrument was designed to measure wind velocities using Doppler shifts of O_2 rotational lines in the 550 nm to 770 nm region, and contained a triple FPI system providing 0.05 cm^{-1} resolution (equivalent to $\lambda/\Delta\lambda \approx 3 \times 10^5$).

Imaging Fourier transform spectroscopy

Principles of FT spectroscopy

The essential features of a Fourier transform spectrometer (FTS) are illustrated in Figure 18.5, which shows the layout of a Michelson interferometer. In a Fourier transform spectrometer such an interferometer is employed as an autocorrelation spectrometer — one which compares the signal to a delayed version of itself over a range of delays. The beam from Port 1 is collimated and split 50:50 by the beam divider. One beam is reflected back by the fixed mirror and the other by the moving mirror. After reflection, the beams again divide at the beam divider and interfere. The beams from the two arms of the interferometer interfere because of the difference in path length of the two optical trains. A typical implementation of the basic Michelson interferometer depicted in Figure 18.5 uses intensity beam dividers and corner-cube reflectors to maintain alignment. On average, half of the power from Port 1 is directed to Port 2 and half goes back to Port 1; likewise, any emission from Port 2 will be divided equally between Port 1 and Port 2. The Michelson interferometer thus has two input and two output ports, but in the basic configuration of Figure 18.5 they are not spatially separated. In the configuration shown the interferogram represents the difference spectrum: Port 1 – Port 2. The Martin-Puplett (M-P) interferometer (Martin 1982; Lambert and Richards 1978)

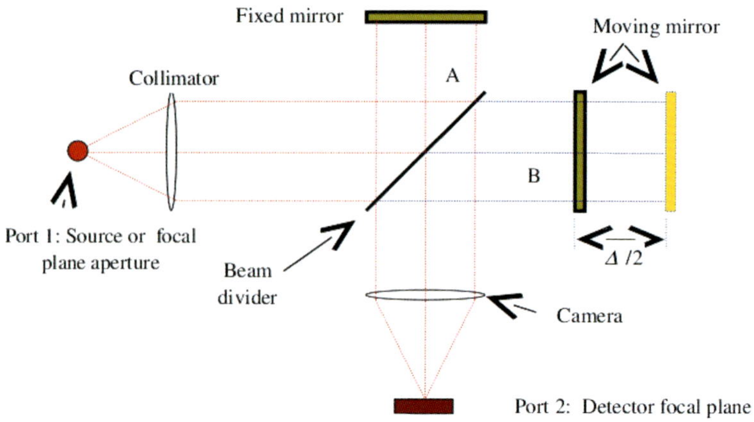

Figure 18.5: Essential features of a Fourier transform spectrometer.

uses polarising beam dividers and roof-top reflectors to achieve separation of the input and output ports (but at a cost of half of the available power).

For an optical path difference $\varDelta$ the phase difference between the interfering beams is

$$\sigma = \frac{2\pi\varDelta}{\lambda} = 2\pi\, k\varDelta \quad , \tag{18.18}$$

where the wavenumber k is $1/\lambda$.

For an intensity beam divider with input amplitude A_0 (intensity $I_0 \propto A_0^2$), the amplitude of the reflected and transmitted beams is $A_0/\sqrt{2}$, and on the second interaction, the amplitude is reduced by a further factor of $\sqrt{2}$ to $A_0/2$. The amplitude in the output port is therefore

$$A_{\mathrm{out}} = \frac{A_0}{2} + \frac{A_0}{2}\mathrm{e}^{\mathrm{i}\,2\pi\,k\varDelta} \quad . \tag{18.19}$$

The corresponding intensity as a function of optical path difference is

$$I_{\mathrm{out}}(\varDelta) = A_0 A_0^* = \frac{A_0^2}{4}[1+\mathrm{e}^{\mathrm{i}\,2\pi\,k\varDelta}][1+\mathrm{e}^{-\mathrm{i}\,2\pi\,k\varDelta}] = \frac{I_0}{2}[1+\cos(2\pi\,k\varDelta)] \quad , \tag{18.20}$$

which is recorded by the detector as the interferogram. For a monochromatic input, the interferogram is a cosine wave. For an arbitrary spectral input $S(k)$, the interferogram is the sum of the relevant monochromatic components:

$$I_{\mathrm{out}}(\varDelta) = \int_0^\infty \frac{S(k)}{2}[1+\cos(2\pi\,k\varDelta)]\mathrm{d}k \quad . \tag{18.21}$$

Application of the Fourier integral theorem gives:

$$S(k) = 4\int_{-\infty}^{\infty}[I_{\mathrm{out}}(\varDelta) - \frac{1}{2}\,I_{\mathrm{out}}(0)]\cos(2\pi\,k\varDelta)\mathrm{d}\varDelta \quad . \tag{18.22}$$

The inverse Fourier transform of the interferogram thus gives the input spectrum. Perfect reproduction of the spectrum would require the optical path difference range to be infinite. With a real interferometer, the interferogram will be measured over a finite range of path difference. If the range is $\pm\Delta_{\max}$, the interferogram is truncated by a rectangular function equal to unity between these limits and zero outside:

$$\begin{aligned} T &= 1 \quad \text{for} \quad -\Delta_{\max} < \Delta < \Delta_{\max} \\ &= 0 \quad \text{otherwise} \quad . \end{aligned} \tag{18.23}$$

The Fourier transform of this truncation function represents the measured spectrum for a monochromatic input, i.e., the instrument response function, $R(k, \Delta_{\max})$:

$$R(k, \Delta_{\max}) = 2\int_{-\Delta_{\max}}^{\Delta_{\max}} \cos(2\pi\, k\Delta)\mathrm{d}\Delta = 2\Delta\left[\frac{\sin(2\pi\, k\Delta)}{2\pi\, k\Delta}\right] \quad . \tag{18.24}$$

Figure 18.6 shows the shape of this function. The spectral resolution is characterised by its FWHM, given by

$$\delta\, k = \frac{1.21}{2\,\Delta} \quad . \tag{18.25}$$

The secondary features in the instrument response function are due to the sharp cut-off in $T(\Delta)$ and can be suppressed in the course of data processing by multiplying the interferogram by a suitable apodisation function. A standard method, termed linear apodisation, is simply to taper linearly the amplitude of the fringes to zero at the maximum path difference. The new effective instrument response function can then be determined by Fourier transforming this triangular response, which gives

$$R(k, \Delta_{\max}) = \left[\frac{\sin(\pi\, k\Delta)}{\pi\, k\Delta}\right]^2 2\Delta \quad . \tag{18.26}$$

This new effective instrument response is also shown for comparison in Figure 18.6. It is immediately apparent that the suppression of the sidelobes comes at some cost to the spectral resolution via broadening of the main peak by a factor of $\approx\sqrt{2}$. An additional feature of any real interferometer is that the interferogram is not sampled continuously but at discrete intervals $\mathrm{d}\Delta$. According to Nyquist's sampling theorem, the corresponding highest frequency that can be measured unambiguously, known as the Nyquist frequency, is

$$k_{\max} = \frac{1}{2\,\mathrm{d}\Delta} \quad . \tag{18.27}$$

Higher frequencies in the input spectrum are aliased into the measured spectral band (0 to $k_{\max}$), contributing to false features overlaying the spectrum. In contrast to the FPI, a simple low-pass optical filter can be used to remove frequencies above the Nyquist frequency. A popular variant of the FT spectrometer is a scanning instrument, in which the mirror is translated continuously at constant speed so that the optical fringes appear as low-frequency audio components in the detector output. Nyquist filtering in this case can be achieved conveniently by appropriate electrical filtering in the detector readout.

As with the FPI, the performance of a Fourier transform interferometer is influenced by the parallelism of the beam. Departures from parallelism, which are

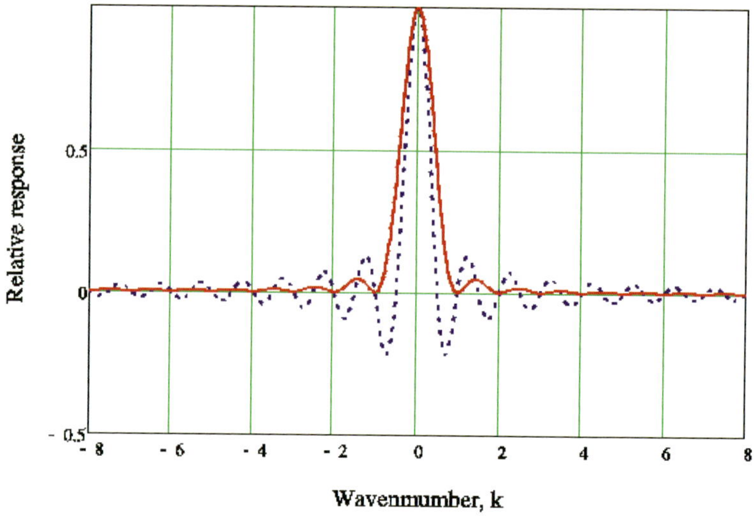

Figure 18.6: Normalised FTS instrument response function for $\Delta_{\max} = 1$, unapodised (blue dotted line) and with linear apodisation (red solid line).

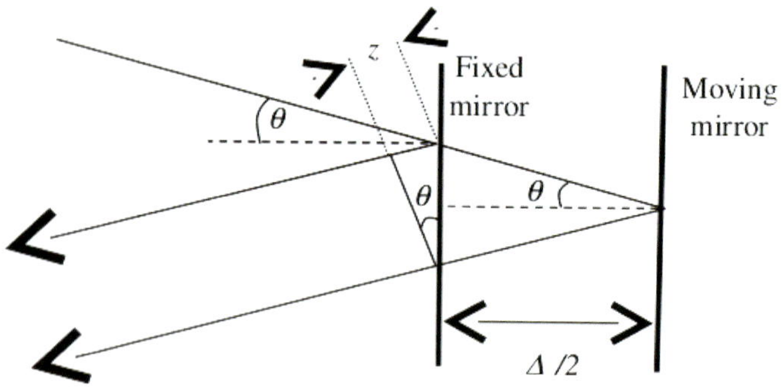

Figure 18.7: Path difference for off-axis rays.

inevitable when viewing an extended object, result in phase shifts which cause distortions to the interferogram when the beams interfere. The geometry for an off-axis angle θ is depicted in Figure 18.7, from which one can show that the optical path difference z between the two rays at the off-axis angle θ is given by $z = \Delta \cos\theta$.

The interferogram (Equation 18.20) is thus modified to

$$I_{\text{out}}(\Delta) = \frac{I_0}{2}[1 + \cos(2\pi\, k \Delta \cos\theta)] \quad . \tag{18.28}$$

Whereas on-axis rays produce a spectral feature at wavenumber k, the off-axis rays produce a feature at $k \cos\theta$. The corresponding resolution limit is

$$\delta k = k(1 - \cos\theta) \approx \frac{k\,\theta^2}{2} \tag{18.29}$$

for small θ. The corresponding limit on resolving power is

$$\frac{\delta k}{k} = \frac{\theta^2}{2} = \frac{\Omega}{2\pi} \quad , \tag{18.30}$$

where Ω is the beam solid angle.

It should be noted that in comparison to the FPI the efficiency of the FTS is very high because there is no loss in order-sorting filtering, and the inherent relative efficiency of the beam division can approach 100 %.

Advantages and disadvantages of FTS instruments

The main advantages of FTS instruments are

1. broad instantaneous spectral coverage, often allowing many spectral features to be observed simultaneously;
2. well-defined peak of the response function and the ability to match the response function to the scientific objectives by choice of a suitable apodisation function;
3. suitability for direct imaging at high spectral resolution with a compact instrument;
4. adjustable spectral resolution (dictated by the choice of maximum optical path difference);
5. high overall optical efficiency.

The main disadvantage compared to monochromator-based spectrometers such as the grating or Fabry–Perot is that in background-limited regime the wide instantaneous bandwidth contributes additional photon noise, leading to lower sensitivity as the price paid for broad wavelength coverage.

Examples of space-borne FTS instruments

FTS instruments are particularly useful in applications that require broad wavelength coverage, and they have been widely used in astronomy and Earth-observing satellite missions. A detailed review of remote sensing instruments is given by Persky (1995). Here we provide brief details of some instruments recently flown and currently planned.

The Composite Infrared Spectrometer (CIRS, Flasar et al 2004) is an instrument on board the *Cassini* satellite mission to the Saturnian system. CIRS contains two FT spectrometers using the same actuator, covering 7 µm to 17 µm and 17 µm to 1000 µm and providing spectral resolution between (0.5 and 15.5) cm^{-1}. The long-wavelength band is covered by a Martin-Puplett interferometer, using 2 µm spacing metal-grid polarisers and two thermopiles as detectors, each with the same 0.4 mrad field of view (FOV). The spectrometer for the shorter wavelength band mid-infrared interferometer is a Michelson with a KBr intensity beam-splitter and corner-cube retro-reflectors, and two 1×10 linear photodetector arrays, with each

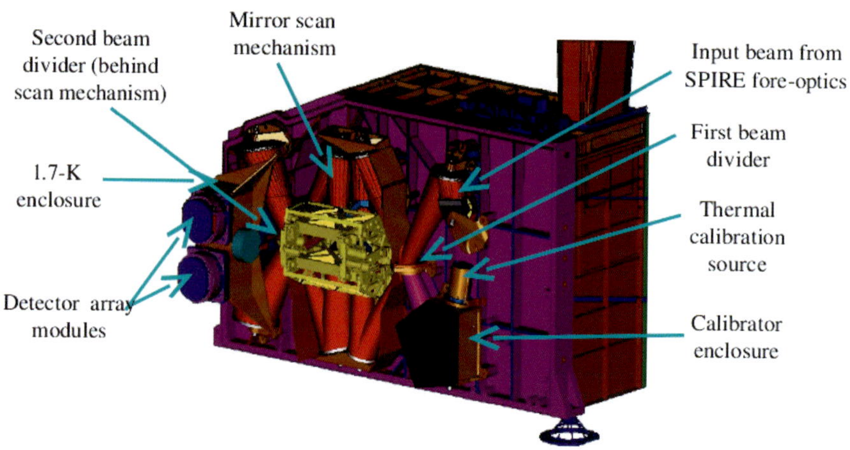

Figure 18.8: Opto-mechanical layout of the *Herschel*-SPIRE Fourier Transform Spectrometer.

detector having a 0.3 mrad FOV. CIRS has observed the atmospheres of Jupiter, Saturn and Titan, characterising temperature, gas composition, and atmospheric aerosols with vertical and horizontal resolution from tropospheric to stratospheric altitudes. The Tropospheric Emission Spectrometer (TES, Beer et al 2004) is an imaging FTS operating on NASA's *EOS* satellite. TES covers wavelengths between 3.2 µm and 15.4 µm and carries out global measurements of O_3, H_2O, CO, CH_4 and other important species as a function of altitude in the Earth's troposphere. The TES optics feature back-to-back corner-cube reflectors with a common translator mechanism to vary the optical path difference (OPD). One input port views the atmosphere while the other views an internal calibration source. The optics are cooled to 180 K to reduce the thermal background on the detectors and to provide high contrast between the Earth's atmosphere and the instrument background. Two focal planes are located at the two output ports and are each split into two channels by dichroic beam dividers. Four 1×16 arrays of photovoltaic detectors operate at 65 K and view spatially equivalent pixels.

The Spectral and Photometric Imaging Receiver (SPIRE) instrument (Griffin et al 2008) is currently flying on the *Herschel Space Observatory*, launched by ESA on 14 May 2009. SPIRE contains a three-band sub-millimetre camera and an imaging FTS which covers 194 µm to 672 µm. The FTS (Figure 18.8) has an approximately circular field of view of 2.6′ diameter and spectral resolution adjustable between 0.042 cm^{-1} and 2 cm^{-1} ($\lambda/\Delta\lambda = 20$ to 1000 at 250 µm). To minimise the background on the detectors, the entire SPIRE instrument is cooled, with temperature levels of ≈ 5 K for much of the optics, 2 K for the final optics, and 0.3 K for the bolometric detector arrays. The FTS uses two broadband intensity beam splitters in a Mach–Zehnder configuration, with a single back-to-back scanning roof-top fold mirror serving both interferometer arms. All four ports are independently accessible as in the M-P polarising FTS, but the throughput is a factor of two higher than for the M-P as none of the incoming radiation is rejected, and there is no sensitivity to the polarisation of the incident radiation. A thermal calibration source at the second input port allows the background power from the telescope to be matched. The amplitude of the interferogram central maximum is proportional to the difference in the radiant power from the two ports, so this allows the large telescope background to be nulled, reducing the dynamic range requirements for the detector sampling. Two detector arrays are placed in the two output ports to accommodate the whole wavelength range in two sections with overlapping bands. The SPIRE FTS is used to characterise the physics and chemistry of interstellar dust and gas in our own and nearby galaxies.

An FTS spectrometer is also proposed for *Herschel*'s successor, the Japanese Space Infrared telescope for Cosmology and Astrophysics (*SPICA*) mission. The SAFARI instrument (Swinyard et al 2008) for *SPICA* is based on an adaptation of the SPIRE FTS design, to operate in the 30 µm to 200 µm range with much improved sensitivity provided by the cold (4 K) telescope and the use of improved far infrared detectors. Future far infrared interferometer mission concepts such as *SPECS* (Harwit et al 2007), *SPIRIT* (Leisawitz et al 2007) and *FIRI* (Helmich and Ivison 2009), propose to carry out high angular resolution imaging spectroscopy using a double Fourier spatial and spectral interferometric technique (Mariotti and Ridgeway 1988; Elias et al 2007).

Bibliography

Ade PAR, Pisano G, Tucker C, Weaver C (2006) A review of metal mesh filters. Proc SPIE 6275, 62750U (15 pp.)

Beer R, Glavich TA, Rider DM (2001) Tropospheric emission spectrometer for the Earth Observing System's *Aura* satellite. Appl Opt 40:2356–2367

Born M, Wolf E (1999) Principles of Optics (7th edition). Cambridge University Press.

Benford DJ, Moseley SH, Stacey GJ (plus two authors) (2003) Far-Infrared Imaging Spectroscopy with SAFIRE on SOFIA. Proc SPIE 4857:105–114

Bradford CM, Stacey GJ, Swain MR (plus six authors) (2002) SPIFI: A Direct-detection Imaging Spectrometer for Submillimeter Wavelengths. Appl Opt 41:2561–2574

Clegg PE, Ade PAR, Armand C (plus 60 authors) (1996) The *ISO* Long Wavelength Spectrometer. Astron Astrophys 315:L38–L42

Davis GR, Furniss I, Towlson W (plus seven authors) (1995) Design and performance of cryogenic, scanning Fabry–Perot interferometers for the Long-Wavelength Spectrometer on the Infrared Space Observatory. Appl Opt 34:92–107

Elias NM, Harwit M, Leisawitz D, Rinehart SA (2007) The mathematics of double-Fourier interferometers. Astrophys J 657:1178–2000

de Graauw T, Haser LN, Beintema DA (plus 57 authors) (1996) Observing with the *ISO* Short-Wavelength Spectrometer. Astron Astrophys 315:L49–L54

Flasar FM, Kunde VG, Abbas MM (plus 43 authors) (2004) Exploring the Saturn system in the thermal infrared: The Composite Infrared Spectrometer. Space Sci Rev 115:169–297

Harwit M, Leisawitz D, Rinehart SA (2007) A kilometer baseline far-infrared/submillimeter interferometer in space. New Astronomy Reviews 50:228–234

Hays PB, Abreu VJ, Dobbs ME (plus three authors) (1993) The High Resolution Doppler Imager on the Upper Atmosphere Research Satellite. J Geophys Res 98:10713–10723

Hecht E (1987) Optics (2nd edition), Addison Wesley

Helmich FP, Ivison RJ (2009) FIRI – a far-infrared interferometer. Experimental Astronomy 23:245–276

Lambert DK, Richards PL (1978) Martin-Puplett interferometer: an analysis. Appl Opt 17:1595–1602

Leisawitz D, Baker C, Barger A (plus 52 authors) (2007) The space infrared interferometric telescope (SPIRIT): High-resolution imaging and spectroscopy in the far-infrared. Adv Space Res 40:689–703

Mariotti J-M, Ridgway ST (1988) Double-Fourier spatio-spectral interferometry: combining high spectral and high spatial interferometry in the near infrared. Astron Astrophys 195:350–363

Martin DH (1982) Polarizing (Martin-Puplett) interferometric spectrometers for the near- and submillimeter spectra. In Button KJ (ed.), Infared and Millimeter Waves, 6, Systems and Components, Ch. 2, Academic Press, New York 65–148

Persky MJ (1995) A review of spaceborne infrared Fourier transform spectrometers for remote sensing. Rev Sci Instrum 66:4763–4797

Poglitsch A, Beeman JW, Geis N (plus seven authors) (1991) The MPE/UCB far-infrared imaging Fabry–Perot interferometer (FIFI). Int J of Infrared and Millimeter Waves 12:859–884

Ulrich R (1967) Far-infrared properties of metallic mesh and its complementary structure. Infrared Physics 7:37–55

— 19 —

Imaging Michelson interferometers

Alan M. Title[I]

Abstract

The imaging solid wide-field Michelson (SWFM) interferometer has a performance superior to that of a wide-field birefringent element because of its smaller angle sensitivity and its relative insensitivity to temperature. For these reasons SWFM filter systems were chosen as the imaging spectrometers for the Michelson Doppler Imager on *SOHO* and for the Helioseismic and Magnetic Imager on the *SDO*. Theory and practice in space are described for SWFM filters.

Advantages of Michelson imaging techniques

Spectral imaging has been a basic tool for solar and stellar observations. The techniques used to isolate spectral features include coloured filters, thin film all-deposited dielectric filters (DF), Fabry–Perot (FP) interferometers both solid and air spaced, birefringent filters (BF), and Michelson interferometers (MI). The spectrally narrowest DF (0.3 nm to 0.5 nm) can isolate individual spectral lines or clean continuum, if the spectral region under study is not too complex. FP, BF, and MI systems can be made sufficiently narrow to sample within spectral lines, and all these devices can be made tuneable. For narrow-band spectral imaging there are design trades that favor one filtering system over another. The MI in its wide-field configuration (WFM) and especially its solid wide-field (SWFM) configuration has significant advantages in étendue, the solid angle aperture product, compared to FPs, MIs, BFs and even wide-fielded BFs (WFBF).

Etendue is an invariant of an optical system. All interferometers have angle sensitivity; the wavelength of maximum transmission is a function of incident angle. As the spectral resolution and/or the aperture of the telescope increases, the angle sensitivity of the spectral isolation device becomes more critical. The étendue advantage of the SWFM configuration compared to the FP, MI, or WFBF occurs because the angle sensitivity has a fourth-order rather a than second order dependence. In principle the angle dependence can be reduced to any even power with a Michelson arm design of sufficient complexity. Temperature and wavelength sensitivity also can be greatly reduced by proper selection of the arm materials. In their polarized variant SWFMs can be very simply and repeatably tuned. These

[I]LMSAL—Lockheed Martin Solar and Astrophysics Laboratory, Palo Alto CA, USA

capabilities are especially important for helioseismology where some of the physical measurements are inferred from sequences of velocity measurements that have durations of months to years.

Another solid Michelson variant uses arm materials that have both the same physical length and index of refraction at the design wavelength. This configuration is called an Equal Index Solid Michelson (EISM). In the EISM the transmission change with wavelength is caused by differences in the dispersions of the glasses in the two arms.

Because there are not an infinite number of optical glasses it is unlikely that a pure EISM design would be available at a desired wavelength, but a simple variation on the design can be made to operate in essentially any wavelength throughout the visible domain.

Theory of the wide-field Michelson

Evans (1949) suggested that an "analog birefringent crystal" could be made using an MI with a polarizing beam splitter. Mertz (1965) introduced the idea of a wide-field Michelson (WFM) in the early 1960s. In the 1970s it was realized that the analog birefringent crystal concept could be extended further by modifying the WFM to create an SWFM (Title and Ramsey 1980).

A normal MI is a two-beam interferometer. A beam splitter divides an incoming state of light; the two beams are sent along optical paths of different optical length, and then recombined by the beam splitter. The optical path-length difference causes the spectral modulation of the combined beam. In the SWFM (see Figure 19.1) the optical path difference remains, but the ray paths of the output beams are nearly identical.

Here a polarizing beam splitter divides an incoming beam polarized at 45° to the plane of incidence of the beam splitter. The polarizing beam splitter reflects the light that is polarized perpendicular to the plane of incidence (s wave) toward arm 1 and transmits the light parallel to the plane of incidence (p wave) to arm 2. Quarterwave plates in front of the mirrors in the two arms rotate the planes of polarization by 90° so that the beam splitter now reflects the return light (now s wave) from arm 2 and transmits reflected the light (now p wave) from arm 1. As a result the output from the beam splitter is a pair of orthogonally polarized beams with a relative time delay caused by the difference in optical path length of the two arms. Because the beams came from a single polarized state they can be made to interfere by being recombined by a quarterwave plate and an exit polarizer. The phase of the interference and hence the spectral locations of the transmission peaks can be selected by rotating an exit polarizer. Alternately a halfwave plate can be rotated between the exit quarterwave plate and a fixed polarizer.

Insight into wide fielding the WFM is gained by a thought experiment with an air-spaced MI. Assume an observer looks into the exit face of the interferometer at a source located in front of the input face. Observed would be two images of the source caused by the mirror reflections in arms 1 and 2. The spacing of the images would be twice the difference in the distances of mirrors 1 and 2 from the beam splitter. Suppose that mirror 1 is farther away from the beam splitter than

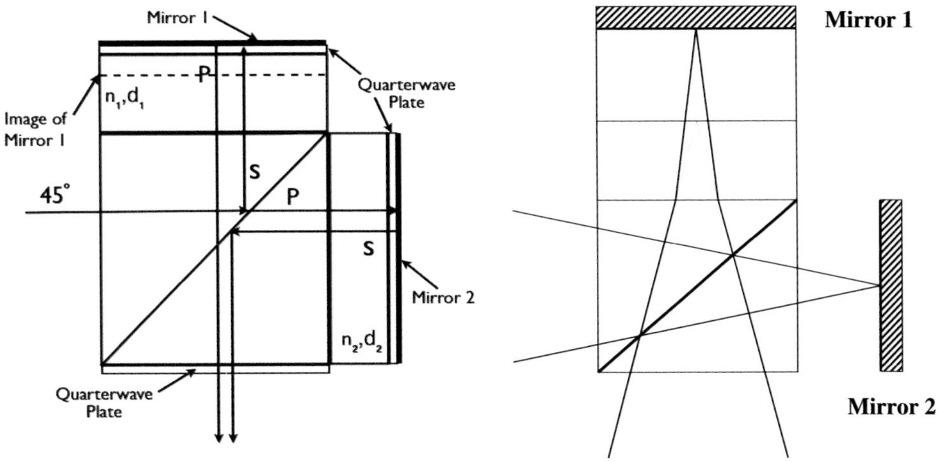

Figure 19.1: Left: Optical schematic of a polarizing SWFM interferometer. Right: The course of light rays in a Michelson interferometer with a glass block in one arm shows why the the apparent position of mirror 1 moves towards the observer, when a glass block is inserted: the thicker line follows the actual course of an oblique light ray, while the thinner line shows its apparent course. (For simplicity the beam splitter is assumed to have the same refractive index as air.)

mirror 2. Now introduce a glass block into arm 1 before mirror 1. A glass block causes the apparent position of the mirror 1 to move toward the observer. When there are blocks in both arms the condition for optical coincidence becomes

$$\frac{d_1}{n_1} = \frac{d_2}{n_2} \quad , \tag{19.1}$$

where the d_i and n_i, $i = 1, 2$, are the geometrical path lengths and indices of refraction, respectively (cf., right-hand part of Figure 19.1). But while the apparent interfering sources are coincident there remains a normal-incidence path difference

$$\delta(0) = 2\, d_1 \frac{(n_1^2 - n_2^2)}{n_1} \quad . \tag{19.2}$$

The angle-dependent path difference is

$$\delta(\theta) = 2 \left(n_1\, d_1 \sqrt{1 - \frac{\sin^2(\theta)}{n_1^2}} - n_2\, d_2 \sqrt{1 - \frac{\sin^2(\theta)}{n_2^2}} \right) \quad , \tag{19.3}$$

where θ is the angle of incidence. Using condition 19.1, assuming $n_1 > n_2$, and expanding Equation 19.3 in a power series in θ, then to fourth order the path difference becomes

$$\delta(\theta) = 2\, d_1 \frac{(n_1^2 - n_2^2)}{n_1} \left(1 + \frac{\theta^4}{8\, n_1^2\, n_2^2} \right) \quad . \tag{19.4}$$

The wide-field condition, Equation 19.1, causes the second-order term to be zero.

For comparison, a solid FP interferometer with a spacer of thickness d and index n has a path-length angle dependence of

$$\delta_{\mathrm{FP}}(\theta) = 2\,d\,n\,(1+\frac{\theta^2}{2\,n^2}) \quad . \tag{19.5}$$

The wavelength shift of the transmission maxima relative to the design wavelength is just the angle sensitivity. A 0.1 Å shift at 5000 Å with a FP and SWFM are caused by incident angles of 0.544° and 7.89°, respectively, and the corresponding maximum f/ratios are $f/52.7$ and $f/7.89$, respectively, where the FP has $n = 1.5$ and the WFM has $n_1 = 1.5$ and $n_2 = 1$. The corresponding values for a 1 Å shift are 1.72° and 14.03° and $f/16.67$ and $f/2.04$, respectively.

The phase difference introduced by the interferometer is just

$$\phi(\theta) = \frac{2\,\pi\,\delta(\theta)}{\lambda} \quad . \tag{19.6}$$

The temperature sensitivity is obtained from the relationship

$$\frac{\partial\phi}{\partial\lambda}\Delta\lambda = \frac{\partial\phi}{\partial T}\Delta T \quad . \tag{19.7}$$

With some algebra the temperature sensitivity can be shown to be

$$\begin{aligned} \frac{\Delta\lambda}{\Delta T} &= -\lambda\left[n_1^2\left(\frac{\frac{\partial d_1}{\partial T}}{d_1}+\frac{\frac{\partial n_1}{\partial T}}{n_1}\right)-n_2^2\left(\frac{\frac{\partial d_2}{\partial T}}{d_2}+\frac{\frac{\partial n_2}{\partial T}}{n_2}\right)\right] \\ &\quad \left[1-n_1^2\frac{\frac{\partial n_1}{\partial\lambda}-\frac{\partial n_2}{\partial\lambda}\left(\frac{n_2}{n_1}\right)^2}{n_1^2-n_2^2}\right]^{-1} \quad . \end{aligned} \tag{19.8}$$

This expression is zero if

$$n_1^2\left(\frac{\frac{\partial d_1}{\partial T}}{d_1}+\frac{\frac{\partial n_1}{\partial T}}{n_1}\right) = n_2^2\left(\frac{\frac{\partial d_2}{\partial T}}{d_2}+\frac{\frac{\partial n_2}{\partial T}}{n_2}\right) \quad . \tag{19.9}$$

That is, if the product of the square of the index of refraction of the arm material times the sum of the coefficients of linear expansion and index change with temperature are equal in the two arms, the wavelength change with temperature will be eliminated. This criterion can be partially satisfied with available glasses with the result that the temperature sensitivity of the transmission peaks is reduced near the design wavelength. One class of solutions has a glass block in one arm and the other spaced by a material with an appropriate coefficient of linear expansion. Note that the term $1/(n_1^2 - n_2^2)$ in the denominator increases the temperature sensitivity as n_2 approaches n_1, so it is desirable to have as great an index difference as possible between n_1 and n_2.

Extensions of the field of view

Higher-order wide fielding

The approach taken above to widening the field of view of the SWFM was to make the location of the apparent sources coincident. This causes the second-order term in the expansion of the angle sensitivity of the path difference to vanish. Viewed more mathematically, the path-difference equation was expanded in a power series and the conditions were determined for which the coefficient of the second-order term vanishes. This methodology can be generalized to eliminate all of the higher order terms in the angle sensitivity. One can write down a general solution for the arm lengths and optical path difference for an interferometer with one block in arm 1 and n blocks in arm 2. A system with n blocks in arm 2 will have the first angle-dependent term of order $2(n+1)$. The arm lengths and the total retardation, n_E are

$$d_i = \frac{\prod_{k=2}^{n}(n_1^2 - n_k^2)}{\prod_{k=2}^{n}(n_i^2 - n_k^2)} \quad k \neq i \quad , \tag{19.10}$$

$$n_E = 2\frac{\prod_{k=2}^{n}(n_1^2 - n_k^2)}{n_1^{n+1}} \quad . \tag{19.11}$$

There are always solutions to conditions 19.10 and 19.11. If a d_i is negative it means the block is in arm 1 rather than arm 2.

It may appear to the reader that the simplest version of the SWFM is good enough, but there can be advantages for more complex systems when it is desired to stabilize temperature sensitivity and well as the temperature and wavelength range of the wide-field condition. These properties have not been investigated in this chapter.

Equi-index Michelsons (EIMs)

It was argued above that wide fielding occurs because the apparent sources can be made to appear coincident. This clearly occurs when both arms of the interferometer have glass blocks of equal thickness d and equal indices of refraction at a design wavelength. However, if the glasses in the two arms have different dispersions, modulation will still occur. Let

$$\begin{aligned} n_1 &= n_0 + \alpha_1(\lambda - \lambda_0) \\ n_2 &= n_0 + \alpha_2(\lambda - \lambda_0) \quad , \end{aligned} \tag{19.12}$$

where α_1 and α_2 are the dispersions of the glasses in arms 1 and 2, respectively. With these definitions the phase difference ϕ as a function of wavelength and angle can be written as

$$\begin{aligned} \phi(\theta) = \frac{2\pi d}{\lambda} \Bigg\{ & [n_0 + \alpha_1(\lambda - \lambda_0)]\sqrt{1 - \frac{\sin^2\theta}{[n_0 + \alpha_1(\lambda - \lambda_0)]^2}} \\ & -[n_0 + \alpha_2(\lambda - \lambda_0)]\sqrt{1 - \frac{\sin^2\theta}{[n_0 + \alpha_2(\lambda - \lambda_0)]^2}} \Bigg\} \end{aligned} \tag{19.13}$$

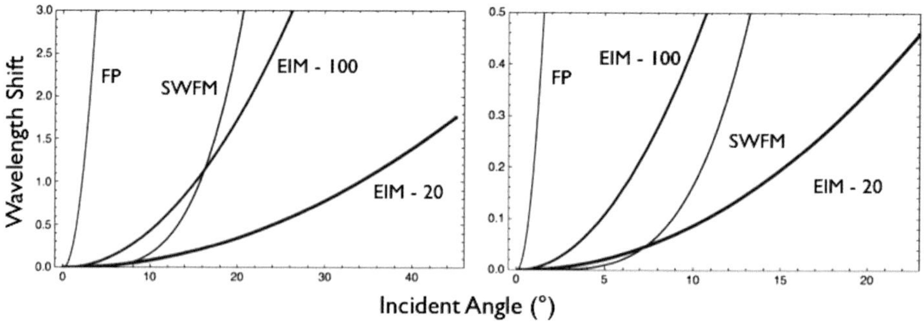

Figure 19.2: Comparison of FP, SWFM, and EIM operated at 20 and 100 times the $\Delta\lambda_{\text{FWHM}}$ at the design wavelength versus incident angle. The plot on the left shows the angle shift range from 0 to 3 Å, while the plot on the right shows the range 0 to 0.5 Å. An $f/1$ system has an incident angle of 22.5°.

At $\lambda = \lambda_0$, the phase difference is zero and the transmission $\tau = \cos^2 \frac{\phi}{2}$ is one. The transmission falls to zero for a phase shift of π. Half intensity occurs at phase shift $\phi = -\frac{\pi}{2}$ and $\phi = +\frac{\pi}{2}$, therefore the full width at half maximum $\Delta\lambda_{\text{FWHM}}$ corresponds to $\phi = \pi$.

$$\Delta\lambda_{\text{FWHM}} = \frac{\lambda}{4\,d\,(\alpha_1 - \alpha_2)} \quad . \tag{19.14}$$

Note that the $\Delta\lambda_{\text{FWHM}}$ does not depend on the index of refraction, but only on the dispersions. The Schott glasses sf57D and lasf41 have dispersions of -3.17×10^{-5} and -1.65×10^{-5}, respectively, at 5000 Å and both their indices are near 1.87. Assuming for the moment that the indices are equal, then the lengths of the arms for a 0.1 Å and 1 Å $\Delta\lambda_{\text{FWHM}}$ are 8.2 cm and 0.82 cm, respectively. There are lasf glasses that have similar dispersions, but slightly different indices, so a near match is possible throughout the visible regime. An exact equi-index condition can be achieved anywhere in the visible with three glasses, say sf57D in one arm and the proper thicknesses of lasf31 and lasf41 in the other.

The wavelength shift with angle at a spectral distance of $N \times \Delta\lambda_{\text{FWHM}}$ is

$$\Delta\lambda(N) = -N \times \Delta\lambda_{\text{FWHM}}\left(\frac{\theta^2}{2\,n^2}\right) \quad . \tag{19.15}$$

Relation 19.15 implies that there is a considerable decrease in angle sensitivity at some distance away from the equi-index point. Figure 19.2 illustrates this and compares the angle sensitivities of an SWFM and FP, with indices of 1.87, to EIMs operated at N times the $\Delta\lambda_{\text{FWHM}}$ for the cases of N equals 20 and 100. The figure indicates that a 1 Å $\Delta\lambda_{\text{FWHM}}$ EIM operated at $\pm$ 20 Å of the equi-index point has a shift in wavelength of less than half its $\Delta\lambda_{\text{FWHM}}$ for incident angles of less than 25°, i.e., it would shift slightly less than 0.5 Å in an $f/1$ beam. Similarly operated as far as $\pm$ 100 Å from the equi-index point, the maximum shift in an $f/2.5$ beam would be 0.5 Å.

Design considerations for space applications

Temperature control

The most important design concern for temperature control systems is how smoothly the temperature is controlled. This depends on the quality of the proportional control, the width of the dead band, and how the various components of the interferometer respond to temperature. The ideal approach for a tuneable system is to isolate the interferometers from a slowly changing environment in an enclosure that has substantial thermal inertia. Temperature monitors on or close to the interferometer can then be used to measure the temperature, and tuning can compensate any measured temperature change.

As discussed above, significant reduction of the temperature sensitivity of the interferometers can be achieved by a selection of materials that mitigate against temperature changes. In such designs care must be taken that the compensating materials do not have significant differences in their response to temperature changes. Shown in Figure 19.3 is the measured performance of a WFB element made of calcite and ammonium dihydrogen phosphate (ADP). The temperature sensitivity is less than 0.01 $\text{Å}\,\text{K}^{-1}$.

SWFMs have been used as the narrow spectral isolation system in both the Michelson Doppler Imager (MDI, Scherrer et al 1995) on *SOHO* and the Heliospheric and Magnetic Imager (HMI, HMI proposal 2002) for *SDO*. Both of these interferometers have one arm made of BK7G18, which is the same material as the beam splitter. This design detail eliminates the effects of differential thermal expansion between that arm and the beam splitter. In MDI and HMI the other arm is located by a spacer with the property that its thermal expansion compensates the product of the square of the index of refraction times the coefficients of thermal expansion and index of refraction of the BK7G18 (Equation 19.10). The MDI and HMI interferometers differ in that the MDI spacer is made of copper while the HMI spacer is calcium fluoride, which is an optical quality material.

There are two significant advantages to a calcium fluoride design. The first is thermal. Copper responds much faster to changes in the environment than the BK7G18, which forms the bulk of the interferometer. Therefore, when power is expended in the oven enclosure, because of the much more rapid response of the copper, there is a shift in the path difference in the interferometer that remains until thermal equilibrium is again achieved. This problem is much less severe with the calcium fluoride spacer.

The measured temperature sensitivity of the HMI SWFM is 0.003 $\text{Å}\,\text{K}^{-1}$. At 6173 Å, the operating wavelength for Doppler measurements, this corresponds to 150.7 $\text{m}\,\text{s}^{-1}\text{K}^{-1}$. The HMI interferometers are in an oven that controls the temperature stability to better than 0.01 $\text{K}\,\text{h}^{-1}$, which corresponds to a velocity stability of better than 1.51 $\text{m}\,\text{s}^{-1}\text{h}^{-1}$. This is a truly remarkable stability of about $5 \times 10^{-9}\,\text{h}^{-1}$. The actual velocity stability can only be determined in orbit, but it is expected to be about an order of magnitude better.

The second advantage of the fluoride is its transparency. Since the spacer length determines the mirror parallelism, its thickness needs to be uniform to $\lambda/50$, which is difficult to achieve with an opaque material like copper. The simplest method for

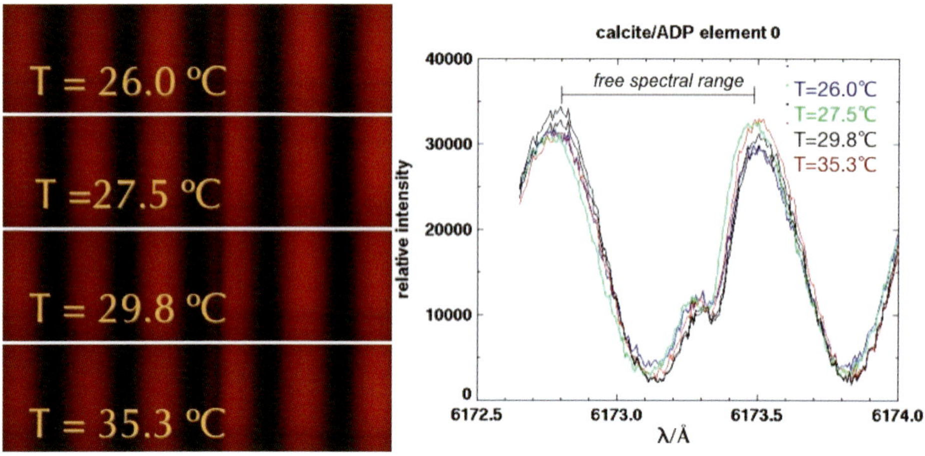

Figure 19.3: The left hand panel shows photographs of the spectral modulation of a solar spectrum by a calcite/ADP WFBF element. The dark line near the letter C is the spectral line Fe 6173 Å. The panel on the right shows the intensity modulation of the pattern on the left.

producing the spacer would be to manufacture a flat plate to the design flatness uniformity and then to core it to make an annular ring. Unfortunately, an annular spacer does not work because the spacer will not have the same thermal coefficient of expansion as the beam splitter when the condition for thermal insensitivity (Equation 19.10) is satisfied. Since the thermal expansion coefficient of the spacer is different from that of the beam splitter and mirror, any temperature change of the interferometer will bend the mirror mounted on the spacer. To eliminate bending the ring must be broken into a series of posts — a "Stonehenge" design. Again it is much easier to carry out the slicing of the angular ring with a transparent material. Shown in Figure 19.4 is the layout of the optical components of the HMI SWFMs.

Tuning implementation

An SWFM with an air-spaced arm has several options for tuning. The simplest approach conceptually is to make the spacer layer length electrically adjustable. This could be accomplished by techniques similar to those used by capacitance stabilized piezo-electrically controlled FPs. Internal spacer techniques all require maintaining the parallelism of the air-spaced mirror to $\lambda/50$ with respect to the beam-splitter cube. More seriously, the spacing must remain stable to 0.56 Å h^{-1}, and the tuning control must be repeatable to better than 0.1 Å to match the stability achieved by MDI and HMI. This is difficult to achieve in a space application.

The tuning method used by MDI and HMI is implemented by rotation of a halfwave plate between the SWFM and the exit polarizer. In both MDI and HMI, an exit quarterwave plate mounted on the beam splitter converts the elliptically polarized light before the quarterwave plate to linearly polarized light at angles between 0° and 180°. The final HMI SWFM has a free spectral range of 0.172 Å

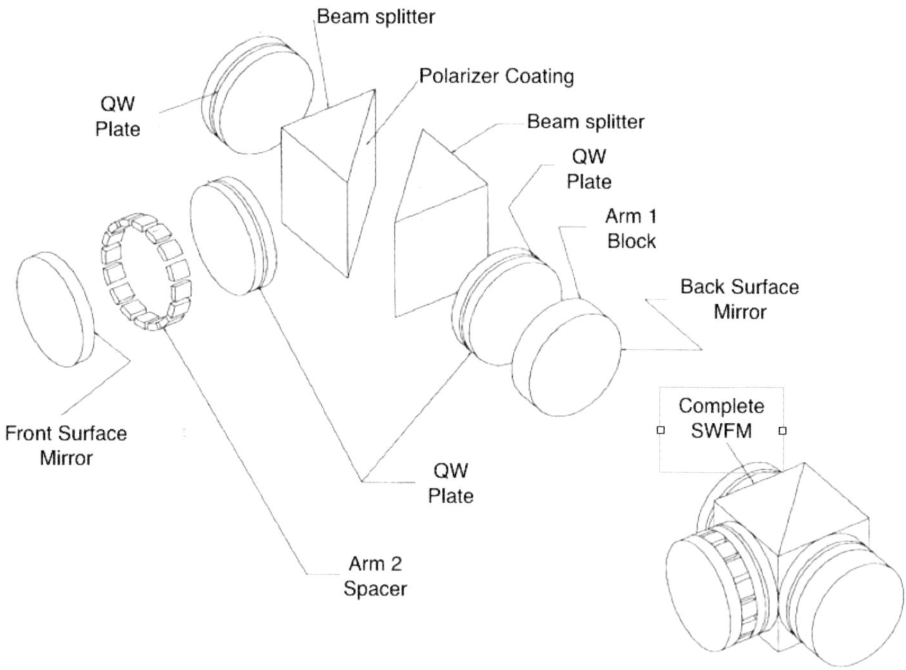

Figure 19.4: Components of the HMI SWFM.

that is scanned by a 180° rotation of a polarizer. In the MDI and HMI instruments, instead of rotating a polarizer, a halfwave plate is rotated between the fixed quarterwave and a fixed-exit polarizer. A halfwave plate rotates a plane of polarization by twice the angle between the waveplate and the polarization direction. That is, 90° of rotation tunes over a free spectral range. The halfwave plates are mounted in the center of hollow-core DC torque motors. There are no internal or external gears. Fixed magnets in the armature determine armature positioning. The measured repeatability of the motor positions is $\pm 5''$ or an equivalent velocity repeatability of $0.5\ \mathrm{m\,s^{-1}}$.

Optical coatings

The all air-glass surfaces of the SWFM need to be anti-reflection (AR) coated to reduce the possible secondary interference modulation. This is relatively straightforward to achieve with modern coating systems. However, care must be taken that the air-glass coatings do not bend the optics. Much more critical is the design and execution of the beam-splitter coating. The most serious problems result from the phase changes in reflection and transmission with the angle of the s- and p-wave light. Less of a problem is the balance of the reflection and transmission amplitudes. Achieving a relative balance to 5 % only results in a small loss in throughput.

Spacecraft instruments

The MDI instrument on the *SOHO* spacecraft and the recently delivered HMI instrument for the *SDO* are both designed to make maps of the magnetic and velocity field of the entire Sun. The velocity maps are constructed from images made at a sequence of spectral positions, tuning positions of the interferometers, in a magnetically sensitive iron line, while the magnetic maps are constructed from a combination of spectral images that are made in an array of prepared polarization states at a set of spectral positions. For cost and telemetry reasons, MDI only takes images in circularly polarized light. This just allows measurement of the line-of-sight magnetic field, while HMI collects images in both circular and linear polarizations that allow construction of the full vector magnetic field.

Both MDI and HMI use a fixed temperature compensated WFB filter followed by two tuneable SWFMs. In the case of HMI the final element of the WFB is also tuned. The third tuning element is required for HMI because *SDO* will fly in a geosynchronous orbit where the magnitude of the variation in the velocity of the spacecraft along the Earth-Sun line is too large to be compensated by tuning only two interferometers. MDI flies in an orbit around L1 and the velocity along the Earth-Sun line is much smaller. Hence only the two SWFMs are tuned. Besides the addition of a more complex polarization analyzer the major difference between MDI and HMI is in the image detectors that are 1024×1024 and 4096×4096 CCDs, respectively. The larger detectors in HMI allow acoustic tomography measurements over the full solar disk. These larger arrays require a 16-fold increase of data rate in order to downlink the science data at essentially the same cadence for the two instruments. The cadence requirement occurs from the necessity to adequately sample the solar five-minute oscillations. HMI has two focal planes mainly for redundancy.

Shown in Figure 19.5 is a layout of the major optical and mechanical components of the HMI instrument. In operation, incident sunlight strikes a front window that only transmits in a 50 Å band centered on 6173 Å. The remaining light is focused by the primary and secondary lenses through a beam splitter that transmits a small fraction of the light to the limb-sensor focal plane, while the majority of the light is reflected toward the filter section. On the way to that focal plane the light passes through a pair of filter wheels that contain focus blocks, lenses, and polarization selection waveplates. The waveplates in conjunction with the polarizer in front of the Lyot constitute the polarization analyzer. Before the limb-sensor focal plane the light reflects off the image stabilization mirror. This mirror is mounted on piezo-electric transducers that are in a closed-loop servo system controlled by the solar limb sensor in the sensor focal plane. A telecentric lens in the main beam collimates the image of the objective, so that the ray bundles to each point in the focal plane are identical through the filter system. The telecentric lens, the Lyot filter, the two SWFMs, and a lens that reimages the solar image on the detector focal planes are in the well-isolated precision temperature controller enclosure. Following the reimaging lens are beam splitter, shutters, and the two camera focal planes. In order to achieve the precision measurements required for helioseismology, both MDI and HMI were designed to be calibrated. For that reason the focus is controlled by a set of focus blocks that displace the focus in discrete steps rather

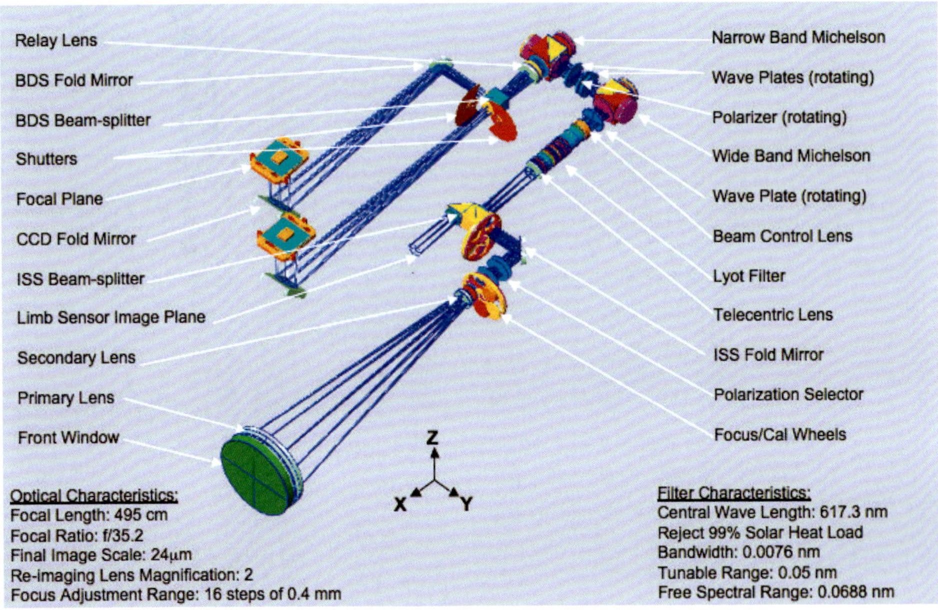

Figure 19.5: Optical schematic of the HMI showing the location of the major components of the optical package.

than continuously. The tuning waveplates that control the measuring wavelengths are relatively coarse, but very reliable. Before flight the optical distortions in the focal plane were carefully measured as a function of the temperature of the entire package.

Conclusions

The SWFM has been demonstrated to be a robust spectral imaging system for use in space. The MDI system on *SOHO* survived loss of solar pointing that caused the instrument to drop below 190 K for several months. After the recovery of the pointing the filter system functioned as before the event and it is still functioning perfectly after more than 10 years of operation. The HMI interferometers are of much higher quality and were manufactured more quickly both because of the use of calcium fluoride spacers and the improvements in optical polishing technology. It will be interesting to follow the future development of solid interferometer systems for spectral regions in the ultraviolet and infrared. Other areas of possible development are beam splitters, polarizers, and retardation plates made using micro-lithographic technology. These devices can be less angle sensitive, large, and operate in a broader spectral domain.

Solid Michelson interferometers are in use for solar and Earth-observing spacecraft. In addition, it is likely that future applications will be found on astronomical spacecraft.

Bibliography

Evans JW (1949) The birefringent filter. JOSA 39:229–242

HMI proposal (2002) http://hmi.stanford.edu/proposal_April_2002/HMI_view_proposal_public.pdf

Mertz L (1965) Transformations in optics. Wiley, New York

Scherrer PH, Bogart RS, Bush RI (plus ten authors) (1995) The Solar Oscillations Investigation – Michelson Doppler Imager. Sol Phys 162:129–188

Title AM, Ramsey HE (1980) Improvements in birefringent filters: 6. Analog birefringent elements. Appl Opt 19:2046–2058

IV. Detectors

— 20 —

Detector types used in space

Anuschka Pauluhn[I]

Abstract

The main characteristics of present and future detector types for space applications discussed in this volume are summarised in the form of a table. The order follows wavelength range. For reasons of brevity abbreviations are only explained in the following chapters. Similarly, only the most important features and capabilities of the detectors in question are listed; again details are found in the corresponding chapters.

General remarks

The technology of photo-electric detectors has gone through huge developments in the course of the space age. Imaging detectors with read-outs suitable for space use have become available during the past four decades and are standard today. Systems are now available that can not only determine the location of the detected photon, but also its time of arrival and energy.

The need for high voltage for the operation of many early photon detectors in space has caused problems, because of corona and arc discharges. This problem has now been overcome with experience in handling high voltage in space on the one hand; on the other hand most solid-state detectors are operating at low voltages.

Cryogenics is far from being simple for ground-based laboratories, even more so in space. Efficient and long-lasting cooling devices are an important part of detector development.

Last but not least, after all development and design has been done, the requirements of cleanliness and calibration remain important throughout; the latter including a continuing performance monitoring in orbit.

In Table 20.1 we present an overview of the most important features of the detector systems, whose operating principles, construction details and performance characteristics are described in Chapters 21 through 32 (Smith 2010, Timothy 2010, Waltham 2010, Holland 2010, Schühle 2010, Schühle and Hochedez 2010, Martin and Verhoeve 2010, Porter 2010, Eaton 2010, Raab 2010, Wild 2010, Fröhlich 2010).

[I]PSI—Paul Scherrer Institut, Villigen, Switzerland

Wavelength range	Type	Characteristics/capability	Performance	Requirements	Chapter
γ-rays, X-rays, $1\ \mathrm{GeV} \geq E > 10\ \mathrm{keV}$	Scintillators	Solid-state detectors based on (depending on energy range) photoelectric absorption, Compton scattering, pair production	Photon counting, can be pixellated	Background can be problematic	21
X-rays	X-ray calorimeters	Energy measurements for single photons by sensing temperature rise in absorber material	Photon counting, high efficiencies; high resolving power; pixellated focal plane detector; simultaneous spectral and spatial resolution	Choice of thermistor technology is crucial; cooling to < 0.1 K required; larger arrays needed	28
X-rays to NIR	MCP	Microchannel plates; combination with discrete anode arrays, analogue readout arrays or digital readout arrays (e.g., MAMA)	Photon counting, time tagging and TOF mass spectrometry possible	No cooling at high energies, but high voltage and high vacuum required; radiation tolerant	22
X-rays 0.1 keV to 100 keV, UV – VIS	CCD, CMOS	Solid state image sensors; large pixel arrays, each with its own readout electronics (microprocessor chips); sensor and electronics integrated	Fast readout systems; shutterless operation possible; low voltages (1.8 to 5) V for CMOS; energy discrimination at X-ray energies	Compact, rugged; Si very responsive to visible light; for CCDs cooling to -60 °C to -100 °C to avoid dark current; CCDs susceptible to radiation damage; CMOS relatively radiation hard	23, 24
UV – VIS	ICCD, IAPS	MCP intensifier + imaging chip; selection of photocathode determines spectral range	Large dynamic range (8 orders of magnitude); photon coun-ting or integrating mode	No cooling necessary	25
UV	BOLD, solar-blind detectors	Wide band gap (≥ 3.4 eV) semiconductors	Imaging arrays; bump-bonding of WBGM on readout circuit	Low dark current; room temperature operation possible	26

Table 20.1: Summary of wavelength ranges and principal detector classes.

Wavelength range	Type	Characteristics/capability	Performance	Requirements	Chapter
X-rays – submillimetre	Superconducting tunnel junctions (STJ)	Generation of "free" charge carriers by breaking Cooper pairs in a superconducting material by the absorbed radiation	Quasi-particle processes, and thus faster than thermal processes; high energy resolution	Readout is demanding; need stable, low voltage; operation temp. 10 % T_c, i.e., 0.1 K to 1 K	27
IR – submillimetre	Bolometers	Energy measurements by sensing temperature rise in absorber material	Imaging arrays in focal plane; several temp. sensors: thermistors, pyroelectric and ferroelectric sensors (mostly in uncooled arrays); NTD germanium thermistors, TES	Uncooled operation for Sun/Earth/planet observation; cooling (0.1 K to 0.3 K) for other astronomical objects	29
NIR 1 µm to 5 µm	Photodiodes	Based on junction between oppositely doped zones on a semiconductor; wavelength cutoff depending on supercond. material	Available as large arrays, with low-noise, low-power readout circuits	Need cooling to 10 K to 40 K	30
MIR 5 µm to 40 µm	Photoconductors	Single crystal, bulk-type, Si-based; or blocked-impurity-band (BIB)	Non-linear response under certain operating conditions; BIB exhibit lower noise	Cooling required (BIB: 3 K to 6 K)	30
FIR 40 µm to 200 µm	Photoconductors	Bulk-type Ge-based; or mechanically stressed	Bump bonding not possible for stressed arrays; non-linear effects; costly manufacturing	Cooling required (1 K to 2 K)	30
FIR >40 µm, submillimetre range	Coherent detectors, heterodyne systems	Mixer + local oscillator + spectrometer	High resolution ($> 10^6$); efficient signal coupling and stability of the devices is critical	Cooling needed for HEB and SIS mixers, not for Schottky mixers	31
TSI, integration over all wavelengths	Radiometers	Conversion of radiation into thermal energy, measured by an electrically calibrated thermal flux meter	Geometry of the components important: cavities for absorption of the radiation with "precision apertures" (defining the flux-collecting area)	Long-term behaviour of instruments (exposure-dependent and exposure-independent) needs to be modelled and corrected	32

Table 20.1: Summary of wavelength ranges and principal detector classes (continued).

Bibliography

Eaton HAC (2010) Infrared imaging bolometers. ISSI SR-009:477–483

Fröhlich C (2010) Solar radiometry. ISSI SR-009:525–540

Holland A (2010) X-ray CCDs. ISSI SR-009:409–418

Martin DDE, Verhoeve P (2010) Superconducting tunnel junctions. ISSI SR-009:441–457

Porter FS (2010) X-ray calorimeters. ISSI SR-009:459–475

Raab W (2010) Semiconductors for low energies: incoherent infrared/submillimetre detectors. ISSI SR-009:485–501

Schühle U (2010) Intensified solid state sensor cameras: ICCD and IAPS. ISSI SR-009:419–428

Schühle U, Hochedez J-F (2010) Solar-blind UV detectors based on wide band gap semiconductors. ISSI SR-009:429–439

Timothy JGT (2010) Microchannel plates for photon detection and imaging in space. ISSI SR-009:365–391

Smith DM (2010) Hard X-ray and γ-ray detectors. ISSI SR-009:345–364

Waltham N (2010) CCD and CMOS sensors. ISSI SR-009:391–408

Wild W (2010) Coherent far-infrared / submillimeter detectors. ISSI SR-009:503–523

— 21 —

Hard X-ray and γ-ray detectors

David M. Smith[I]

Abstract

The detection of photons above 10 keV up to gigaelectronvolt energies is challenging due to the penetrating nature of the radiation, which can require large detector volumes, resulting in correspondingly high background. In this energy range, most detectors in space are either scintillators or solid-state detectors. The choice of detector technology depends on the energy range of interest, expected levels of signal and background, required energy and spatial resolution, particle environment on orbit, and other factors. This section covers the materials and configurations commonly used from 10 keV to >1 GeV.

Introduction

Most high-energy detectors in space are based on scintillators or solid-state detectors. Scintillators are the older technology and are generally used where large detector volumes and lower cost are paramount. Solid-state diode detectors, e.g., germanium, silicon, and cadmium telluride (CdTe) or cadmium zinc telluride ($Cd_{1-x}Zn_xTe$ or CZT), generally have better energy resolution than scintillators and a lower energy threshold, and can be more easily pixellated for fine spatial resolution if that is required. They are more expensive than scintillators and more difficult to produce, package and read out in large volumes. For all these materials, photons are detected when their energy is transferred to electrons via photoelectric absorption, Compton scattering, or pair production (which also produces positrons, of course). The high-energy particles come to a stop in the detector volume, producing ionization that is detected by varying methods. I will review the most commonly used solid-state and scintillator materials and detector configurations. Considerable work has also been done worldwide on gas and liquid detectors and on newer or less common semiconductors and scintillators. For a much more detailed discussion of a wider range of detectors, as well as a treatment of general considerations in photon counting and electronics, see the excellent textbook by Knoll (2000).

[I]Physics Department and Santa Cruz Institute for Particle Physics, University of California, Santa Cruz, USA

Configurations and energy regimes

Thin and monolithic detectors

The optimum configuration and material for a detector depend most strongly on the energy range of the photons to be observed. Figure 21.1 shows the cross-sections for photoelectric absorption, Compton scattering, and pair production for photons by elements commonly used in detectors: silicon and germanium (in solid-state detectors) and iodine and bismuth (in common scintillators). Simple efficiency calculations based on cross-sections can assist with instrument design, particularly when photoelectric interactions are dominant, but Monte Carlo simulation is the most powerful and flexible tool. It can be used to model the response to source and background radiation and to incident particles other than photons as well. The packages most commonly used are GEANT3 and GEANT4 (GEometry ANd Tracking), originally developed at the European Organization for Nuclear Research (CERN) for accelerator applications (CERN 1993; Agostinelli et al 2003). Since the cross-section for photoelectric absorption is large at low energies, low-energy detectors can be quite thin. Photoelectric absorption is also a strong function of atomic number, so that, for example, a silicon detector 1 mm thick will absorb > 50 % of X-rays up to 23 keV, while a 1 mm CdTe detector will do the same up to 110 keV. At these low energies, high spatial resolution is often desired when there is an imaging system using focusing optics or a coded mask (see Chapter 12, Hurford 2010). This can be accomplished with multiple small detectors, by pixellating the electrodes of solid-state detectors, or by using multiple or position-sensitive phototubes to read out a scintillator (the "Anger camera" configuration, after inventor Hal Oscar Anger).

At energies of more than about 300 keV, photoelectric cross-sections are small even at high atomic number, and detectors must be made large enough that photons can Compton scatter in the detector and still be photoelectrically absorbed afterwards. Even though the Compton cross-section is nearly independent of atomic number, a high atomic number is still critical for stopping the dowscattered photon before it escapes the detector carrying off some of its energy. A low atomic number can be desirable for the scattering plane of a Compton telescope or for a detector or shield designed to stop charged particles or X-rays and pass γ-rays through. In many cases the optimum solution for maximizing sensitivity will be to have separate detectors for low energies (thin) and high energies (thick). Since most cosmic sources have falling energy spectra, high-energy detectors will generally need larger area than low-energy detectors in order to reach comparable sensitivity. Even large monolithic detectors can serve as elements for a coarse imaging system when placed in a large array. *INTEGRAL* provides two good examples. The spectrometer on *INTEGRAL* (SPI, Vedrenne et al 1998) has coaxial germanium detectors with a characteristic size of 7 cm serving as pixels below a large coded mask (see Chapter 12), and the imager on board the *INTEGRAL* spacecraft (IBIS) includes thick fingers of CsI serving as pixels beneath a finer mask than SPI's (Ubertini et al 2003). The large germanium detectors on *RHESSI* (Smith et al 2002) sit below rotation modulation collimators (see Chapter 12) that do not require position sensitivity.

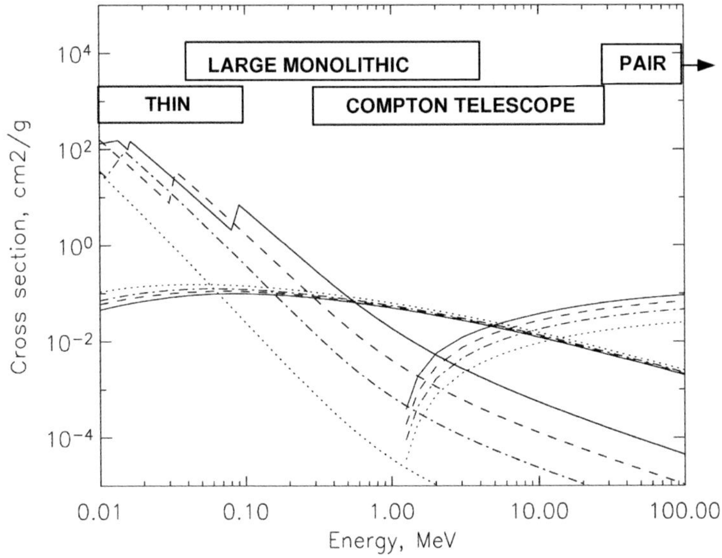

Figure 21.1: Cross-sections for photoelectric absorption (falling), pair production (rising above 1 MeV), and Compton scattering (flat), data from Berger et al (1998). Dotted line: silicon. Dash/dot line: germanium. Dashed line: iodine. Solid line: bismuth. The cross-sections of cadmium and tellurium (i.e., CdTe and CZT detectors) are similar to iodine. The approximate energy regimes for basic detector configurations are shown (thin detectors, large monolithic detectors, Compton telescopes, and pair-tracking telescopes). The pair-tracking regime extends beyond the plot to hundreds of gigaelectronvolts.

Compton and pair-tracking telescopes

At megaelectronvolt and gigaelectronvolt energies, the physics of the photon interactions in matter can be exploited to reject background and determine the direction of the incoming photon. In the range between about 0.1 MeV and 20 MeV large-volume detectors with position sensitivity in three dimensions can record the entire sequence of Compton interactions in the detector volume (Figure 21.2a). This allows the direction of the incoming photon to be reconstructed and background to be rejected effectively (see Chapter 11, Schönfelder and Kanbach 2010). At energies of 10 MeV and higher, where pair production is the dominant photon interaction with matter, a pair-conversion tracking system can be used. For example, the Large Area Telescope (LAT, Atwood et al 2007) on *Fermi (GLAST)* consists of alternating thin layers of passive tungsten and active silicon strip detectors (Figure 21.2b). Pair production takes place in the tungsten, since the high atomic number gives a high cross-section (Figure 21.1). The electron/positron pair has high enough energy to penetrate multiple W/Si layers; the penetration depth gives the initial γ-ray energy, while the position-sensitive detectors allow the track to be extrapolated backwards to give the arrival direction of the γ-ray. The highest-energy pairs that

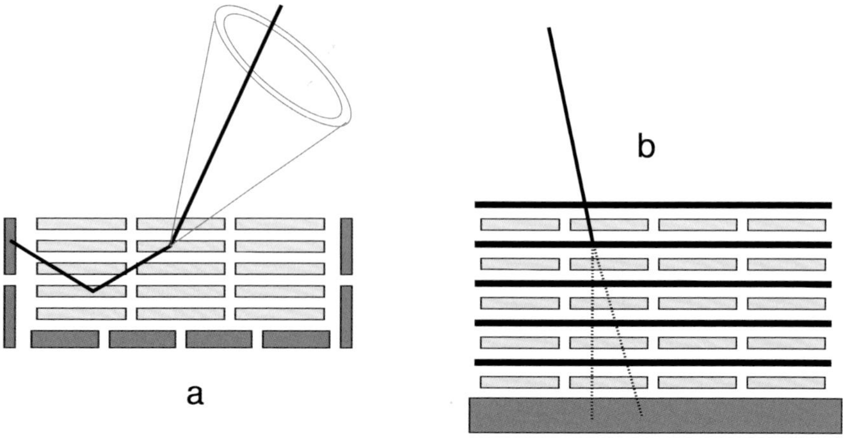

Figure 21.2: Design concepts for (a) a modern Compton telescope and (b) a pair telescope. The energies and angles between interactions in the Compton telescope trace the incident photon to an annulus on the sky. Here the telescope is shown with low-Z scattering detectors and high-Z absorbing detectors (Bloser et al 2002; Zych et al 2006; Takahashi et al 2006), although single-medium instruments are also being designed and built (Boggs and Chang 2007; Kurfess et al 2004; Aprile et al 1998). See Chapter 11 for more details. The pair telescope represents the *Fermi (GLAST)* design (Atwood et al 2007), with passive tungsten layers to convert the γ into an e^+/e^- pair, silicon detectors to track the pair, and a heavy calorimeter to absorb the remaining energy. Relativistic beaming of the pair and high spatial resolution allow reconstruction of the incident photon's direction.

penetrate the tracker are stopped in a calorimeter. At gigaelectronvolt energies, this extrapolation can be very precise. The tracking detectors are not required to measure deposited energy.

Detector materials

Scintillation detectors

Scintillators can be produced in large, monolithic volumes and in a variety of shapes. They can have low to high atomic numbers (and therefore stopping power), ranging from plastic scintillators at the low end to bismuth germanate ($Bi_4Ge_3O_{12}$ or BGO) at the high end. Plastic scintillators can be doped with high-Z atoms, like lead, to improve their stopping power. In inorganic scintillators, ionization produces free electrons that can move around the crystal until they fall back into the valence band. In *activated* crystals, such as NaI(Tl) and CsI(Na), the trace activator element provides a fast route to the valence band via intermediate states. Since the amount of activator is small, the crystal remains transparent to the scintillation photon emitted when the activator's excited state decays. In unactivated crystals such as BGO, one ion of the pure crystal (Bi^{3+}) provides the scintillation photons, with a large enough shift between its emission and absorption frequencies that the crystal is still transparent. In organic scintillators (plastic and liquid), large molecules are excited by the passing energetic particles; the scintillation is produced when they relax to their ground state. In all cases, scintillation light can be collected and multiplied by a sensor such as a photomultiplier tube (PMT), photodiode, or microchannel plate. Scintillation light can be reflected many times before being collected, so the PMT(s) need not have a direct line of sight to every part of the detector. In fact, uniformity of light collection — and therefore energy resolution — is sometimes improved by treating the surfaces of the best-viewed parts of the detector so that they reflect scintillation light poorly.

In space applications, extremely high-energy deposits, up to many gigaelectron-volts, can occur in a detector due to the passage of a cosmic-ray iron nucleus (or other heavy element) or the spallation of a nucleus in the detector by any cosmic ray. NaI (Fishman and Austin 1977) and CsI (Hurley 1978) scintillators display phosphorescence in which a fraction of the light is emitted over a much longer time (hundreds of milliseconds) than the primary fluorescence. Thus, when a particularly large energy deposit occurs, the crystal can glow rather brightly for up to a second; the effect on the data depends on the design of the electronics.

Pulse-shape analysis, whether by analogue electronics or via flash digitization of the PMT signal, has multiple uses for scintillators. If two different scintillators have very different light decay times, they can be read out by a single phototube if they are sandwiched together, with the energy deposited in each still separable in the PMT signal. This can be used as a form of active shielding to veto charged particles or photons that interact in both scintillators. This configuration is called a "phoswich" (phosphor sandwich). Pulse-shape analysis can also be used to distinguish neutron and cosmic-ray ion interactions from the interactions of photons or electrons in CsI (Biggerstaff et al 1961) and plastic and liquid scintillators (Morris et al 1992).

Table 21.1 summarizes some properties of the scintillators most commonly used in space, as well as two particularly promising lanthanum halide scintillators that have recently become available. These new materials have good stopping power and excellent relative energy resolution (3 % to 4 % FWHM at 662 keV versus about 7 % for the industry standard NaI). They have a moderate internal radioactivity giving a count rate of $\approx 1.8\ \mathrm{cm^{-3}\,s^{-1}}$, mostly from ^{138}La (Owens et al 2007). This is of the same order as the background from other sources that an unshielded detector would receive in low Earth orbit, but could dominate the background if the detector is well shielded (see Page 356).

Semiconductor detectors

In semiconductor detectors, the electrons and holes excited into the conduction band by the passage of energetic particles are swept toward opposite electrodes on the detector surface by an applied electric field. The image charges induced on one or both electrodes, as they change with the movement of electrons and holes in the crystal, provide a small current pulse. This pulse is generally read by a charge-sensitive (integrating) preamplifier, followed by a shaping amplifier.

Table 21.1: Properties of some scintillators.

Material	Light yield (photons/keV)	Decay time (ns)	Peak λ (nm)	Density (g/cm^3)	Max. Z	Notes
NaI(Tl)	38	230	415	3.67	53	*a*
CsI(Na)	39	460, 4180	420	4.51	55	*b*
CsI(Tl)	65	680, 3340	540	4.51	55	*b*
BGO	8.2	300	480	7.13	83	*c*
GSO	9.0	56, 400	440	6.71	64	*d*
BC–408	10.6	2.1	425	1.03	6	*e*
$LaBr_3$(Ce)	63	16	380	5.29	57	*f*
$LaCl_3$(Ce)	49	28	350	3.85	57	*g*

a Good energy resolution. Hygroscopic; phosphorescent; susceptible to thermal shock.
b Slightly hygroscopic; phosphorescent. Denser, less brittle than NaI. Pulse-shape discrimination of particle types is possible.
c Excellent stopping power; inferior energy resolution; easily machined; non-hygroscopic.
d Gd_2SiO_5; non-hygroscopic. Used in the Hard X-ray Detector (HXD) on *Suzaku* (Takahashi et al 2007).
e A commonly used plastic.
f New material; some internal background from radioactivity. Best energy resolution, good stopping power.
g Similar to $LaBr_3$, resolution and density not quite as high. Large crystals were developed earlier.
Data are taken from Tables 8.1 and 8.3 of Knoll (2000) and from Bicron (2007).

Semiconductor detectors are capable of much better energy resolution than scintillators, since the collection of electrons in the conduction band is much more complete than the conversion to scintillation light and light collection in scintillators. The noise performance of the electronics must be excellent, however, if the natural resolution of the detector is to be approached, i.e., the limit due to counting statistics of the liberated electron/hole pairs (including the Fano factor; Knoll 2000). In the preamplifier, noise currents are converted to voltage noise proportional to the detector capacitance (Spieler 2005); thus detectors with large volumes can be expected to show poorer resolutions. Resolution can be preserved at large volumes if the electrode configuration has intrinsically low capacitance. Examples of such configurations are silicon drift detectors (Rehak et al 1985) and germanium LO-AX^{TM} (ORTEC 2008) and drift (Luke 1988) detectors, which have one large and one small, more pointlike, electrode. Pixellating the anode and reading out each pixel as a separate detector also results in low capacitance and excellent noise performance. Some of the most common semiconductor detector and electrode configurations are sketched in Figure 21.3.

Large energy deposits from cosmic rays in semiconductor detectors do not carry the risk of long-duration detector response that they do in scintillators. It is very

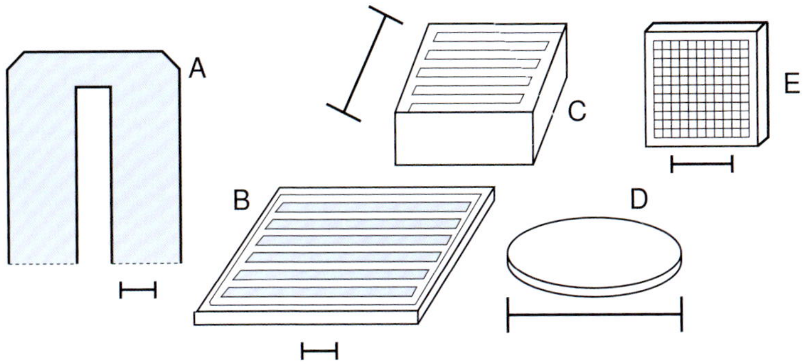

Figure 21.3: Common electrode configurations for solid-state detectors. Each is shown with a 1 cm marker to indicate the scale of a typical detector. A) Coaxial germanium detector (cross-section of a cylindrically symmetrical shape). Electrodes are on the inner bore and the outer surface; the insulating surface is at the back (dashed line). B) Silicon strip detector (typically 300 µm thick, 1 mm strips). C) Monolithic CZT detector with coplanar (interleaved) electrode to null the signal from the holes and improve energy resolution (Luke 1994). D) Simple silicon p-i-n diode with plane electrodes at top and bottom. E) CZT pixel detector. The pixel and strip detectors are shown with guard rings around the edge, used to capture any leakage current on the side surfaces and keep it away from the electronics chains reading out the volume of the detector. Pixel and strip detectors have been made from all the common semiconductor materials (Ge, Si, CZT/CdTe).

important, however, to test the response of the *electronics* to huge energy deposits. Many designs can become paralyzed for a significant amount of time by a multi-gigaelectronvolt energy deposit, or else produce false counts by triggering on ringing. Interactions this energetic do not occur in the laboratory, where the only cosmic-ray particles are muons, so they ought to be simulated either at an accelerator or with a pulser before any electronics design is declared ready for space flight. An extensive treatment of semiconductor detectors and their electronics is given by Spieler (2005).

Germanium

Germanium detectors are preeminent for spectroscopy in the range from hundreds of kiloelectronvolts to a few megaelectronvolts. Germanium crystals can be grown in large volumes at extremely high purity, with single detectors up to 4 kg (Sangsingkeow 1999). High purity guarantees that both electrons and holes can move untrapped through the whole crystal volume, and that the detector volume can be depleted of charge carriers due to impurities by a manageable applied field of $\approx$ (500 to 1000) V/cm. The small bandgap gives good counting statistics for the liberated electron/hole pairs, but requires low operating temperatures — below 130 K, and preferably much lower — to prevent thermal excitation into the conduction band and a large leakage current (Knoll 2000). Energy resolution of ≈ 0.3 % FWHM at 662 keV can be achieved with a good detector and optimized electronics. If high resolution is scientifically important, germanium should be considered. High resolution is important any time weak lines are being observed, even if the exact profile of the line to be observed is *not* needed, and even if it is *not* necessary to separate nearby lines. Since γ-ray observations in space are often dominated by background, a good energy resolution reduces the amount of background against which the signal of a γ-ray line is to be detected, greatly increasing sensitivity.

Operation of germanium detectors in space is challenging due to the need to keep them cold. Passive cooling is possible if a large area and solid angle of radiator can always be presented to deep space (avoiding the Sun, and, in low Earth orbit, the Earth as well). This technique was used for the Transient Gamma-Ray Spectrometer (TGRS) on the *Wind* spacecraft (Seifert et al 1996) and the Gamma-Ray Spectrometer on *Mars Odyssey* (Boynton et al 2004). Cryogens can also be used, but they have a large mass and limit the mission life time; the germanium spectrometer on *HEAO-3* ran out of cryogen after a pioneering 154 day mission (Mahoney et al 1980). Some recent instruments have relied on Stirling-cycle mechanical coolers (Vedrenne et al 1998; Smith et al 2002; Goldsten et al 2007). The design of the radiator for waste heat is still critical in that case, but the requirements are not as severe as for passive cooling. *RHESSI*, for example, often has the Earth nearly filling the field of view of its radiator for part of its orbit. Cryocoolers can be very expensive to qualify for space flight and this should not be underestimated in mission planning.

All the germanium detectors mentioned above were in the closed-end coaxial configuration (Figure 21.3A), with an outer contact on the sides and across the front face of the detector, an inner contact lining a bore that goes most of the way through the crystal, and an intrinsic (insulating) surface on the back face. This configuration is a compromise between large volume, low capacitance (compared to two flat electrodes on either side of a comparable crystal) and the lowest possible distance between the electrodes (to keep the depletion voltage manageable). Thick germanium strip detectors are also being developed for Compton telescopes and other applications that require position sensitivity, and have already flown on the Nuclear Compton Telescope balloon payload (Boggs and Chang 2007). The x and y positions are measured by localizing the charge collection to individual strips on each electrode (the strips on one side run perpendicular to those on the other), while the z position is measured by the relative arrival times of the electrons and holes at their respective electrodes (Amman and Luke 2000).

In addition to cooling, the other particular challenge for germanium detectors in space is radiation damage. Defects in the crystal lattice caused by nuclear interactions of protons and neutrons create sites that can trap holes as they drift through the crystal. Electrons in the conduction band are not comparably affected. Since germanium detectors are designed to use both the electron and hole signals, hole trapping reduces energy resolution. Because protons lose energy rapidly by ionization, they must have high energy to penetrate the layers of passive material around the detector (e.g., the cryostat) and penetrate beyond the outer surface of the crystal. Depending on the detector or cryostat configuration, the lower limit on relevant proton energy is of the order of 100 MeV. Neutrons, on the other hand, can penetrate the full volume of the crystal regardless of their energy, and are relevant even at a few megaelectronvolts (Leleux et al 2003). Protons below 100 MeV can

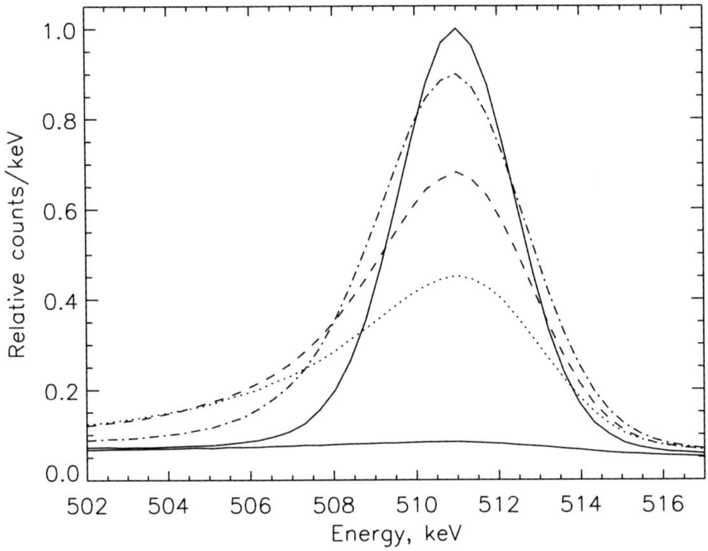

Figure 21.4: Effect of radiation damage on γ-ray spectroscopic performance of a coaxial germanium detector. The 511 keV background line from positron annihilation is shown in data from the *RHESSI* satellite taken from 2002 to 2007. The symmetrical, narrow peak is from the start of the mission. The next two lines (dot-dash and dashed) show the effect of moderate to severe damage on the line resolution due to hole trapping. The last two lines (dotted and the nearly flat solid line at the bottom) show the loss of effective area at very severe levels of damage due to volumes in the crystal that are no longer depleted (active). At this point, the *RHESSI* detectors were annealed. In general, annealing would be performed at a much earlier stage of damage.

convert to neutrons via spallation in spacecraft materials and therefore still cause damage.

Strategies to reduce the effect of radiation damage include choice of orbit, operating procedures, detector geometry, shielding, and annealing. In low Earth orbit, most radiation damage will come from radiation belt protons seen during passage through the South Atlantic Anomaly (SAA). This can be avoided if the orbit is equatorial (inclination less than about 10°). This is the most benign orbit available, since the magnetosphere also protects the instrument from solar energetic particles. In high Earth orbit or interplanetary space, a large solar energetic particle event can inflict severe damage in a short time (Pirard et al 2007; Owens et al 2007). Galactic cosmic rays and their secondary neutrons are otherwise the dominant source of damage for all orbits, but those below a cutoff energy are blocked by Earth's magnetosphere for an equatorial orbit. An orbit that spends much of its time in the heart of the radiation belts has by far the highest dose of all, and would certainly prohibit the use of germanium. The Space Environment Information System

(SPENVIS) webpage (Heynderickx et al 2004) is an extremely valuable resource for estimating the irradiation by radiation-belt and solar protons in various orbits.

A good choice of detector geometry can limit the severity of the effect of radiation damage by limiting the amount of germanium that the holes must traverse. In the coaxial configuration, most of the volume is near the outside of the detector. Thus, by applying negative high voltage (HV) to the outer contact, the holes are made to take the shorter path for the majority of interactions (Pehl et al 1979). The result is a line shape with a sharp peak and a long tail (due to the few interactions near the bore). This is shown in Figure 21.4. This polarity provides good uniformity of field within the crystal for slightly n-type material. The opposite polarity (used with p-type material) will show broadening of the line much earlier and more severely as the holes migrate all the way to the central bore. Shielding the detectors can be very effective at blocking solar and SAA protons, but not cosmic rays, which are much more energetic and penetrating. Very often, shielding is also desired to reduce background (see Page 356). Keeping the detectors very cold reduces the amount of trapping for a given radiation dose as long as the detector is never warmed up (Brückner et al 1991). Raising the HV on a damaged detector, when possible, can reduce the effect of damage somewhat (Kurczynski et al 1999). If large currents are passed through a damaged crystal, many of the hole traps will fill, and the effect of radiation damage will be much less severe for a few minutes, until these traps empty again. This occurs when a spacecraft passes through the SAA. The opposite effect occurs for damaged n-type germanium detectors when the HV is turned off: when it is turned back on, there are temporarily *more* unfilled traps than in equilibrium, and the resolution will be degraded until an equilibrium between detrapping and hole production is reached on the same timescale of minutes (Koenen et al 1995).

Even if all these factors are taken into account in design, virtually any germanium detector that is *not* in a low Earth, equatorial orbit will have to be annealed. When the crystal is heated to well above operating temperatures, many of the damage sites become de-activated (not repaired, since the annealing temperatures are far too low to actually move atoms around in the lattice). The mechanism is not well understood. The literature includes many small-scale experiments that do not give a good, overall formula for estimating the efficacy of the annealing process given detector type, damage history, and annealing temperature and duration. Temperatures of 50 °C to 100 °C and durations of days to weeks are typical of operations in space (Lonjou et al 2005; Boynton et al 2004); when in doubt, the longer and warmer, the better (Brückner et al 1991). Some annealing does take place at room temperature (Raudorf et al 1984) but this takes longer to be effective, and cannot eliminate trapping completely.

Silicon

Silicon can be grown in large volumes but not to as high a purity as germanium, and is therefore harder to deplete. Silicon detectors are therefore generally thin (typically 300 μm), and used for purposes where that is appropriate. The thickest silicon detectors (up to ≈ 1 cm) are made from slightly p-type material and have lithium ions drifted through the crystal bulk to compensate the intrinsic impurities, a technique formerly used for germanium before high-purity material became available. Small, simple planar p-i-n detectors are often used for X-ray detection up to a few tens of kiloelectronvolts, as in the top detector layer of the hard X-ray instrument on the *Suzaku* spacecraft (Takahashi et al 2007). Small Si drift detectors (SDDs), in which the field is shaped to lead the electrons to a small collecting contact, show improved resolution over p-i-n detectors due to their smaller capacitance (Rehak et al 1985) and can also provide position sensitivity when the drift time is measured in the electronics. Large, thin Si strip detectors (single or double sided) can be used when position resolution is important but energy resolution and low-energy threshold are not, such as in the silicon tracker on *Fermi*, where the requirement is to register the passage of high-energy electrons. Thick Si strip detectors have been proposed for a Compton telescope operating in a mode where a final photoelectric absorption is not necessary (Kurfess et al 2004). Si detectors benefit from cooling to reduce leakage current, but at a more modest level than germanium (temperatures of −20 °C to 0 °C). This can be accomplished by a careful passive cooling design or the use of simple thermoelectric (Peltier) coolers.

Cadmium telluride and cadmium zinc telluride

Cadmium telluride (CdTe) and cadmium zinc telluride ($Cd_{1-x}Zn_xTe$, CZT) detectors offer two advantages relative to germanium: they can be operated at room temperature and they have better photoelectric stopping power. It is difficult to grow large crystals of high quality, and the largest detectors available are (1 to 4) cm^3 (Chen et al 2008). Efficient detection in the megaelectronvolt range therefore requires a three-dimensional array of detectors (Natalucci et al 2008) to take the place of a single large germanium coaxial detector, with very careful control of passive material within the detector volume to prevent undetected Compton scatters. When energy response greater than 30 keV is needed but it is not necessary to go above a few hundred kiloelectronvolts, a single layer of CZT/CdTe detectors is often the best choice. If pixels of a few millimetres or larger are desired, an array can be made out of individual detectors, as was done for the Burst Alert Telescope (BAT, Barthelmy et al 2005) on the *Swift* mission (CZT) and the front detector layer of the IBIS imager on *INTEGRAL* (CdTe, Lebrun et al 1996). For smaller pixels, it can be advantageous to pixellate the electrode on one side of a larger detector and read signals out of each pixel. Not only does this provide greater position resolution with smaller gaps, the small pixels have very low capacitance and excellent energy resolution. Pixellated CZT detectors have been used on two hard X-ray focusing balloon payloads: the High Energy Focusing Telescope (HEFT, Bolotnikov et al 2001) and the INternational Focusing Optics Collaboration for μCrab Sensitivity (InFOCμS) (Tueller et al 2005). For InFOCμS, signal traces were routed away from the detector to the ASIC electronics, while for HEFT the preamplifiers were put onto the ASIC with the same spacing as the detector pixels and bump-bonded directly to the detector. The HEFT detector technology is being adapted for *NuSTAR*, an upcoming NASA small explorer satellite using focusing optics (Harrison et al 2005).

CZT and CdTe (and other compound semiconductors) differ from germanium and silicon in that holes are much less mobile than electrons, and suffer trapping even in crystals that are not radiation-damaged. Thus the best energy resolution is obtained when only the electrons contribute to the energy signal. This is not possible for a simple planar configuration (a rectangular detector with plane electrodes on opposite sides), but there are a number of ways to improve the situation. Coplanar grid (Luke 1994; Amman and Luke 1997) and pseudo-Frisch grid (McGregor et al 1998) electrode configurations can cancel most of the contribution of the holes to the signal for thick (1 cm or 2 cm) single detectors, resulting in an electron-only signal that recovers the excellent energy resolution of a very thin detector. If a pixellated detector is desired for spatial resolution or low capacitance anyway, pixellating the anode also ensures that the electrons contribute most of the detected signal as they get very near the anode (the "small pixel effect"). For a thick CZT or Cd/Te detector, it is also possible to measure the depth of the interaction within the crystal by measuring the risetime of the current pulse and use this information to correct the energy measurement either by analogue or digital techniques downstream (Richter and Siffert 1992). This technique was used for *INTEGRAL*/IBIS (Lebrun et al 1996).

General considerations

Background

Since most γ-ray detectors in space are not detecting focused radiation, count rates are often dominated by background rather than signal. Bright transient events such as cosmic γ-ray bursts, solar flares, and terrestrial γ-ray flashes are often source-dominated, but for most applications both *reducing* and *estimating* background levels is important. For an unshielded detector in low Earth orbit, the dominant sources of continuum background are the cosmic diffuse radiation (below about 150 keV) and the "albedo" glow of γ-rays from the Earth's atmosphere due to interactions of cosmic rays (above 150 keV) (Dean et al 2003; Gehrels 1992). From a few hundred kiloelectronvolts to a few megaelectronvolts, radioactivity in detectors themselves and, to a lesser extent, in passive materials nearby can be a significant source of both line and continuum background. Radioactivity can be natural (e.g., from ^{40}K and the daughters of ^{238}U), or induced by cosmic rays, solar or radiation-belt protons, or neutrons created in the spacecraft or in the Earth's atmosphere. Induced radioactivity can be prompt when short-lived nuclear states have been excited, or have a half-life from seconds to years. Above a few megaelectronvolts, the dominant component is likely to be from minimum-ionizing cosmic rays clipping the corners of the detector. Recently, there has been a great deal of effort put into refining tools for simulating the expected backgrounds (Dean et al 2003; Weidenspointner et al 2005) (see Figure 21.5).

Reducing background should first be approached by selecting the right detector thickness (no thicker than necessary) and material (one that is not intrinsically radioactive and does not become badly so when exposed to cosmic rays on orbit). But if the instrument is not meant to observe the entire sky, there should usually also be some shielding. Below $\approx$ 100 keV, passive shielding can be adequate. It is often arranged in a "graded" configuration, with a high-Z material like lead or

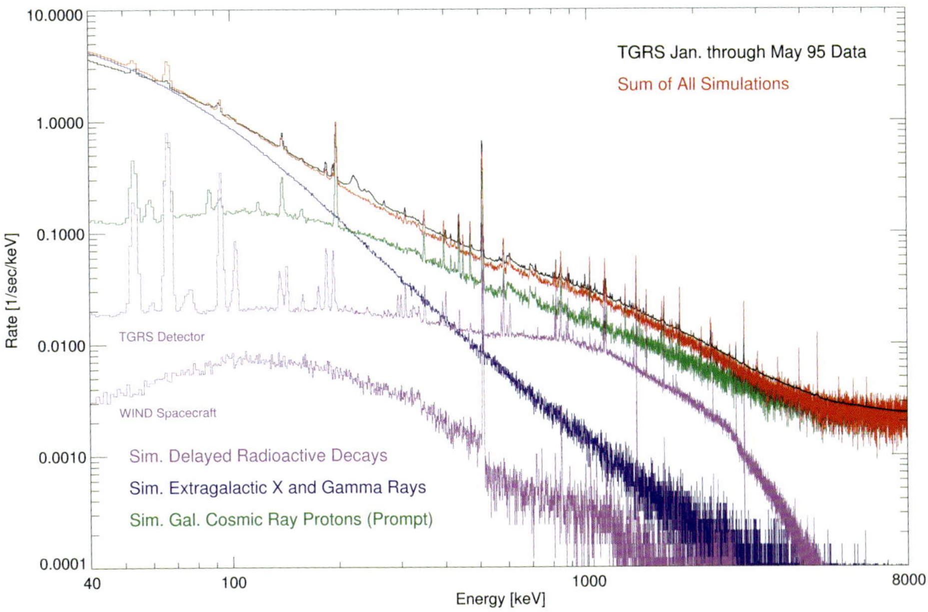

Figure 21.5: State-of-the-art *a priori* modelling of γ-ray instrumental background: data from *Wind*/TGRS and a simulation of all important background components using MGGPOD (Weidenspointner et al 2005).

tungsten on the outside followed by one or two layers of lower-Z material, each of which is meant to stop K-shell X-rays from the previous layer before they reach the detector. At Compton-dominated energies (see Figure 21.1), thick passive material would be necessary to stop incoming γ-rays. In space, however, such a shield can actually create more background than it stops, due to reprocessing of incident cosmic rays into neutrons, γ-rays, and multiple charged particles. Thus graded-Z and other passive shields shouldn't be more than a few millimetres thick. However, *active* shielding with several centimetres of inorganic scintillator can be very effective. In this case, cosmic rays are vetoed along with their daughter particles produced in the shield, and only a single interaction is necessary to veto a background photon, even if it then interacts in the central detector. An active shield will also veto photons from the target that interact in the central detector but scatter out of it ("Compton shield" mode). Even active shields produce background via neutron production (Naya et al 1996). A study for *INTEGRAL*/SPI found that 5 cm of BGO was the optimum shield thickness for its orbit (Diallo et al 1996). At the highest energies, a thin, active plastic shield can veto the prompt components due to cosmic rays: clipping of the detector by the cosmic rays themselves and prompt nuclear de-excitations in passive materials near the detectors. But it should be established that these background components will be important in the energy range of interest before choosing to use a plastic veto. The ability to veto charged particles that are *not* cosmic rays is desirable for an orbit outside the Earth's magnetosphere (for solar particles) or a low Earth orbit that goes to high magnetic latitudes (for precipitating outer-belt electrons and, for nearly polar orbits, solar particles as well).

Estimating background is less important for detectors in imaging configurations (e.g., coded mask, rotating grid, or Compton telescope) that have ways to reject background based on incident direction. In these cases, it is enough to predict the background accurately enough to have confidence in the instrument's sensitivity. But for non-imaging detectors, it may be necessary to know the fractional background to 1 % or better to study faint sources. This cannot be done via *a priori* modelling, if only because cosmic-ray fluxes fluctuate much more than this. Instead, background is subtracted by finding a period of time when the source is not visible but the background is expected to match that during the observation. For highly collimated instruments, this is best accomplished by "chopping" between the source position and an empty field nearby, as was done with the Oriented Scintillation Spectrometer Experiment (OSSE) on *CGRO* and the High-Energy X-ray Timing Experiment (HEXTE) on *RXTE* (Johnson et al 1993; Rothschild et al 1998). For uncollimated or wide-field instruments viewing a transient event like a cosmic γ-ray burst or a solar flare, time intervals just before and after the event, or (in low Earth orbit) $\pm$ 15 orbits ($\approx$ one day) away often provide an excellent background measurement. The case of a non-chopping instrument measuring a non-transient source is the most difficult. A variety of techniques can be mixed, combining observations and modelling, including the use of the Earth as an occulter and the generation of background databases incorporating large amounts of data from throughout the mission. Such a database can be used to extract the dependence of background on parameters such as orbital position and cosmic-ray flux (Jahoda et al 2006; Smith et al 1996).

Livetime

Instrumental livetime is generally of concern only when background is not — i.e., when very bright cosmic, solar or terrestrial transients are of primary interest. In these cases, all stages of the signal chain should be analyzed to make sure that the highest expected count rate can be recorded. The intrinsic response time of the detector material (scintillation light decay or electron and hole drift times in a solid state detector) may be a consideration in detector choice, but only if pulse shaping times and throughput in the rest of the electronics can be designed to keep up with the detector's capability. Large detectors can be pixellated or replaced with many small ones to reduce deadtime, at the expense of an increase in the number of electronics chains needed. The deadtime caused by an active shield veto should always be estimated, particularly when the instrument is very large or will be studying bright transients. When scattering between detector elements is part of the source detection (such as in a Compton telescope or γ-ray polarimeter), the frequency with which two independent background or source interactions will fall within the instrument's coincidence time window by chance and be mistaken for a scatter should always be calculated.

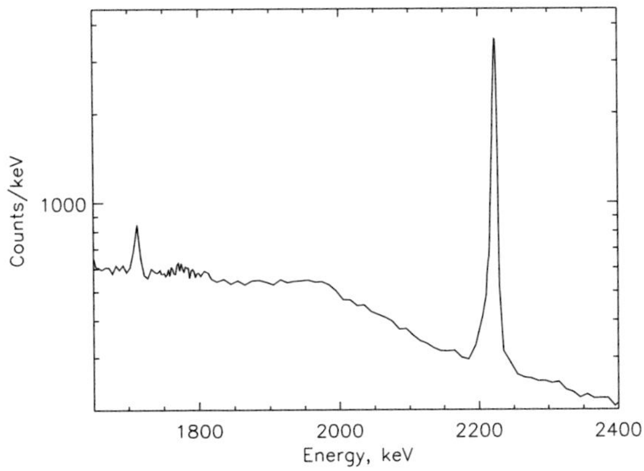

Figure 21.6: *RHESSI* spectrum of the solar flare of 28 October 2003. The emission in this energy range is dominated by the response to the 2.223 MeV line, which includes the photopeak, the Compton continuum from photons that scatter nearly 180° out of the detector (cutting off around 2 MeV), and the first 511 keV escape peak (1.712 MeV). There is an underlying, falling continuum due to other flare components as well.

Spectral response

Lastly, it is important to understand the energy response of any detector design due to the physics of the high-energy photon interactions. Incomplete collection of the incident photon energy is critical for both line and continuum spectroscopy, but is most obvious to the eye when a narrow line is being observed. At energies where Compton scattering becomes important, a Compton continuum below the incident energy appears in the spectrum due to scattering either into or out of the detector. At low energies (within a factor of ≈ 2 from the K-edge of the detector material), a K-shell X-ray escape peak appears since absorption occurs very close to the surface. At megaelectronvolt energies, two escape peaks appear corresponding to the escape of one or both 511 keV photons following pair production and annihilation of the positron. Figure 21.6 shows the photopeak line, Compton backscatter feature, and first 511 keV escape peak from a *RHESSI* observation of the 2.223 MeV line from a solar flare, from neutron capture by a proton producing deuterium. These effects — combined with the blurring effect of finite energy resolution — combine to make up the "response matrix", the function that maps the input spectrum of incoming photons to the output spectrum of detector counts. Good stopping power (high atomic number) and an active Compton shield can keep the diagonal components of this matrix dominant, making interpretation of the spectrum easier. It is not possible to unambiguously invert a nondiagonal instrument response matrix and calculate a unique photon spectrum given the observed count spectrum. The usual practice is to convolve models of the expected spectrum with the response matrix and compare the results to the observed count spectrum. Monte Carlo tools such as GEANT should be used to model and predict the response matrix for any new design.

Outlook

Germanium (or, at hard X-ray energies, the other solid state detectors) provides high enough energy resolution for most astrophysical purposes > 10 keV. BGO provides stopping power and large volume up to the mass limit that a launch vehicle can reasonably haul. So the current challenges in detector development are combining high resolution with high volume, and adding fine, 2- or 3-dimensional spatial resolution for Compton telescopes, coded masks, or imagers. Germanium strip detectors provide an appealing compromise of very high spatial and energy resolution with moderate detector volume and stopping power. The key development issues are the difficulty, power and expense of cooling and minimizing the amount of passive material in and around the array, material needed for thermal control as well as mechanical and electrical connections. Where very high spatial and spectral resolution are not required, the new lanthanum halide scintillators are becoming a popular option in proposals for space instruments. They have spectral resolution better than other scintillators and very good stopping power. They are appealing if their intrinsic background can be tolerated. Recently, high-performance Anger camera prototypes have been developed using SDDs to read out a large scintillator (Lechner et al 2008; Labanti et al 2008); SDDs have also been coupled to a physically pixellated scintillator (Yamaoka et al 2007). These technologies may provide an interesting alternative to pixellated semiconductors for moderate to high spatial resolution. Development is also always in progress on other technologies that are promising but still present technical challenges, such as liquid xenon (Aprile et al 1998; 2008), semiconductors such as TlBr (Chen et al 2008; Hitomi et al 2008) and HgI_2 (van den Berg et al 2007), and scintillators such as SrI_2 (Wilson et al 2008; Hawrami et al 2008) and BaI_2 (Cherepy et al 2007).

Acknowledgement

I would like to thank Mark McConnell for very helpful suggestions on the manuscript of this chapter.

Bibliography

Agostinelli S, Allison J, Amako K (plus 124 authors) (2003) GEANT4—a simulation toolkit. Nuc Inst Meth A 506:250–303

Amman M, Luke PN (2000) Three-dimensional position sensing and field shaping in orthogonal-strip germanium gamma-ray detectors. Nuc Inst Meth A 452:155–166

Aprile E, Egorov V, Giboni KL (plus eight authors) (1998) The electronics read out and data acquisition system for a liquid xenon time projection chamber as a balloon-borne Compton telescope. Nuc Inst Meth A 412:425–436

Aprile E, Curioni A, Giboni KL (plus three authors) (1998) Compton imaging of MeV gamma-rays with the Liquid Xenon Gamma-Ray Imaging Telescope (LXeGRIT). Nuc Inst Meth A 593:414–425

Atwood WB, Bagagli R, Baldini L (plus 58 authors) (2007) Design and initial tests of the tracker-converter of the Gamma-ray Large Area Space Telescope. Astroparticle Phys 28:422–434

Barthelmy SD, Barbier M, Cummings JR (plus 12 authors) (2005) The Burst Alert Telescope (BAT) on the *SWIFT* Midex Mission. Space Sci Rev 120:143–164

Berger MJ, Hubbell JH, Seltzer SM (plus four authors) (1998) XCOM: Photon Cross Sections Database, NIST Standard Reference Database 8 (XGAM) (online)

Bicron (Saint Gobain Crystals) (2007) BrilLanCe^{TM} 350 ($LaCl_3$(Ce)) and BrilLanCe^{TM} 380 ($LaCl_3$(Ce)) data sheets, http://www.detectors.saint-gobain.com/

Biggerstaff JA, Becker RL, McEllistrem MT (1961) Charged particle discrimination in a CsI(T1) detector. Nuc Inst Meth 10:327–332

Bloser PF, Andritschke R, Kanbach G (plus three authors) (2002) The MEGA advanced Compton telescope project. New Astron Rev 46:611–616

Boggs S, Kurfess J, Ryan J (plus 30 authors) (2006) The Advanced Compton Telescope. Proc SPIE 6266:662624-1–15

Boggs S, Chang Y-H for the NCT collaboration (2007) The Nuclear Compton Telescope (NCT): Scientific goals and expected sensitivity. Adv Space Res 40:1281–1287

Bolotnikov AE, Cook WR, Boggs SE (plus two authors) (2001) Development of high spectral resolution CdZnTe pixel detectors for astronomical hard X-ray telescopes. Nuc Inst Meth A 458:585–592

Boynton WV, Feldman WC, Mitrofanov IG (plus 26 authors) (2004) The Mars Odyssey gamma-ray spectrometer instrument suite. Space Sci Rev 110:37–83

Brückner J, Koerfer M, Waenke H (plus six authors) (1991) Proton-induced radiation damage in germanium detectors. IEEE Trans Nuc Sci 38:209–217

CERN Application Software Group (1993) Detector description and simulation tool. CERN Program Library Long Writeup W5013

Chen H, Awadalla SA, Iniewski K (plus 17 authors) (2008) Characterization of large cadmium zinc telluride crystals grown by traveling heater method. J Appl Phys 103:014903-1–5

Cherepy NJ, Hull G, Niedermayr TR (plus nine authors) (2007) Barium iodide single-crystal scintillator detectors. Proc SPIE 6706:670616-1–7

Dean AJ, Bird AJ, Diallo N (plus five authors) (2003) The modelling of background noise in astronomical gamma-ray telescopes. Space Sci Rev 105:285–376

Diallo N, Collin M, Bersillon O (plus three authors) (1996) Impacts of the shield thickness on the induced background. Proc SPIE 2806:545–550

Fishman GJ, Austin RW (1977) Large-area multi-crystal NaI(Tl) detectors for X-ray and gamma-ray astronomy. Nuc Inst Meth 140:193–196

Gehrels N (1992) Instrumental background in gamma-ray spectrometers flown in low Earth orbit. Nuc Inst Meth A 315:513–528

Goldsten JO, Rhodes EA, Boynton WV (plus 21 authors) (2007) The MESSENGER gamma-ray and neutron spectrometer. Space Sci Rev 131:339–391

Harrison F, Christensen FE, Craig W (plus ten authors) (2005) Development of the HEFT and *NuSTAR* focusing telescopes. Experimental Astron 20:131–137

Hawrami R, Groza M, Cui Y (plus four authors) (2008) SrI_2: a novel scintillator crystal for nuclear isotope identifiers. Proc SPIE 7079:70790Y-1–11

Heynderickx D, Quaghebeur B, Wera J (plus two authors) (2004) New radiation environment and effects models in the European Space Agency's Space Environment Information System (SPENVIS). Space Weather 2:S10S03

Hitomi K, Kikuchi Y, Shoji T, Ishii K (2008) Evaluation of TlBr detectors with Tl electrodes. Proc SPIE 7079:70790J-1–7

Hurford GJ (2010) X-ray imaging with collimators, masks and grids. ISSI SR-009:223–234

Hurley K (1978) Phosphorescence in CsI crystals. Astron Astrophys 69:313–319

Jahoda K, Markwardt CB, Radeva Y (plus five authors) (2006) Calibration of the Rossi X-Ray Timing Explorer Proportional Counter Array. Astrophys J Suppl 163:401–423

Johnson WN, Kinzer RL, Kurfess JD (plus seven authors) (1993) The Oriented Scintillation Spectrometer Experiment — Instrument description. Astrophys J Suppl 86:693–712

Koenen M, Kinzer RL, Kurfess JD (plus two authors) (1995) Radiation damage in large-volume n- and p-type high-purity germanium detectors irradiated by 1.5 GeV protons. IEEE Trans Nuc Sci 42:653–658

Knoll GK (2000) Radiation detection and measurement. 3rd Edition. John Wiley & Sons, Inc, Hoboken

Kurczynski P, Pehl RH, Hull EL (plus 14 authors) (1999) Long-term radiation damage to a spaceborne germanium spectrometer. Nuc Inst Meth A 431:141–147

Kurfess JD, Johnson WN, Kroeger RA (plus four authors) (2004) An advanced Compton telescope based on thick, position-sensitive solid-state detectors. New Astron Rev 48:293–298

Labanti C, Marisaldi M, Fuschino F (plus four authors) (2008) Position sensitive X- and gamma-ray scintillator detector for new space telescopes. Proc SPIE 7021:702116-1–9

Lebrun F, Blondel C, Fondeur I (plus three authors) (1996) ISGRI: a CdTe array imager for *INTEGRAL*. Proc SPIE 2806:258–268

Lechner P, Eckhard R, Fiorini C (plus six authors) (2008) Hard X-ray and gamma-ray imaging and spectroscopy using scintillators coupled to silicon drift detectors. Proc SPIE 7021:702111-1–10

Lei F, Dean AJ, Hills GL (1997) Compton polarimetry in gamma-ray astronomy, Space Sci Rev 82:309–388

Leleux P, Albernhe F, Borrel V (plus 15 authors) (2003) Neutron-induced nuclear reactions and degradation in germanium detectors. Astron Astrophys 411:L85–90

Lonjou V, Roques JP, von Ballmoos P (plus five authors) (2005) Characterization of the in-flight degradation of the *INTEGRAL*/SPI detectors. Nuc Inst Meth A 554:320–330

Luke PN (1988) Low noise germanium radial drift detector. Nuc Inst Meth A 271:567–570

Luke PN (1994) Single-polarity charge sensing in ionization detectors using coplanar electrodes. Appl Phys Lett 65:2884–2886

Mahoney WA, Ling JC, Jacobson AS, Tapphorn RM (1980) The *HEAO-3* gamma-ray spectrometer. Nuc Inst Meth 178:363–381

McGregor DS, He Z, Seifert HA (plus two authors) (1998) Single charge carrier type sensing with a parallel strip pseudo-Frisch-grid CdZnTe semiconductor radiation detector. Appl Phys Lett 72:792–794

Morris DJ, Aarts H, Bennett K (plus seven authors) (1992) Neutron induced background in the COMPTEL detector on the Gamma-Ray Observatory. NASCP 3137:102–108

Natalucci L, Alvarez JM, Barriere N (plus 13 authors) (2008) CdZnTe detector for hard X-ray and low energy gamma-ray focusing telescope. Proc SPIE 7011:70111S-1–10

Naya JE, Jean P, Bockholt J (plus three authors) (1996) The neutron spectrum inside the shielding of balloon-borne Ge spectrometers. Nuc Inst Meth A 368:832–846

Owens A, Brandenburg S, Buis E-J (plus six authors) (2007) An assessment of radiation damage in space-based germanium detectors due to solar proton events. Nuc Inst Meth A 583:285–301

ORTEC division of AMETEK, Inc. (2008) LO-AX™ Low-Energy Photon Detector datasheet, http://www.ortec-online.com/

Pehl RH, Madden NW, Elliott JH (plus three authors) (1979) Radiation damage resistance of reverse electrode Ge coaxial detectors. IEEE Trans Nuc Sci NS 26:321–323

Pirard B, Cabrera J, d'Uston C (plus four authors) (2007) Solar proton damage in high-purity germanium detectors. Nuc Inst Meth A 572:698–707

Raudorf T, Trammell RC, Wagner S, Pehl RH (1984) Performance of reverse electrode HPGe coaxial detectors after light damage by fast neutrons. IEEE Trans Nuc Sci NS 31:253–257

Rehak P, Gatti E, Longoni A (plus five authors) (1985) Semiconductor drift chambers for position and energy measurements. Nuc Inst Meth A 235:224–234

Richter M, Siffert P (2002), High resolution gamma-ray spectroscopy with CdTe detector systems. Nuc Inst Meth A 322:529–537

Rothschild RE, Blanco PR, Gruber DE (plus six authors) (1998) In-flight performance of the high energy X-ray timing experiment on the Rossi X-ray Timing Explorer. Astrophys J 496:538–549

Sangsingkeow P (1999) Recent developments in HPGe material and detectors for gamma-ray spectroscopy. Proc SPIE 3768:204–211

Schönfelder V, Aarts H, Bennett K (plus 27 authors) (1993) Instrument description and performance of the imaging gamma-ray telescope COMPTEL aboard the Compton Gamma-Ray Observatory. Astrophys J Suppl 86:657–692

Schönfelder V, Kanbach G (2010) Imaging through Compton scattering and pair creation. ISSI SR-009:207–222

Seifert H, Teegarden BJ , Cline TL (plus seven authors) (1996) Gamma-ray observations with the Transient Gamma-Ray Spectrometer (TGRS. Astron Astrophys Suppl 120:653–656

Smith DM, Leventhal M, Cavallo R (plus three authors) (1996) All-sky search for transient souces near 0.5 MeV with the Burst and Transient Source Experiment (BATSE). Astrophys J 471:783–795

Smith DM, Lin RP, Turin P (plus 17 authors) (2002) The *RHESSI* spectrometer. Sol Phys 210:33–60

Spieler H (2005) Semiconductor Detector Systems. Oxford University Press, Oxford, New York

Takahashi T, Mitsuda K, Kunieda H (2006) The *NeXT* mission. Proc SPIE 6266:62660D-1–12

Takahashi T, Abe K, Endo M (plus 66 authors) (2007) Hard X-Ray Detector (HXD) on board *Suzaku*. Pub Ast Soc Japan 59:35–51

Tueller J, Krimm HA, Okajima T (plus 12 authors) (2005) InFOCμS hard X-ray imaging telescope. Exp Astron 20:121–129

Ubertini P, Lebrun F, Di Cocco G (plus 17 authors) (2003) IBIS: The Imager on-board *INTEGRAL*. Astron Astrophys 411:L131–139

van den Berg L, Bastian LF, Zhang F (plus two authors) (2007) Fabrication and performance of mercuric iodide pixellated detectors. Proc SPIE 6706:67060J-1–7

Vedrenne G, Kandel B, Albernhe F (plus 25 authors) (1998) The SPI spectrometer for the *INTEGRAL* mission. Physica Scripta 77:35–38

Weidenspointner G, Harris MJ, Sturner S (plus two authors) (2005) MGGPOD: a Monte Carlo Suite for Modeling Instrumental Line and Continuum Backgrounds in Gamma-Ray Astronomy. Astrophys J Suppl 156:69–91

Wilson, CM, van Loef EV, Glodo J (plus seven authors) (2008) Strontium iodide scintillators for high energy resolution gamma ray spectroscopy. Proc SPIE 7079:707917-1–7

Yamaoka K, Arai Y, Doshida T (plus ten authors) (2007) Development of a gamma-ray burst detector based on silicon drift detectors and scintillators. Proc SPIE 6686:66860K-1–9

Zych AD, O'Neill TJ, Bhattacharya D (plus four authors) (2006) TIGRE prototype gamma-ray balloon instrument. Proc SPIE 6319:631919-1–10

— 22 —

Microchannel plates for photon detection and imaging in space

J. GETHYN TIMOTHY[I]

Abstract

This chapter describes the development of the continuous-dynode electron multiplier and its evolution into the microchannel plate (MCP). The fabrication procedures for MCPs are described and the performance characteristics of MCPs and MCP stacks are enumerated. The configurations and performances of imaging MCP electronic readout systems currently in use in space instruments are described in detail. The unique capabilities of MCP electronic systems for fast timing and for time-tag imaging are listed. Finally, the prospects for the development of the next generation of MCPs are briefly outlined.

Background

The conventional photomultiplier tube (PMT), invented in the 1930s, employs a photocathode to convert the detected photon to a photoelectron, and a discrete dynode electron multiplier to amplify the charge of the single photoelectron to the level that can be recorded by commercial electron circuits. Still in wide use today, a number of dynode configurations, e.g., as shown in Figure 22.1, have been developed for PMTs. The discrete dynode electron multiplier is not optimized for space observations. First, the dynode structure has to be ruggedized in order to survive the launch environment. Second, the complex arrangement of the high-voltage dynode electronics does not permit compact packaging. Third, and most important, most of the efficient secondary electron emitting materials used to coat the dynodes cannot survive exposure to air. This prevents the use of the multiplier in an open structure configuration in the windowless region at extreme ultraviolet (EUV) wavelengths below about 116 nm. One exception was the use of a multiplier with high work function beryllium-copper dynodes to study EUV radiation at wavelengths between 10 nm and 80 nm (Piore et al 1952).

The concept of a continuous-dynode electron multiplier was first proposed by Farnsworth (1930), but the fabrication of practical devices was delayed until advances in the understanding of the processes of secondary electron emission and

[I]Nightsen, Inc., Tiverton RI, USA

(a)

(b)

(c)

(d)

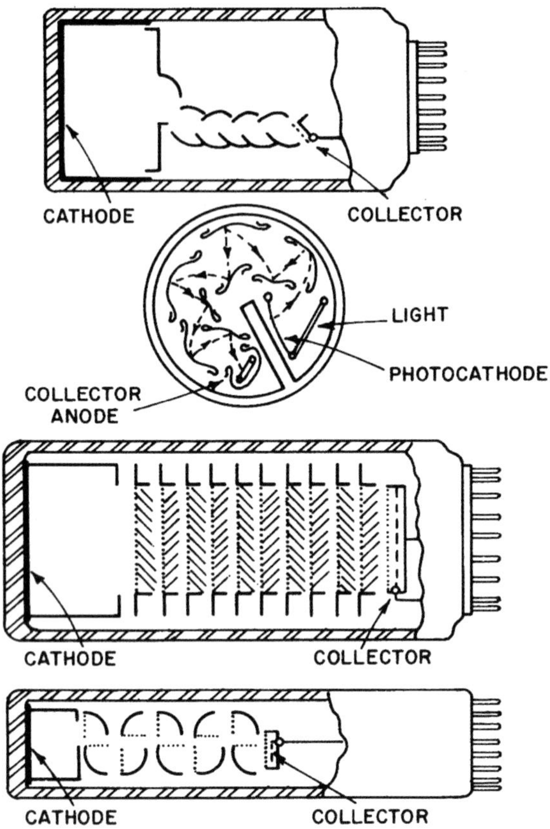

Figure 22.1: Dynode configurations for PMTs: a) linear focused; b) squirrel cage; c) venetian blind; d) box-and-grid (from Young 1974).

in the development of high surface resistance glasses had been made in the 1960s. The first practical continuous-dynode multiplier was the Bendix magnetic electron multiplier (MEM) (Heroux and Hinteregger 1960; Timothy et al 1967). Macar et al (1970) have given a detailed analysis of the electron multiplication process in the MEM.

As shown in Timothy et al (1967), the MEM consists of two glass strips with baked-on high-resistance tin oxide and antimony coatings surrounded by a series of magnets. An electrostatic field is established along the strips by four high-voltage connections. A photoelectron, formed on the opaque photocathode, is accelerated in a trochoidal path along the dynode strip, causing the emission of secondary electrons at each impact. Gains of order 10^6 were obtained with applied potentials of the order of 2000 V, and the MEM operating characteristics remained stable with exposure to the ambient atmosphere. Detection efficiencies of up to 5 % at 60 nm were obtained with opaque tungsten photocathodes.

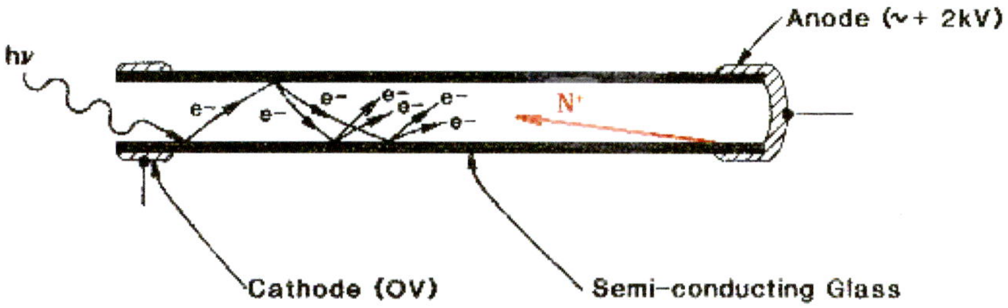

Figure 22.2: Schematic of the CEM.

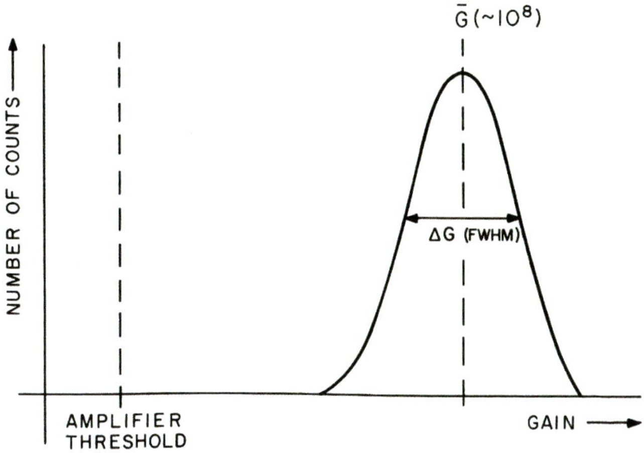

Figure 22.3: Schematic of saturated output pulse height distribution for a curved-channel CEM.

A major advance in the development of continuous-dynode electron multipliers took place in the 1960s with the production of high-resistance surfaces on lead glasses through the process of high-temperature $\approx 300\,^{\circ}$C to $\approx 450\,^{\circ}$C reduction in a hydrogen atmosphere (Blodgett 1951). This led to the independent development of the continuous channel electron multiplier (CEM) by Oschepkov et al (1960) in the Soviet Union, by Goodrich and Wiley (1962) at the Bendix Research Laboratories in the USA, and by Adams and Manley (1966) at the Mullard Laboratories in the United Kingdom and at the Laboratoires d'Electronique et de Physique Appliquée (LEP) in France (both part of the Philips Corporation). The mode-of-operation of the CEM is shown in Figure 22.2. The multiplier is operated under vacuum with a high voltage established along the channel. A high-energy photon or charged particle striking the wall of the channel releases an electron with some initial energy that is accelerated along the channel axis, drifting across to strike the wall with sufficient energy to release secondary electrons. This process is repeated many times with a final output pulse containing up to 10^8 electrons. As the voltage along the channel is increased, the energy of the electrons striking the wall will increase, but

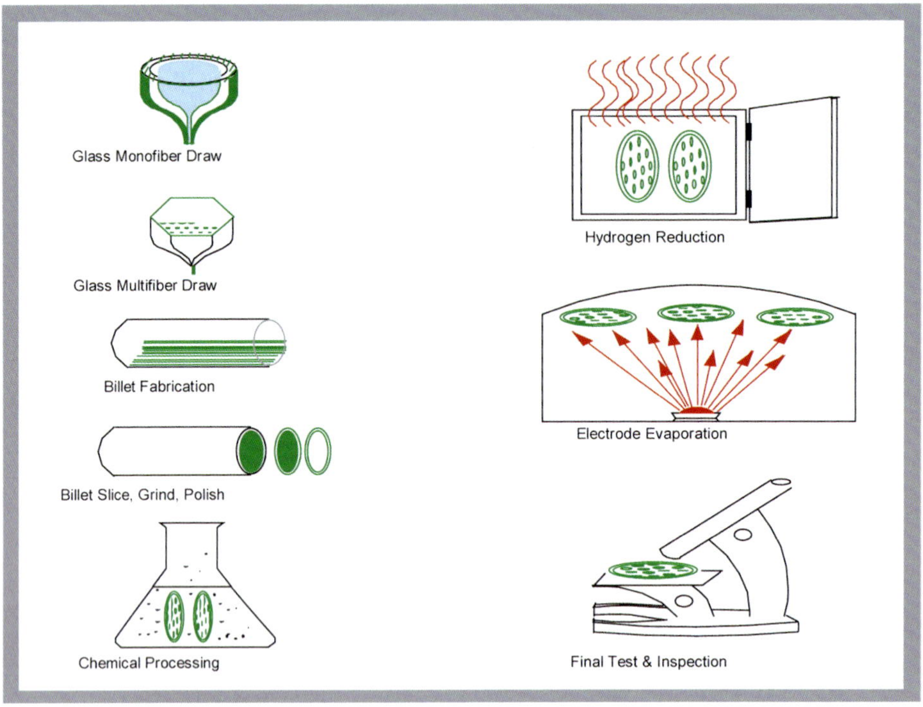

Figure 22.4: Process for fabricating MCPs using etchable core glass (from Laprade et al 1997).

the total number of impacts will decrease. It can thus be expected that the electron gain (amplification) will increase to a maximum value and then start to decrease. Adams and Manley (1966) derived an expression showing the dependence of the gain on the applied voltage as:

$$G = \left(\frac{K\,V_0^2}{4\,V\,\alpha^2}\right)^{4V\alpha^2/V_0} \quad , \tag{22.1}$$

where G is the gain, V_0 is the energy gained by an electron traversing the applied potential difference, V is the initial energy of the secondary electron, α is the length-to-diameter ratio of the channel, K is a constant from the relation $\partial \approx KV_C$, with ∂ the secondary emission coefficient and V_C the collision energy.

However, the experimental gain curve does not show a maximum, as predicted by the model, but continues to much higher gain levels and then starts to saturate. This is because, first, the secondary electrons are not all emitted orthogonally, and, second and more important, the gain is enhanced by ion feedback. As shown in the schematic in Figure 22.2, the voltage that accelerates the electrons down the channel can also accelerate a positive ion, caused by the impact of an electron with a residual gas molecule, back to the input where it can impact the wall and restart the gain process. The level of ion feedback depends both on the ambient pressure and the level of the applied voltage.

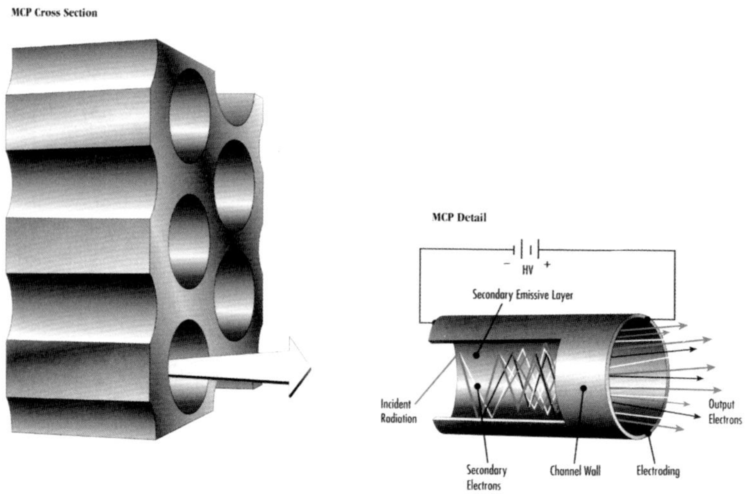

Figure 22.5: Schematic showing mode-of-operation of the MCP (from GEOC, now Photonis USA).

The feedback of positive ions can be prevented by curving the channel, forcing the ions to impact the wall in a short distance compared with the total channel length. Evans (1965) calculated the degree of curvature required to effectively suppress ion feedback and permit stable operation of a CEM at gain levels high enough to permit single photoelectron detection. Eschard and Manley (1971) have reviewed the differences in the performance characteristics of the different types of CEM. The curved-channel CEM is effectively immune to ion feedback when operated under high vacuum at pressures less than about 1.3×10^{-5} hPa. Further, when operated at high voltages (> 2 kV), the distribution of output pulses changes from a negative exponential form to the quasi-Gaussian saturated form shown in Figure 22.3. Evans (1965) has suggested that the saturation is caused by the distortion of the potential gradient within the channel by the deposition of positive charge on the walls at the output end. A CEM with an internal channel diameter of order 1.0 mm and a length-to-diameter ratio of order 100:1 can produce a modal gain $> 10^8$ and a relative resolution (ΔG or R) in the range 35 % to 60 %. The CEM provides a stable photometric response when operated in gain saturation (Timothy and Lapson (1974), and has been used with great success in the open-structure mode to study EUV radiation on a number of sounding rockets and orbital space missions (see, for example, Timothy et al 1975 and Reeves et al 1977).

Microchannel plates

It is important to note from Equation 22.1 that the fundamental electrical characteristics of the CEM depend on the length-to-diameter ratio α of the channel and not on the absolute physical dimensions. The size of the channel can thus be reduced to the limit set by the available glass technology. Further, many channels

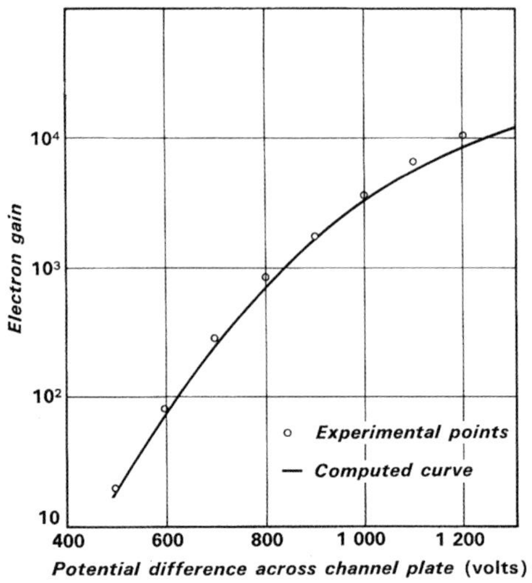

Figure 22.6: Measured and computed gain as functions of voltage for an MCP with a channel length-to-diameter ratio of 40:1 (from Guest 1971).

can be bonded together to produce a detector with an image recording capability. The first microchannel plates (MCP) were assembled by bonding together about five thousand CEMs with ≈ 100 µm channel diameters at the Bendix Research Laboratories (Wiley and Hendee 1962).[1] These early efforts were rapidly replaced by the application of fiber draw techniques, as described by Washington et al (1971).

The first processes utilized the drawing of hollow fibers, or metal core fibers. Subsequently, the process was modified to utilize the drawing of fibers with etchable glass cores. This is the process used to this day in Europe, the US (Wiza 1979), and in the rest of the world. The process sequence is shown in Figure 22.4. A two-part billet, several centimetres in diameter, consisting of an etchable glass core surrounded by a proprietary lead glass cladding is drawn down to a single fiber. About 3×10^3 fibers are then fused together to form a hexagonal multi-fiber billet. This billet is then drawn down again to form hexagonal multi-fibers. These multi-fibers are then fused together to create the MCP boule or billet. The MCP wafers are then produced from this boule by slicing, grinding and polishing. The next steps are the etching out of the core glass and chemical processing of the resulting lead glass interior surfaces of the channels. The semi-conducting surfaces of the channels are then produced by heating the MCP wafer in a reducing atmosphere of hydrogen to yield a very thin free lead layer. Finally the input and output faces of the MCP wafer are electroded, typically with nichrome or inconel, to provide

[1] The progression of MCP manufacturers in Europe and the USA requires clarification in order to avoid confusion. In Europe, Philips closed the Mullard MCP fabrication facility in the UK in 1988 and moved MCP fabrication to Photonis in Brive, France. Photonis became independent of Philips in 1998. In the USA, Bendix Research Laboratories in Ann Arbor, Michigan transferred to Galileo Electro-Optics Corporation (GEOC), Sturbridge, Massachusetts in 1972. In 1999 GEOC was purchased by Burle Industries and became Burle Electro Optics. In 2005 Burle Electro Optics was purchased by Photonis and is currently Photonis USA.

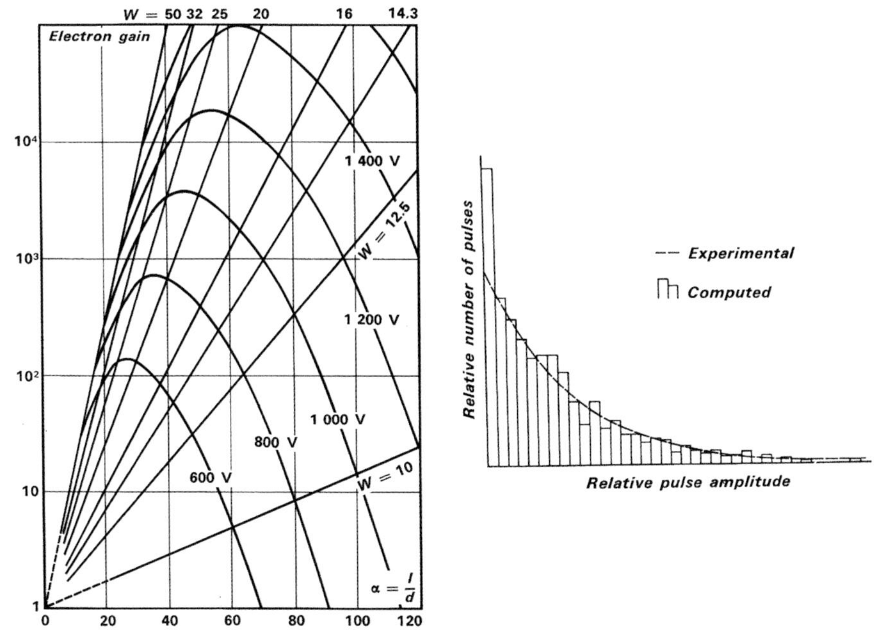

Figure 22.7: a) Universal gain curves for an MCP. $W = V/\alpha$ (from Guest 1971). b) Measured and computed output pulse height distributions for a straight-channel MCP operated at low gain (from Guest 1971).

electrical contacts. The basic structure and mode-of-operation of the MCP is shown in Figure 22.5. Great care must be taken with the fusion of the hexagonal multi-fibers in order to avoid channel distortions at the interfaces. Distortion of the channels at the multi-fiber interfaces can change the length-to-diameter ratio and the gain at a given voltage. This leads to a hexagonal pattern in the flat-field response known as "chicken wire". Recent advances in reducing multi-fiber edge effects have been reviewed by Siegmund et al (2007). Most MCPs have a planar format, although MCPs with a steep curvature to match the focal planes of optical instruments have also been successfully fabricated (see, for example, Siegmund et al 1990).

MCP performance characteristics

The expression for the CEM gain (Equation 22.1) has been further refined by Guest (1971) in a computer model for a channel whose performance is not modified by space charge and wall charging effects, i.e., at gain less than 10^5. This model is in good agreement with the experimental data, as shown in Figure 22.6. Most importantly, the Guest model allows the calculation of a set of universal gain curves for an MCP, as shown in Figure 22.7a. It can be seen that there is an optimum combination of applied voltage V and length-to-diameter ratio α where

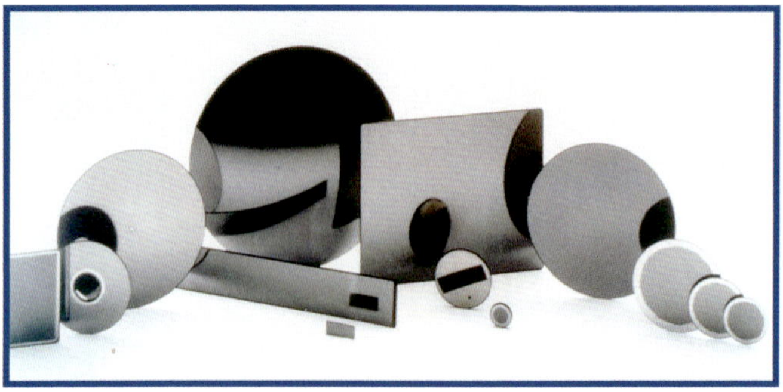

Figure 22.8: Different MCP configurations (from the Burle Electro-Optics, now Photonis USA, catalog). Diameters of the circular MCPs range from ≈ 18 mm to ≈ 125 mm.

slight variations in the length-to-diameter ratio do not significantly affect the gain. Specifically for the optimum length-to-diameter ratio, the ratio W is:

$$W = \frac{V}{\alpha} \approx 22.5 \quad . \tag{22.2}$$

Channel length-to-diameter ratios in the range 40:1 to 60:1 are accordingly used in image intensifier tubes at operating voltages between 1000 V and 1200 V (see, for example, Csorba 1980). At these voltages the MCP produces a negative exponential output pulse height distribution, with most of the pulses at low amplitudes, as shown in Figure 22.7b. This is clearly not optimum for a stable photometric response as a slight reduction in the gain will cause a number of pulses to be lost below the threshold of the electronics, significantly reducing the detective quantum efficiency (DQE). The development of MCPs has been inexorably linked with the development of military night-vision image tubes. Currently MCPs are available from a number of vendors with channel diameters ranging from 25 µm down to 2 µm. Typically the open-area ratio of the channels is of the order of 60 % of the area of the plate. The diameters of circular-format MCPs range from ≈ 18 mm to ≈ 150 mm. MCPs can be obtained in rectangular formats, with holes in the centre for ion detection applications, and with active areas configured to match specific readout systems, moreover with solid rims or rimless (see Figure 22.8). Catalogs of MCP configurations can be obtained from, amongst others, www.photonis.com, www.burle.com, and www.sales.hamamatsu.com. The best MCPs for high-resolution imaging are the 18 mm, 25 mm, and 40 mm formats, since these are fabricated using boule fusion techniques rather than the block fusion techniques employed for larger formats. Typically, the channels in these MCPs have bias angles with respect to the input face of the plate in the range 0° to 20°.

The maximum gain that can be realized without significant ion feedback in a straight-channel MCP is of the order 10^3 to 10^4. These gain levels are not high enough for electronic readout systems. In order to obtain higher gains without ion feedback, alternative MCP configurations must be employed. Three types of high-gain MCP configurations are currently in use, as shown in Figure 22.9. The first type is the "chevron" MCP stack shown in Figure 22.9a. The performance

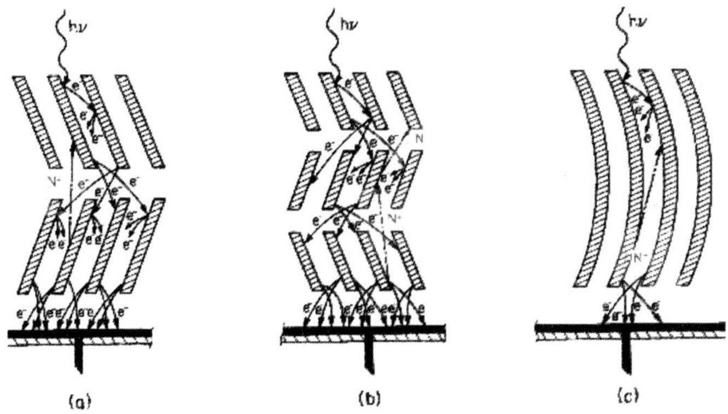

Figure 22.9: High-gain MCP configurations. a) Chevron MCP stack. b) Z-plate MCP stack. c) Curved-channel C-plate MCP.

characteristics of this configuration have been reported by Wiza (1979). In the chevron MCP stack two straight-channel MCPs with channel bias angles typically in the range from 8° to 10° are mounted in sequence with the bias angles set so that positive ions are trapped at the interface between the two MCPs. An improvement on the chevron MCP stack is the "Z-plate" MCP stack shown in Figure 22.9b. In the "Z-stack" three MCPs with matched resistances are mounted in a butt-faced configuration with the positive ions trapped at the two MCP interfaces (see, for example, Siegmund et al 1994). The Z-stack provides a significantly better reduction of ion feedback than the chevron MCP stack and is the most-used high-gain MCP configuration at this time.

The third type of high-gain MCP configuration is the curved-channel MCP or C-plate MCP shown in Figure 22.9c. The channels in this MCP are curved in a manner analogous to that of the curved-channel CEM. Washington (1971) stated that for a curvature sufficient to mask the output 30 % of the channel would be required to effectively suppress ion feedback to the channel input. The first curved-channel MCPs with the output end strongly curved ("J-plates") were fabricated by LEP in France and Mullard in the UK (see, for example, Boutot et al 1974). Typical length-to-diameter ratios for C-plate MCPs lie in the range 80:1 to 120:1. A single C-plate MCP can be operated at high gain $> 10^6$.

C-plate MCPs with, first, J-sections, and later C-sections were successfully fabricated in the US by GEOC starting in the early 1970s (Timothy and Bybee 1977a; Timothy 1981).

Typical relative resolutions for chevron MCP stacks range from ≈ 120 % to ≈ 60 % at gains around 10^7 and for Z-plate MCP stacks from ≈ 60 % to ≈ 35 % at gains around 10^8. A single C-plate MCP produces resolutions ranging from ≈ 50 % to ≈ 35 % at gains around 10^6. Examples of the saturated output pulse height distributions from a Z-stack MCP and a C-plate MCP are shown in Figure 22.10. The single C-plate MCP is the mechanically most stable of the three configurations, and produces the least spreading of the output charge cloud. It also produces the lowest level of ion feedback. However, C-plate MCPs have proved extremely difficult

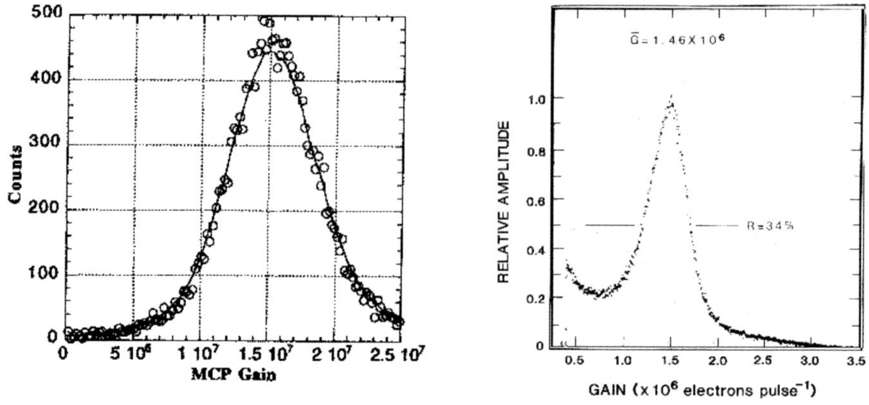

Figure 22.10: High-gain MCP output pulse-height distributions. a) Z-plate MCP stack (from Siegmund et al 1994). b) C-plate MCP (from Timothy 1981).

to manufacture. Corbett et al (1992) have described some of the more critical problems. The result is that C-plates have only been produced for space missions with adequate resources, such as the *HST*, and C-plates are not commercially available at this time. The best high-gain MCP configuration available at this time is the Z-stack (Siegmund et al 1995).

Image intensifier tubes

Thousands of MCPs are produced each month for use in image intensifier tubes for military and civilian night-vision applications. Csorba (1985a) has described the different types of image intensifier tube developed to date. The development of the second generation (Gen II) wafer tube (see Figure 22.11), employing an MCP to amplify the internal electron current, revolutionized passive night vision. Earlier tube configurations (Gen 0 and Gen I), which employed a photocathode and accelerated the photoelectron directly to a phosphor, yielded amplifications of the order of 1000. In contrast, the Gen II wafer tube with an MCP provides amplifications of the order of 10 000 to 20 000.

The emphasis in night vision is on the response to red and near-infrared wavelengths, since starlight has a predominantly red spectrum. For many years photocathode materials were developed empirically by trial and error, as described by Sommer (1980). The Cs-Na-K-Sb trialkali (S-20) multialkali photocathode was found to have the best red response and is typically used in the Gen II tubes. In the 1970s negative-affinity photocathodes were developed from physical principles (Spicer 1975). In particular, the GaAs photocathode, doped with extremely thin layers of Cs and O, was found to have an extremely high response at red and near-infrared wavelengths. This photocathode is used in the state-of-the-art Gen III image tube that produces amplifications of the order 30 000 to 50 000.

Other photocathode materials can be employed in the image intensifier tubes for scientific applications. At shorter wavelengths the S-11 bialkali photocathode

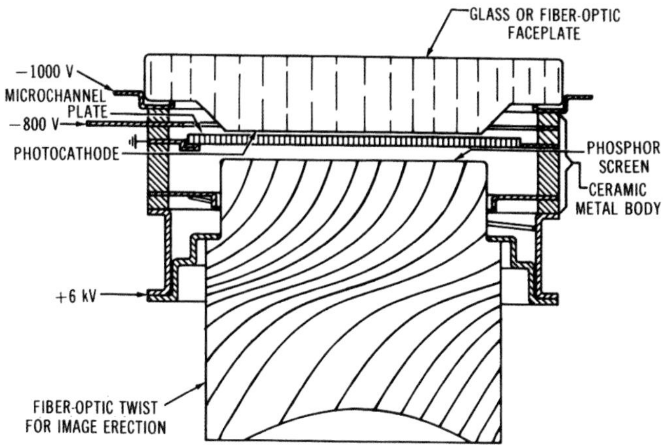

Figure 22.11: Structure of Generation II night-vision wafer tube employing an MCP intensifier (from Csorba 1985a).

provides a strong blue response, and at vacuum ultraviolet (VUV) wavelengths the Cs-Te photocathode provides a strong response down to about 170 nm, with a high degree of solar blindness at wavelengths longer than about 350 nm. All of these semi-transparent photocathode materials must be remotely processed in order not to contaminate the MCP with Cs, and must be mounted on a window in proximity focus with the MCP input.

At wavelengths below about 170 nm, high work function opaque photocathode materials are available which can be deposited directly on the front face of the MCP. Opaque CsI deposited on the front face of the MCP provides a strong response at wavelengths below about 180 nm. However, CsI is hygroscopic and the quantum efficiency degrades on exposure to air. The DQEs of image tubes with MgF_2 windows and semi-transparent Cs-Te and opaque CsI photocathodes are shown in Figure 22.12.

At shorter wavelengths higher work function photocathode materials are available which are stable on exposure to air. Materials such as opaque MgF_2, KBr, RbBr, and CsBr, deposited on the front face of an MCP provide high DQEs at wavelengths progressively shorter than about 150 nm (Siegmund et al 1987; Tremsin and Siegmund 2005).

The output phosphor screen in a typical night-vision tube is a P-20, which produces a green image to which the human eye is most responsive. Other phosphor screens are available, as described by Csorba (1985b).

Electronic readout systems for the MCP image intensifier tubes are required for use in space. The modifications to the tubes required to use CCD or CMOS-APS electronic readout systems are described in Chapter 25 (Schühle 2010). Direct electronic readout systems for high-gain MCPs are described in the next section of this chapter.

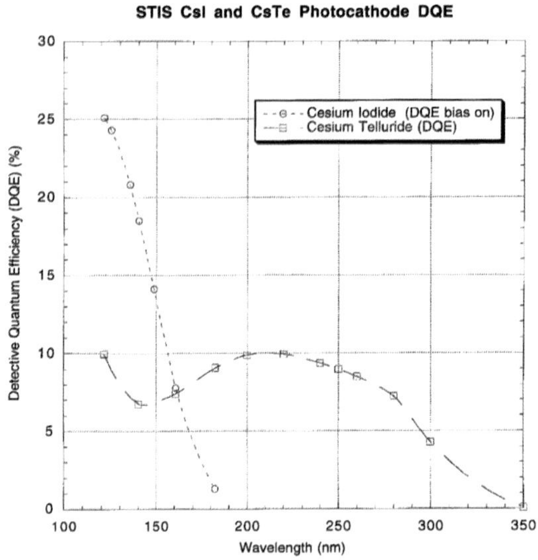

Figure 22.12: DQEs of image tubes developed for the Space Telescope Imaging Spectrograph (STIS) with MgF_2 windows and, respectively, a semi-transparent Cs-Te photocathode and an opaque CsI photocathode.

Electronic readout systems

Three classes of electronic readout systems are available for use with high-gain MCPs. The first class is the discrete anode arrays in which each pixel electrode is connected to its own amplifier/discriminator and counting circuit. The second class is the analog readout arrays, which employ charge-division or timing-discrimination techniques to determine the coordinates of the detected photon. The third class is the digital readout arrays, which employ specially configured anodes to determine the coordinates of the detected photon.

In the discrete anode readout arrays the spatial location and resolution are defined by the geometry of the pixel electrodes, as shown in Figure 22.13. As an example, a special double linear array was developed for the spectrometer on the *SPARTAN 201* mission (Kohl et al 1994), as shown in Figure 22.14. Linear arrays of 1×160 pixels and area arrays of 10×10 pixels have been used with 25 mm format and 40 mm format high-gain MCPs. Discrete anode MCP photomultiplier tubes are commercially available (see, for example, www.burle.com). There are local and global dynamic range limitations with all three classes of electronic readout systems for high-gain MCPs. The local dynamic range is set by the output count rate capability of the MCP. The modal gain will start to fall when the avalanche current reaches about 10 % to 20 % of the wall current, as shown in Figure 22.15. Reducing the resistance of the channels can increase the dynamic range. A C-plate MCP with the characteristics of that shown in Figure 22.15, but with a resistance of only 26 MΩ, produced a linear resistance to output count rates greater than

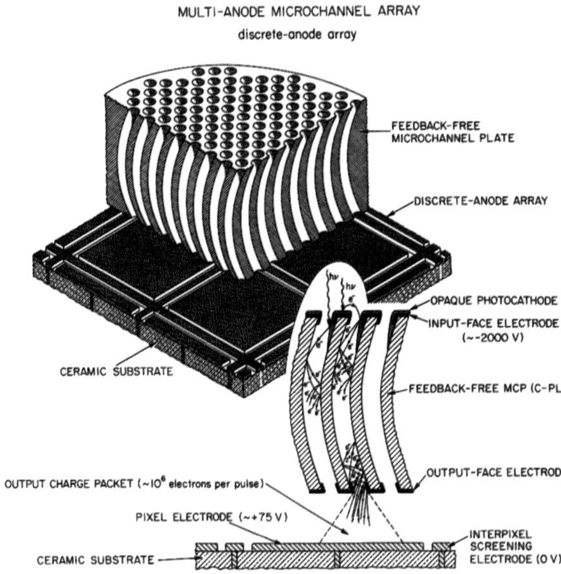

Figure 22.13: Schematic of a discrete anode MCP readout array. A C-plate is shown but the array can also employ chevron or Z-stack MCPs (from Timothy and Bybee 1977b).

10^5 $\mathrm{mm^{-2}\,s^{-1}}$. Small-pore MCPs will typically have the largest dynamic range, albeit at slightly lower gain levels.

Similar gain reductions are observed with chevron and Z-stack MCPs, but are mitigated by charge spreading into adjacent channels in the second and third MCPs. This however leads to a scene-dependent dynamic range reduction when observing bright objects covering many channels in the first MCP, because of the overlapping charge clouds. Wilhelm et al (1997) have discussed this effect in the context of observing strong solar emission lines with the SUMER instrument on the *SOHO* mission (see Chapter 5, Timothy et al 2010). There is a limit to the conductivity of an MCP, since the lead glass has a negative temperature coefficient of resistance (see Timothy 1981). Too low a resistance will result in thermal runaway from ohmic heating, leading to the destruction of the MCP.

The number of pixels in a discrete anode array is limited to several hundred both by the connector technology of the tubes and the complexity and power requirements of the associated electronics. Much larger array formats are required for imaging and spectroscopy, particularly in space at wavelengths below ≈ 300 nm. A series of analog and digital systems for encoding the position of the detected photon have been developed to meet this requirement.

The first of the analog systems is the resistive anode encoder (RAE) or RANICON (Lampton and Paresce 1974; Wiza 1979). The readout array is a resistive anode with readouts at either end for a linear array, and at each corner for an area array. The position of the detected photon is determined by the proportion of the charge collected at each end, or by pulse rise time discrimination. In order to avoid the inherent distortions in the two-dimensional RAE, a circular arc-terminated

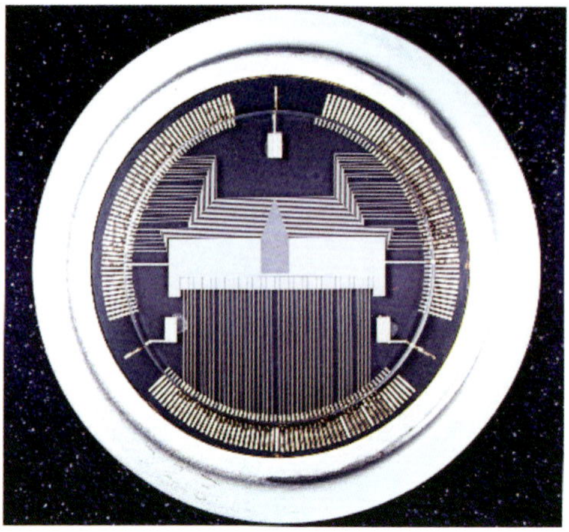

Figure 22.14: Discrete linear anode arrays of 1×48 pixels and 1×42 pixels used to measure the solar H Ly α line profile and intensity (see Kohl et al 1994). The 1×48 array is ≈ 3.3 mm long and the 1×42 array is ≈ 24 mm long.

resistive anode is employed, as described by Lampton and Carlson (1979). Early RANICONs yielded formats of $\approx 100 \times 100$ pixels, pixel dimensions of order 100 μm or better, and total count rates of a few hundred counts per second with MCP gains in excess of about 5×10^6. Later, PHOTICON tubes with bialkali, trialkali, or GaAs photocathodes, and a stack of five MCPs produced formats of 1024×1024 pixels, pixel dimensions of $\approx 70\ \mu m^2$, and total count rates of a few hundred kilohertz (Clampin and Paresce 1989). The RAE was used as the soft X-ray detector (Mason et al 1984) on the *EXOSAT* mission (de Korte et al 1981).

An analog readout array that eliminates the inherent resistive noise in the RANICON readout array is the wedge-and-strip array (WSA) (Martin et al 1981). A four-electrode WSA is shown in Figure 22.16. Three and five electrode patterns can also be used. Siegmund et al (1986) have described the operating characteristics of the WSA readout systems. Large-format WSA detectors with more than 512 $\times$ 512 pixels have been successfully used on the *EUVE* mission (see Chapter 5 of this volume). Siegmund et al (1984) have described the configuration of these detectors, and Vallerga et al (1994) have described the long-term on-orbit performance. Siegmund et al (1985) have described the performance characteristics of the large-format Z-stack MCPs used with these detectors.

A variation of the WSA, namely the one-dimensional spiral anode (SPAN) detector (Breeveld et al 1992) was used in the CDS instrument on the *SOHO* mission (see Chapter 5 of this volume).

A two-dimensional crossed-wire anode array has been developed for the *Einstein* (Giacconi et al 1979) and *AXAF* (now *Chandra*) (Weisskopf et al 2002) X-ray observatories, and has also been used on the *ROSAT* mission (Trümper 1984). Schematics of configuration of the High Resolution Imager (HRI), and of one dimension of the decoding scheme are shown in Figure 22.17 (Kellogg et al 1976). The performances of the HRI on *ROSAT* and on *Chandra* have been given, respectively, by Zombeck et al (1995), and by Murray et al (2000).

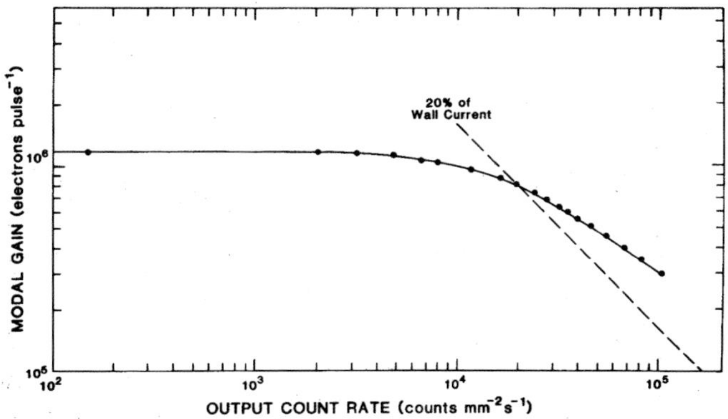

Figure 22.15: Modal gain as a function of count rate per unit area for a C-plate MCP with 25 µm channels, active area 25 mm in diameter and a resistance of 300 MΩ (from Timothy 1981).

The characteristics of the *Chandra* HRI are a format of 1024 × 1024 pixels, a spatial resolution ≈ 20 µm FWHM (set by the MCP pore size), and a total output count rate > 150 s^{-1}. The most sophisticated analog MCP readout systems available today are the double delay line (DDL) and crossed delay line (XDL) (Siegmund et al 1993 and 1994), and the crossed strip (XS) detector (Siegmund et al 2001). XDL detectors with formats of 360 × 1024 pixels were flown in the UVCS and SUMER instruments on the *SOHO* mission (Siegmund et al 1994). Pixel dimensions were ≈ 25 µm FWHM and the global counting rates were stable to > 4 × 10^5 s^{-1}. XDL arrays with formats of ≈ 2200 × 2200 resolution elements are flying on the *GALEX* mission (Siegmund et al 1999).

Larger-format XDL arrays, optimized for spectroscopy, with ≈ 9000 × 200 resolution elements have been flown on the *FUSE* mission. These arrays employ longer helical DDL anodes deposited on two sides of a substrate (see Siegmund et al 1997). An upgraded version of this detector, the COS instrument, was installed in the *HST* in May 2009 (McPhate et al 2000).

The advantage of the XS array over the DDL and XDL arrays is the ability to operate at MCP gain levels ≈ 5×10^6 rather than ≈ 10^8, albeit with low global count rates (≈ 10 kHz). New readout techniques with the potential for global count rates ≈ 1 MHz, have recently been described by Tremsin et al (2006). The XS detector should have the first flight test in the J-PEX sounding rocket instrument (Berendse et al 2006). The cross strip (XS) anode array employs a coarse (≈ 0.5 mm) multilayer metal and ceramic pattern that encodes event positions by direct sensing of the charge on each strip and determination of the charge cloud centroid for each event (see Figure 22.19). The first class of digital readout arrays is the coded anode converter (CODACON). The simplified schematic of a one-dimensional CODACON readout array is shown in Figure 22.20. A charge pulse on a charge spreader causes charge to be induced in the code tracks. Differential amplifiers connected to each bit track pair determine which side of a pair has the most charge. The CODACON thus employs a 2^n encoding scheme where n is the number of code track pairs.

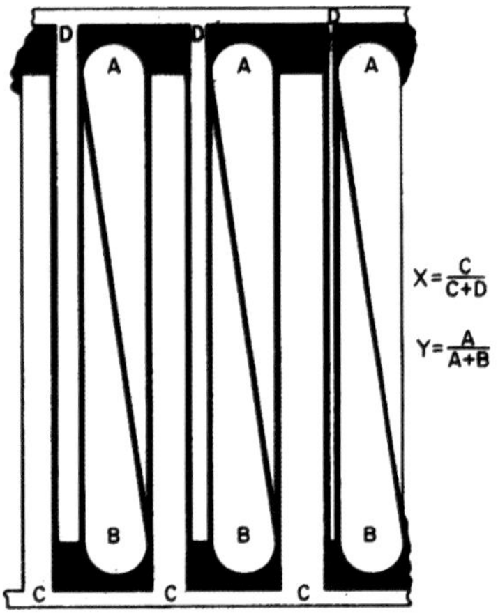

Figure 22.16: Four-electrode anode pattern for a WSA (from Martin et al 1981).

Both one- and two-dimensional CODACONs have been built employing versions of Gray's binary code, as described by Lawrence and McClintock (1996) and the references therein. A 64 × 1024 pixel CODACON with a global count rate of 10^5 s^{-1} has been flown in the Ultraviolet Imaging Spectrograph on the *Cassini* mission to Saturn (McClintock et al 1993).

A light-weight ultraviolet imaging spectrograph has been developed for remote sensing of planetary atmospheres, based on the *Cassini* instrument and using a light-weight CODACON tube assembly together with FPGA electronics (McClintock and Lawrence 1996).

The second class of digital readout arrays is the multi-anode microchannel array (MAMA). The MAMA employs $2(n+1)$ electrodes to uniquely define $n(n+2)$ position locations in each axis. Thus $32+34$ electrodes uniquely define 32×34 or 1088 position locations. Details of the MAMA encoding scheme have been given by Slater et al (1990) and by Timothy (1994). The schematics of a coincidence-anode MAMA detector are shown in Figure 22.21. The simplified block diagram of a two-dimensional system is shown in Figure 22.21a, and the detection and encoding of multi-fold events, where the charge pulse stimulates two or more electrodes (the position of an event on a single electrode cannot be determined), is shown in Figure 22.21b.

The ability to discriminate between the detection of events on odd and even numbers of anode electrodes permits the spatial resolution to be doubled, as described by Kasle and Morgan (1991) and by Kasle and Horch (1992). The imaging MAMA detector designed for the STIS instrument on *HST* (see Figure 22.22) can accordingly be used as a 1024 × 1024 pixel array, or as a high-resolution 2048 × 2048 array. The spatial resolution is set by the pore size of the channels in the single C-plate MCP, namely 12 µm diameter channels on 15 µm centres.

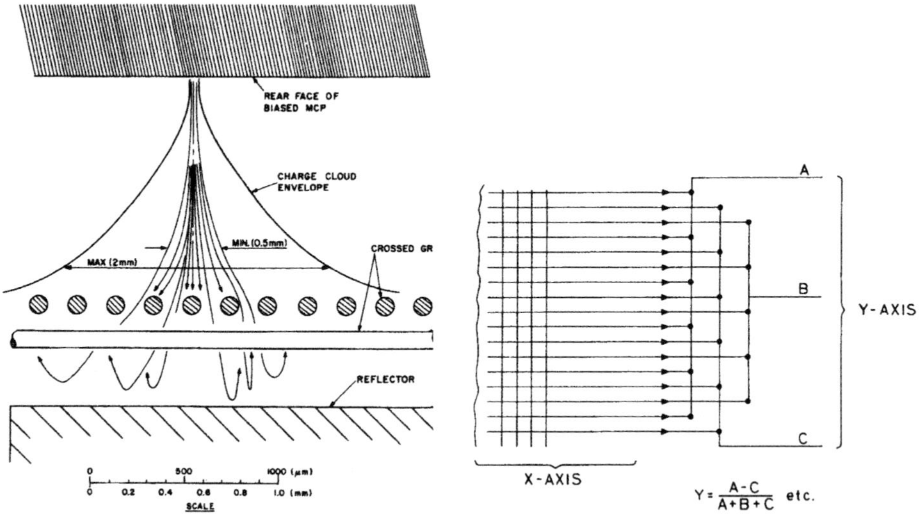

Figure 22.17: Schematics of the HRI readout system (from Kellogg et al 1976).
a) Configuration of the two-dimensional array. b) Schematic of one dimension of the position encoding technique.

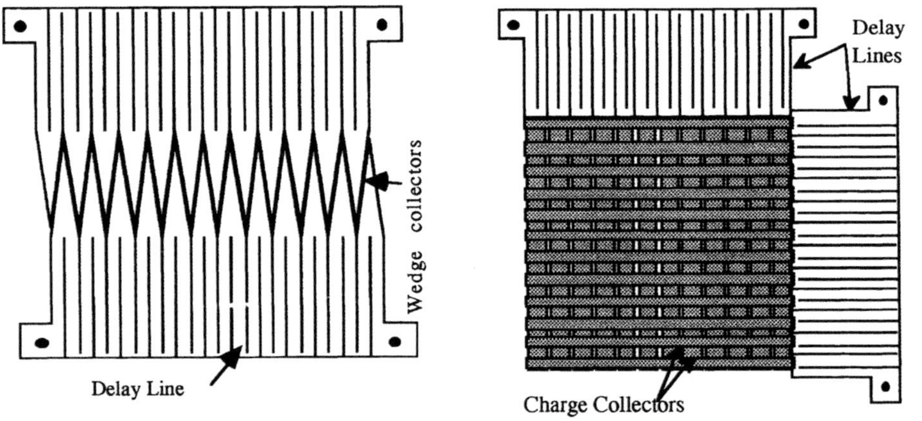

Figure 22.18: Schematics of DDL and XDL arrays (from Siegmund et al 1993). a) DDL anode with wedge charge collectors and external delay lines. b) XDL with external delay lines and crossed finger charge collectors.

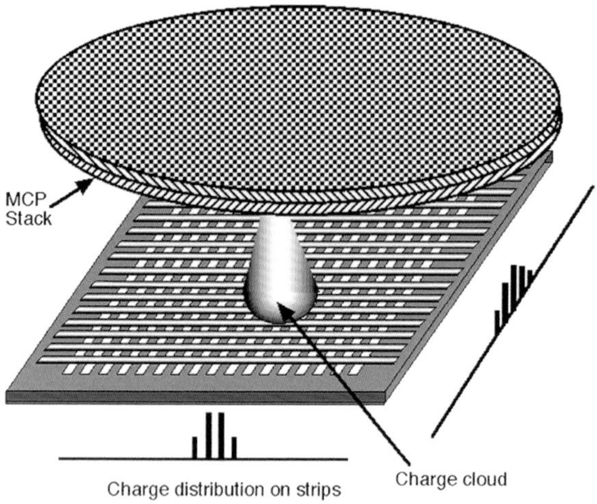

Figure 22.19: Schematic of XS anode array (from Siegmund et al 2001).

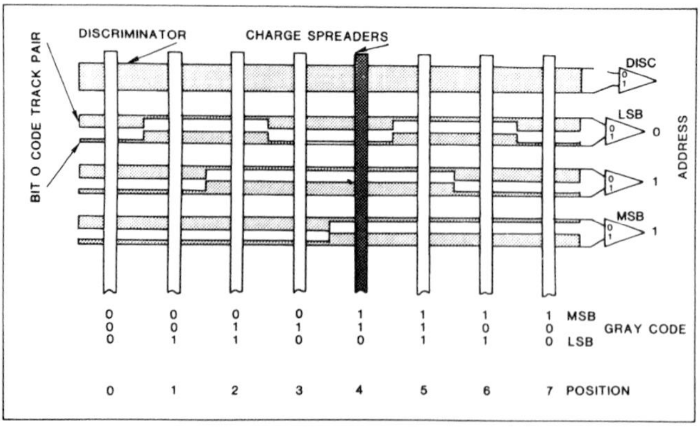

Figure 22.20: Simplified example of an eight-channel one-dimensional CODACON array (from McClintock et al 1982).

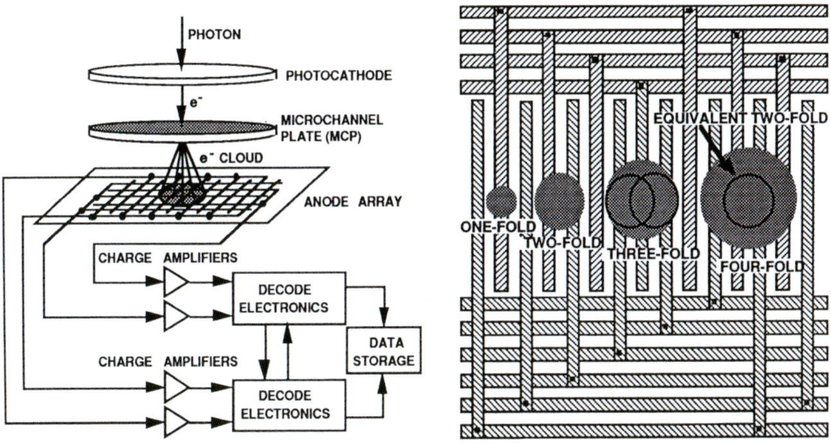

Figure 22.21: MAMA detector system schematics (from Kasle and Morgan 1991). Left: System block diagram. Right: One-dimensional fine-fine encoding scheme showing detection of multi-fold events.

Figure 22.22: 1024 × 1024 pixel (2048 × 2048 pixel high resolution) MAMA array with 25 µm by 25 µm pixels (12.5 µm by 12.5 µm pixels high resolution) for the STIS instrument on *HST*.

Two MAMA tubes are used in STIS; an NUV tube with a semi-transparent Cs-Te photocathode on an MgF_2 window, and an FUV tube with an opaque CsI photocathode and an MgF_2 window (see Figure 22.22). Details of the performances of these detectors have been given by Kimble et al (1998 and 1999). The detectors worked flawlessly from launch in 1997 until the STIS electronics failure in 2004. Further, the flight spare FUV MAMA is currently in use in the solar-blind channel of the ACS instrument on *HST*, and the flight spare NUV MAMA has been installed in the COS instrument on-orbit in *HST* in May 2009 (see Chapter 5, Timothy et al 2010).

Fast timing and time-tag imaging

The transit time of electrons in a CEM or MCP depends linearly on the channel length, and the transit time jitter is proportional to the transit time for a given applied voltage and length-to-diameter ratio (see Wiza 1979). CEMs and conventional PMTs typically produce pulses having a FWHM in the range from 5 ns to 30 ns. By comparison MCPs with channel diameters less than 12 µm produce pulses having an FWHM less than 1 ns. For example, a chevron MCP with 5 µm diameter channels produces a pulse width of order 0.5 ns (Laprade et al 1996). This gives the MCP unique capabilities for fast timing applications, including time of flight (TOF) mass spectrometry (see, for example, Roman et al 2008). The result of this characteristic is that all of the electronic readout systems described here can record both the coordinates and the time of arrival of each detected photon to an accuracy set by the pulse-pair resolution of the electronics. This capability, which is unique to the MCP detectors, has a number of applications in space science. First, the integration time of the binned and time-tagged photons can be selected after the observations have been completed. Second, with a position reference, blurring of the image due to motion can be corrected. Third, frequency analyses of observed sources can be determined, and for pulsed sources the spectra of the different pulses can be selected and analyzed. The examples given here are for the MAMA detector systems, but are equally applicable to all of the described systems. An example of the laboratory reconstruction of a moving image is shown in Figure 22.23.

This technique was used to correct the jitter and gyro drift of a sounding rocket during FUV imagery of the galaxy NGC 6240 using a 256×1024 pixel MAMA detector system, by using star images within the field of view as references (Smith et al 1992). The raw and corrected images can be seen in Morgan et al (1990). An excellent example of the ability to select the optimum integration time with time-tag imaging is speckle interferometry, as described by Horch et al (1992). Some of the results of this technique are described by Horch et al (1999). The ultimate application for this technique is the ability to determine the spectra of the different pulses from a pulsed source. A specific example of this is the recording of the FUV spectra of the Crab pulsar pulses over the wavelength range from 160 nm to 320 nm using STIS (Gull et al 1998), as shown in Figure 22.24.

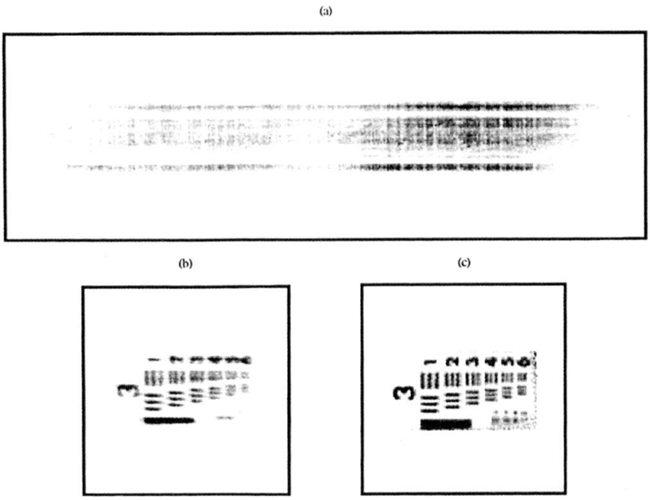

Figure 22.23: Image reconstruction of a moving image (from Giaretta et al 1993). a) Integrated image made from the raw time-tagged data of a portion of an Air Force test target while it was moving across the detector. b) Reconstructed image made by dividing the time-tagged data of the moving target into 0.1 s frames and centroiding each frame before co-adding. c) Flat-fielded image of the stationary Air Force target.

The future

Over 30 years ago conventional PMTs were considered, at best, obsolescent, and more likely obsolete. Yet today, thousands of these devices are in use. Similarly today, the major emphasis is on solid-state devices such as the CCDs and CMOS arrays (see Chapter 23, Waltham 2010). Yet the unique capabilities of MCPs and MCP stacks, plus their vast use in night-vision applications and particle detection (Wüest et al 2007), suggest that science-quality lead glass MCPs will be available for a long time to come.

The principal development needs are, first, to replace the lead glass, which does not have optimum thermal properties, and, second, to develop efficient solar-blind photocathodes such as GaN and diamond. The development of silicon-based MCPs using lithographic techniques, which eliminate the multi-fiber errors has been in progress for some time (see, for example, Siegmund et al 2004). However it has not yet proved possible to obtain adequate gain from these MCPs. In parallel, the development of MCPs fabricated from aluminum oxide is also under way (Drobychev et al 2006).

Whatever the results of these developments, there will always be a requirement in UV and X-ray astrophysics instruments for solar-blind, photon-counting imaging detector systems, which are radiation tolerant and do not require cryogenic cooling.

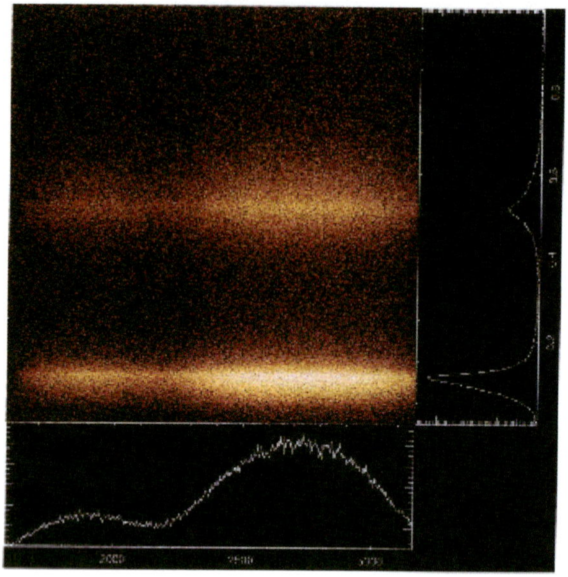

Figure 22.24: Crab pulsar profile spectra from 160 nm to 320 nm (from Gull et al 1998). The data do not indicate any wavelength dependence of the pulse profile.

Notes

The references to commercial organizations in this chapter are for technical purposes only and do not represent any endorsement on the part of the author, the editors, or the publisher.

Bibliography

Adams J, Manley BW (1966) The mechanism of channel electron multiplication. IEEE Trans Nucl Sci NS-13:88–99

Berendse FB, Cruddace RG, Kowalski MP (plus eight authors) (2006) The Joint Astrophysical Plasmadynamic Experiment extreme ultraviolet spectrometer: resolving power. Proc SPIE 6266:62660V-1–62660V-11

Blodgett KB (1951) Surface conductivity of lead silicate glass after hydrogen treatment. J Am Ceram Soc 34:14–27

Boutot JP, Eschard G, Polaert R, Duchenois V (1974) A microchannel plate with curved channels: an improvement in gain, relative variance and ion noise for channel plate tubes. Proc of Sixth Symposium on Photo-electronic Image Devices, London, pp 103–111

Breeveld AA, Edgar ML, Smith A (plus two authors) (1992) A SPAN MCP detector for the *SOHO* Coronal Diagnostics Spectrometer. Rev Sci Instrum 63:673–676

Clampin M, Paresce F (1989) Photon-counting imaging with a GaAs photocathode: evaluation of the Red-RANICON for astronomical imaging. Astron Astrophys 225:578–584

Corbett MB, Feller WB, Laprade BN (plus four authors) (1992) Development Efforts to Improve Curved-channel Microchannel Plates. Proc SPIE 1764:240–251

Csorba IP (1980) Current gain parameters of microchannel plates. Appl Opt 19:3863–3866

Csorba IP (1985a) Phosphor screens, in image tubes. Howard W Sams & Co, Inc, (Publishing Subsidiary of ITT), 29–42

Csorba IP (1985b) Image intensifier tube types, in Image Tubes, Howard W Sams & Co, Inc, (Publishing Subsidiary of ITT), 174–193

De Korte PAJ, Bleeker JAM, den Boggende AJF (plus six authors) (1981) The X-ray imaging telescopes on *EXOSAT*. Space Sci Rev 30:495–511

Drobychev G, Barysevich A, Delendik K (plus four authors) (2006) Development of micro-channel plates on the basis of aluminum oxide. Nucl Instr Meth A 567:290–293

Evans DS (1965) Low energy charged-particle detection using the continuous-channel electron multiplier. Rev Sci Instrum 36:375–382

Eschard G, Manley BW (1971) Principles and characteristics of channel electron multipliers. Acta Electronica 14:19–39

Farnsworth PT (1930) Electron multiplier, US Patent number 1 969 399

Giacconi R, Branduardi G, Briel U (plus 28 authors) The *Einstein (HEAO 2)* X-ray observatory. Astrophys J 230:540–550

Giaretta G, Horch E, Timothy JG, Heanue JF (1993) Time-tag photon detection with the MAMA detector system. Proc SPIE 1952:316–328

Goodrich GW, Wiley WC (1962) Continuous channel electron multiplier. Rev Sci Instrum 33:761–762

Guest AJ (1971) A computer model of channel multiplier plate performance. Acta Electronica 14:79–97

Gull TR, Lindler DJ, Crenshaw DM (plus eight authors) (1998) Space telescope imaging spectrograph near-ultraviolet time-tagged spectra of the Crab pulsar. Astrophys J 495:L51–54

Heroux L, Hinteregger HE (1960) Resistance strip magnetic electron multiplier for the extreme ultraviolet. Rev Sci Instrum 31:280–286

Horch E, Morgan JS, Giaretta G, Kasle DB (1992), A new speckle interferometry system for the MAMA detector. Pub Astron Soc Pacific 104:939–948

Horch E, Ninkov Z, van Altena WF (plus three authors) (1999) Speckle observations of binary stars with the WIYN telescope I. Measures during 1997. Astronom J 117:548–561

Kasle DB, Morgan JS (1991) High resolution decoding of multi-anode microchannel array detectors. Proc SPIE 1549:52–58

Kasle DB, Horch EP (1992) Performance of high resolution decoding with multi-anode microchannel array detectors. Proc SPIE 1764:202–216

Kellogg E, Henry P, Murray S, Van Speybroek L (1976), High-resolution imaging X-ray detector. Rev Sci Instrum 47:282–290

Kimble RA, Woodgate BE, Bowers C (plus 50 authors) (1998) The on-orbit performance of the Space Telescope Imaging Spectrograph, Proc SPIE 3356:188–202

Kimble RA, Abraham J, Argebright VS (plus 15 authors) (1999) In-flight performance of the MAMA detectors on the Space Telescope Imaging Spectrograph. Proc SPIE 3764:209–225

Kohl JL, Gardner LD, Strachan L, Hassler DM (1994) Ultraviolet spectroscopy of the extended solar corona during the *SPARTAN* 201 Mission. Space Sci Rev 70:253–261

Lampton M, Paresce F (1974) The Ranicon: A resistive anode image converter. Rev Sci Instrum 45:1098–1105

Lampton M, Carlson CW (1979) Low-distortion resistive anodes for two-dimensional position-sensitive MCP systems. Rev Sci Instrum 50:1093–1097

Laprade B, Dykstra M, Langevin F (1996) The development of an ultra small pore microchannel plate for space sciences applications. (1996) Proc SPIE 2808:72–85

Laprade B, Cochran RC, Langevin F, Dykstra MW (1997) Characterization of an ultra small pore microchannel plate. Proc SPIE 3173:474–485

Lawrence GM, McClintock WE (1996) Compact ultraviolet imaging microchannel plate detectors using CODed Anode Convertor (CODACON) readout systems. Proc SPIE 2831: 104–111

Macar PJ, Rechavi J, Huber MCE, Reeves EM (1970) Solar-blind photoelectric detection systems for satellite applications. Appl Opt 9:581–593

Martin C, Jelinsky P, Lampton M, Malina RF (1981) Wedge-and-strip anodes for centroid position sensitive photon and particle detectors. Rev Sci Instrum 52:1067–1074

Mason IM, Braduardi-Raymont G, Culhane JL (plus three authors) The *EXOSAT* imaging X-ray detectors. (1984) IEEE Trans Nucl Sci NS-31:795–800

McClintock WE, Barth CA, Steele RE (plus two authors) (1982) Rocket-borne instrument with a high resolution microchannel plate detector for planetary UV astronomy. Appl Opt 21:3071–3079

McClintock WE, Lawrence GM, Kohnert RA, Esposito LW (1993) Optical design of the Ultraviolet Imaging Spectrograph for the *Cassini* mission to Saturn. Opt Engineering 32:3038–3046

McClintock WE, Lawrence GM (1996) Low-mass, Low-power Ultraviolet Telescope-Imaging Spectrograph for Planetary Atmosphere Remote Sensing. Proc SPIE 2807:256–266

McPhate JB, Siegmund OH, Gaines G (plus two authors) (2000) The Cosmic Origins Spectrograph FUV detector. Proc SPIE 4139:25–33

Morgan JS, Timothy JG, Smith AM (plus two authors) (1990) High time-resolution imaging with the MAMA detector systems, Proc SPIE 1235:347–357

Murray SS, Austin G, Chappell J (plus seven authors) In-flight performance of the *Chandra* High Resolution Camera. Proc SPIE 4012:68–80

Oschepkov PK, Skvortsov BN, Osanov BA, Sibrikov I (1960) Application of a continuous secondary electron multiplication for amplifying small currents (Translation). Pribory Tekh Eksper 4:89–91

Piore ER, Harvey GG, Gyorgy EM, Kingston RH (1952) A high vacuum recording spectrograph for the study of radiation from solids in the 100–800 Å range. Rev Sci Instrum 23:8–12

Reeves EM, Huber MCE, Timothy JG (1977) Extreme UV spectroheliometer on the Apollo Telescope Mount. Appl Opt 16:837–857

Roman PA, Brickerhoff WB, Getty SA (plus five authors) (2008) A miniature MEMS and NEMS enabled time-of-flight mass spectrometer for investigations in planetary science. Proc SPIE 6959:6959G-1–13

Schühle U (2010) Intensified solid state sensor cameras: ICCD and IAPS. ISSI SR-009:419–428

Siegmund OHW, Malina RF, Coburn K, Werthimer D (1984) Microchannel plate EUV detectors for the Extreme Ultraviolet Explorer. IEEE Trans Nucl Sci NS-31:776–779

Siegmund OHW, Coburn K, Malina RF (1985) Investigation of large format microchannel plate Z configurations. IEEE Trans Nucl Sci NS-32:443–447

Siegmund OHW, Lampton M, Bixler J (plus two authors) (1986) Operational characteristics of wedge and strip readout systems. IEEE Trans Nucl Sci 33:724–727

Siegmund OHW, Everman E, Vallerga JV (plus two authors) (1987) Ultraviolet quantum detection efficiency of potassium bromide as an opaque photocathode applied to microchannel plates. Appl Opt 26:3607–3614

Siegmund OHW, Cully SL, Gaines GA (plus three authors) (1990) Highly curved microchannel plates, EUV, X-ray and Gamma-ray instrumentation for astronomy. Proc SPIE 1344:346–354

Siegmund OHW, Gummin M, Stock J (plus three authors) (1993) High resolution monolithic delay line readout techniques for two dimensional microchannel plate detectors. Proc SPIE 2006:176–186

Siegmund OHW, Gummin M, Stock J (plus ten authors) (1997) Performance of the double delay line microchannel plate detectors for the Far Ultraviolet Spectroscopic Explorer. Proc SPIE 3114:283–294

Siegmund OHW, Stock JM, Marsh DR (plus 17 authors) (1994) Delay line detectors for the UVCS and SUMER instruments on the *SOHO* satellite, EUV, X-ray, and Gamma-ray instrumentation for astronomy. Proc SPIE 2280:89–100

Siegmund OHW, Gummin MA, Sasseen T (plus eight authors) (1995) Microchannel Plates for the UVCS and SUMER instruments on the *SOHO* satellite. Proc SPIE 2518:344–355

Siegmund OHW, Jelinsky P, Jelinsky S (plus six authors) (1999) High-resolution cross delay line detectors for the GALEX mission. Proc SPIE 3765:429–440

Siegmund OHW, Tremsin AS, Vallerga JV, Hull J (2001) Cross strip imaging anodes for microchannel plate detectors. IEEE Trans Nucl Sci 48:430–434

Siegmund OHW, Vallerga JV, McPhate J, Tremsin AS (2004) Next generation microchannel plate detector technologies for UV Astronomy. Proc SPIE 5488:789–800

Siegmund OHW, Vallerga J, Tremsin A, McPhate J (2007) Microchannel plates: Recent advances in performance. Proc SPIE 6686, web reference 66860W

Slater DC, Timothy JG, Morgan JS, Kasle DB (1990) Imaging MAMA Detector Systems. Proc SPIE 1243:35–49

Smith AM, Hill RS, Vrba FJ, Timothy JG (1992) Far-ultraviolet MAMA detector imagery and emission-line CCD imagery of NGC 6240. Astrophys J 391:L81–84

Sommer AH, Photoemissive Materials (1980) Robert E Krieger Publishing Company, Huntington, New York

Spicer WE (1975) Negative electron affinity photocathode research and technology. IEEE Electron Devices 21:57–59

Timothy AF, Timothy JG, Willmore AP (1967) The performance of open structure photomultipliers in the 1100-Å to 250-Å wavelength region. Appl Opt 6:1319–1326

Timothy JG, Bybee RL (1977a) Preliminary results with microchannel plates employing curved microchannels to inhibit ion feedback. Rev Sci Instrum 48:292–299

Timothy JG, Bybee RL (1977b), Multi-anode microchannel arrays. Proc SPIE 116:24–32

Timothy JG, Lapson LB (1974) Use of channel electron multipliers as secondary standard detectors at EUV wavelengths. Appl Opt 13:1417–1430

Timothy JG, Reeves EM, Chambers RM (plus two authors) (1975) A sounding rocket spectroheliometer for photometric studies at extreme ultraviolet wavelengths. Space Sci Instrum 1:23–49

Timothy JG (1981) Curved-channel microchannel array plates. Rev Sci Instrum 52:1131–1142

Timothy JG (1994) Recent advances with the MAMA detector systems. Proc SPIE 2278:134–137

Timothy JG, Wilhelm K, Xia L (2010) The extra-terrestrial vacuum-ultraviolet wavelength range. ISSI SR-009:89–112

Tremsin AS, Siegmund OHW (2005) The quantum efficiency and stability of UV and soft X-ray photocathodes. Proc SPIE 5920:1–13

Tremsin AS, Siegmund OHW, Vallerga JV, Hull JS (2006) Novel high resolution readout for UV and X-ray photon counting detectors with microchannel plates. Proc SPIE 6276:627616-1–627616-11

Vallerga JV, Eckert M, Sirk M (plus two authors) Long-term orbital performance of the MCP detectors aboard the Extreme Ultraviolet Explorer. Proc SPIE 2280:57–68

Waltham N (2010) CCD and CMOS sensors. ISSI SR-009:391–408

Washington D (1971) Improvements in or relating to Electron Multipliers. UK Patent 1352732

Washington D, Duchenois V, Polaert R, Beasley RM (1971) Technology of channel plate manufacture. Acta Electronica 14:201–224

Weisskopf MC, Brinkman B, Canizares C (plus three authors) (2002) An overview of the performance and scientific results from the *Chandra* X-Ray observatory. Publ Astron Soc Pacific 114:1–24

Wiley WC, Hendee CF (1962) Electron multipliers utilizing continuous strip surfaces. IRE Trans Nucl Sci NS-9:103–106

Wilhelm K, Lemaire P, Curdt W (plus 16 authors) (1997) First Results of the SUMER Telescope and Spectrometer on *SOHO*; I. Spectra and Spectroradiometry. Sol Phys 170:75–104

Wiza JL (1979) Microchannel plate detectors. Nucl Instrum Meth 162:587–601

Wüest M, Evans DS, McFadden JP (plus seven authors) (2007) Review of instruments. ISSI SR-007:11–101

Young AT (1974) Photomultipliers: Their cause and cure, in Methods of Experimental Physics 12 (ed. N Carleton) Academic Press, New York and London, 1–94

Zombeck MV, David LP, Harnden FR Jr, Kearns K (1995) Orbital performance of the high resolution imager on *ROSAT*. Proc SPIE 2518:304–320

— 23 —

CCD and CMOS sensors

NICK WALTHAM[I]

Abstract

The charge-coupled device (CCD) has been developed primarily as a compact image sensor for consumer and industrial markets, but is now also the preeminent visible and ultraviolet wavelength image sensor in many fields of scientific research including space-science and both Earth and planetary remote sensing. Today's scientific or science-grade CCD will strive to maximise pixel count, focal plane coverage, photon detection efficiency over the broadest spectral range and signal dynamic range whilst maintaining the lowest possible readout noise. The relatively recent emergence of complementary metal oxide semiconductor (CMOS) image sensor technology is arguably the most important development in solid-state imaging since the invention of the CCD. CMOS technology enables the integration on a single silicon chip of a large array of photodiode pixels alongside all of the ancillary electronics needed to address the array and digitise the resulting analogue video signal. Compared to the CCD, CMOS promises a more compact, lower mass, lower power and potentially more radiation tolerant camera.

The charge-coupled device

The concept of the charge-coupled device (CCD) emerged from the search for a silicon-based electrical equivalent of the magnetic bubble memory (Boyle and Smith 1970). In its simplest implementation, the CCD structure consists of a series of closely spaced electrodes separated from an underlying semiconductor substrate by a thin insulating oxide layer (Figure 23.1a). When a bias voltage is applied to an electrode, a depletion region is formed in the semiconductor immediately beneath it. The depletion region is in effect a potential well which can store an electrical charge packet. By pulsing the electrodes in an appropriate sequence the potential well, and hence its charge packet, can be transferred through the semiconductor (Figure 23.1b). A shift register can be formed by adding circuits for the insertion and detection of charge packets.

Although the CCD was originally conceived as a device to store digital information, it was evident that because a potential well could store variable quantities of charge it could also convey analogue signals. As a result, the CCD concept has

[I]Space Science and Technology Department, Rutherford Appleton Laboratory, Harwell Science and Innovation Campus, Didcot, UK

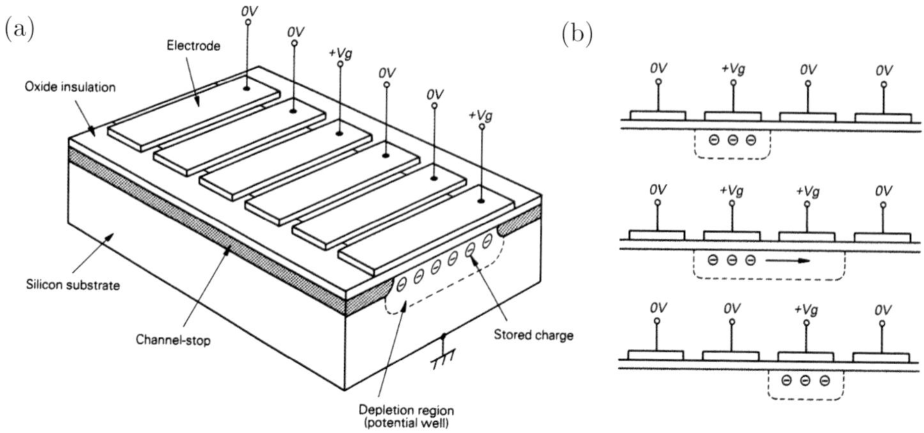

Figure 23.1: (a) Charge-coupled device structure. (b) Charge transfer operation.

been applied to analogue signal processing functions including simple delay lines and transverse filters. Without question, however, the greatest impact of the CCD has been in the application of solid-state image sensors. Today the CCD appears throughout the consumer, industrial, medical, security and scientific imaging sectors. Only the relatively recent emergence of CMOS sensors has challenged the CCD in the consumer markets of mobile telephones and digital cameras where miniaturisation and lower power consumption are key requirements. This begs the question of why the CCD has become the image sensor of choice for nearly all visible light imaging systems and in particular scientific applications. There already exist a number of texts that describe the concepts, physics and operation of the basic CCD structure, including its variations and refinements for particular applications — see for example: Séquin and Tompsett (1975), Beynon and Lamb (1980), Janesick (2001) and Holst and Lomheim (2007). Only a brief overview can be presented here, summarising the types of arrays and technologies that have been developed and optimised for scientific imaging applications and in particular space instrumentation.

Prior to the development of the CCD, electronic imaging relied mostly on the use of camera tubes. As a solid-state device the CCD offered the immediate attractions of compactness, ruggedness and low-voltage operation. Of importance is that silicon is very responsive to visible light. At the peak of its sensitivity a back-illuminated CCD is capable of absorbing and sensing very nearly all incident photons. With appropriate fabrication and optimisation, silicon can also respond well to extreme ultraviolet light and soft X-rays. In its simplest form, the basic structure of a CCD image sensor is formed from an array of electrodes running orthogonally to a series of isolated charge transfer channels (Figure 23.2). The electrodes are usually connected together in groups of two, three or four phases. Biasing of the electrode phases creates an array of isolated potential wells, or pixels, that collect the photon-generated electrons. Following an exposure, the imaging area electrodes

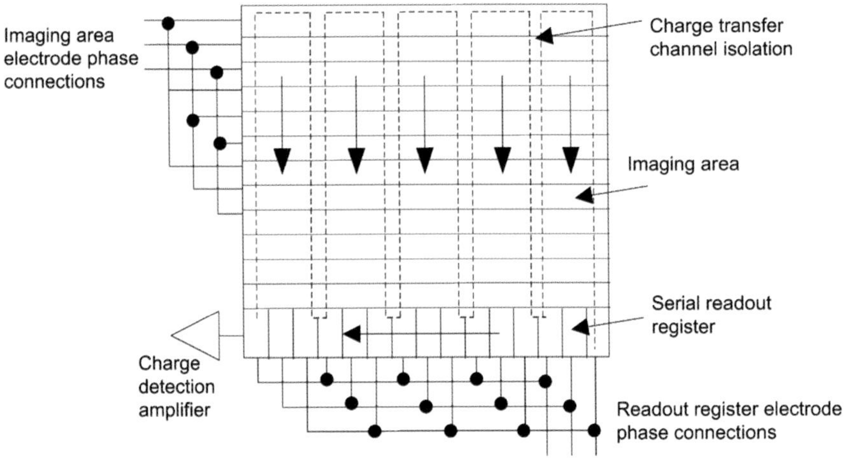

Figure 23.2: Full-frame CCD array architecture.

are pulsed or "clocked" to transfer the integrated image charge pattern down the array one line at a time. The lowest line is transferred into a serial readout register that runs orthogonally to the imaging area transfer channels or columns. This register is clocked separately and at a higher rate allowing each pixel to be read out sequentially through a charge detection amplifier. Once the complete line has been read out the imaging area electrodes can be clocked again so that the next image line is transferred into the readout register. This sequence is repeated until all image lines have been read.

The CCD architecture described above is commonly referred to as a full-frame CCD array, the most popular for scientific imaging applications in space and ground-based astronomy. However, an obvious limitation is that the image will be smeared if the CCD remains exposed while being read out, particularly if the frame readout time is a significant fraction of the exposure time. A common solution is to incorporate a mechanical shutter in front of the CCD so that an integrated image can be read out in darkness. An alternative solution is to add a light-shielded storage array to which the integrated charge pattern can be transferred rapidly at the end of the exposure period. The storage array can then be read out while a new image accumulates in the imaging array. There are two basic formats: the frame-transfer CCD (Figure 23.3a) and the interline-transfer CCD (Figure 23.3b).

In the frame-transfer CCD the storage array is added beneath the imaging array. This effectively doubles the size of the silicon chip and is unattractive in commercial cost-sensitive applications. In the interline-transfer CCD the storage array is incorporated as light-shielded columns adjacent to the imaging columns. The weakness in this approach is that the sensitivity is effectively halved. Unsurprisingly, this architecture has seen little interest in the scientific community but rapidly gained popularity in the digital camera and camcorder markets. Two further CCD architectures applicable to Earth and planetary remote sensing are the linear CCD (Figure 23.4a), and the time-delay-integration (TDI) CCD (Figure 23.5a). The linear CCD consists of a row of photodiodes from which photon-generated charge is

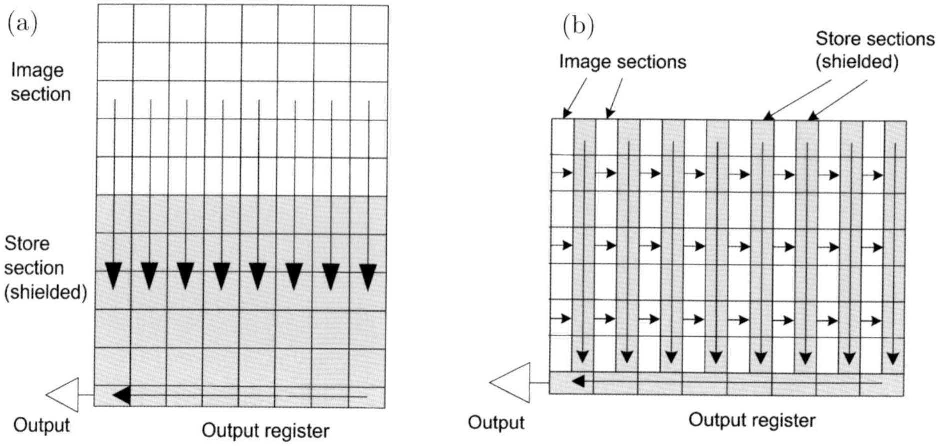

Figure 23.3: (a) Frame-transfer CCD architecture. (b) Interline-transfer CCD architecture.

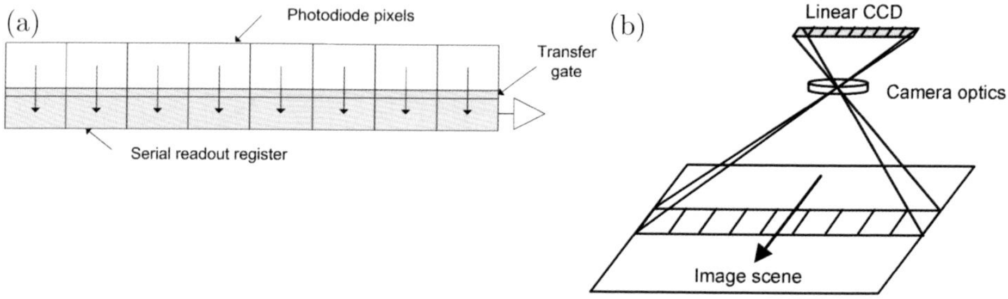

Figure 23.4: (a) The linear CCD. (b) Push-broom imaging.

transferred into a serial readout register. The acquisition of a two-dimensional image relies on the sensor being scanned across the target scene. Linear CCDs with 12 000 pixels or more are readily available. The technique of push-broom imaging is used to acquire very high resolution imagery (Figure 23.4b).

The TDI CCD is advantageous when there is insufficient illumination to obtain a useful signal from a conventional linear array. Its architecture is essentially the same as a full-frame CCD array except that the x-y image format is typically very much larger in x than in y (Figure 23.5a). To avoid image smear, push-broom imaging requires the integrating CCD charge packets to be clocked down the array at the same rate as the projected image scans over the array (Figure 23.5b). The TDI provides gain because the target scene is now integrated over the number of lines in the array. TDI CCDs have been used successfully to obtain very high resolution imagery of Mars (Ebben et al 2007).

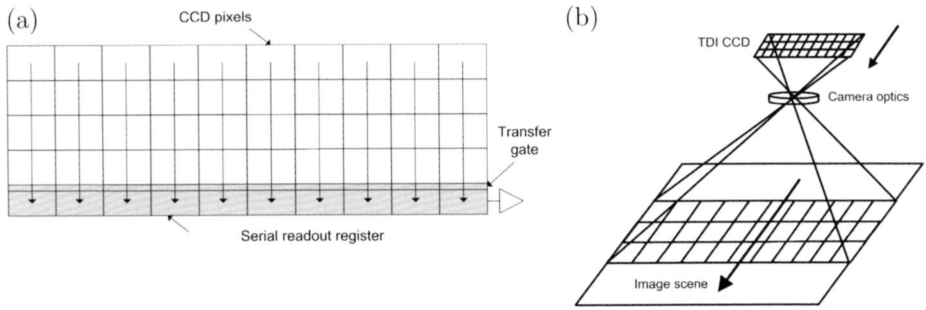

Figure 23.5: (a) The TDI CCD. (b) TDI CCD push-broom imaging.

The scientific CCD

CCD fabrication technologies have undergone continuous development to improve performance. The development of large-format CCD arrays for scientific applications was driven largely by the aspirations of the ground-based astronomical community and the requirement for a solid-state image sensor to replace the vidicon camera tube in space applications. In the late 1970s a number of astronomical research groups developed camera systems using the first commercially available sensors from Fairchild — see for example: Marcus et al (1979) and Leach et al (1980). At the same time the CCD was being proposed for what were to become NASA's *HST* and the *Galileo* mission to Jupiter. The earliest sensors, for example the Fairchild CCD201, had just 100 pixels × 100 pixels. RCA rose to the challenge of designing a CCD with 320 pixels × 512 pixels, the minimum needed to meet the requirements of the US TV standard. Texas Instruments collaborated with JPL under NASA sponsorship to develop arrays of 400 pixels × 400 pixels and later 800 pixels × 800 pixels — the dawn of the scientific CCD. This quest for a large-area, high-sensitivity and low-noise solid state image sensor resulted in many of the developments now seen in today's scientific or science-grade CCD catalogue. The scientific CCD typically has several million pixels, high quantum efficiency over a broad spectral range, low readout noise and wide dynamic range.

Charge-transfer efficiency

The first significant development in CCD fabrication addressed the poor charge-transfer efficiency (CTE) of the simple surface-channel CCD architecture illustrated in Figure 23.1a. Interface "traps" at the silicon-silicon dioxide interface were found to absorb and release charge with differing time constants, resulting in the smearing of charge packets being transferred through them. The CTE was found to be so poor as to make large-format image sensors impractical. The problem was overcome with the introduction of the buried-channel CCD in which an additional layer of n-type silicon is formed at the top of the p-type substrate and immediately below the silicon dioxide insulating layer. The n-type silicon forces the potential wells to form deeper within the substrate and away from the traps at the silicon-silicon dioxide

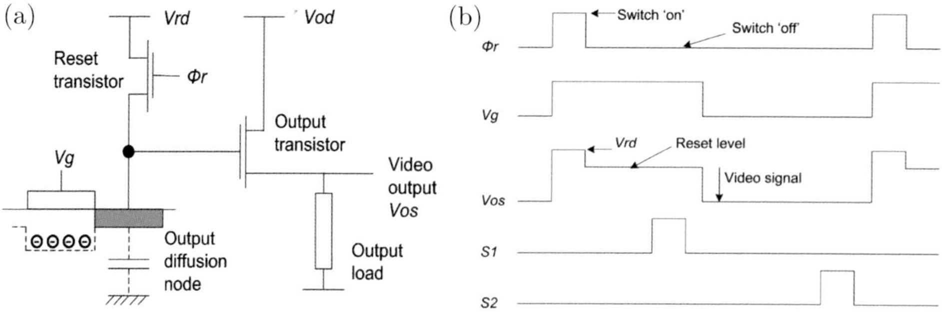

Figure 23.6: (a) CCD charge detection amplifier. (b) Correlated double sampling.

interface. Signal charge can now be transferred through the substrate without it coming into contact with the traps. Transfer efficiencies better than 99.999 % and as much as 99.9999 % are now achieved, enabling thousands of transfers without significant signal loss or smear.

Readout noise

Another important and on-going development has been to reduce readout noise. The output circuit of a CCD is a charge detection amplifier consisting of an output diffusion which has an associated parasitic node capacitance, a reset transistor to recharge the node capacitance and an output transistor operating as a source-follower to sense voltage variations on the node (Figure 23.6a). Pixel readout requires that the output node capacitance is pre-charged to a "reset" potential ($V_{\rm rd}$) by pulsing the reset transistor "on" and "off" with a reset clock ($\emptyset_{\rm r}$) as illustrated in Figure 23.6b. In practice there is a small discharge of the output node resulting from feed-through of the reset clock as it falls low. Signal charge from the last gate of the serial readout register ($V_{\rm g}$) is then transferred onto the output node resulting in further discharge. The output transistor presents a video output signal ($V_{\rm os}$). As the reset clock falls low, Johnson noise from the channel resistance of the reset transistor is frozen on the output node capacitance (C) and results in RMS pixel-to-pixel reset voltage fluctuations of $\sqrt{k_{\rm B}T/C}$. This noise is commonly referred to as $k_{\rm B}TC$ noise, or reset noise, and would be the dominant source of CCD readout noise without a cancellation technique known as correlated double sampling (CDS). CDS acquires two samples of the CCD output, one before ($S1$), and one after ($S2$) the signal charge has been transferred onto the output node (Figure 23.6b). Subtraction of the two samples cancels the $k_{\rm B}TC$ noise as it is remains "correlated" or unchanged between the two samples. After reset-noise cancellation, transistor noise in the source-follower output transistor becomes the dominant noise source. The resultant readout noise will depend on the signal measurement bandwidth and thus the CCD readout rate. Higher readout rates require greater bandwidth in the readout electronics and result in higher readout noise. The best of today's CCDs achieve a readout noise of $\approx$ 6 electrons RMS when read at $\approx$ 1 MHz, or as little as $\approx$ 2 electrons RMS when read at $\approx$ 100 kHz. A compromise is required

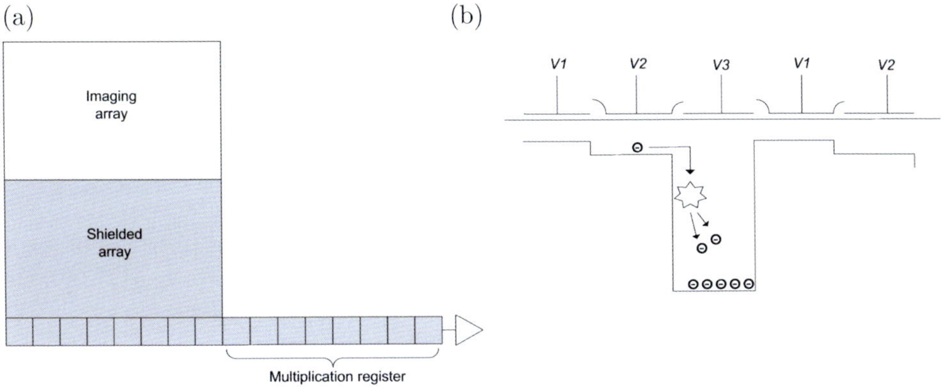

Figure 23.7: (a) The electron multiplication CCD. (b) Avalanche breakdown gain.

between a low readout rate to minimise readout noise and a high readout rate to minimise the overall frame readout time. High frame readout rate is clearly desirable to maximise observing efficiency. One solution to the dilemma is to add additional readout amplifiers to the CCD array such that a number of sub-sections can be read in parallel. Low readout rate for minimal noise is now possible without compromising observing efficiency. An added benefit is that the additional readout amplifiers provide a degree of redundancy in the case of a failure of a single readout amplifier. A disadvantage, however, is that additional calibration of the data is required to correct for gain mismatch between the multiple readout amplifiers.

Electron multiplication readouts

To reduce readout noise still further, two CCD manufacturers have developed electron multiplication readouts to provide on-chip gain (Hynecek 2001). A multiplication register of typically several hundred stages is added to the end of the serial readout register (Figure 23.7a). One electrode phase in the multiplication register is clocked at significantly higher voltage (typically up to 40 V in amplitude) which enables a probability of avalanche multiplication (Figure 23.7b). The word "probability" is key as the net gain of a single stage might only be ≈ 1.01 but after 500 stages the net gain is 1.01^{500} or ≈ 145. Because the gain is applied before the charge detection output amplifier, the effective readout noise is the output noise divided by the gain. This enables very low readout noise to be maintained at the high readout rates of TV cameras (≈ 10 MHz).

The electron multiplication CCD provides a solid-state equivalent of an intensified camera tube and with sufficient gain a solid-state photon counting detector. Not to be overlooked is an additional noise, introduced by the discrete nature of the multiplication process, which effectively increases the shot noise by a factor $\sqrt{2}$ (Robbins and Hadwen 2003). The net gain is highly dependent on the amplitude of the high-voltage clock and so appropriate control of its stability is necessary. Despite these and further issues concerned with spurious charge and ageing of the gain stage, the electron multiplication CCD provides a new and powerful tool for low-light imaging.

Spectral range

A third continuing development of the CCD has been the challenge to maximise its sensitivity over the widest spectral range possible. The first CCDs were "front-illuminated" requiring the incident photons to penetrate through semi-transparent polysilicon electrodes to reach the underlying substrate. Unfortunately, the absorption depth of silicon is such that a large fraction of the light is absorbed within the electrodes, particularly at the shorter "blue" wavelengths. On the other hand, there is a greater chance that the longer "red" and near-infrared wavelength photons will penetrate deeply into the silicon substrate below the depletion regions. Electrons from these photons are free to drift laterally in a field-free region with the result that image resolution is degraded or blurred. A solution is to manufacture the CCD on epitaxial silicon which consists of a thin layer of the nominally-doped silicon, typically 10 µm to 20 µm thick, on top of a very highly-doped bulk substrate. Electrons created in the bulk substrate are likely to recombine before they drift to the depletion regions. Image resolution is maintained but at the cost of sensitivity to "red" wavelengths. To solve the lack of sensitivity to "blue" wavelengths researchers and manufacturers started to develop the thinned "back-illuminated" CCD. This relies on the bulk of the silicon substrate being removed by chemical etching, and the CCD being illuminated through the back surface rather than through the polysilicon electrodes. In principle, the sensitivity, or quantum efficiency (QE), is very much increased. However, a bare untreated silicon surface will contain a high density of recombination centres or trapping sites. Any photoelectrons which make contact with this surface will re-combine with holes and be lost. Surface passivation is therefore necessary to reap the full benefits of the thinning and to maximise the QE, particularly for the shorter "blue" wavelengths which are absorbed close to this surface. An early method, referred to as back-surface charging, relied on the CCD being held in a vacuum and subjected to a flooded exposure of ultraviolet light. This caused the native oxide to hold a negative charge that repelled photo-generated electrons away and towards the depletion regions. Experiments demonstrated very high QE but the process was unstable and required periodic recharging. A more reliable and stable method, referred to as back-surface doping, relies on a thin layer of dopant, typically boron, being incorporated in the back surface to form a p^+-layer (Figure 23.8a). The change of doping concentration gives rise to a small potential step that repels photo-generated electrons towards the depletion regions (Figure 23.8b).

Dopant is generally introduced by ion implantation but does not become active unless in substitutional lattice sites. The lattice needs to reach a high temperature for this to occur but, because the thinned substrate will not tolerate furnace temperatures, only the thin doped surface layer is heated, using a very short pulse from an ultraviolet laser which heats the surface silicon to $\approx$ 1500 °C whilst increasing the temperature of the remaining silicon by only 5 °C to 10 °C. With the addition of an anti-reflection coating, excellent QE can now be obtained over a broad spectral range. The advantages of rear-illumination are clear from the characteristic QE curves of front- and rear-illuminated CCDs illustrated in Figure 23.9. However, multiple internal reflections within the thinned silicon substrate can give rise to interference fringing, typically at wavelengths in excess of $\approx$ 750 nm to

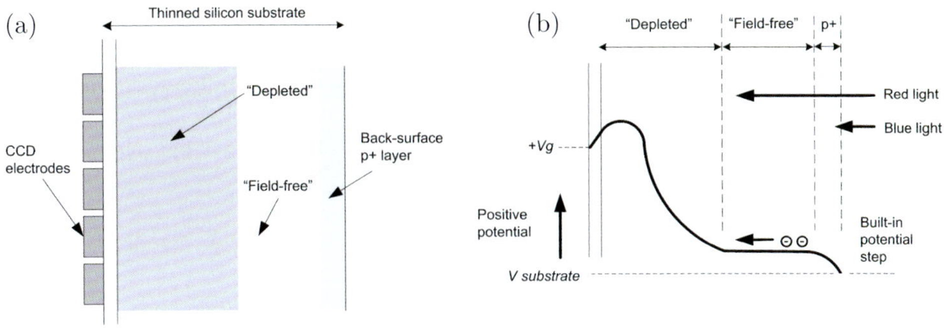

Figure 23.8: (a) The thinned back-illuminated CCD. (b) Potential profile of the CCD.

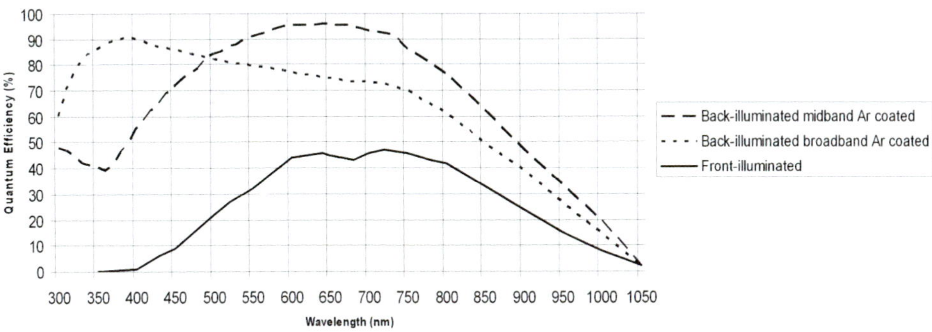

Figure 23.9: Quantum efficiency curves for front- and back-illuminated CCDs.

800 nm, and rising in amplitude to ≈ 20 % at longer wavelengths. An example frame is reproduced in Figure 23.10. Their calibration and removal in subsequent image processing can prove problematic in some applications such as narrow-band imaging and spectroscopy. One method of suppressing the fringing is to apply an anti-reflection coating peaked at the fringing wavelengths that minimises internal reflections. A second method is to etch a $\lambda/4$ (≈ 80 nm) deep groove over each half pixel so any fringing effect over the two half pixels is complementary.

Focal plane coverage

The final significant developments of the science-grade CCD addressed the requirement for greater focal plane coverage than provided by the first CCDs which were limited by the maximum reticle size of the manufacturing photolithography to ≈ 20 mm $\times$ 20 mm. The technique of "stitching", as illustrated in Figure 23.11, was developed to allow a large-format CCD to be constructed by precision lithography from a selection of reticles, each containing a different part of the overall CCD design. CCD size is now limited only by the size of the silicon wafer and

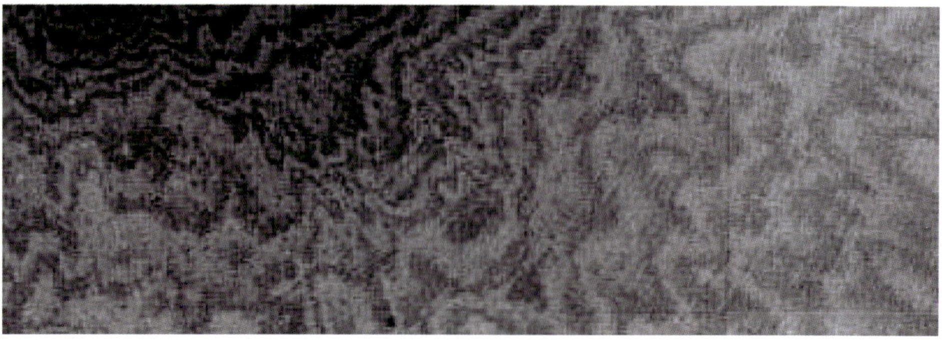

Figure 23.10: Interference fringing in a thinned back-illuminated CCD.

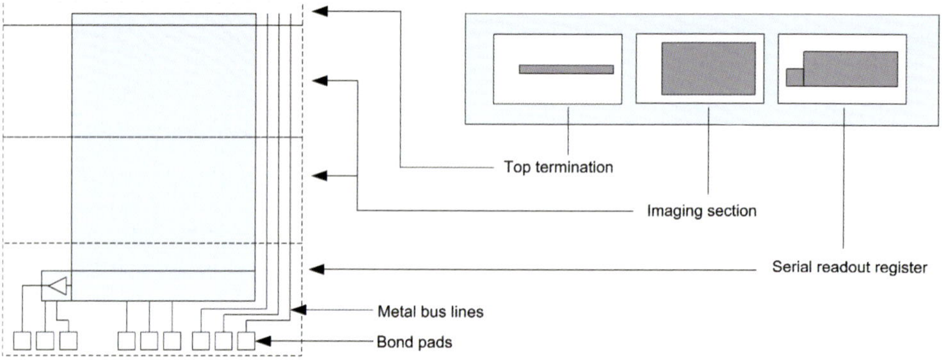

Figure 23.11: Fabrication of large-format CCDs by stitching.

the more practical aspects of defect-free yield and hence cost. For even greater coverage, an obvious solution is to populate the focal plane with more than one CCD, accepting that there will be gaps between the chips due to the chip packaging and necessary electrical connections. To minimise these gaps, CCDs have been designed with minimal non-imaging on-chip circuitry and chip-carrier packaging along one or more edges such that two or more chips can be closely butted up against each other. Gaps of only ≈ 0.2 mm to 0.5 mm are achievable. By way of example, photographs of two *HST* Wide Field Camera-3 (WFC3) CCDs, first on a five-inch (12.7 cm) silicon wafer and then butted up together within the focal plane assembly, are reproduced in Figure 23.12. The basic CCD design is a stitched array of ≈ 4096 pixels $\times$ 2048 pixels with 15 µm pixel pitch such that the two chips butted together provide an overall coverage of ≈ 4096 pixels $\times$ 4096 pixels or ≈ 61.4 mm $\times$ 61.4 mm.

Dark current

Depending on the application, an important CCD characteristic is its leakage current, otherwise known as dark current. All semiconductors suffer from leakage currents that result from electrons having sufficient thermal energy to break free

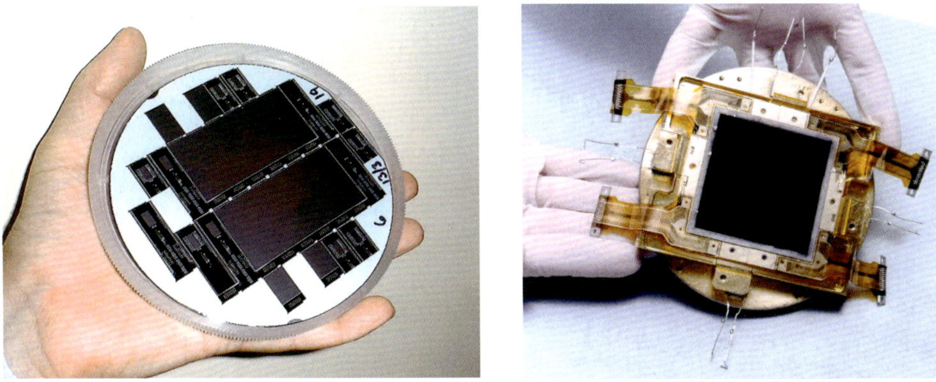

Figure 23.12: *HST* WFC3 CCDs on a five-inch wafer and integrated in the focal plane assembly (© e2v technologies plc and Ball Aerospace & Technologies Corp.).

from the lattice. The surface traps at the silicon-silicon dioxide interface are largely responsible for the dark current generated in a CCD. At 20 °C, 1 nA cm^{-2} is typical, but this decreases by a factor of two for roughly every 7 °C to 8 °C reduction in operating temperature. Dark current has associated shot noise and a relative statistical pixel-to-pixel non-uniformity of roughly 3 % to 10 % RMS. At 20 °C, dark current may well be the dominant contributor to the overall readout noise of a CCD, even in moderate frame rate applications like TV sensors. However, with sufficient cooling, typically in the range −60 °C to −100 °C, dark current can be reduced to negligible levels even for the hour-long exposures often required for astronomy. Liquid nitrogen has been popular for ground-based applications, whereas passive radiators have been used to cool sensors in space instrumentation.

An important consideration is the need for cleanliness because contamination, including materials out-gassing from the immediate surroundings, can condense on the cold CCD (Kimble et al 1994). The issue is particularly serious for CCDs imaging in the extreme ultraviolet as even the smallest layer of surface contamination can significantly reduce the detector's sensitivity. One solution to the problem is to provide a cold-trap on which contamination is condensed before it reaches the CCD. Another solution is to add a decontamination heater to the CCD package to allow periodic warming and out-gassing of contaminants. One operating technique and a further development have enabled CCDs to be operated with significantly reduced dark current. Biasing the CCD such that the low level of the CCD clocks is significantly below the substrate potential causes holes to accumulate at the silicon surface, fill surface traps, and so suppress the dominant surface dark signal generation. The holes effectively change the surface from n-type to p-type and the silicon can be considered as "inverted". The technique of "dither-clocking" relies on a confined signal charge being periodically shifted back and forth between adjacent electrodes and ensures that all of the silicon is periodically inverted. The dark current suppression is a function of operating temperature and the frequency of the "dither" and can result in more than two orders of magnitude of reduction. A further development was to add an implant under one of the electrode phases to define fixed potential wells under which charge could be collected with the whole of the surface inverted. Dither clocking was no longer necessary. This CCD variant is referred to as a multi-pinned phase (MPP) CCD, or sometimes as inverted-mode operation (IMO). Disadvantages of the MPP CCD are firstly a reduction in the charge storage capacity, or full-well capacity, of roughly 20 % and secondly a significantly increased dark signal non-uniformity.

Dynamic range

Finally, two of the most important metrics of a science-grade CCD are its full-well capacity and readout noise. Together these determine the dynamic range of the CCD with $\approx 1 \times 10^5$ or more being obtainable from today's best devices. Associated with these parameters is the relative linearity, which is typically 0.5 % to 1 % and dominated by the characteristics of the charge detection amplifier. Full-well capacity is a measure of the maximum signal charge that a pixel can store and transfer, and is a function of the isolated electron storage area which is determined by both pixel size and the number of electrode phases used to define the pixel. A 4-phase pixel holds and transfers charge beneath half of the pixel area whilst a 3-phase pixel can only work with one third of the pixel area. A 2-phase pixel operates beneath half of the pixel area but the storage density is limited by a greatly reduced potential well depth. By way of example, a 12 µm × 12 µm pixel will typically store ≈ 60 000 electrons (2-phase), ≈ 100 000 electrons (3-phase) and ≈ 200 000 electrons (4-phase). A larger 24 µm × 24 µm 4-phase pixel could hold more than 10^6 electrons.

This necessarily brief introduction to the science-grade CCD concludes by directing the reader to the literature for more in-depth discussion, recommending Janesick (2001) in particular as a comprehensive source.

The CCD in space

The first CCDs to be flown in space were on the two Russian *Vega* probes launched in 1984 to image the nucleus of comet Halley in 1986. ESA followed with CCD cameras on board the *Giotto* spacecraft, also to comet Halley. The development of the CCD for space applications can be traced back to 1974 when, under the sponsorship of NASA, JPL undertook a programme to develop large-format CCD arrays for the *Galileo* mission to Jupiter, launched in 1989, and the *HST*, launched in 1990. Today the CCD is the preeminent visible- and ultraviolet-wavelength image sensor in space science and in both Earth and planetary remote sensing.

Radiation damage

One of the most important issues for operation in the space environment is radiation damage. Radiation changes the operational and performance characteristics of CCDs, and the primary concerns are ionisation, displacement damage and transient effects.

Ionising radiation results in the accumulation of trapped charge in the CCD's oxide and the generation of traps at the silicon-silicon dioxide interface. The trapped charge results in changes to the effective bias voltages applied to the CCD and are referred to as flat-band voltage shifts. The accumulation of radiation is quantified by a parameter called the total ionising dose (TID). A biased CCD will typically exhibit a flat-band voltage shift of ≈ 0.08 V krad^{-1} (Si).[1] Shifts of up to ≈ 2 V can usually be tolerated by careful optimisation of the CCD bias voltages prior to launch and/or by enabling software-controlled adjustments to be made during the mission. The second effect of ionising radiation is a significant increase in the surface dark current. Besides operating at a reduced temperature, inverted-mode operation, either by dither-clocking or the use of an MPP CCD, will generate holes that will suppress the additional dark charge.

Displacement damage results from energetic particles, for example protons and neutrons, which collide with the silicon atoms and displace them from their lattice sites, creating vacancy-interstitial pairs. Many of these recombine but some vacancies can link with phosphorous atoms to form a trap (e-centre) that degrades CTE. Particle-induced lattice damage can also give rise to an increase in dark current and its non-uniformity, and introduce a new type of pixel noise referred to as "random telegraph signals" (RTS). A good review of the subject can be found in Hopkinson et al (1996). The observable degradation in CTE is very much application dependent but is particularly serious in the field of spectroscopic X-ray astronomy. Signal charge trapped by a proton-induced defect is likely to be later released into trailing pixels with resulting image smear. This arises because the time for an empty trap to capture charge is very short whereas the time constant of the release is very much longer. The release time constant is exponentially temperature dependent and so a complex interdependency exists between operating temperature and the readout rate of the CCD. The observable CTE degradation is also a strong function of the general background signal present on the CCD with the smearing of small charge packets being more pronounced in the absence of a background signal to fill the traps. CTE is often assessed by the degree of smearing observed in X-ray events from an Fe^{55} source (Janesick 2001). Events will normally be confined to one or two pixels but will become increasingly smeared with increasing radiation-induced CTE degradation (Figure 23.13). The same effect has been observed in the images of star fields on *HST* (Kimble et al 2000) and has significant implications for the accuracy of any radiometric calibration. The second result of displacement damage is an increase in the contribution of dark current and its non-uniformity from the bulk silicon. Individual pixels may become localised regions of high dark current, often referred to as "hot pixels", and the number can be expected to increase during the mission life. Some pixels exhibit RTS noise, an unstable increase in dark current which jumps randomly between well-defined levels (Hopkins and Hopkinson 1993, 1995). The final aspect of displacement damage to address is the finding that a good proportion ≈ 60 % to 85 % of CTE damage can be repaired by annealing. This involves periodically heating the CCD to ≈ 120 °C, or higher, for tens of hours at a time (Holland 1991). Similar studies have also demonstrated

[1] 0.008 V Gy^{-1}

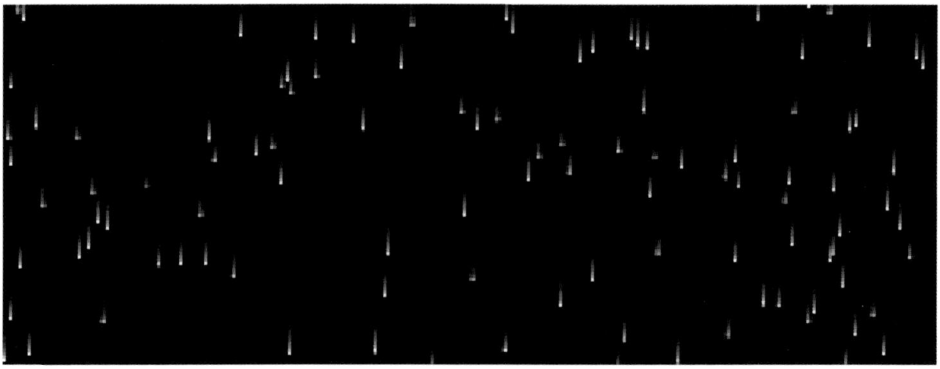

Figure 23.13: Smearing of Fe^{55} X-ray events after CTE degradation in an irradiated CCD.

partial annealing of hot-pixel damage at elevated temperatures (Holland et al 1990; Holland 1991).

Transient radiation effects occur when a particle, for example a cosmic ray, passes through the active volume of the CCD. Ionisation creates charge along the particle's path and results in a track of charge that may traverse many pixels. Although the events are transient and cause no lasting damage, they can result in significant noise within an image. The identification and rejection of cosmic-ray events in subsequent image processing requires that the observer acquires two images of the same field and accepts only those features that are common to both images.

Protective measures

The importance of the CCD in space instrumentation has led to considerable research being undertaken to understand radiation damage, how best to protect against it and the development of radiation-hardening technologies. The first defence has been to ensure that the CCD is optimally shielded, typically with up to $\approx$ 25 mm of aluminium. Significantly more shielding worsens matters because incident high-energy particles create an excess of secondary events from the shielding itself. The effects of ionising radiation can also be reduced by manufacturing the CCD with a modified gate dielectric layer but has the penalty of some loss in production yield. Minimisation of CTE degradation has seen the development of the supplementary buried channel or notch. The concept is to confine the signal charge to a narrow channel, and thus a smaller volume, such that it is exposed to a reduced number of trapping sites. This can be effective for the transfer of small signals but the advantage is lost once the signal spills out of the notch and into the wider channel. Finally, although CCD manufacturing has focused almost exclusively on the buried n-channel CCD, p-channel devices are now being explored for their greater resilience to CTE degradation, principally because of the absence of phosphorous atoms with which to form e-centres (Hopkinson 1999).

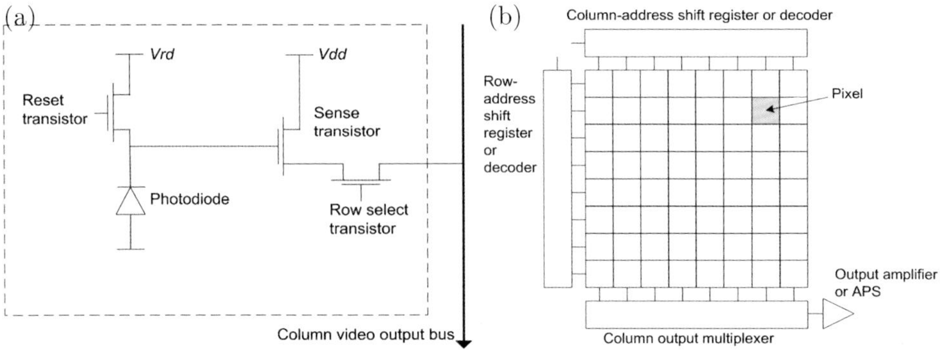

Figure 23.14: (a) 3-Transistor CMOS Active Pixel Sensor. (b) Array architecture.

CMOS

The emergence of CMOS image sensor technology is arguably the most important development in solid-state imaging since the invention of the CCD. CMOS image sensors, of which today most can be referred to as CMOS Active Pixel Sensors (APS), exploit the same silicon chip technology used in microprocessor systems. The attraction is mostly due to the nature of CMOS technology in that many millions of transistors can be integrated on a single silicon circuit. This presents the opportunity to integrate a large array of pixels, each with its own photodiode and readout transistors, alongside all of the ancillary electronics needed to address the array, buffer the analogue video signal and even digitise it ready for processing, storage or display.

CMOS sensors, like CMOS integrated circuits, operate at considerably lower voltages than CCDs, typically between 1.8 V and 5 V depending on process form and geometry. The large-scale functional integration and low-voltage operation are clearly very attractive in consumer markets that demand compactness, long battery life and low production cost. As a consequence, many digital cameras and mobile telephones now exploit CMOS rather than CCD sensors.

The question arises as to how CMOS compares to CCD technology for scientific imaging applications and in particular space instrumentation. CMOS sensors are designed in many architectural forms and a good introduction can be found in Holst and Lomheim (2007). In its simplest form the CMOS APS pixel consists of a photodiode and three transistors: one to pre-charge the photodiode, one to sense the signal voltage on the photodiode and one to select the row (Figure 23.14a). A pixel array will typically be x-y addressed by shift registers (Figure 23.14b) although some implementations employ address decoders to enable random pixel access. The APS pixel array is accessed one row at a time by enabling all the row-select transistors within a single row of pixels. At the bottom of the array, the individual pixels within the row are selected and read out column-by-column. The video signal may be fed through a multiplexer to an analogue output amplifier, a serial analogue-to-digital converter (ADC) or an array of column-parallel ADCs.

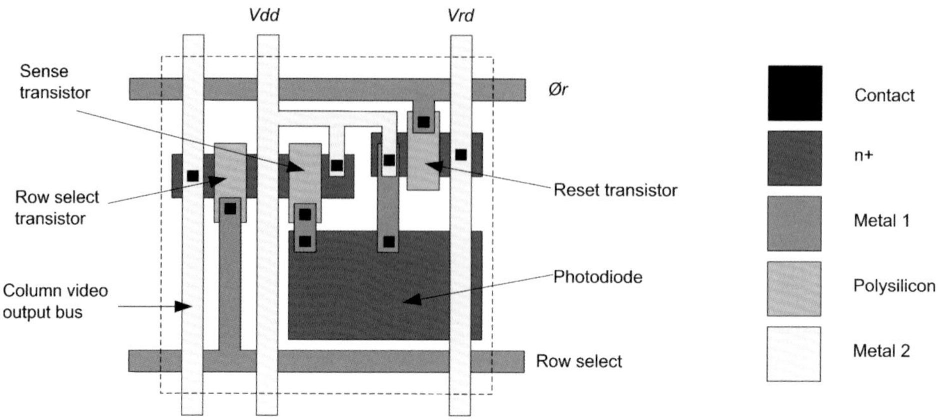

Figure 23.15: Layout of a typical CMOS pixel.

A disadvantage of the basic three-transistor pixel is that it is subject to k_BTC noise, which is typically the dominant source of readout noise. CDS has been found to work with varying degrees of success by storing a "reset" frame off-chip prior to an exposure and later subtracting it from the "signal" readout frame. To enable on-chip CDS, researchers and manufacturers have developed four-transistor pixels in which the new transistor is used as a gate between the photodiode and the sense transistor. The concept is that the output node can be sampled after reset and before signal from the photodiode is switched onto the output node. This appears analogous to the CCD output amplifier circuit but requires the photodiode to be "pinned" at appropriate potential. This ensures that all the signal charge is transferred to the output node, thus avoiding image lag, and that the k_BTC noise remains correlated between the two CDS samples. The disadvantage of the pinned photodiode is that the pinning significantly reduces the charge storage capacity of the pixel and thus also the usable linear dynamic range.

Quantum efficiency

The quest for a scientific CMOS sensor is today still in its infancy. The quantum efficiency of front-illuminated CMOS sensors is compromised by the in-pixel electronics and aluminium bus tracks reducing the "fill factor", a measure of the fraction of the pixel's area that is actually sensitive to light. The layout of a typical CMOS pixel is shown in Figure 23.15. In this example, the photodiode occupies only 19 % of the pixel area (as defined by the dotted line) but will typically yield a fill factor $\approx$ 30 %. The losses arise from the reflection of light from the aluminium bus lines and photon-induced electrons in the substrate being absorbed and lost within the in-pixel transistor electronics. An obvious solution to the problem is to thin and back-illuminate the sensor and several research groups have now demonstrated back-illuminated CMOS sensors that achieve quantum efficiencies in line with their CCD counterparts (Waltham et al 2007; Janesick et al 2007).

Readout noise and dynamic range

A second area of research is to minimise readout noise and maximise charge storage capacity and linearity. A limitation is the linear voltage swing that can be obtained outside the transistor threshold regions in modern low-voltage CMOS processes. The linear dynamic range of today's best CMOS sensors is ≈ 5000, considerably less than a CCD. Several approaches to overcoming the problem are being investigated including the concept of pixels that deliberately behave in a non-linear fashion and sensors that allow individual pixels to have varying exposure periods. More complex five- and six-transistor pixels are also being investigated in the pursuit of increased dynamic range (Janesick et al 2006, 2007). We can anticipate further progress in the future as researchers adapt to exploiting the advantages of CMOS technology rather than attempting to emulate the elegance of the CCD.

CMOS in space

CMOS sensors are already used in space, having applications in satellite bus instrumentation such as star trackers and inspection cameras. CMOS is yet to have a significant impact in scientific payloads for which the CCD remains dominant. The principal advantages of CMOS over the CCD for space instrumentation are compactness, low mass, low power and radiation hardness. The CCD remains unchallenged in dynamic range and photometric accuracy. The effects of radiation in CMOS and CCD sensors are similar in that both suffer from ionising radiation and displacement damage. However, the key advantage of the CMOS APS is that there is no degradation of CTE. A new effect, unseen in the CCD, is a susceptibility to single-event latch-up (SEL). SEL is a potentially destructive condition, triggered by an energetic particle, in which parasitic circuit elements form the equivalent of a silicon-controlled rectifier. The result is a large and potentially damaging increase in supply current that can only be recovered from by temporarily removing power. CMOS sensors will also exhibit the other effects of displacement damage such as hot pixels and RTS noise. CMOS sensors will undoubtedly play an increasing role in space instrumentation, but today they are most likely to appear in those applications that have extreme size, mass or power constraints, applications which require complex operational modes such as random pixel access, or in those missions for which the radiation damage to a CCD would prove insurmountable.

Bibliography

Beynon JDE, Lamb DR (1980) Charge-coupled devices and their applications. McGraw-Hill

Boyle WS, Smith GE (1970) Charge-coupled semiconductor devices. Bell Syst Tech J 49:587

Ebben TH, Bergstrom J, Spuhler P, Delamere A, Gallagher D (2007) Mission to Mars: the HiRISE camera on-board MRO. Proc SPIE 6690:6690B-1–14

Janesick J (2001) Scientific Charge-Coupled Devices. Proc SPIE Press Book, ISBN 9780819436986

Janesick J, Andrews J, Elliott T (2006) Fundamental performance differences between CMOS and CCD imagers: Part I, Proc SPIE 6276:62760M-1–19

Janesick J, Andrews J, Tower J, Grygon M, Elliott T, Cheng J, Lesser M, Pinter J (2007) Fundamental performance differences between CMOS and CCD imagers; Part II. Proc SPIE 6690:669003-1–23

Holland AD (1991) Annealing of proton-induced displacement damage in CCDs for space use, Proc. of 10th Symposium on Photoelectronic Image Devices, IoP, 33–40

Holland A, Abbey A, Lumb D, McCarthy K (1990) Proton damage effects in EEV charge coupled devices. Proc SPIE 1344:378–395

Holst GC, Lomheim TS (2007) CMOS/CCD Sensors and Camera Systems, Bellingham, Wash.; The International Society for Optical Engineering; JCD Publishing

Hopkins IH, Hopkinson GR (1993) Random Telegraph Signals from Proton-irradiated CCDs. IEEE Trans Nuc Sci 40:1567–1574

Hopkins IH, Hopkinson GR (1995) Further measurements of random telegraph signals from proton-irradiated CCDs. IEEE Trans Nuc Sci 42:2074–2081

Hopkinson GR (1999) Proton damage effects on p-channel CCDs, IEEE Trans Nuc Sci 46:1790–1796

Hopkinson GR, Dale CJ, Marshall PW (1996) Proton effects in charge-coupled devices, IEEE Trans Nuc Sci 43:614–627

Hynecek J (2001) Impactron – a new solid state image intensifier. IEEE Trans on Electron Devices, 48:2238–2241

Kimble R, Brown L, Fowler W (plus 11 authors) (1994) CCD detector for the Space Telescope Imaging Spectrograph. Proc SPIE 2282:169–181

Kimble RA, Goudfrooij P, Gilliland RL (2000) Radiation damage effects on the CCD detector of STIS, UV, optical and IR space telescopes. Proc SPIE 4013:532–544

Leach RW, Schild RE, Gursky H (plus three authors) (1980) Description, performance and calibration of a charge-coupled device camera. PASP 92:233–245

Marcus S, Nelson R, Lynds R (1979) Preliminary evaluation of a Fairchild CCD-211 and a new camera system. Proc SPIE 172:207–231

Robbins MS, Hadwen BJ (2003) The noise performance of electron multiplying charge-coupled devices. IEEE Trans on Electron Devices 50:1227–1232

Séquin CH, Tompsett MF (1975) Charge transfer devices. Academic Press, NY

Waltham NR, Prydderch M, Mapson-Menard, H (plus three authors) (2007) Development of a thinned back-illuminated CMOS active pixel sensor for extreme ultraviolet spectroscopy and imaging in space science. Proc SPIE 6690:669007-1–14

— 24 —

X-ray CCDs

Andrew Holland[I]

Abstract

In recent years, the silicon charge coupled device (CCD) has found use in many applications in space science providing imaging solutions to a variety of instrument needs. Whilst the majority of imaging applications using CCDs are in the visible region, the ability of silicon to perform efficient detection of X-rays in the 0.1 keV to 10 keV band enables their use in, for example, X-ray astronomy. Since the early 1990s several instruments have been flown using X-ray-optimised CCDs as both non-imaging and imaging detectors at the focal planes of X-ray telescopes. This chapter reviews the use of the silicon CCD for direct detection in the X-ray region and describes some successful applications, together with radiation damage effects when used in the space environment.

X-ray interactions in silicon CCDs

High-energy photons in the X-ray band can be detected and imaged using silicon CCDs. Figure 24.1 depicts the X-ray absorption coefficient σ_{I} as a function of photon energy E_γ. Since the detection thicknesses are measured in $\approx$ 10 µm or $\approx$ 30 µm, silicon can efficiently stop X-rays in the energy range 100 eV to 10 keV, where the photoelectric effect dominates. At these energies, the photon interacts with bound electrons in the silicon atom and ejects an energetic photo-electron. The excited atom then returns to its ground state through a series of Auger and fluorescence processes, whilst the energetic photo-electron releases energy by ionisation (Fraser et al 1994). The whole process results in the creation of an electron cloud around the initial X-ray absorption site, with the mean number of free electrons generated equal to $N_{\mathrm{e}} = E_\gamma/\omega$, where E_γ is the photon energy and ω is the mean energy required to liberate one electron-hole pair. In silicon at room temperature, $\omega = 3.65$ eV. Thus a 6 keV photon would create typically 1640 electrons per X-ray interaction (as compared to only one electron per optical photon interaction).

[I]Planetary and Space Science Research Institute, Open University, Walton Hall, Milton Keynes, UK

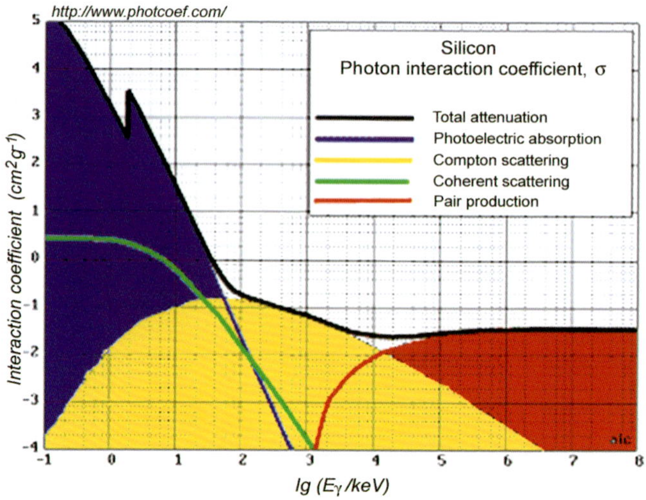

Figure 24.1: Absorption coefficient vs. photon energy in silicon. The dominant energy range for direct absorption in silicon is between 100 eV and 10 keV where the photoelectric effect is important.

The X-rays are penetrating and can be absorbed throughout the active structure in the CCD. Figure 24.2 depicts the various layers which affect signal generation in a CCD, and shows the two illumination methods of front-illumination and back-illumination. As described in Chapter 23 (Waltham 2010), the CCD is initially fabricated on a high-purity epitaxial silicon layer which is grown on top of a substrate. The electrodes plus gate oxide and nitride layers are electrically inactive and form a "dead" layer. In the front-illuminated structure, X-rays impinge on the CCD from the front/top as indicated in Figure 24.2.

The X-rays may be absorbed in either the depletion layer, the field-free layer, or the substrate. As described earlier, some devices are back-illuminated, where the substrate and some of the field-free layer are removed through a combination of mechanical plus chemical means. In most back-illuminated devices, e.g., for optical applications, a finite field-free layer is retained (around 5 μm to 10 μm thick). For X-ray applications, however, the device can be thinned into the depletion layer, which results in a fully-depleted structure. Figure 24.3 shows how the signal, generated in each of the depletion and field-free regions in the front-illuminated structure, spreads prior to being stored at the pixel potential minimum just below the electrode surface. In general, the signal arising from X-ray interactions in the depletion region is rapidly collected and is stored in single pixels, while that from the field-free region undergoes lateral diffusion during collection, producing larger "event" sizes, and is collected over several pixels. The image in Figure 24.3 shows discrete X-ray photon interactions in a CCD in what is referred to as "photon counting mode". In this image the single and multi-pixel events are clearly identified.

The detection efficiency of the X-ray CCD is therefore built up from several discrete components. For low-energy photons, absorption occurs in the dead layers above the X-ray sensitive volume, while the limit to high-energy detection is affected

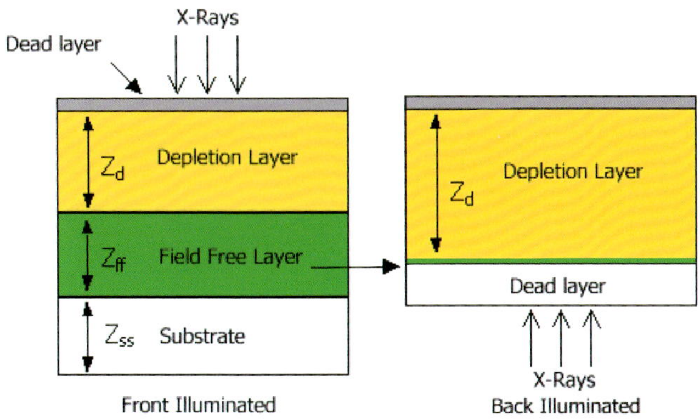

Figure 24.2: Detection and "dead" layers both for both, front and back-illuminated structures.

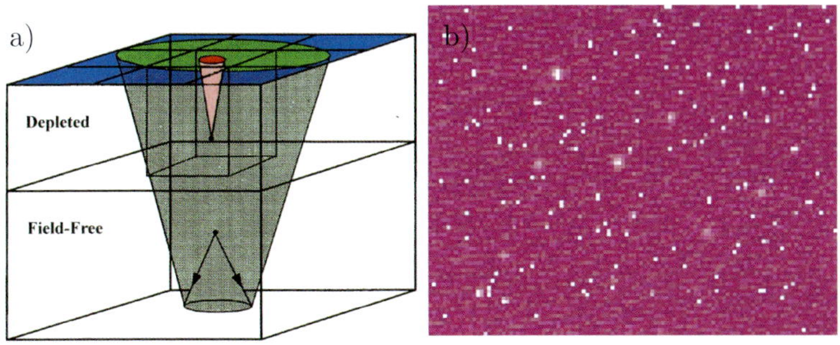

Figure 24.3: a) Sketch showing charge spreading from the point interactions. b) Image of discrete photon detections from these layers demonstrating the effect of charge spreading.

by the absorption of photons in the active layer. In the front-illuminated device this can be typically equivalent to ≈ 1 µm of silicon, while in the back-illuminated device the real dead layer can be as little as 50 nm. Standard CCDs have depletion depths of between 3 µm and 10 µm, and for X-ray applications "deep depletion" devices constructed on high-purity silicon can provide depletion depths of 30 µm to 300 µm, depending upon silicon type and CCD design and bias. Care must therefore be exercised when specifying CCDs for specific applications to ensure that the best detection efficiency and CCD type (front/back-illumination) is obtained.

Figure 24.4 gives the quantum efficiency (QE) as a function of photon energy for several types of CCD construction, ranging from a standard commercial TV-type sensor, to a special fully-depleted, 300 µm thick back-illuminated type. Factors such as price, yield and availability might be traded against detection efficiency and energy range.

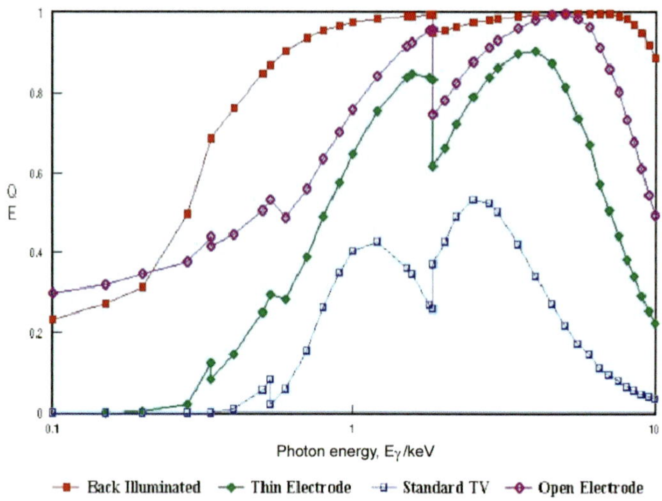

Figure 24.4: Comparison of different types of CCD construction and their resulting quantum efficiency.

Noise sources and X-ray energy resolution

The detection sensitivity and energy resolution are dependent upon the system noise, leakage current, etc. At high readout frequencies, in the absence of other degrading factors, the on-chip amplifier white noise σ_{w} dominates and $\sigma_{\mathrm{w}} \propto \sqrt{f}$, where f is the readout frequency (or more accurately the electronic bandwidth of the system). As the readout speed decreases, the noise profile settles at the output node's $1/f$ noise floor, where readout noise is independent of readout speed. This $1/f$ noise floor for scientific type CCDs can be between two to three electrons RMS. However, for low readout frequencies, the integrated leakage current can dominate, and Figure 24.5a shows the additional noise contributions arising from leakage current at a number of temperatures, for both normal and inverted, or MPP (multi-pinned phase, cf., Chapter 23, Waltham 2010), mode operation. For many scientific space applications where noise is important, the devices are operated at temperatures around -100 °C to ensure that the contribution from leakage current is negligible.

In X-ray photon counting spectroscopy applications, whilst the number of photo-generated electrons is E_γ/ω, the variance on this electron number is not given by the normal Poisson statistics, but is smaller due to the correlation between the energy-recovery processes (Fraser et al 1994). In this instance the variance on the signal generated is given by $F\,E_\gamma/\omega$, where $F = 0.115$ is the Fano factor in silicon (Fraser et al 1994). The total X-ray energy resolution as a function of energy may be expressed as

$$\Delta E_{\mathrm{FWHM}} = 2.35\,\omega\sqrt{\sigma^2 + F\frac{E_\gamma}{\omega}} \quad . \tag{24.1}$$

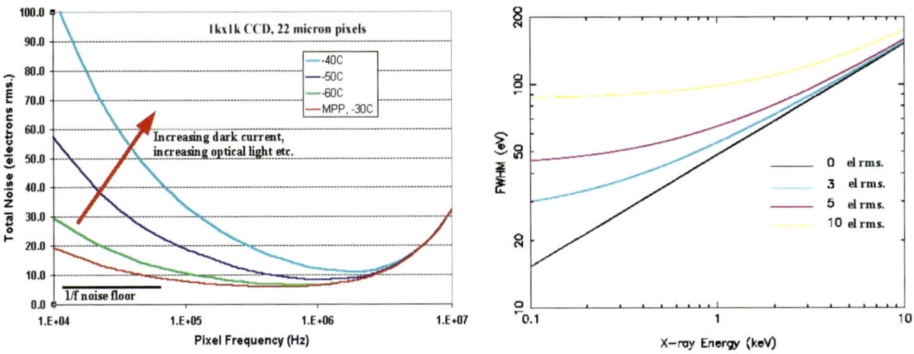

Figure 24.5: a) Total system noise as a function of readout speed for different operational temperatures. b) X-ray FWHM resolution vs. energy for several readout noise values.

This relationship is given in Figure 24.5b as a function of X-ray energy for a number of different readout-noise values. Since the efficient detection of photons starts at energies above $\approx$ 200 eV to 300 eV, system noise of three to five electrons RMS is satisfactory for most applications of this type (e.g., *XMM-Newton* EPIC MOS cameras operate with $\approx$ 4.5 electrons RMS). The figure also demonstrates that an energy resolution of only 200 eV is required to resolve the elemental lines from the elements heavier than oxygen, which is an important parameter for instruments for planetary science investigating elemental composition.

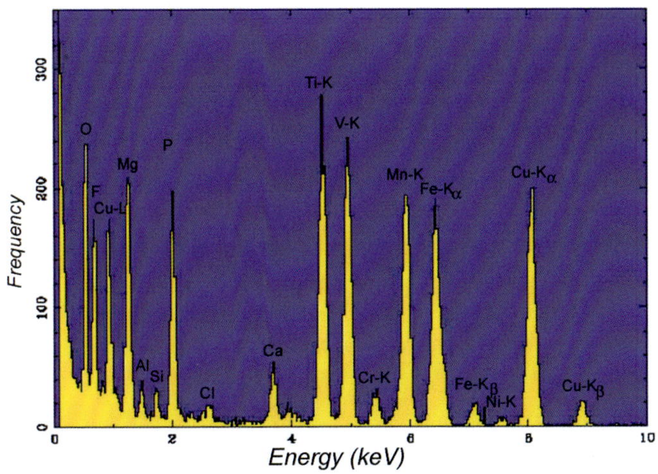

Figure 24.6: X-ray spectrum from a variety of elemental fluorescence lines between oxygen at 0.5 keV and copper at 8 keV and 9 keV, demonstrating the energy resolving power of the CCD in X-ray photon counting mode.

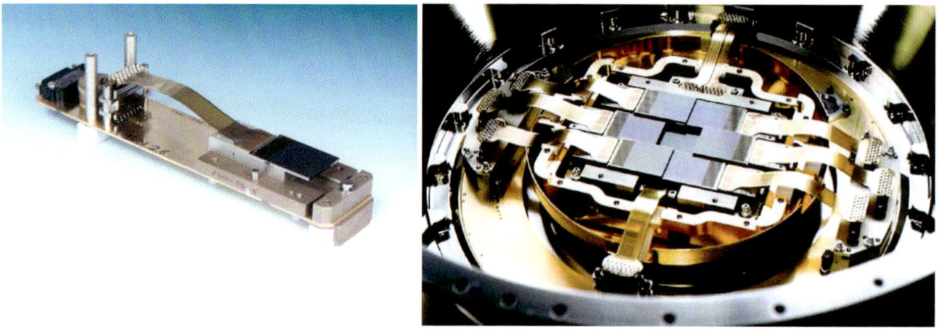

Figure 24.7: CCD22 from e2v designed for the EPIC cameras on *XMM-Newton* together with one of the flight focal plane arrays which consists of an array of seven such detectors.

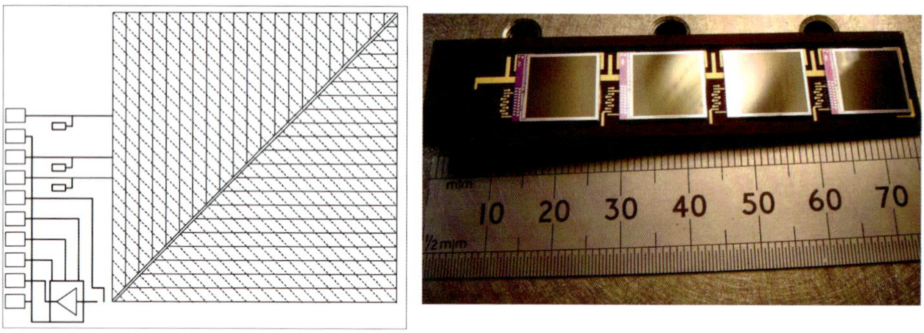

Figure 24.8: Schematic showing the SCD layout, together with a photograph of the SCD module with four detectors for the D-CIXS instrument.

X-ray CCD arrays

Figure 24.7 shows a single flight CCD developed by e2v technologies and Leicester University for the EPIC instrument on ESA's *XMM-Newton* spacecraft (Turner et al 2001). The two EPIC MOS CCD cameras, at the focal surfaces of X-ray grazing-incidence optics, comprise arrays of seven such CCDs in a focal plane array. The devices are optimised for single photon X-ray spectroscopy in the 3 eV to 10 keV band, and are of the "open electrode" format for enhanced soft X-ray detection efficiency (Holland 1996). Due to the large depth of focus of the X-ray optic, the CCDs can be overlapped to minimise the dead space in the array.

The swept charge device (SCD)

A variant detector type based on CCD technology has been generated which dispenses with the spatial (imaging) information and which is designed as a photon counting X-ray spectrometer. The "swept charge device" (SCD) is a relatively simple CCD, having $\approx$ 10 interconnections, compared to between 20 to 30 for

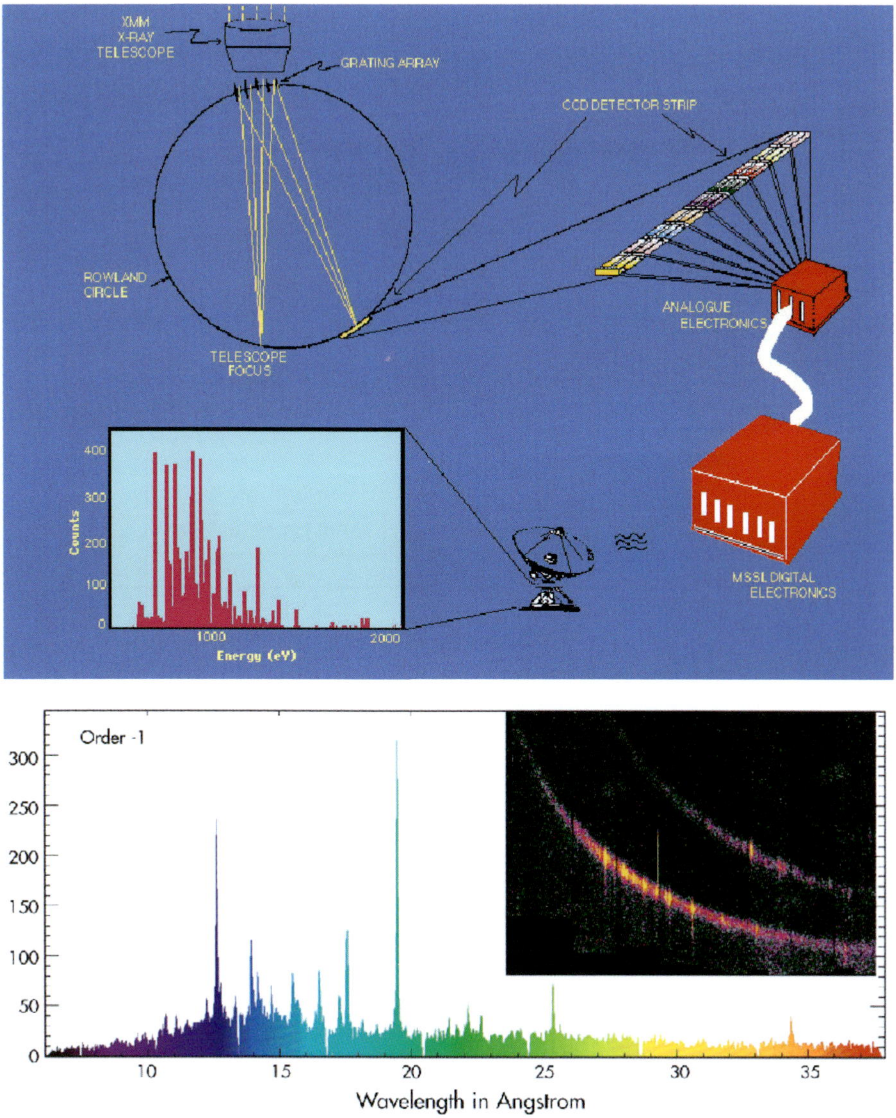

Figure 24.9: Upper: Dispersed X-ray spectrometer system used on ESA's *XMM-Newton* where gratings are used to disperse X-rays for readout by an array of CCDs.
Lower: First-light results from the RGS instrument (courtesy of ESA).

a normal CCD. The device is continuously read out and is similar in operation to a linear CCD though with large area. The continuous readout maintains a low leakage current by "dither mode clocking", and X-ray spectroscopy can be obtained at much higher temperatures than normal. The original SCD, which possessed an area of ≈ 1 cm^2, could produce Fano-limited spectroscopy at temperatures as

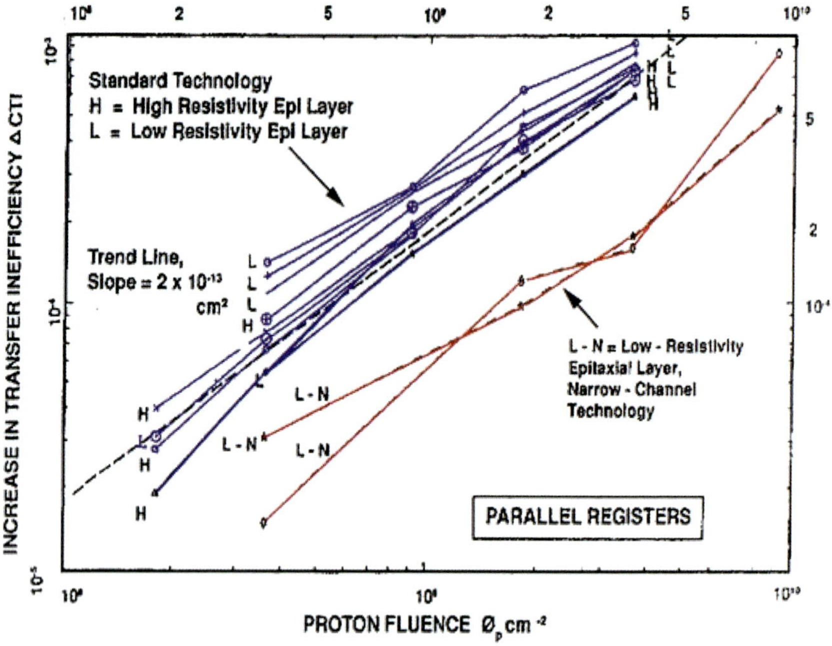

Figure 24.10: Increase in parallel CTI for e2v's standard CCD62 as a function of 10 MeV proton fluence, together with results for an experimental narrow channel CCD for improved hardness.

high as −10 °C, while newer generations of the device can yield good spectroscopy even at +10 °C for detection areas of 4 cm^2. The figure also shows a module containing four original-type SCD detectors which was used for lunar elemental X-ray fluorescence mapping in the D-CIXS instrument on ESA's *Smart-1* spacecraft (Grande 2003), and has been launched in October 2008 on the C1XS spectrometer of India's *Chandrayaan-1* lunar orbiter (Howe et al 2009).

Dispersed X-ray spectroscopy

An alternative method for X-ray spectroscopy is through the use of an energy-dispersive medium such as a Bragg crystal or diffraction grating. Figure 24.9 shows the configuration adopted for the reflection grating spectrometer (RGS) instrument on *XMM-Newton*, where a grating is used behind the X-ray optic to disperse the X-rays (Brinkman 2001). In this approach the spatial information is essentially lost, and the technique is mainly intended for the brighter sources. The dispersed X-rays come to a focus on the Rowland circle of the grating, and a linear array of nine CCDs is used to read out the spectrum, where linear position equates to X-ray energy. The original "first light" results from the instrument are also shown in the figure. These demonstrate the improved energy resolution that can be achieved using the dispersed spectrum method.

Space radiation damage

The space environment and its impact on CCDs was discussed in Chapter 23 (Waltham 2010). For a thorough description of radiation effects see Holmes-Siedle (2002). In the X-ray photon-counting applications, the most degrading effect is that of the high-energy protons, originating from the Sun or from the trapped radiation belts and South Atlantic Anomaly (SAA) (Holland 1992). These protons create trap sites which capture electrons from the X-ray signals and can release the charge when the X-ray signal has been removed by readout. The release-time constants are dependent upon operating temperature and trap species, but the signal can be released in a range between nanoseconds to minutes after trapping. In the X-ray spectroscopy application, where the energy resolution is well defined through the system noise and Fano statistical variation, such carrier removal through trapping can significantly degrade the FWHM energy resolution of the CCD.

The hardness to radiation damage can be improved by restricting the charge storage and transport within the CCD during readout. Figure 24.10 shows the increase in CTI (charge transfer inefficiency) for a CCD against fluence for 10 MeV protons, showing results for standard off-the-shelf CCDs together with experimental devices intended to improve the radiation-transport properties.

Future developments

Current and future developments in X-ray direct-detection CCDs are targeted at improving the readout speed, and hence the X-ray throughput of the systems, to enable their use with X-ray optics having larger effective area. In addition, the size of CCDs can now be much larger, e.g., up to 6 cm $\times$ 6 cm so either single detectors can replace earlier arrays, or the focal-plane array size can be much larger. The high-energy detection efficiency arises through the use of high-purity bulk silicon, such that detection of photons out to 20 keV can be achieved with a relative detection efficiency of 90 % at 10 keV. Finally the CCD technology is being increasingly combined with application specific integrated circuit (ASIC) technology so that much more of the drive and readout function of the electronics is included at the CCD module level. This is especially important for large mosaic arrays intended for future telescopes (both optical and X-ray). The improvement of radiation hardness in CCDs is an ongoing development, and measures such as thin-gate dielectrics, narrow-channel technology and p-channel technology are being used to improve the performance of these sensors for use in space.

Bibliography

Brinkman AC, Behar E, Güdel M (plus 15 authors) (2001) First light measurements with the *XMM-Newton* reflection grating spectrometers: Evidence for an inverse first ionisation potential effect and anomalous Ne abundance in the Coronae of HR 1099. Astron Astrophys 365:L324–L328

Fraser GW, Abbey AF, Holland AD (plus three authors) (1994) The X-ray energy response of silicon: (A) Theory. Nucl Instrum Methods Phys Res Section A 350:368–378

Grande M, Browning R, Waltham N (plus 45 authors) (2003) The D-CIXS X-ray mapping spectrometer on *SMART-1*. Planet Space Sci 51:427–433

Holland AD, Turner MJL, Abbey AF, Pool P (1996) The MOS CCDs for the EPIC on *XMM*. Proc SPIE 2808:414–420

Holland AD (1992) The effect of bulk traps in proton-irradiated EEV CCDs. Nucl Instrum Methods Phys Res Section A 326:335–343

Holland AD (2003) New Developments in CCD and pixel arrays. Nucl Instrum Methods Phys Res Section A 513:308–312

Holmes-Siedle A, Adams L (2002) Handbook of Radiation Effects. Oxford Press, ISBN 0-19-850733-X

Howe CJ, Parker D, Maddison B (plus thirteen authors) (2009) *Chandrayaan-1* X-ray Spectrometer (C1XS) – Instrument design & technical details. Planet Space Sci 57:735–743

Turner MJL, Abbey A, Arnaud M (plus 60 authors) (2001) The European photonimaging camera on *XMM-Newton*: The MOS cameras. Astron Astrophys 365:L27–L35

Waltham N (2010) CCD and CMOS sensors. ISSI SR-009:391–408

— 25 —

Intensified solid state sensor cameras: ICCD and IAPS

Udo Schühle[I]

Abstract

We describe the general design of intensified charge coupled devices and intensified active pixel sensors: cameras using microchannel plate intensifiers in combination with imaging arrays, like CCDs or CMOS-APS sensors. Several design options are compared and the capabilities and shortcomings of these devices will be highlighted and discussed. We describe, in particular, the properties of the intensifiers, phosphor anodes, and coupling schemes. The coupling between the intensifier and the image sensor is a special technological step that adds to this type of camera a great amount of flexibility and design options.

Introduction

An intensified charge coupled device (ICCD) or an intensified active pixel sensor (IAPS) camera is just a special form of a microchannel plate (MCP) detector system where the image encoding system uses a CCD or a CMOS-APS sensor chip for building a digital image. The MCP-intensifier in this case is using a phosphor anode to convert the electron pulse into visible light, which will be accumulated in the imaging sensor. Like all MCP detectors, they can be used in a large spectral range from the visible to the extreme ultraviolet where secondary electron emission from a suitable photocathode material is efficient. But after the development of thinned, back-side illuminated CCDs for the EUV range, MCP intensifiers have more been used in the particular photo-emission regime in the VUV, where materials are not transparent enough to allow penetration of the radiation into the bulk of the sensor where charge carriers can be collected.

The use of MCP intensifiers with an imaging chip increases not only the complexity to the camera design but it also adds important flexibility:

- the intensifier can be made more sensitive in a preferred spectral range by selection of the photocathode material,
- by switching the high voltage of the MCP, the intensifier can be used as an electronic shutter,

[I]MPS—Max-Planck-Institut für Sonnensystemforschung, Katlenburg-Lindau, Germany

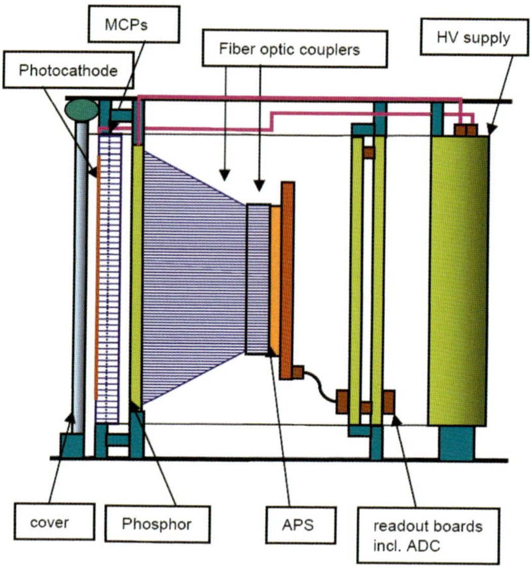

Figure 25.1: General scheme of ICCD / IAPS cameras. The size of each individual component is not to scale and may be selected according to availability and the performance goals of the assembly.

- the size of the focal plane, and thus the image scale of the optical system, becomes independent of the imaging chip size, as it can be adapted by the optical magnification of the coupling system,
- the intensifier can be made blind in visible light without an additional filter,
- the amplification of the intensifier, given by the high voltage of the intensifier and the phosphor conversion factor, can be adjusted within a large range to the signal strength,
- it allows the image sensor to be used in current accumulation mode or in single-photon counting mode.

In addition, in photon counting mode the signal can be discriminated against the dark noise. This allows operation at room temperatures without the need for cooling the sensor – an important advantage in space instruments, as it saves weight and reduces the risk of contamination build-up on the sensor.

On the other hand, the use of MCP intensifiers implies carrying along a high-voltage power supply, and it adds to the camera the drawbacks generally associated with MCP detectors, e.g., the limited resolution given by the multi-fibre bundles and corresponding flatfield structures. In addition, there may be distortions of the image related with the microchannel plates, the optical coupling between the intensifier and the image sensor, or the electric field between them. To minimize such complications, the components must be carefully selected and well matched with each other.

General design of ICCDs and IAPSs

An intensified CCD or APS camera, in general, consists of a channel plate intensifier, a CCD or APS imaging chip with corresponding readout electronics, and an optical coupling between them. The general scheme is depicted in Figure 25.1.

The intensifier may be under vacuum, covered with a UV-transparent window, or open-faced (and therefore operable only under vacuum), depending on the wavelength range. Below 110 nm no window can be used. The photocathode plane actually represents the focal plane of the optical system. It may be defined by the front surface of the channel plate, or it may be a semi-transparent coating deposited on the inner side of the window. In an open-faced system, an additional photocathode coating may be deposited as a thick, opaque layer on the front face of the channel plate.

Depending on the gain factor needed, there may be one single MCP or a stack of two or three MCPs to produce electron pulses that are accelerated by a high voltage across a "proximity gap" onto the anode. The anode has a special coating. The main portion is a phosphor converting the electron pulse into a flash of visible light. In general, the sensitivity of the image sensor is high at this wavelength, such that a single channel plate-phosphor combination creates pulses strong enough for all types of readout schemes of the camera, but in some cases a double or triple stack of MCPs has been used. In any case, the light at the phosphor must be transferred by optical means to the sensor. At this stage, the image can be magnified or demagnified, to adapt to the sensor size, or even attenuated if needed to adapt the output of the intensifier to the sensor's full-well capacity.

MCP intensifiers and photocathodes

The intensifiers used for ICCD/IAPS are the same MCP tubes as described in Chapter 22 (Timothy 2010). The only difference is the anode containing the phosphor coating described further below. In brief, the intensifier tube consists of a ceramic-metal housing with metal contacts for applying the high voltage to the MCPs of approximately 1 kV per MCP for electron amplification and the phosphor anode. An electric field of several kilovolt per milimetre is needed for the photoconversion and to avoid charge spreading in the proximity gap, the space between the MCP output and the anode. The input side of the intensifier may be a bare MCP or may have a coating of the appropriate material to increase the detection efficiency by enhancing secondary electron creation (photo effect) in the desired wavelength range. In the VUV spectral range the photocathodes are mostly made of alkali halides, e.g., CsI, KBr, CsBr, KI, RbBr, etc. (Siegmund 1999), materials with a large band gap (between valence and conduction bands) to suppress emission at longer wavelengths. These materials are also sufficiently stable under normal environmental conditions if high humidity can be avoided. For wavelengths longer

Figure 25.2: Intensifier housing with fibre optic output window. Left: inside-view with the phosphor screen and the pads for placing the 25 mm circular MCPs. Right: the fibre-optic output window at the back side of the tube.

than 110 nm, materials that are not stable under normal conditions, like multi-alkali combinations, can be used with a window-sealed housing under vacuum.

The phosphor screen anode

The electron pulses leaving the MCP output side are accelerated by a high potential onto a proximity-focused phosphor screen deposited on a fibre optic face plate that is used as the intensifier output window. The value of the electric field (up to 6 kV/mm) depends on the size of the gap and determines the amount of transverse spreading of the charge cloud and the intensity of the phosphor flash. The amount of transverse charge spreading that is acceptable depends on the resolution of the sensor reading the image: it may be much larger when single-photon events are being detected with position-encoding (centroiding) the charge cloud after detection. A typical intensifier tube with a fibre optic bundle coated with a phosphor screen is shown in Figure 25.2. The choice of the screen material depends on the spectrum of the luminous emission in relation to the wavelength with the highest efficiency of the sensor, the conversion efficiency of the phosphor material, and the response (decay) time of the light flash.

Typical phosphor screens have a broad spectral output in the near-UV, blue, or green spectral range with conversion efficiencies between 10^2 and 10^3 photons per electron (depending on the accelerating voltage). The decay times (to 1/10 of the maximum intensity) of common phosphors may be between 100 ns and 10 ms. The decay time of the luminescence should be related to the normal readout time of the image sensor. While CCDs in general use a shutter for timing the exposure, and the photon events are captured during the exposure time and then transferred to the readout amplifiers, CMOS-APS sensors run shutterless by reading out line by line

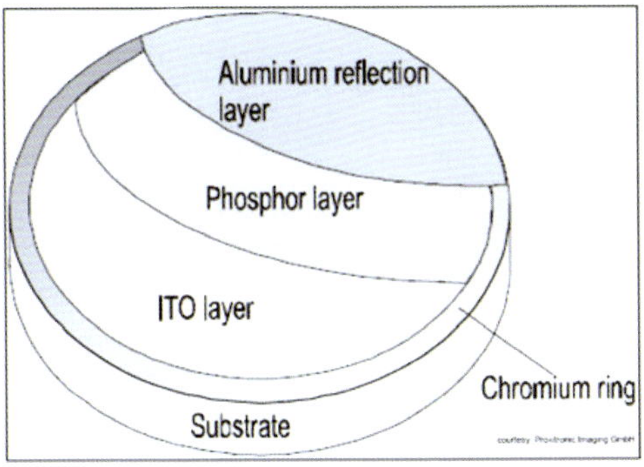

Figure 25.3: Design of the phosphor-coated output window of an intensifier.

and the exposure time of each line may be chosen much shorter than the readout time of the whole image. In this operational mode the lines are exposed sequentially and if the phosphor decay time is comparable to the line exposure time, the events spreading over several lines will not be detected correctly. In this case the decay time should be large compared to the line exposure time. On the other hand, if the exposure time is long (for example, longer than 0.1 s for a 1024×1024 pixel sensor), then the phosphor decay time is much shorter and no such interference occurs.

To maintain a homogeneous and stable electric field inside the proximity gap, the MCP output face and the phosphor screen must have conductive coatings. A thin metal layer (e.g., 50 nm of aluminium) on top of the phosphor may serve several purposes: it provides the homogeneous conductive coating, it increases the efficiency by reflecting the luminescence light back towards the sensor, and it prevents visible light from passing through the intensifier, thus making it visible-blind. Figure 25.3 shows the schematic design of an intensifier output window.

Coupling schemes

The image produced on the phosphor screen must be relayed to the imaging sensor device (CCD or APS) by an optical transfer. The simplest way (but rarely used) is to place the sensor in close proximity to the phosphor screen. Another way is to use an optical lens to transfer the image onto the plane of the sensor. In this case, only a transparent window is used as substrate for the screen and the sensor may as well have a protective window. One advantage of this optical transfer is the option of using the magnification of the lens for adaptation of the image scale to the size of the sensor. This scheme has been used successfully in instruments flown on space missions.

A more direct way of image transfer can be achieved by fibre optic coupling where the phosphor screen is applied to a fibre optic face plate. A fibre optic face plate is a coherent multi-fibre plate, which acts as a zero-depth window, transferring an image pixel by pixel (fibre by fibre) from one face of the plate to the other. Standard fibre sizes are between (3 and 25) µm. This fibre optic coupler can be glued directly onto the image sensor, providing 1:1 imaging from the intensifier

Figure 25.4: Fibre optic tapers as advertised by the manufacturer (Schott North America, Inc.).

to the sensor, or they can be made as a taper, magnifying (or reducing) within a range of magnifications given by the manufacturer. The common magnification range is up to 4:1 and sizes up to 75 mm diameter at the large end can be made. The geometric distortion of such a taper is negligible compared to the distortion introduced by the intensifier. Figure 25.4 shows some fibre optic tapers that are commercially available.

A prerequisite for the direct coupling of the sensor with the fibre optic is that the sensor comes with a removable cover window. Although removing the window from the sensor is possible in principle, it is a difficult process. The fibre optic coupler can be glued directly onto the active area of the sensor by optical cement. However, in most circumstances the sensor has to be able to go through extreme temperature cycles, which makes a hard coupling impractical because of stress developed by mismatch of the thermal expansion coefficients (CTE) of the different materials. This can be solved by leaving a small gap of approximately 5 μm between the sensor and the coupler. A loss of contrast resulting from this gap can be avoided by filling the gap with oil of high refractive index. Of course, this coupling must then be supported mechanically around the interface by epoxy.

It may be important to adjust the orientation of the sensor with respect to the intensifier within tight tolerances. In such a case, it is more practical to first couple the sensor to another fibre optic face plate or taper, as described, which can afterwards be connected with the intensifier output window. Figure 25.5 shows, as an example, the engineering model of an IAPS camera where this coupling scheme has been used. It is a camera of the Rapid Acquisition Imaging Spectrograph Experiment (RAISE), an instrument (Hassler et al 2004) to be launched by a NASA sounding rocket for a spectroscopic observation of the Sun.

Sensors and their readout schemes

The image produced by the intensifier on the phosphor screen is detected by an imaging array sensitive to the light of the phosphor. Imaging sensors that have been

Figure 25.5: An engineering model of a camera with intensifier and sensor with readout system.

used with intensifiers are charge coupled devices, charge injection devices (CID), or active pixel sensors. As we have seen in Chapter 23 (Waltham 2010), their performance characteristics differ widely, and they will not be further discussed here. The readout scheme depends as well on the sensor type (CCD, CID, or APS) and which operational scheme is supported by the design of the sensor, and this determines the majority of its performance characteristics.

General performance characteristics and limitations

The characteristic of the imaging sensor determines most of the performance of the camera. There are, however, performance limitations that are generally applicable to any of the sensor types. The most important limitation is related to the dynamic range of the entire camera system, which may be due to the intensifier or the sensor (or both). However, the combination of intensifier with the imaging sensor provides us with a variety of design flexibilities to adapt the camera to a range of incoming photon fluxes, corresponding to a vast dynamic range of about eight orders of magnitude.

The light spots of the intensifier are generally larger than a sensor pixel, and the size and intensity may vary strongly depending on the operating voltage of the intensifier. In principle, the photon-electron-photon conversion of the intensifier may be such that a single photon can produce a light spot that fills the full-well charge capacity of the image sensor. This obviously leads to a *local dynamic range* limitation caused by the amplification of the intensifier. The dynamic range is, first of all, determined by the resolution (depth) of the sensor electronic digitization. For an optimal adaptation of the intensifier with the image sensor, the detected incoming photon should be converted into one digital unit (DN).

For high photon input rates the intensifier has to be operated at very low gain, in order to adjust the intensifier such that saturation is avoided within the given exposure time. For very high input rates it may be necessary to reduce the MCP high voltage to a very low gain setting. Since the phosphor also has a high conversion efficiency (typically 100 photons per electron) it may be useful, in order to reduce the brightness of the light pulse, to implement an attenuator within the fibre optic coupler behind the phosphor screen. In this way, the charge created per photon in the image sensor (converted to DN per photon) can be adjusted without compromising the quantum detection efficiency.

For low photon input rates, on the other hand, the high amplification of the intensifier provides the possibility to operate the camera in photon-counting mode. In this mode, the MCPs are used at a higher gain producing a spot of light for each photon event. If the spot of light is large enough, the event positions may be determined within the image sensor by finding the centre of gravity of the charge signal level of adjacent pixels. The event locations are then passed on to memory and binned into an image. Such a centroiding scheme achieves a much better accuracy than the pixel resolution of the sensor (Vallerga et al 1997). In addition, before finding the centroid of each event, the signals can be discriminated against dark noise and single-pixel noise events caused by, e.g., cosmic rays or "hot" pixels. As a result the camera is free of dark noise and can be used at extremely low input signals. At the same time, since dark noise is discriminated, the sensor can be operated without the need of cryogenic cooling.

Past and future intensified cameras for space missions

Soon after CCDs became available and space-qualified for scientific camera systems, efforts started to use them with microchannel plate intensifiers to increase the sensitivity in certain wavelength ranges. The first intensified cameras were developed in the late 1970s for the *Spacelab 1* mission. Further developments of ICCD cameras for space missions followed during the 1980s (Torr et al 1986). Due to the high efficiency of CCDs in the visible, they can be operated with intensifiers at very low gain resulting in a high count rate capability. Such a camera was built for the Coronal Diagnostic Spectrometer (CDS) (Thompson 1999), which has been operating almost continuously for more than 13 years on the *SOHO* spacecraft. The camera uses a single-MCP intensifier for the spectral bands 31 nm to 38 nm and 51 nm to 63 nm. Although the electrons are accelerated onto the phosphor anode on a fibre optic plate, a single lens focuses the image from the fibre optic window onto the CCD (1024 × 1024 pixel format). The camera is operated in "normal" mode (not photon counting) with passive cooling to reduce the dark noise.

An ICCD camera (with 2048 × 2048 pixel format) was built for the Tunable X-ray Imager (TXI) rocket experiment for the wavelength range of 17.1 nm to 21.1 nm that was flown in June 2001 (Golub et al 2002). The camera used a phosphor coated fibre optic taper *in front of* the MCP intensifier and another phosphor coated fibre optic plate as anode for coupling with the CCD. The intensifier, in this case, amplified the image of the phosphor and operated as a shutter by switching the high voltage of the MCPs.

A camera with curved intensifier was built for the Wideband Imaging Camera (Mende et al 2000) of the Imager for Magnetopause-to-Aurora Global Exploration (*IMAGE*) mission of NASA. The instrument is sensitive in the spectral region from 140 nm to 190 nm with a caesium iodide photocathode. The MCP is curved to accommodate the focal surface of the concentric optics of the camera. The phosphor anode of the intensifier is deposited on a concave side of the fibre optic output window, which is then coupled to the CCD with a fibre optic taper.

In conjunction with the fast readout of the image sensor, the intensifier enables the counting and position centroiding of each photon event, making this type of camera favourable for observations with low photon flux. Thus, cameras have been developed that can operate in normal (current integrating) mode and in photon-counting mode (Vallerga et al 1995).

Recently, intensified CCD cameras have been built for the UV Monitoring Telescope of ESA's *XMM-Newton* mission (Mason et al 2001), and a similar camera for the *Swift* mission (Gehrels 2005) in the Ultraviolet and Spectroscopic Telescope (UVOT) instrument. Both cameras cover a very large spectral range from 170 nm to 600 nm using an S20 photocathode on the inside of the entrance window and a fibre optic taper coupling with the CCD.

In an attempt to avoid the shortcoming of CCDs, mainly the transfer of change along pixel rows, and aiming at a high-speed readout system, an intensifier was coupled with a CID (Charge Induction Device) that allows the addressing of individual pixels and a fast readout of sub-arrays. Such a camera was built for the Optical Monitor on the *XMM-Newton* mission (Morrissey et al 1998).

With the same reasoning, CMOS-APS sensors are now more often chosen as the image sensor for space cameras. Since these devices avoid charge transfer, they are more resistant to degradation by space radiation. In addition, the random accessibility of the image area and the flexible readout modes (snapshot or rolling shutter) that avoid the need for a mechanical shutter, make this type of sensor attractive for space applications. Since their sensitivity in the VUV is still low compared to microchannel plate detectors, they will, in this wavelength range, preferably be used with intensifiers. With the availability of space-qualified APS sensor devices (of 1024 × 1024 pixel format), cameras with intensifiers are now being built for future missions. The first camera of this type was flown successfully in November 2007 on the Solar EUV Normal-Incidence Spectrometer (EUNIS) rocket experiment (Thomas and Davila 2001). This camera uses three APS sensors (with 9 μm pixel size) connected with three fibre optic couplers to one fibre optic output window of the intensifier. In this way, a large spectral range in the focal plane can be covered by the large area of the MCP intensifier, to be imaged by the small size of three active pixel sensors. Another IAPS camera is to be flown in 2010 on the rocket spectrograph RAISE (Hassler et al 2004). For this instrument, two cameras have been designed for high count rates with a single MCP operating at relatively low gain and a STAR-1000 APS sensor (Cypress Inc.) with 1024 × 1024 pixels that can be read out at a cadence of 10 Hz. To enhance the sensitivity in the large wavelength range of the spectrograph (between 60 nm and 156 nm), the

MCP of one camera has been coated with potassium bromide and the other with caesium iodide.

Bibliography

Gehrels N (2004) The *Swift* X-ray burst mission. New Astron Rev 48:431–435

Golub L, Deluca E, Hamilton P (plus four authors) (2002) A photometric imaging solar telescope, tunable in the extreme ultraviolet, utilizing multilayer X-ray optics. Rev Sci Instrum 73:1908–1913

Hassler DM, DeForest CE, Slater D (plus four authors) (2004) Future observations with RAISE, 35th COSPAR Scientific Assembly, Paris, France:2280

Mason KO, Breeveld A, Much R (plus 16 authors) (2001) The *XMM-Newton* Optical/UV Monitor Telescope. Astron Astrophys 365 1:L36–L44

Mende SB, Heetderks HD, Frey HU (plus 12 authors) (2003) Far ultraviolet imaging from the *IMAGE* spacecraft: 2 Wideband FUV imaging. Space Sci Rev 91:271–285

Morrissey P, Norton T, Kimble R (1998) A photon counting intensified CID detector for space astronomy. Proc SPIE 3356:1036–1045

Siegmund OHW (1999) Microchannel plate detector technologies for next generation UV instruments. ASP Conf Ser 164:374–391

Thomas RT, Davila JM (2001) EUNIS: a Solar EUV Normal-Incidence Spectrometer. Proc SPIE 4498:161–172

Thompson WT (1999) UV detectors aboard *SOHO*. Proc SPIE 3764:196–208

Timothy JG (2010) Microchannel plates for photon detection and imaging in space. ISSI SR-009:365–391

Torr MR, Torr GT, Baum R, Spielmaker R (1986) Intensified-CCD focal plane detector for space applications: a second generation. Appl Opt 25:2768–2777

Vallerga J, Jelinsky P, Siegmund OHW (1995) Initial results from a photon counting intensified CCD development. Proc SPIE 2518:410–421

Vallerga J, Siegmund OHW, Dalcomo J, Jelinsky P (1997) High-resolution (< 10 μm) photon-counting intensified CCD. Proc SPIE 3019:156–167

Waltham N (2010) CCD and CMOS sensors. ISSI SR-009:391–408

— 26 —

Solar-blind UV detectors based on wide band gap semiconductors

Udo Schühle[I] and Jean-François Hochedez[II]

Abstract

Solid-state photon detectors based on semiconductors other than silicon are not yet considered mature technology but their current development opens new possibilities, also for space observations. Such devices are especially attractive for ultraviolet radiation detection, as semiconductor materials with band gaps larger than that of silicon can be produced and used as "visible-blind" or "solar-blind" detectors that are not affected by daylight. Here we evaluate the advantages of such detectors compared to silicon-based devices and report on the semiconductor detectors that have been fabricated in recent years with materials having large band-gap energies. We describe the most common pixel designs and characterize their general properties.

Wide band gap materials and solar blindness

The present chapter addresses the prospects provided by sensors based on wide band gap materials (henceforth WBG materials or WBGM) as the "active" semiconducting volume. The latter corresponds usually to a solid-state layer where the photons create electron/hole pairs, subsequently collected to produce the signal. WBG materials have also been applied to photo-emissive detectors, e.g., as photocathode coatings on micro-channel plate detectors (Siegmund et al 2003), but in this chapter, we restrict ourselves to solid-state ultraviolet (UV) detection devices made of WBG semiconductors where charge creation and charge collection are taking place inside the material.

One advantage of detectors made of WBG materials is that they can be made "solar-blind". However, solar blindness is a relatively ill-defined concept because it would depend on typical outdoor lighting conditions varying with local time and the Earth's atmosphere. "Visible-blind" UV detectors are required to have a higher responsivity in the UV range than in lower energy ranges. The cut-off of a visible-blind detector is thus at a wavelength shorter than 400 nm. Detectors with a cut-off below 280 nm could be defined as solar-blind because they respond

[I]MPS—Max-Planck-Institut für Sonnensystemforschung, Katlenburg-Lindau, Germany
[II]ROB—Royal Observatory of Belgium, Brussels

only to UV radiation with wavelengths shorter than the solar radiation that can penetrate the atmosphere of the Earth. They produce no measurable signal if they are exposed to normal outdoor lighting. In space, the concept of solar blindness would need to be redefined since there is no atmosphere left between the detector and the Sun. It would be appealing to define solar blindness as insensitivity to the solar black-body spectrum. Solar-blind detectors should hence not respond to the FUV, nor photons of lower energy (200 nm to 1220 nm), because the photospheric black-body radiation still prevails in those ranges (cf., Figure 26.1). However, WBG material sensors cannot offer such ideal properties. Real-world solar-blind detectors are sensitive in the VUV (10 nm to 200 nm) or in the XUV (1 nm to 30 nm) ranges, but as little as possible to the MUV (200 nm to 300 nm) and higher wavelengths. To illustrate the above, in Figure 26.1 we relate the quantum efficiency of a diamond photodiode with the solar spectral irradiance as seen from space and from Earth at sea level.

Which are the WBG materials? In case there is no impurity or defect, the minimum photon energy needed to create an electron-hole pair as a signal in the conduction band is associated with the band gap of the material. As the energy of the band gap is thus equal to the energy of the cut-off, we consider the gap between the valence and the conduction bands of the semiconductor to be "wide" if it enables visible-blind detection, that is, if the gap is above ≈ 3.1 eV (400 nm). The WBG materials of interest are C* (diamond), $Al_xGa_{1-x}N$ with all possible stoichiometries from GaN to AlN, BN, ZnS, ZnO, ZnSe, and combinations as, for example, $Al_{1-x}B_xN$. SiC will not be dealt with as it falls short with indirect band gaps at 3.25 eV (4H-SiC), 3.0 eV (6H-SiC), and 2.2 eV (3C-SiC). We focus in the following on GaN (3.4 eV), diamond (5.5 eV), AlN (6.2 eV), and cubic boron nitride (cBN, 6.4 eV).

Actual WBG materials are not ideal by any means. They usually exhibit many crystallographic imperfections influencing the mobility of charge carriers. This requires the artificial synthesis of WBG material layers, because natural substrates are rare and actually inappropriate without certain doping profiles. Although the various techniques for growing epitaxial layers have made immense strides in the recent years, diamond, cBN, AlN, and even GaN must still be acknowledged as relying on new/emerging technology.

Radiation hardness, dark current, and more

A consequence of the large band gap is the reduction of the dark current, permitting room temperature operations. In addition to the improved noise performance, this is a great advantage in terms of sparing the resources (mass and power) needed to cool a subsystem in space. Moreover, it brings the additional benefit of avoiding a cold surface that would trap residual molecules. Not only water can deposit on the cold surface; it is well known that hydrocarbons outgassing from organic materials used in space instrumentation may deposit on the cold detector, and polymerize under UV illumination, leading to irreversible opaque absorbing layers (Schühle 2003). It is therefore highly beneficial to operate UV sensors in warm conditions.

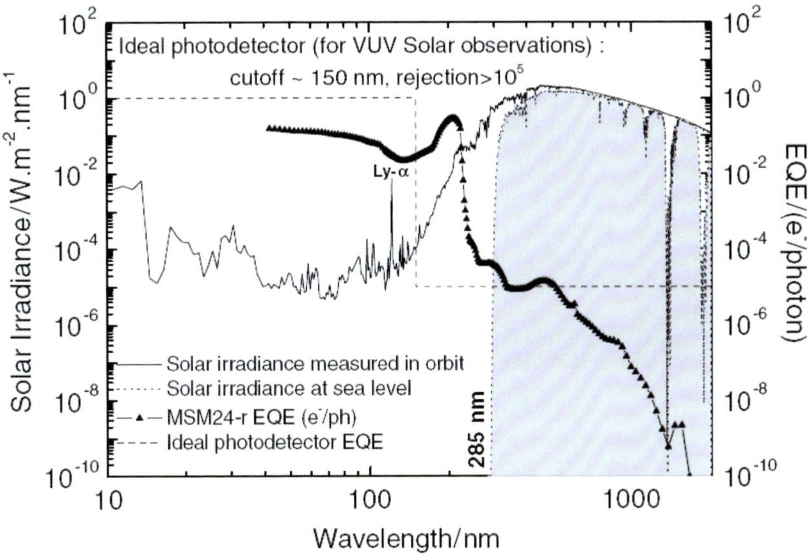

Figure 26.1: The solar spectral irradiance measured in space and from Earth (left ordinate axis) compared with the quantum efficiency (right ordinate axis) of an ideal WBG detector for solar observations and of a real diamond MSM detector (BenMoussa et al 2008b).

An additional asset for space applications is radiation hardness (see, e.g., Rahman et al 2004), which is of paramount importance for the lifetime of missions and instruments (Hochedez et al 2003; Schühle et al 2004). The mobility of charge carriers in WBG materials can be very high (C*, GaN). It provides a fast response of detectors and enables high-cadence observations, including new photon-counting readout schemes. To summarize, in Table 26.1 we list the main differences between silicon-based and WBGM-based devices.

Table 26.1: Differences between silicon- and prospective WBGM-based devices.

Silicon-based detectors	WBG-based detectors
Dark current at room temp., need cooling	Lower dark current
Cold trap of contaminants	Less contamination (long-term stable)
Degradation by displacement radiation	Rad-hardness (increased lifetime)
QE evolving and inhomogeneous	QE stable and better flatfield
Minimal pixel size ≈ 5 µm	Sustain higher electrical fields
Need blocking filters to suppress visible	Need fewer blocking filters

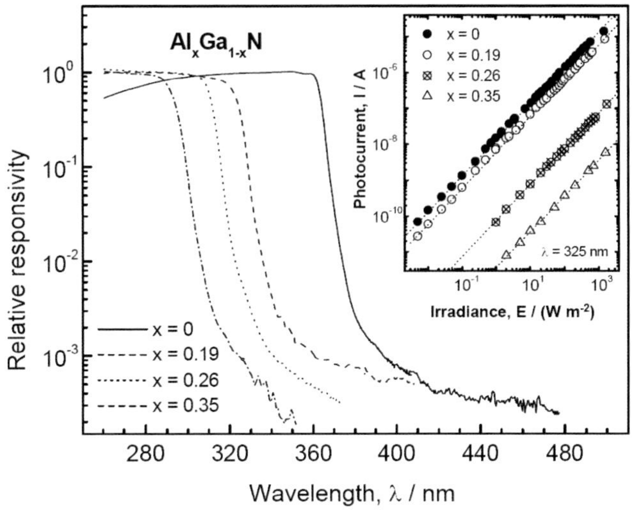

Figure 26.2: Relative responsivity of AlGaN Schottky diodes with different aluminium content at their threshold energies (after Muñoz et al 2001). The insert shows their linearity over three decades of irradiance.

Solutions for WBG materialbased sensors

Most of the WBGM-based devices built so far have been designed specifically for visible-blind UV detection applications. In recent years different designs have been developed, from single-pixel photodiode-type detectors to imaging arrays, based on various WBG materials. The "Blind to Optical Light Detectors" (BOLD) project has been initiated in 1999 to design and develop diamond, nitride, and other WBGM–based detectors, especially for astrophysics (Hochedez et al 2000). In 2010, the BOLD team expects to characterize a novel AlGaN-based imaging prototype. In practice, the material can be selected according to the specific requirements of the application.

WBG materials usually have a large penetration depth above the wavelength threshold, and a shallow penetration depth below. This leads to optimal charge collection and therefore to high efficiency in a passband around the band gap. The detector itself serves thereby as a spectral filter. By selecting the material according to the wavelength of application, the detector is most efficient in that given spectral region (see, e.g., BenMoussa et al 2006a and references therein). Devices built from III-V nitrides are promising for a wide range of UV applications, and technological progress has been made with GaN and AlGaN devices (Pearton et al 1999; Monroy et al 2003; Aslam et al 2005). With the band-gap energies of 3.4 eV of GaN and 6.2 eV of AlN it is possible to build UV detectors with precisely selected responsivity thresholds between these energies by selecting the fraction of aluminium in the $Al_xGa_{1-x}N$ alloy material (Muñoz et al 2001; Sandvik et al 2001; Tripathy et al

2007). Figure 26.2 shows as example the relative responsivity of AlGaN devices (Schottky-diodes) with different aluminium content at their threshold energies.

Diamond is another promising WBG material. Both single-crystal and polycrystalline thin-film diamond devices have been explored as photoconductive detectors (McKeag and Jackman 1998). Its band-gap energy of 5.5 eV corresponds to a threshold wavelength of 225 nm. It has been identified as a promising material for solar-blind UV detectors in space environments.

A possible alternative to diamond is boron nitride (BN) with a band-gap energy of 6.4 eV. Recently, metal-semiconductor-metal (MSM) photoconductors based on cubic BN films have been produced (Soltani et al 2008). The main obstacle in developing UV detectors with WBGM is in producing high-quality material by growing high-purity semiconductor thin film layers with minimal defects and dislocations on the substrate material. Major problems relate to impurities, vacancies, lattice discontinuities, phase boundaries, lattice mismatch with the substrate, and appropriate doping with n- and p-type materials. Many techniques for epitaxial growth are being investigated but their description is beyond the scope of this book.

Pixel architectures – a brief comparative assessment

There are different types of semiconductor detectors, the most important are: photoconductors, MSM diodes, Schottky barrier and p-i-n photodiodes. The schematic structures are depicted in Figure 26.3, using AlGaN as the active WBG semiconductor material. The following subsections provide a brief comparison of performances and limitations. For more details we refer to the review by Razeghi and Rogalski (1996).

Photoconductors: Photoconductors consist of a semiconductor layer with two ohmic contacts and a voltage bias applied between them. Under illumination, the conductance of the semiconductor changes with the intensity of the incident radiation. The current is mainly due to majority charge carriers since they are free to flow across the ohmic contacts. However, the majority carrier current depends on the presence of the minority carriers. The minority carriers pile up at one of the contacts, where they cause additional injection of majority carriers (current continuity and charge conservation) until the minority carriers recombine. This effect can cause large "photoconductive" *gain*, which depends primarily on the ratio of the minority carrier lifetime to the majority carrier transit time. Long carrier lifetimes therefore cause large gain, but also a slow response time. The photoconductive gain is one of the good features of photoconductors. An undesirable attribute of photoconductors is a relatively high dark current. However, photoconductors are simple and useful for those applications where speed and dark current are not particularly important.

MSM photodiodes: The MSM diode is made of two Schottky barrier diodes back-to-back. The entire semiconductor layer between two metal electrodes becomes fully depleted under sufficient bias. The geometric parameters of a conventional MSM photodetector pixel element determine its performance. The device consists of alternating metal contacts deposited on the semiconductor active layer. MSM structures represent a very simple photodetector design. Their main limi-

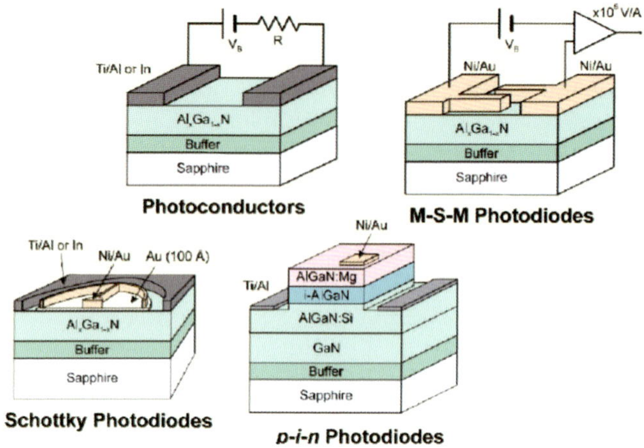

Figure 26.3: Schematic structure of different semiconductor photodetectors (courtesy of E. Muñoz).

tations include relatively large dark current (since a bias voltage is required) and somewhat lower external quantum efficiency due to electrode shadowing.[1]

Schottky diodes: Schottky diodes are a good alternative to MSM detectors due to the fact that these detectors can operate in photovoltaic mode, i.e., without bias voltage (and no resulting dark current), and therefore with high signal/noise ratio. A Schottky diode requires an ohmic contact, i.e., a conductive layer like, n-type AlGaN.

Pin photodiodes: The most common form of photodiode used is the pin diode which consists of a p-type, highly doped, transparent contact layer on top of an undoped absorbing layer and an n-type, highly doped contact layer on the bottom. Since there are very few charge carriers in the intrinsic region, the space charge region reaches completely through from the p to the n region. This results in detectors with several important advantages: they are fast, stable, linear and low-noise. The primary disadvantage of the pin diodes is that with moderate doping concentrations in the conducting regions for low resistance, the resulting depletion region is quite thin. This causes two problems:
— low efficiency when little absorption occurs in a thin depletion region, and
— relatively high capacitance which decreases device speed.

[1]For interdigitated fingers, Au or Al ($<$10 nm) can be used as EUV semi-transparent contact. Three different approaches to improve the responsivity of MSM detectors are: size reduction towards sub-micrometre fingers, use of semitransparent electrodes, or asymmetric electrodes. All these solutions provide better detector characteristics in terms of dark current, UV/visible contrast, linearity, and responsivity. The shadowing of the active area by the interdigitated contacts is the main drawback of the MSM detector, reducing their efficiency. The spacing-to-width ratio of the fingers is the major parameter that determines the detector's reponsivity.

A general rule is that only carriers generated within the depletion region are *efficiently and rapidly* collected as photocurrent. The goal is to create a diode with a wide depletion region.[2]

In summary, the preferred type of photosensitive diode for a given application depends on its inherent properties and the choice may depend on various constraints, not only related to performance but also to ease of the fabrication process: photoconductors and Schottky diodes can be built with co-planar structures, making their production simpler than pin diodes which consist of different layers (sandwich-type) with different doping.

In general, photoconductive devices have a high sensitivity and may profit from internal photoelectric gain (in constant voltage operation mode). Since they need a bias voltage, they have low impedance but usually a higher background (dark signal) than Schottky or pin diodes, in which the charge is created by photo effect against an internal potential barrier. The latter provides the advantages of bias-free operation, low background signal, fast response, and better signal stability.

Responsivity

The responsivities reported are quite high just above the band-gap energy, reaching values up to 0.2 AW^{-1}. For example: Parish et al (1999) report on AlGaN pin detectors on laterally epitaxially overgrown GaN with a dark current level of 10 $nAcm^{-2}$ at -5 V bias voltage and a peak responsivity of 0.05 AW^{-1} at 285 nm. Motogaito et al (2001) realized a responsivity of 0.15 AW^{-1} with GaN Schottky diodes. Pau et al (2004) have reported responsitivities of GaN Schottky diodes of 0.03 AW^{-1} and of MSM photoconductors of 0.04 AW^{-1}. The responsivity depends also on the separation of the charge-collecting metal electrodes on the semiconductor surface. The smaller the distance between the metal electrodes, the better is the charge collection. Malinowski et al (2008) show GaN devices (threshold at 370 nm) with a peak responsivity of 0.162 AW^{-1} at 360 nm with a rejection of longer wavelengths of almost four orders of magnitude towards 400 nm. High responsivity can be obtained with avalanche photodiodes using their internal gain with reverse-biased operation. Tut et al (2007) report an internal gain of 1560 with AlGaN photodiodes and a peak responsivity of 0.13 AW^{-1} at 272 nm.

While the efficiency of the devices is highest just above the threshold, at shorter wavelengths, where the penetration depth decreases, the detection of charge carriers is decreased by processes such as surface recombination and photoemission losses (Motogaito et al 2001, BenMoussa et al 2008a), a complication that needs to be carefully addressed in the design of detectors for the VUV range.

Photodetectors based on diamond material could recently progress due to success in producing high-quality thin films with sufficiently low defects and concentrations of n-type and p-type dopants (e.g., Nesladek 2005). The first diamond

[2]The material quality is crucial since the optical performance in the EUV is strongly influenced by the surface characteristics (generation of e-h pairs near the surface, dead layer, oxide). A high-quality active layer is required and the thickness for total absorption should be optimised for low defects, cracks-free, low roughness, and good structural homogeneity. The doping level of the active layer is also one of the most critical parameters and should be below 10^{16} cm^{-3} in order to achieve a depletion reach-through region for reasonable electrode gap spacing.

Figure 26.4: Diamond MSM photodiode of the LYRA EUV radiometer aboard *PROBA2* (courtesy A. BenMoussa).

detectors that will be flown on a space mission will be the detectors of the solar XUV-VUV radiometer LYRA aboard the *PROBA2* satellite (see Figure 26.4, Hochedez et al 2006; BenMoussa et al 2006). LYRA will carry several large area (4.6 mm diameter) diamond diodes which have been especially fabricated for this mission. They have a sufficiently uniform response over their active area and reach a peak responsivity of 0.048 AW^{-1} at 210 nm (BenMoussa et al 2008b).

Imaging arrays and hybridization

We distinguish two technologies for the fabrication of imaging arrays: first, deposition of the active material for charge creation directly on the signal readout integrated circuit (ROIC) and second, hybridization, where the WBG semiconductor array and the silicon-based readout circuitry are developed and fabricated independently to the same pixel scale (pitch) and then bonded to each other. The two arrays are combined by indium bump bonding, which involves flipping one array on the other (flip-chip integration). The definition of indium bumps is presently limiting the miniaturization of pixel size and pitch. Figure 26.5 shows a detail of the indium bump structure processed on a sensor array.

Using this technique, focal plane arrays (FPA) of AlGaN detectors have been produced by several groups, e.g., with 256×256 pixels (Lamarre et al 2001; Aslam et al 2005) and 320×256 pixels (Long et al 2002; McClintock et al 2005). These devices employ deposition of the array structure on a transparent substrate and illumination from the back side (back-illumination) through the substrate. For shorter wavelengths than the transmission cut-off of the substrate material,however, — in the VUV range — front-side illumination of the device is required. Further in the EUV (below $\approx$ 30 nm) back-side illumination is again possible with a device that is back-side thinned, making it transparent to the EUV and X-ray radiation ranges.

By deposing the WBG material directly on the readout circuit, the monolithic array fabrication is less technology intensive in principle, since the fabrication can be carried out on one wafer and the flip-chip bonding process is not required. This

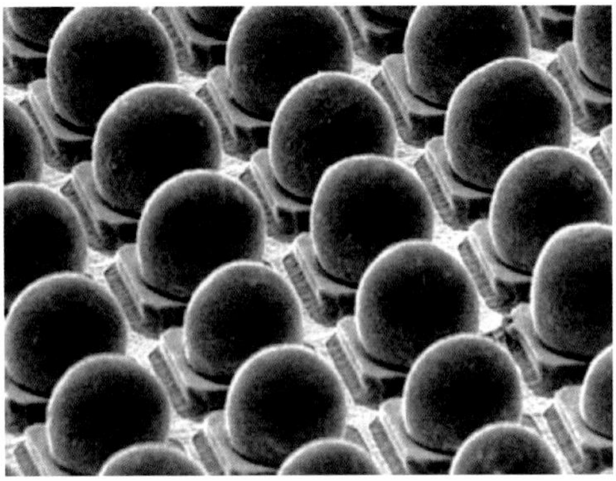

Figure 26.5: Scanning electron microscope (SEM) picture of a detail of a 256 × 256-element sensor array realized in III-V material after the indium bump processing. The pixel pitch is 25 µm (courtesy of IMEC, Belgium).

technique has been applied with infrared-sensitive materials and may offer larger yields of ultraviolet detector arrays in the future.

It is the conviction of the authors that future scientific discoveries rely on new detector developments. The presented progress with WBGMs bears out the prospects for enhanced performance of future missions.

Acknowledgements

We gratefully acknowledge the contributions to this work by A. BenMoussa and by other colleagues of the BOLD projects.

Bibliography

Aslam S, Yana F, Pugel DE (plus six authors) (2005) Development of ultra-high sensitivity wide-band gap UV-EUV detectors at NASA Goddard Space Flight Center. Proc SPIE 5901:59011J-1–12

BenMoussa A, Schühle U, Scholze F (plus ten authors) (2006a) Radiometric characteristics of new diamond PIN photodiodes. Meas Sci Technol 17: 913–917

BenMoussa A, Hochedez JF, Schühle U (plus 13 authors) (2006b) Diamond detectors for LYRA, the solar VUV radiometer on board PROBA2. Diamond Rel Mat 15:802–806

BenMoussa A, Hochedez JF, Dahal R (plus seven authors) (2008a) Characterization of AlN metal-semiconductor-metal diodes in the spectral range of 44–360 nm: Photoemission assessments. Appl Phys Lett 92:0221081–0221083

BenMoussa A, Soltani A, Haenen K (plus eight authors) (2008b) New developments on diamond photodetector for VUV solar observations. Semicond Sci Technol 23:0350261–0350267

Hochedez JF, Verwichte E, Bergonzo P (plus eight authors) (2000) Future diamond UV imagers for solar physics. Phys Stat Sol A 181:141–149

Hochedez JF, Schühle U, Pau JL (plus 23 authors) (2003) New UV detectors for solar observations. Proc SPIE 4853:419–426

Lamarre P, Hairston A, Tobin S (plus 13 authors) (2001) AlGaN p-i-n photodiode arrays for solar-blind applications. Phys Stat Sol A 188:289–292

Long JP, Varadaraajan S, Matthews J, Schetzina JF (2002) UV detectors and focal plane array imagers based on AlGaN p-i-n photodiodes. Opto-Electr Rev 10:251–260

Malinowski P, John J, Lorenz A (and 13 authors) (2008) AlGaN photodetectors for applications in the extreme ultraviolet (EUV) wavelength range. Proc SPIE 7003A: 70030N1–70030N8

McClintock R, Mayes K, Yasan A (plus three authors) (2005) 320 × 256 solar-blind focal plane arrays based on $Al_xGa_{1-x}N$. Appl Phys Lett 86:0111171–0111173

McKeag RD, Jackman RB (1998) Diamond photodetectors for UV laser-based applications. Proc SPIE 3484:182–194

Monroy E, Omnès F, Calle F (2003) Wide-band gap semiconductor ultraviolet photodetectors. Semiconductor Science and Technology 18:R33–R51

Motogaito A, Ohta K, Hiramatsu K (plus four authors) (2001) Characterization of GaN based UV-VUV detectors in the range 3.4–25 eV by using synchrotron radiation. Phys Stat Sol A 188:337-340

Muñoz E, Monroy E, Pau JL (plus three authors) (2001) III nitrides and UV detection. J Phys Condens Matter 13:7115–7137

Nesladek M (2005) Conventional n-type doping in diamond: state of the art and recent progress. Semicond Sci Technol 20:R19–R27

Parish G, Keller S, Kozodoy P (plus seven authors) (1999) High-performance (Al,Ga)N-based solar-blind ultraviolet p-i-n detectors on laterally epitaxially overgrown GaN. Appl Phys Lett 75:247–249

Pau JL, Rivera C, Muñoz E (plus six authors) (2004) Response of ultra-low dislocation density GaN photodetectors in the near- and vacuum-ultraviolet. J Appl Phys 95:8275–8279

Pearton SJ, Zolper JC, Shul RJ, Ren F (1999) GaN: Processing, defects, and devices. J Appl Phys 86:1–78

Rahman M, Al-Ajili A, Bates R (plus 14 authors) (2004) Super-radiation hard detector technologies: 3-D and widegap detectors. IEEE Trans Nucl Sci:2256–2261

Razeghi M, Rogalski A (1996) Semiconductor ultraviolet detectors. J Appl Phys 79:7433–7473

Sandvik P, Mi K, Shahedipour F (plus four authors) (2001) $Al_xGa_{1-x}N$ for solar-blind UV detectors. J Crystal Growth 231:366–370

Schühle U (2003) Cleanliness and calibration stability of UV instruments on *SOHO*. Proc SPIE 4853:88–97

Schühle U, Hochedez JF, Pau JL (plus 18 authors) (2004) Development of imaging arrays for solar UV observations based on wide band gap materials. Proc SPIE 5171:231–238

Siegmund OHW, Tremsin A, Martin A (plus three authors) (2003) GaN photocathodes for UV detection and imaging. Proc SPIE 5164:134–143

Soltani A, Barkad HA, Mattalah M (plus nine authors) (2008) 193 nm deep-ultraviolet solar-blind cubic boron nitride based photodetectors. Appl Phys Lett 92:0535011–0535013

Tripathi N, Grandusky JR, Jindal V, (plus two authors) (2007) AlGaN based tunable hyperspectral detector. Appl Phys Lett 90:2311031–2311033

Tut T, Gokkavas M, Inal A, Ozbay E (2007) $Al_xGa_{1-x}N$-based avalanche photodiodes with high reproducible avalanche gain. Appl Phys Lett 90:1635061–1635063

— 27 —

Superconducting tunnel junctions

DIDIER D.E. MARTIN[I] AND PETER VERHOEVE[I]

Abstract

Superconducting tunnel junctions (STJ) are a class of cryogenic detectors that rely on the generation of free charge carriers by breaking Cooper pairs in a superconducting material with the use of absorbed photon energy. In an STJ, consisting of two superconducting films separated by a thin insulating barrier, the charge carriers can be detected through the tunnel-current pulse they produce if the STJ is under a finite voltage bias. The number of charge carriers generated is proportional to the energy of the absorbed photon, and, depending on the material of choice, ranges from several hundreds to a few thousand per electronvolt of photon energy. This allows STJs to be used as photon-counting detectors with intrinsic energy resolution over a wide energy band from the near infrared to well into the X-ray band. The operating temperature is typically at 10 % of the critical temperature T_c of the superconducting material and may range from 0.1 K to 1 K. Although they have not been deployed in space applications yet, they could well be envisaged as spectrometers with an energy-resolving power of several hundreds in the soft X-ray range, or as highly efficient order-sorting detectors in UV grating spectrographs. STJs can be used simultaneously as absorbers and read-out elements, or alternatively, if a larger sensitive area is required, two or more can be attached as read-out elements to a separate absorber. The state of the art performance comprises energy resolutions of 2 eV to 11 eV in the soft X-ray energy range from 0.4 keV to 5.9 keV, and of 0.1 eV to 0.2 eV in the near-infrared and visible range from 0.5 eV to 5 eV. Imaging arrays of >100 pixels of close-packed STJs have been made and operated in ground-based astronomical applications.

Introduction

Superconducting tunnel junctions (STJ) are a family of cryogenic sensors. As we shall describe in the next paragraph, they rely on the creation of 'free' charge carriers by the absorbed radiation. The energy required to generate these excitations is typically three orders of magnitude lower than that in classical semiconductors. This makes them extremely sensitive to radiation from the sub-millimetre to the X-ray wavelengths, in principle allowing single-photon detection with intrinsic energy

[I]Advanced Studies and Technology Preparation Division, ESA-ESTEC, The Netherlands

resolving power. STJs operate on a different principle than microcalorimeters and consequently have their own strengths and weaknesses. Although neither of them have been flown in space yet, it is expected that each will have their own domain of application. Advantages of STJs over microcalorimeters can be summarized as follows:

- STJs are directly grown on solid substrates (e.g., sapphire) and do not require complex micromachining processes.
- For a given energy resolution, operating temperatures can be substantially higher (e.g., 300 mK for STJs vs. 50 mK for transition-edge sensors, TES).
- Quasiparticle relaxation processes are typically faster than the thermal processes on which calorimeters rely. This makes STJs faster detectors and implies they could be used with higher photon-counting rates.

STJs or superconductor-insulator-superconductor (SIS) junctions were first used by Giaever (1960) to study quasiparticle densities in Sn. In 1969 Wood and White (1969) observed the first α-particle induced pulses from a biased STJ made of Sn electrodes. A wider interest was generated through the efforts of Kurakado and Mazaki (1980) to further develop Sn STJs, which resulted in the first X-ray detections with a photon energy, E, of 5.9 keV in 1986 (Twerenbold 1986; Kraus et al 1986), still with Sn STJs. Sn proved to be unstable under thermal cycling and was replaced by materials like Nb, Al, and Ta. Since then, many other groups have been developing STJs for such different applications as particle and nuclear physics, astrophysics, material analysis, and mass spectroscopy.

Principle of operation of STJs as radiation detectors

In its simplest form an STJ consists of two superconducting layers, acting as radiation absorbers, separated by a thin insulating barrier (see Figure 27.1, left). Below the critical temperature T_c the electrons in a superconductor are organised in Cooper pairs, with a binding energy 2Δ with Δ the superconductor's energy gap which is inversely proportional to T_c. At finite temperature, a fraction of the Cooper pairs is broken into excited states, called quasi-particles (QP), through the continuous pair breaking by thermal phonons with energy $> 2\Delta$ and the recombination of QPs with the emission of a phonon. The rates for these processes are material dependent and tend to get slower for lower-T_c superconductors (Kaplan et al 1976). A semiconductor representation of the energy-level diagram with electron-like and hole-like quasiparticles for a biased STJ at $T > 0$ is shown in Figure 27.1, middle. The charge carriers associated with QPs can tunnel quantum-mechanically across the insulating barrier. Four different tunnel processes are possible, two of which are depicted in Figure 27.2 for the case of a symmetric STJ under a bias voltage U_b. The other two (3 and 4) are the reverse processes of these. Processes 1 and 3 involve the direct transfer of a quasiparticle from left to right and vice versa, respectively, and correspond to an electron transferred in the same direction. Processes 2 and

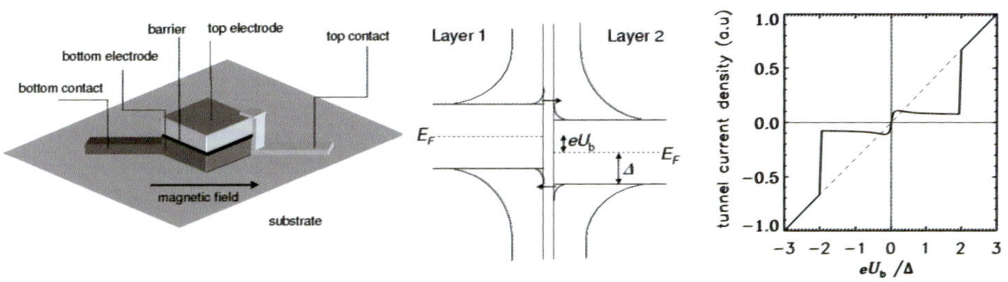

Figure 27.1: Left: Schematic representation of a superconducting tunnel junction. Middle: the energy level diagram in the semiconductor representation for an STJ, voltage biased at U_b. Right: The current-voltage characteristic for this structure.

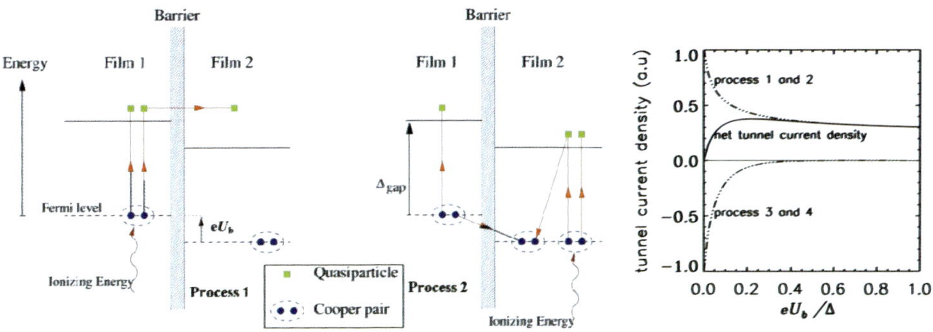

Figure 27.2: Left: Schematic representation of the possible tunnel processes that can occur in an STJ. Processes 3 and 4 (not indicated) are the exact opposite of 1 and 2. The vertical scale represents energy. Right: Resulting current-voltage characteristic for an STJ at $T > 0$, in arbitrary units.

4 involve the additional annihilation of a Cooper pair in one film and creation in the other film. Effectively in process 2 a QP is transferred from right to left, but an electron is still going from left to right, as in process 1. At sufficient high bias voltage, the occupancy of the higher energy levels is sufficiently low to leave only processes 1 and 2 significantly contributing to the tunnel current (see Figure 27.2, right). The resulting *I*-*V* curve is schematically shown in Figure 27.1, right. For bias voltages $U_b < 2\Delta/e$ (the sub-gap region), the current is weakly dependent on bias voltage and scales with the temperature dependent quasiparticle density as $n(T) \propto T^{\frac{1}{2}} e^{-\frac{\Delta}{k_B T}}$, with k_B the Boltzmann constant. Above $2\Delta/e$ the device exhibits the normal resistance of the barrier (represented by the dashed line).

The absorption of a photon with energy E in a superconductor is followed by a series of fast processes involving both electrons and phonons (Kozorezov et al 2000, and references therein), in which the photon energy is converted into quasiparticles in excess of the thermal density by the breaking of Cooper pairs. For typical transition metals the time scale of this conversion process ranges from nanoseconds (niobium) to microseconds (hafnium). The average number of excess carriers is $N_0(E) = E/(1.7\Delta)$ (Kurakado 1982; Rando 1992). Thus, in a superconductor such as tantalum (Δ=0.7 meV), the initial mean number of free charge carriers created is $N_0(E) \approx 10^3$ per electronvolt of photon energy. The excess quasiparticle population produced through photo-absorption in one of the electrodes of a voltage-biased STJ will be detected as a tunnel-current pulse with amplitude proportional to the incoming photon's energy. Absorptions in either electrode of the STJ will give rise to the same polarity current pulse, and moreover, quasiparticles that have been transferred to the other electrode by one of the two processes 1 or 2 described above, can continue to contribute to the signal by 'backtunnelling' (Gray 1978) as long as they are available in the electrode volume. On average each quasiparticle will contribute $<n>$ times to the signal before it is lost from the system through recombination or trapping. Hence the mean number of *detected* charge carriers $N = <n> N_0$. The detector responsivity R is referred to as the number of tunnelled electrons per electronvolt of photon energy: $R = <n> /(1.7\Delta)$. For a symmetric STJ the value of $<n> \approx \Gamma_{\mathrm{tun}}/\Gamma_{\mathrm{loss}}$, with Γ_{tun} the tunnel rate across the barrier and Γ_{loss} the loss rate in the system. The tunnel rate for bias voltages $k_{\mathrm{B}}T \ll eU_{\mathrm{b}} < 2\Delta$ can be approximated as (de Korte et al 1992):

$$\Gamma_{\mathrm{tun}} = \frac{1}{4\, e^2 \rho_n N(0) d} \sqrt{\frac{(\Delta + e\,U_{\mathrm{b}})^2}{(\Delta + e\,U_{\mathrm{b}})^2 - \Delta^2}} \quad , \tag{27.1}$$

with ρ_n the resistivity of the barrier, $N(0)$ the normal-state single spin density of state of the superconductor, d the electrode thickness. Typical values for Γ_{tun} in a Ta STJ with $d = 100$ nm and $\rho_n = 2 \times 10^{-6}\ \Omega\,\mathrm{cm}^2$, are of the order of $10^6\ \mathrm{s}^{-1}$. The loss rate Γ_{loss} is ultimately limited by quasiparticle recombination and can be orders of magnitude slower than the tunnel rate. In practice however, other processes like diffusion of the QPs out of the electrode volume or trapping and recombination in defects at surfaces, edges or bulk impurities will dominate the loss rate. Multiple tunnelling can be enhanced by including layers of a lower-gap material on either side of the barrier, such that quasiparticles are collected and confined close to the barrier and the tunnel rate is enhanced (Booth 1987), and away from possible loss sites such as surfaces. Typically $<n>$ can vary widely from 1 to 100 and depends on the size and nature of the STJ. Such 'trapping' layers allow the separation of radiation-absorbing volume and tunnelling volume, which is particularly useful in the case of high-energy radiation for which absorber thicknesses of the order of 1 μm may be required to provide sufficient detection efficiency. Modelling of the time-dependent response to photon absorption by STJs involves a description of quasiparticle tunnel and loss processes, including their interactions with phonons. Traditionally such models start from the Rothwarf-Taylor balance equations for quasiparticles and phonons (Rothwarf and Taylor 1967), which assumes all quasiparticles to be and remain at the superconducting gap energy. More elaborate models now allow for a more realistic energy distribution of quasiparticles, as well as detailed descriptions of quasiparticle loss mechanisms (traps) that can explain the observed dependence of STJ responsivity on external parameters such as bias voltage and temperature (Kozorezov et al 2008). However, the physical nature and origin of these loss mechanisms, which determine the quasiparticle loss rates, and therewith the detector response time, are still largely unknown.

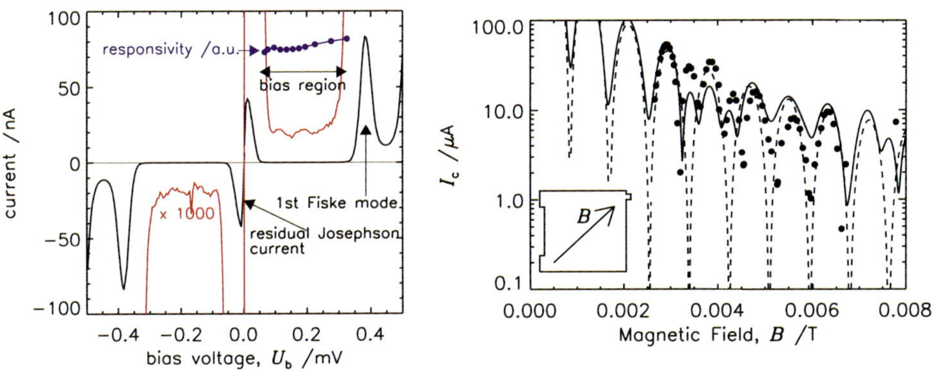

Figure 27.3: Left: Detail of the I-V curve of a 25 μm × 25 μm Ta-based STJ. The blue dots indicate the dependence of the device responsivity on bias voltage. Right: Measured (dots) Josephson current amplitude versus applied magnetic field strength. The dashed and solid lines are predicted curves for a perfectly square STJ and a more realistic shape as shown in the inset, respectively.

In a practical application the STJ will be operated at $T \approx 0.1\, T_c$, sufficiently low to avoid any significant population of thermally excited quasiparticles. In this case the subgap current will be dominated by any residual leakage current of the barrier. In addition, Cooper-pair tunnelling gives rise to the zero-voltage DC-Josephson current (Josephson 1962), and to Fiske steps (Kulik 1965) induced by AC-Josephson currents and occurring at non-zero bias voltages which scale inversely with the lateral size of the STJ. This is illustrated in Figure 27.3 (left), which shows the relevant part of the I-V curve of a 25 μm × 25 μm sized Ta-based STJ. The amplitudes of DC-Josephson current and Fiske steps can be suppressed by the application of a magnetic field parallel to the plane of the barrier (see Figure 27.1). The dependence on magnetic field strength is periodic, with a B^{-n} envelope where n depends on the shape of the STJ (Peterson 1991). Irregularities in the shape of the STJ can distort the periodicity of the B-field dependence. This is shown in Figure 27.3 (right) which compares the measured DC-Josephson current amplitude with predicted curves for a perfectly square STJ and for a more realistic shape. It is important to suppress the Josephson current and the Fiske modes sufficiently to allow stable voltage biasing. It is therefore important that the shapes of individual STJs in a detector array are as identical as possible, so as to enable simultaneous suppression for all STJs with a single uniform magnetic field value. It also implies that there is a practical limit to the size of the STJs, since the decreasing separation of DC-Josephson current and Fiske steps with increasing STJ size will make stable biasing prohibitively difficult.

One way to circumvent this size limitation is to separate the potentially large absorber area from the much smaller tunnel area, a concept first introduced by Kraus et al (1989). In its simplest incarnation such a distributed read-out imaging device (DROID) consists of a linear absorber with an STJ for read-out at either end (see Figure 27.4). Quasiparticles generated by photons in the absorber will diffuse

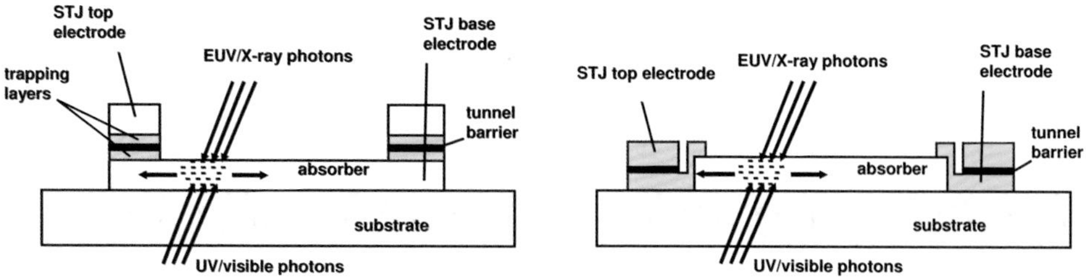

Figure 27.4: Schematic of a 1-dimensional DROID. Left: Stacked configuration: STJs and absorber form an integral structure. Right: Lateral configuration: the STJs are adjacent to the absorber and may be of a different superconducting material.

into the STJs where they are detected. The sum of the two signal amplitudes is a measure for the energy of the incident photon, and the ratio of the signal amplitudes correlates with the position of absorption along the absorber. By choosing the STJ a lower-gap superconductor material or by incorporating lower-gap trapping layers, the quasiparticles can be confined in the STJs, which enhances the signal amplitudes and yields better position resolution. This concept can be expanded to two-dimensional imaging by using a square absorber with four STJs.

Alternatively, the STJs could be made on a dielectric or electrically isolated absorber. Phonons generated by photons hitting the absorber can diffuse into the STJs and break Cooper pairs. If a large number of STJs connected in series is used to cover a large fraction of the backside area of the absorber (Booth et al 1993), a high phonon collection efficiency is possible, in combination with good energy resolution.

The electronic signal chain required to read out STJs is similar to that of many photon counting detectors and always consists of six main components: device biasing; low-noise amplification; signal conditioning; sampling and conversion to the digital domain; digital processing and storage. With the advent of fast and accurate analogue-to-digital converters (ADC) as well as powerful digital processors, the boundary between the analogue and digital domains is rapidly moving closer to the pre-amplification stage. Nevertheless, the first four components of the electronics signal chain are always present. The most critical part is the first stage, amplification and biasing. Since STJs present a fairly high dynamic impedance, the pre-amplifier usually consists of a (room-temperature) charge-sensitive amplifier with a J-FET as input device, although SQUIDs have also been used for this purpose (Frank et al 1996). The feedback circuit, a parallel combination of integrating capacitor and resistor, is tailored to the expected signal amplitudes. In some instances a parallel active device can also be used to facilitate biasing and diagnosis (Martin et al 2003). STJs need to be biased at a precise, very stable and low voltage (typically a few 100 μV). Standard discrete components do not offer this capability. Also, any circuit placed at the input node will affect the noise performance. For this reason, an electrometer operational amplifier is often used. It samples the input (DC) voltage and through a DC feedback-loop will regulate the input node's voltage. The inverting input of the OpAmp can be connected to

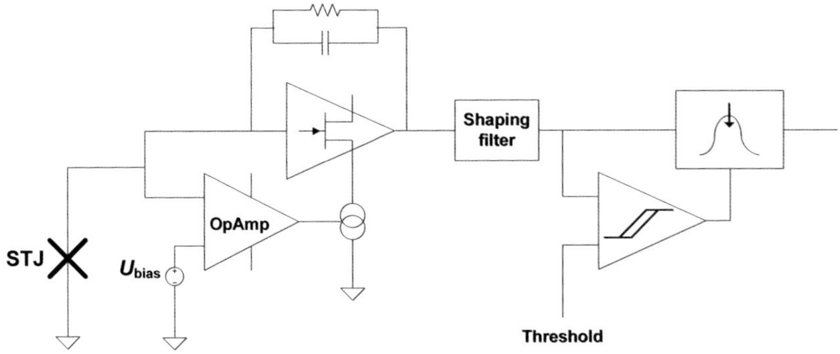

Figure 27.5: Typical STJ readout chain. Coupling is DC to a charge sensitive amplifier.

a digital-to-analogue converter (DAC) to allow software control of the bias voltage (Figure 27.5).

The limiting energy resolution (Fano limit) of an STJ is determined by the variance on the initial number of quasiparticles produced after photo-absorption $N_0(E)$ and given by:

$$\delta E_{\mathrm{Fano}} = 2.355\sqrt{1.7EF\Delta} \quad , \tag{27.2}$$

with $F \approx 0.2$ the Fano factor (Kurakado 1982; Rando 1992). In addition, statistical fluctuations in the tunnelling process will contribute to the energy resolution (Goldie et al 1994), resulting in the so-called tunnel-limited resolution, which is given by:

$$\delta E_{\mathrm{tunnel}} = 2.355\sqrt{1.7E(F+G)\Delta} \quad . \tag{27.3}$$

For a symmetrical STJ, $G = 1 + <n>^{-1}$, which approaches unity for high-responsivity devices. This holds for the case that the entire signal pulse is measured. The tunnel-noise contribution can sometimes be significantly reduced by a proper choice of the electronic integration time (Hiller et al 2001; Verhoeve et al 2002). It should be noted that Fano-limited energy resolution is only achieved if each of the initially generated charge carriers tunnels exactly once across the barrier. In practice, such a detector is difficult to realise and, moreover, its responsivity would be low ($<n>= 1$), which compromises the signal-to-noise ratio for low-energy photons. The predicted tunnel-limited resolution as a function of photon energy for Nb, Ta, Re, Al and Mo is shown in Figure 27.6.

Further degradation of the energy resolution may occur due to:

- electronic noise related to the impedance and the residual leakage current of the STJ and read-out electronics (δE_{el}, independent of E) (Jochum et al 1994)

- spatial non-uniformity in the responsivity of the detector ($\delta E_{\text{non-unif}} \propto E$): typically the signals induced by photons absorbed near the edges of the detector or close to the electrical contacts could be lower than for the centre of the detector (Verhoeve et al 1998). Alternatively, the presence of interfaces

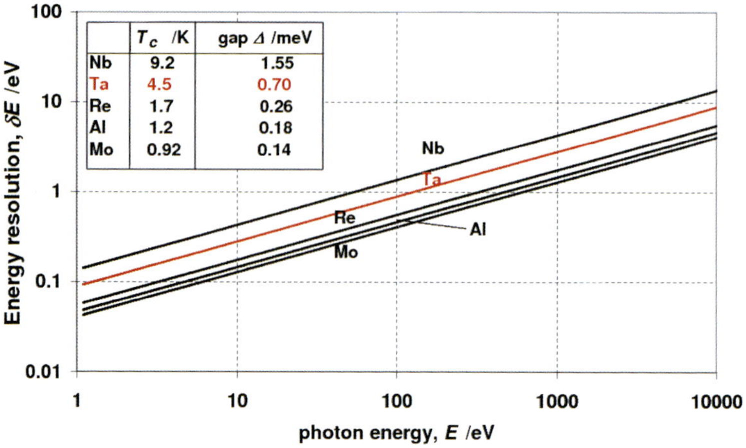

Figure 27.6: Tunnel-limited energy resolution as a function of photon energy for STJs made of five different elemental superconductors.

(e.g., substrate-bottom electrode) may give rise to variations in the initial number of quasiparticles due to loss of phonons during the energy conversion process, depending on the absorption depth in the detector (Kozorezov et al 2007)

- pile-up effects and DC-currents induced by low-energy background photons (e.g., IR) emitted by relatively warm parts of the experimental set-up (le Grand et al 1997)

- incomplete cooling of quasiparticles down to the device's energy gap before tunnelling may give rise to increased statistical broadening ('cancellation noise' $\delta E_{\text{canc}} \propto E^{\frac{1}{2}}$) (Segall 1991).

At X-ray energies, the contribution of spatial non-uniformities in the detector's response is usually found non-negligible or even dominating, whereas tunnel noise and electronic noise are most dominating in the visible. Cancellation noise and related electronic noise components are particularly relevant at X-ray energies in devices with low energy gap, where thermalization of the quasiparticles may be slow compared to the tunnel rate.

Fabrication of STJs

The fabrication of STJs is typically done on a bulk substrate (sapphire or Si wafers) and starts with the deposition of the superconducting multilayer under ultra-high vacuum using DC-magnetron sputtering in an argon atmosphere. In the case of a crystalline substrate like sapphire the base layer (e.g., Ta) may be grown epitaxially at high temperature. Growth rate and substrate temperature are usually optimized for the highest residual resistivity ratio (RRR) values, while the

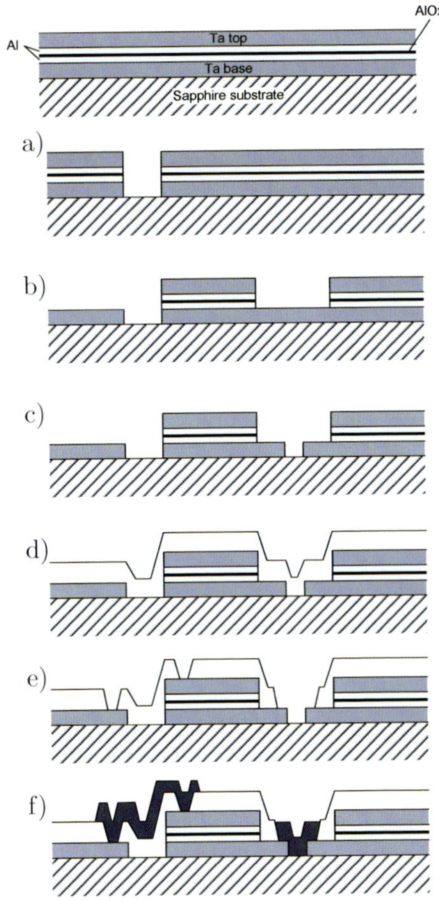

Figure 27.7: Typical fabrication process steps for Ta/Al STJs.
Initial Ta-Al-Al oxide-Al-Ta multilayer on sapphire substrate.
a) Base etch defines the STJs and leads using photolithography and plasma or wet chemical etch.
b) Mesa etch removes the top layers and oxide barrier to form the leads. This is achieved using the same process as for the base etch, but stopping at the base film.
c) Snip which is an etch of the base film, in preparation of the Nb plug deposition. It interrupts the contact in the base film between pixels. This step can also be achieved in the base etch.
d) Passivation is done by reactively sputtering silicon from a high purity target in an O_2 atmosphere.
e) Vias are etched in the SiO_2 to allow for top contacts and bridges between base contacts. This is done with CHF_3 reactive ion etching.
f) Niobium is deposited, and can be patterned by lift-off technique for top and base film contacts.

argon pressure is set for least film stress. The growth proceeds with the deposition of an aluminium layer, part of which is subsequently oxidized under an oxygen atmosphere to form a thin insulating tunnel barrier. In order to achieve a very transmissive but defect-free barrier it is of utmost importance to ensure that the underlying films are as flat as possible. Top aluminium is then deposited under the same conditions as for the base Al film, followed by the top material (e.g., Ta). The top layers are grown at temperatures below 120 °C in order to preserve the quality of the tunnel barrier. Photo-lithographic techniques are used to pattern any structure into the multilayer, which may eventually be diced into detector chips. A typical sequence of processing steps to obtain Ta/Al STJs in arrays for optical applications (see Figure 27.9) is shown in Figure 27.7.

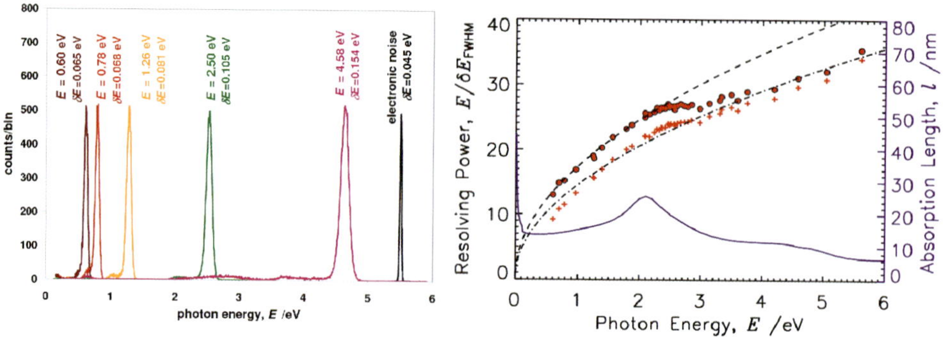

Figure 27.8: Left: Composite spectrum of the response of a 30 µm × 30 µm Ta/Al STJ to monochromatic illumination at five different photon energies from near-IR to UV. Right: Measured (red dots) and intrinsic (red + signs) energy resolving power as a function of energy for the same STJ.

State-of-the-art performance

UV-Vis–near-IR performance

Continuous efforts to improve the performance led to the first single photon detection with Nb-based STJs in the UV to visible wavelength range in 1996 (Peacock 1996). Useful spectroscopic capabilities were demonstrated as soon as the Nb was replaced by Ta, which yields much higher responsivity due to longer quasi-particle life times. Today the best single Ta-based STJs are performing very close to the predicted limit, with a resolving power $E/\delta E \approx 23$ at $E = 2.5$ eV including noise contributions from the read-out electronics $<$ 0.05 eV (Martin et al 2006a). Figure 27.8 shows some spectra in the wavelength range 270 nm to 2060 nm (E = 0.6 eV to 4.6 eV) obtained with a 30 µm × 30 µm STJ consisting of 100 nm thick Ta electrodes with 30 nm thick Al trapping layers. The baseline noise level for this detector suggests that photon counting should be possible up to wavelengths as long as $\lambda \approx 10$ µm. The right-hand side of Figure 27.8 shows the measured resolving power $E/\delta E$ as a function of photon wavelength, together with the intrinsic resolving power which is obtained after correction for the electronic noise contribution. The dashed lines represent the predicted energy resolution from Equation 27.3 for this device with $G = 0.8$ and $G = 0.5$, indicating a performance well within the predicted tunnel limit of $G = 1$.

The detection efficiency of STJs in this wavelength range is only limited by reflection losses at the metal surfaces, provided the thickness is of the order of 100 nm. A Ta-based STJ thus has a relative detection efficiency of ≈ 70 % for wavelengths $\lambda \approx 150$ nm to 600 nm, rolling off to 20 % at $\lambda \approx 1000$ nm and ≈ 5 % at $\lambda \approx 2000$ nm. The combination of this detection efficiency with high time resolution, spectroscopic resolution and imaging in a single non-dispersive instrument allows for niche applications in UV-visible astronomy. Examples of possible targets are eclipsing white dwarfs (eclipse-timing and -mapping), pulsating white dwarfs (studying the wavelength dependence of the pulsations in order to study the inter-

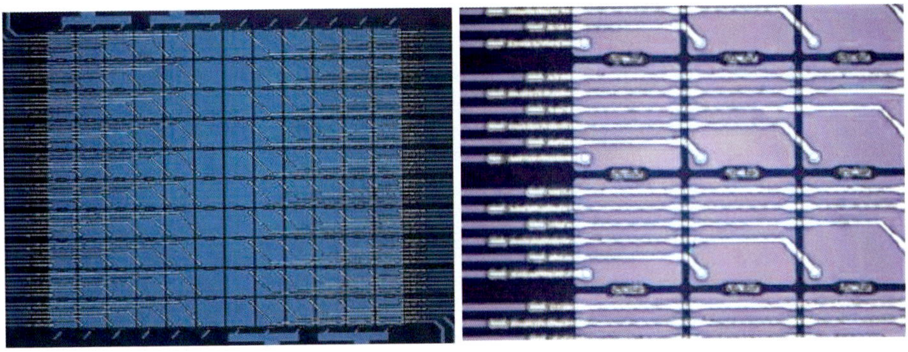

Figure 27.9: Left: 10×12 pixel S-Cam3 array of Ta/Al STJs. Note that illumination is from the backside through the sapphire substrate. Right: Detail, showing the interpixel gaps (< 2 μm) and the Nb contacts and interconnects between the base electrodes of pixels in columns. Groups of three columns share a common return wire.

nal structure), fast variability properties of black hole candidates, neutron stars and pulsars. Alternatively, the combination of low-resolution spectroscopy and imaging allows for the simultaneous measurement of redshifts of a number of objects within the field of view. ESA/ESTEC has been developing a demonstrator instrument, S-Cam, for ground-based astronomy, based on Ta/Al STJs. Since its first deployment in 1999 at the 4.2 m William Herschel Telescope at La Palma (Spain), it has evolved from a 6×6 pixel camera with a wavelength resolving power of $\lambda/\delta\lambda \approx 5$ at $\lambda = 500$ nm, to a 10×12 pixel camera with $\lambda/\delta\lambda \approx 15$ that is regularly used at ESA's 1 m optical ground station telescope (Tenerife, Spain) (Martin et al 2006b). S-Cam provides a maximum relative efficiency of ≈ 30 % in a wavelength band of λ = 340 nm to 740 nm and a time resolution of 1 μs. The maximum useful count rate is ≈ 5000 s^{-1} per pixel, limited by the signal decay time of ≈ 20 μs. The detector is operated at T = 285 mK, provided by a cryogenic chain consisting of ^{3}He and ^{4}He sorption coolers and a liquid helium bath. Figure 27.9 shows the current detector array with 35 μm × 35 μm sized pixels and 88 % fill factor. The detector is illuminated through the transparent sapphire substrate, such that the wiring can be routed across the pixels without penalty in detection efficiency. Each pixel is connected to a dedicated read-out chain consisting of a charge-sensitive preamplifier (at ambient temperature) and programmable finite impulse response filters for pulse shaping.

Such limited arrays can already provide good observational conditions for point sources, including correction for background photons, but larger arrays are needed for the imaging of extended or multiple objects. The scalability of the pixelated arrays described above appears to be limited: the accommodation of the dense contact wiring is a problem, the readout electronics grow proportionally, and the thermal load from the increasing number of signal wires will exceed the available cooling. A couple of alternatives can alleviate this challenge. In the matrix readout

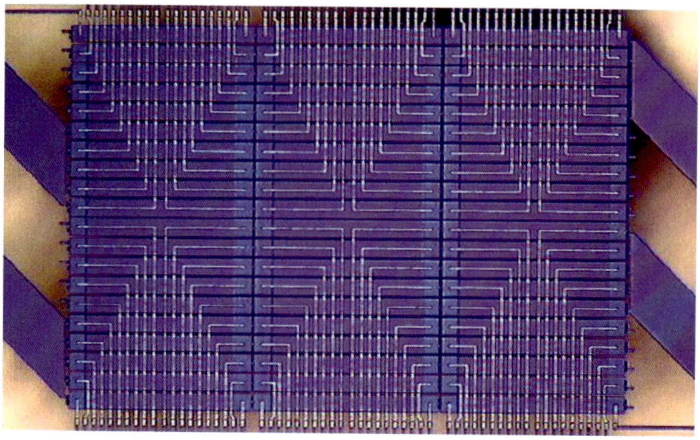

Figure 27.10: 3×20 array of 1-dimensional DROIDs (360 µm×30 µm, Ta absorber, Ta/Al readout STJs).

mode (Martin et al 2000), arrays of pixel detectors are interconnected by their top and base electrodes in rows and columns. Each row and column is then connected to an amplifier. Simultaneous event detection in a row and column identifies the photon absorption location. DROIDs as described above provide another elegant way of increasing the total detection area. Several groups have demonstrated the feasibility of this concept for UV-visible applications, all with Ta absorbers and Ta/Al or Al readout STJs (Verhoeve et al 2000; Wilson et al 2000; Jerjen et al 2006). Figure 27.10 shows an example of a 3×20 element array of 1-D DROIDs, with a total of 120 STJs and the equivalent of 660 pixels.

X-ray performance

As the advantage of cryogenic detectors over semiconductor detectors lies in the predicted much better energy resolution, most of the STJ development work has focussed on trying to realize this prediction in the X-ray energy regime. For practical reasons, a radioactive ^{55}Fe sample emitting 5.9 keV X-rays is most commonly used as a photon source, and the measured energy resolution achieved at this energy has become a figure of merit. A variety of superconducting materials has been used to fabricate single STJs on bulk substrates, most of them yielding $\delta E < 50$ eV at $E = 5.9$ keV. Noticeable examples are $\delta E = 49$ eV with Sn (Rothmund and Zehnder 1988), $\delta E = 29$ eV with Nb (Frank 1996), and $\delta E = 21$ eV with Ta (Brammertz et al 2001). The best results reported to date have been achieved with Al: $\delta E = 12$ eV was achieved with an Al STJ on a Si_3N_4 membrane (Angloher et al 2001). Figure 27.11, left, shows the corresponding spectrum, with a double set of lines, characteristic for thin single STJs in which top and base electrode are both sensitive. The Si_3N_4 membrane plays an important role in reducing the number of phonon induced events from the substrate. The right-hand side of Figure 27.11 shows a further improvement on this result with $\delta E = 10.8$ eV (Huber

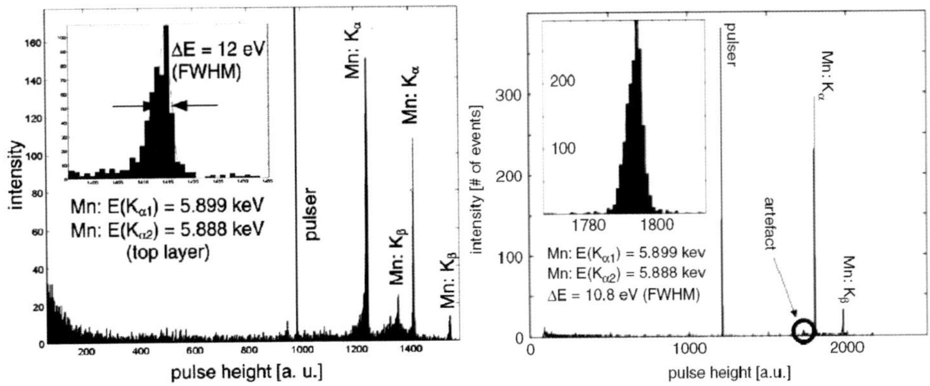

Figure 27.11: Left: Spectral response for ^{55}Fe illumination of a 100 μm ×100 μm Al STJ on a Si_3N_4 membrane (from Angloher 2001). Right: Same for a similar STJ with a 1.3 μm thick Pb absorber on top (from Huber 2004).

et al 2004). In this case the spectral response has been much improved by using a phonon-coupled Pb absorber on top of the Al STJ.

Very promising results have also been obtained with DROIDs: $\delta E = 13$ eV with a 100 μm ×200 μm, 570 nm thick Ta absorber and Al readout STJs in the lateral configuration of Figure 27.4, right (Li et al 2002), and $\delta E = 2.4$ eV at $E = 500$ eV with a 20 μm ×100 μm, 100 nm thick Ta absorber and Ta/Al readout STJs in the stacked configuration of Figure 27.4, left (Den Hartog et al 2002). Designs for arrays of STJs for X-ray astronomy have not been reported to date; they will be inherently more complicated than those presented above, since the detectors need to be illuminated from above. The routing and dimensions of the contact wiring will therefore have to be organized in such a way that a minimum of incident light is obscured by it.

STJ operation in space

In order to operate these sensitive detectors, a suitable environment needs to be created. The operating temperature needs to be chosen such as to minimize bias currents due to thermal quasiparticle generation. As a rule of thumb, this requires the operating temperature to be about one tenth of the superconducting material's critical temperature. For Ta/Al STJs, this is $\approx$300 mK. Such temperatures can be achieved conveniently using as a last cooling stage ^{4}He–^{3}He sorption coolers. Such coolers are completely sealed units that do not require any moving parts and can be easily controlled electrically by a set of heaters. Special care needs to be taken to minimize heat leaks to warmer parts of the instrument. Two conduction paths are usually present: the mechanical and electrical interfaces. As the number of pixels to be read out increases, the latter will become the dominant factor. The cold finger on which the detector is attached is usually suspended with Kevlar strings. Electrical connections are made using superconducting Nb/Ti wires. These wires, often assembled in looms, are thermally anchored at each available intermediate temperature stage to intercept the heat load from the higher temperature parts. In addition to the conduction path, also the radiative path needs to be optimized. All available temperature stages need to be optically closed from each other to avoid black-body IR radiation leakage onto the detector assembly, increasing the heat load to the cooling system, but also to avoid flooding the detector with unwanted photons. For the useful incoming radiation to reach the detector, appropriate filters need to be inserted in the optical path. For NIR-Vis-UV applications, this can be achieved using various IR blocking glasses (Martin et al 2006b) or plastics (Bay et al 2006), while thin Al foils are used for X-ray applications. For an instrument, and particularly a space instrument based on these detectors, particular care needs to be taken against electromagnetic interference. As STJs are usually read out by room-temperature charge-sensitive amplifiers, the connecting wires running to the detector need to be shielded, which always remains a challenge as thermal decoupling between cooling stages still needs to be ensured. In addition, the wires need to be mechanically anchored so as to minimize charge displacements due to vibrations. For a space mission, the design needs to be tolerant against cosmic rays. STJs seem to be intrinsically radiation tolerant, although more measurements are needed to corroborate preliminary results. The radiation tolerance could be attributable to the small interaction volumes (very thin insulator barrier). Nevertheless, the system needs to discriminate these events as they will often be recorded. Every cosmic-ray interaction with the substrate will generate a high concentration of energetic phonons that can reach the STJ array. Usually, a large number of pixels will therefore simultaneously register an event, which can be used to discriminate them against true photons (by anti-coincidence). To date, two space applications have been considered. In the late 1990s, a proposal was submitted to NASA for an optical/UV spectro-imager based on an array of STJs. The unique capability of these sensors allows simultaneous imaging with low energy resolution and high timing accuracy. Alternatively, the intrinsic energy resolving power can be put to use as an order sorting in a high-resolution spectrometer (Jakobsen 1999). The second application is related to X-ray detection. The next generation X-ray observatories (Parmar et al 2006; Stahle et al 1999) will require very sensitive imaging spectrometers. These cryogenic sensors are particularly well suited for X-ray energies up to $\approx$3 keV. Their intrinsic energy resolution ($\approx$1 eV to 2 eV below 1 keV) alleviates the need for inefficient dispersive gratings.

Bibliography

Angloher G, Hettl P, Huber M (plus five authors) (2001) Energy resolution of 12 eV at 5.9 keV from Al-superconducting tunnel junction detectors. J Appl Phys 89:1425–1429

Bay TJ, Burney JA, Barral J (plus seven authors) (2006) The optical imaging TES detector array: considerations for a cryogenic imaging instrument. Nucl Instrum Meth A 559:506–508

Booth NE (1987) Quasiparticle trapping and the quasiparticle multiplier. Appl Phys Lett 50:293–295

Booth NE, Brink PL, Gaitskell RJ (plus four authors) (1993) Superconducting tunnel junctions and quasiparticle trapping. J Low Temp Phys 93:521–532

Brammertz G, Verhoeve P, Peacock A (plus four authors) (2001) Development of practical soft X-ray spectrometers. IEEE Trans Appl Supercon 11:828–831

Frank M, Mears CA, Labov SE (plus five authors) (1996) High-resolution X-ray detectors with high-speed SQUID readout of superconducting tunnel junctions. Nucl Instr Meth A 370:41–43

Giaever I (1960) Energy gap in superconductors measured by electron tunnelling. Phys Rev Lett 5:147–148

Goldie DJ, Brink PL, Patel C (plus two authors) (1994) The statistical noise due to tunneling in superconducting tunnel junction detectors. Appl Phys Lett 64:3169–3171

Gray KE (1978) A superconducting transistor. Appl Phys Lett 32:392–395

le Grand JB, Mears CA, Labov SE (plus four authors) (1997) Infrared photon induced noise in detectors for photons and particles. Proceedings of Low Temperature Detectors 7, Munich, Germany. published by MPI Physik, pp 106–107 ISBN 3-00-002266-X

Den Hartog RH, Kozorezov A, Martin D (plus five authors) (2002) Large format distributed readout imaging devices for X-ray imaging spectroscopy. AIP Conf Proc 605:11–14

Hiller LJ, van den Berg ML, Labov SE (2001) Multiple tunnelling noise in superconducting tunnel junctions from partial current integration. Appl Phys Lett 79:4441–4443

Huber M, Angloher G, Hollerith C (plus four authors) (2004) Superconducting tunnel junctions as detectors for high-resolution X-ray spectroscopy. X-Ray Spectrom 33:253–255

Jakobsen P (1999) Superconducting tunnel junction detectors for optical and UV astronomy. ASP Conf Series 164:397–404

Jerjen I, Kirk E, Schmid E, Zehnder A (2006) Distributed readout detectors using superconducting tunnel junctions. Nucl Instr Meth A 559:497–499

Jochum J, Kraus H, Gutsche M (plus three authors) (1994) Electronic noise of superconducting tunnel junction detectors. Nucl Instr Meth A, 338:458–466

Josephson BD (1962) Possible new effects in superconducting tunnelling. Phys Letters 1:251–253

Kaplan SB, Chi CC, Langenberg DN (plus three authors) (1976) Quasiparticle and phonon lifetimes in superconductors. Phys Rev B, 14:4854–4873

de Korte PAJ, van den Berg ML, Bruijn MP (plus five authors) (1992) Superconductive tunnel junctions for X-ray spectroscopy. Proc SPIE 1743:24–35

Kozorezov AG, Volkov AF, Wigmore JK (plus three authors) (2000) Quasiparticle-phonon downconversion in nonequilibrium superconductors. Phys Rev B 61:11807–11819

Kozorezov AG, Wigmore JK, Martin D (plus two authors) (2007) Electron energy down-conversion in thin superconducting films. Phys Rev B 75:094513

Kozorezov AG, Hijmering RA, Brammertz G (plus six authors) (2008) Dynamics of non-equilibrium quasiparticles in narrow-gap superconducting tunnel junctions. Phys Rev B 77:14501–14511

Kraus H, Peterreins Th, Pröbst F (plus four authors) (1986) High-resolution X-ray detection with superconducting tunnel functions. Europhys Lett 1:161–166

Kraus H, von Feilitsch F, Jochum J (plus three authors) (1989) Quasiparticle trapping in a superconductive detector system exhibiting high energy and position resolution. Physics Lett B 231:195–202

Kulik IO (1965) Theory of steps of voltage-current characteristic of Josephson tunnel current. JETP Lett 2:84–87

Kurakado M, Mazaki H (1980) Quasiparticle excitation in a superconducting tunnel junction by α particles. Phys Rev B 22:168–173

Kurakado M (1982) Possibility of high resolution detectors using superconducting tunnel junctions. Nucl Instr Meth 196:275–277

Li L, Frunzio L, Wilson CM (plus four authors) (2002) Spatial uniformity of single photon 1-D imaging detectors using superconducting tunnel junctions. AIP Conf Proc 605:145–148

Martin DDE, Peacock A, Verhoeve P (plus two authors) (2000) Matrix readout scheme for superconducting tunnel junction arrays. Rev Sci Instr 71:3543–3551

Martin DDE, Verhoeve P, den Hartog RH (plus five authors) (2003) 12×10 pixels superconducting tunnel junction array based spectro-photometer for optical astronomy. Proc SPIE 4841:805–814

Martin DDE, Verhoeve P, Peacock A (plus four authors) (2006a) Resolution limitation due to phonon losses in superconducting tunnel junctions. Appl Phys Lett 88:123510

Martin DDE, Verhoeve P, Oosterbroek T (plus three authors) (2006b) Accurate time-resolved optical photospectroscopy with superconducting tunnel junction arrays. Proc SPIE 6269:62960O

Parmar AN, Arnaud M, Barcons X (plus 14 authors) (2007), *XEUS* - the X-ray evolving universe spectroscopy mission. Proc SPIE 6266:62661R

Peacock A, Verhoeve P, Rando N (plus nine authors) (1996) Single optical photon detection with a superconducting tunnel junction. Nature 381:135–137

Peterson RL (1991) Sidelobe suppression in small Josephson junctions. Cryogenics 31:132–135

Rando N, Peacock A, van Dordrecht A (plus six authors) (1992) The properties of niobium superconducting tunneling junctions as X-ray detectors. Nucl Instr Meth A 313:173–195

Rothwarf A, Taylor BN (1967) Measurement of recombination lifetimes in superconductors. Phys Rev Lett 19:27–30

Rothmund W, Zehnder A (1988) In: Barone A (eds) Superconductive Particle Detectors World Scientific, Singapore, p 52, ISBN: 9971506114

Segall K, Wilson C, Frunzio L (plus six authors) (2000) Noise mechanisms in superconducting tunnel-junction detectors. Appl Phys Lett 76:3998–4000

Stahle CK, Bandler S, Barbee TW (plus 12 authors) (1999) Toward a 2-eV microcalorimeter X-ray spectrometer for *Constellation-X*. Proc SPIE 3765:82–93

Twerenbold D (1986) Giaever-type superconducting tunnelling junctions as high-resolution X-ray detectors. Europhys Lett 1:209–214

Verhoeve P, Rando N, Peacock A (plus two authors) (1998) High-resolution X-ray spectra measured using tantalum superconducting tunnel junctions. Appl Phys Lett 72:3359–3361

Verhoeve P, den Hartog R, Martin D (plus three authors) (2000) Development of distributed readout imaging detectors based on superconducting tunnel junctions for UV/optical astronomy. Proc SPIE 4008:683–694

Verhoeve P, den Hartog R, Kozorezov A (plus four authors) (2002) Time dependence of tunnel statistics and the energy resolution of superconducting tunnel junctions. J Appl Phys 92:6072–6081

Wilson CM, Segall K, Frunzio L (plus six authors) (2000) Optical/UV single-photon imaging spectrometers using superconducting tunnel junctions. Nuclear Instr Methods in Phys Research A 444:449–452

Wood GH, White BL (1969) Pulses induced in tunnelling currents between superconductors by alpha-particle bombardment. Appl Phys Lett 15:237–239

— 28 —

X-ray calorimeters

F. Scott Porter[I]

Abstract

X-ray calorimeter instruments for astrophysics have seen rapid development since they were invented in 1984. The prime instrument on all currently planned X-ray spectroscopic observatories is based on calorimeter technology. This relatively simple detection concept that senses the energy of an incident photon by measuring the temperature rise of an absorber material at very low temperatures can form the basis of a very high-performance, non-dispersive spectrometer. State-of-the-art calorimeter instruments have resolving powers of over 3000, large simultaneous bandpasses, and near unit efficiency. This coupled with the intrinsic imaging capability of a pixilated X-ray calorimeter array, allows true spectral-spatial instruments to be constructed. This chapter briefly reviews the detection scheme, the state of the art in X-ray calorimeter instruments and the future outlook for this technology.

Introduction

An X-ray calorimeter is a conceptually simple device that provides simultaneous spatial and spectral observations for X-ray astrophysics. Current X-ray observatories use charge coupled devices (CCD) and precision-focusing optics to achieve high spatial resolution but with limited inherent spectral resolution. High-resolution spectroscopy is achieved with dispersive optics and position-sensitive detectors. Dispersive optics are additionally constrained by relatively low efficiency and a limited bandpass. Moreover, the performance degrades for objects larger than the point-spread function of the optics. In contrast, an X-ray calorimeter provides a broad-band, high-resolution spectrometer with inherent imaging capability where every pixel in a pixilated focal plane is a high resolution spectrometer. In addition, the performance of a calorimeter is not affected by the spatial extent of the source and is insensitive to polarization.

As in all branches of astrophysics, spectroscopy is the key to isolating the physical interactions occurring in a celestial source. X-ray spectroscopy, in particular, yields information on the composition, temperature, ionization equilibrium, density, and bulk and turbulent motion of the X-ray emitting plasmas that make up the bulk

[I]NASA Goddard Space Flight Center, Greenbelt, USA

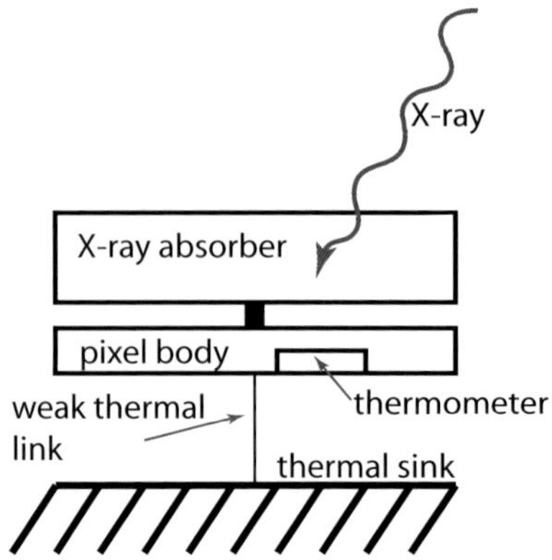

Figure 28.1: A schematic view of a calorimeter detector. An incoming X-ray is photo-absorbed in the absorber and the resultant equilibrium thermal signal is registered with a sensitive thermometer. The detector is thermally reset to the heat-sink temperature through a weak thermal link.

of the observable matter in the Universe. Specifically, X-ray spectroscopy probes, for example, accretion around black holes and active galactic nuclei, shock-heated plasmas in stellar winds and supernova remnants, mergers of clusters of galaxies, and the decay processes in γ-ray bursts. In addition, X-ray back-lighting and photo-ionization provide key diagnostics of the intervening material between the observer and cosmic X-ray sources. With current X-ray observatories, high-resolution spectroscopy is limited to a handful of the brightest sources and only over a limited spectral range concentrated below 2 keV. Most major X-ray observatories planned for the next decade including *Astro-H* and *IXO*[1], include a calorimeter system as one of the primary focal-plane instruments. In this chapter, we briefly review the operating concepts behind a calorimeter instrument, the current state of the art in detector systems, current and future spaceflight instruments, and prospects for future large-scale imaging spectrometers. A complete development of calorimeter theory is given in Mather (1982), Moseley et al (1984), and McCammon (2005).

The basic concept of a calorimeter is shown in Figure 28.1, consisting of an X-ray absorber, a sensitive thermometer, and a heat sink held at low temperature, typically below 0.1 K. Figure 28.2 shows a spaceflight detector with its components labelled. Incident X-rays are photo-absorbed in the X-ray absorber at temperature T, releasing a primary photo-electron and one or more Auger electrons. The liberated electrons then thermalize into the available energy reservoirs in the solid-state system at a higher temperature T'. Phonon-states in an insulator, phonon and electronic states in a metal, and phonon and quasi-particle states in a superconductor dominate the thermalization channels, although more exotic reservoirs can

[1] *IXO* is the merger of the U.S. *Constellation-X* and the European *XEUS* missions.

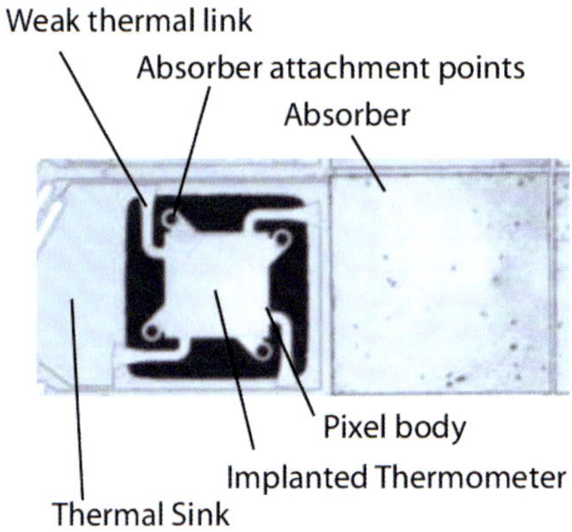

Figure 28.2: An XRS spaceflight calorimeter detector (cf., Page 468) without the absorber (left) and after the 8 μm thick HgTe absorber is attached (right) (Stahle et al 2004). The absorbers are 624 μm × 624 μm.

also contribute. The change in temperature as a function of energy is then simply governed by the heat capacity of the system as

$$\Delta T = T' - T \sim \frac{E}{C(\bar{T})} \quad , \tag{28.1}$$

where E is the energy of the incident photon, C is the heat capacity of all contributing states in the absorber and $\bar{T}$ designates the mean value of T. A thermometer placed in good thermal contact with the absorber then senses the change in temperature and thus the energy of the incident photon through the simple relationship 28.1. Finally, a weak thermal link to the heat sink returns the detector to its base temperature for the next photon interaction. An example of the thermal response for a detector from a 6 keV X-ray is shown in Figure 28.3. An arbitrarily sensitive thermometer with infinite dynamic range coupled to an arbitrarily small heat capacity has no upper limit to the spectral resolving power. As we shall see, practical considerations and specific instrument requirements generally limit the performance of the detector system.

Calorimeters are intrinsically imaging in a pixelated detector. For example, Figure 28.4 shows a calorimeter array built for the X-ray Quantum Calorimeter (XQC) sounding rocket with 36 detector elements. Coupled with focusing X-ray optics, the detector channels are completely independent, both spatially and spectrally, yielding an X-ray image where each imaging element is an independent high-resolution spectrometer. Larger imaging arrays, including the 4096 element *IXO* detector, are currently in development, and even larger megapixel arrays are planned for the future. As a spatial-spectral instrument, the calorimeter detector is similar and complementary to an X-ray CCD. The spectral resolution of an X-ray CCD is fundamentally limited by the counting statistics of charges formed in the cascade from an X-ray liberated photoelectron to carriers at the band-gap of the material. This gives them moderate spectroscopic resolving power of about 50 at 6 keV. On the

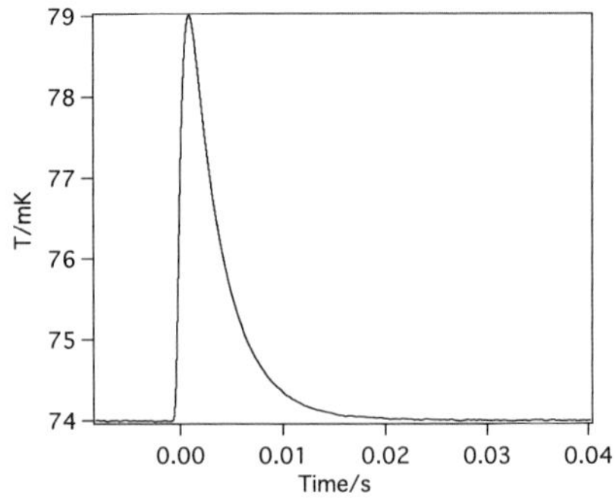

Figure 28.3: The thermal signature of an individual X-ray photon as measured by the semiconductor calorimeter shown in Figure 28.2. The incident 6 keV X-ray produces a 5 mK temperature rise in the detector before it relaxes to its equilibrium state through the weak thermal link to the heat sink.

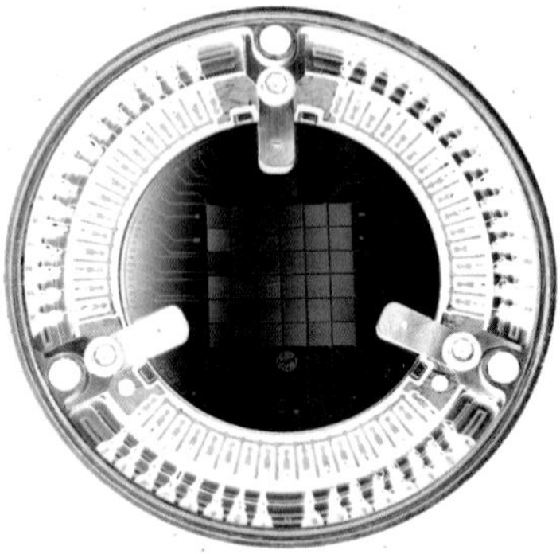

Figure 28.4: The 36 pixel calorimeter array for the X-ray Quantum Calorimeter (XQC) sounding rocket instrument. Each X-ray absorber is 2 mm $\times$2 mm $\times$ 0.8 µm (McCammon et al 2008).

other hand, CCDs scale to very large imaging arrays because one can easily transfer the X-ray initiated charge cloud from the interaction site to collection nodes on the exterior of the array. For calorimeters the situation is inverted. The spectral resolving power has no fundamental limit, and practical detectors have achieved resolving powers of 3000 at 6 keV, but the array size is limited by our ability to fabricate detector arrays with individually wired single or small groups of pixels. No analogue to the charge transfer mechanism in CCDs has been invented for a thermal detector. An example of the relative performance of a calorimeter array is shown in Figure 28.5 for a simulated observation of M82 along with a simulated CCD observation of the same object. A real-world example is shown in Figure 28.6 where a plasma of He-like and H-like iron was observed at the Electron Beam Ion Trap (EBIT) facility at the Lawrence Livermore National Laboratory using the EBIT Calorimeter Spectrometer (ECS) instrument (Porter et al 2008a). In prac-

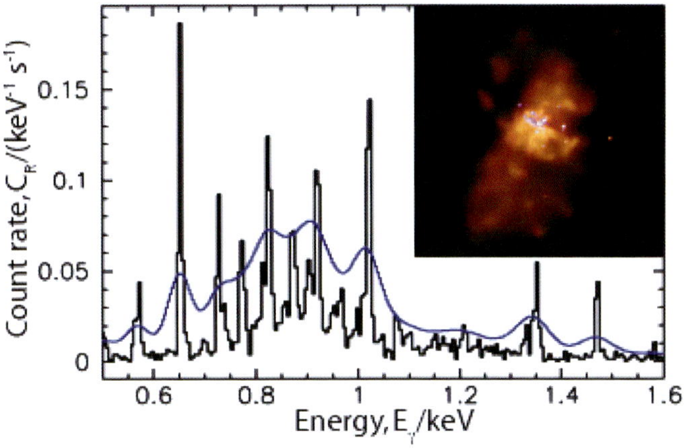

Figure 28.5: A simulated spectrum of the starburst galaxy M82 (inset, Credit: NASA/CXC/SAO/PSU/CMU) using the *Astro-H*/SXS calorimeter array. A simulated lower resolution spectrum for an X-ray CCD instrument is also shown (R. Smith, personal communication, 2008).

tical detectors, the performance of a calorimeter is limited by both extrinsic noise such as amplifier and photon shot noise, and intrinsic noise such as Johnson noise and thermodynamic noise. Thermodynamic noise, which is the thermal noise associated with energy exchange across the weak thermal link shown in Figure 28.1, fundamentally limits the resolving power of the detector. For an optimized detector this is given for $\alpha \gg 1$ as:

$$\Delta E \propto \sqrt{\frac{k_B T_0 C_0}{\alpha}} \quad , \tag{28.2}$$

where T_0 is the temperature of the heat sink, C_0 is the heat capacity at that temperature, k_B is Boltzmann's constant, and α is a measure of the thermistor responsivity (Moseley et al 1984; McCammon 2005). High spectroscopic resolution is generally achieved by running at very low bath temperatures, usually < 100 mK, with small, low heat-capacity systems, and high-responsivity thermometers. In designing a practical detector, other noise sources in the system are minimized with respect to the thermodynamic noise to achieve the highest possible spectral resolving power.

Calorimeter detectors

Over the last two decades, this field has seen rapid development in the United States, Europe and Japan. Current calorimeter research concentrates on three main components of the technology: the thermistor design, the absorber design, and the readout system.

X-ray calorimeter detectors can encompass a huge performance phase space that includes spectral resolving power, counting rates, detector size, and dynamic range (bandpass). Optimizing for a specific set of instrument requirements is a highly

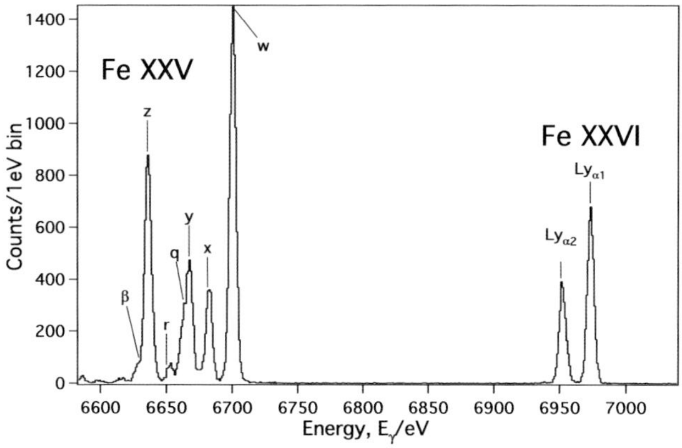

Figure 28.6: The spectrum emitted by highly charged H-like and He-like iron ions, measured with the ECS. The ECS is a production spectrometer based on the *Suzaku*/XRS calorimeter instrument (Porter et al 2008a).

coupled problem, and most of the optimization phase space is unexplored. Two examples of very different optimizations illustrate this point. Figure 28.4 shows the detector array for the X-ray Quantum Calorimeter (XQC) sounding rocket experiment. It is a 36-pixel device with very large (2 mm $\times$ 2 mm) pixels and an energy resolution of 7 eV at 1.5 keV. The detector is optimized for low energies up to about 1 keV. However, the large size of the absorber limits the thickness and thus the bandpass over which it has high quantum efficiency. In contrast, high-energy detectors with thicker, more opaque absorbers, have been optimized for energies up to 100 keV with energy resolutions of around 25 eV at 60 keV (Doriese et al 2007; Porter et al 2008b). The high-energy detector is optimized for dynamic range at the expense of both energy resolution and speed. The low-energy detector was optimized differently, emphasizing detector area and energy resolution over dynamic range. Many other optimizations are possible including detectors for the visible and UV range with a dynamic range of a few electronvolts (see, for example, Romani et al 2001), *alpha*-particle detectors with a dynamic range of several million electronvolts (Horansky et al 2008), and high-cadence solar X-ray detectors with small but very fast pixels. Detector optimization, however, is highly dependent on the underlying calorimeter technology where some component choices lend themselves more easily to certain applications. Calorimeters are usually classified by the choice of thermistor technology, which generally drives the design of the rest of the detector array and the instrument as a whole. Table 28.1 gives an overview of the major thermistor technologies and how they cascade into a few of the other system parameters.

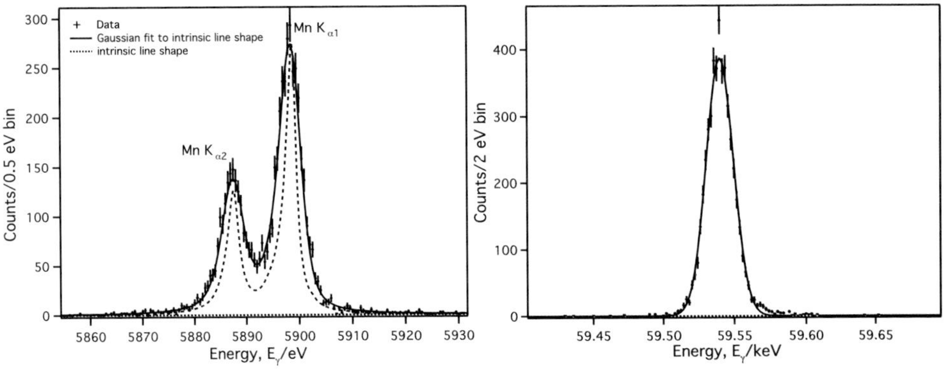

Figure 28.7: Left: A Mn K_α spectrum measured with a semiconductor calorimeter array giving 3.2 eV FWHM spectral resolution at 5.9 keV. The X-ray absorbers were 400 µm × 400 µm × 8 µm of HgTe with 95 % quantum efficiency at 6 keV (Porter et al 2006). Right: A spectrum of the 60 keV emission line from ^{241}Am measured with a high-energy semiconductor calorimeter array giving 22 eV FWHM spectral resolution at 60 keV. The X-ray absorbers were 500 µm × 500 µm × 100 µm of HgTe with 32 % quantum efficiency at 60 keV (Porter et al 2008b).

Detectors based on semiconductor thermistors

Detectors based on semiconductor thermistors (McCammon 2005b) are the most mature and have been optimized for a wide variety of applications. Semiconductor thermistors have relatively low responsivity, usually specified as:

$$\alpha \equiv \frac{\mathrm{d}\lg\frac{R}{\Omega}}{\mathrm{d}\lg\frac{T}{\mathrm{K}}} \quad , \tag{28.3}$$

where R is the resistance, T is the temperature, and α is generally less than 10 for a semiconductor thermistor. However, with a careful choice of absorber material, extremely high-performance detectors can be constructed. Figure 28.7 shows the spectrum of a ^{55}Fe radioactive source using a 400 µm × 400 µm detector with 95 % quantum efficiency at 6 keV giving a spectral resolving power of almost 2000. A spectrum using a device optimized for high energies is also shown in Figure 28.7 giving a resolving power of over 2000 at 60 keV in an array of 500 µm × 500 µm pixels with 32 % quantum efficiency at 60 keV. Semiconductor thermistors are generally read out by cooled junction field effect transistors (JFET) (Porter et al 1999) and are operated at high impedance to minimize the contributions of amplifier noise. The high impedance of the detectors and the use of JFET amplifiers severely limits the size of the focal plane arrays. Most semiconductor detector arrays are less than 100 pixels, although the HAWC infrared bolometer array for the SOFIA airborne observatory has 384 cryogenic JFET readout channels (Harper 2004), and a 256-pixel X-ray detector is being developed at GSFC.

Table 28.1: Classes of X-ray calorimeters by thermistor technology, for semiconductor, Transition Edge Sensor (TES), Metallic Magnetic Calorimeter (MMC), and Microwave Kinetic Inductance Detector (MKID) detector schemes. MKIDs, while not strictly equilibrium thermal detectors, show huge promise for future megapixel detector arrays.

	Semiconductor	TES	MMC	MKID
Physical effect	Mott hopping	normal-superconducting transition	magnetic susceptibility	quasi-particle modulated impedance
Readout	JFET	SQUID	SQUID	microwave resonator
Temperature, T/mK	50 to 100	50 to 100	50 to 100	50 to 100
Pixel size[a], s/mm	0.5 to 2.0	0.05 to 1.0	0.05 to 0.5	0.2 strip
Multiplexable	not yet	yes	yes	yes
Best $\Delta E/\mathrm{eV}$ at 6 keV	3.2	2.1	2.7	62
Bandpass[a], E/eV	20 to 10^5	1 to 5×10^6	100 to 5×10^6	30 to 6×10^5

[a] Demonstrated

Transition-edge sensors

Newer thermistor technologies allow for easier access to other regions of the detector optimization space and expand the range of accessible absorber materials. Specifically, transition-edge sensor (TES) calorimeters are the leading candidate for the *IXO* observatory that is expected to be launched around 2018 – 2020. Transition-edge thermistors use a superconductor that is biased in the middle of a very narrow normal-superconducting transition. A TES thermistor has a very high responsivity with an α between 100 and 1000, but over a fairly narrow temperature range, generally less than 1 mK for temperatures near 100 mK. From Equation 28.2, one can see that a high α can lead to increased energy resolution. However, a high α also leads to a much greater range of allowable heat capacities, and thus, in principle, a greater range of acceptable absorber materials. TES thermistors are voltage biased, where the power dissipated in the thermistor heats the device into its transition. This is a stable bias point for a thermal detector since the absorption of energy increases the resistance of the device, thus decreasing the Joule power dissipation and keeping the TES near its bias point. In operation, the device is biased to a temperature well above the bath temperature, generally $\approx$ 100 mK for a 50 mK bath temperature. The large temperature gradient and the decrease in bias power during an X-ray event vastly improve the detector recovery time. This has been termed "extreme electro-thermal feedback". A similar, though smaller, effect occurs with semiconductor thermistors. TES detector systems have achieved

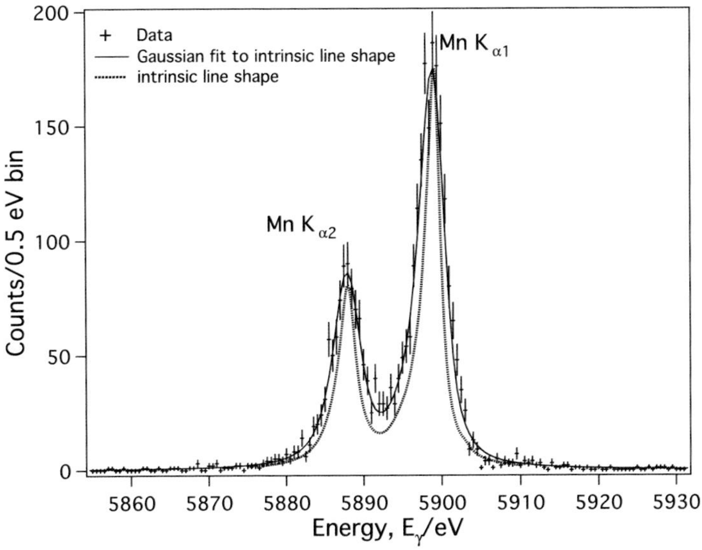

Figure 28.8: A Mn K_α spectrum measured with a TES calorimeter array giving 2.4 eV FWHM spectral resolution at 5.9 keV. The X-ray absorbers were 250 μm × 250 μm × 5.3 μm of Au with 99 % quantum efficiency at 6 keV (Iyomoto et al 2008).

excellent energy resolution across a wide range of optimizations. An example is shown in Figure 28.8 for a 250 μm × 250 μm pixel that gives a resolving power of almost 3000 at 6 keV (Iyomoto et al 2008).

Semiconductor and TES calorimeters have received the most development funding for the longest period of time and are thus the most technologically evolved thermistor technologies. Newer systems using magnetic susceptibility thermistors (Fleischmann et al 2005) and kinetic inductance thermistors (Mazin et al 2008) are again broadening the phase space for detector optimization. Both of these technologies may lend themselves to very large detector arrays, approaching 10^6 pixels.

Read-out technology

The thermistor chosen for the detector system drives the readout technology that ultimately drives the complexity and scalability of the instrument. Semiconductors are well matched to JFET front-end amplifiers that must be placed in close proximity to the detector array to prevent microphonic pickup noise in the wiring to the high-impedance detectors. A typical semiconductor readout circuit, with its complex thermal staging, is shown in Figure 28.9. TES detectors, in contrast, are low-impedance devices that are well matched to a current-sensing superconducting quantum interference device (SQUID) amplifier. SQUIDs are high-sensitivity, low-power amplifiers that operate at cryogenic temperatures and are well suited

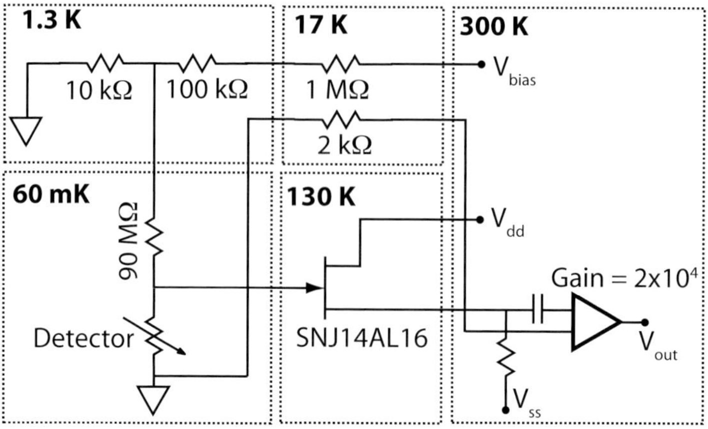

Figure 28.9: A standard readout circuit for a semiconductor calorimeter detector. The detector is current biased through a large load resistor and read-out with a cryogenic JFET source follower that is placed in close proximity to the detector. This specific circuit is from the *Suzaku*/XRS instrument (Kelly et al 2007).

to calorimeter instruments. An important advantage of SQUID readout systems is that several high-density readout systems (Kiviranta et al 2002; Reintsema et al 2003; Irwin et al 2006) have been developed. These readout systems combine multiple pixels onto a single readout node, significantly reducing the wiring complexity of a detector array. For example, Figure 28.10 shows a schematic for a SQUID-based time division multiplexer (TDM) developed at the National Institute of Standards and Technology (NIST) that implements row-column readout. In the TDM system, front-end SQUIDs on the focal plane sample a column of detector channels using a single output amplifier. This is akin to a typical commercial laboratory digitizer where a number of inputs are multiplexed to a single amplifier and analogue-to-digital converter. The result is a major reduction in the wiring complexity of an $N \times N$ array, where the wiring scale goes from $O(N^2)$ to $O(N)$ vastly simplifying the instrument design. A 4 column by 32 row NIST TDM multiplexer system with a NASA/GSFC 8×8 TES detector array is shown in Figure 28.11.

Spaceflight calorimeter instruments

X-ray calorimeters have a relatively short spaceflight history but the number of planned near-term instruments is quite large. These are summarized in Table 28.2. Here we briefly describe the XQC and *Suzaku*/XRS instruments as examples of adapting a complex calorimeter detector systems to a space environment. In the next section, we discuss future missions including *Astro-H* and *IXO* where mission requirements define the majority of current calorimeter research.

The XQC instrument was the first calorimeter instrument in space and has flown four times on a suborbital sounding rocket since 1995. The instrument is designed without focusing optics to measure the detailed spectral structure of the

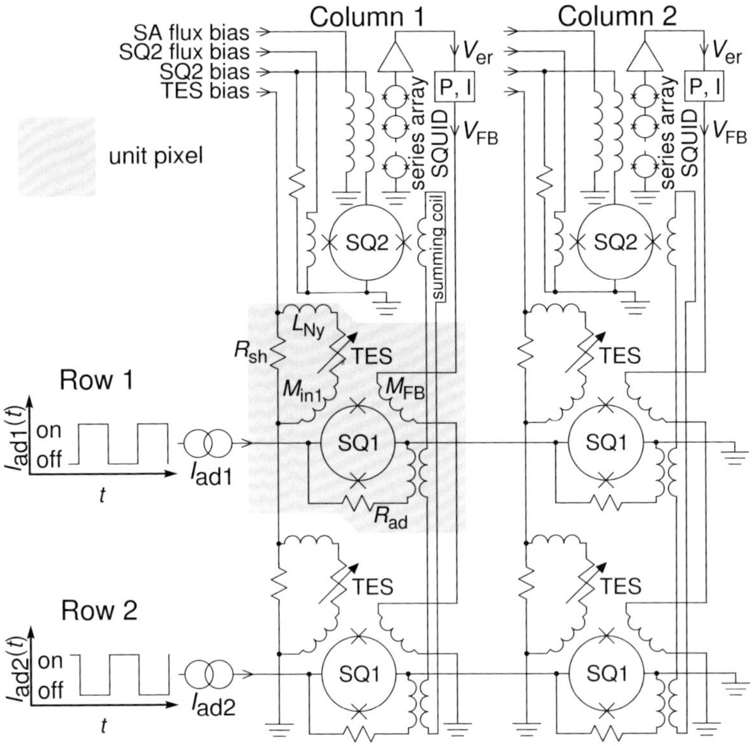

Figure 28.10: A time division multiplexer (TDM) readout circuit for a TES calorimeter instrument (Doriese et al 2007). The TDM is a row-column system where front-end SQUIDs are commonly addressed for each row with the outputs for all the SQUIDs in a column combined on to a single output amplifier. This decreases the wiring complexity from $O(N^2)$ to $O(N)$ for an $N \times N$ detector array. A 1024 pixel array could be read out with a 32 column $\times$ 32 row multiplexer.

soft X-ray background. The results have shown that the ubiquitous soft X-ray emission that dominates the X-ray sky is largely composed of individual emission lines that are almost certainly due to a superposition of geospheric and heliospheric charge exchange emission and line emission from the local interstellar medium (McCammon et al 2002). The XQC instrument is a good model of a spaceborne calorimeter since it encompasses all aspects of an orbital instrument except a long duration cryogenics system, but with the added complexity of very high vibration loads, and the requirement to achieve operational equilibrium within 60 s of launch (Porter et al 2000).

The XQC uses an implanted silicon semiconductor detector array that operates at 50 mK with an individual JFET amplifier per channel operated at 130 K. The cryostat uses 4 L of liquid helium pumped to 1.6 K to precool an adiabatic demagnetization refrigerator (ADR) that uses a 50 g ferric ammonium alum (FAA) salt pill in a 4 T superconducting magnet to reach 50 mK. The most recent flight (April, 2008) used the new 36 pixel detector array shown in Figure 28.4 (McCammon et al

Figure 28.11: A test fixture for a TDM multiplexer and TES calorimeter array. The TDM is a NIST 4 column by 32 row multiplexer (Doriese et al 2007) with a NASA/GSFC 8×8 TES calorimeter array in the centre (Iyomoto et al 2008). The array pixels are 250 µm × 250 µm each.

2008). The XQC detector array is monolithically fabricated using micromachined silicon with the semiconductor thermistors formed in-situ by ion-implantation. The absorbers consist of HgTe, epitaxially grown to a thickness of 0.8 µm on a silicon backing layer which is then epoxied to spacer posts on the pixel. The new XQC array uses very large 2 mm × 2 mm pixels to give a total collecting area of 1.44 cm^2 and has a resolution of 7 eV FWHM at 1.5 keV during ground testing.

The XRS2 instrument on the *Suzaku* observatory uses technology very similar to the XQC but was developed for long-term operation in space (Kelley et al 2007). The original XRS instrument was launched on the *Astro-E* observatory in 1999 but did not reach orbit due to a failure of the first stage booster. The XRS instrument on *Suzaku* was launched in 2005 and worked flawlessly until the cryogens were prematurely exhausted due to an instrument accommodation error on the spacecraft five weeks after launch. So far, the XRS2 is our only experience with an orbital calorimeter instrument, successfully demonstrating many of the supporting technologies that will be used on future calorimeter instruments.

The XRS2 used a 6×6 calorimeter array fabricated using the same process as the array in the XQC instrument but with a different optimization (Stahle et al 2004). The pixel size of the XRS is smaller at 0.625 mm × 0.625 mm but the HgTe absorber material is ten times thicker. This extends the bandpass of the instrument to above 10 keV with high quantum efficiency and yields an energy resolution, on orbit, of 7 eV FWHM at 6 keV (Kelley et al 2007). The XRS used a four-stage cooler to reach its 60 mK operating temperature consisting of a 100 K mechanical cryocooler, 120 L of solid Neon at 17 K, 32 L of space-pumped superfluid helium at 1.3 K, and an ADR at 60 mK. The ADR is relatively large with 920 g of FAA in a 2 T superconducting magnet, but it contains no moving parts and has a hold time, on orbit, of ≈ 36 h with a 1 h recharge. An important legacy of the XRS2

Table 28.2: Current and planned space-borne calorimeter instruments. Launch dates and performance of future missions are best estimates from publicly accessible data.

Mission	Launch	Technology	Array size (pixels)	Pixel area, a/mm^2	Bandpass, E/keV	Resolution (FWHM), $\Delta E/\text{eV}$
XQC1	1995 – 1999	semic.	2×18	0.5×2.0	0.05 to 3	8 @ 1 keV
XQC2	2008	semic.	6×6	2.0×2.0	0.05 to 3	7 @ 1.5 keV
Astro-E	1999	semic.	2×18	0.25×1.0	0.3 to 10	11 @ 6 keV
Suzaku	2005	semic.	6×6	0.63×0.63	0.3 to 10	7 @ 6 keV[a]
Astro-H	2014	semic.	6×6	0.82×0.82	0.3 to 10	4 @ 6 keV
Micro-X	2011	TES	11×11	0.5×0.5	0.1 to 3	2 @ 2 keV
IXO	2020	TES	64×64	0.25×0.25	0.3 to 12	2 @ 6 keV

[a] On-orbit performance.

instrument is the onboard data-processing system that analyzed each X-ray event using an "optimal" digital filter to estimate the X-ray energy (Boyce et al 1999). Only the processed event list was telemetered to the ground. This is necessary to reduce the very large intrinsic data rate of the instrument (6.3 Mbit/s for XRS, $\approx$6.4 Gbit/s for *IXO*) to the $<$ 100 kbit/s average data rate typically available on orbit. Future calorimeter missions will likely use a variant of this system.

The XRS detector system continues to be an important technology for X-ray astrophysics. A ground-based version of the instrument has been operating almost continuously at the EBIT at Lawrence Livermore National Laboratory since 2000, making fundamental measurements of astrophysical plasmas in the laboratory (Porter et al 2000, 2008b). In addition, a similar detector system with improved performance will be flown on the *Astro-H* observatory in 2014.

Future spaceflight instruments

Beyond the near-term *Astro-H* observatory mentioned in the previous section, most calorimeter development is targeted for the *IXO* mission. This large orbiting observatory has a long ($>$ 20 m) focal length with precision X-ray optics that place severe constraints on the focal-plane instruments. For example, oversampling the $5''$ point-spread-function of the *IXO* telescope requires $5''$ pixels that are less than 416 µm × 416 µm in size. To cover the required $5'$ field of view (FOV) then requires a 4096-element imaging array. The instrument's FOV is extremely important, providing crucial spatial information for diffuse objects, and critical background information for point sources. For example, the FOVs for *Suzaku*/XRS and *IXO*/XMS (X-ray Microcalorimeter Spectrometer) are shown in Figure 28.12 in comparison to a large, diffuse galaxy cluster. To achieve these large array sizes, calorimeter development has centered on monolithically fabricated TES arrays with multiplexed readouts.

Semiconductor calorimeters are well developed, space-flight qualified, and are likely to achieve the spectral resolution requirements for the *IXO* mission. How-

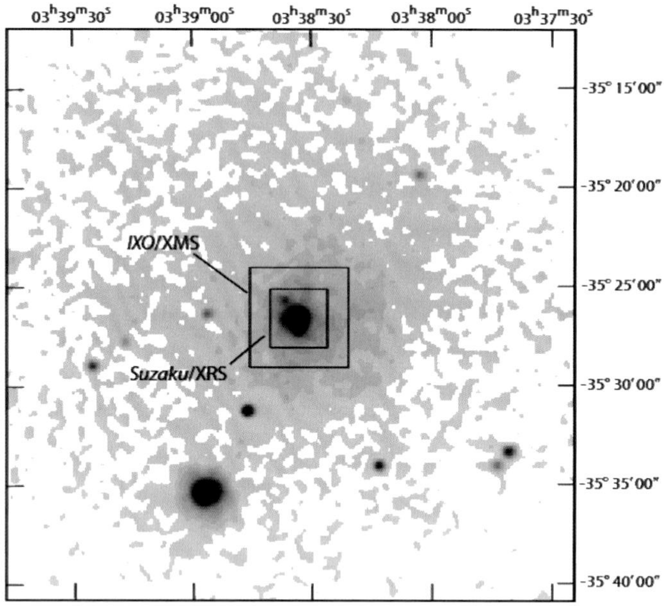

Figure 28.12: An X-ray image of the Fornax galaxy cluster observed with *ROSAT*. The FOVs for the *Suzaku*/XRS and *IXO*/XMS are overlaid on the image. One can see that even the very large calorimeter array for *IXO* yields only a relatively small FOV.

ever, the arraying and readout technologies do not scale well beyond a few hundred channels. TES detectors are now being fabricated in fairly large arrays, currently 8×8 (Iyomoto et al 2008) and will scale to 32×32 in the next year. These arrays are being instrumented successfully with multiplexed readouts, with one such example shown in Figure 28.11. Importantly, small-scale science instruments using this technology are already being constructed. These include the EURECA TES synchrotron experiment, shown in Figure 28.13, that will be a test-bed for candidate *IXO* detectors and frequency-domain multiplexing (de Korte et al 2008), the Transition-Edge Microcalorimeter Spectrometer (TEMS) that will be deployed at EBIT (Porter et al 2008a), and the *Micro-X* sounding rocket experiment which will observe the Puppis A supernova remnant with a 11×11 multiplexed TES array in 2011 (Figueroa-Feliciano et al 2008).

Outlook

X-ray calorimeter technology has progressed substantially since its inception in 1984 with ground and first-generation spaceflight instruments already in operation. Near-term missions such as *Astro-H* will use improved versions of these instruments and the technology is well advanced for the next-generation instrument that will be deployed on the *IXO* observatory. These missions will provide exceptional spec-

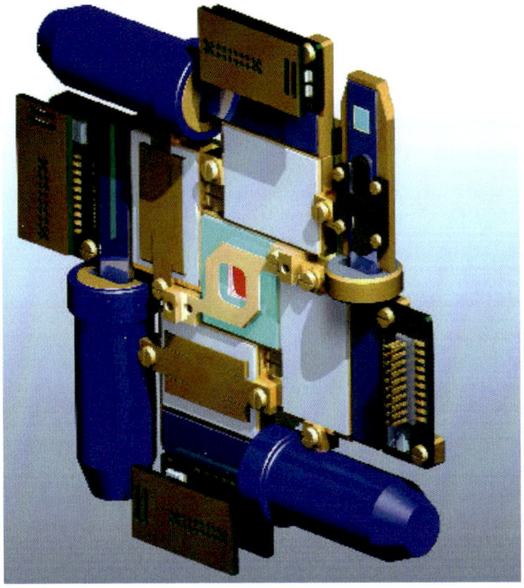

Figure 28.13: A solid model of the European Calorimeter Array (EURECA) focal plane. EURECA is a European-Japanese test bed for calorimeter detectors for *IXO*. The EURECA focal plane consists of a 5×5 TES calorimeter array and a frequency division multiplexer system. In the solid model, the detector array is in the middle and is surrounded by four filter boards for channel isolation. The four high-bandwidth SQUID output amplifiers are located in the corners (de Korte et al 2008; L. Gottardi, personal communication 2008).

trometers with very high sensitivity, enabling routine, high-resolution spectroscopic observations of nearly every class of X-ray emitting object.

Future missions such as the *Generation-X* (Windhorst et al 2006) and *MAXIM* (Gendreau et al 2004) mission concepts will require much larger detector arrays to match their precision ($< 0.1''$) optics with a reasonable field of view. Scaling the current arraying and readout technologies to this level will be challenging at best. The most likely scenario for achieving megapixel calorimeter arrays is to adopt detector elements that are spatially sensitive and also to adopt a readout technology that can pack a very high density of detector channels per readout node. Position-sensitive TES detectors have been under development for several years (Figueroa-Feliciano et al 2006; Smith et al 2006), and have shown that for a modest degradation in detector performance, sub-pixel position information can be encoded in the detector response. In addition, microwave SQUID circuits are being developed to pack large numbers of TES detectors on to a single high-bandwidth readout node (Irwin et al 2006).

Newer calorimeter technologies may also play a large role in achieving megapixel scale arrays. Magnetic calorimeters are dissipation-less systems that substantially improve the very serious problem of thermal management on a large detector array. Kinetic inductor detectors have both sub-pixel position resolution and are read out in a high-Q microwave resonance circuit that can pack several thousand channels onto a single readout node.

The future of X-ray calorimeter spectrometers is exciting, with a number of near-term missions ready to move from development into implementation. The long-term future of X-ray astrophysics, however, will require precision X-ray optics and matching imaging spectrometers with large fields of view and very high spectral resolution. The challenge will be to find clever new ways to package and instrument calorimeter arrays on this much larger scale.

Acknowledgements

The author wishes to thank the X-ray detector development groups at the Goddard Space Flight Center, the National Institute of Standards and Technology, and the Netherlands Institute for Space Research for providing material, and the National Aeronautics and Space Administration for financial support for this work.

Bibliography

Boyce KR, Audley MD, Baker RG (plus eight authors) (1999) Design and performance of the *Astro-E*/XRS signal processing system. Proc SPIE 3765:741–750

de Korte P, Anquita J, Bakker F (plus 39 authors) (2008) EURECA — European-Japanese microcalorimeter array. J Low Temp Phys 151:733–739

Doriese WB, Ullom JN, Beall JA (plus 14 authors) (2007) 14-pixel, multiplexed array of gamma-ray microcalorimeters with 47 eV energy resolution at 103 keV. Appl Phys Lett 90:193508-1–3

Figueroa-Feliciano E, Bandler SR, Boyce K (plus nine authors) (2006) Expanding the *Constellation-X* field of view with position-sensitive X-ray microcalorimeters. Proc SPIE 6276:627615-1–8

Figueroa-Feliciano E, Bandler SR, Bautz M (plus 25 authors) (2008) Micro-X: Mission overview and science goals. J Low Temp Phys 151:740–745

Gendreau KC, Cash WC, Gorenstein P (plus three authors) (2004) *MAXIM*: the black hole imager. Proc SPIE 5488:394–402

Harper DA, Bartels AE, Casey SC (plus 20 authors) (2004) Development of the HAWC far-infrared camera for SOFIA. Proc SPIE 5492:1064–1073

Horansky RD, Ullom JN, Beall JA (plus seven authors) (2008) Analysis of nuclear material by alpha spectroscopy with a transition-edge microcalorimeter. J Low Temp Phys 151:1067–1073

Irwin KD, Beall JA, Doriese WB (plus nine authors) (2006) Microwave SQUID multiplexers for low-temperature detectors. NIMPA 559:802–804

Iyomoto N, Bandler SR, Brekosky RP (plus nine authors) (2008) Close-packed arrays of transition-edge X-ray microcalorimeters with high spectral resolution at 5.9 keV. Appl Phys Lett 92:013508-1–3

Kelley RL, Mitsuda K, Allen CA (plus 41 authors) (2007) The *Suzaku* high resolution X-ray spectrometer. PASJ 59:77–112

Kiviranta M, Seppa H, van der Kuur J, de Korte P (2002) SQUID-based readout schemes for microcalorimeter arrays. AIP Conf Proc 605:295–300

Mather JC (1982) Bolometer noise: nonequilibrium theory. Appl Opt 21:1125–1129
Mazin BA, Eckart ME, Bumble B (plus four authors) (2008) Optical/UV and X-ray microwave kinetic inductance strip detectors. J Low Temp Phys 151:537–543
McCammon D, Almy R, Apodaca E (plus 17 authors) (2002) A high spectral resolution observation of the soft X-ray diffuse background with thermal detectors. Astrophys J 576:188–203
McCammon D (2005a) Thermal equilibrium calorimeters—An introduction. In: Enss C (ed) Cryogenic particle detection, pp 1–34. Springer, Berlin Heidelberg New York
McCammon D (2005b) Semiconductor thermistors. In: Enss C (ed) Cryogenic particle detection, pp 35–61. Springer, Berlin Heidelberg New York
McCammon D, Barger K, Brandl DE (plus nine authors) (2008) The X-ray quantum calorimeter sounding rocket experiment: Improvements for the next flight. J Low Temp Phys 151:715–720
Moseley SH, Mather JC, McCammon D (1984) Thermal detectors as X-ray spectrometers. J Appl Phys 56:1257–1262
Porter FS, Audley MD, Brokosky R (plus 13 authors) (1999) The detector assembly and the ultralow-temperature refrigerator for XRS. Proc SPIE 3765:729–740
Porter FS, Almy R, Apodaca E (plus seven authors) (2000) The XQC microcalorimeter sounding rocket: A stable LTD platform 30 seconds after rocket motor burnout. NIMPA 444:220–223
Porter FS, Audley MD, Beiersdorfer P (plus nine authors) (2000) Laboratory astrophysics using a spare XRS microcalorimeter. Proc SPIE 4140:407–418
Porter FS, Kelley RL, Kilbourne CA (2006) High resolution X-ray microcalorimeters. NIMPA 559:436–438
Porter FS, Beiersdorfer P, Brown GV (plus eight authors) (2008a) The EBIT calorimeter spectrometer: A new, permanent user facility at the LLNL EBIT. J Low Temp Phys 151:1061–1066
Porter FS, Beck BR, Beiersdorfer P (plus nine authors) (2008b) The XRS microcalorimeter spectrometer at the Livermore electron beam ion trap. Can J Phys 86:231–240
Reintsema CD, Beyer J, Nam SW (plus seven authors) (2003) Prototype system for superconducting quantum interference device multiplexing of large-format transition-edge sensor arrays. Rev Sci Inst 74:4500–4508
Romani RW, Miller AJ, Cabrera B (plus two authors) (2001) Phase-resolved Crab studies with a cryogenic transition-edge sensor spectrophotometer. Astrophys J 563:221–228
Smith SJ, Whitford CH, Fraser GW (2006) Signal processing for distributed readout using TESs. NIMPA 559:814–816
Stahle CK, Allen CA, Boyce KR (plus 16 authors) (2004) The next-generation microcalorimeter array of XRS on *Astro-E2*. NIMPA 520:466–468
Windhorst RA, Cameron RA, Brissenden RJ (plus ten authors) (2006) *Generation-X*: An X-ray observatory designed to observe first light objects. New Astron Rev 50:121–126

— 29 —

Infrared imaging bolometers

Harry A. C. Eaton[I]

Abstract

Infrared imaging bolometers have found great utility in space imaging applications. Bolometers measure incident energy by observing a temperature change caused when a material absorbs photons. Because there is great flexibility in choosing the absorbing material characteristics, bolometers can readily be designed to have narrow or broad-band characteristics over a wide range of wavelengths. Imaging bolometers are simply formed from arrays of individual bolometers in the focal plane of a telescope. Uncooled bolometers with high pixel counts are applied for studying the Sun and planets, while imaging bolometers cooled to very low temperatures but having only a few pixels have been designed for deep-space imaging applications, typically in the very-long wavelength to sub-millimetre region. Ultra-low temperature imaging bolometers have achieved detectivities better than the best photo-current detectors in the far-infrared region.

Introduction

Bolometers are capable of detecting radiant energy over a very broad spectral range but have found their greatest utility in the infrared range. Most recent developments of imaging bolometers have focused on the long-wave IR (LWIR) range from 8 µm to 13 µm because the atmosphere is fairly transparent in this range, which also encompasses the peak emission from warm objects on Earth. While these developments are driven by commercial and military market interest the resulting technologies are quite usefully adapted to space applications. Many commercial imaging bolometers that are designed for the LWIR can be adapted for use in short-wave (SWIR, 1 µm to 3 µm) and midwave (MWIR, 3 µm to 8 µm) ranges. With somewhat greater difficulty they can be adapted for far infrared or very broad-band applications as well (Foukal and Libonate 2001). For faint or cold astronomical targets, cryogenically cooled very long-wave IR (VLWIR) and sub-millimetre bolometer arrays have been developed, and these have recently been adapted for use in space.

[I]Johns Hopkins University Applied Physics Laboratory, Laurel MD, USA

History

In the 1980s the Defense Advanced Research Project Agency (DARPA) funded researchers at Honeywell International and Texas Instruments to develop small uncooled infrared imagers using micro-bolometer arrays. The results of this research were declassified in 1992 and have led to an explosive growth in both research on bolometer-based imagers as well as commercially available bolometer focal-plane arrays (FPA). The first uncooled micro-bolometer FPA was flown in space in 1997 (Spinhirne et al 2000). Since that time there have been many space missions incorporating micro-bolometer FPAs, most of which have been for Earth and planetary observations.

The first cryogenically cooled IR bolometer array to fly in space was the FIRP instrument on board the Japanese *IRTS* satellite flown in 1995 (Hirao et al 1996). Since then major advancements in low-temperature bolometers have resulted in detectivities surpassing those of photon-detector technologies.

Bolometer fundamentals

In its simplest form, a bolometer consists of an energy-absorbing material connected to a thermometer with a weak thermal connection to a constant-temperature heat sink. Incident radiant energy heats the absorber, raising its temperature, which is measured by the thermometer. A wide range of materials can be used as the energy absorber, allowing bolometer designers significant freedom to control the operating spectral range, including broad and narrow tuning ranges in wavelength, λ. Figure 29.1 shows the schematic structure of a bolometer element. The absorber/thermometer's heat capacity is C_{p} while G_{c} represents the equivalent thermal conductance from the element to the heat sink including (linearized approximations of) any radiative and convective components. Solving the heat-balance equation (ignoring any self-heating of the thermometer) gives the rise in the detector element temperature

$$\Delta T = \int \frac{\varepsilon(\lambda) P(\lambda)}{\sqrt{G_{\mathrm{c}}^2 + \omega^2 C_{\mathrm{p}}^2}} \, \mathrm{d}\lambda \quad , \tag{29.1}$$

where ε is the emissivity of the absorber, P is the power incident on the absorber element, and ω is the frequency of the incident power envelope. Since low G_{c} improves the bolometer performance, steps are taken to minimize it including using low-emissivity materials on the detector back side (facing the heat sink), evacuated packages and long, thin support and connecting elements with low thermal conductivity. Values for G_{c} of 10^7 W/K are typical for micro-bolometer arrays. The speed of response of a bolometer is governed by the bolometer time constant

$$\tau_{\mathrm{b}} = \frac{C_{\mathrm{p}}}{G_{\mathrm{c}}} \quad , \tag{29.2}$$

and is much slower than most photon detectors. For micro-machined structures found in typical bolometer FPAs, the heat capacity is low enough for response times of the order of a few milliseconds.

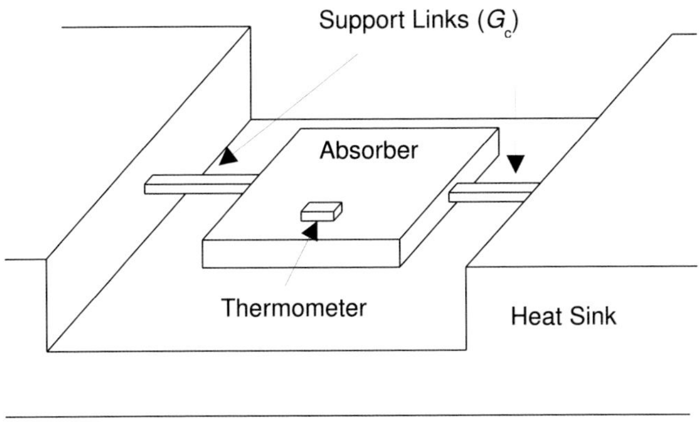

Figure 29.1: A schematic representation of a bolometer element. Incident energy from above heats the absorber element.

The maximum bolometer detectivity is limited by random fluctuations in the radiant power exchange between the detector and scene and is given by (Rogalski 2002)

$$D_{\mathrm{b}}^{*} = \sqrt{\frac{\varepsilon}{8\, k_{\mathrm{B}}\, \sigma\, (T_{\mathrm{d}}^{5} + T_{\mathrm{s}}^{5})}} \quad , \tag{29.3}$$

where k_{B} is Boltzmann's constant, σ is the Stefan-Boltzmann constant, T_{d} is the detector temperature, and T_{s} is the scene temperature. This formula provides an upper bound for bolometer detectivity. In practice additional noise sources in the thermometer and phonon noise of the detector element reduce the detectivity. As can be seen from Equation 29.3, it is desirable that the detector be colder than the scene temperature, but there are diminishing returns for additional cooling. In practice this means that observations of warm targets such as the Sun, Earth and planets can easily be accomplished with detectors near room temperature, but astrophysical observations must been done with detectors cooled to a few kelvins or less.

Uncooled bolometer arrays

The primary figure of merit for uncooled FPA performance is the noise equivalent temperature difference (NETD, ΔT_{NE}). NETD is the temperature difference across a black-body source that produces a signal difference equal to the rms detector noise. It is given by:

$$\Delta T_{\mathrm{NE}} = \frac{4}{\pi} \frac{F^2}{D^* \, \frac{\partial R}{\partial T}} \sqrt{\frac{B}{A}} \quad , \tag{29.4}$$

where F is the ratio of the focal length of the optics to its aperture diameter, $\frac{\partial R}{\partial T}$ is the derivative of the black-body radiance with respect to temperature at the wavelength of operation, D^* is the actual pixel detectivity, B is the noise bandwidth of the measurement, and A is the area of the detector pixel. NETD is dependent on integration time (frame rate) because the detector noise is a function of the measurement bandwidth. The area dependence of the NETD makes the filling factor of the array an important parameter to optimize. NETDs as low as 23 mK have been reported for micro-bolometer arrays having 50 μm × 50 μm pixel size with $F/1$ optics operating at room temperature at 60 frames per second (Howard 2000). For cryogenically cooled detectors the NETD is not reported because the temperature derivative of radiance is very small at cold temperatures.

Thermometers

Various technologies are available to measure the temperature rise of the detector element. These include temperature-dependent resistance (thermistors), pyroelectric and ferroelectric effects, thermopiles, and mechanical cantilevers. Each technique has various advantages and disadvantages, especially for FPAs. Temperature-dependent resistive sensors are dominant in the market at present and have the advantages of monolithic array fabrication (with resulting low cost) and very good detectivity; drawbacks include complex element trimming circuitry and high $1/f$ noise. Pyroelectric and ferroelectric detectors sense changes in temperature and therefore require a chopper to modulate the signal, but are not affected by $1/f$ noise and are minimally affected by static offset variations across the array. Pyroelectric materials also require hybrid assembly techniques that raise the G_c and C_p values, but they do have the advantage that the hybrid structure can be coated with materials that are also not amenable to monolithic processing such as gold-black. Thermopiles (Kanno et al 1994) allow high linearity and negligible $1/f$ noise, but suffer from low signal levels with resulting poor detectivities and NEDT. Mechanical cantilever sensors operate at DC and theoretically have better noise performance (Datskos et al 2004), but require bulky and complex readout systems.

Temperature-dependent resistance sensors (thermistors)

Vanadium oxide (VO_x) is the dominant temperature-sensitive resistance material used for uncooled monolithic IR FPAs, although many FPAs use amorphous silicon. A very large amount of research and technology development has gone into perfecting monolithic FPAs based on these technologies. Much of the improvements have focused on improving spatial NETD problems through on-array correction techniques. The growing trend towards higher pixel counts has resulted in smaller pixel sizes, yet reductions in circuit noise have kept pace to maintain similar NETD (Murphy 2002). Resistance-type micro-bolometer FPAs have been used in several Earth-observation and planetary space missions (Murphy et al 2000; Corlay and Arnolfo 2001; Lancaster et al 2003; Pope et al 2006).

Pyroelectric sensors

Among the uncooled IR FPAs the pyroelectric FPAs gained favour rapidly. This was primarily due to the superior uniformity and resulting good image quality. Another advantage to such capacitive temperature sensors is that there is no self-heating of the temperature sensing element. The main drawback to the technology is the hybrid fabrication technique required to integrate the sensor into the electronic readout, even though this was actually an advantage for an application to solar astronomy because it allowed a relatively simple means to coat the detector with a thin film of gold-black to establish a broad, flat spectral response. The solar bolometric imager (SBI) project used a camera modified in this way to make the first flat full-spectrum images of the Sun from a stratospheric balloon platform in 2003 (http://sd-www.jhuapl.edu/SBI/) (Bernasconi et al 2004). Such images allow the investigation of the contribution to total solar irradiance (TSI, cf., Fröhlich 2010) by areas of the Sun outside of active regions (Foukal and Bernasconi 2008). A similar instrument is proposed for a small explorer (*SMEX*) space mission in 2012.

Cryogenic bolometer arrays

Compact closed-cycle cooling systems and bolometer-array fabrication have both advanced significantly in recent years, paving the way for space-based far-infrared and sub-millimetre telescopes with unprecedented performance.

Two different types of thermometer are most often used with cryogenically cooled bolometer arrays: neutron-transmutation-doped (NTD) germanium thermistors and superconducting transition-edge sensors (TES). Both are usually operated at sensor temperatures ranging from 100 mK to 300 mK. NTD sensors are usually interfaced to conventional metal oxide semiconductor field-effect transistor (MOSFET) amplifiers and biased with an AC signal in order to avoid the amplifier $1/f$ noise and offsets. The MOSFET amplifiers are operated at much warmer temperatures than the sensors (typically ≈ 70 K). TES sensors are usually interfaced to a superconducting quantum interference device (SQUID) amplifier, which is effectively an ultra-low noise ammeter that can operate at the same temperature as the TES thermometer. Both types of temperature sensor exhibit self-heating that is significant compared to the intended incident radiation. The sensor self-heating leads to an electrothermal feedback that reduces the sensor time constant and decreases the temperature change from incident radiation (Irwin 1995). In the case of TES sensors the electrothermal feedback can be very large, reducing the time constant by a factor of 100 or more and providing excellent linearity and dynamic range due to the small temperature excursions. NTD thermistors have a much smaller but non-negligible electrothermal feedback. Both NTD and TES sensors can be interfaced to the telescope as either bare elements or coupled through a feedhorn into an integrating cavity. There are significant advantages and disadvantages to each approach. Advantages of the feedhorn include increased signal gain, good stray light and interference rejection, which can be very important for sensitive bolometers (Yvon et al 2008), smaller numbers of sensors, and ability to control the field of view. The main disadvantage is that the focal plane is under-sampled because the size of the horns spaces the pixels farther apart than the Nyquist distance. This means that the telescope must sweep or jiggle across the sky to form a complete image. With bare detector elements the full image can be sampled at once but it is much more difficult to shield for stray light and interference. Bare

Figure 29.2: The NTD detector FPA used for the BLAST instrument. Fine wire grids inside the small circular areas are tuned to absorb the element's target wavelength (350 μm). Pixel spacing is 1.75 mm to accommodate a $2\,f\,\lambda$ feedhorn for $F/5$ optics. Photo courtesy of Jamie Bock, JPL.

focal planes need about 16 times more detectors and must have lower phonon and Johnson noise since the received signal (and thus photon noise) is smaller. The advantage of full-image sampling of course is simpler telescope pointing and faster image acquisition.

Current space missions using NTD bolometer arrays include *Planck*-HFI (http://public.planck.fr, Yun et al 2003) and *Herschel*-SPIRE (http://sci.esa.int/science-e/www/area/index.cfm?fareaid=16, Nguyen et al 2004). Several near-space observations have been made with NTD bolometer arrays on the BLAST stratospheric balloon telescope (http://blastexperiment.info, Devlin et al 2004). Figure 29.2 shows the FPA elements used for the BLAST instrument. Proposed space missions using TES bolometer arrays include *SPIRIT* (Benford et al 2007) and *SPICA* (http://www.ir.isas.jaxa.jp/SPICA, Kenyon et al 2006). The *SPIRIT* mission will examine dust disks of distant stars to study planetary formation. *SPICA* will be used for detailed mapping of objects such as protostars and ultra luminous infrared galaxies.

Bibliography

Benford DJ, Rinehart SA, Leisawitz DT, Hyde TT (2007) Cryogenic far-infrared detectors for the Space Infrared Interferometric Telescope (*SPIRIT*). Proc SPIE 6687:66870-1–12

Bernasconi PN, Eaton HAC, Foukal P, Rust D (2004) The solar bolometric imager. Adv Space Res 33:1746–1754

Corlay G, Arnolfo M-C, Bret-Dibat T (2001) Microbolometer in space: IASI and PICASSO-CENA. Act Astronautica 48:299–303

Datskos PG, Lavrik NV, Rajic S (2004) Performance of uncooled microcantilever thermal detectors. Rev Sci Instrum 75:1134–1146

Devlin M, Ade PAR, Aretxaga I (plus 24 authors) (2004) The Balloon-borne Large Aperture Submillimeter Telescope (BLAST). Proc SPIE 5498:42–54

Foukal P, Libonate S (2001) Total-light imager with flat spectral response for solar photometric measurements. Appl Opt 40:1138–1146

Foukal P, Bernasconi PN (2008) Do photospheric brightness structures outside magnetic flux tubes contribute to solar luminosity variation? Sol Phys 248:1–15

Fröhlich C (2010) Solar radiometry. ISSI SR-009:525–540

Hirao T, Matsumoto T, Sato S (plus four authors) (1996) Flight performance of the Far-InfraRed Phometer (FIRP). Proc SPIE 2817:276–285

Howard PE, Clarke JE, Bradley MG (plus two authors) (2000) Progress in uncooled focal plane sensor technology at Boeing. Proc SPIE 4130:168–174

Irwin KD (1995) An application of electrothermal feedback for high resolution cryogenic particle detection. Appl Phys Lett 66:1998–2000

Kanno T, Saga M, Matsumoto S (plus 11 authors) (1994) Uncooled infrared focal plane array having 128×128 thermopile detector elements. Proc SPIE 2269:450–459

Kenyon M, Day PK, Bradford CM (plus two authors) (2006) Background-limited membrane isolated TES bolometers for far-IR/submillimeter direct detection spectroscopy. Nuc Instrum Meth Phys Res A 559:456–458

Lancaster R, Skillman D, Welch W (plus two authors) (2003) The Economical Microbolometer-Based Environmental Radiometer Satellite (*EMBERSAT*) designed for forest fire detection and monitoring. Int Workshop on Thermal Detectors, Washington, DC, TDW2003:2–25

Murphy D, Finch R, Kenneday A, Wyles J (plus eight authors (2000) Multi-spectral uncooled microbolometer sensors for the Mars 2001 Orbiter THEMIS instrument. IEEE Aero Conf Proc 3:151–163

Murphy DF, Ray M, Wyles R (plus ten authors) (2002) High-sensitivity 25 μm microbolometer FPAs. Proc SPIE 4721:99–110

Nguyen HT, Bock JJ, Ringold P (plus 11 authors) (2004) A report on the laboratory performance of the spectroscopic detector arrays for SPIRE/HSO. Proc SPIE 5498:196–207

Pope T, Bergeron A, Bourqui P (plus eight authors) (2006) Linear microbolometer arrays for space and terrestrial imaging. Proc SPIE 6206:62061P-1–9

Rogalski A (2002) Comparison of photon and thermal detector performance. In: Handbook of Infra-Red Detection Technologies Vol. 1. (New York: Elsevier), 5–81

Spinhirne J, Scott V, Lancaster R (plus two authors) (2000) Performance and results from a space borne, uncooled microbolometer array spectral radiometric imager. Aero Conf Proc IEEE 3:125–134

Yun M, Beeman JW, Bhatia R (plus eight authors) (2003) Bolometric detectors for the Planck surveyor. Proc SPIE 4855:136–147

Yvon D, Panh J, Landé J (plus eight authors) (2008) My bolometer is running a fever, or why very low noise performance requires global design of the apparatus. J Low Temp Phys 151:448–458

— 30 —

Semiconductors for low energies: incoherent infrared/sub-millimetre detectors

Walfried Raab[I]

Abstract

Semiconductor-based detectors have been used to detect infrared radiation since the early days of space astronomy, and are now the workhorse of infrared space missions. Whilst their leading position in space applications is starting to be challenged by the promise of large-format, high-performance bolometer arrays, semiconductor devices are currently the most sensitive detector type at infrared wavelengths and will undoubtedly continue to play an important role. This chapter gives an overview of the state-of-the-art semiconductor-based detector types and the materials commonly used to cover the different wavelengths, providing insight into the key performance parameters and system trade-offs. Important considerations for the operation of semiconductor-based detectors and the most common read-out techniques and interconnection schemes used are also discussed. The chapter is rounded off by a brief overview of the most promising technologies currently under development in this waveband.

Introduction

The electronic structure of the atoms in a semiconductor crystal causes the formation of a band structure with a valence band filled with electrons and (at low temperatures) an unfilled conduction band, separated by the band-gap energy. Intrinsic photoconduction can be observed in very high-purity semiconductors such as silicon, germanium, HgCdTe or InSb in which absorbed photons excite electrons from the valence band into the conduction band. The long-wavelength limit for intrinsic photoconductivity is set by the wavelength at which the energy of the photon equals the band-gap energy. Extrinsic semiconductors are formed by adding a small amount of an impurity in a process called doping. Impurities with a larger number of valence electrons than the semiconductor form a donor band with an energy just below that of the conduction band of the semiconductor. Semiconductors doped in this way are called n-type, because electrons in the donor band can

[I]MPE—Max-Planck-Institut für extraterrestrische Forschung, Garching, Germany

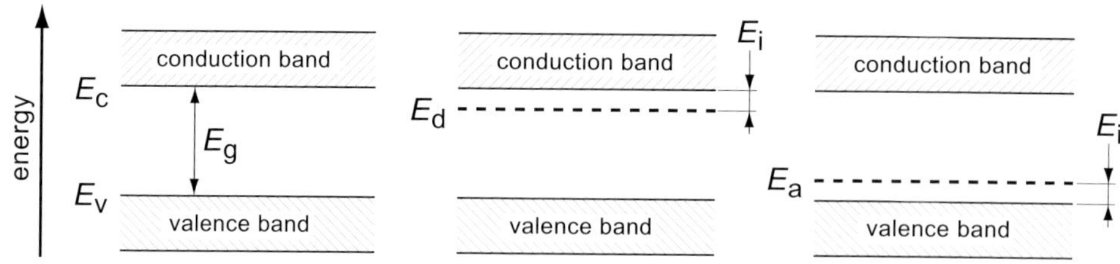

Figure 30.1: Illustration of the band structures of intrinsic (left) and extrinsic semiconductors (middle: n-doping, right: p-doping).

easily be excited into the conduction band so that the dominant charge carriers are electrons. In a similar way an acceptor band is formed just above the valence band by impurities with a smaller number of valence electrons. These semiconductors are called p-type, because the dominant charge carriers are holes left in the conduction band after excitation of electrons into the acceptor band. A simplified illustration of the band structures of various semiconductor types is shown in Figure 30.1. Doped semiconductors are described by the notation *semiconductor:dopant* so, for example, Ge:Ga identifies a germanium semiconductor with gallium as the major impurity.

There are two principal types of semiconductor-based detectors used in the infrared and sub-millimetre waveband: photoconductors and photodiodes. In the following sections the underlying physical principles behind the two detectors will be briefly introduced and examples of the implementation across the wavebands will be given. For a much more detailed discussion of semiconductor detectors, and photodetectors in general, we recommend the textbook by Rieke (2003) and the review article Rieke (2007).

Detectors across the infrared/sub-millimetre waveband

Near-infrared detectors (1 µm to 5 µm)

The leading detector technology for the wavelength range between 1 µm and 5 µm is the photodiode. A photodiode is based on a junction between oppositely doped zones in a sample of semiconductor. The formation of a pn-junction is illustrated in Figure 30.2. The Fermi level in the n-type material is higher relative to that of the p-type material because of the relative electron excess in the n-type material. After contact of the two materials, electrons in the vicinity of the junction will diffuse from the n-type into the p-type material until the Fermi levels are equal throughout the entire semiconductor. This electron flow will produce a space charge region with a negative charge in the p-type and a positive charge in the n-type material and a voltage across the junction that counteracts the flow of charge carriers. An equilibrium is established at the voltage U_0, known as the contact potential. The space charge region is then almost completely depleted of free charge carriers. Photodiodes are operated with reverse bias, i.e., with an external

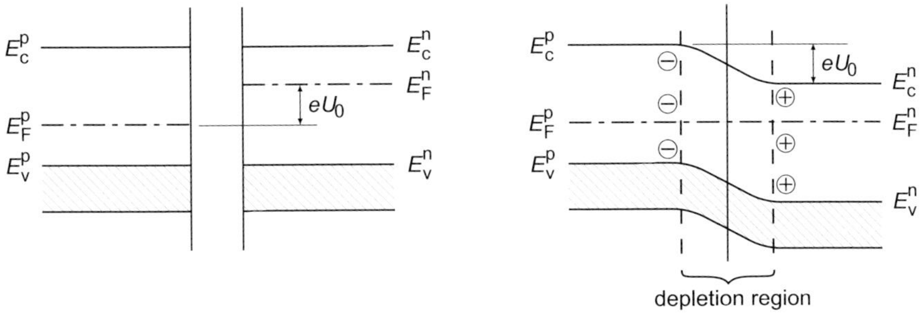

Figure 30.2: Formation of a pn-junction used for photodiodes. Shown on the left is the band structure before, and on the right after the (hypothetical) contact.

voltage adding to the contact potential. The external voltage increases the potential across the junction, increasing the width of the depletion region and consequently the resistance of the device. Charge carriers produced by photoexcitation in, or near, the depletion region are driven across the junction by the electric field, thus carrying a net current. To avoid excess dark current, the device must be cooled to sufficiently low temperatures that thermal excitation is negligible.

Designed for military and remote sensing applications, photodiode arrays are well-developed and commercially available in large format arrays of up to 2048×2048 pixels with integrated low-noise, low-power read-out circuits. Practically all near-infrared arrays used in space missions to date have been based on InSb or HgCdTe semiconductors. While the band gap of InSb equals 0.18 eV (at 300 K) corresponding to a cut-off wavelength of 6.8 μm, the band gap of HgCdTe can be tuned by varying the relative concentration of Hg and Cd. The compound HgTe is a metallic conductor ($E_g < 0$) while CdTe is a semiconductor with a band-gap energy of 1.55 eV, which corresponds to a cut-off wavelength of 0.8 μm. The band-gap energy of $Hg_{1-x}Cd_xTe$ varies monotonically with x between these values. By carefully choosing the constituents of HgCdTe, band gap tailored photodiodes with cut-off wavelengths anywhere between ≈ 0.8 μm and ≈ 15 μm can be made (D'Souza et al 2000).

Molecular beam epitaxy (MBE) is now the leading technology for growing device grade HgCdTe structures because of the precision control over growth parameters and excellent reproducibility (D'Souza et al 2000; Garnett et al 2004). HgCdTe or InSb diode arrays processed in epitaxially grown films are directly attached to silicon read-out integrating circuits (ROICs) using the technique of indium bump bonding (see Page 496) to form a hybrid focal plane array.

Space-astronomy instruments using HgCdTe or InSb photodiode detector arrays include: DIRBE on *COBE* (single-pixel InSb), Nicmos on *HST* (multiple 256×256 pixel HgCdTe), IRAC on *Spitzer* (256×256 pixel InSb), IRC on the Japanese *Akari* (512×412 pixel InSb) and NIRSpec and NIRCam for *JWST* (multiple 2000×2000 pixel HgCdTe arrays).

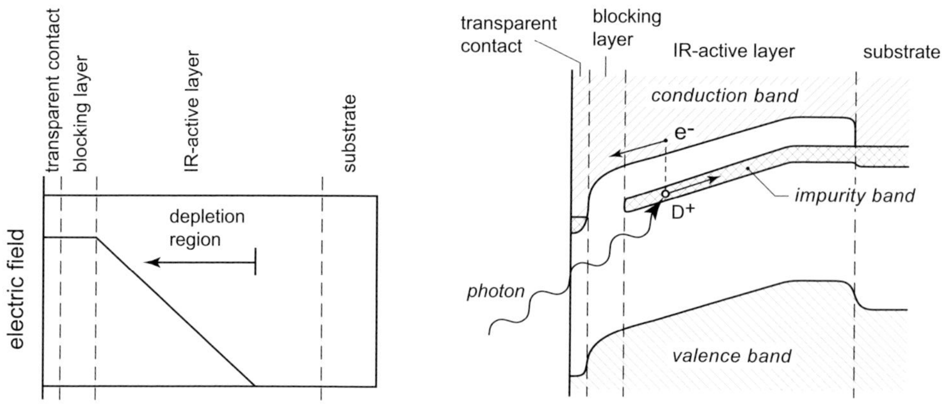

Figure 30.3: Cross section, electric field (left) and band structure (right) of a blocked impurity band detector.

Mid-infrared detectors (5 µm to 40 µm)

The first mid-infrared detectors used in spaceborne missions were single crystal, bulk-type photoconductors. A wide variety of extrinsic, silicon-based semiconductors sensitive in the wavelength range between 2.5 µm and 26 µm were used in the *IRAS* and *ISO* missions, including Si:Ga, Si:B, Si:Sb, and Si:As.

One of the major problems in the design of extrinsic photoconductors is that the doping concentration is driven by conflicting requirements: the doping concentration needs to be as high as possible to get high photon-absorption coefficients (the doping concentration is limited by hopping conduction induced by direct transfer of charge carriers from one impurity site to the next in heavily-doped semiconductors). In contrast a low doping concentration is also desirable to achieve a low electrical conductivity which in turn reduces Johnson noise. An elegant way around this dilemma is the use of separate layers for photon absorption and carrier conduction, allowing the optimization of optical and electrical properties of the photodetector independently. A device following this idea is the so-called blocked impurity band (BIB) detector (Petroff and Stapelbroek 1986; Stetson et al 1986; Szmulowicz and Madarsz 1987; Herter 1994).

The two separate layers of a BIB device are a heavily doped absorbing layer and a thin, high-purity blocking layer, epitaxially grown on a degenerately doped substrate which also serves as a contact layer. In the case of a front-illuminated device described in Figure 30.3, a transparent contact is implanted into the blocking layer.

The absorbing, infrared-active layer can be doped to a level at which hopping conduction would be unacceptably high in a bulk-type detector because the blocking layer interrupts the impurity band, effectively stopping charge transfer due to hopping. A positive bias voltage is applied to the transparent contact to drive out free charge carriers from the active layer around the interface to the blocking layer, generating an electrical field across the depletion region. When a photon with an energy greater than the photoionisation threshold is absorbed in the depletion region of the infrared-active layer, it excites an electron from the impurity band into the conduction band. The resulting electron-hole pair is swept out by the electric field towards the respective contact and, since the conduction bands of the blocking layer and the active layer are continuous, the excited electron can pass the blocking layer and reach the transparent contact. The design of BIB detectors offers a number of advantages over conventional extrinsic photoconductors (Rieke 2003; Herter 1994): the high absorption coefficient of the absorbing layer means that detectors with comparatively small active volumes can be made, providing low susceptibility to cosmic rays (Page 497) without compromising quantum efficiency (Page 491). Also, due to the heavy doping of the active layer, the impurity band increases in width, therefore effectively decreasing the energy gap between the impurity band and the conduction band. As a result, BIB detectors typically offer spectral responses extended towards longer wavelengths compared to bulk-type photoconductors with the same dopant. Ultimately BIB devices also provide better noise performance compared to conventional photoconductors: the positive and negative charge carriers separate after generation and therefore no recombination occurs in the depletion region. Since generation and recombination of the charge carriers in this case are statistically independent processes, only a single noise event is associated with the detection of each photon, leading to a decrease in detector noise by a factor of $\sqrt{2}$ compared to conventional photoconductors.

The technically most mature mid-infrared BIB detector arrays of today are based on Si:As and Si:Sb. Single-pixel Si:Ga BIB detectors were only used in the DIRBE instrument on board *COBE*. Si:As arrays are typically used between 2 µm and 28 µm, while Si:Sb arrays cover the wavelength range between 7 µm and 40 µm, albeit with a lower quantum efficiency than Si:As below $\approx$25 µm. Astronomical space instruments using Si:As and Si:Sb BIB arrays include: SWS on *ISO* (12 pixel Si:As), IRAC, IRS and MIPS on *Spitzer* (up to 256×256 pixel Si:As, 128×128 pixel Si:Sb), IRC on *Akari* (256×256 pixel Si:As) and MIRI on *JWST* (1000×1000 pixel Si:As).

Far-infrared detectors (40 µm to 200 µm)

Radiation in this waveband is strongly absorbed by the water-vapour bands in the Earth's atmosphere, rendering these wavelengths unattractive for military remote sensing applications. Consequently, advances in array performance in the far-infrared have relied on research and development programmes in comparatively small research institutes and universities. As a result, space-grade far-infrared photoconductor arrays are considerably less developed than their shortwave counterparts and, in general, not available commercially.

Far-infrared photoconductors are typically bulk-type, extrinsic, germanium-based semiconductors. Although other doping elements have been explored, Ge:Be and Ge:Ga are the only photoconductors used in space astronomy missions observing at wavelengths longwards of $\approx$40 µm. Ge:Be has a band-gap energy of 24 meV corresponding to a cut-off wavelength of $\approx$52 µm. Ge:Be photoconductors were used in the *ISO* spectrometers, however they suffered from comparatively low quantum efficiencies, so further development of the material was stopped. The material of choice for far-infrared photoconductors today is Ge:Ga. With a band-gap energy

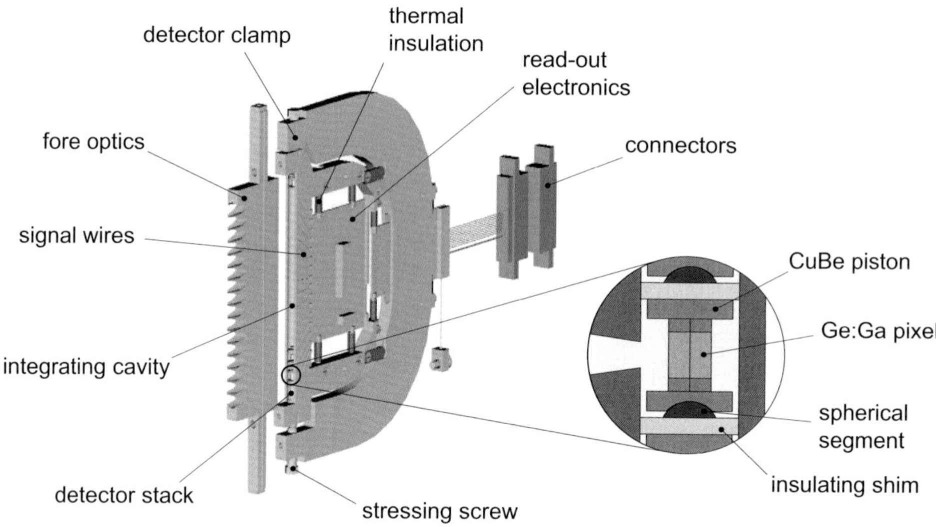

Figure 30.4: Design of a stressed Ge:Ga photoconductor module used in the *Herschel*-PACS instrument. The small insert shows the assembly of the individual components in the pixel stack.

of 11 meV, Ge:Ga detectors are sensitive to wavelengths of up to ≈ 115 µm. Unlike detector arrays for shorter wavelengths, which are manufactured using lithographic processes, Ge:Ga arrays are manufactured by hand: individual Ge:Ga pixels or linear strips of pixels are typically mounted next to one another to form far-infrared photoconductor arrays. The resulting mechanical assembly procedures are relatively complex and thus limit the available array size, so that the largest Ge:Ga array manufactured to date is the 32×32 pixel used in the 55 µm to 95 µm waveband of *Spitzer*-MIPS (Young et al 1998). Recent developments have enabled the fabrication of monolithic 2D arrays, with pixels formed using lithographic and wet-etching processes (Fujiwara et al 2003). Monolithic Ge:Ga arrays can be contacted directly to Si read-out circuits using In-bump bonds (see Page 496) to form direct hybrid arrays.

Applying mechanical stress to a p-type semiconductor can modify the donor binding energy and consequently the cut-off wavelength. The effect is particularly pronounced in semiconductors with a diamond-like lattice (e.g., germanium) stressed along the [100] axis (Price 1961). Application of uniaxial stress to a Ge:Ga crystal reduces the donor binding energy to about 5 meV, effectively extending the responsivity out to ≈ 220 µm (Kazanskii et al 1977).

Stressed arrays are manufactured by stacking individual pixels and applying mechanical stress by a spring-loaded stressing screw. The design of stressed arrays is even more complex than that of unstressed arrays, and consequently the manufacture of large-format stressed Ge:Ga arrays is extremely time consuming and expensive. The largest stressed Ge:Ga array built to date is the 16×25 pixel array used in the PACS instrument on the *Herschel* satellite. Figure 30.4 shows the design of a 16 pixel stressed Ge:Ga sub-module of PACS; 25 of these modules are used to realise the 16×25 pixel detector array. Shown in the exploded view is the complex design of the stressed stack, illustrating the arrangement of pixels, stressing pistons, insulating shims and spherical segments used to compensate uneven and non-parallel pixel surfaces.

The doping concentration in extrinsic bulk photoconductors must be kept low to avoid unwanted conduction modes (see Page 486) and excess noise leading to a comparatively low absorption coefficient in the device. Absorption coefficients are typically about three orders of magnitude lower than those measured for direct absorption in intrinsic photoconductors. Consequently, extrinsic photoconductors need to be large to yield useful quantum efficiencies. Typical dimensions of individual pixels are of the order of 1 mm making them particularly vulnerable to cosmic-ray damage (Page 497).

Photoconductors are widely used for far-infrared wavelengths on space-astronomy missions. Instruments using these detectors include: LRS and CPC on *IRAS* (Ge:Ga), DIRBE on *COBE* (single-pixel Ge:Ga), SWS, LWS and ISOPhot on *ISO* (various detectors including: 12 pixel Ge:Be, 3×3 pixel Ge:Ga, 2×2 pixel stressed Ge:Ga), MIPS on *Spitzer* (32×32 pixel Ge:Ga, 2×20 pixel stressed Ge:Ga), FIS on *Akari* (up to 20×3 pixel monolithic Ge:Ga, up to 15×3 pixel stressed Ge:Ga) and PACS on *Herschel* (16×25 pixel Ge:Ge, 16×25 pixel stressed Ge:Ge).

Detector performance

General parameters

The properties and hence the performance of detectors and detector arrays can be described by a number of key parameters. A basic property of any detector is its wavelength-dependent quantum efficiency, $\eta(\lambda)$, defined to be the ratio of total photon flux incident on the detector and the flux that is absorbed by the detector to create charge carriers. Together with the photoconductive gain, G, which is the ratio of carrier lifetime to carrier transit time to one of the electrodes, the product $\eta(\lambda)\,G$ is thus the probability that a photon incident on the detector produces a charge carrier which also reaches an electrode. Another important figure of merit of a detector is its responsivity, defined as the electrical current output divided by the incident photon power. The responsivity of photoconductors as well as photodiodes at a given wavelength is proportional to the product $\eta(\lambda)\,\lambda$. The long-wavelength limit for both the responsivity and the quantum efficiency is the cut-off wavelength, corresponding to the band-gap energy of the detector material. Responsivity and quantum efficiency are therefore important figures to define the wavelength range over which a detector can be used. Figure 30.5 shows typical quantum efficiencies of various detectors discussed on Page 486 to Page 491. It is interesting to note the presence of a marked gap in the wavelength coverage around 40 µm; no semiconductor-based detector array is currently available that covers this wavelength range with useful quantum efficiency. Table 30.1 and Table 30.2 list the performance of selected state-of-the-art semiconductor-based infrared detectors. In addition to the properties discussed above, the tables also quote values for the read noise (see Page 495), the dark current (being the signal produced by the detector in

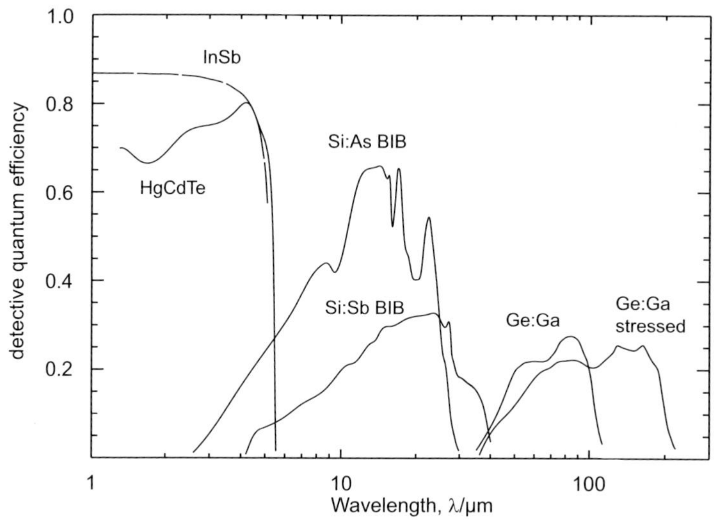

Figure 30.5: Typical quantum efficiencies, $\eta(\lambda)$, of semiconductors used in the wavelength range between 1 μm and 200 μm. InSb: 1024×1024 pixel test array from NOAO (Hoffman et al 1998), HgCdTe: 1024×1024 pixel MBE grown array without anti-reflection coating (Garnett et al 2004), Si:As and Si:Sb: 256×256 pixel BIB arrays used in the FORECAST instrument (Adams et al 2004), Ge:Ga and stressed Ge:Ga: 16×25 pixel arrays used in the *Herschel*-PACS instrument.

absence of any illumination) and the typical detector operating temperature. The low excitation energies of the semiconductor materials used to fabricate IR detectors place constraints on the operating temperatures of the detectors: detectors are typically cooled to cryogenic temperatures to reduce thermal excitation of charge carriers, which in turn would increase the detector dark current. (For more detailed information on the complex subject of dark current and its statistical manifestation in various detector technologies see Rieke (2003).) The table entry for germanium-based far-infrared arrays also includes the noise equivalent power (NEP), defined as the signal power that gives an RMS signal-to-noise ratio of unity in a system that has an electronic bandpass of 1 Hz.

The relative importance of each of the performance parameters given in Table 30.1 and Table 30.2 depends critically on the specific scientific goals, operating waveband, and details of type and design of the instrument in which the detectors are to be used. In certain cases, a particular parameter may even become irrelevant: for example, the read noise, or NEP, of a detector can become insignificant under "background-limited conditions", as under such conditions the noise arising from the photon statistics of the background radiation (e.g., thermal emission from optical components like the telescope, or astronomical sources like zodiacal light) itself is dominant. Other parameters such as the array format are extremely important when observing spatially extended sources or undertaking large-area surveys, but are of little importance when studying unresolved, point-like sources. In general, instrument designers will define and optimize the "observing efficiency" of

Table 30.1: Performance of state-of-the-art semiconductor-based infrared arrays. N_e denotes the number of electrons. InSb values from: Hoffman et al (1998); Pipher et al (2000); Fazio et al (2004), HgCdTe from: Garnett et al (2004); Figer et al (2004); Rauscher and Ressler (2005), Si:As BIB from: McMurray et al (2000); Fazio et al (2004), Si:Sb BIB from: Van Cleve et al (1995); Roellig et al (1998); Houck et al (2004). Read-noise values are given for 200 s frame time except: HgCdTe: 1000 s, Si:As 8.0 μm: 50 s, and Si:Sb: 32 s.

	InSb	HgCdTe	Si:As	Si:Sb
Type	photodiode	photodiode	BIB	BIB
Mission	*Spitzer*	*JWST*	*Spitzer*	*Spitzer*
Instrument	IRAC	test array	IRAC	IRS
Wavelength range, λ/μm	3.6/4.5	0.6 to 5	5.8/8.0	14 to 40
Array format, pixel	256^2	2048^2	256^2	128^2
Pixel size, s/μm	30	18	30	75
Operating temperature, T/K	15	38	6	3.8 to 5
Quantum efficiency, η	0.87	>0.8	0.45/0.7	0.4
Read noise, N_e	8.1/6.8	<10	6.6/13	30
Dark current, N_e/s	0.2	<0.005	<1/3.8	30

their instrument, such that a given science goal can be achieved within a minimum amount of observing time. A careful and detailed study of the trade-offs between all parameters is therefore required during the design phase of any instrument.

Photoconductor transient response

Extrinsic bulk-type photoconductors exhibit non-linear behaviour for changes in illumination under certain operating conditions. Figure 30.6 shows the typical response of a photoconductor to a step in the illumination. A fraction of the signal called the fast component rises almost instantaneously after the illumination step, while the rest of the signal reaches the new equilibrium level with a much slower time constant. The time constant of the slow component and the ratio of fast to slow component depend on many different operational parameters. The most important one of these is the level of background radiation, but the size of the illumination step, the uniformity of the illumination across the pixel, the dimensions of the photoconductor, and the detailed physical properties of the semiconductor material are also critical. The time constant of the slow component increases with decreasing background radiation and can reach several minutes or even hours at the low signal levels typical of space-astronomy missions (Williams 1969; Haegel et al 1999). The transient response is associated with the relatively slow adjustment of the electric field near the injecting contact (the contact which replaces charge carriers absorbed by the other contact) to the new equilibrium conditions after the illumination change. After the step in the illumination, charge carriers are swept out of the semiconductor leaving behind an excess of ionized impurities and thus a net space charge which perturbs the electric field in the crystal. Neutralization of the space charge takes place near the injecting contact, with a dielectric relaxation time of $\tau_p = \varepsilon\,\rho$, where ρ and ε are the resistivity and the dielectric

Table 30.2: Performance of state-of-the-art germanium-based far-infrared arrays. N_e again denotes the number of electrons. Ge:Ga values from: Young et al (1998); Rieke et al (2004), monolithic Ge:Ga from: Kawada et al (2004); Shirahata et al (2004), and stressed Ge:Ga from: Poglitsch et al (2004).

	Ge:Ga	Monolithic Ge:Ga	Stressed Ge:Ga
Mission	*Spitzer*	*Akari*	*Herschel*
Instrument	MIPS	FIS	PACS
Wavelength range, λ/μm	50 to 100	50 to 110	110 to 210
Array format, pixel	32×32	20×3	16×25
Pixel size, s/mm	0.75	0.5	3.6
Operating temperature, T/K	1.5	2.2	1.8
Quantum efficiency	>0.2	0.4	0.3
Read noise, N_e	≈100		
NEP, W/Hz$^{1/2}$		7.5×10^{-17}	8.9×10^{-18}
Dark current, N_e/s	<200	1000	$\approx10^5$

constant of the bulk semiconductor, respectively. Since bulk photoconductors have extremely high resistivity (of order of 10^{12} Ω m under low illumination), the dielectric relaxation time quickly reaches the large values discussed above. Extrinsic bulk photoconductors can also exhibit a behaviour known as the hook anomaly, also shown in Figure 30.6. The hook anomaly arises from non-uniform illumination of the detector, in particular of the region near the injecting contact (Haegel et al 2001).

These non-linearities of the photoconductors can cause a strong time-dependent response which is extremely undesirable in observing modes such as scanning, or in applications such as Fourier-transform spectroscopy. To obtain reliable data, a comprehensive model for the response of the detectors is required. Such a model exists for silicon-based photoconductors (Schubert et al 1994; Schubert and Fouks 1995; Schubert et al 1995), but whilst some progress has been reported (Coulais et al 2005; Jeong et al 2004), no such model exists for germanium-based photoconductors.

Read-out techniques

The performance of any detector heavily depends on the quality of the read-out electronics. One fundamental difficulty in reading out semiconductor devices such as photoconductors or photodiodes lies in their intrinsically high impedance. First-stage electronics therefore need to be capable of detecting the extremely small resulting currents with a very high degree of linearity, and without adding additional noise. Very high input impedance, ultra low-noise amplifiers built with field effect transistors (FET) are typically used for read-out circuits.

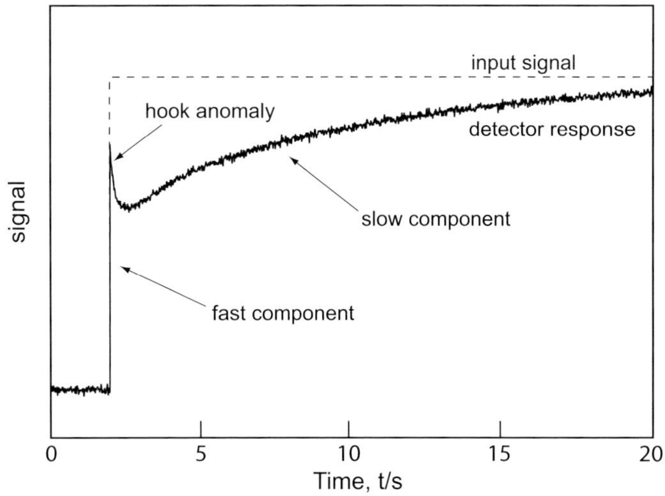

Figure 30.6: Response of a photoconductor to a step in the illumination (Haegel et al 2001).

Read-out circuits are often designed as integrating amplifiers. The output signal of these devices is proportional to a charge collected in an integrating capacitor. In the case of semiconductor detectors, this charge is modulated by the detector impedance which, in turn, is a direct function of the photon flux. The charge collected in the capacitor can be cleared by closing a switch, short circuiting the integrating capacitor and resetting the amplifier. The uncertainty in the measurement of the collected charge is typically expressed in electrons and is referred to as the read noise of the amplifier.

A simple integrating amplifier built with a single FET as a source-follower is shown in the left panel of Figure 30.7. In this design the charge is directly deposited into the input capacitor C_i of the FET, so that the FET is used both to collect the charge and to produce a relatively low impedance output signal which is readily matched to the subsequent electronic circuits. Inherent to the design of source-follower integrating amplifiers is a reduction of the detector bias as charge accumulates in the integrating capacitor. Source-follower amplifiers are therefore best suited for use with detectors with a weak bias dependence of their responsivity. This limitation can be partially overcome using a direct injection amplifier. Here an additional FET is placed between the detector and the source-follower which effectively isolates the integrating capacitor and its voltage variations from the detector. An integrating circuit that maintains the detector bias voltage and provides an intrinsically linear output is the capacitive transimpedance amplifier (CTIA). The layout of a CTIA is shown in the right hand panel of Figure 30.7. By design, the ideal operational amplifier in feedback mode maintains the input node at zero volts, therefore keeping the bias voltage across the detector constant. Unlike the simple integrating amplifier, the CTIA accumulates the charge in an external integrating capacitor C_f. It is important for the linearity of the output signal that

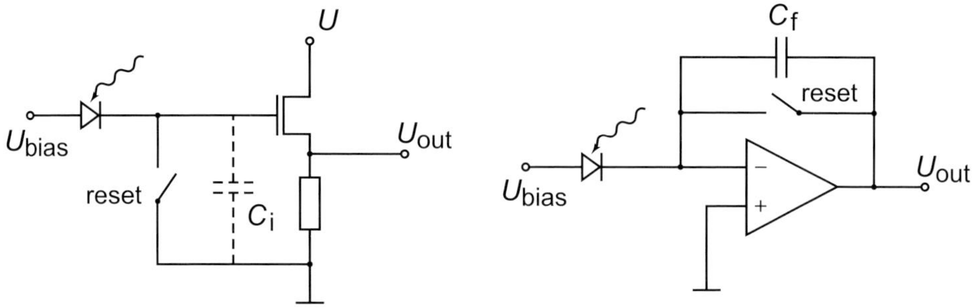

Figure 30.7: Basic read-out diagrams used to read out semiconductor-based detectors. Left: Simple integrating amplifier. Right: Capacitive feedback amplifier (CTIA).

the capacitance of the integrating capacitor is voltage independent. Values for the feedback capacitor need to be relatively small to provide a large output signal, minimizing the influence of the amplifier electronics noise on the signal-to-noise ratio. Again, a reset switch is used to discharge the integrating capacitor and reset the integration process.

A second consequence of the high impedance of photoconductors and photodiodes is that it becomes essential to use short wires to connect the detectors and the read-out electronics. In practice, this means that the first-stage read-out electronics need to be as close to the detectors as possible and thus must be able to operate at the same or similar cryogenic temperature. To date, complex low-noise integrated circuits are only available in silicon, making the manufacture of detectors and their read-out on a common substrate impossible. An elegant way to connect detector pixels to the read-out electronics is provided by the technique of indium bump bonding, also referred to as hybrid integration. A cross section of an indium bump-bonded hybrid array is shown in Figure 30.8. An array of detectors is connected to an evenly spaced grid of read-out electronics by squeezing indium spheres between the detector and the silicon read-out. The easily deformed indium welds to metal contact pads on the detectors and the read-out elements, providing electrical contact and mechanical stability. Indium bump bonding is commonly used for building infrared semiconductor arrays; almost all detectors discussed in this chapter are manufactured as hybrid arrays using this technique. The indium bump-bond technique, however, can not be used for stressed arrays because the typical large size of stressed Ge:Ga arrays leads to a very high differential thermal contraction between the detector material and the silicon read-out. Instead, the pixels of stressed arrays are connected to the input channels of the read-out circuits by individual wires as is the case in the stressed Ge:Ga module used in the *Herschel*-PACS instrument illustrated in Figure 30.4. The individual routing of signal wires is a time-consuming process further adding to the complexity and manufacturing cost of these devices.

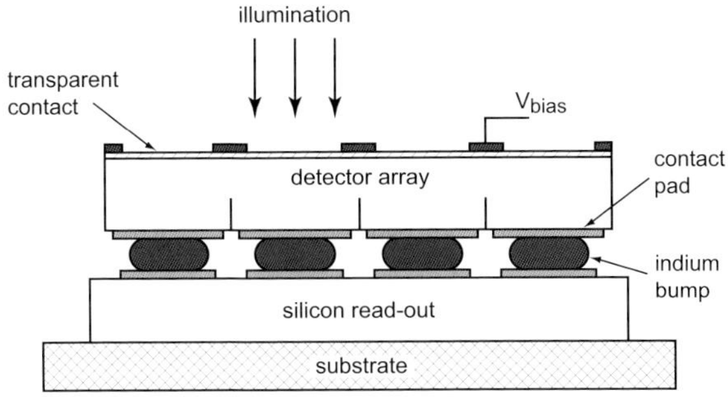

Figure 30.8: Indium bump-bond technique. Detectors and read-out electronics are bonded by indium balls of a size similar to the size of the individual pixels.

Cosmic-ray effects

Ionizing radiation, most notably high-energy ions of the solar wind, creates large numbers of free electrons when passing through the sensitive volume of a detector. These electrons can severely affect the performance of semiconductor-based detectors for periods long enough to corrupt a considerable fraction of the data recorded. The effects of ionizing radiation are complex and depend in detail on the radiation environment, the type and amount of shielding provided by the satellite structure, and the type of detectors and their operating conditions. Few general rules for the operation of semiconductors in space missions exist and extensive irradiation tests before launch are required in order to identify and address the cosmic-ray effects in the individual mission environment.

The inherently large volume of bulk-type far-infrared detectors such as Ge:Ga means that the probability of interaction with cosmic-ray particles is much higher than that for the low-volume pixels of photodiode or BIB detector arrays. Consequently these detectors are much more susceptible to cosmic-ray damage/effects than commercial near- and mid-infrared detectors. Far-infrared arrays usually also have far fewer pixels and so a significant fraction of an array can be affected by a single cosmic-ray hit. Figure 30.9 illustrates some of the cosmic-ray effects observed in Ge:Ga photoconductors. The two main effects directly visible in the individual integration ramps are sudden steps in the integration ramp known as "glitches", and an increase of the subsequent ramp slope leading to a persistent increase in detector responsivity. Models for the increase in responsivity can be found in the literature (Patrashin et al 1997, 1999). The increase of the ramp slope continues with subsequent particle hits leading to a long-term drift of the detector responsivity as shown in the right panel of Figure 30.9. The responsivity eventually reaches a plateau when the rates of partial curing of the radiation damage by infrared radiation and further radiation damage are in equilibrium. As the responsivity increases during radiation exposure, the detectors can also enter a non-linear operating regime where excited charge carriers are accelerated to energies sufficient to create a charge-carrier avalanche. The integration ramps then display a huge number of glitches that are no longer related to cosmic rays, but to the charge-carrier avalanche events. This mode of operation, called spiking, critically depends on the

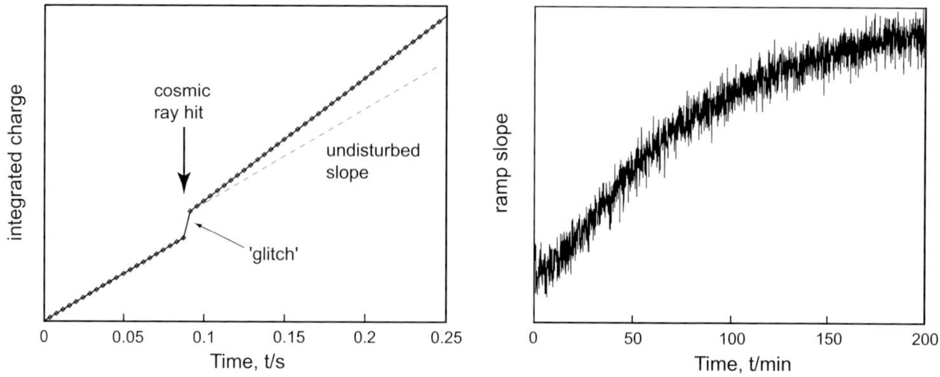

Figure 30.9: Illustration of cosmic-ray effects. Left: A sudden step in an integration ramp after a cosmic-ray hit, and the subsequent change of the integration ramp slope. Right: Change of the integration ramp slope under constant cosmic-ray intensity. The ramp slope eventually reaches a plateau.

applied bias voltage. To avoid this regime, photoconductors operated in space require significantly reduced bias voltages compared to the optimal values derived in radiation free laboratory measurements. Drifts in the detector responsivity can be addressed by on and off source chopping, a common technique in the infrared to subtract background radiation. Rapid chopping is preferred when detectors show significant responsivity drifts.

Various methods for reducing radiation effects have been developed. The most effective of these include the temporary increase of the bias voltage beyond the break down voltage of the semiconductor (bias boost), moderate increase of the detector temperature (thermal annealing), and irradiation of the detectors with infrared radiation (flashing). Thermal annealing and infrared flashing have proven to be the most effective and are usually preferred over boosting the bias voltage which often can not fully restore the detector responsivity.

Future developments

The extremely low thermal background radiation of future large cryogenic infrared space missions certainly will require more sensitive detectors than currently available, with higher quantum efficiencies, better noise performance and lower dark current — independent of the detector technology used. In the case of the semiconductor-based detectors discussed in this chapter, development is focused primarily on materials research aimed at extending the wavelength coverage beyond 200 μm, and covering the responsivity gap around 40 μm (compare Figure 30.5). Larger arrays are also highly desirable, in particular for the longer wavelengths. However, the mechanical complexity and manufacturing cost quickly place limits on the application of stressed Ge:Ga technology at the long wavelength end of the detector material discussed on Page 486 to Page 491. From the discussion on BIB devices on Page 488 to Page 489 it is clear that the development of BIB devices for the far infrared is highly desirable in order to transfer the advantages of BIB detectors into this wavelength regime. Such research is currently carried out by various groups in Europe, the United States, and Japan (Reichertz et al 2005; Watanabe et al 2004).

One particularly interesting semiconductor is GaAs because of the very low excitation energy levels formed by n-type dopants such as sulfur, silicon, selenium, or tellurium. The excitation energy for all these dopants lies at around 6 meV, corresponding to a cut-off wavelength of ≈ 210 µm. At the high doping concentration of the absorption layer typical for BIB devices, a broad impurity band forms (see Page 488), so that a wavelength response out to ≈ 330 µm is expected for these devices. A GaAs BIB detector would therefore not require the complex stressing mechanisms of Ge:Ga which would then enable the manufacture of single-chip arrays with large numbers of pixels. To date, however, GaAs BIB detectors have yet to realize their full potential. The main hurdle is the requisite purity of the material. The unintentional doping of the blocking layer due to impurity atoms needs to be less than 10^{13} cm^{-3}, while the acceptor concentration in the absorption layer needs to be at least an order of magnitude lower to allow the formation of a depletion region of sufficient width. Establishing the required sharp transition of doping concentration at the blocking-/absorption-layer interface presents an additional challenge in the manufacturing of GaAs BIB detectors.

Other promising semiconductor materials for BIB manufacture are based on doped germanium, in particular Ge:Ga and Ge:B. Research on these materials is focused on closing the wavelength gap in the responsivity of conventional semiconductors around 40 µm. Germanium-based devices face the same problems as GaAs in terms of material purity and dopant concentration control. A new technique using ion implantation of boron into high-purity germanium, forming a so-called IBIB (ion-implanted blocked impurity band) device has been reported recently (Beeman et al 2007). The basic development step using this technique is the fabrication of deep implantations to form absorption layers wide enough for significant photon absorption.

Bibliography

Adams JD, Herter TL, Keller LD (plus four authors) (2004) Testing of mid-infrared detector arrays for FORECAST. Proc SPIE 5499:442–451

Beeman JW, Goyal S, Reichertz LA, Haller EE (2007) Ion-implanted Ge:B far-infrared blocked-impurity-band detectors. Infrared Phys Techn 51:60–65

Coulais A, Malaizé J, Giovannelli J-F (plus six authors) (2005) Non-linear transient models and transient corrections methods for IR low-background photo-detectors. ASP Conf Ser 314:566–569

D'Souza A, Wijewarnasuriya PS, Poksheva JG (2000) HgCdTe infrared detectors. In: AGU Perera, HC Liu (ed) Handbook of Thin Film Devices 2:1–25 Academic Press, San Diego, London

Fazio GG, Hora JL, Allen LE (plus 62 authors) (2004) The Infrared Array Camera (IRAC) for the *Spitzer* Space Telescope. Astrophys J Suppl Ser 154:10–17

Figer DF, Rauscher BJ, Regan MW (plus four authors) (2004) Independent testing of *JWST* detector prototypes. Proc SPIE 5167:270–301

Fujiwara M, Hirao T, Kawada M (plus five authors) (2003) Development of a gallium-doped germanium far-infrared photoconductor direct hybrid two-dimensional array. Appl Opt 42:2166–2173

Garnett JD, Zandian M, DeWames RE (plus twelve authors) (2004) Performance of 5 micron, molecular beam epitaxy HgCdTe sensor chip assemblies (SCAs) for the *NGST* mission and ground-based astronomy. In: P Amico, JW Beletic, JE Beletic (eds) Scientific detectors for astronomy, pp 59–79, Kluwer Academic Publishers, Dordrecht

Haegel NM, Simoes JC, White AM, Beeman JW (1999) Transient behavior of infrared photoconductors: application of a numerical model. Appl Opt 38:1910–1919

Haegel NM, Schwartz WR, Zinter J (plus two authors) (2001) Origin of the Hook effect in extrinsic photoconductors. Appl Opt 40:5748–5754

Herter T (1994) IBC arrays: present and future prospects. In: I McLean (ed) Infrared Astronomy with Arrays, pp 409–417, Kluwer, Netherlands

Hoffman AW, Ando AD, Estrada JD (plus 8 authors) (1998) Near IR arrays for ground-based and space-based astronomy. Proc SPIE 3354:24–29

Houck JR, Roellig TL, Van Cleve J (plus 32 authors) (2004) The infrared spectrograph on the *Spitzer* Space Telescope. Proc SPIE 5487:62–76

Jeong WS, Pak S, Lee HM (plus 10 authors) (2004) *ASTRO-F*/FIS observing simulation including detector characteristics. Adv Space Res 34:573–577

Kawada M, Shibai H, Kaneda H, Nakagawa T (2004) Far-Infrared Surveyor: design, operation, and performance. Proc SPIE 5487:359–368

Kazanskii AG, Richards PL, Haller EE (1977) Far-infrared photoconductivity of unaxially stressed germanium. Appl Phys Lett 31:496–497

McMurray RE, Johnson RR, McCreight CR (plus eleven authors) (2000) Si:As IBC array performance for SIRTF/IRAC. Proc SPIE 4131:62–69

Patrashin M, Fouks B, Groezinger U (plus two authors) (1997) Residual conductivity of stressed Ge:Ga photoconductors after low-dose gamma irradiation. J Appl Phys 82:1450–1453

Patrashin M, Hiromoto N, Fouks B (plus two authors) (1999) Low-dose radiation effects in extrinsic photoconductors. J Appl Phys. 86:3797–3803

Petroff MD, Stapelbroek MG (1986) Blocked impurity band detectors. United States Patent No. 4568960, filed October 1980, granted 4 February 1986

Pipher JL, Forrest WJ, Glaccum WJ (plus eleven authors) (2000) InSb arrays for IRAC (InfraRed Array Camera) on *SIRTF* (Space Infrared Telescope Facility). Proc SPIE 4131:7–12

Poglitsch A, Waelkens C, Bauer OH (plus ten authors) (2004) The photodetector array camera and spectrometer (PACS) for the *Herschel* Space Observatory. Proc SPIE 5487:425–43

Price PJ (1961) Strain dependence of the acceptor binding energy in diamond-type semiconductors. Phys Rev 124:713–716

Rauscher BJ, Ressler ME (2005) The James Webb Space Telescope and its infrared detectors. Exp Astron 19:149–162

Reichertz LA, Cardozo BL, Beeman JW (plus six authors) (2005) First results on GaAs blocked impurity band (BIB) structures for far-infrared detector arrays. Proc SPIE 5883:164–171

Rieke G (2003) Detection of Light. From the Ultraviolet to the Submillimeter. Second edition. Cambridge University Press, Cambridge

Rieke GH, Young ET, Engelbracht CW (plus 40 authors) (2004) The Multiband Imaging Photometer for *Spitzer* (MIPS). Astrophys J Suppl Ser 154:25–29

Rieke GH (2007) Infrared detector arrays for astronomy. Ann Rev Astron Astrophys 45:77–115

Roellig TL, Houck JR, Van Cleve J (plus six authors) (1998) Infrared spectrograph for the Space Infrared Telescope Facility (*SIRTF*). Proc SPIE 3354:1192–1203

Schubert J, Roth G, Wolf J (plus two authors) (1994) Correction and curing of in-orbit-induced nonideal behavior of ISOPHOTS's photodetectors. Proc SPIE 2268:283–294

Schubert J, Fouks BI (1995) Precise theoretical description of photoresponse for detectors of ISOPHOT's Si:Ga array. Proc SPIE 2475:487–498

Schubert J, Fouks BI, Lemke D, Wolf J (1995) Transient response of ISOPHOT Si:Ga infrared photodetectors: experimental results and application of the theory of nonstationary processes. Proc SPIE 2553:461–469

Shirahata M, Matsuura S, Makiuti S (plus nine authors) (2004) Pre-flight performance measurements of a monolithic Ge:Ga array detector for the far-infrared surveyor on board *ASTRO-F*. Proc SPIE 5487:369–380

Stetson SB, Reynolds DB, Stapelbroek MG, Stermer RL (1986) Design and performance of blocked-impurity-band detector focal plane arrays. Proc SPIE 686:48–65

Szmulowicz F, Madarsz FL (1987) Blocked impurity band detectors: an analytical model—figures of merit. J Appl Phys 62:2533–2540

Van Cleve J, Herter T, Butturini R (plus four authors) (1995) Evaluation of Si:As and Si:Sb blocked impurity band detectors for *SIRTF* and *WIRE*. Proc SPIE 2553:502–513

Watanabe K, Murakami H, Ohata T (plus two authors) (2004) Development of GaAs photoconductors for far-infrared/submillimeter astronomy. Proc SPIE 5498:637–646

Williams RL (1969) Response characteristics of extrinsic photoconductors. J Appl Phys 40:184–192

Young ET, Davies JT, Thompson CL (plus seven authors) (1998) Far-infrared imaging array for *SIRTF*. Proc SPIE 3354:57–65

— 31 —

Coherent far-infrared / submillimetre detectors

WOLFGANG WILD[I]

Abstract

Coherent detectors in the far-infrared and submillimetre spectral ranges provide high to extremely high spectral resolution ($> 10^6$). Heterodyne techniques and components have been developed up to terahertz frequencies and moved from ground-based to space applications. The quality of a coherent (heterodyne) receiver depends critically on the signal coupling, the mixer and the mixing element, the local oscillator, and the spectrometer. An overview of the state of the art of far-infrared/submillimetre heterodyne receivers is given. The main developments for the different receiver components are described with the emphasis on their use in past, present and future space missions.

Introduction

Coherent (heterodyne) detectors play an important role for the detection of photons in the far-infrared/submillimetre regime, in particular for high spectral resolution observations and interferometry. In Chapter 14 (Swinyard and Wild 2010) we have introduced the basic working principle of a heterodyne system and its sub-components. The sensitivity of a heterodyne receiver strongly depends on the following factors of which the mixer plays a prominent role.

- Efficient coupling of the signal radiation to the detector element.
- Quality of the mixer and the non-linear mixing device.
- Power level, frequency linewidth and stability of the local oscillator.
- Availability of suitable backend spectrometer technology.

In the following sections we will briefly address each of these components.

[I]SRON Netherlands Institute for Space Research and European Southern Observatory, Garching, Germany

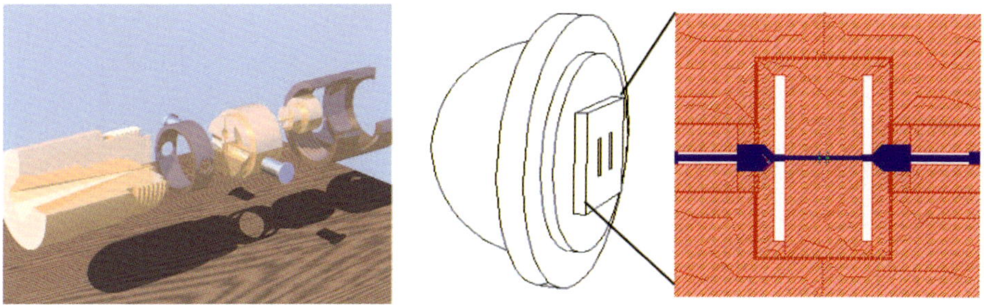

Figure 31.1: Left: Typical waveguide mixer configuration with a feedhorn. The detector element is placed at the end of the waveguide. Right: Quasi-optical coupling with double-slot antenna and lens. The detector element is in the centre between the two vertical linear antenna slots.

Signal coupling

The efficient coupling of the RF signal to the mixing device is important to achieve good sensitivity and is commonly done either via feedhorns and a waveguide or quasi-optically, i.e., via a planar antenna structure and a lens. Figure 31.1 shows the two schemes. The optical design in the submillimetre and far-infrared regime uses the formalism of "quasi-optics" which takes into account the finite diameter of a submillimetre beam (measured in wavelengths) and the corresponding diffraction and propagation effects. These effects can be rather different from geometrical optics and need to be taken into account when designing a submillimetre optical system. See, e.g., Goldsmith (1997) and Lesurf (1990) for an in-depth treatment of quasi-optics. Signal coupling using waveguide structures is often preferred since they are well understood and modelled and provide clean beams and high efficiency. However, at higher frequencies they become increasingly difficult to manufacture — for example, a rectangular waveguide at 2 THz has dimensions of the order of 65 µm × 130 µm. On the other hand, this difficulty is increasingly overcome as (micro-) machining techniques are further refined, and feedhorn structures up to 2.5 THz have been fabricated (Figure 31.2). While quasi-optical structures are easier to manufacture, they provide a very fast beam (low f-ratio) which is difficult to match to the telescope optics, and the detector element needs to be extremely well aligned ($\ll$ 1 µm) with respect to the lens at high frequencies. For a low-noise superconductor-insulator-superconductor (SIS, see Page 505) mixer, the signal and local-oscillator (LO, see Page 512) radiation need to be coupled efficiently to the small SIS mixing element. In a waveguide mixer, this is done by placing the SIS junction together with coupling and impedance matching structures on a rectangular substrate which is mounted across the waveguide at the end of a feedhorn. The radiation is picked up by a waveguide probe on the chip, and a microstrip transmission line couples it to the mixing element.The stripline is designed in such a way that it transmits the low-frequency IF (intermediate-frequency) signal to the substrate end (where electrical contact is made to the IF amplifier) and blocks transmission of the high-frequency astronomical and LO signals. Figure 31.3 shows an example

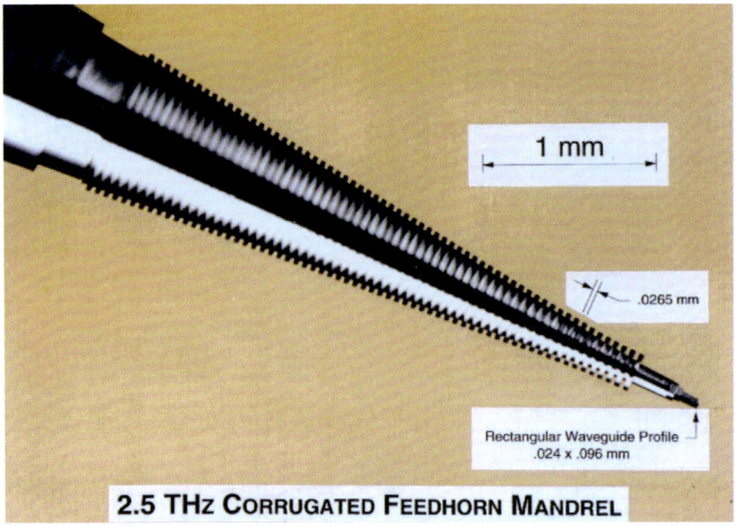

Figure 31.2: Example of state-of-the-art micro-machining: a mandrel for a corrugated feedhorn at 2.5 THz with structures as small as 26 μm. (Courtesy B. Ellison, RAL).

of an SIS junction geometry on a substrate. Both types of signal coupling – waveguide and quasi-optical – are employed in the HIFI instrument for *Herschel*. Below 1.3 THz waveguide mixers are used, and above 1.3 THz quasi-optical coupling has been chosen.

Mixers

The choice of the non-linear mixing device is crucial for the receiver sensitivity, and several technologies are used in the submillimetre and far-infrared regime: SIS junctions, hot-electron-bolometer (HEB) devices, and Schottky diodes. Table 31.1 lists their key characteristics. It is expected that with further progress in space cryocooler technology, future heterodyne space missions such as a far-infared space interferometer (Wild et al 2006) can fully exploit the potential of the very sensitive superconducting mixers at submillimetre and far-infrared frequencies.

SIS mixers

SIS mixers provide unsurpassed sensitivity approaching the quantum limit in the submillimetre regime up to $\approx$ 1.3 THz. The non-linear mixing element is an SIS tunnel junction formed by two superconductors (typically Nb) separated by a very thin (few nm) insulating layer. Figure 31.4 shows the cross section of a typical SIS junction. The insulator is so thin that charged particles can tunnel quantum-mechanically through the barrier. The area of the junction needs to be small to keep the junction capacitance low (the capacitance acts increasingly as a short

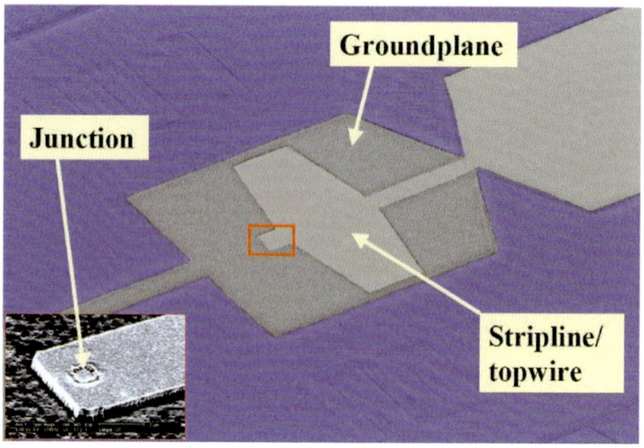

Figure 31.3: Geometry of the central part of an SIS-mixer chip. The stripline/topwire of width 56 µm on top of the groundplane (60 µm width) forms a radiation probe which couples to the SIS junction. The insert shows an SEM photograph of an SIS junction of dimensions 1 µm by 1 µm. (Courtesy C. Lodewijk, TU Delft).

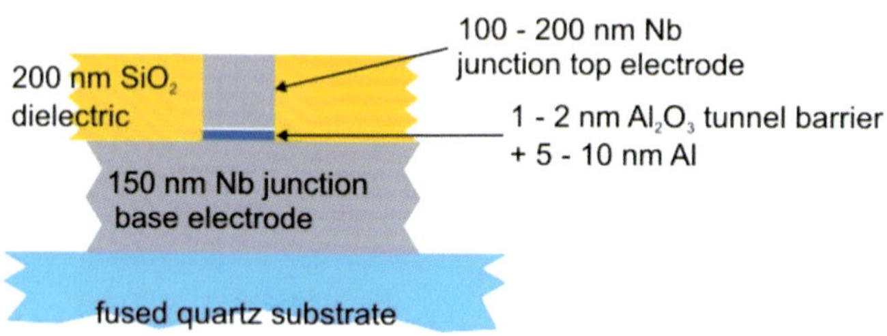

Figure 31.4: Schematic cross section of a typical SIS tunnel junction which is formed by two layers of superconducting Nb separated by a very thin insulator (Al_2O_3 in this case).

circuit at high frequencies). A typical junction size for a submillimetre SIS device is 1 µm × 1 µm or smaller.

In a simple (and necessarily incomplete) picture, the I-V curve and basic detection mechanism of the SIS mixer can be understood with the help of the semiconductor picture of an SIS junction (Figures 31.5 and 31.6). The filled and empty energy states of a superconductor are separated by the band gap 2Δ, i.e., the energy needed to break up a Cooper pair. Above the energy gap, only so-called quasi-particles (essentially single electrons) can exist. For bias voltages less than the gap voltage, $V < 2\Delta/e$, and without incident radiation, these single electrons cannot tunnel across the insulation layer since no free energy states are available on the other side of the barrier. At a voltage of $V_g = 2\Delta/e$ (the gap voltage) quasi-particles at the top of the filled band are able to tunnel into empty states on the other side of the barrier causing a sharp current rise due to the singularity of the density of states $D(\epsilon)$ close to the band gap (Figure 31.5). For voltages

Table 31.1: Key characteristics of far-infrared/submillimetre coherent detector (mixer) technologies.

	SIS	HEB	Schottky
RF range	up to $\approx$ 1.3 THz	(1.3 to 5) THz	up to 3 THz
IF bandwidth	large (> 8 GHz)	small (< 4 GHz)	large ($\gg$ 8 GHz)
Sensitivity[a] $/T_{\rm min}$	excellent: $\approx$ 2 to 6	medium: $\approx$ 8 to 10	poor: $\approx$ 20 to 40
LO power requirement	low: $\approx$ 1 µW	very low: $\leq$ 1 µW	high: $\approx$ 1 mW
Operating temperature	$\leq$ 4 K	$\leq$ 4 K	70 to 300 K
Astronomical space missions	*Herschel*-HIFI, future scientific space missions	*Herschel*-HIFI, future scientific space missions	*SWAS*, *ODIN*, *Rosetta*-MIRO

[a] in units of $T_{\rm min} = h\nu/k_{\rm B}$, the minimum achievable noise temperature for a single sideband receiver, with h the Planck constant, ν the receiver input frequency and $k_{\rm B}$ the Boltzmann constant (cf., Chapter 14, Swinyard and Wild 2010).

greater than the gap voltage the SIS junction behaves essentially like a resistor. SIS junctions can detect radiation through the process of the photon-assisted tunnelling of quasi-particles, and the SIS I-V curve is modified in the presence of radiation. This is schematically shown in Figure 31.6. If radiation is applied to the SIS junction, a quasi-particle can tunnel through the insulation barrier even for a bias voltage below the gap voltage, i.e., $V < 2\Delta/e$, if the extra energy of an incident and absorbed photon is sufficient to reach an empty state on the other side of the barrier (hence the name photon-assisted tunneling). In other words, in the presence of electromagnetic radiation and a bias voltage $V_{\rm bias}$ across the junction, an electron absorbing a photon of energy $h\,\nu$ can tunnel across the barrier when its total energy $E = e\,V_{\rm bias} + h\,\nu$ is equal to or larger than the energy gap 2Δ. In this case the resulting I-V curve shows photon steps, and the most efficient mixing is achieved when the SIS junction is biased at the first photon step just below the gap voltage. The voltage width of the photon step corresponds to the photon energy, i.e., $\Delta V_{\rm photon} = h\,\nu/e$. This very simplified picture can only illustrate the general principle of the SIS mixer. A complete quantum theory for tunnel-junctions mixers was developed by Tucker (1979) and Tucker and Feldman (1985). For somewhat easier reading on SIS tunnel junctions and mixers, see one of the many PhD theses on the subject, e.g., Jackson (2005) or Teipen (2006).

Different materials have been used for SIS junctions and striplines/ground planes. Up to $\approx$ 700 GHz, Nb-Al_2O_3-Nb junctions with Nb striplines on Nb can be used. Above 700 GHz, losses in the Nb superconductor result in increasingly poorer mixer sensitivity, and NbTiN-Al_2O_3-Nb or NbTiN-AlN-NbTi junctions with Al on NbTiN microstrip lines are more suitable for low-noise terahertz SIS mixers. SIS mixing devices with these material combinations are used in *Herschel*-HIFI (Delorme et al 2005; Justen et al 2004; Jackson et al 2006; Karpov et al 2004). Figure 14.13 in Chapter 14 (Swinyard and Wild 2010) shows the sensitivities for the HIFI flight mixers.

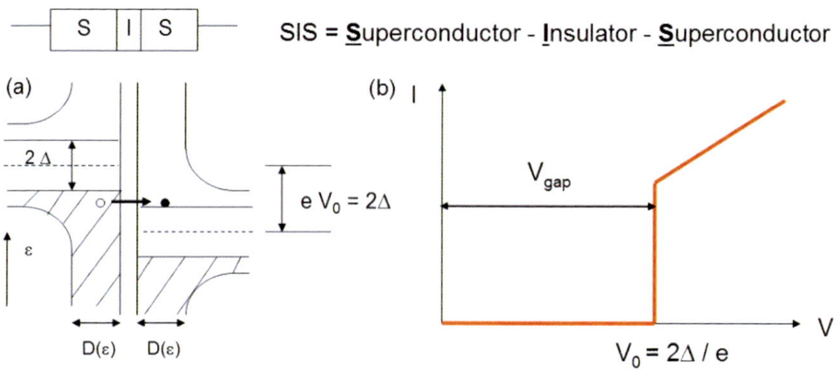

Figure 31.5: a) Semiconductor picture of an SIS tunnel junction with a bias voltage $V_0 = 2\Delta/e$ applied and no incident radiation. A quasi-particle can tunnel through the insulator barrier. b) Resulting I-V curve of an ideal SIS junction with no local oscillator power applied. Note the strong I-V non-linearity.

HEB mixers

Hot-electron-bolometer mixers are the most sensitive coherent detectors for frequencies above $\approx$ 1.3 THz and up to $\approx$ 6 THz. They are thus complementary to SIS mixers which work well up to $\approx$ 1.3 THz. Bolometers are thermal detectors which can also be used as mixers if local oscillator radiation is applied. A far-infrared HEB mixer device consists of a very thin bridge of superconducting material (typically NbN) with submicron dimensions on a silicon substrate. Incident radiation heats the electrons which then cool through phonon coupling to the substrate (Figure 31.7). For efficient detection of radiation the device is coupled to a suitable antenna structure. Figure 31.8 shows an HEB device and its coupling to a twin-slot antenna.

In contrast to the rapid technological progress in the development of HEB mixers and the achieved results, the physical processes leading to heterodyne mixing in HEBs are not as well understood as, e.g., for SIS or Schottky mixers. At present the results can be best understood with the hotspot mixing model (Wilms Floet et al 1999, Merkel et al 2000) explaining the HEB heterodyne mixing process in terms of an electronic hotspot, the length of which oscillates at the intermediate frequency. Figure 31.9 illustrates the basic idea.

The combination of the incident RF signal and LO power on the HEB microbridge causes a time-dependent heating of the electron gas in the superconductor where the small difference in frequency results in a beating of the power at the intermediate frequency (Figure 31.9a). The incident LO power (which is much higher than the signal power) in combination with a DC bias current will create a resistive state when the critical current density of the superconductor is exceeded and heating above the critical temperature T_c occurs in the bridge centre (Figure 31.9b). The contacts remain superconducting because the current density is much lower there and the bath temperature T_b to which the HEB is coupled is well below the critical temperature. The resulting electronic hotspot (i.e., the normal conducting

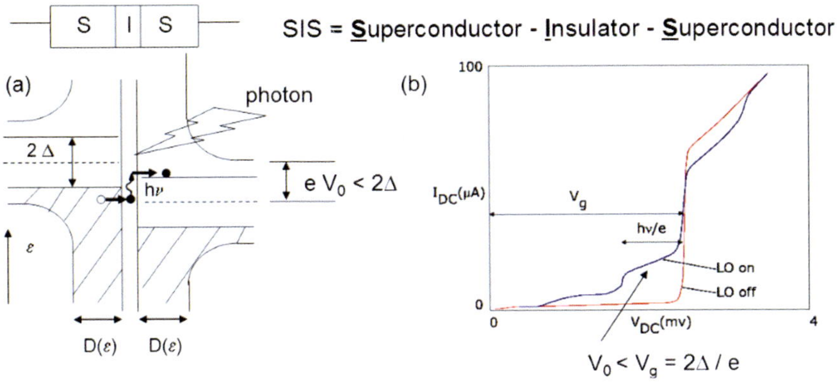

Figure 31.6: a) SIS tunnel junction with a bias voltage $V_0 < 2\Delta/e$ applied and with incident radiation. A quasi-particle can tunnel through the insulator barrier absorbing the energy $h\nu$ of an incident photon. b) Resulting I-V curve of a real SIS junction with local oscillator power on and off. The incident radiation causes photon steps of width $h\nu/e$.

region in the microbridge) is due to the combined heating of the DC and incident LO power. Its length (and consequently the microbridge resistance) varies according to the variation in total power which occurs at the IF. Thus, the response of the HEB is caused by a normal conducting hotspot the length of which oscillates at the IF. This process is called hotspot mixing. The real situation in HEB mixers is more complex but the simplified hotspot mixing model allows to understand a number of HEB characteristics including a qualitative explanation of the mixer gain and noise temperature. HEB mixers for the far-infrared range are not yet as mature as SIS or Schottky mixers despite their rapid development. It can be expected that the ongoing research efforts – to a large extent motivated by current and future astronomical space missions – will result in further progress in the areas of HEB device theory, sensitivity, IF bandwidth, signal-coupling schemes, and number of pixels. The past years have seen a significant improvement of sensitivity which was largely driven by the *Herschel*-HIFI space project. Consequently, any given information here can only reflect the present state-of-the-art as of 2010 which may change quite rapidly.

As of 2010, best sensitivities of $\approx$ (6 to 10) $h\nu/k_B$ over the frequency range 0.6 THz to 5.3 THz have been achieved (Figure 31.10) making HEB mixers the most sensitive coherent detectors above $\approx$ 1.3 THz. The IF bandwidth is relatively small (in particular in comparison with SIS and Schottky mixers) due to the fact that HEBs are intrinsically thermal devices. Typically $\approx$ (3.5 to 5) GHz of IF noise bandwidth have been realized to date.

HEB mixers have their first application in a space mission in the two highest frequency bands (1410 GHz – 1703 GHz, and 1626 GHz – 1910 GHz) of the HIFI instrument on board ESA's *Herschel*. Figure 31.11 shows a HIFI 1.6 THz HEB mixer (Cherednichenko et al 2002, 2008) with quasi-optical coupling via a silicon lens and a double slot antenna. Apart from quasi-optical HEB mixers at terahertz

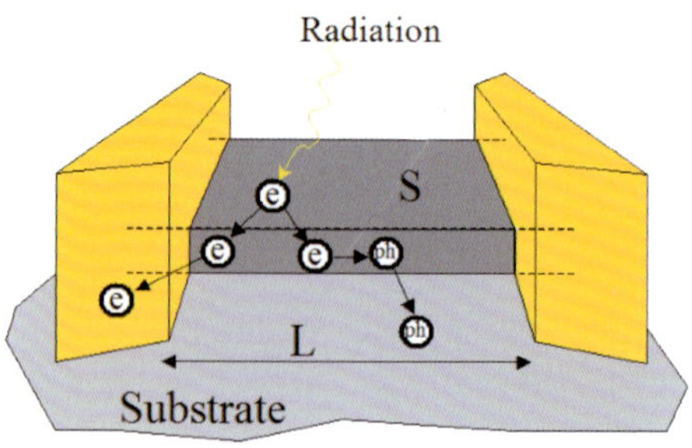

Figure 31.7: General principle of an HEB device. The electrons in a small superconducting bridge are heated through incident radiation and cool through phonon coupling to the substrate.

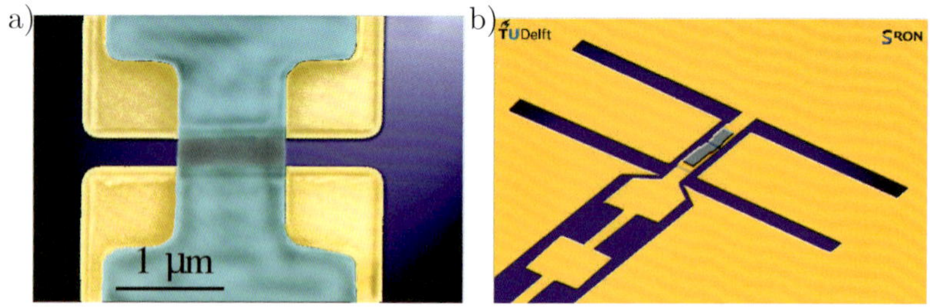

Figure 31.8: a) Photograph of an HEB device consisting of a thin submicron-size superconducting NbN bridge across gold contact pads. b) Double-slot antenna structure for terahertz frequencies with the HEB detector device at the centre. (Courtesy J.R. Gao, SRON/TU Delft).

frequencies, waveguide mixers with the HEB device on a membrane have been developed (see, e.g., Munoz et al 2006).

Schottky mixers

Schottky mixers use the non-linear I-V characteristic of a Schottky diode for frequency down-conversion. A Schottky diode is a metal-semiconductor (usually GaAs at terahertz frequencies) junction as opposed to a semiconductor-semiconductor junction in a normal diode. The current in a Schottky diode is carried by electrons only, making them very fast devices. The response speed is basically limited by the junction capacitance which can be made as low as fractions of a picofarad and enables operating frequencies well into the terahertz range. The characteristic of very fast response times make Schottky diodes suitable for (sub-) millimetre and far-infrared mixing (see Crowe et al 1992 for an excellent discussion of the Schottky diode physics and engineering at that time). Schottky mixers for submillimetre astronomy have been built in one of two technical implementations: whisker-contacted diodes and more recently planar diodes (Figure 31.12). Whisker

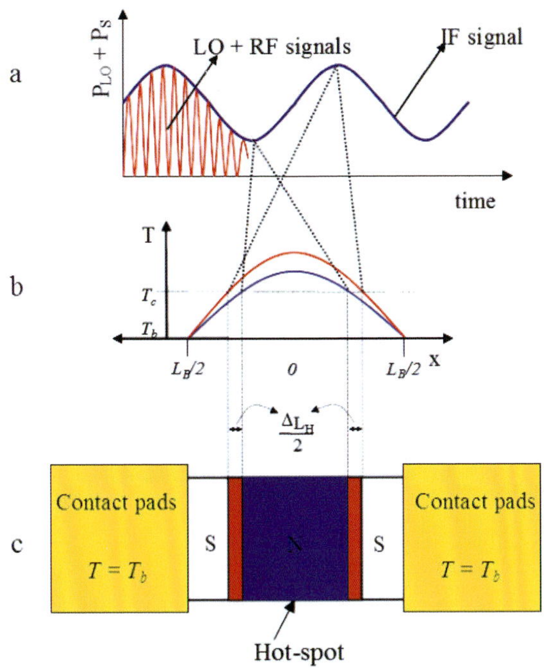

Figure 31.9: Schematic of the hotspot model for hot-electron-bolometer heterodyne mixing. a) Absorbed LO and RF signal power in the superconducting microbridge as a function of time. The slight difference in their frequencies results in a beating of the power at the IF. b) Resulting temperature profile in the microbridge for two different points in time and power levels. c) Top view of the device. As a consequence of the modulation of the dissipated power, the length of the hotspot, and thus the resistance of the microbridge, oscillates at the IF. The N and S regions refer to the normal conducting (hotspot) and superconducting parts, respectively. (Courtesy J.R. Gao, SRON/TU Delft. Text after Wilms Floet et al 1999).

contacts have been realized within waveguides or with a long wire as antenna in a corner-cube reflector (although the latter type has not been used in space due to its fragility). Planar diodes in general are less fragile, can be handled and space qualified more easily, and are therefore replacing whisker-contacted Schottky diodes.

For astronomical applications submillimetre receivers using Schottky mixers have the general disadvantage of being significantly less sensitive than those using SIS or HEB mixers (by factors of typically 10 to 20 and even up to 30). On the other hand, a major advantage of Schottky mixers is the fact that they can be operated over a wide range of temperatures including room temperature although they are often cooled to lower temperatures to improve the noise performance. This makes them the prime choice of submillimetre/far-infrared coherent detectors in applications where active cryogenic cooling is too costly, undesirable, or simply impossible. This is the case for example on Earth-observation missions or planetary missions

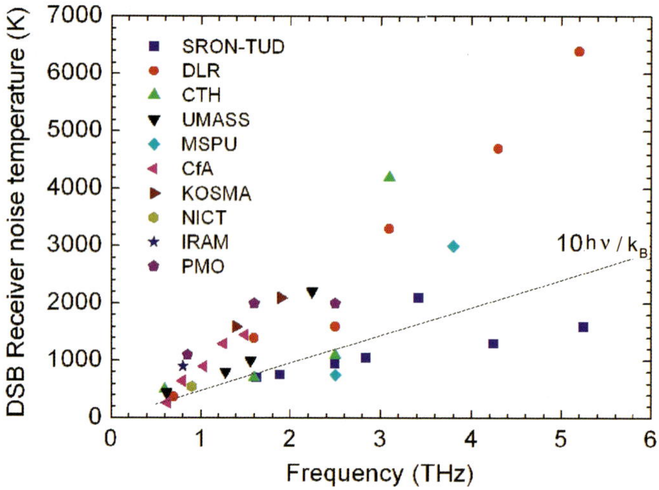

Figure 31.10: HEB mixer sensitivities from 0.6 THz to 5.3 THz achieved until 2010. Best results are ≈ (6 to 10) $T_{\min}(=h\,\nu/k_{\mathrm{B}})$. (Compilation by J.R. Gao, SRON/TU Delft).

where payload mass and power restrictions make it difficult to use 4 K detectors. In addition, Schottky mixers provide a large IF bandwidth which for some applications (e.g., atmospheric science) is crucial. For astronomical missions, however, where instrument sensitivity is a major priority, the use of SIS and HEB mixers is preferred unless operational and spacecraft limitations enforce the utilization of Schottky mixers.

Heterodyne receivers using submillimetre Schottky mixers have been used in the *SWAS*, *ODIN* and *Rosetta*-MIRO astronomical space missions, launched in 1998, 2001, and 2004, respectively. NASA's *SWAS* was the first astronomical heterodyne space mission and carried two receivers passively cooled to about 175 K and operating around selected frequencies corresponding to rotational transitions of the indicated molecule or atom: 487 GHz (O_2), 492 GHz (C I), 551 GHz (^{13}CO), 548 GHz ($H_2^{18}O$), and 557 GHz ($H_2^{16}O$). For a detailed description of *SWAS* see Melnick et al (2000) and Tolls et al (2004). *ODIN*—a Swedish-led dual-purpose astronomy/aeronomy mission—has tunable receivers and allows observation at the frequency bands 118.25 GHz to 119.25 GHz, 486.1 GHz to 503.9 GHz, and 541.0 GHz to 580.4 GHz. Nordh et al (2003) and Frisk et al (2003) give a description of *ODIN* and its radiometer. *Rosetta*-MIRO (Gulkis et al 2007) operates at centre frequencies of 190 GHz and 562 GHz to study water and other molecules from a comet.

Local oscillators

Every coherent heterodyne receiver system contains a local oscillator (LO). The LO produces a monochromatic reference signal which is combined with the astro-

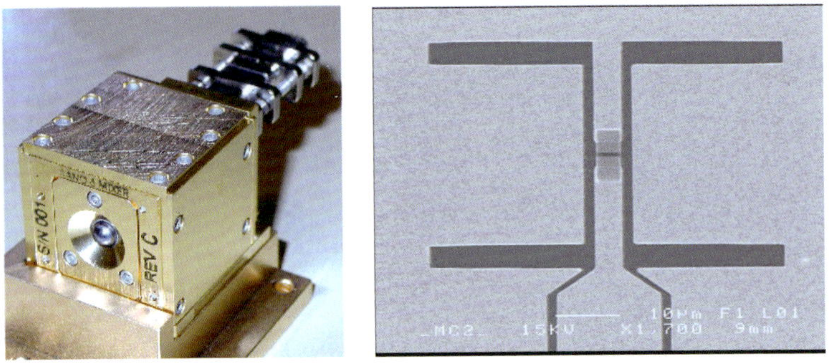

Figure 31.11: Left: Quasi-optical HEB mixer for HIFI Band 6 using a silicon lens for signal coupling to a planar antenna behind the lens. Right: SEM photograph of the planar antenna structure consisting of a double-slot antenna with an overall horizontal slot length of 56 µm and a vertical slot separation of 32 µm. The HEB mixing device is in the centre. (Courtesy S. Cherednichenko, Chalmers University of Technology).

Figure 31.12: Example of a planar Schottky diode as used in Schottky mixers for submillimetre frequencies. (Courtesy B. Ellison, RAL).

nomical signal in the mixer to generate the lower-frequency IF signal. In general, a useful LO needs to meet the following requirements:

1. **Sufficient output power to operate the mixer.** Depending on the type of mixer (SIS, HEB or Schottky) and the frequency of operation, the required power ranges from less than a microwatt (for HEB mixers) to several milliwatts (for Schottky mixers). For frequencies in the terahertz range, even producing a few microwatts of output power can pose a significant technical challenge.

2. **Narrow linewidth.** Since the IF output signal is the combination of the LO and sky signals, a narrow linewidth of the LO signal is required to have the IF signal as a true (only frequency-shifted, not line-broadened) image of the sky signal.

Table 31.2: Key characteristics of terahertz local-oscillator technologies for space.

	Frequency range	Output power	Relative bandwidth	Efficiency (DC power)	Mass	Remarks
Multiplier chain	$\leq$2 THz (3 THz ?)	nW to mW	$\approx$ 10 % to 15 %	low	small	on HIFI
Gas laser	$\gg$ 1 THz	high	very narrow	low	large	discrete lines
QCL	(1.2 to 4.7) THz	$\approx$ mW	few percent	ok?	small	cryogenic, maturing fast

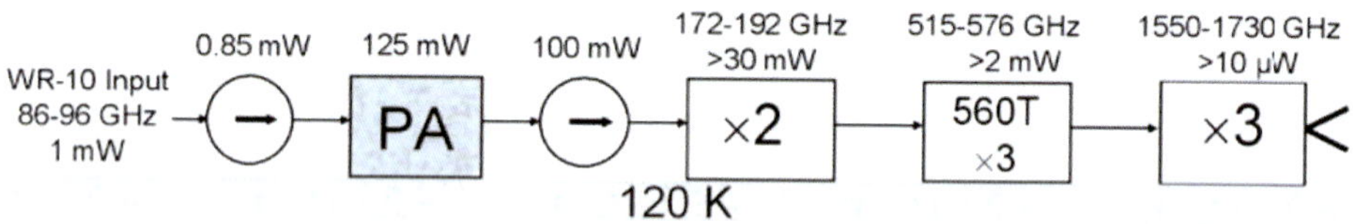

Figure 31.13: Principle of a multiplier-chain local oscillator. A low-frequency signal around 90 GHz is amplified with a power amplifier (PA) and multiplied in various stages to generate a high-frequency output signal (from Maestrini et al 2005).

3. **Extremely high stability of frequency and output power.** This is required to obtain a stable IF output signal.

4. **Tunability over a wide range of frequencies.** If a receiver is used for the observation of a wide range of input frequencies, a tunable LO is required. Alternatively, a fixed tuned LO can be used for a receiver operating at a single input frequency.

5. **Low phase noise.** Achieving ultimate receiver sensitivity poses tight constraints on the maximum allowable LO phase noise.

6. **Low spurious signal content.** Spurious LO signals (such as higher-order harmonics) create spurious output signals of the heterodyne receiver because the mixer combines the input frequencies in the IF. A clean and unique output therefore requires a spectrally clean LO.

7. **Compliance with available on-board resources,** such as electrical input power, mass, size and cooling are important for a system to be used on a spacecraft.

The most important LO technologies for space missions in the submillimetre and far-infrared range are multiplier-chain LOs, gas lasers and quantum-cascade lasers (QCL). Table 31.2 lists their key characteristics.

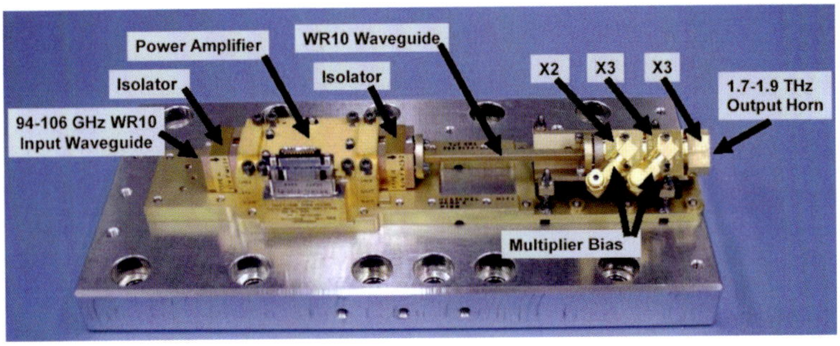

Figure 31.14: Photograph of a 1.7 THz to 1.9 THz local-oscillator chain for *Herschel*-HIFI as described by Maestrini et al (2004). (Image from Ward et al 2005.)

Multiplier-chain local oscillators

Multiplier chains driven by high-power millimetre-wave sources have been developed for millimetre and submillimetre applications and recently up to 2 THz for the LO system of *Herschel*-HIFI (see Pearson et al 2000 for a description). Maestrini et al (2008) give a concise overview of terahertz multipliers. Figure 31.13 shows the general principle. A (high-power) low-frequency signal is amplified with a millimetre-power amplifier (e.g., Ferber et al 2002) and then multiplied in several multiplier stages to obtain a (low-power) high-frequency output signal. The basis for frequency multiplication is the non-linear characteristic of a (whisker-contacted or planar) Schottky diode generating higher harmonics. By proper design of the multiplier certain harmonics are suppressed while others are favoured. The efficiency of the multiplier decreases strongly with increasing frequency, and the overall efficiency of multiplier chains can be as low as a fraction of a percent.

Figure 31.14 shows a 1.7 THz to 1.9 THz mulitplier chain developed for *Herschel*-HIFI, and Figure 31.15 shows the achieved output powers for HIFI multiplier chains from 1.1 THz to 1.9 THz. HIFI carries 14 multiplier chains with different multiplication schemes to reach all desired output frequencies for a limited range of input frequencies.

Gas and quantum-cascade lasers

Far-infrared (FIR) gas lasers are used in the laboratory for terahertz applications and mixer development. A space-qualified version of a CO_2-pumped FIR laser at 2.5 THz has been used in the MLS instrument on the Earth-atmosphere *Aura* mission. However, for many applications, and in particular for future space missions, smaller and lighter LO sources are required. Extremely promising is the development of terahertz quantum-cascade lasers (QCL) (Köhler et al 2002) which has opened up the possibility for heterodyne receivers operating as high as 6 THz ($\lambda = 50$ µm) and beyond. QCLs have first been demonstrated at infrared wavelengths (4.2 µm, Faist et al 1994) and since then progressed to ever longer wavelengths. By 2006, wavelength 150 µm (4 THz) had been reached (Worrall et al 2006), and as of 2010, the longest wavelength achieved is 250 µm (1.2 THz) by

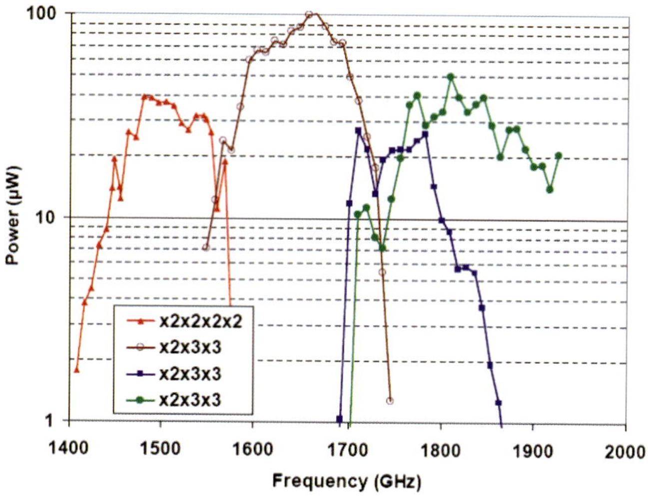

Figure 31.15: Output power of four local-oscillator chains from 1.4 THz to 1.9 THz developed for *Herschel*-HIFI (Ward et al 2005).

Walther et al (2007). QCLs are engineered structures where the laser frequency can in principle be "custom designed" and is not fixed by material properties (as, e.g., in gas lasers). The lasing action of a QCL is due to intersubband transitions in the conduction band of GaAs/AlGaAs heterostructures as illustrated in Figure 31.16. Electrons injected in the injector regions cause light emission in the active regions by transitions between two energy levels whose energy difference depends on the properties of the quantum wells created by the heterostructure. In this way, electrons can "cascade" down the different regions and lasing can be achieved. QCLs are being developed rapidly and have great potential for use in many fields (see, e.g., Williams 2007 for a review). Recently, QCLs have been used for the first time as local oscillators in heterodyne systems, showing their potential for future applications (Gao et al 2005; Hübers et al 2005). QCLs have been demonstrated in CW operation in the range from (1.2 to 4.7) THz with sufficient output power to pump terahertz mixers (including Schottky mixers). For a practical terahertz system, issues like phase-locking, beam quality, tuning range, and stability are being investigated and/or improved by several groups (e.g., Betz et al 2005; Baryshev et al 2006; Adam et al 2006; Kooi et al 2006). To overcome the tuning range limitation, integrated systems with many QCLs could be developed covering several (adjacent) frequencies with the same HEB mixer.

Spectrometer backends

The spectrometer backend analyzes the low-frequency IF signal from the frequency down-conversion process and produces an output spectrum which ideally contains the same information as the high-frequency astronomical signal. The spectral resolution of a heterodyne system is in fact determined by the spectrometer.

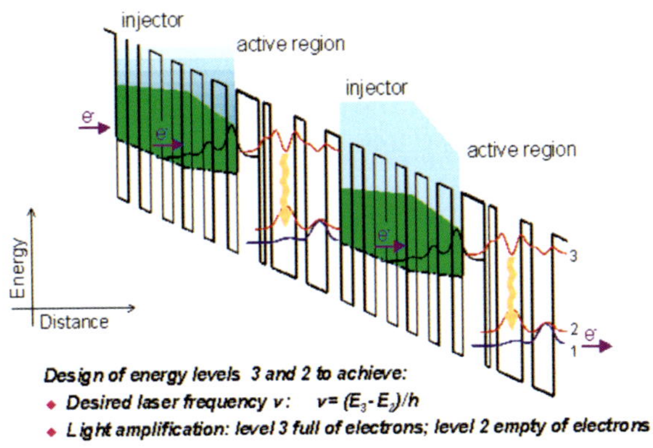

Figure 31.16: Principle of a quantum-cascade laser. (Adapted from Faist et al 1996.)

Table 31.3: Key characteristics of spectrometer backends built for heterodyne space missions.

	SWAS	*ODIN*		*Herschel*-HIFI	
Type	AOS	AOS	AC	AOS[a]	AC[b]
Bandwidth	1.4 GHz	1.0 GHz	(96 to 800) MHz	4.0 GHz	(235 to 470) MHz
Spectral resolution	1.4 MHz	1 MHz	(0.15 to 1.1) MHz	1.1 MHz	(0.125 to 1.0) MHz
Size	9.8 dm^3	4.5 dm^3	1.2 dm^3	20.9 dm^3	14.1 dm^3
Mass	7.5 kg	5.9 kg	1 kg	12.3 kg	12.4 kg
Power consumption	5.5 W	19.5 W	18 W	28 W	65 W
Built by	University of Cologne	Meudon, CESR, LAS	Omnisys Instruments	University of Cologne	CESR Toulouse, Obs. Bordeaux

[a] Values are for one unit. HIFI-*Herschel* has two AOS units.
[b] Values are for one unit. HIFI-*Herschel* has two AC units.

There are several types of spectrometer backends which are used in space: the acousto-optical spectrometer (AOS), the autocorrelator (AC), and probably for the future the fast-Fourier-transform spectrometer (FFTS). Table 31.3 lists the key parameters. Filterbanks are routinely used with ground-based observatories, but due to their size, mass and power requirements they have not been used on space missions and will not be considered here. The AOS and AC have space heritage and the FFTS is a very promising backend for future heterodyne space missions. We will briefly explain the basic principles and main characteristics of these three spectrometer types.

Acousto-optical spectrometer

An acousto-optical spectrometer (AOS) uses the diffraction of light by acoustic waves. Figure 31.17 shows a schematic. The central element is the Bragg cell which is an acousto-optical modulator. It uses the acousto-optic effect to diffract light with

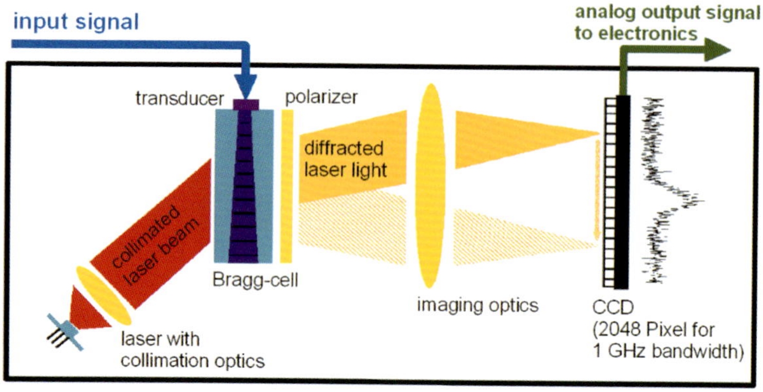

Figure 31.17: Principle of an acousto-optical spectrometer (AOS). A collimated laser beam is diffracted in the Bragg cell according to the spectral content of the input signal and imaged onto a linear CCD array. (Courtesy R. Schieder, University of Cologne.)

sound waves. The (IF) input signal in the gigahertz frequency range is converted to acoustic waves inside the Bragg cell crystal with a piezo-electric transducer attached to it. The acoustic waves in the crystal cause periodic variations of the density and the refractive index and form a phase grating. This phase grating diffracts an incident collimated beam of laser light according to the spectral content of the input signal. The diffraction angle depends on the frequency of the acoustic wave. The angular dispersion and intensity of the diffracted light represent the frequency and amplitude components of the input signal since the acoustic wave spectrum in the crystal is a true image of the input signal spectrum. The diffracted laser light (i.e., the spectrum) is recorded with a linear CCD array which is located in the focal plane of suitable imaging optics.

AOSs are suitable spectrometers for space applications due to their relatively simple design, wide bandwidth, good frequency resolution and low power and mass (Schieder et al 2002). They were used in the *SWAS* mission providing 1.4 GHz of bandwidth (Frerik et al 1999) and on the *ODIN* satellite providing 1.0 GHz of bandwidth (Lecacheux et al 1998). For HIFI-*Herschel*, two AOSs with 4 GHz bandwidth each, i.e., a total of 8 GHz bandwidth, are used (Schieder et al 2000). Figure 31.18 shows the optical unit.

Autocorrelator and fast-Fourier-transform spectrometer

The autocorrelator (AC) and fast-Fourier-transform spectrometer (FFTS) are both digital backends using the Fourier transform of the IF signal. Figure 31.19 shows the principle. There are two ways to obtain the power spectrum from the time domain signal (i.e., the IF signal from the heterodyne receiver). The theoretical basis is the Wiener-Khinchin-theorem of the Fourier transform which states that the spectrum $S(\omega)$ of a signal is the Fourier transform of its autocorrelation function $R(\tau)$. An AC first calculates the autocorrelation function and obtains the spectrum

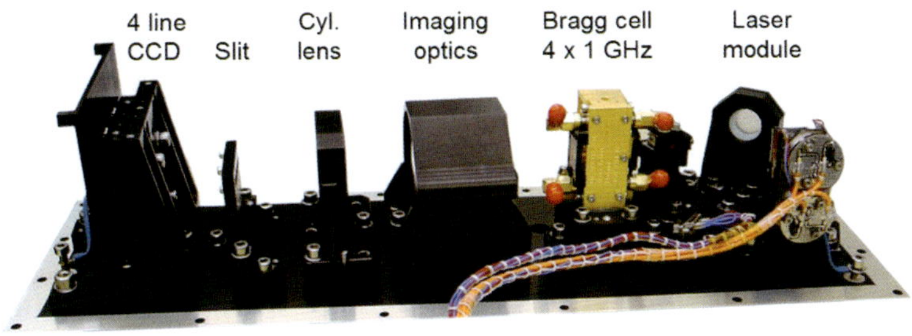

Figure 31.18: Acousto-optical spectrometer (AOS) for *Herschel*-HIFI providing a bandwith of 4 GHz and a spectral resolution of 1.1 MHz (Schieder et al 2000).

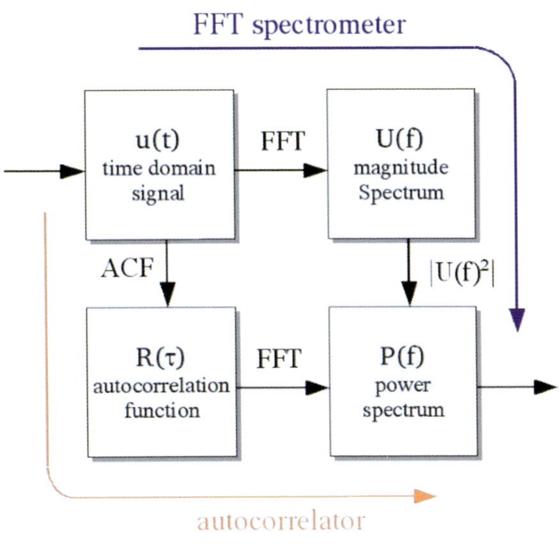

Figure 31.19: Visualization of the Wiener-Khinchin-theorem. The spectrum of a time-domain signal can be obtained via two routes. (Courtesy B. Klein, Max-Planck-Institut für Radioastronomie.)

via a Fourier transform. The autocorrelation function in a digital AC is computed by multiplication of the periodically time-shifted signals, followed by a summation of these. If this is done with 1 or 2 bits, simple logical elements can be used. ACs provide smaller bandwidth than AOSs but can be reconfigured by adapting the spectral resolution and bandwidth for a particular observation (within limits set by the AC hardware, e.g., the sampling speed and number of channels). ACs have flown on *ODIN* (Emrich 1997) and are part of the *Herschel*-HIFI instrument (Caïs et al 1998; Belgacem et al 1998). The second possibility of obtaining the spectrum of a time domain signal is via direct fast-Fourier-transform (FFT) as shown in Figure 31.19. An FFTS directly digitizes the IF signal and computes the Fourier transform. Until recently this method was not very practical due to the high demand on sampler speed and digital processing power. However, advances in these areas (in particular using FPGAs, i.e., field-programmable gate arrays, for signal processing) have made it possible to build FFTSs with large bandwidth and many resolution channels (Klein et al 2006a). An FFTS with a bandwidth of 2 × 1 GHz and a frequency resolution of about 100 kHz is in use at the ground-based APEX observatory (Klein et al 2006b). The rapid progress and commercial availability of key components (in particular the ADCs and FPGAs) make more powerful FFTSs also for space applications possible. For example, a single-board FFTS with 1.5 GHz bandwidth, 8000 channels, 20 W power consumption and a mass of 0.5 kg has recently been developed for radio-astronomy applications (Klein 2007), and a single-board spectrometer with 2.5 GHz bandwidth and 32 000 channels is the next step. These parameters and the rapid progress make the FFTS a very interesting backend for future space missions. It should be noted, however, that the current FFTS technology would have to be adapted to the space environment concerning radiation hardness and electronic component cooling.

Bibliography

Adam AJL, Kašalynas I, Hovenier JN (plus seven authors) (2006) Beam patterns of terahertz quantum cascade lasers with subwavelength cavity dimensions. Appl Phys Lett 88:151–105

Baryshev A, Hovenier JN, Adam AJL (plus seven authors) (2006) Phase locking and spectral linewidth of a two-mode terahertz quantum cascade laser. Appl Phys Lett 89:031115-1–3

Belgacem M, Ravera L, Caux E, Caïs P, Cros A (2003) The high resolution versatile digital spectrometer of HIFI-HSO. New Astronomy 9:43–50

Betz AL, Boreiko RT, Williams BS (plus three authors) (2005) Frequency and phase-lock control of a 3 THz quantum cascade laser. Opt Lett 30:1837–1839

Caïs P, Ravera L, Lagrange D (plus three authors) (1998) First spectra from a new, wide band, hybrid digital spectrometer for the FIRST-HIFI instrument. International Journal of Infrared and Millimeter Waves 19 (11):1471–1487

Cherednichenko S, Drakinskiy V, Berg T, Khosropanah P, Kollberg E (2008) Hot-electron bolometer terahertz mixers for the *Herschel* Space Observatory. Rev Sci Instr 79:034501-1–10

Cherednichenko S, Kroug M, Khosropanah P (plus eight authors) (2002) 1.6 THz heterodyne receiver for the far infrared space telescope. Physica C 372–376:427–431

Crowe TW, Mattauch RJ, Röser HP (plus three authors) (1992) GaAs Schottky diodes for THz mixing applications. Proc IEEE 80(11):1827–1841

Delorme Y, Salez M, Lecomte B (plus six authors) (2005) Space qualified SIS mixers for *Herschel* Space Observatory's HIFI band-1 instrument. Proc 16th Int Symp on Space THz Technology, Göteborg, Sweden; May 2-4,2005, 444–448

Emrich A (1997) Autocorrelation spectrometers for space borne (sub)millimetre astronomy. ESA SP-401:361

Faist J, Capasso F, Sirtori C (plus five authors) (1996) High power mid-infrared ($\lambda \approx 5$ µm) quantum cascade lasers operating above room temperature. Appl Phys Lett 68:3680–3682

Faist J, Capasso F, Sivco DL (plus three authors) (1994) Quantum Cascade Lasers, Science 264:553–556

Ferber RR, Gaier TC, Pearson JC (plus six authors) (2002) W-band power amplifier development for the *Herschel* HIFI instrument. Proc SPIE 4855:468–479

Frerick J, Klumb M, Schieder R, Tolls V, Winnewisser G (1999) *SWAS*-AOS: The first acousto-optical spectrometer in space. Proc SPIE 3759:170–179

Frisk U, Hagström M, Ala-Laurinaho J (plus 48 authors) (2003) The *Odin* satellite I. Radiometer design and test. Astronom Astrophys Lett 402:L27–L34

Gao JR, Hovenier JN, Yang ZQ (plus ten authors) (2005) Terahertz heterodyne receiver based on a quantum cascade laser and a superconducting bolometer. Appl Phys Lett 86:244104–244107

Goldsmith PF (1997) Quasioptical Systems: Gaussian Beam Quasioptical Propagation and Applications. Wiley-IEEE Press, ISBN 978-0-7803-3439-7

Gulkis S, Frerking M, Crovisier J (plus 30 authors) (2007) MIRO: Microwave Instrument for Rosetta Orbiter. Space Sci Rev 128:561–597

Hübers HW, Pavlov S, Semenov A (plus six authors) (2005) Terahertz quantum cascade laser as local oscillator in a heterodyne receiver. Opt Express, 13:5890–5896

Jackson BD (2005) NbTiN-based THz SIS mixers for the *Herschel* Space Observatory. PhD thesis, Casimir PhD Series Delft-Leiden, the Netherlands. ISBN 90-8593-001-4

Jackson BD, de Lange G, Zijlstra T (plus four authors) (2006) Low-noise 0.80.96-and 0.961.12-THz superconductor-insulator-superconductor mixers for the *Herschel* space observatory. IEEE Trans Microwave Theory and Techniques 54:547–558

Justen M, Schultz M, Tils T (plus five authors) (2004) SIS flight mixers for band 2 of the HIFI instrument of the *Herschel* Space Observatory. Proc 12th Int Conf on Terahertz Electronics, Conf digest 27:437–438

Karpov A, Miller D, Rice FR, Stern JA (plus three authors) (2004) Low-noise SIS mixer for far-infrared radio astronomy. Proc SPIE 5498:616-621

Klein B (2007), personal communication

Klein B, Philipp SD, Güsten R, Krämer I, Samtleben D (2006a) A new generation of spectrometers for radio astronomy: Fast Fourier Transform Spectrometer. Proc SPIE 6275:627511-1–12

Klein B, Philipp SD, Krämer I (plus three authors) (2006b) The APEX digital Fast Fourier Transform Spectrometer. Astron Astrophys Lett 454:L29L32

Köhler R, Tredicucci A, Beltram F (plus six authors) (2002) Terahertz semiconductor-heterostructure laser. Nature 417:156–159 (2002)

Kooi JW, Baselmans JJA, Baryshev A (plus six authors) (2006) Stability of heterodyne terahertz receivers. J Appl Phys 100:064904-1–9

Lecacheux A, Rosolen C, Michet D, Clerc V (1998) Space-qualified wideband and ultrawideband acousto-optical spectrometers for millimeter and submillimetre radio astronomy. Proc SPIE 3357:519–532

Lesurf JCG (1990) Millimetre-wave Optics, Devices and Systems. Institute of Physics Publishing (UK), ISBN 0-85274-129-4

Maestrini A, Ward J, Chattopadhyay G (plus two authors) (2008) Terahertz sources based on frequency multiplication and their applications. FREQUENZ Journal of RF-Engineering and Telecommunications 2008/5:118–122, ISSN 0016-1136

Maestrini A, Ward J, Gill J, Javadi H (plus five authors) (2004) A 1.7- 1.9 THz local oscillator source. IEEE Microwave and Wireless Comp Lett 14, 253–255

Maestrini A, Ward JS, Javadi H (plus five authors) (2005) Local Oscillator Chain for 1.55 to 1.75 THz with 100 μW Peak Power. IEEE Microwave and Wireless Components Letters 15(12):871–873

Mehdi I (2004) THz local oscillator technology. Proc SPIE 5498:103–112

Melnick G, Stauffer JR, Ashby MLN (plus 17 authors) (2000) The Submillimeter Wave Astronomy Satellite: Science Objectives and Instrument Description. Astrophys J Lett 539:L77-L85

Merkel HF, Khosropanah P, Wilms Floet D (plus two authors) (2000) Conversion gain and fluctuation noise of phonon-cooled hot electron bolometer in hot spot regime. IEEE MTT-S 48:690–699

Munoz PP, Bedorf S, Brandt M (plus three authors) (2006) THz waveguide mixers with NbTiN HEBs on silicon nitride membranes. IEEE Microwave and Wireless Components Lett 16 (11):606–608

Nordh HL, von Schéele F, Frisk U (plus 15 authors) (2003) The Odin orbital observatory. Astronom Astrophys Lett 402:L21–L25

Pearson JC, Guesten R, Klein T, Whyborn ND (2000) Local oscillator system for the heterodyne instrument for FIRST (HIFI). Proc SPIE 4013:264–274

Schieder R, Siebertz O, Gal C (plus six authors) (2002) Toward very large bandwidth with acousto-optical spectrometers. Proc SPIE 4855:313–324

Schieder R, Siebertz O, Schloeder F (plus four authors) (2000) Wide-band spectrometer for HIFI-FIRST. Proc SPIE 4013:313–324

Swinyard B, Wild W (2010) Far-infrared imaging and spectroscopic instrumentation. ISSI SR-009:241–261

Teipen RTA (2006) Design and analysis of a broadband SIS-mixer for the Heterodyne Instrument for the Far Infrared (HIFI) on the *Herschel* Space Observatory. PhD thesis, University of Cologne, Germany

Tolls V, Melnick GJ, Ashby MLN (plus 15 authors) (2004) Submillimeter Wave Astronomy Satellite performance on the ground and in orbit. Astrophys J Suppl 152:137–162

Tucker JR (1979) Quantum limited detection in tunnel junction mixers. IEEE J Quantum Electron QE-15(11):1234–1258

Tucker JR, Feldman MJ (1985) Quantum detection at millimeter wavelengths. Rev Mod Phys 57(4):1055–1113

Walther C, Fischer M, Scalari G (plus three authors) (2007) Quantum cascade lasers operating from 1.2 to 1.6 THz. Appl Phys Lett 91:131122-1–3

Ward J, Schlecht E, Chattopadhyay G (plus five authors) (2005) Local oscillators from 1.4 to 1.9 THz Proc 16th International Symposium on Space Terahertz Technology, Göteborg, Sweden http://www.mc2.chalmers.se/mc2/conferences/ISSTT/ISSTT2005-proceedings.pdf

Wild W, de Graauw Th, Helmich F, Cernicharo J (plus 13 authors) (2006) ESPRIT: a space interferometer concept for the far-infrared. Proc SPIE 6265:62651Z

Williams BS (2007) Terahertz quantum-cascade lasers. Nature Photonics 1:517–525

Wilms Floet D, Miedema E, Klapwijk TM, Gao JR (1999) Hotspot mixing: A framework for heterodyne mixing in superconducting hot electron bolometers. Appl Phys Lett 74:433–435

Worrall C, Alton J, Houghton M (plus four authors) (2006) Continuous wave operation of a superlattice quantum cascade laser emitting at 2 THz. Opt Expr 14:171–178

— 32 —

Solar radiometry

Claus Fröhlich[I]

Abstract

The classical radiometry for total solar irradiance (TSI) measurements is described using examples of the four types of radiometers currently used in space. The design, characterization and operation of these radiometers are described. Besides the instrumental characteristics determining the measurement uncertainties, an important issue is possible long-term changes of the radiometers exposed to solar irradiance — especially in the EUV — and the space environment. A model for the degradation has been developed which can explain the behaviour of most radiometers in space. The TSI record since 1978 from different platforms and radiometers can be combined in a composite time series which demonstrates that although the assumed uncertainty of the present state-of-the-art radiometers is insufficient, their short- and long-term precision is good enough to produce a reliable time series of TSI over almost 30 years.

Introduction

The term solar radiometry is generally used to refer to measurements of the 'solar constant', the total solar irradiance (TSI), integrated over all wavelengths and reduced to the mean Sun-Earth distance, 1 ua; it is an observation of the Sun as a star. The Smithsonian Astrophysical Observatory initiated a ground-based programme for the determination of solar irradiance variability already in 1902, but they were not able to distinguish between solar and atmospheric effects (see, e.g., Abbot 1952; Aldrich and Hoover 1954). It was not so much a problem of solar radiometry — accurate pyrheliometers have been known since the late 19th century (e.g., Fröhlich 1991) — but of the atmospheric extinction. Measurements from high-flying aircrafts and balloons and from outside the atmosphere, i.e., from rockets and satellites, started in the sixties and yielded first results about the inconstant solar constant, see, e.g., Drummond et al (1968); Plamondon (1969); Willson (1972) and for a review Fröhlich (1977). However, a reliable record of TSI started only in November 1978, when *NIMBUS7* was launched with an electrically calibrated radiometer (ECR) on board. All the measurements from the radiometers in space

[I]Physikalisch-Meteorologisches Observatorium Davos, World Radiation Center, Switzerland

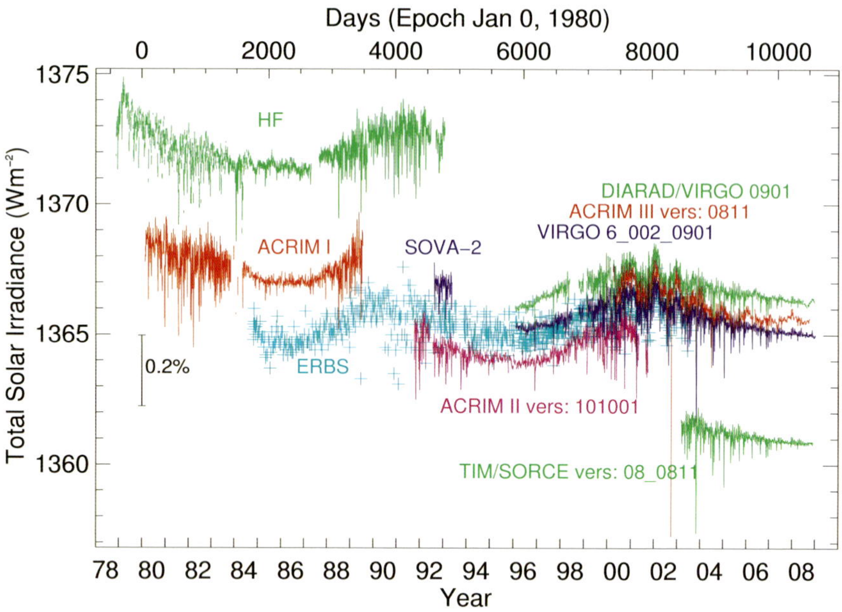

Figure 32.1: Measurements of TSI since November 1978 are plotted as originally published. During the solar activity minimum in 1986 the three data sets ranged from (1364. . . 1371) $\mathrm{W\,m^{-2}}$, whereas during the next minimum in 1996 the range was already reduced to about 2 $\mathrm{W\,m^{-2}}$. At that time, the measurements were within their stated relative uncertainties of the order of $\pm 0.1\,\%$ to $\pm 0.2\,\%$ and it was generally agreed that the characterization of the ECRs had improved, especially the determination of the aperture area. With the advent of the results from TIM on *SORCE* the community has been faced with a serious problem: TIM is measuring almost 5 $\mathrm{W\,m^{-2}}$ lower, and an accuracy of nearly ten times better than that of the classical radiometers is stated.

since then are shown in Figure 32.1, which illustrates the evolution of the state-of-the-art and the improvement in understanding this aspect of solar radiometry.

In the following we will describe the classical radiometry for TSI measurements using examples of the four types of radiometers currently used in space. Besides the instrumental characteristics determining the uncertainties, an important issue is possible long-term changes of the radiometers exposed to solar irradiance in the space environment. A short discussion of the resulting composites — attempts to combine all the existing time series into a contiguous one — will conclude 'Solar radiometry'.

Principles and characterization of solar radiometers

Solar radiometers are based on the conversion of radiation to thermal energy which is measured by an electrically calibrated thermal flux meter. Cavities are used to improve the absorption of solar radiation. They have an aperture, called a

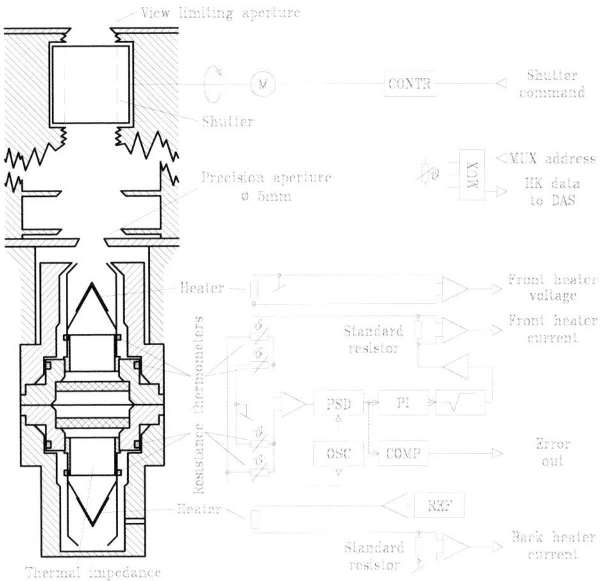

Figure 32.2: Schematic diagram of the PMO6V radiometer with its control electronics. The shutter behind the view limiting aperture is a drum which can be closed by turning it by 90°. The electronics control the active mode operation in which the heat flux (temperature difference across the thermal impedance of the front cavity) is maintained constant during the illuminated and reference phases.

'precision' aperture, which determines the flux-defining area, and a shutter, which opens and closes alternately while the thermal flux to the heat sink is maintained constant; this is called the active mode of operation, hence the name active cavity radiometer (ACR). Four types are currently used in space: ACRIM-III on *ACRIMSat* (Willson 1979, 2001), PMO6V and DIARAD within VIRGO on *SOHO* (Brusa and Fröhlich 1986; Crommelynck et al 1987; Fröhlich et al 1995, 1997) and TIM on *SORCE* (Kopp and Lawrence 2005; Kopp et al 2005b). These references may be consulted for details of the construction of the radiometers, of their characterization and on how the uncertainties are determined. Before we discuss the differences and similarities of the different approaches, we introduce the principle of solar radiometers with the PMO6V radiometer as an example. Figure 32.2 from Fröhlich et al (1995) shows a cut through the radiometer and a block diagram of the control and measurement electronics. The front cavity is used to measure the radiation and the rear facing one is the compensating part of the differential heat flux meter; in this configuration the back cavity cannot be used for radiation measurements. In contrast DIARAD and TIM have all cavities side by side which allows radiation measurements with any of them alternatively. The active mode operation with a shutter open-closed cycle of 60 s/60 s is realized with the control circuit consisting of a Wheatstone bridge with the four wire-wound thermometers, a phase-sensitive detector (PSD) for the error signal, a proportional-integral (PI) control and a square-root circuit controlling the heater power in the front cavity. The value is

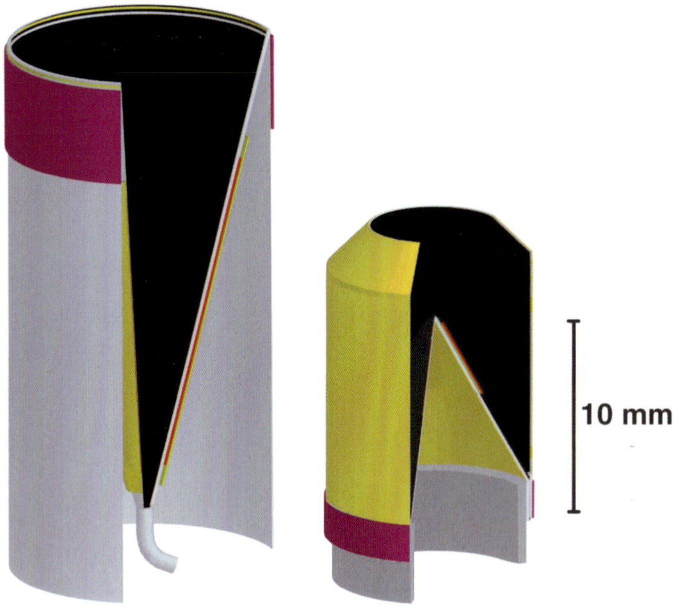

Figure 32.3: The cavities of ACRIM-III (left) and PMO6-V (right). The electrical heaters (red) are wire-wound around the outside of the cone for ACRIM and a flexible printed circuit of Constantan glued to the inverted cone for PMO6V. The temperature sensors (pink) are wire wound around the warm end of the thermal resistor, which is made of silver (ACRIM, light gray as the silver cone) and stainless steel (PMO6V, dark gray); both are soldered to the heat sink.

set by the amount of power in the back cavity (REF). The electrical power in the front cavity is measured as voltage drop over the heater and a standard resistor.

First we will discuss the cavities and related effects of the non-equivalence between electrical and radiative heating. Figure 32.3 shows the cavities of ACRIM-III and PMO6-V. The cone angle of ACRIM is 30° and the one of PMO6 60°; the black paint is in both cases specular. Ideally a ray parallel to the optical axis would undergo six and five reflections for ACRIM and PMO6, respectively, before it leaves the cavity. This would yield a reflectivity of the order of a few 10^{-6}. The measured values for ACRIM and PMO6, however, are of the order of 1.2×10^{-4} and 3.0×10^{-4}, respectively (Willson 1979; Brusa and Fröhlich 1986). The latter is larger because there is no means to reduce the flat part of the tip of the cone of the PMO6 radiometers, whereas ACRIM prevents direct reflections from the bottom with the curved light trap. The thermal resistor of ACRIM is a silver tube soldered to the cone and to the heat sink at the bottom; the one of PMO6 is a stainless steel tube from the inverted cone to the heat sink. A thin copper wire (pink) wound around and glued to the warm end of the thermal resistor serves as the temperature sensor for both radiometers. The electrical heater (red) is wire wound around the cone of ACRIM and covered with aluminized mylar to minimize radiative losses to the outside. The one of PMO6 is a flexible printed circuit made from an etched 5 μm Constantan foil forming a 90 Ω heater bonded to a 20 μm Kapton foil which is directly glued to the front side of the cone and then covered by the black paint. In both radiometers the heater covers the same area as is illuminated on-axis through the 'precision' aperture, an indispensable measure to minimize the non-equivalence. Each radiometer has two receivers in a back-to-back configuration, one for solar measurements and the other as reference of the differential heat-flow meter. The temperature sensor windings of both cavities are elements of a Wheatstone bridge, the output of which regulates the electrical energy dissipated in the primary cavity in such a way that the temperature difference over the thermal resistor is maintained constant. In the case of PMO6V the reference cavity is heated with the same electrical power as used during the reference phase of the operational cavity so that the bridge output can be nulled and does not need to be offset to set the temperature difference.

The effects producing the non-equivalence of these two radiometers are quite different: due to the inverted cone of PMO6 the first reflected radiation does leave the electrically heated part, and produces an extra temperature gradient along the outer shield that is not present in the electrically heated case. In the case of ACRIM some of the electrical heater power may be lost to the outside; in vacuum the effect is probably small due to the radiation shield wrapped around the cone, but it is unknown and uncorrected. In air, however, it is quite large as ground comparisons have shown. For PMO6 the effect is described in Brusa and Fröhlich (1986) and determined from the difference in responsivity in air and vacuum with the argument that the extra losses of the shield in vacuum and thus the non-equivalence are regarded as negligible due to the gold plating of the outside of the shield. This relative correction is rather high with values between 1.5×10^{-3} and 4.0×10^{-3}. The results shown in Table II of Brusa and Fröhlich (1986) indicate, however, that these corrections are well determined, as they reduce the relative standard deviation of the comparison to PMO2 — one of the reference radiometers of the World Radiometric Reference (WRR) — from 1.12×10^{-3} to 0.58×10^{-3}. Moreover, the relative change of the TSI measurement is 2.3×10^{-3}, which corresponds roughly to the average of all the non-equivalence corrections. This turns out to be the most important correction for the PMO6 radiometers with a contribution to the overall 3σ uncertainty of more than one third. For space applications of the PMO6 radiometers this correction is only used to transfer the WRR to space with the results of ground comparison performed in air. In vacuum a corresponding correction is neglected because the losses are only about 3 % to 5 % of those in air, yielding a small correction of less than 10^{-4}.

Figure 32.4 shows the cavities of DIARAD and TIM. The DIARAD cavity has a flat bottom and uses a diffuse paint. The cone angle of TIM is 30° and the black is not a paint, but an etched nickel phosphorous (NiP) layer deposited inside the silver cone. The measured reflectivities for solar radiation of the cavities are about 2.0×10^{-4} for TIM and 2.5×10^{-4} for DIARAD, which is compatible with the enhancement of such geometries under the assumption of diffuse reflections. In order to avoid direct reflections from the bottom of the cone of TIM a thin and pointed tungsten needle is inserted at the bottom which reflects the radiation back to the cone. The heater of TIM covers the illuminated area of nominally 0.5 cm^2. For

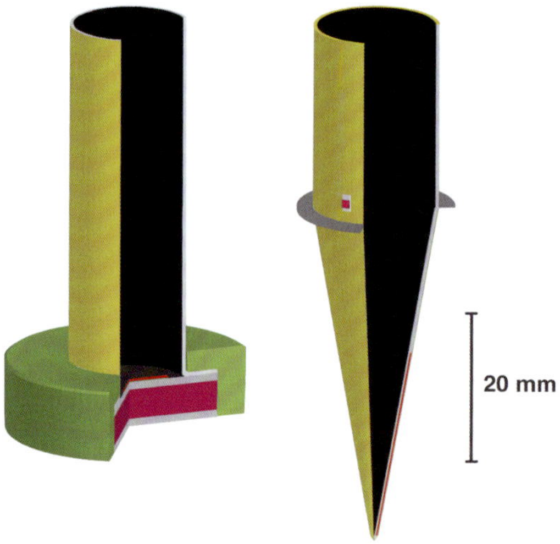

Figure 32.4: The cavities of DIARAD (left) and TIM (right). The electrical heaters (red) are wire-wound around the outside of the cone and encapsulated for TIM, and a flexible printed circuit of Constantan glued to the upper face of the heat flow device (pink) for DIARAD, which is bonded to the heat sink with an indium foil. Three thermistors (pink) soldered to diamond flakes measure the temperature of the cavity for TIM. The thermal resistor to the heat sink is shown schematically as a ring (dark gray) of stainless steel; in reality it is spoked.

DIARAD the heater covers a circular area of 0.95 cm^2 of which only 0.5 cm^2 or slightly more than 50 % is illuminated by the Sun. The radiometers developed at Institut Royal Météorologique de Belgique (IRMB) were the first to arrange both the operational and reference cavities side by side which allows to use both alternatively also for solar measurements — hence the name Dual Irradiance Absolute RADiometer. A further important advantage of this arrangement is that it provides a very similar thermal environment to both receivers, which substantially reduces the sensitivity to changes of temperature and related gradients by, e.g., eclipses during a low Earth orbit, producing thermal waves from the front to the back. In TIM this arrangement is extended to two pairs which allows now simultaneous measurements and comparison of each of the two detector pairs.

The effects of the non-equivalence for DIARAD and TIM are quite different. For DIARAD the non-equivalence is due to the difference in the area which is irradiated and electrically heated and due to radiation reflected from the bottom to the sidewall, a similar effect as for the PMO6 cavities. With results from mapping the sensitivity over the bottom of the cavity and along the sidewall with a laser beam, a corresponding correction — called *efficiency* (Crommelynck 1989) — is estimated. In vacuum, the relative correction amounts to 1.3×10^{-4} for DIARAD on VIRGO (Crommelynck and Dewitte 2005).

The accurate knowledge of the size of the 'precision' aperture is obviously very important – it defines the area in which the solar radiation is collected. Figure 32.1 illustrates the importance very nicely as the differences in the early measurements were mainly due to the uncertainty of the aperture area. Both, the manufacturing of precise and round apertures with short or zero land (cylindrical part of the aperture) and the measurement techniques, have been substantially improved during recent years and allow now for relative uncertainties as low as $(5 \text{ to } 10) \times 10^{-5}$ (Johnson et al 2003; Litorja et al 2007) for apertures with diameters of 5 mm to 8 mm. The apertures of ACRIM, PMO6V and DIARAD are made of stainless steel and turned or milled to shape. The one of DIARAD is covered with a thin layer of evaporated nickel, the surfaces of the others are as manufactured. The TIM aperture is made from nickel-covered aluminium which is diamond-turned to shape. In the classical radiometers the 'precision' apertures are placed directly in front of the cavities, and a further aperture at about 100 mm to 140 mm from the 'precision' aperture limits the field of view to approximately 5° full angle with an angle of 1° between the opening of the apertures, called slope. This is compatible with the view-limiting geometry of pyrheliometers used on ground for solar radiometry and allows direct comparison with, e.g., the World Radiometric Reference (see, e.g., Fröhlich 1991), maintained at PMOD/WRC in Davos. TIM has the aperture geometry inverted with the 'precision' aperture (8 mm) at the front of the instrument and the viewing angle is limited by the diameter of the cavity entrance (16 mm at 104 mm behind the 'precision' aperture). This obviously needs a large cavity with a rather long time constant, which is, however, no problem for TIM as it is not operated in a quasi-steady-state mode as the classical radiometers.

From the beginning of solar radiometry in the sixties one was concerned about possible heating of the 'precision' aperture directly in front of the cavity which would emit additional IR radiation into the cavity. This IR radiation is proportional to the solar irradiance and hence results in an increase of responsivity. In the early seventies tests with electrical heaters were performed (see, e.g., Geist 1972), but no conclusive results were obtained, mainly because the emphasis was more on how much the overall temperature of the aperture increased than on the detailed temperature distribution especially at the edge of the aperture. This edge has a short enough thermal time constant for the temperature to rise and fall during the shuttered operation and thus to contribute to the signal in the cavity, whereas an overall temperature increase of the aperture does not influence the measurement. This effect is very difficult to determine experimentally. The problem came back with the detection of the early increase of the responsivity of the PMO6 radiometer in VIRGO (Fröhlich et al 1997), which can only be explained by a steadily increasing aperture heating due to, e.g., a blackening of the aperture due to the strong solar UV radiation from, e.g., Ly α during the early exposure of the radiometer. This interpretation was confirmed by inspection of the apertures of the PMO6 type radiometers of SOVA-2 which were in space for more than one year during the *EURECA* mission and then retrieved. The same effect is also present in the HF, ACRIMs and ERBE radiometers as shown by Fröhlich (2003, 2006). So, the aperture heating seems to be a substantial contribution to a yet uncorrected effect of the classical solar radiometers. Due to the nickel plating of the DIARAD aperture the early increase effect could be prevented, but this does not mean that the aperture heating does not also increase the sensitivity of DIARAD. For TIM the influence of aperture heating is negligible because of the very small view angle of the aperture from the cavity. Optical power measurements (with under-filled apertures) with PMO6 radiometers and a spare TIM and comparison with the cryogenic radiation scale at NPL (Romero et al 1991, 1996) and NIST (Kopp and Rice 2007), respectively, show agreement of their responsivity ratio to within $< 0.1\,\%$. Thus, the difference in absolute values between the classical radiometers and TIM is most likely related to the arrangement of the apertures and more specifically to aperture heating of the radiometers with the aperture directly in front of the cavity. Consequently, a detailed study of this effect in classical radiometers is needed in order to determine it accurately.

Diffraction is another effect which needs to be included in the corrections. This correction depends on the relative arrangement of the two apertures. If the larger aperture is in front of the radiometer, its diffraction adds radiation through the 'precision' aperture due to Babinet's theorem. In the other case, the 'precision' aperture in front diffracts radiation out of the view-limiting aperture, reducing the received radiation. In other words, the amount diffracted out is greater than the amount diffracted in. Thus, the aperture arrangement of the classical radiometers leads to a relative increase of the measured radiation by $(0.2 \text{ to } 1.3) \times 10^{-3}$ depending on the size of their apertures and the distance between them, and reduces it for TIM by 4.2×10^{-4} (Brusa and Fröhlich 1986; Kopp et al 2005a; Shirley 1998). These corrections are calculated from theory which is based on exactly co-aligned and circular apertures; nevertheless, their relative uncertainties seem to be at the level of a few percent.

Another important aspect of solar radiometry is the influence of the immediate thermal environment seen by the detector. Most important is the difference in IR radiation received by the cavity during the shutter open and closed phases, respectively. Temperature sensors in the front part of the radiometers and on the shutter, together with thermal models, are used to estimate and correct these effects. An obvious test for these estimates is measurement of the radiation from deep space. Other effects for the characterization like stray light and lead[1] heating have been described in some detail (Willson 1979; Brusa and Fröhlich 1986; Crommelynck and Dewitte 2005; Kopp and Lawrence 2005) and will not be discussed further. Problems related to the electrical power measurements and their uncertainties are also not discussed here, but details from the VIRGO experiment as an example can be found in Fröhlich et al (1995, 1997).

Operation of solar radiometers

Classical radiometers are operated in a quasi-stationary mode with shutter cycles of 60 s to 90 s. At the end of each phase the electrical power is read and the irradiance S evaluated according to

$$S = \frac{C}{A}(P_{\text{closed}} - P_{\text{open}}) \quad , \tag{32.1}$$

with A the area of the 'precision' aperture and C the total correction factor as determined by characterization. P_{closed} and P_{open} are the closed and open electrical power readings. This operation relies on a rather short time constant which is

[1] i.e., electrical connections

improved with the overall gain of the servo loop which is typically between three and five. So, an open-loop or natural 1/e time constant of about 20 s is reduced to 4 s to 7 s. Normally, there is not only one measurement before the end of a phase, but a few, so that the servo-loop characteristics can also be checked in flight. In contrast to this classical active-mode operation TIM uses phase-sensitive detection at the fundamental shutter period. This avoids many problems with effects at higher frequencies. For the behaviour of the non-equivalence in the classical case one needs to take into account all contributions from frequencies of up to at least ten times the shutter frequency because a full square wave has to be reconstructed. The TIM radiometer is operated in the same way as the classical radiometers, that is, the electrical power is always adjusted so that the temperature difference — or the heat flux — remains constant. The electrical power is derived from a constant voltage source with pulse-width modulation. The evaluation of the phase-sensitive signal is somewhat more complicated because most terms are complex phasor components representing the amplitude and phase of sinusoidal variations (indicated as bold-type symbols). With A for the aperture area and α for the absorptivity of the cavity, the irradiance S is evaluated from the time series of the fraction $\mathbf{D}$ of the electrical power P_0 as

$$S = \frac{P_0}{\alpha A} \operatorname{Re}\left[-\frac{\mathbf{Z}_{\mathrm{el}}}{\mathbf{Z}_{\mathrm{rad}}} \frac{1}{\mathbf{T}} \left(\mathbf{D} + \frac{\mathbf{D} - \mathbf{F}}{\mathbf{G}} \right) \right] \quad , \tag{32.2}$$

which corrects also for the complex servo system gain $\mathbf{G}$, the shutter timing $\mathbf{T}$, and the equivalence as ratio of the corresponding impedances $\mathbf{Z}_{\mathrm{el}}/\mathbf{Z}_{\mathrm{rad}}$ and the applied feed-forward values $\mathbf{F}$ at the shutter frequency. This latter effect accelerates the control system with the information of the last shutter-open values fed forward. The time series of the instantaneous fraction of power fed to the cavity $\mathbf{D}$ is generated from a pulse-width modulator controlled by a digital signal processor maintaining the thermal balance of the cavity. This signal is sampled at 100 Hz and transmitted to ground as numbers from 0 to 64 000, representing the fraction $q = D/64\,000$ of P_0. The advantage of the transmission of $\mathbf{D}$ is obviously that one can also analyse other frequency terms for diagnostics by choosing a different $\mathbf{T}$. The final S is convoluted with a Gaussian filter of about 400 s width. This means that fluctuations due to, e.g., p modes cannot be resolved, but for the objective of TSI for Sun-Earth connections this seems justified. Due to the fact that the result is a true integral the aliases due to such higher-frequency variations are suppressed anyway.

The value of TSI as determined from the above algorithms includes all known instrumental effects. It has to be normalized to 1 ua, the mean Sun-Earth distance, by multiplying the measured value by $(r_{\mathrm{E}}/\mathrm{ua})^2$, the actual value of the Sun-instrument distance being r_{E}. The relative correction varies between $\pm 3\,\%$ and can be done to a relative uncertainty of at least 1.0×10^{-7} from the ephemeris of the spacecraft. A further correction is needed for the Doppler effect due to the radial velocity between the source and the radiometer. The correction is discussed in Chapter 2 (Wilhelm and Fröhlich 2010) and is proportional to $(1 - 2v_r/c_0)$ with v_r being positive away from the Sun. For a low-orbit satellite this correction can amount to up to $\pm 6 \times 10^{-5}$ with radial velocities of up to ± 9 km s^{-1}.

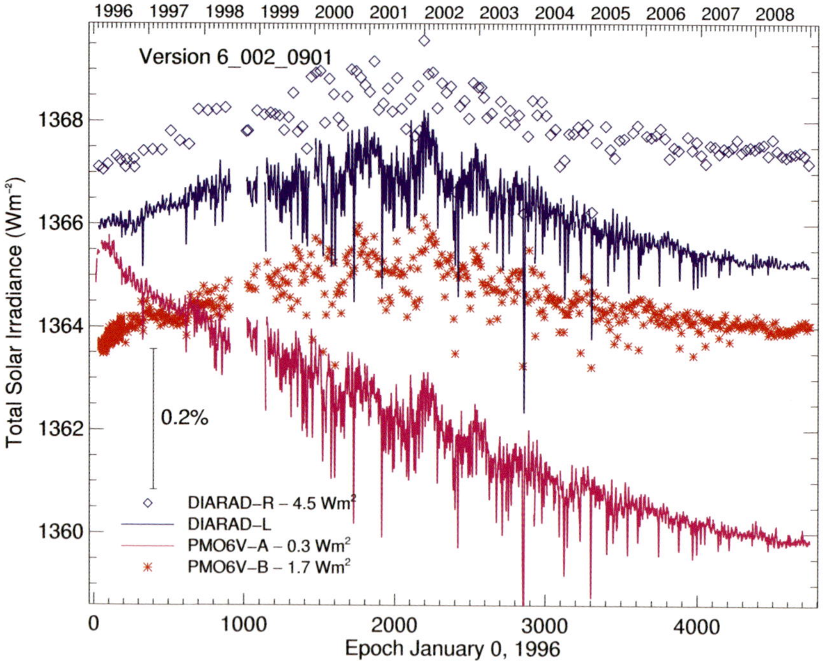

Figure 32.5: Level-1 data of the two radiometers on VIRGO: DIARAD-L and R and PMO6V-A and B. Note the difference in the amount of degradation of PMO6V-A relative to DIARAD-L and the early increase of the PMO6V-A and B.

Long-term behaviour of solar radiometers in space

From Figure 32.5 it is obvious that the long-term behaviours of the four VIRGO radiometers differ substantially from each other, and important corrections are needed to deduce a reliable TSI from these data. Already at this stage of the evaluation the different long-term behaviour of the operational PMO6V and DIARAD is very obvious. Also prominent is the early increase of the PMO6V radiometers during the first weeks of exposure. From the comparison of a back-up instrument of the same type with much less exposure to solar radiation, changes due to exposure to the Sun can be determined. But these data are sparse and we need a reliable way to interpolate between the reference measurements in order to enable continuous correction of the operational radiometer. One may use fitting of polynomials of higher degrees (Willson and Hudson 1991; Willson and Helizon 2005) or one can use some other methods, e.g., running means (Dewitte et al 2004a). A much better way is to use a model which also helps to understand the physical mechanisms. Such a model is based on a hyperbolic function (e.g., Fröhlich and Anklin 2000; Fröhlich and Finsterle 2001; Fröhlich 2006) which is the solution of the differential equation describing the 'siliconizing' of a quartz window exposed to UV radiation, that is, a change of the optical properties due to the formation of silicon at the surface with a subsequent decrease of the response of the underlying quartz to radiation exposure. The time-dependent sensitivity change $\Delta S(t)$ with t_{exp} for the exposure time can be described as

$$\Delta S(t) = a \left[\left(1 + \frac{1}{b\,\tau_{\mathrm{C}}} \int_0^{t_{\mathrm{exp}}} (\lambda\, m(t) + 1)\, \mathrm{d}t \right)^{-b} - 1 \right] \tag{32.3}$$

with a, λ, b and $1/(b\,\tau_{\mathrm{C}})$ as adjustable parameters. The parameter b has been included in τ_{C}, which then corresponds to a $1/e$ time constant. This transforms the hyperbolic function to an exponential for large b, reducing the number of parameters:

$$\Delta S(t) = a \left[\exp\left(-\frac{1}{\tau_{\mathrm{C}}} \int_0^{t_{\mathrm{exp}}} (\lambda\, m(t) + 1)\, \mathrm{d}t \right) - 1 \right]. \tag{32.4}$$

For $b < 20$ the differences are substantial and the hyperbolic function should be used. Only if b starts to increase during the fitting process, the hyperbolic function can be replaced with the exponential one. The integral corresponds to the dose received during the exposure time, and for $m(t)$ the Mg II index is used as surrogate for the UV radiation, normalized to -0.5 to 0.5. The fitted parameter λ provides the information about the wavelength responsible for the effect as it is proportional to the cycle variability of the corresponding wavelength, and 2λ corresponds to the cycle amplitude of that radiation. Thus, for, e.g., Ly α with a solar cycle variation of (0.06 to 0.10) mW m^{-2} (Rottman et al 2004), $\lambda = (1/2)(0.1/0.06) = 0.83$. This analysis not only explains the dose dependence of the changes, but also provides information about the physical mechanisms behind it.

In the case of VIRGO this analysis works only for the PMO6V radiometers, mainly because the changes of DIARAD are a mixture of exposure-dependent and also non-exposure-dependent changes. For an illustration of the method, the behaviour of ACRIM-I is shown in Figure 32.6 and the hyperbolic function of the degradation explains the behaviour quite well. The correction of the early increase is not shown here, for details of this correction see Figure 2 of Fröhlich (2006). It shows also how important the dose is, and that neither the exposure time alone nor a simple polynomial fit is sufficient — especially if the activity level is changing during the period analysed.

For the PMO6V on VIRGO this analysis is complicated by the fact that also the back-up instrument was exposed more than normally due to the change in operations (Fröhlich et al 1997). Thus, it shows a significant early increase which has to be corrected before the back-up can be used for the determination of the exposure-dependent changes of PMO6V-A. The principles of these dose-dependent corrections are described in Fröhlich (2003), however, for version 5. Since then the exposure-dependent changes of PMO6V have been revised for version 6 as well as the corresponding non-exposure-dependent changes of DIARAD. This has improved the internal consistency of the corrections substantially. The most recent VIRGO results (version 6.2, updated to end of January 2009) are shown in Figure 32.1; they are the basis for the PMOD composite. Also included in Figure 32.1 is the DIARAD time series which is determined from the DIARAD-L and R alone (updated from Dewitte et al 2004a). Their analysis assumes arbitrarily that there is no change over the *SOHO* 'vacations' (i.e., the loss of Sun-pointing attitude

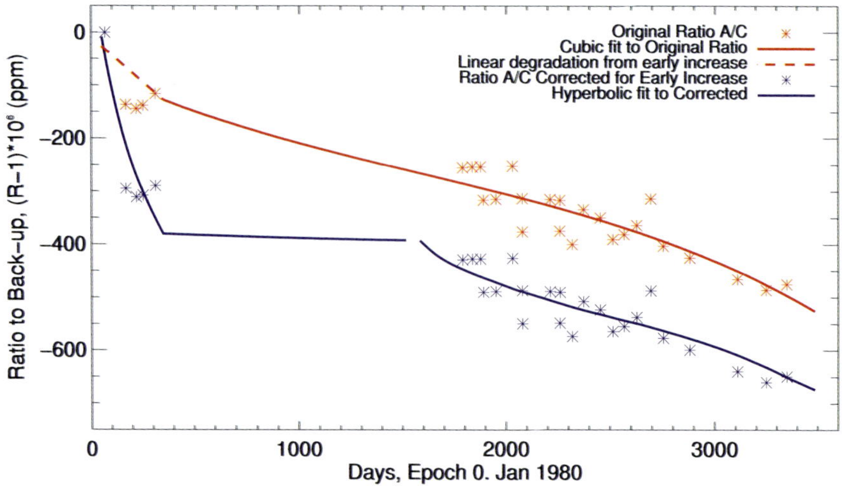

Figure 32.6: The original ratios of sensor A to sensor C (red symbols) of the ACRIM-I experiment on *SMM* are from Willson and Hudson (1991). The cubic fit corresponds to the correction originally applied and the dashed line includes the linear fit found by fitting the early increase (not shown). The blue symbols are corrected for the early increase and then fitted with a hyperbolic function. Note the dose-dependent change over the period of the spin mode (roughly after day 2700 on the time axis) and after the solar minimum of 1986 (after day 2100). Adapted form Figure 3 of Fröhlich (2006).

from June to September 1998) which is not confirmed by comparison with, e.g., ACRIM-II and ERBE. Besides this obvious misinterpretation, the overall behaviour shows a much stronger decrease over solar cycle 23 than any other time series during this period. This is due to the neglect of the non-exposure-dependent increase of sensitivity which cannot be assessed by analysing the operational and the back-up radiometer alone. It can only be detected by comparison with an independent radiometer, e.g., PMO6V. Albeit this effect was not anticipated before launch, the reason for having two different radiometers within VIRGO was: "Although the designs of both radiometers are based on the same principle, the physical realization is different" (from Fröhlich et al 1995). The residuals of the comparison of DIARAD and PMO6V, both corrected for exposure-dependent changes (level-1.8), can now be fitted to an exponential function which describes the long-term sensitivity change. Comparisons with ACRIM-II or ERBE instead of PMO6V yield essentially the same results, indicating that it is indeed a sensitivity change of DIARAD. A similar effect is also found for the HF as shown in Fröhlich (2006). So, these two radiometers show an increase of sensitivity which is independent of exposure to radiation and fortunately no such effect is present in ACRIM, ERBE and PMO6 radiometers. Thus, the DIARAD/VIRGO data as published by Dewitte et al (2004a) and Dewitte et al (2004b) are confusing and are not correctly representing the irradiance from the Sun during solar cycle 23.

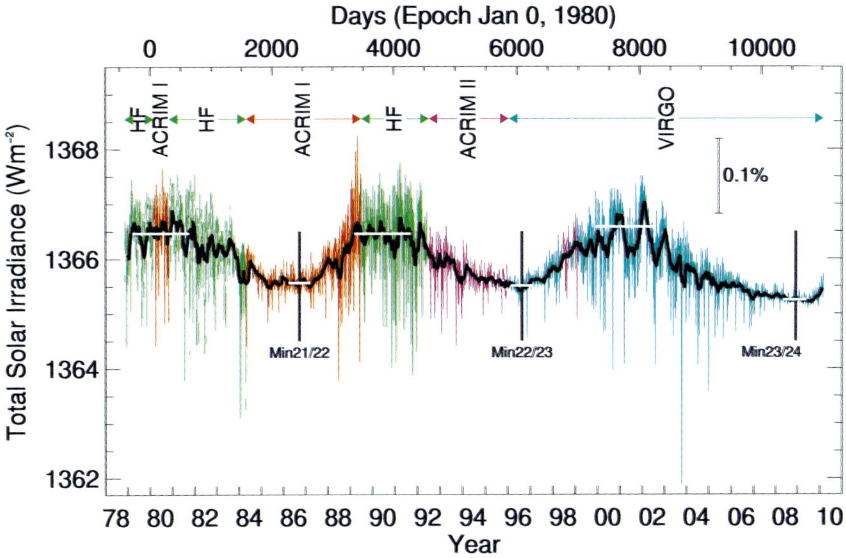

Figure 32.7: Shown is the PMOD composite (version 41_62_1004), updated to end of March 2010. The minima values as one-year averages amount to (1365.46, 1365.51, 1365.25) W m^{-2} with an average of 1365.44 W m^{-2} and the amplitudes of three cycles as difference of the three-year average at maximum to the adjacent minima to (0.928, 0.919, 1.038) W m^{-2}. In contrast to cycles 21 and 22 there is a substantial trend over cycle 23 with a difference between the two adjacent minima amounting to a change of 26 % relative to the amplitude of this cycle (see, e.g., Fröhlich 2009), which is a very interesting and important new result.

Results and discussion of solar radiometry

With the VIRGO TSI data for solar cycle 23 and the data from HF, ACRIM-I and -II for the period before 1996 we can construct a composite TSI as described in Fröhlich (2006), and in Figure 32.7 the so-called PMOD composite is shown as an example. There are two other composites from the ACRIM (Willson and Mordvinov 2003) and IRMB (Dewitte et al 2004b) teams. The differences between these and the PMOD composite have been discussed extensively in Fröhlich (2006, 2009) and Lockwood and Fröhlich (2007, 2008), where also detailed information about the reasons for the differences can be found. Figure 32.7 shows very interesting results of solar variability during the last three solar cycles. Most importantly, the most recent minimum is lower than the previous ones, and the length of this last solar cycle is also much longer than the previous ones.

The results show that overlapping measurements from different platforms enable to construct a reliable TSI record, demonstrating that the relative precision of the individual radiometers is much higher than their uncertainty. The result also shows that only a detailed understanding of the degradation helps to correct the data over longer periods than the 11-year solar cycle. Comparison with simultaneous TSI measurements from ACRIM-II and III and TIM/*SORCE* indicate that

VIRGO may overestimate this trend (Fröhlich 2009), but a relative change of at least 18 % (instead of the 26 % stated in Figure 32.7) can be confirmed. With a long-term stability of $\pm 3.5 \times 10^{-5}$ per decade (Fröhlich 2004) and $\pm 5.0 \times 10^{-5}$ per decade (Fröhlich 2009) the presently observed low value of TSI is highly significant but different from what is observed in *spectral* irradiance (see Figure 2.3 of Chapter 2) which shows essentially no change between minima. This means that there are different physical mechanisms responsible for the trend in TSI and the cycle variation of TSI and SSI. The former may be due to a long-term change in the photospheric temperature which only marginally influences UV radiation of the Sun.

Another still unsolved problem of solar radiometry is the absolute value of TSI which differs by more than 4 $\mathrm{W\,m^{-2}}$ between the classical radiometers and TIM/*SORCE* (see Figure 32.1).

Acknowledgements

The author would like to thank Greg Kopp, LASP at University of Colorodo, Steven Dewitte and Andre Chevalier, Institut Royale Météorologique de Belgique, Richard Willson, Columbus University, for many helpful discussions about their radiometers. And last but not least all these results would not have possible without the continuing support of the Swiss National Science Foundation, and the *SOHO* and VIRGO Teams, which is gratefully acknowledged. *SOHO* is a cooperative mission of ESA and NASA.

Bibliography

Abbot C (1952) Periodicities in the solar-constant measures. Smithsonian MiscColl 117(10):1–31

Aldrich L, Hoover W (1954) Annals of the Astrophysical Observatory of the Smithsonian Institution, vol 7, Smithsonian Institution, Washington, DC, U.S.A., chap 7: Statistical Studies of the Solar-Constant Record, pp 165–168

Brusa RW, Fröhlich C (1986) Absolute radiometers (PMO6) and their experimental characterization. Appl Opt 25:4173–4180

Crommelynck D (1989) Factors limiting the accuracy of absolute radiometry. In: New Developments and Applications in Optical Radiometry (ed N Fox) Inst Phys Conf Ser 92, London, pp 19–24

Crommelynck D, Dewitte S (2005) The DIARAD type instruments: Principles and error estimates, presented at the "Total Solar Irradiance Workshop" at NIST, Gaithersberg, Md, USA, 18-20 July 2005

Crommelynck DA, Brusa RW, Domingo V (1987) Results of the solar constant experiment onboard Spacelab 1. Sol Phys 107:1–9

Dewitte S, Crommelynck D, Joukoff A (2004a) Total solar irradiance observations from DIARAD/VIRGO. J Geophys Res 109:A02102

Dewitte S, Crommelynck D, Mekaoui S, Joukoff A (2004b) Measurement and uncertainty of the long-term total solar irradiance trend. Sol Phys 224:209–216

Drummond AJ, Hickey JR, Scholes WJ (1968) New value for the solar constant of radiation. Nature 218:259–261, DOI 10.1038/218259b0

Fröhlich C (1977) Contemporary measures of the solar constant. In: The Solar Output and its Variation (ed. OR White), Colorado Associated Univ. Press, Boulder, pp 93–109

Fröhlich C (1991) History of solar radiometry and the World Radiometric Reference. Metrologia 28:111–115

Fröhlich C (2003) Long-term behaviour of space radiometers. Metrologia 40:60–65

Fröhlich C (2004) Solar irradiance variability. In: Geophysical Monograph 141: Solar Variability and its Effect on Climate, American Geophysical Union, Washington DC, USA, chap 2: Solar Energy Flux Variations, pp 97–110

Fröhlich C (2006) Solar irradiance variability since 1978: Revision of the PMOD composite during solar cycle 21. Space Sci Rev 125:53–65

Fröhlich C (2009) Total solar irradiance variability: What have we learned about its variability from the record of the last three solar cycles? In: Climate and Weather of the Sun-Earth System(CAWSES): Selected Papers from the 2007 Kyoto Symposium, October, 23-27 2007, (eds T Tsuda, R Fujii, K Shibata, M Geller) Terra Publishing, Setagaya-ku, Tokyo, Japan, 217–230

Fröhlich C, Anklin M (2000) Uncertainty of total solar irradiance: An assessment of the last 20 years of space radiometry. Metrologia 37:387–391

Fröhlich C, Finsterle W (2001) VIRGO radiometry and total solar irradiance 1996-2000 revised. ASP Conf Ser 203:105–110

Fröhlich C, Romero J, Roth H (plus 21 authors) (1995) VIRGO: Experiment for helioseismology and solar irradiance monitoring. Sol Phys 162:101–128

Fröhlich C, Crommelynck D, Wehrli C (plus 7 authors) (1997) In-flight performances of VIRGO solar irradiance instruments on SOHO. Sol Phys 175:267–286

Geist J (1972) Fundamental principles of absolute radiometry and the philosophy of this NBS program (1968-1971). In: NBS Technical Note 594-1, U.S. Department of Commerce, National Bureau of Standards, Gaithersburg, Md, USA

Johnson BC, Litorja M, Butler JJ (2003) Preliminary results of aperture-area comparison for exo-atmospheric solar irradiance. Proc SPIE 5151:454–462

Kopp G, Lawrence G (2005) The Total Irradiance Monitor (TIM): Instrument design. Sol Phys 230:91–109

Kopp G, Rice J (2007) Optical power comparison of TIM at NIST, personal communication

Kopp G, Heuerman K, Lawrence G (2005a) The Total Irradiance Monitor (TIM): Instrument calibration. Sol Phys 230:111–127

Kopp G, Lawrence G, Rottman G (2005b) The Total Irradiance Monitor (TIM): Science results. Sol Phys 230:129–139

Litorja M, Johnson BC, Fowler J (2007) Area measurements of apertures for exo-atmospheric solar irradiance for JPL. Proc SPIE 6677:667–708

Lockwood M, Fröhlich C (2007) Recent oppositely directed trends in solar climate forcing and the global mean surface air temperature. Proc R Soc A 463:2447–2460

Lockwood M, Fröhlich C (2008) Recent oppositely-directed trends in solar climate forcings and the global mean surface air temperature: II. different reconstructions of the total solar irradiance variation and dependence on response timescale. Proc R Soc A 464:1367–1385

Plamondon J (1969) The Mariner Mars 1969 temperature control flux monitor. JPL Space Prog Summary 37-59, Vol.III(37-59):162–168

Romero J, Fox N, Fröhlich C (1991) First comparison of the solar and SI radiometric scale. Metrologia 28:125–128

Romero J, Fox N, Fröhlich C (1996) Improved comparison of the World Radiometric Reference and SI radiometric scale. Metrologia 32:523–524

Rottman G, Floyd L, Viereck R (2004) Measurement of the solar ultraviolet irradiance.Geophysical Monograph 141:111–126

Shirley EL (1998) Revised formulas for diffraction effects with point and extended sources. Appl Opt 37:6581–6590

Wilhelm K, Fröhlich C (2010) Photons—from source to detector. ISSI SR-009:23–54

Willson R, Helizon R (2005) Active cavity radiometer irradiance monitors: S.I. measurements uncertainties, presented at the "Total Solar Irradiance Workshop" at NIST, Gaithersberg, Md, USA, 18-20 July 2005

Willson RC (1972) Experimental comparisons of the International Pyrheliometric Scale with the absolute radiation scale. Nature 239:208–209

Willson RC (1979) Active cavity radiometer type iv. Appl Opt 18:179–188

Willson RC (2001) The ACRIMSAT/ACRIM III Experiment: Extending the precision, long-term total solar irradiance climate database. The Earth Observer 13:14–17

Willson RC, Hudson HS (1991) The Sun's luminosity over a complete solar cycle. Nature 351:42–44

Willson RC, Mordvinov AV (2003) Secular total solar irradiance trend during solar cycles 21-23. Geophys Res Lett 30:1199

V. Polarimetry

— 33 —

Stokes polarimetry of the Zeeman and Hanle effects

Jan O. Stenflo[I]

Abstract

Magnetic fields are responsible for almost all variability in the Universe on intermediate time scales. The information on the magnetic fields is encoded in the polarization of the radiation from the Sun and stars through the Zeeman and Hanle effects. Stokes polarimetry is the observational tool that gives us access to this information and allows us to determine the structure and evolution of the fields. Space-based observations are needed for two main reasons: (1) To allow high and stable angular resolution over a large field of view. (2) To get access to the vacuum ultraviolet (VUV), which contains information on the magnetic fields in the corona and the chromosphere-corona transition region. VUV polarimetry has considerable potential but has been much neglected in the past.

Dimensionality and trade-offs in information space

From scalar to 4-vector measurements

Spectro-polarimetry enhances the dimensionality of information space from a scalar problem (radiance) in the case of ordinary spectrometry to a 4-D (polarization vector) problem. The different components of the Stokes vector (which we will define later, see Page 545) carry different and complementary types of information that are not contained in the intensity spectrum, in particular information on magnetic fields in the radiating medium.

Trade-offs in parameter space

For the design of observing programmes it is helpful to consider another 4-D parameter space, namely that spanned by the angular, spectral, and temporal resolutions, together with the radiometric and polarimetric accuracies. These four parameters cannot be optimized simultaneously, even with the largest conceivable telescopes; trade-offs are always needed (cf., Stenflo 2001). The radiance of a star does not depend on its distance, only on its effective temperature. Therefore the

[I]Institute of Astronomy, ETH Zurich, CH-8093 Zurich, Switzerland

proximity of the Sun does not help much when we study resolution elements close to the telescope diffraction limit; we quickly run out of photons.

In Stokes polarimetry it has become possible to eliminate systematic effects such that the polarimetric accuracy is only constrained by the quantum limit of the photon Poisson statistics, down to levels of 5×10^{-6} in the degree of polarization (Stenflo 2004a). For a precision of 10^{-5} we need to collect 10^{10} photo-electrons in the detector, or, when accounting for all the efficiencies and optical transmissions of the telescope system, of the order of 10^{12} photons per resolution element. This has routinely been achieved in combination with high spectral resolution, but at the expense of the angular and temporal resolutions.

In most solar magnetic field work priority has been given to angular resolution at the expense of polarimetric precision. Then the temporal resolution has to be sufficiently high, since smaller structures evolve faster, and the integration time should not be longer than the evolutionary time scale. To resolve 100 km on the Sun ($0.14''$) the integration time should be shorter than 10 s, the approximate time it takes for a sound wave to cross the resolution element. The photon flux then limits the relative polarization accuracy in resolved spectral lines to at best 0.1 %. This is fully sufficient for most applications of the longitudinal Zeeman effect (line-of-sight magnetograms in the form of circular-polarization maps), but it is marginal for the transverse Zeeman effect and insufficient for the Hanle effect.

Physical origins of polarization

Polarization is associated with broken symmetries. The spatial symmetry may be broken by the presence of a magnetic field. This is the origin of the Zeeman effect. The symmetry can also be broken by anisotropic scattering. The presence of magnetic fields then leads to polarization phenomena that are covered by the term "Hanle effect". By Zeeman-effect polarization we generally refer to polarization effects that are not dependent on coherent scattering processes.

The Zeeman and Hanle effects are sensitive to magnetic fields in very different parameter ranges and therefore provide mutually highly complementary diagnostic information. While the Zeeman effect has been applied in astrophysical contexts for a century, since Hale used it to discover magnetic fields in sunspots (Hale 1908), applications of the Hanle effect are still in their infancy.

Stokes vector imaging

Why the Stokes formalism is needed

The polarization of an electromagnetic wave is determined by the oscillations of the two orthogonal electric vectors E_{x} and E_{y} in an x-y coordinate system in a plane transverse to the wave propagation. E_{x} and E_{y} can be combined into a complex 2-vector, the *Jones vector*, which is characterized by four parameters (two amplitudes and two phases). A Jones vector however always represents light that is fully elliptically polarized. It is incapable of describing partially polarized radiation, i.e., a natural beam of light.

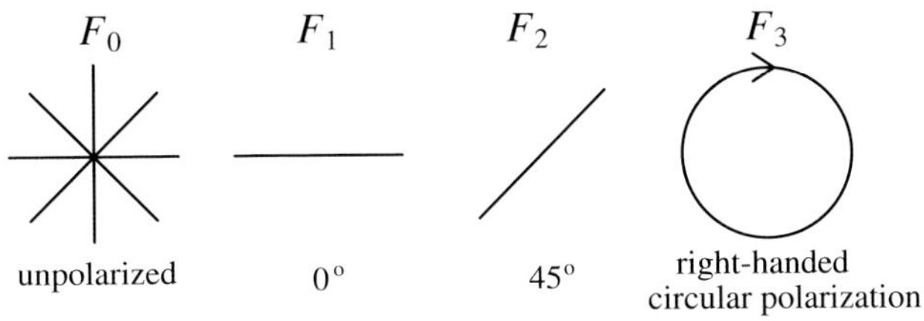

Figure 33.1: Idealized polarizing filters used in the operational definition of the Stokes parameters.

Each wave train or photon is always fully elliptically polarized. Partial polarization is produced by the process of incoherent superposition of an ensemble of wave packets with arbitrary phase relations. To properly describe the polarization state of such an ensemble we need the Stokes formalism (cf., Stenflo 1994).

The most convenient way of defining the *Stokes vector* $\boldsymbol{I}$, which consists of the four Stokes parameters I, Q, U, V, or S_k, $k = 0, 1, 2, 3$, i.e.,

$$\boldsymbol{I} = \begin{pmatrix} S_0 \\ S_1 \\ S_2 \\ S_3 \end{pmatrix} \equiv \begin{pmatrix} I \\ Q \\ U \\ V \end{pmatrix}, \tag{33.1}$$

is in terms of the four idealized filters F_k shown in Figure 33.1. F_0 represents empty space, F_1 and F_2 transmit linear polarization with the electric vector at position angles 0 and 45°, respectively, while F_3 transmits right-handed circular polarization. Let I_k be the radiance of the beam behind each filter. Then

$$I_k = \frac{1}{2}(S_0 + S_k)\,. \tag{33.2}$$

The physical meaning of the Stokes parameters S_k can be understood as follows: If we replace filters F_k with ones that transmit the orthogonal polarization state, then Q, U, and V change sign. From this it follows that Stokes I represents the intensity, Stokes Q the intensity difference between horizontal and vertical linear polarization, Stokes U the intensity difference between linear polarizations at $+45°$ and $-45°$, Stokes V the intensity difference between right- and left-handed circular polarizations.

In the Stokes formalism the effects of a medium can be described in terms of a 4×4 matrix, the *Mueller matrix*, which operates on the incoming Stokes vector to produce the output Stokes vector. The medium can be a stellar atmosphere, telescope optics, a train of polarization optics (modulators, wave plates, polarizers), or all of this combined. It can be treated as a "black box" described by a single Mueller matrix $\boldsymbol{M}$. If the ith component inside the "black box" has Mueller matrix $\boldsymbol{M}_i$, then the total matrix is formed by matrix multiplications:

$$\boldsymbol{M} = \boldsymbol{M}_n \boldsymbol{M}_{n-1} \ldots \boldsymbol{M}_2 \boldsymbol{M}_1\,. \tag{33.3}$$

The ordering of the matrices is essential, from right to left when following the direction of the light beam. Calibration is done by inserting known polarizers in front of the system (cf., Stenflo 1994).

Polarization modulation and demodulation

To measure polarization we need to form differences between orthogonal polarization states. Beam splitters may be used to record the orthogonal states simultaneously on different portions of a detector. This is not convenient for simultaneous recording of all of the four Stokes parameters, moreover the different pixel sensitivities in the different parts of the detector (the gain table) limit the polarization accuracy. High-precision vector polarimetry therefore uses temporal modulation of the various polarization states. In ground-based observations one needs to modulate faster than the atmospheric seeing fluctuations ($\gtrsim 1\,\text{kHz}$), but this requirement can be much relaxed in space-based observations. Mechanical modulation (rotating retarder plate) may introduce beam wobble and influence of optical inhomogeneities. Better precision is achieved with electro-optical modulation, which can be done with piezoelastic modulators (PEM), or with nematic or ferro-electric liquid crystals. PEMs have the great advantage that basically any optical material may be used, like lithium fluoride, which has good transmittance in the vacuum ultraviolet. The main disadvantage is the high, sinusoidal modulation frequency (typically 50 kHz), which may not be changed, since PEMs are resonant devices. The compatibility problem that this causes when PEMs are used in combination with the slow read-out of large CCD detectors has been elegantly solved with the ZIMPOL (Zurich Imaging Polarimeter) technology (Povel 1995, 2001; Gandorfer et al 2005), which in the future might become superseded by CMOS technology. For space applications the use of rotating wave plates or nematic liquid crystals in the visible, and rotating wave plates in the vacuum ultraviolet, may be the appropriate solution.

Instrumental polarization

Since the telescope optics are generally polarizing they corrupt the Stokes vector that we want to determine. Therefore the optics for the polarization analysis (the modulation package, preceded by the polarization calibration optics) should be placed as early as possible in the optical train. The part of the telescope that is in front of the polarization calibration optics should produce as little polarization as possible. This is the case if this part of the optical system is axially symmetric. If significant instrumental polarization is unavoidable in the telescope design, then it is much more manageable if it is constant in time. Although instrumental polarization may in principle be calibrated, modelled, and removed in the data reduction, large instrumental polarization should be optically compensated for as much as possible to avoid nasty effects of detector non-linearities, which generate spurious polarization signals that are next to impossible to calibrate or model.

Figure 33.2: Line-of-sight component of the magnetic field recorded by the *Hinode* satellite on 13 December 2006 during a proton flare. Bright and dark areas represent magnetic fields directed towards and away from the observer. *Hinode* is a Japanese mission developed, launched and operated by ISAS/JAXA, in partnership with NAOJ, NASA and STFC (UK). Additional operational support is provided by ESA and NSC (Norway).

Spectral and spatial information space

Polarimetry may be combined with any system for spectral selection. The data cube consists of the two spatial coordinates on the Sun, x and y, and the wavelength coordinate λ. Since the detector is a 2-D device, scanning is needed to cover a 3-D portion of the data cube. With (stigmatic) spectrographs one gets images in (x, λ) space for each given y, and then needs to stepwise scan in y to build up an image of the Sun. With narrow-band filters one gets (x, y) images for each given λ, and needs to stepwise tune the filter in λ to build up line-profile information.

One interesting alternative that combines these two features is the *solar chromatograph* (Stenflo 1973b), which gives monochromatic (x, y) images, but such that the wavelength λ varies linearly with y. Therefore line profile information is simultaneously present in each image, although it is convolved with the spatial coordinate. This solution is based on the concept of subtractive double dispersion and has so far only been systematically implemented in the MSDP instrument of the French THEMIS telescope on Tenerife (Mein 2002).

Advantages of observations above Earth's atmosphere

Space-based Stokes vector polarimetry has two main advantages: (1) It gives superior and stable angular resolution over a large field of view. (2) It gives access to the extraterrestrial/vacuum ultraviolet (VUV) part of the spectrum. The stable angular resolution is needed to explore the evolution of the small-scale magnetic fields, to understand magneto-convection, dynamo processes and the underlying mechanisms of solar and stellar activity. Access to the VUV is needed to explore the magnetic fields in the solar transition region and corona. This is of profound importance for understanding the dynamics and the heating processes of the outer solar atmosphere, which ultimately control the space weather and terrestrial effects.

At the time of writing the highest angular resolution in solar observations is still achieved from the ground thanks to the application of adaptive optics: the images from the Swedish Solar Telescope SST on La Palma reach the diffraction limit of $0.1''$ for the 1 m telescope. Adaptive-optics correction, however, only works over a small field of view, which in future may be improved upon by more complex multi-conjugate adaptive optics systems (Berkefeld et al 2001). The quality of the adaptive-optics correction improves when the atmospheric seeing is better, but since the seeing is almost always highly variable on time scales of minutes or less, it is next to impossible to obtain evolutionary sequences that can compete with space-based observations. In contrast, the spacecraft *Hinode* achieves an angular resolution of $0.2''$ over the whole large field of view (cf., Figure 33.2), and the resolution remains the same over the course of the mission, thus allowing evolutionary sequences of extraordinary quality and length (Kosugi et al 2007).

For observations in the visible, space-based instruments of modest spatial resolution still have great advantages. An example is the Michelson Doppler Imager (MDI) on the *SOHO* spacecraft (Scherrer et al 1995; Title 2010). In spite of its relatively low spatial resolution of $4''$ it has been of tremendous value, since it has given us a continuous time series that covers more than one solar activity cycle with high-quality full-disk magnetograms that all have the same resolution.

Zeeman-effect observations

Magnetic-field evolution

Although the magnetic-field evolution occurs on all scales, the key to the understanding of solar magnetism, basic dynamo processes, and the underlying mechanisms of stellar activity may be found in the smallest scales. Some of the most fundamental still unanswered questions concern the emergence, decay, and removal of the photospheric magnetic flux. All the emerged magnetic flux has to be removed on the solar-cycle time scale and be in statistical equilibrium with the emergence rate, otherwise the photosphere would quickly get choked with undisposed magnetic flux. We know that as we go from active regions to ephemeral regions and still smaller scales, the emergence rate increases dramatically, but we do not know how one can get rid of the flux at such tremendous rates. We need to determine the relative contributions of the following three alternative removal mechanisms: (i) Cancellation of opposite polarities (reconnection). (ii) Flux retraction (reprocessing in the convection zone). (iii) Flux expulsion (and the role of coronal mass ejections). In spite of decades of hard work we still know practically nothing about the relative roles of these processes.

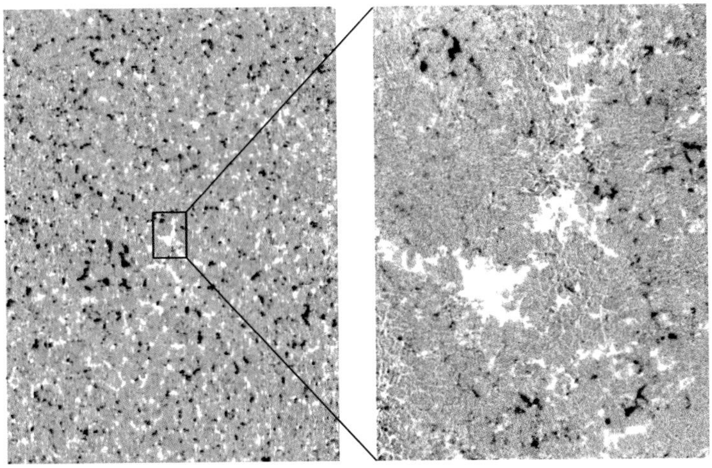

Figure 33.3: Illustration of the fractal-like pattern of quiet-Sun magnetic fields. The rectangular area covered by the left map (from Kitt Peak) is about 15 % of the area of the solar disk, while the map to the right (from the Swedish La Palma telescope) covers an area that is 100 times smaller. The white and black areas correspond to magnetic flux of positive and negative polarities, separated by grey voids of seemingly no flux. Analysis of Hanle-effect observations (cf., Page 550) of atomic and molecular lines have shown that these grey regions are actually not voids at all, but are teeming with turbulent magnetic fields that carry a significant magnetic energy density. Since these turbulent fields are tangled with mixed polarities on very small scales, they are invisible to observations of the Zeeman effect, but they are revealed by observing the Hanle effect (Stenflo 2004b).

Small-scale structuring

Observations with improved angular resolution in combination with indirect diagnostics have shown that the magnetic structuring extends from the global scales towards the diffusion scales far beyond the telescope resolution. There is a remarkable degree of self-similarity over the various scales, suggesting a fractal-like nature (cf., Figure 33.3). The scales that we are beginning to resolve (the dimensions of the photon mean free path and the pressure scale height) are of critical importance for a physical understanding of the scale spectrum of magneto-convection. Since however the structuring continues far into the unresolved domain, we always need to complement the direct, resolved observations with indirect techniques to extract statistical information about the unresolved structures, like the line-ratio method (Stenflo 1973a). These indirect methods are now being extended through applications of the Hanle effect.

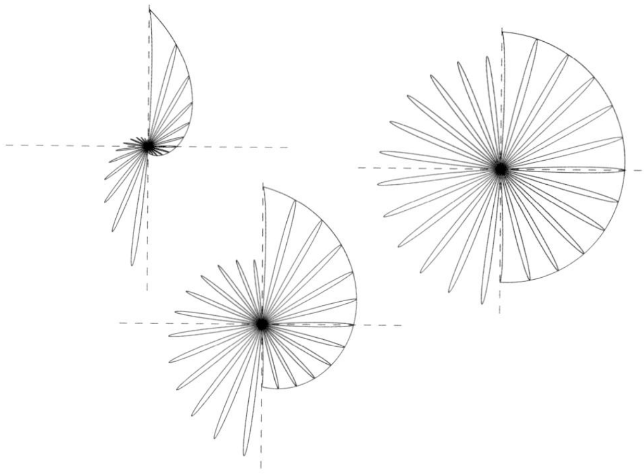

Figure 33.4: Trajectory of a damped classical oscillator, illustrating the Hanle-effect depolarization and rotation of the plane of polarization when the magnetic field is oriented along the line of sight. The three diagrams represent different values of the field strength, which increases from left to right.

Hanle-effect observations

Classical and quantum descriptions of the Hanle effect

In contrast to the Zeeman-effect polarization the Hanle effect is a coherency phenomenon that only occurs in coherent scattering processes. Magnetic fields remove the degeneracy of the radiatively excited and coherently superposed magnetic substates and thereby cause partial decoherence that leaves a signature in the polarization of the scattered radiation.

In a classical description the damped dipole oscillations that are induced by the incident radiation precess in the presence of a magnetic field, as pictured in Figure 33.4. In the illustrated case it is assumed that vertical oscillations are induced by the excitation process, and that we observe the emitted radiation along the magnetic field direction. The trajectory of the damped oscillator forms a rosette pattern that becomes more isotropic when the field is stronger. The emitted polarization is obtained from the Fourier transform of the rosette pattern. In the absence of magnetic fields the scattered radiation would be linearly polarized in the vertical direction (in the illustrated case), but as the field strength is increased, the plane of polarization rotates, and the amount of polarization is reduced.

If the classical equation for the dipole oscillation is decomposed in the three Cartesian coordinate equations, the component equations are coupled to each other due to the $\boldsymbol{v} \times \boldsymbol{B}$ term of the Lorentz force. If we however decompose in terms of complex spherical vectors, the component equations decouple (cf., Stenflo 1994). These three components correspond to the $\Delta m = 0, \pm 1$ transitions (the π and σ components) in the quantum-mechanical picture. Since they oscillate with slightly different frequencies due to the Larmor precession (which is the source of the Zeeman splitting), the damped oscillators gradually get out of phase, which leads to partial decoherence depending on the strength of the magnetic field.

Observational signatures of the Hanle effect

The Hanle effect leaves its imprints in the linear polarization. Its two main signatures are depolarization and rotation of the plane of polarization when the scattering geometry resembles 90° scattering. Figure 33.5 shows an example of the qualitatively different signatures of the Hanle and Zeeman effects. The Hanle effect also occurs in forward scattering, where it generates linear polarization in the presence of transverse magnetic fields (Trujillo Bueno 2001).

Preconditions for the Hanle effect to be observable at all are (i) that coherent scattering plays a significant role in the formation of the spectral line, and (ii) that the scattering polarization has observable amplitude. Condition (i) favours strong resonance lines, which anyway dominate the spectrum from the chromosphere-corona transition region. Condition (ii) requires that the incident radiation field of the scattering process is significantly anisotropic. In the vacuum ultraviolet the radiance contrasts of the solar structures are much larger than in the visible, which implies that the local anisotropies of the radiation field are large. The expected local fluctuations of the scattering polarization are therefore expected to be large as well. In the visible the contrasts are much smaller, with the consequence that the anisotropy due to the limb darkening becomes more important than the local radiance fluctuations. This global anisotropy determines the polarization scale of the "Second Solar Spectrum" (Stenflo and Keller 1997), the highly-structured spectrum in linear polarization that is exclusively due to coherent scattering.

Due to the scattering geometry when limb darkening is the source of the anisotropy, the polarization amplitude increases monotonically as we approach the limb. Therefore, in the visible part of the spectrum, the best conditions for the observability of the Hanle effect are found in a zone near (but inside) the solar limb, or in prominences above the limb. The situation is somewhat different in the vacuum ultraviolet, since the anisotropies are more local than global. Therefore the observability should be less restricted to a limb zone. For structures that are rather high above the solar surface the global anisotropy would however dominate, if the total illumination from the underlying solar disk occupies a solid angle that is significantly smaller than a half sphere (2π).

Complementarity of the Zeeman and Hanle effects

While the Zeeman-effect polarization becomes measurable when the Zeeman splitting is comparable to the Doppler width of the spectral line used, the sensitivity range for the Hanle effect is where the Zeeman splitting is comparable to the damping width or inverse life time of the excited atomic or molecular state. Since the damping width is typically two orders of magnitude smaller than the Doppler width, the Hanle effect is sensitive to correspondingly weaker fields. This makes it particularly suited for diagnostics of magnetic fields in the transition region and corona, where the fields are relatively weak due to the rapid expansion with height of the many strong, photospheric flux concentrations. In addition, since the Zeeman

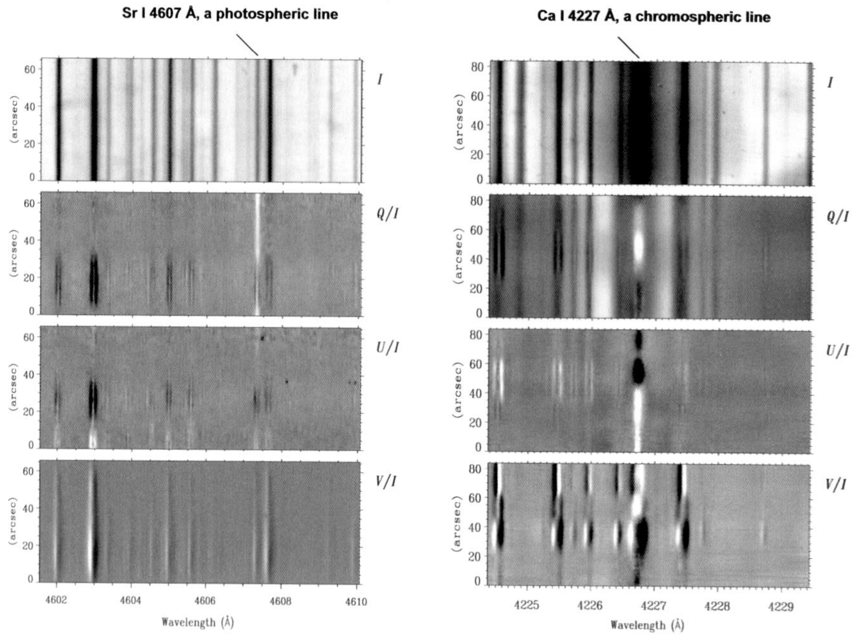

Figure 33.5: Examples of Hanle-effect signatures in Stokes spectra (the intensity I and the three fractional polarizations Q/I, U/I, and V/I) recorded with ground-based instrumentation, illustrating the difference between the Hanle and Zeeman effects. The photospheric Sr I 4607 Å line and the chromospheric Ca I 4227 Å line exhibit strong linear polarization due to coherent scattering, which in the presence of magnetic fields gets modified by the Hanle effect (seen as spatial variations of Q/I and U/I along the slit). The surrounding spectral lines display the usual transverse Zeeman effect in the linear polarization (Stokes Q/I and U/I), and the longitudinal Zeeman effect in the circular polarization (Stokes V/I). The recordings were made with the Zurich Imaging Polarimeter (ZIMPOL) at the McMath-Pierce facility (Kitt Peak).

splitting scales with λ^2, the signatures of the Zeeman effect are tiny in the vacuum ultraviolet.

Another fundamental area where the Zeeman and Hanle effects are highly complementary concerns the spatially unresolved domain of magneto-convection. The Zeeman effect is blind to spatially unresolved turbulent fields with zero net magnetic flux, in contrast to the Hanle effect that has different symmetry properties (the "sign" of the Hanle depolarization is independent of field polarity).

Observational strategy and interpretational issues

The Hanle-effect signatures depend on the anisotropies of the radiation field, on the scattering geometry, and on optical-depth effects in the medium, in addition to the magnetic field. To disentangle these effects it helps to apply a strategy of using combinations of spectral lines that are affected similarly by the non-magnetic factors, but which differ in their magnetic sensitivities (in analogy with the line-ratio technique for Zeeman-effect observations). A further problem is that the Hanle effect in principle delivers only two observables (amount of depolarization and rotation of the plane of polarization), while the magnetic field vector has three spatial components. For a unique interpretation one therefore needs additional observational or modelling constraints.

The longitudinal Zeeman effect has the great advantage that it allows circular polarization maps to be directly interpreted as maps of the line-of-sight component of the magnetic flux density. In contrast, the magnetic-field information that is carried by the Hanle-effect signatures is rather convolved with other factors in a way that prevents magnetic maps from being extracted without additional information. Still, the Hanle effect provides information about parameter domains that are not accessible by other means, but which are needed for understanding the fundamental physical processes on the Sun.

Stokes polarimetry in the vacuum ultraviolet

Advantages of the VUV region

The VUV and X-ray regions of the solar spectrum are full of resonance lines that are formed in the chromosphere-corona transition region or above. Observations in such lines allow us to diagnose the physics of the outer solar atmosphere where the coronal heating takes place, the solar wind is driven, and the space weather is generated. Crude information on coronal magnetic fields can be obtained with radio-astronomical techniques, forbidden-line scattering polarization in the visible, and (more recently) by recordings of the longitudinal Zeeman effect of forbidden coronal lines in the near infrared. Spectro-polarimetry in the vacuum ultraviolet remains a promising but largely unexploited area with considerable potential.

The VUV region down to about 1050 Å is of special interest, since transmission optics like lithium fluoride can be used throughout this range, which makes it feasible to apply retarders and modulators for complete Stokes polarimetry. At shorter wavelengths one may still determine the linear polarization by using the partially polarizing properties of oblique reflections, but circular polarization appears to be out of reach. This may however not be such a serious limitation, since the Hanle effect has its signatures in the linear polarization, while the circular polarization is a property of the Zeeman effect.

Due to its λ^2 dependence the Zeeman effect will be ineffective at short wavelengths. In the VUV its observability will mainly be limited to sunspots. Since the Hanle effect is sensitive to considerably weaker fields than the Zeeman effect, we expect that it will play the leading diagnostic role in the VUV and below. Coronal magnetic fields have field strengths in the Hanle sensitivity range for allowed line transitions. All the allowed coronal and transition-region lines are in the EUV or soft X-ray region, which can only be accessed from space.

The Hanle effect can also be used to diagnose the expanding envelopes of hot stars, but also here the relevant spectral lines are in the VUV, inaccessible from ground. Since magnetic fields play a key role for the physics of stellar transition

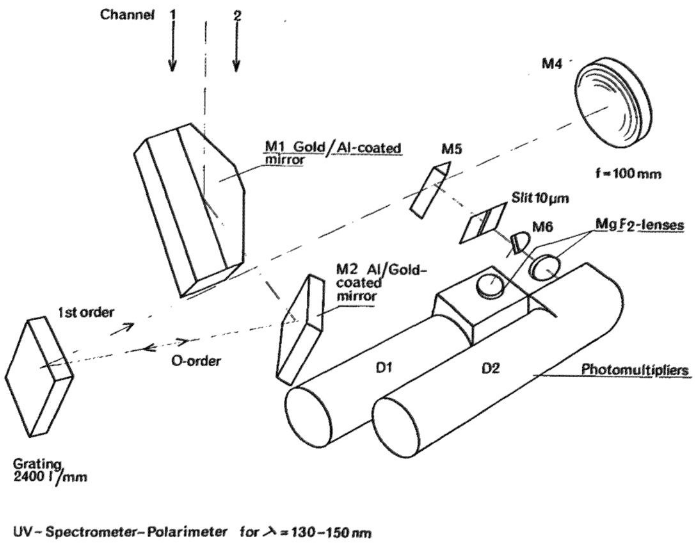

Figure 33.6: Optical scheme of the first space-based spectro-polarimeter for the recording of scattering polarization in the vacuum ultraviolet (Stenflo et al 1976). It was flown on *Intercosmos 16* in 1976.

regions, coronae, and stellar winds, we need space-based observations of the Hanle effect to diagnose these domains.

Choice of polarization optics

Very little has been done in the area of VUV polarimetry, it still represents almost virgin territory. For the detection of linear polarization one needs to make a trade-off between polarization efficiency and transmission. Thus Brewster-angle reflection on uncoated dielectric surfaces gives complete polarization, but the reflectivity is generally low. Coated reflective surfaces give good throughput and varying degree of polarization, depending on the choice of coating. A good choice is to use 60° reflection (near the Brewster angle) on a gold-coated mirror, which gives a degree of polarization of about 70 %. In contrast, an aluminium-coated mirror gives about 5 % polarization or less.

Magnesium fluoride (MgF_2) is birefringent and may be used as a retarder down to 1150 Å. It can also be used as the material for a polarizing Wollaston beam splitter. A rotating MgF_2 retarder plate was used by the UVSP instrument (Woodgate et al 1980) on the *SMM* satellite to record the circular polarization due to the Zeeman effect in the C IV 1548.2 Å line above a sunspot (Henze et al 1982), and to record the scattering polarization across the Mg II k and h lines near the solar limb (Henze and Stenflo 1987).

The first attempt to record scattering polarization in the VUV was made in 1976 with a Swedish-built spectro-polarimeter on the Soviet satellite *Intercosmos 16* (Stenflo et al 1976, 1980). As shown in Figure 33.6, most of the polarization analysis took place already at the first, oblique plane mirror, which was divided in two halves, one coated with gold, the other with aluminium. The beams from the two halves were sent to two different photomultipliers. The ratio between the signals from the two detectors could be normalized to unity at the unpolarized disk centre and would then differ from unity if other parts of the solar disk were polarized.

Gratings produce partial linear polarization and may therefore serve the role of polarization analyzer. This brings polarimetric capabilities to instruments that have not been specifically designed for polarimetry. Not only the UVSP instrument on *SMM* but also SUMER on *SOHO* made use of this property. In the case of SUMER the rotation of the whole *SOHO* spacecraft was used to detect the linear O VI 1032 Å line polarization in the corona, through modulation of the signal with spacecraft roll angle (Hassler et al 1997; Raouafi et al 1999).

For future instruments it should be possible to develop piezoelastic modulators from lithium fluoride (LiF). Thereby one should be able to design a polarimetric system that could record the full Stokes vector down to a wavelength of 1050 Å.

Solar and non-solar opportunities

While Stokes polarimetry in the VUV remains a seriously neglected area in the planning of space missions, a few concrete projects are in an advanced stage. Thus polarimetric instrumentation has been planned to be used in the *ASCE* mission (Gardner et al 2003; Romoli et al 2003) to follow up on the initial results from the *SOHO* mission in the coronal EUV lines. The linear polarization in these lines carries information on the acceleration of the solar wind through the Doppler dimming effect (Fineschi 2001) as well as on coronal magnetic fields through the Hanle effect.

Another mission, developed for launch as a sounding rocket payload, is the Solar Ultraviolet Magnetograph Investigation (SUMI, West et al 2000), which plans to measure magnetic fields in the solar transition region by recording the polarization caused by the Zeeman effect in the C IV 1548.2 Å and 1550.8 Å lines and in the Mg II k and h lines at 2795 Å and 2803 Å.

In the area of stellar physics the astronomy department at the University of Wisconsin in Madison has taken the lead in applying EUV scattering polarization and the Hanle effect to constrain the geometry, dynamics, and magnetic fields in hot, expanding stellar envelopes with P Cygni type spectral lines (Cassinelli and Nordsieck 2001; Ignace et al 2004). For this purpose they have developed a sounding rocket payload, the Far-Ultraviolet SpectroPolarimeter (FUSP) for high-precision spectro-polarimetry in the range 1050 Å to 1500 Å. The polarization analysis is done with a stressed lithium fluoride rotating wave plate, which is followed by a diamond Brewster-angle mirror (Nordsieck et al 2003).

Bibliography

Berkefeld T, Soltau D, von der Lühe O (2006) Multi-conjugate solar adaptive optics with the VTT and GREGOR. Proc SPIE 6272 627205:1–9

Cassinelli JP, Nordsieck KH, Ignace R (2001) Determination of magnetic fields in the winds from hot stars using the Hanle effect. ASP Conf Ser 248:409–414
Fineschi S (2001) Space-based instrumentation for magnetic field studies of solar and stellar atmospheres. Astron Soc Pacific Conf Ser 248:597–606
Gandorfer AM, Povel HP, Steiner P (plus six authors) (2005) Solar polarimetry in the near UV with the Zurich Imaging Polarimeter ZIMPOL II. Astron Astrophys 422:703–708
Gardner LD, Kohl JL, Daigneau PS (plus eight authors) (2003) The Advanced Spectroscopic and Coronagraphic Explorer: science payload design concept. Proc SPIE 4843:1–7
Hale GE (1908) On the probable existence of a magnetic field in Sun-spots. Astrophys J 28:315–343
Hassler DM, Lemaire P, Longval Y (1997) Polarization sensitivity of the SUMER instrument on *SOHO*. Appl Opt 36:353–359
Henze W, Stenflo JO (1987) Polarimetry in the Mg II H and K lines. Sol Phys 111:243–254
Henze W Jr, Tandberg-Hanssen E, Hagyard MJ (plus seven authors) (1982) Observations of the longitudinal magnetic field in the transition region and photosphere of a sunspot. Sol Phys 81:231–244
Ignace R, Nordsieck KH, Cassinelli JP (2004) The Hanle effect as a diagnostic of magnetic fields in stellar envelopes. IV. Application to polarized P Cygni wind lines. Astrophys J 609:1018–1034
Kosugi T, Matsuzaki K, Sakao T (plus twenty-two authors) (2007) The *Hinode (Solar-B)* mission: An overview. Sol Phys 243:3–17
Mein P (2002) The MSDP of THEMIS: Capabilities, first results and prospects. Astron Astrophys 381:271–278
Nordsieck KH, Jaehnig KP, Burgh EB (plus three authors) (2003) Instrumentation for high-resolution spectropolarimetry in the visible and far-ultraviolet. Proc SPIE 4843:170–179
Povel H (1995) Imaging Stokes polarimetry with piezoelastic modulators and charge-coupled-device image sensors. Opt Eng 34:1870–1878
Povel H (2001) Ground-based instrumentation for solar magnetic field studies, with special emphasis on the Zurich Imaging Polarimeters. Astron Soc Pacific Conf Ser 248:543–552
Raouafi N-E, Lemaire P, Sahal-Bréchot S (1999) Detection of the O VI 103.2 nm line polarization by the SUMER spectrometer on the *SOHO* spacecraft. Astron Astrophys 345:999–1005
Romoli M, Fineschi S, Uslenghi M (plus seven authors) (2003) The ASCE EUV polarimeter. Mem Soc Astron Italiana 74:835–838
Scherrer PH, Bogart, RS, Bush RI (plus ten authors) (1995) The Solar Oscillations Investigation – Michelson Doppler Imager. Sol Phys 162:129–188
Stenflo JO (1973a) Magnetic-field structure of the photospheric network. Sol Phys 32:41–63
Stenflo JO (1973b) Solar chromatograph. Appl Opt 12:805–809
Stenflo JO (1994) Solar Magnetic Fields — Polarized Radiation Diagnostics. Kluwer, Dordrecht, 385 pp

Stenflo JO (2001) Limitations and opportunities for the diagnostics of solar and stellar magnetic fields. Astron Soc Pacific Conf Ser 248:639–650

Stenflo JO (2004a) The new world of scattering physics seen by high-precision imaging polarimetry. Rev Mod Astron 17:269–296

Stenflo JO (2004b) Hidden magnetism. Nature 430:304–305

Stenflo JO, Keller CU (1997) The second solar spectrum. A new window for diagnostics of the Sun. Astron Astrophys 321:927–934

Stenflo JO, Holzreuter R (2003) Distribution of magnetic fields at scales beyond the spatial resolution limit. Astron Soc Pacific Conf Ser 286:169–176

Stenflo JO, Biverot H, Stenmark L (1976) Ultraviolet polarimeter to record resonance-line polarization in the solar spectrum around 130 – 150 nm. Appl Opt 15:1188–1198

Stenflo JO, Dravins D, Wihlborg N (plus five authors) (1980) Search for spectral line polarization in the solar vacuum ultraviolet. Sol Phys 66:13–19

Title AM (2010) Michelson interferometers. ISSI SR-009:327–338

Trujillo Bueno J (2001) Atomic polarization and the Hanle effect. Astron Soc Pacific Conf Ser 236:161–195

West EA, Porter JG, Davis JM (plus four authors) (2000) Overview of the solar Ultraviolet Magnetograph Investigation. Proc SPIE 4139:350–361

Woodgate BE, Brandt JC, Kalet MW (plus seven authors) (1980) The Ultraviolet Spectrometer and Polarimeter on the Solar Maximum Mission. Sol Phys 65:73–90

— 34 —

Polarimetry at high energies

Wojtek Hajdas[I] and Estela Suarez-Garcia[II]

Abstract

Using polarimetry as a tool to study high-energy photons was until now rather uncommon. Recent satellite measurements proved the high power of polarization observables and initiated a rapid development of new instruments. We describe the physical mechanisms leading to photon polarization, and discuss the processes used for its detection at energies from few kiloelectronvolts X-rays up to mega-electronvolts γ-rays. Past attempts and instrumentation applied for polarization measurements are summarized and reviewed with respect to experimental difficulties. We also present current and planned polarimeters and the utilization of novel, energy specific, detection technologies. Finally, we discuss the importance of proper modelling and calibration of the polarimeter, and describe several methods used to characterize it.

Introduction

The polarization data at X- and γ-ray photon energies remain until now very scarce. Several past attempts to measure polarization concentrated mainly on observations of the Sun, Crab nebula, and a few X-ray binaries. They resulted in a single detection of a relative polarization of 19 % from the Crab (Weisskopf et al 1978), and hardly conclusive results from solar flares (Suarez-Garcia et al 2006; Boggs et al 2006). On the other hand, the lower-energy regimes of the electromagnetic spectrum routinely benefit from polarization observations. Such disparity is largely caused by great difficulties related with both construction of the high-energy polarimeters and conducting of the measurements in space. Polarized high-energy photons can be produced by the following physical processes: Compton and inverse Compton scattering, magneto- and electrostatic bremsstrahlung, and photon splitting in extreme magnetic fields (Kallman 2004). As polarization signatures are clearly distinctive, studying them can provide essential information on both the emission mechanism and the source geometry with its size, uniformity or even structure of the magnetic field. In addition, polarization can sample the extreme physics, basic symmetries and invariants, as well as the nature of gravity (Fan 2007). Therefore it has a great potential in answering the fundamental questions

[I]PSI—Paul Scherrer Institut, Villigen, Switzerland
[II]DPNC & ISDC, University of Geneva, Switzerland

on the physics of the Universe. The list of targets for polarization observation includes black holes, pulsars, X-ray binaries and flare emitters and exotic objects, such as γ-ray bursts (GRBs), soft gamma repeaters and magnetars (Lazzati 2006). Polarization measurements aim to determine two variables: the polarization degree and the polarization direction of the incoming photon flux (Lei et al 1997).

Measuring principles

The selection of detection techniques applied to determine the level of the linear polarization of photons is strongly related to their energies. Thomson scattering, photo-electric effect and Bragg reflection are utilized at energies of about a few kiloelectronvolts to a few tens of kiloelectronvolts. For the Bragg-reflection technique, the incoming photons undergo a constructive interference during reflection off the crystal at the glancing angle (θ_B) (see Equation 2.50 in Chapter 2, Wilhelm and Fröhlich 2010). The maximum reflectivity occurs for photons having their electric vectors parallel to the crystal planes, while it equals zero if the direction of electric vectors is normal. The Bragg-crystal method, although highly efficient as a polarization analyzer, can suffer from systematic effects and a narrow energy range strictly selected by the Bragg law. Another technique in the same energy regime is based on the photo-electric effect. It analyzes the angular distribution of electrons emitted from the atom shell after absorption of the photon. The cross section depends on the azimuth angle ξ between the photon electric vector and the direction of the electron emission. The formula for electrons ejected from the K-shell (Ramsey et al 1994) is:

$$\frac{\mathrm{d}\sigma}{\mathrm{d}\Omega} \approx \frac{\sin^2\theta_\mathrm{e}\cos\xi}{(1-\beta\cos\theta_\mathrm{e})^4} \quad , \tag{34.1}$$

where θ_e is the electron ejection angle and β is its velocity divided by the speed of light. In order to maximize their efficiency, polarimeters based on the photo-effect utilize high-Z materials as detectors and work in the energy range where the photo-absorption cross section is the highest. The Thomson-scattering technique is also utilized at energies below ≈ 30 keV. The process is equivalent to the Compton effect in the classic limit in which photon energy is much smaller than the mass of the electron. The azimuth distribution of the Compton-scattered photon depends on the initial direction of its polarization vector and is given by the Klein-Nishina cross section (Heitler 1954):

$$\frac{\mathrm{d}\sigma}{\mathrm{d}\Omega} = \frac{r_0^2}{2}\left(\frac{E^{'}}{E}\right)^2\left(\frac{E}{E^{'}} + \frac{E^{'}}{E} - 2\sin^2\theta\cos^2\eta\right) \quad , \tag{34.2}$$

where r_0 is the classical electron radius, θ is the scattering angle and η is the azimuth angle between the scattering plane and the direction of the photon polarization vector, and E and $E^{'}$ are the initial and final photon energies, respectively. The cross section in the classic limit is obtained for $E^{'} \cong E$, i.e., photon energy conservation.

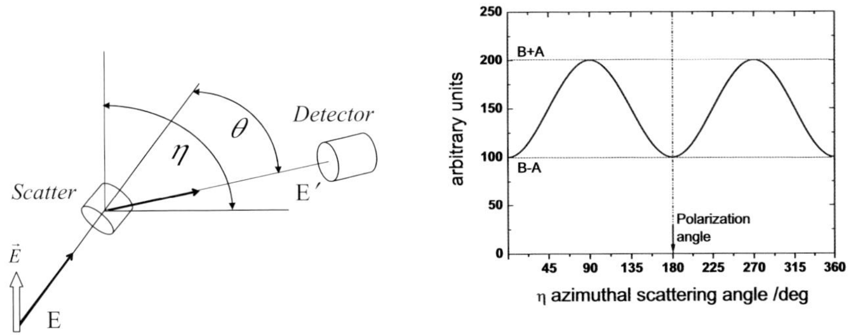

Figure 34.1: Left: Measurement of the linear polarization of γ-rays. $\vec{E}$ is the photon polarization (electric) vector. Right: Ideal modulation curve pattern from the Compton scattering of polarized photons.

The most widely-used technique to measure the photon polarization at energies from tens of kiloelectronvolts up to several megaelectronvolts is the Compton scattering. The measurement principle is described in Figure 34.1. Polarimeters based on the Compton effect usually select low-Z materials as scatterers in order to maximize the probability of Compton scattering over the photo-electric absorption.

At even higher energies of tens of megaelectronvolts the electron-positron pair production becomes a dominating process in interactions of photons with matter. The cross section for pair production with a polarized photon can be written as (Mattox et al 1990):

$$\sigma(\delta) = \frac{\sigma_0}{2\pi}\left(1 + \Pi R \cos 2\delta\right) \quad , \tag{34.3}$$

where σ_0 is the total cross section for pair production, δ is the angle between the electron-positron plane and the incident direction of the photon electric vector. Π is the initial polarization degree of the photons and R specifies the asymmetry ratio for the pair-creation process. To date, there are no working polarimeters based on the pair production. Their typical construction makes use of layers of heavy-element materials maximizing the probability of pair production. They are followed by position-sensitive detectors that measure the electron and positron tracks.

For each of the above processes, the angular distribution of the measured observable, either photon or electron, can be described by a cosine function of the azimuth angle, $C(\eta)$, (McConnell et al 2002). The modulation amplitude A and the offset B of the cosine function are linked to the polarization level of the incident photon flux. The phase of the modulation curve, η_0, corresponds to the direction of the photon polarization vector (see Figure 34.1).

$$C(\eta) = A\cos\left[2(\eta - \eta_0) + \pi/2\right] + B \quad . \tag{34.4}$$

The main experimental challenge is to measure this distribution with the highest possible precision. Polarimetric capabilities of any instrument based on the measurement techniques described above are determined by a single parameter called modulation factor. The modulation factor is extracted from the parameters A and B from Equation 34.4 as:

$$\mu_{\Pi} = \frac{A}{B} \quad . \tag{34.5}$$

It describes the response of the polarimeter to a photon flux with a polarization degree of Π. Given an instrumental modulation factor for fully polarized photons, μ_{100}, the degree of polarization Π of the incoming photons is equal to:

$$\Pi = \frac{\mu_{\Pi}}{\mu_{100}} \tag{34.6}$$

The polarization degree is defined as a positive quantity with non-Gaussian statistics spanned between 0 and 1. For the source flux S_{f} and background rate B_{r}, the minimum detectable polarization Π_{minD} at the significance level n_{σ} is expressed as (Novick 1975):

$$\Pi_{\mathrm{minD}} = \frac{n_{\sigma}}{\mu_{100}\, S_{\mathrm{f}}\, \varepsilon\, A_{\mathrm{p}}} \sqrt{\frac{S_{\mathrm{f}}\, \varepsilon\, A_{\mathrm{p}} + B_{\mathrm{r}}}{T}} \quad , \tag{34.7}$$

where A_{p} is the detection area, ε the detection efficiency, and T the observing time. The product $\varepsilon A_{\mathrm{p}}$ is also known as the effective area A_{eff} of the detector. Another quantity used to characterize an operational performance of the polarimeter is its figure of merit defined as a product of A_{eff} and μ_{100} divided by a square root of the instrument background rate.

Existing polarimeters

The choice of the polarimeter detection technique, the optimization of its parameters and the observation methodology depend both on the objectives of the project and on the constraints imposed by the satellite. This section illustrates the topic with several representative examples of existing polarimeters.

SXRP and SPN-R

The Stellar X-Ray Polarimeter SXRP (Ferocci et al 1994) was designed for the *SRG* mission that has not been flown yet. It is a hybrid consisting of the Bragg graphite crystal and the low-Z lithium scatterer surrounded by the imaging proportional counters (Figure 34.2). The instrument covers the energy range from 2 keV to 20 keV. To reduce systematic effects the polarimeter rotates around its optical axis. An extra anticoincidence shield minimizes the background counting rate. The SXRP has very high values of μ_{100} and A_{eff}: 99 % and 10 cm^2 for the Bragg and 71 % and 65 cm^2 for the Thomson processes, respectively. The polarimeter is well understood with a thorough laboratory calibration and modelling. Its flight model is still waiting for the next flight opportunity. A similar Thomson polarimeter called SPN-R flew aboard the *Coronas-F* satellite. Its scatterer is made of beryllium and is surrounded by three pairs of the CsI(Na) detectors. The active area of SPN-R A_{eff} is 1 cm^2 only. Over five years, between 2001 and 2005, the instrument observed

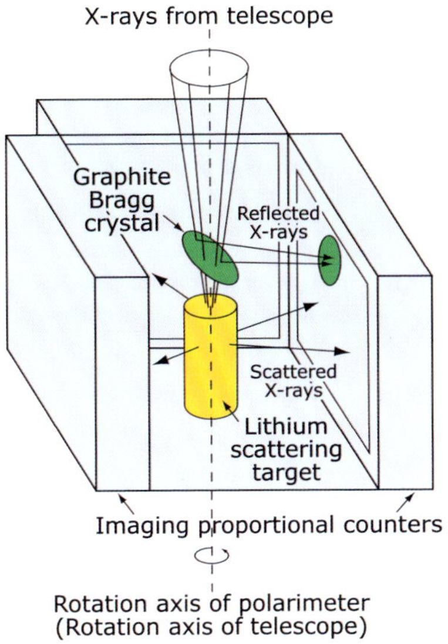

Figure 34.2: A diagram of the SXRP rotating polarimeter with the Bragg crystal and lithium scatterer.

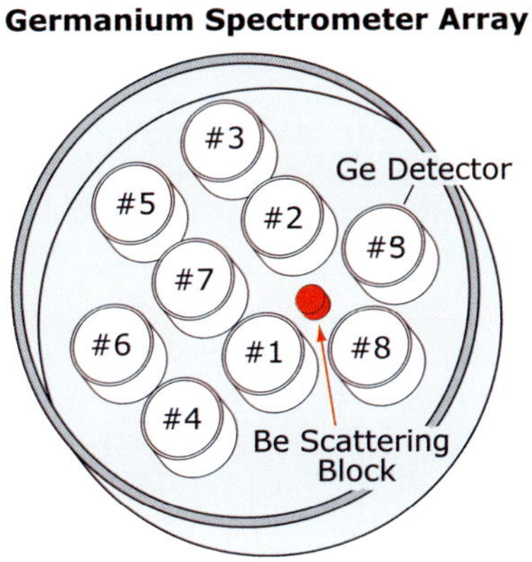

Figure 34.3: The *RHESSI* spectrometer array with nine Ge-detectors and beryllium scatterer.

tens of solar flares. For only eight of them it was possible to provide just the upper polarization limits: $\Pi_{3\sigma} < 8$ % to 40 %.

COMPTEL and BATSE

The large area imaging Compton Telescope COMPTEL (Schönfelder et al 1993) and the Burst And Transient Source Experiment BATSE (Fishman et al 1992) were flown on board the NASA *CGRO* mission. COMPTEL operated at energies from 0.75 MeV to 30 MeV and consisted of seven low-Z scintillators as scatterers (liquid NE213, A = 4188 cm^2) followed by fourteen high-Z detectors (NaI(Tl), A = 8620 cm^2) placed 1.5 m below the scatterer plane (see also Chapter 11, Schönfelder and Kanbach 2010, for a more detailed description of the instrument). Its field of view (FOV) was equal to 1 sr but its $A_{\rm eff}$ was smaller than 20 cm^2 and μ_{100} below 8 %. The instrument was not optimized for polarimetry and the analysis of several GRBs revealed very low statistics and serious systematic effects. Even polarimetric observations of the Crab nebula turned out to be impossible. Slightly more successful was the other instrument – BATSE. It had eight large area NaI(Tl) detectors (A = 2000 cm^2 each) placed at the spacecraft corners. It operated between 40 keV and 600 keV and had a full 4π FOV. BATSE was not a polarimeter but had good directional capabilities to localize GRBs. Polarization detection was possible using the Earth albedo scattering. After tedious analysis supported by complex Monte Carlo modelling, only two GRBs out of the about 3000 in the BATSE catalogue were analyzed. For both bursts the results gave the lower limit of the polarization: $\Pi > 30$ % and 50 % (Willis et al 2005).

RHESSI

RHESSI (Lin et al 2002) is a NASA-*SMEX* mission for imaging the Sun at energies between 3 keV and 20 MeV. Its spectrometer, made of nine germanium detectors, is located in the plane perpendicular to the satellite rotation axis. Although not primarily designed as polarimeter, *RHESSI* can be used as such in two different modes: passive and active. In the passive mode (30 keV to 80 keV) one measures polarization studying the photons that scatter in the Be-block (see Figure 34.3) and stop in the Ge-detectors surrounding it (McConnell et al 2002, 2004). In the active mode (100 keV to 2 MeV), the polarization information is extracted from the photons scattered between two neighbouring Ge-detectors. As a polarimeter, *RHESSI* has a wide FOV, but a moderate μ_{100} and a small A_{eff} (1 cm^2). In 2003, a result obtained using the active mode was published claiming a very high polarization ($\Pi = 80$ %) from GRB021206 (Coburn and Boggs 2003). Although later analysis found this result rather controversial (Rutledge and Fox 2004; Wigger et al 2004), it stimulated the theoretical and experimental astrophysics communities, and it accelerated the development of hard X-ray polarimetry. To date, *RHESSI* has been used to study the polarization of eight solar flares (2 % $< \Pi <$ 54 % with a relative standard uncertainty from 5 % to 26 %) (Boggs et al 2006; Suarez-Garcia et al 2006). *RHESSI* polarimetric capabilities are limited by the high levels of Earth-scattered photons and accidental background.

SPI and IBIS on *INTEGRAL*

INTEGRAL is an ESA satellite launched in 2002. Its two main instruments, the SPI spectrometer and the IBIS imager, were not designed for polarization detection, but have such capabilities (Lei et al 1997). The spectrometer has 19 large hexagonal Ge-detectors covering energies from 20 keV to 8 MeV. Its μ_{100} is moderate (20 %) but its large A_{eff} reaches 50 cm^2. To date the SPI instrument could measure polarization of only one GRB (GRB041219a). Two groups performed independent analysis detecting very high polarization degrees of $\Pi = 60$ % (McGlynn et al 2007) and $\Pi = 98$ % (Kalemci et al 2007) with relative uncertainties of 31 % and 33 %, respectively. One should note that although the results are mutually consistent, the level of uncertainty is very large. With respect to observations of Crab (routinely used for the energy and flux calibration), calculations predict the Π_{minD} to be on the level of a few percent for 10^6 s long measurements. The first polarization data from several weeks of Crab observations resulted, as in case of the GRB, in very high values of polarization. In the energy range from 100 keV to 1 MeV the polarization level was found to be $\Pi = 46$ % with a relative uncertainty of 10 % (Dean et al 2008).

The IBIS imager has two detection layers with CdTl and CsI detector arrays. These detectors can be combined to work as a Compton telescope with $A_{\text{eff}} \cong$ 100 cm^2. Despite the IBIS large μ_{100} (30 %), no polarization data have yet been published. The major problem of the *INTEGRAL* polarimetry is its lack of proper calibration of its modulation factors and a large uncertainty of the systematic and instrumental effects.

New polarimeters

As shown, until now X- and γ-ray polarimetry has been largely neglected and attempts to measure polarization have suffered from various obstacles. The difficulties in the instrument construction and in carrying out the observations came in parallel with beliefs that the expected polarization levels should be quite low. These subjects have been recently challenged by the latest results from several instruments on satellites and by the new theoretical models. It caused a rapid development of novel polarimeters and the perfection of measurement techniques making use of the latest advances in the detection technology (McConnell and Bloser 2006). Modern gas micropattern detectors are used at low energies by PIXIE (Costa et al 2001), Mu-PIC or 1DPSPC (Nakajima et al 2004). Also classic types like Thomson and crystal polarimeters are further developed and improved like XPE and PLEXAS (Marshall et al 1998). The large scintillator arrays are core to the design in the classic Compton energy range for instruments like GRAPE (McConnell et al 1999), POGO (Larsson and Pearce 2004) and POLAR (Produit et al 2005). For polarimetry at even higher energies, either Si or Ge microstrip detectors are applied as for MEGA (Kanbach et al 2001), TIGRE (O'Neill et al 1996) and NCT (Boggs et al 2001). Novel semiconductor pixel detectors (CdTl, CdZnTl) are planned for the SGD and CIPHER polarimeters (Curado da Silva et al 2003) and a time projection chamber filled with liquid xenon is adopted for the high-energy polarimetry in the LXeGRIT detector (Aprile et al 2002). For all of the above instruments, big emphasis is given to maximizing the values of A_{eff} and μ_{100} and optimizing the signal-to-noise ratio. Current developments and modern detection techniques are described below for three energy regions. For this purpose a few novel polarimeters are selected as typical examples.

Energies around 10 keV – PIXIE (*IXO*)

New polarimetric techniques based on the photo-effect have recently been developed to study polarization at lower energies. An example of such an instrument is the Pixel Imaging Experiment (PIXIE) under design for the ESA *IXO* (formerly *XEUS*) mission (Costa et al 2001). PIXIE measures the tracks of electrons ejected from the atom after absorption of the photon. From the angular distributions of the tracks one can deduce photon polarization. At energies around 10 keV the length of the electron track in matter is very short. Thus, its precise visualization requires micropattern gas detectors as a photon absorption medium. The image of the ionization pattern is made using gas electron multipliers (GEMs) directly coupled to the readout electronics (ASIC, application specific integrated circuit, see Figure 34.4). In the typical operational energy range between 2 keV and 12 keV, the μ_{100} reaches values of about 50 %. The total area of the GEM is slightly below 2 cm^2 while its pixel size is of the order 50 µm. The thickness of the gas chamber is about 10 mm only. This implies that for efficient operation the detector must be mounted behind a large photon concentrator, e.g., an *IXO* mirror with 3 m^2 effective area. Further challenges are related with a precise modelling and advanced reconstruction techniques of the complex 3D electron tracks. Despite such obstacles

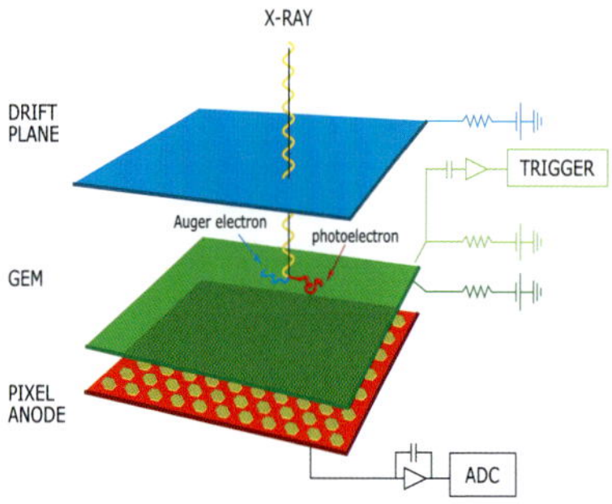

Figure 34.4: PIXIE polarimeter concept with a gas micropattern detector GEM directly connected to the readout ASIC.

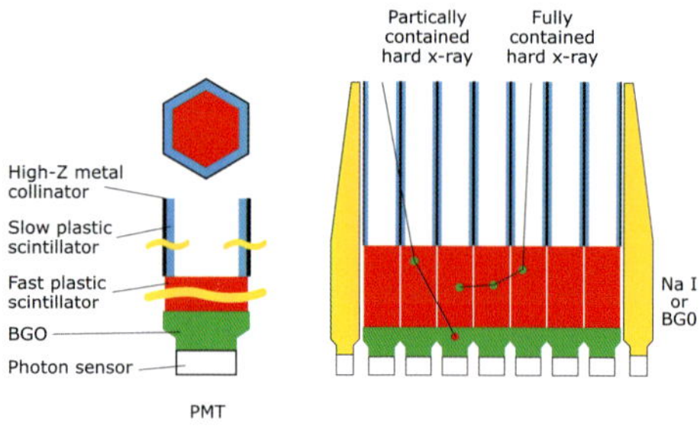

Figure 34.5: Concept of the POGO pointing instrument made of a well-type array of phoswich detectors.

the prototype was constructed and its polarimetric potential has been successfully demonstrated.

Energies around 100 keV – GRAPE, POLAR and POGO

Not only the energies but also the field of view that a polarimeter has to cover depends on the sources that one would like to study. For point-like objects like the Crab nebula a pointing instrument with a narrow field of view is preferable as its background levels are smaller. In the case of transient sources, like γ-ray bursts, that can appear in any position of the sky the field of view must be as large as allowed by external constraints, e.g., by the satellite or Earth shadow.

When a large FOV is required at energies around 100 keV, the classical design consists of an array of low-Z plastic scatterers where photons experience Compton scattering, and high-Z absorbers where they undergo photopeak reaction. This is the concept of the Gamma Ray Polarimeter Experiment GRAPE (McConnell et al 1999), constituted of a 64×64 array of plastic and BGO scintillators (see Figure 34.6). An alternative design is being developed for POLAR (Produit et al 2005), where all elements are low-Z plastic scintillators and Compton scattering is the main effect. In this case a photon usually produces several energy depositions and the two largest ones are selected to draw the modulation curve. Despite the difference in their designs, both GRAPE and POLAR detect photons in the range 50 keV to 500 keV with high $\mu_{100} \approx 40$ % to 60 % and large $A_{\rm eff} \approx 100$ cm^2 to 400 cm^2. Both instruments have also very good off-axis performance that makes them perfect for GRB polarimetry. The demo model of GRAPE has already been tested in a balloon flight and the proposal for a new space mission dedicated to γ-ray polarimetry is under evaluation by NASA.

The Polarized Gamma-ray Observer POGO (Larsson and Pearce 2004) is a polarimeter designed to study point-like sources. It is constituted of ca. 400 well-type phoswich counters with fast and slow scintillators placed on top of BGOs (see Figure 34.5). Its energy range is 25 to 200 keV and it has a small FOV $\approx 2.2^\circ \times 2.2^\circ$. Large $A_{\rm tot} \approx 1$ m^2 and $\mu_{100} \approx 25$ %, and a good signal-to-noise ratio maximize POGO's figure of merit. The instrument is suitable for long-duration

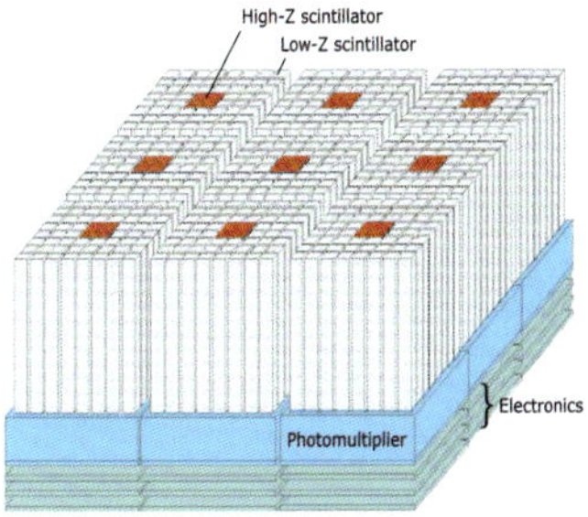

Figure 34.6: A diagram of the GRAPE polarimeter with high-Z absorbing scintillators surrounded by an array of the low-Z scattering ones.

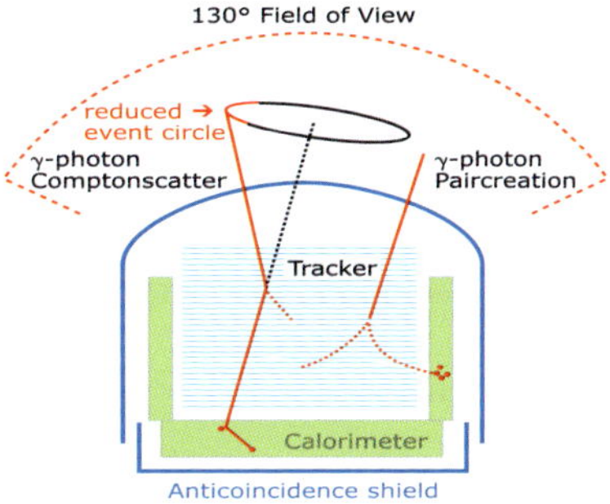

Figure 34.7: Si-microstrip tracker and arrays of CsI(Tl) absorbers build the MEGA instrument. The 4π anticoincidence shield reduces the charged-particle background.

balloon flights with pointing capabilities. Within 6 h it can reach a 6 % level of Π_{minD} for 100 mCrab strong sources. A small version of POGO has been flown in a balloon, and a larger version will fly in the near future.

Energy around and above 1000 keV – MEGA

Astrophysical γ-ray polarimetry at energies where electron-positron pair production starts to dominate is to date fully untouched. Negative analysis of data from both the *COS-B* and the EGRET on *CGRO* instruments in the past as well as a limited sensitivity of the present *Fermi (GLAST)* mission indicate certain difficulties. In order to succeed, the future instruments have to assure very high detection efficiency and superior tracking precision. Both parameters are intensely optimized for the Medium Energy Gamma-Ray Astronomy Experiment MEGA (Kanbach et al 2001). This instrument consists of a tracker made of 32 layers of large area (36 cm × 36 cm), double-sided Si-microstrip detectors and a calorimeter made of the three-dimensional CsI(Tl)/PIN diode arrays as shown in Figure 34.7. MEGA operates between 0.4 MeV and 50 MeV although measurements of polarization (at least with the Compton scattering) are feasible only up to 5 MeV. The predicted μ_{100} values for the Compton mode span between 19 % at 0.7 MeV and 3 % at 5 MeV. They were confirmed with the prototype version of MEGA (scaled down to about 10 % of the full instrument volume) during calibration runs at the synchrotron. The instrument has a large field of view of 130° (full cone) and is equipped on the sides and on top with an anticoincidence shielding. Its moderate overall dimensions of 1.2 m width and 1.3 m height make it an ideal polarimeter for both a satellite mission and a long-duration balloon flight.

Modelling and calibration

The modulation factor derived from calculations with the Compton cross-section formula overestimates in most cases the instrumental one. The reason for this lies in the extreme difficulty of including all polarimeter aspects into the analytical calculation. Therefore, the determination of the polarimeter modulation factor (μ_{100}) and its effective area (A_{eff}) must rely on both Monte Carlo simulations and laboratory calibrations.

Polarization-sensitive codes of photon tracking, e.g., GEANT4 (CERN 2007), EGSnrc (NRC/SLAC 2007) or MCNP (LANL 2007), are necessary for construction of the instrument mass-model. Together with an accurate description of the geometry and materials that constitute the instrument and its surroundings (spacecraft, balloon, ...), it is also important to introduce into the simulation the numerous sources of background that will affect the polarimeter in flight. The cosmic ray particles, the cosmic X-ray background, and the photons that arrive at the polarimeter either after being scattered in the materials of the spacecraft or in the Earth's atmosphere are only some of the most significant examples of background that have to be considered.

More advanced studies of the signal conversion in the detector can also be performed using recent software developments. For example, with the optical photon package of GEANT4 one can generate and track optical photons through the scintillator down to the photomultiplier including their absorption as well as reflection from the walls and the coating. Further implementation of the sensitive detector volumes and particle or γ-ray hits is also possible.

Independently of the simulation packages chosen, it is always necessary to verify the Monte Carlo results with proper laboratory tests and calibration procedures. These tests have to be done in a wide range of energies paying special attention to the extremes of the spectral range designed for the polarimeter. The laboratory calibration of μ_{100}, its angular variability as well as the effective area, detection thresholds and counting rate dependence, etc., are essential for the polarimeter's in-flight success. Studies must be done with both polarized and unpolarized beams, to properly verify the *zero polarization* response of the instrument and discover possible systematic effects causing false polarization patterns or responses. Synchrotron radiation is commonly used to obtain fully polarized photon fluxes. For somewhat lower levels of polarization a simple laboratory option is to build a setup where γ-rays from a radioactive source (e.g., Cs^{137}) are applied to the polarimeter after being Compton scattered in some material. The Compton process will partially polarize the reflected photons, with the maximum polarization level for the scattering angles around 90°. In addition, it is recommended to test the polarimeter in particle beams to study the influence of the background particles (e.g., protons from radiation belts and cosmic rays or neutrons from the nuclear reactions).

Both simulations and laboratory verification help in the elimination and correction of systematic effects and spurious signals. For example sensitivity variations between individual channels of the photomultiplier can be diminished after adjustment of their discriminator thresholds. For those effects which are inherent characteristics of the detector, a specific data analysis technique must be applied. An example is the pixellation effect, inherent to any pixellated geometry of the polarimeter. It can be eliminated by randomizing the position of the hit inside each pixel, or by applying the Decoupled Ring Technique (DRT) from Lei et al (1997), where further descriptions of data analysis techniques useful in different situations can be found.

Summary

During recent years X- and γ-ray polarimetry experienced considerable progress. Innovative detection technologies, like gas tracking detectors, pixellated semiconductor or large scintillator arrays, greatly influence the design of novel polarimeters. Lessons learnt from the existing instruments brought further improvements especially in reduction of various background sources and systematic effects. Modern simulation tools and accelerator-based calibration methods are now routinely applied to enhance the performance. In addition one also observes big theoretical advances with polarization used as a probe of various physical processes and mechanisms in the Universe. A number of original polarimeters are currently under development worldwide. They will cover the whole energy range from hard X-rays up to medium-energy γ-rays. Some instruments are already approved for future missions while others are in the application phase. Long awaited, high scientific return provided by X- and γ-ray polarization measurements can be expected within the next decade.

Bibliography

Aprile E, Curioni A, Giboni KL (plus seven authors) (2002) The LXeGRIT Compton telescope prototype: current status and future prospects. Proc SPIE 4851:140–152

Boggs SE, Jean P, Lin RP (plus nine authors) (2001) The nuclear Compton telescope: a balloon-borne soft γ-ray spectrometer polarimeter and imager. AIP Conf Proc 587:877–881

Boggs SE, Coburn W, Kalemci E (2006) Gamma-ray polarimetry of two X-class solar flares. Astrophys J 638:1129–1139

Coburn W, Boggs SE (2003) Polarization of the prompt γ-ray emission from the γ-ray burst of 6 December 2002. Nature 423:415–417

Costa E, Soffitta P, Bellazzini R (plus three authors) (2001) An efficient photoelectric X-ray polarimeter for the study of black holes and neutron stars. Nature 411:662–665

Curado da Silva RM, Caroli E, Stephen JB, Siffert P (2003) Polarimeter telescope concept for hard X-ray astronomy. Experim Astron 15:45–65

Dean AJ, Clark DJ, Stephen JB (plus seven authors) (2008) Polarized gamma-ray emission from the Crab. Science 321:1183–1185

EGSnrc NRC webpage (2007): http://www.irs.inms.nrc.ca/EGSnrc/EGSnrc.html

Fan Y-Z, Wei D-M, Xu D (2007) γ-ray burst ultraviolet/optical afterglow polarimetry as a probe of quantum gravity. MNRAS 376:1857–1860

Ferocci M, Costa E, Matt G (plus two authors) (1994) X-ray scattering polarimetry with scintillating fibers of different materials. Proc SPIE 2283:275–287

Fishman GJ, Meegan CA, Wilson RB (plus two authors) (1992) The BATSE experiment on the Compton Gamma Ray Observatory: Status and some early results. NASA CP-3137:26–34

Geant4 CERN webpage (2007): http://geant4.web.cern.ch/geant4

Heitler W (1954) Quantum theory of radiation. 3rd edition, Oxford University Press, Oxford

Kalemci T, Boggs SE, Kouveliotou C (plus two authors) (2007) Search for polarization from the prompt gamma-ray emission of GRB 041219a with SPI on *INTEGRAL*. Astrophys J Suppl 169:75–82

Kallman T (2004) Astrophysical motivation for X-ray polarimetry. Adv Space Res 34:2673–2677

Kanbach G and the MEGA collaboration (2001) MEGA-A next generation mission in medium energy gamma-ray astronomy. AIP Conf Proc 587:887

MCNP webpage: http://mcnp-green.lanl.gov/index.html (2007) LANL

Lazzati D (2006) Polarization in the prompt emission of gamma-ray bursts and their afterglows. New J Phys 8:131 (8 pages)

Larsson S, Pearce M (2004) POGO: The polarised gamma-ray observer. NIMP A 525:148–152

Lei F, Dean AJ, Hills GL (1997) Compton polarimetry in gamma-ray astronomy. Space Sci Rev 82:309–388

Lin, RP, Dennis BR, Hurford GJ (plus 63 authors) (2002) The Reuven Ramaty High-Energy Solar Spectroscopic Imager (*RHESSI*). Sol Phys 210:3–32

Mattox JR, Mayer-Hasselwander HA, Strong AW (1990) Analysis of the COS B data for evidence of linear polarization of VELA pulsar gamma rays. Astrophys J 363:270–273

Marshall HL, Corbet R, Takeshima T (plus six authors) (1998) RXTE observations of gamma-ray burst afterglows. Bull Am Astr Soc 30:762

McConnell ML, Ryan JM, Smith DM (plus two authors) (2002) *RHESSI* as a hard X-ray polarimeter. Sol Phys 210:125–142

McConnell ML, Macri JR, McClish M, Ryan JM (1999) Recent laboratory tests of a hard X-ray solar flare polarimeter. Proc SPIE 3764:70–78

McConnell ML, Smith DM, Emslie AG (plus three authors) (2004) Hard X-ray solar flare polarimetry with *RHESSI*. Adv Space Res 34:462–466

McConnell ML, Bloser PF (2006) Status and future prospects for γ-ray polarimetry. Chinese J Astron Astrophys Suppl 6:237–246

McGlynn S, Clark DJ, Dean AJ (plus six authors) (2007) Polarisation studies of the prompt gamma-ray emission from GRB 041219a using the spectrometer aboard *INTEGRAL*. Astron Astrophys 466:895–904

Nakajima M, Mihara T, Tamagawa T (plus two authors) (2004) Proportional counter with 7 µm carbon fiber and detection of X-ray polarization. X-ray Polarimetry Workshop, SLAC, Abstract under http://www-conf.slac.stanford.edu/xray_polar/reg/absdetail.asp?absID=48

Novick R (1975) Stellar and solar X-ray polarimetry. Space Sci Rev 18:389–408

O'Neill TJ, Akyuez A, Bhattacharya D (plus six authors) (1996) Tracking, imaging and polarimeter properties of the TIGRE instrument. Astron Astrophys 120:661–664

Produit N, Barao F, Deluit S (plus seven authors) (2005) POLAR, a compact detector for gamma-ray bursts photon polarization measurements. NIMP A 550:616–625

Ramsey BD, Austin RA, Decher R (1994) Instrumentation for X-ray astronomy. Space Sci Rev 69:139–204

Rutledge RE, Fox DB (2004) Re-analysis of polarization in the γ-ray flux of GRB 021206. MNRAS 350:1288–1300

Schönfelder V, Aarts H, Bennett K (plus 27 authors) (1993) Instrument description and performance of the Imaging Gamma-Ray Telescope COMPTEL aboard the Compton Gamma-Ray Observatory. Astrophys J Suppl 86:657–692

Schönfelder V, Kanbach G (2010) Imaging through Compton scattering and pair creation. ISSI SR-009:207–222

Suarez-Garcia E, Hajdas W, Wigger C (plus four authors) (2006) X-ray polarization of solar flares measured with *RHESSI*. Sol Phys 239:149–172

Weisskopf MC, Silver EH, Kestenbaum HL (plus two authors) (1978) A precision measurement of the X-ray polarization of the Crab Nebula without pulsar contamination. Astrophys J 220:L117–L121

Wigger C, Hajdas W, Arzner K (plus two authors) (2004) Gamma-ray burst polarization: Limits from *RHESSI* measurements. Astrophys J 613:1088–1100

Wilhelm K, Fröhlich C (2010) Photons—from source to detector. ISSI SR-009:23–54

Willis DR, Barlow EJ, Bird AJ (plus six authors) (2005) Evidence of polarisation in the prompt gamma-ray emission from GRB 930131 and GRB 960924 Astron Astrophys 439:245–253

— 35 —

Polarization measurements of the Cosmic Microwave Background

ETTORE CARRETTI[I] AND CYRILLE ROSSET[II]

Abstract

The polarization of the Cosmic Microwave Background (CMB) brings key information of early stages of the Universe such as the epoch of re-ionization and the inflation time. Its faint signal asks for both high sensitivity and high polarization purity which requires cutting-edge and specifically designed instruments. In this chapter we review the instruments used in CMB space experiments to date and give an overview of the next developments aimed at detection of the elusive B-mode, which will permit us to look back in time to inflation.

CMB polarization

The Cosmic Microwave Background radiation (CMB) is the relic emission of the Big Bang and carries the signature of the primeval Universe conditions. The study of its temperature anisotropies by spacecraft like *COBE*, *WMAP* and now *Planck* has been enabling high-precision measurements of cosmological quantities like the density of the Universe, its expansion rate, and the composition of the matter–energy which fills it (Spergel et al 2007).

The polarization is the next CMB "holy grail". One of its two components (the E-mode) allows the study of the re-ionization history of the Universe and the formation of first stars and galaxies. The other component (B-mode) looks even further back in time. It is the signature of the gravitational-waves background emitted by the inflation and enables a direct probe of this epoch.

The CMB polarization is weak and a tiny fraction of the temperature anisotropies (Figure 35.1). The anisotropy of the E-mode is about 0.3 μK on large angular scales (≈ 1 % of the anisotropies). Only first detections have been carried out so far by *WMAP* (Page et al 2007) and a few sub-orbital experiments (e.g., *DASI* and *QUaD*, see Pryke et al 2009, and references therein), while significantly more is expected from *Planck*. The B-mode is even fainter and only upper limits exist so

[I]Istituto Nazionale di Astrofisica – Istituto di Radioastronomia, Bologna, Italy
present address: CSIRO – Australia Telescope National Facility, Parkes, NSW, Australia
[II]Laboratoire de l'Accélérateur Linéaire, Université Paris Sud-11, Orsay, France
present address: Laboratoire Astroparticules et Cosmologie, Université Paris Diderot, France

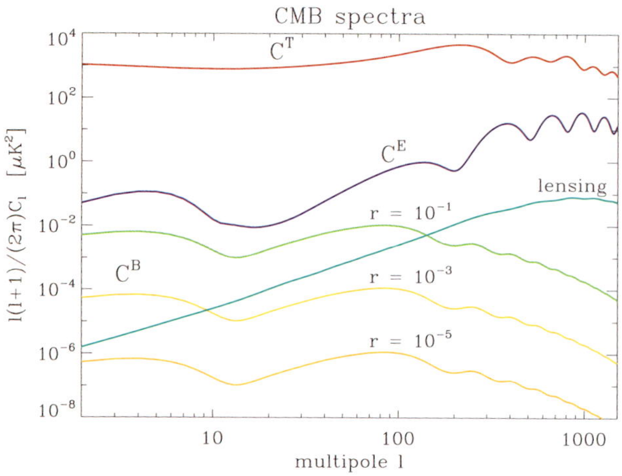

Figure 35.1: CMB angular power spectra assuming the *concordance* cosmological model after *WMAP* data (Spergel et al 2007): temperature (C^{T}), E-mode (C^{E}), and B-mode (C^{B}). The latter is reported for three cases of the tensor-to-scalar perturbation power ratio r, which measures the power of the primordial gravitational wave background. The contribution to C^{B} by gravitational lensing is also reported. The CMB has an intrinsic limit of $\Delta r \approx 10^{-5}$, corresponding to a B-mode anisotropy of $\approx$ 1 nK.

far. The emission range of interest runs from the present upper limit of $\approx$ 0.2 μK down to the CMB intrinsic limit of $\approx$1 nK, four orders of magnitude lower than the anisotropies of the total CMB emission (Amarie et al 2005). These unusual conditions of both low signal and polarization fraction require specifically designed experiments: low-noise detectors to reach the required sensitivity; high polarization purity to minimize leakages from the unpolarized component; a highly stable environment for high thermal stability and minimisation of spillover pollution. The latter requires a quiet space environment like the Sun-Earth L2 Lagrange point, where the telescope can be efficiently shielded from the Sun, Earth, and Moon, ensuring stable conditions. In fact, the best CMB measurements have been carried out from space so far.

The tiny signal also makes relevant the contamination by foreground emissions, mainly of the Galaxy. Their minimum occurs at 70 GHz to 100 GHz, which sets the best CMB frequency range, where radio and bolometric technologies overlap. In addition, the need to trace these foregrounds asks for observations at both lower and higher frequencies. Consequently, both these two technologies are required, and a wide frequency range (from 30 GHz to 300 GHz) must be covered.

In this paper we review radio and bolometric receivers used in CMB space experiments. Then we will introduce the new ideas under study to match the challenge posed by the faint B-mode, i.e., the aim of the next-generation CMB space missions.

Correlation receivers

Radiometric receivers are based on direct amplification at the frequency of the signal. Usually, low-noise amplifiers (LNA) of high electron-mobility transistors (HEMT) are used because of their high sensitivity especially if cooled down to cryogenic temperatures (typically 20 K or 80 K).

On the other hand, HEMT-LNAs are affected by gain instabilities. This adds an extra term to the white noise, called $1/f$ noise after its power spectrum (Wollack 1995). Its contribution increases with the integration time τ instead of decreasing as $\tau^{-1/2}$, and thus prevents any benefit of long exposures. A key parameter is the knee frequency, i.e., the frequency at which the 1/f noise component equals the white noise. This defines the longest useful integration time ($\tau_{\rm max} \approx 1/f_{\rm knee}$).

Total-power receivers feature $f_{\rm knee}$ in the range from 100 Hz to 1 kHz (at 100 GHz) resulting in too short integration times. Correlation receivers therefore are needed to correlate signals amplified by different LNAs and largely reduce the common mode and, in turn, the fluctuations induced by gain instabilities (Cortiglioni and Carretti 2006). In this case, $f_{\rm knee}$ typically ranges from 0.01 Hz to 0.1 Hz, which can keep the instrument stable for tens of seconds. This is sufficient for a space experiment. In fact, with an appropriate scanning strategy (e.g., by spinning and precessing the spacecraft) it is possible to modulate the signal on time scales from 1 min to 2 min, which enables data-reduction software to remove instability effects on longer times (cf., for example, Delabrouille 1998; Sbarra et al 2003). Correlation receivers can come in several flavours. In the following we will describe those of the last generation of space experiments: *Planck*–LFI, *WMAP*, and SPOrt.

CMB polarization observations

***Planck*-LFI.** ESA's *Planck* has been launched on 14 May 2009. The Low Frequency Instrument (*Planck*-LFI) consists of radiometers to cover three frequency bands from 30 GHz to 70 GHz and is expected to significantly improve the polarization results of *WMAP* (see Page 578).

Planck is primarily devoted to CMB temperature anisotropies and measures the total power of the two linear polarizations $|E_{\rm x}|^2$, $|E_{\rm y}|^2$. The sensitivity to both $|E_{\rm x}|^2$ and $|E_{\rm y}|^2$ gives it polarization capabilities. Stokes Q is measured by the difference[1]

$$Q = \frac{|E_{\rm x}|^2 - |E_{\rm y}|^2}{2} \quad . \tag{35.1}$$

Stokes U is obtained by a second receiver spun by 45°, as two receivers are needed to get Q and U.

The total-power measurement is obtained by a pseudo-correlation (which ensures the needed stability) based on the difference between the sky signal (≈ 2.725 K) and a reference-load of comparable intensity: a black body kept at 4 K by a liquid ^{4}He bath. Each linear polarization gathered by the optics is extracted by an orthomode-transducer (OMT) and combined with the reference load signal by a

[1] See Chapter 33 (Stenflo 2010) for the definition of the Stokes parameters.

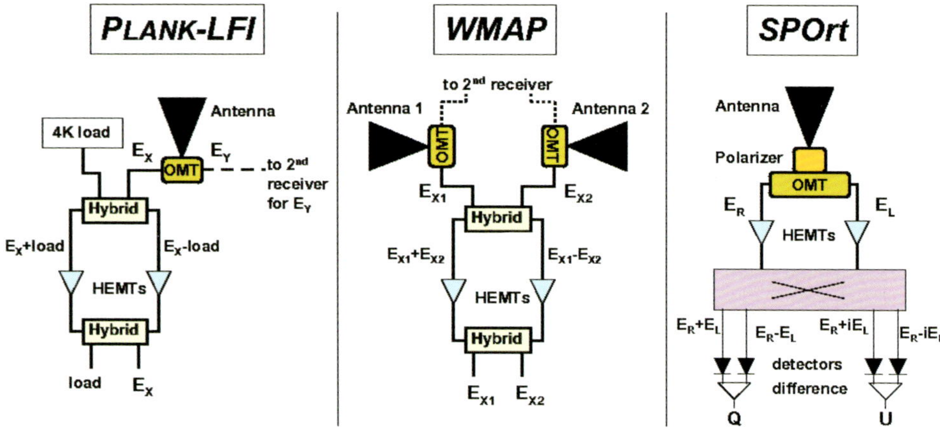

Figure 35.2: Schemes of *Planck*–LFI, *WMAP*, and SPOrt receivers.

180° hybrid (Figure 35.2, left panel). The two outputs, proportional to $E_{\mathrm{x}}^{\mathrm{sky}} + E_{\mathrm{x}}^{\mathrm{load}}$ and $E_{\mathrm{x}}^{\mathrm{sky}} - E_{\mathrm{x}}^{\mathrm{load}}$ (for E_{x} polarization) are then amplified and combined back by a second hybrid to $E_{\mathrm{x}}^{\mathrm{sky}}$ and $E_{\mathrm{x}}^{\mathrm{load}}$. After the detection, sky and load are differenced. The two signals are both amplified by the two LNA chains, and their difference cancels out most of the gain fluctuations (Mennella et al 2004).

WMAP. NASA's *WMAP* (Bennett et al 2003) has been observing since 2001 and covers five bands from 23 GHz to 94 GHz. Besides providing a precise measurement of the CMB anisotropy spectrum, it performed the first detection of the large-scale E-mode polarization (Page et al 2007).

As for *Planck*, its primary goal is the temperature anisotropy, but it can measure the polarization through the difference of $|E_{\mathrm{x}}|^2$ and $|E_{\mathrm{y}}|^2$. *WMAP* uses differential receivers forming the difference of signals collected by two independent telescopes looking at two sky positions spaced by 140°. The receiver scheme is similar to that of *Planck*-LFI, except that the signal from the second telescope is subtracted instead of the reference load (Figure 35.2, mid panel). The other polarization extracted by the two OMTs feeds a second receiver. The basic instrument element thus consists of a pair of radiometers which provide the anisotropy of two pixels 140° spaced for both the two polarizations. The equations to derive Stokes Q and U are given in Page et al (2007).

SPOrt. Unlike the other two instruments, the Sky Polarization Observatory (SPOrt) has been specifically designed for polarization measurements (Cortiglioni et al 2004). It was expected to fly on board the *ISS* and provide sensitivities in polarization comparable to *WMAP*, but it was stopped at the end of Phase B because of the *ISS* schedule uncertainties following the *Columbia* accident. The SPOrt receivers were designed with no compromise with other needs (like the detection of the temperature anisotropies) and the instrumental polarization performances of its prototypes are still the best ever achieved by CMB space experiments (Carretti et al 2004).

The architecture is based on the cross-correlation of the two circularly polarized components ($E_{\rm R}$ and $E_{\rm L}$) which simultaneously gives both the two Stokes parameters Q and U as

$$\begin{aligned} Q &= \mathrm{Re}(E_{\rm R} E_{\rm L}^*) \\ U &= \mathrm{Im}(E_{\rm R} E_{\rm L}^*) \quad , \end{aligned} \tag{35.2}$$

which (in contrast to the 50 % efficiency of both *Planck* and *WMAP*) gives a time efficiency of 100 %.

To realize that, the two linear polarizations $E_{\rm x}$, $E_{\rm y}$ are collected by an antenna (feed horn) while a quarter wave retarder (polarizer) and an orthomode-transducer (OMT) convert them into and extract $E_{\rm R}$ and $E_{\rm L}$ (Figure 35.2, right panel). After amplification, the two polarizations are processed by a correlator to give the four outputs $(E_{\rm R}+E_{\rm L})$, $(E_{\rm R}-E_{\rm L})$, $(E_{\rm R}+\mathrm{i}\,E_{\rm L})$, and $(E_{\rm R}-\mathrm{i}\,E_{\rm L})$, which, once detected and differenced, give Q and U as for Equations 35.2 (Cortiglioni et al 2004).

Instrumental polarization

The low CMB polarization fraction makes the instrumental polarization a major challenge when designing experiments, and several studies have been made to investigate it (e.g., Carretti et al 2001; Franco et al 2003; O'Dea et al 2007)). It is generated by leakages between the two polarizations which give a polarized signal even in the case of a purely unpolarized input. Here we only present the case of the SPOrt architecture, which features best performances (see Cortiglioni and Carretti 2006 for other architectures).

Major leakages occur in the optics, polarizer and OMT, where the two polarizations propagate together. The optics contribution is described by the so-called instrumental polarization beam Π (Carretti et al 2004). This is a complex combination of co- and cross-polar antenna patterns and can be significantly lower than 1 % of the main-beam peak in best devices. It features contributions of opposite sign which makes it negligible on large angular scales, but has to be considered when dealing with instrument resolution scales. The polarizer contribution is given by the differential attenuation between the two polarizations (Carretti et al 2001). Devices developed within the SPOrt programme give a fractional instrumental polarization of $\approx$0.1 %. The OMT is usually the most limiting device. Its contribution is related to the square-root of the cross-talk between the two extracted polarizations, but the exact value is a complex combination of device parameters (Carretti et al 2001). Devices with fractional instrumental polarization of $\approx$ 0.2 % have been achieved, with best performances of 0.02 % at 30 GHz (Carretti et al 2004).

Such levels are sufficient to detect the E-mode and the B-mode of the most optimistic r values. The use of cleaning algorithms will be necessary to detect the B-mode of lowest r.

Receivers based on bolometers

Measuring polarization with bolometers

The increasing sensitivity of bolometers over the years and the large band acceptance (typically 30 % of the central frequency) have made them the detectors of choice for CMB measurements. In recent years, all CMB bolometric experiments have used spider-web bolometers (SWB), developed at JPL. The absorbers of these bolometers have the shape of a spider-web, so that the absorbing part is only around 10 % of the detector surface. This reduces the cosmic-ray glitches in the signal and increases the detector sensitivity by lowering the heat capacity. The noise levels of these detectors reach the theoretical noise limit set by the photon noise of the background radiation. In space the background radiation can be made minimal, as there is no atmospheric emission, and there the photon noise is typically $\approx 40\ \mu\mathrm{K\,s}^{1/2}$.

However, by itself, a bolometer is not sensitive to the polarization state of the incoming radiation, as it absorbs all incoming power. The wish to take advantage of its high sensitivity to measure polarization of the CMB has driven considerable research in the last decade to develop bolometers sensitive to polarization. There are essentially two ways to do this: either by placing a polarizer on the radiation path, to select one polarization state, or by making the bolometer absorb only one direction of linear polarization.

With the first solution, half of the incoming power is reflected back, reducing the efficiency of the device. To make it more efficient, the reflected part must be directed toward a second detector, so that the two polarization states are measured simultaneously: this is done by an OMT, with a polarizer tilted at 45°. However, such a device is bulky as it uses the space of two detectors in the focal plane. The *Archeops* balloon experiment, primarily designed to map the CMB anisotropies, also successfully measured the Galactic dust polarization at 353 GHz (Benoît et al 2004), using three OMTs associated with six spider-web bolometers, revealing highly polarized regions.

The second method has made a breakthrough in recent years with the developments of the polarization-sensitive bolometer (PSB) device (Jones et al 2002). It consists of a pair of silicon nitride micromesh absorbers, separated by 60 μm, each coupled to one direction of linear polarization. This device, combining the advantages of OMT and spider-web bolometers, is much more compact, and the proximity of the two absorbers ensures very similar working conditions (background radiation, spectral bandpass).

The *Planck* High Frequency Instrument

Planck is a third-generation satellite to map the full sky at millimetre wavelength, launched in 2009. The High Frequency Instrument (HFI) carries 32 PSBs at four frequency channels, from 100 GHz to 353 GHz, and 20 spider-web bolometers at five frequencies, from 143 GHz to 857 GHz. Originally, it was designed to map only temperature anisotropies, but given both the theoretical and instrumental developments, it was decided, in the early stage of the project, to replace some spider-web bolometers by PSBs to measure polarization as well. The sensi-

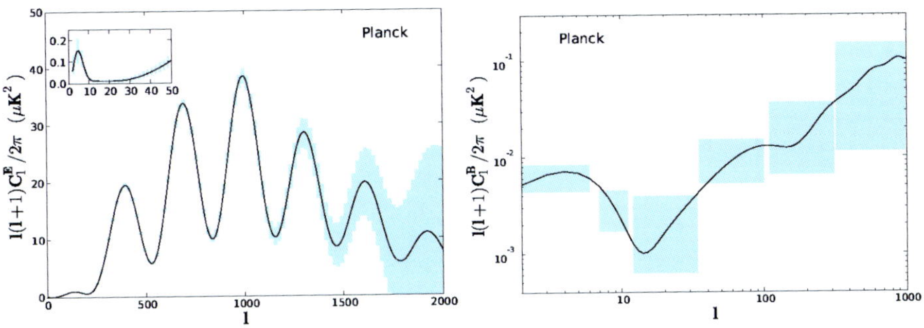

Figure 35.3: Expected E- (left) and B-mode (right) angular power spectra of *Planck* with uncertainty bars. The *WMAP concordance* model has been assumed (Spergel et al 2007) along with tensor-to-scalar perturbation power ratio $r = 0.1$.

tivity expected for the E-mode and B-mode polarization power spectra are shown in Figure 35.3 (it includes both HFI and LFI) .

To make a full-sky map, *Planck* scans the sky along nearly great circles, the axis of which are along the Sun-Earth direction. That way, the full sky is covered after around six months. This scanning strategy allows each of the spider-web bolometers to make a full map. However, this is not true for the PSBs, as each one measures only one direction of polarization (except near the ecliptic poles). Even the combination of the two detectors (the sum and the difference) gives only I and Q Stokes parameters (in a local frame). Therefore, because of the scanning strategy, there is a need to combine at least three detectors to build full-sky I, Q and U maps. This makes the *Planck*-HFI instrument sensitive to various systematic effects, as we shall see in the next section.

The systematic effects

The signal from a single detector of a PSB pair can be written in the following form:

$$m = \frac{g}{2} \int \left\{ (1+\eta) I \tilde{I} + (1-\eta) \left[Q\tilde{Q} + U\tilde{U} \right] \right\} \mathrm{d}\Omega \quad , \tag{35.3}$$

where $\tilde{I}$, $\tilde{Q}$ and $\tilde{U}$, are the beams in term of Stokes parameters, η is the detector cross-polarization and g is the calibration factor. Ideally, a PSB should have $\eta = 0$ and spider-web bolometers $\eta = 1$, but typical values are rather $\eta_{\rm PSB} \simeq 0.05$ and $\eta_{\rm SWB} \simeq 0.95$. Similarly, in ideal conditions, the intensity beam $\tilde{I}$ is Gaussian and symmetric, and $\tilde{Q}$ and $\tilde{U}$ are given by $\tilde{I} \cos 2\psi$ and $\tilde{I} \sin 2\psi$, respectively, where ψ is the angle of the PSB with respect to the sky.

It is clear that if the detectors have different properties, the reconstructed Stokes parameters will be a mix of the sky Stokes parameters. Since the deviation of the polarized anisotropy signal from that of the total CMB radiance is at most 10 %, the most damaging effect is a leakage from I to Q and U. Typical effects that can lead to such leakage are uncertainties in calibration, beam differences or time-constant differences (which lead to different effective beams, see Rosset et al 2007). Other effects, like angle or cross-polarization errors, will mix Q and U, which essentially leads to a leakage from E- to B-mode, since the former is much larger than the latter (O'Dea et al 2007; Hu et al 2003).

For example, a relative fractional calibration uncertainty of 1 % will lead to corresponding uncertainties in Q and U of typically 1 % of I, and hence to uncertainties in power spectra of typically $\Delta C_l^{\mathrm{E}} \approx 10^{-2} C_l^{\mathrm{TE}} + 10^{-4} C_l^{\mathrm{T}}$, where C_l^{TE} is the cross-spectrum between temperature and E-mode. At small angular scales (multipole $l \approx 1000$), where the E-mode is large, the relative uncertainty on the E-power spectrum is typically 1 %. On the other hand, the uncertainty on the B-mode at $l \approx 100$ (typical scale of primordial gravitational waves) is typically 10 times larger than the B-mode spectrum!

This example shows clearly that the measurement of the B-mode will require an exquisite control of systematic effects. Future experiment design must minimize such systematics, and the most important is certainly to avoid using different detectors to build a polarization map.

The future of CMB observation from space

The next step of CMB science is to measure the elusive B-mode and probe the inflation epoch. The CMB community has already moved on: NASA has started *Inflation Probe* with three feasibility studies, while the European community has proposed *B-POL* for the first call of ESA's Cosmic Vision Programme.

As discussed on Page 573, the major challenge of the B-mode is its faintness: its signal is far weaker than the temperature anisotropy and the other polarized component (E-mode). This requires both high sensitivity to detect it and polarization purity to limit contamination by the other two components. The basic idea is to take advantage of the best of each of the two technologies used so far, realizing an instrument with the purity of radiometer front-ends along with bolometer sensitivity.

The simplest architecture is based on front-ends made of a feed horn, collecting the two linear polarizations, and an OMT to separate them as in a radiometer. The signal is then detected by bolometers, instead of being amplified by HEMTs as in a normal radiometer, to take advantage of their better sensitivity. The two detected polarizations must then be subtracted to provide the Stokes parameter Q. A second receiver spun by 45° gets U. A major issue is the need for more detectors to make polarization maps.

An architecture that is a bit more complex uses the circular polarizations like the SPOrt receivers. This requires a polarizer between horn and OMT and a correlation device before the signals are detected with bolometers (Cortiglioni and Carretti 2006). This scheme has the advantage of simultaneously measuring both Q and U with just one unit, with clear advantage in term of calibration. Furthermore, this can reduce by a factor of two the number of receivers necessary to reach the wanted sensitivity, with clear benefit for a space mission in terms of mass and volume reduction. Architectures of this kind are under development for both ground-based and balloon-borne experiments (e.g., CLOVER, North et al 2007; and EBEx, Oxley et al 2004, for the linear polarization scheme) and are the baseline for *B-POL* (both schemes).

Another original architecture, combining advantages of bolometric detectors and interferometry, is also under investigation. It aims at combining the sensitivity of bolometers and low systematic effects of interferometers. The main advantage is to have a very basic optical system (no telescope is needed, feed horns are looking directly at the sky). MBI and BRAIN are two experiments based on this concept (Polenta et al 2007; Timbie et al 2006).

Furthermore, the required sensitivity asks for arrays of thousands of receivers and, in turn, mass production technologies. Studies to produce cheap and high-performance OMTs are ongoing, both in waveguide and planar technology. The latter has evident cost and mass production advantages, but the polarization purity performances are poorer than for the former. Efforts are thus aimed at both improving the performance with the planar technology and reducing the cost of the waveguide devices. The selection will depend on which one will reach the best cost–performance compromise.

Bibliography

Amarie M, Hirata C, Seljak U (2005) Detectability of tensor modes in the presence of foregrounds. Phys Rev D 72:123006-1–7

Bennett CL, Halpern M, Hinshaw G (plus 18 authors) (2003) First-year Wilkinson Microwave Anisotropy Probe (*WMAP*) observations: Preliminary maps and basic results. Astrophys J Suppl Ser 148:1–27

Benoît A, Ade P, Amblard A (plus 65 authors) (2004) First detection of polarization of the submillimetre diffuse galactic dust emission by Archeops. Astron Astrophys 424:571–582

Carretti E, Tascone R, Cortiglioni S (2001) Limits due to instrumental polarization in CMB experiments at microwave wavelengths. New Astron 3:173–187

Carretti E, Cortiglioni S, Sbarra C, Tascone R (2004) Antenna instrumental polarization and its effects on E- and B-modes for CMBP observations. Astron Astrophys 420:437–445

Carretti E, Cortiglioni S, Macculi C (plus 13 authors) (2004) High stability and sensitivity correlation polarimeters for CMB polarization measurements. Proc SPIE 5498:725–734

Cortiglioni S, Bernardi G, Carretti E (plus 28 authors) (2004) The Sky Polarization Observatory. New Astron 9:297–327

Cortiglioni S, Carretti E (2006) Basic considerations about experimental approaches to the B-mode of the CMB polarization. In Cosmic Polarization, ed R. Fabbri, Publisher Research Signpost, Trivandrum, Kerala, India, pp. 67–90

Delabrouille J (1998) Analysis of the accuracy of a destriping method for future cosmic microwave background mapping with the *Planck* surveyor satellite. Astron Astrophys 127:555–567

Franco G, Fosalba P, Tauber JA (2003) Systematic effects in the measurement of polarization by the *Planck* telescope. Astron Astrophys 405:349–366

Hu W, Hedman MM, Zaldarriaga M (2003) Benchmark parameters for CMB polarization experiments. Phys Rev D 67:043004-1–11

Jones WC, Bhatia RS, Bock JJ, Lange AE (2002) A polarization sensitive bolometric detector for observations of the Cosmic Microwave Background. Proc SPIE, Waikaloa, arXiv:astro-ph/0209132v1, 12 p.

Mennella A, Bersanelli M, Cappellini B (plus nine authors) (2004) The Low Frequency Instrument in the ESA *Planck* mission. AIP Conf Proc 703:401–404

North CE, Ade PAR, Audley MD (plus 43 authors) (2007) Clover – Measuring the CMB B-mode polarization. In: Proc Symp Space Terahertz Techn, Caltech, 21-23 March 2007, 6 p.

O'Dea D, Challinor A, Johnson BR (2007) Systematic errors in cosmic microwave background polarization measurements. MNRAS 376:1767–1783

Oxley P, Ade PAR, Baccigalupi C (plus 15 authors) (2004) The EBEx Experiment. Proc SPIE 5543:320–331

Page L, Hinshaw G, Komatsu E (plus 20 authors) (2007) Three-year Wilkinson Microwave Anisotropy Probe (*WMAP*) observations: Polarization analysis. Astrophys J Supl Ser 170:335–376

Pryke C, Ade P, Bock J (plus 28 authors) (2009) Second and third season *QUaD* CMB temperature and polarization power spectra. Astrophys J 692:1247–1270

Polenta G, Ade PAR, Bartlett J (plus 24 authors) (2007) The BRAIN CMB polarization experiment. New Astron Rev 51:256–259

Rosset C, Yurchenko V, Delabrouille J (plus four authors) (2007) Beam mismatch effects in cosmic microwave background polarization measurements. Astron Astrophys 464:405–415

Sbarra C, Carretti E, Cortiglioni S (plus four authors) (2003) An iterative destriping technique for diffuse background polarization data. Astron Astrophys 401:1215–1222

Spergel DN, Bean R, Doré O (plus 19 authors) (2007) Three-year Wilkinson Microwave Anisotropy Probe (*WMAP*) observations: Implications for cosmology. Astrophys J Suppl Ser 170:377–408

Stenflo JO (2010) Stokes polarimetry of the Zeeman and Hanle effects. ISSI SR-009:543–557

Timbie PT, Tucker GS, Ade PAR (plus 15 authors) (2006) The Einstein polarization interferometer for cosmology (EPIC) and the millimeter-wave bolometric interferometer (MBI). New Astron Rev 50:999–1008

Wollack EJ (1995) High-electron-mobility-transistor gain stability and its design implications for wide band millimeter wave receivers. Rev Sci Inst 66:4305–4312

VI. General Techniques

— 36 —

Calibration

MARTIN C.E. HUBER[I], ANUSCHKA PAULUHN[I], J. GETHYN TIMOTHY[II] AND ALEX ZEHNDER[I]

Abstract

Calibrating instruments for photon observations in space involves a number of parameters, most basic among them the pointing accuracy and stability. Wavelength accuracy is important as well. A particularly demanding and complex set of parameters to be determined concerns the responsivity (sometimes also referred to as effective area or detection efficiency) of the telescope-spectrometer combination. The responsivity is a function of wavelength, and for its determination the full set of geometric and spectro-optical properties of the system needs to be quantified. High photon arrival rates may also lead to nonlinearities that have to be assessed. A realistic physical and chemical description of an astronomical object, i.e., the goal of astrophysics, can only be reached with spectroradiometrically calibrated telescopes and spectrometers. We stress that there is no a priori celestial standard. A radiometric calibration must thus assure that observed spectral irradiances (or radiances) are measured in the units of the Système International, which in turn are defined by radiometric standards realised on Earth.

The need for Earth-based radiometric standards

An unambiguous deconvolution of observed signals into meaningful scientific data requires an accurate knowledge of the observing instrument's overall setting and performance. This includes the pointing behaviour, the exact knowledge of the wavelength observed, and the spectral responsivity (including its polarization dependence) — at all times. A proper radiometric[1] calibration of a spectroscopic telescope's responsivity must ultimately be based on a comparison with an Earth-based radiometric standard, because there are no a priori celestial standards. Cook (1994) has discussed the fundamental necessity of defining laboratory standards.

[I]PSI—Paul Scherrer Institut, Villigen, Switzerland

[II]Nightsen, Inc., Tiverton RI, USA

[1]In astronomy, the term photometry is often used when dealing with broadband intensity measurements; and measurements with higher spectral resolution are called spectrophotometry. However, in the terminology of the Système International (SI), photometry refers to intensity determinations that are relevant to human vision. In agreement with the use of the SI throughout this book, we exclusively use the terms radiometry or spectroradiometry.

Traditionally, astronomers have had their own calibration system: magnitudes of celestial objects. When, however, comparing computer-generated stellar atmosphere models with observed stellar spectra had become an astrophysical technique, the observed spectral irradiances had to be known in units of the Système International. Ground-based astronomers, therefore, started in the 1960s to compare the spectral irradiance of the star Vega (α Lyr, the main celestial standard) with that of laboratory standards — mainly with a remote black-body primary standard observed through the same telescope (cf., for example, Oke and Schild 1970, Hayes and Latham 1975).[2] Given this comparison with an Earth-based standard in the visible, the stellar model of α Lyr was then thought to provide a reliable extension into the infrared — until it was found that a dust shell, which had not been part of the stellar model, surrounds Vega, and produces an unexpected bump in its infrared spectral irradiance (Aumann et al 1984).[3]

Users of large astronomical spectroscopic telescopes, such as *HST, XMM-Newton, Chandra, Spitzer, Herschel* (and also users of the large ground-based facilities), as well as the users of Earth-observing instruments such as *Envisat*, may not be aware of the need to connect space-based observations for each wavelength to Earth-based radiometric standards. Radiometric calibration is taken care of for them by the support staff of these facilities: the instrument responsivity or the required metadata are folded into the pipeline of processing, so that data are seen in or can be converted to SI units as a matter of course.

Methods for the radiometric calibration of spectroscopic space telescopes

The methods for radiometric calibration of spectrometric space telescopes may be divided into four categories:
(i) Pre-flight calibration in the laboratory at the component or subsystem levels, or in end-to-end tests (i.e., from the entrance aperture to the detector output) using primary or transfer standards. (ii) Calibration in orbit by visible celestial standards — once these are established through observations or, if necessary, extended to other wavelength ranges by use of theoretical models. (iii) Calibration of an orbiting spectrometric telescope by comparing its observations with simultaneous ones made by a similar "under-flight" instrument on a sub-orbital rocket or the *Space Shuttle*. The latter instrument is then recalibrated in the laboratory after use in space to minimise uncertainties about changes in effective area. (iv) Calibration by use of a primary or transfer standard that is carried into and operated in orbit

[2]The statement re. the Système International should be taken with a grain of salt. Actually Oke and Schild (1970) stated: "... all equipment was designed to permit a determination of the absolute flux of the standard star α Lyr at 5556 Å in units of erg sec^{-1} cm^{-2} Hz^{-1}, since this fundamental parameter is also very uncertainly known." At that time expressing units in the cgs system was still common, and 'absolute flux' was used rather than 'irradiance'. Currently, improved measurements of this kind — striving for (relative) accuracies of 1 % of the absolute responsivity — are being performed in view of the requirements for the proposed *SNAP* mission (Smith et al 2009).

[3]Peterson et al (2006) have recently pointed out additional, albeit more subtle problems with modelling the spectral irradiance of Vega.

as part of the scientific payload.[4] Standards carried into orbit can usually not be used to assure an end-to-end calibration, but can be employed for monitoring the stability of a crucial part of an instrument.

A combination of all four techniques is often used as well. However, in view of (i) the multitude of observing modes and the concomitant, large parameter space that should be covered during a calibration, (ii) concerns about the differences between conditions during ground testing and those of in-orbit operations, and (iii) the inevitable changes in performance over time, instrument models that use well-characterised properties of optical components, coatings, and contaminants have been developed, which allow interpolation between measured benchmark values of instrument performance and prediction of behaviour in time (cf., Rosa 1997; Ballester and Rosa 1997).

Beyond the basic responsivity of an instrument the following must be determined: the plate-scale, the point-spread function (PSF, both on- and off-axis), flat-field maps, straylight and the occurrence of ghost images, radiometric non-linearities owing to pile-up and charge-transfer inefficiency, as well as the timing of the observations. Changes in instrument performance, which may happen between laboratory calibration and orbit, as well as changes occurring during orbital operations, are additional concerns. This leads to the need to monitor the contamination of optical surfaces and fatigue of detectors, and of assessing any effects from eclipses and large solar flares.

Radiometric standards

Ideally, there would be one primary standard[5] for absolute spectral radiometry in all wavelength domains. If we leave out the γ-ray and radio regions, which have their own calibration techniques, four primary standards are still needed to cover the X-ray, vacuum ultraviolet (VUV), visible and infrared wavelength domains: two emission or source standards, viz. black bodies and electron (or positron) storage rings, and two detector standards, namely cryogenic radiometers and double-ionisation chambers.

Black bodies (cf., Kaase et al 1984) are emission standards based on thermodynamics: the radiation from a black body is related to its temperature by the Planck law. The operational temperatures are normally those of melting points of appropriate metals.

[4] An additional concept, viz. having emission transfer standards on the *ISS*, or on a small calibration satellite orbiting the Earth, with the purpose of calibrating co-orbiting satellites in the extreme ultraviolet range has been suggested (Smith et al 1991). This idea has been taken up recently, however, for the case of a satellite observed from the ground; here a satellite-mounted laser helps to provide an accurate assessment of atmospheric absorption (Albert et al 2009).

[5] The criterion for a primary standard is that it gives results directly in SI units without the need for any calibration relative to the quantity being measured. One must be able to write the complete operating equation for the standard without any unknown or empirically determined constants or functions that depend upon the quantity being measured. There is no formal hierarchy of primary standards (Quinn 2000). However, the Bureau International des Poids et Mesures (BIPM, http://www.bipm.org) has the mandate to provide the basis for a single, coherent system of measurements — traceable to the International System of Units (SI) — throughout the world.

Electron (or positron) storage rings (cf., Madden et al 1992) are standards based on electrodynamics. The Schwinger (1949) formula is used to calculate the radiant flux of synchrotron radiation emitted by the accelerated charged particles. Inputs required for the calculation are the energy and current of the stored beam as well as the magnetic induction at the point where the radiation is emitted.

Cryogenic electrical substitution radiometers (ESR, cf., Möstl 1991) are detector standards that are based on accurate current and voltage measurements. The temperature increase of an irradiated cavity, cooled to liquid helium temperatures, is compared by a null method with the temperature increase caused by deposition of accurately measured electrical power into an equivalent cavity. Although corrections for reflections, scattering and diffraction must be made, cryogenic radiometers are accepted as the most accurate radiometric standards (Quinn and Martin 1990).

Rare-gas double-ionisation chambers are detector standards based on the fact that the photo-ionisation yield of rare gases is unity over the wavelength range 30 nm to 102 nm (Samson and Haddad 1974). Ionisation chambers are the only primary standards that have been used in space to date (cf., footnote 6).

Stock et al (1993) discussed the state of agreement between the radiometric scales realised by the first three primary standards mentioned above. They showed that black bodies and cryogenic radiometers produced and detected spectral radiant power with relative standard uncertainties of 0.08 % and 0.007 %, respectively. Thornagel et al (1996) measured emission from an electron storage ring with a cryogenic radiometer and found agreement between the two primary standards within 0.3 % at a photon energy of 15 keV, and 0.08 % for visible radiation.

Storage rings are in general large, complex, and expensive. Cryogenic radiometers and black bodies are less so, but are still inconvenient for routine laboratory applications. Therefore, it is often more practical to use simpler, stable sources and detectors, known as transfer standards, that are themselves calibrated in the laboratory by comparison with the primary standards. Examples include argon arcs (Bridges and Ott 1977), bare metal photoemissive diodes (Saloman 1978), D_2 lamps (Klose et al 1988), tungsten filament incandescent lamps (Waters et al 1988), hollow-cathode (Danzmann et al 1988; Hollandt et al 1994) and Penning (Heise et al 1994) discharges, as well as silicon trap detectors (Goebel et al 1996) and photodiodes (Kuschnerus et al 1998; Canfield et al 1998).

Connecting a transfer standard to the primary standards may require a chain of radiometric comparisons, sometimes also between detector and source standards. Examples are the transportable source standard described by Hollandt et al (2002) and the low light level radiometric calibration spectral source (RCSS) to be used in the calibration of the NIRSpec instrument on the *JWST* (Taubert et al 2009).

Discussing the details of radiometric calibration for the various wavelength domains would lead too far within the context of this book. Accordingly, we will restrict ourselves here to the basic aspects of calibration in order to give a glimpse of what is actually needed to provide scientifically meaningful data from photon-gathering tools in space.

An in-depth discussion of the various aspects of radiometric calibration and inter-calibration using the example of the instruments flown on the *SOHO* mission — in the laboratory, by underflights and inter-comparison of instrument responsivities in orbit, as well as of the strongly related issue of cleanliness — is given in a previous book (Pauluhn et al 2002) published in the same ISSI series. Wilhelm (2010) gave an update on this work by summarising the present state of quantitative solar spectroscopy.

The above-mentioned method of directly comparing an astrophysical object with a primary radiation standard on the ground could so far be extended to space observations only in exceptional cases.[6] Particularly in solar observations, transfer standards have been used successfully.[7] Along with the different techniques that have to be used to calibrate spectrometric space telescopes in different wavelength domains, there come differing accuracies in the diverse parts of the spectrum.

The connection of celestial standards to laboratory standards may be best in the visible and near-infrared wavelength ranges, where one can rely on extensive ground-based work that reduced relative uncertainties to about ± 2 % (Mégessier 1995). Next in claimed accuracy is the ultraviolet range, where white-dwarf models are believed to permit the transfer of calibration from the visible to the ultraviolet with a relative uncertainty of ca. 1 % (cf., e.g., Colina and Bohlin 1994). However, an absolute calibration of a rocket instrument for solar VUV observations, i.e., a calibration based on a comparison with a radiometric laboratory standard (before and after flight), but also including some model assumptions, has an assessed relative accuracy of better than ± 6 % (Didkovsky et al 2009); this is still somewhat worse than the apparent accuracy claimed for *HST* after use of stellar models. For X-ray observations the relative uncertainty of the responsivity (stated as 'absolute effective area') of *Chandra* ranges from 5 % to 20 %.[8] Many efforts are underway to reduce the uncertainties in the infrared, where a factor of two in pre-flight calibration was not uncommon in the case of *ISO* (cf., Lemke et al 2001). As an example, calibration-source developments are in progress for the *JWST* (cf., Taubert et al 2009), which are expected to enable a relative radiometric calibration accuracy in the laboratory of around ± 10 % or better (Hollandt 2009).

Monitoring spectroradiometric response in space

Changes in the effective area of spectrometric telescopes on spacecraft can be caused by exposure to contaminating pre-launch environments and material outgassed from the satellite or the instrument itself. Unless extreme care is exercised, degradation can take particularly dramatic forms for solar telescopes, which are exposed to the harsh electromagnetic emission of the Sun.

It is difficult to detect whether changes in spectroradiometric efficiency have occurred between laboratory measurements and first light in orbit, unless an onboard calibration source that tests the entire system from end to end or reliable celestial standards are available. In order to monitor changes that occur in space,

[6] Carlson et al (1984) and Wienhold et al (2002) describe and mention, respectively, rocket payloads that did include double-ionisation chambers.

[7] A deuterium lamp was used on the Solar Ultraviolet Spectral Irradiance Monitor (SUSIM) on *UARS* (Brueckner et al 1995), and electric substitution radiometers are part of total solar irradiance measurements, as discussed in Chapter 32 (Fröhlich 2010).

[8] http://cxc.harvard.edu/cal/, similar values apply to *XMM-Newton* (cf., http://xmm2.esac.esa.int/external/xmm_sw_cal/calib/). Note also that an International Astronomical Consortium for High Energy Calibration actively pursues the status of inter-calibration between the observatories *RXTE, Chandra, XMM-Newton, Integral, Swift* and *Suzaku* (cf., http://www.iachec.org/).

an appropriate set of non-variable objects is often defined, which are observed at regular intervals. However, this procedure is not without complications for solar and in general for short-wavelength observations, where objects tend to be variable; here on-board calibration systems or under-flights remain the best options.

Calibration as an integral part of a project

As mentioned at the beginning of this chapter, observing astrophysics must be based on the determination of spectral irradiances or spectral radiances. Such measurements require determination of the number, as well as the spectral, spatial and temporal distributions of photons that arrive at an observer's telescope from an object.[9] When combined with atomic and molecular data, spectroradiometrically accurate observations provide the essential means for deriving temperatures, densities, element abundances, ionisation stages, and flow and turbulent velocities in stars. More generally, if we also include data about matter in the condensed state, we can elucidate the physical structure and chemical composition of objects, and the processes that cause them to emit, scatter and absorb radiation and to evolve. Such observations provide inferences not only about plasma, gas, dust and solid bodies but also about the characteristics of interplanetary, interstellar and intergalactic space.

In many past space projects involving spectrometric measurements, conflicts between calibration and other programme goals could not be avoided. Before launch, calibration activities, especially end-to-end spectroradiometric measurements, are frequently the first to be sacrificed, when resources are constrained during the assembly, integration and verification phases. Early in the orbital phase of a mission, when observing time is still limited, science observations have in the past often been given priority over some of the most fundamental calibration work, sometimes with the consequence that the statistical uncertainty of the photon count of science data was much smaller than the uncertainty determined by the accuracy of the instrument parameters used to interpret the data.

In view of future space missions, which promise exquisite, sometimes subpercent accuracy in the measurement of cosmological parameters, stellar masses and ages, parallaxes and proper motions, a rigorous, firmly scheduled effort to achieve a reliable radiometric calibration of spectrometric space telescopes that is traceable to SI standards is well justified. Only in this way will it be possible to derive a physical and chemical description of adequate accuracy.

Bibliography

Albert J, Foster K, Battat J (plus four authors) (2009) Satellite-mounted light sources as photometric calibration standards. eprint arXiv:0908.3343physics

Aumann HH, Beichmann CA, Gillett FC (plus six authors) (1984) Discovery of a shell around alpha Lyrae. Astrophys J 278:L23–L27

[9]In spectral regions, where the photon aspect of electromagnetic radiation is less dominant, 'power' replaces 'photon number per unit time'.

Ballester P, Rosa MR (1997) Modeling echelle spectrographs. Astron Astrophys Suppl 126:563–571

Bridges JM, Ott WR (1977) Vacuum ultraviolet radiometry. III – The argon mini-arc as a new secondary standard of spectral radiance. Appl Opt 16:367–376

Brueckner GE, Floyd LE, Lund PA (plus two authors) (1995) Solar ultraviolet spectral-irradiance observations from the SUSIM-*UARS* experiment. Metrologia 32:661–665

Canfield LR, Vest RE, Korde R (plus two authors) (1998) Absolute silicon photodiodes for 160 nm to 254 nm photons. Metrologia 35:329–334

Carlson RW, Ogawa HS, Judge DL, Phillips E (1984) Absolute measurement of the extreme UV solar flux. Appl Opt 23:2327–2332

Colina L, Bohlin RC (1994) Absolute flux calibration of optical spectrometric standard stars. Astron J 108:1931–1935

Cook AH (1994) The observational foundations of physics. Cambridge University Press, New York

Danzmann K, Günther M, Fischer J (plus two authors) (1988) High current hollow cathode as a radiometric transfer standard source for the extreme vacuum ultraviolet. Appl Opt 27:4947–4951

Didkovsky LV, Judge DL, Wieman SR, McMullin D (2009) Minima of solar cycles 22/23 and 23/24 as seen in *SOHO*/CELIAS/SEM absolute solar EUV flux. Astron Soc Pacific Conf Series (in press)

Fröhlich C (2010) Solar radiometry. ISSI SR-009:525–540

Goebel R, Yilmaz S, Köhler R (1996) Stability under vacuum of silicon trap detectors and their use as transfer instruments in cryogenic radiometry. Appl Opt 35:4404–4407

Hayes DS, Latham DW (1975) A rediscussion of the atmospheric extinction and the absolute spectral-energy distribution of Vega. Astrophys J 197:593–601

Heise C, Kling R, Kock M (plus two authors) (1994) Radiometric characterization of a Penning discharge in the vacuum ultraviolet. Appl Opt 33:5111–5117

Hollandt J (2009) Personal communication

Hollandt J, Kühne M, Wende B (1994) High-current hollow-cathode source as a radiant intensity standard in the 40–125 nm wavelength range. Appl Opt 33:68–74

Hollandt J, Kühne M, Huber MCE, Wende B (2002) Source standards for the radiometric calibration of astronomical instruments in the VUV spectral range traceable to the primary standard BESSY I. ISSI SR-002:51–68

Kaase H, Bischoff K, Metzdorf J (1984) Licht-Forschung 6:29

Klose JZ, Bridges JM, Ott WR (1988) Radiometric calibrations of portable sources in the vacuum ultraviolet. J Res Nat Bur Std 93:21–39

Kuschnerus P, Rabus H, Richter M (plus three authors) (1998) Characterization of photodiodes as transfer detector standards in the 120 nm to 600 nm spectral range. Metrologia 35:355–362

Lemke D, Grözinger U, Schulz B (plus four authors) (2001) The pre-flight calibration of the ISOPHOT instrument. ESA SP-481:57–65

Madden RP, Canfield LR, Furst M (plus two authors) (1992) SURF II: Characteristics, facilities, and plans. Rev Sci Instrum 63:1594–1595

Mégessier C (1995) Accuracy of the astrophysical absolute flux calibrations: visible and near-infrared. Astron Astrophys 296:771–778

Möstl K (1991) in Landolt-Börnstein (New Series) Units & Fundamental Constants in Physics and Chemistry, Subvol. a, Bortfield J & Kramer B eds (Berlin: Springer-Verlag) p 2348

Oke JB, Schild RE (1970) The absolute spectral energy distribution of alpha Lyrae. Astrophys J 161:1015–1023

Peterson DM, Hummel CA, Pauls TA (plus eight authors) (2006) Vega is a rapidly rotating star. Nature 440:896–899

Pauluhn A, Huber MCE, von Steiger R, eds (2002) The Radiometric Calibration of *SOHO*. ISSI SR-002, 387 pp

Quinn TJ (2000) Personal communication

Quinn TJ, Martin JE (1991) Cryogenic radiometry, prospects for further improvements in accuracy. Metrologia 28:155–161

Rosa MR (1997) Physical modeling of scientific instruments. Astron Soc Pacific Conf Ser 125:411–414

Saloman EB (1978) Typical photoefficiency between 20–250 eV of windowless XUV photodiodes with tungsten and anodized aluminum oxide photocathodes. Appl Opt 17:1489–1490

Samson JAR, Haddad GN (1974) Absolute photon-flux measurements in the vacuum ultraviolet. J Opt Soc Am 64:47–54

Schwinger J (1949) On the classical radiation of accelerated electrons. Phys Rev 75:1912–1925

Smith AW, Woodward JT, Jenkins CA (plus two authors) (2009) Absolute flux calibration of stars: calibration of the reference telescope. Metrologia 46: S219–S223

Smith PL, Huber MCE, Parkinson WH (plus two authors) (1991) In-orbit extreme ultraviolet radiometric calibration of satellite instrumentation. In: Extreme Ultraviolet Astronomy, Malina RF, Bowyer S eds, Pergamon: New York pp 390–394

Stock M, Fischer J, Friedrich R (plus four authors) (1993) Present state of the comparison between radiometric scales based on three primary standards. Metrologia 30:439–449

Taubert RD, Monte C, Baltruschat C (plus seven authors) (2009) The spectral photon flux of the radiometric calibration spectral source for the NIRSpec instrument of the James Webb Space Telescope. Metrologia 46:S207–S212

Thornagel R, Fischer J, Friedrich R (plus three authors) (1996) The electron storage ring BESSY as a primary standard source – a radiometric comparison with a cryogenic electrical substitution radiometer in the visible. Metrologia 32:459–462

Waters, WR, Walker JH, Hattenburg AT (1988) The NBS scale of radiance temperature. In: Noncontact Temperature Measurement, ed MC Lee, NASA Washington DC, p 64 (SEE N88-23895 17–29)

Wienhold FG, Anders J, Galuska B (plus seven authors) (2002) The solar package on *ISS*: SOL-ACES. ISSI SR-002:355–360

Wilhelm K (2010) Quantitative solar spectroscopy. Astron Nachrichten 331:502–511

— 37 —

Cryogenics in space

Nicola Rando[I]

Abstract

Cryogenics plays a key role on board space-science missions, with a range of applications, mainly in the domain of astrophysics. Indeed a tremendous progress has been achieved over the last 20 years in cryogenics, with enhanced reliability and simpler operations, thus matching the needs of advanced focal-plane detectors and complex science instrumentation. In this article we provide an overview of recent applications of cryogenics in space, with specific emphasis on science missions. The overview includes an analysis of the impact of cryogenics on the spacecraft system design and of the main technical solutions presently adopted. Critical technology developments and programmatic aspects are also addressed, including specific needs of future science missions and lessons learnt from recent programmes.

Introduction

H. Kamerlingh-Onnes liquefied ^{4}He for the first time in 1908 and discovered superconductivity in 1911. About 100 years after such achievements (Pobell 1996), cryogenics plays a key role on board space-science missions, providing the environment required to perform highly sensitive measurements by suppressing the thermal background radiation and allowing advantage to be taken of the performance of cryogenic detectors.

In the last 20 years several spacecraft have been equipped with cryogenic instrumentation. Among such missions we should mention *IRAS* (launched in 1983), ESA's *ISO* (launched in 1995) (Kessler et al 1996) and, more recently, NASA's *Spitzer* (formerly *SIRTF*, launched in 2006) (Werner 2005) and the Japanese mission *Akari* (IR astronomy mission launched in 2006) (Shibai 2007). New missions involving cryogenics are the ESA missions *Planck* (dedicated to the mapping of the cosmic background radiation) and *Herschel* (far infrared and sub-millimetre observatory), carrying instruments operating at temperatures of 0.1 K and 0.3 K, respectively (Crone et al 2006). In the 10 K to 100 K temperature range, many missions are operational, including military reconnaissance satellites (*Helios*), Earth observation and meteorological satellites (*Meteosat* Second Generation), with IR detectors operating at about 85 K (Cihlar et al 1999). For science missions, we

[I]Science Projects Department, ESA, ESTEC, Noordwijk, The Netherlands

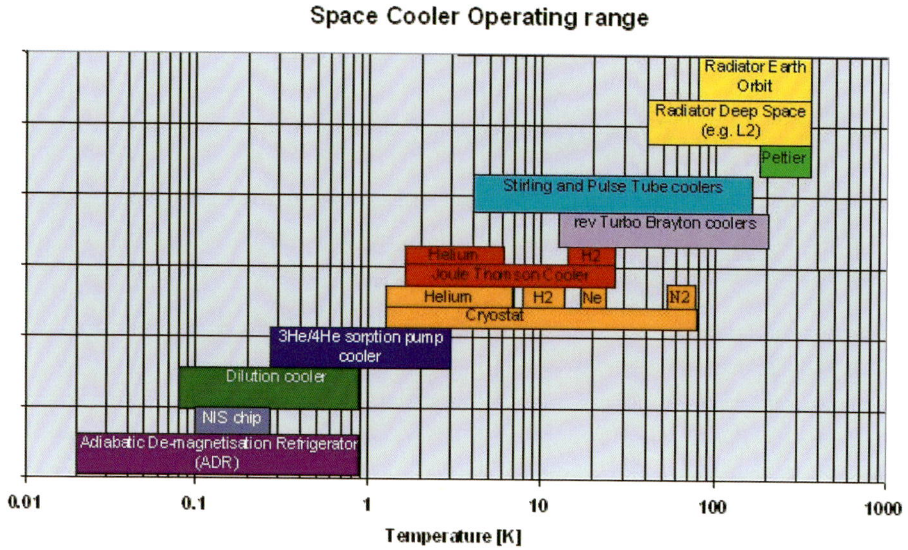

Figure 37.1: Operating temperature range of different space coolers.

should quote the ESA mission *INTEGRAL* (launched in 2002), with Stirling coolers maintaining the detectors of the spectrometer at 80 K (Winkler 2004). Different cryogenic techniques are available depending on requirements, in particular operating temperature and cooling power. Figure 37.1 provides an overview of the different cooling approaches as a function of base temperature. The different cryogenic techniques and the required technology developments will be addressed in the following sections.

Applications of cryogenics in space

Science missions in operations / post-operations

The first cryogenic missions, such as *IRAS*, launched in 1983 (Langford et al 1983), *COBE* (COBE 1992), launched in 1989, and *ISO*, launched in 1995 (Kessler et al 1996), were based on liquid He cryostats, with the bath temperature regulated by adjusting the vapour pressure. Lifetime was correspondingly limited by the amount of cryogen, typically to about 12 to 18 months. More recently, the same approach has been used by *Spitzer* (Figure 37.2, left). *Spitzer*, thanks to an optimised cryogenic system (passive radiation, use of helium gas enthalpy, orbit choice), was designed to provide a minimum lifetime of 2.5 years, using only 360 litres of superfluid He. *WIRE* (launched in 1999 and lost during commissioning), had a two-stage, solid-hydrogen cryostat maintaining the optics below 19 K and the Si:Ga detector array below 7.5 K, with a lifetime of four months (Elliott et al 1994). A hybrid, solid-neon / liquid-helium design was adopted by *ASTRO-E I*

Figure 37.2: Left: *Spitzer* spacecraft during final testing (NASA). Right: *Herschel* spacecraft (artist view – ESA).

and *ASTRO-E II*, renamed *Suzaku* (launched in 2000 and 2005, respectively), two ISAS (Japan) satellites for X-ray astronomy (Ogawara 1998). The *ASTRO-E* spectrometer used calorimeters operating at 65 mK via an adiabatic demagnetisation refrigerator (ADR), hosted in a liquid-helium cryostat and thermally shielded by a solid-neon cooled outer jacket. Predicted lifetime was two years. Finally, the Japanese mission *ASTRO-F* (an IR observatory launched in 2006 and renamed *Akari*) represents a follow-on from the *IRAS* mission, performing an all-sky survey between 6 µm and 180 µm, with a 180 litre He cryostat, working in combination with Stirling coolers (Shibai 2007).

In the temperature range above 50 K, we should mention NICMOS, the Near Infrared Camera and Multi-Object Spectrometer (0.8 µm to 2.5 µm), installed on *HST* in 1997. The detectors were cooled down to 60 K via 120 kg of solid nitrogen, replaced in 2002 by a reverse turbo-Brayton cooler. In this temperature range mechanical coolers (turbo-Brayton, Stirling), replaced the use of cryogens (Thompson et al 1998). In fact, ESA's *INTEGRAL*, dedicated to spectroscopy and imaging between 15 keV and 10 MeV, uses space-qualified Stirling cryo-coolers (Winkler 2004), maintaining the germanium detectors of the spectrometer at 82 K. Similarly NASA's *RHESSI* mission, launched in 2002, also uses a Stirling cooler, maintaining the Ge detectors at 80 K (Hannah et al 2007).

Among the main recent cryogenic missions we should mention ESA's *Planck* and *Herschel*. *Planck*'s objective is to map the temperature anisotropies of the cosmic microwave background over the whole sky (Crone et al 2006). The cryogenic system of *Planck* is based on passive pre-cooling to 50 K to 60 K (telescope), cooling to 20 K via an H_2 Joule-Thomson (JT) cooler (adsorption compressors), cooling to

4 K with a He Joule-Thomson cooler (mechanical compressors) and final cooling (bolometer detectors) to 0.1 K with an open-loop dilution refrigerator (DR). Its nominal mission lifetime is 15 months. The *Herschel* mission (Figure 37.2, right side) is a far-infrared and sub-mm multi-user observatory (85 μm to 600 μm), based on a superfluid helium dewar at 1.65 K (lifetime of 4.5 years) (Pilbratt 2008). The scientific goals will be achieved with three instruments operating at 2 K (heterodyne receiver based on SIS mixers, see Chapter 31, Wild 2010), 1.7 K (photometer based on photo-conductors) and 0.3 K (photo-spectrometer using bolometers and a ^{3}He sorption cooler, respectively). ESA developed the two spacecraft together, and launched them to L2 on a single *Ariane 5* in May 2009.

Science missions under development and under study

An important role will be played by NASA's *JWST* (Gardner et al 2006). This large programme, scheduled for launch in 2013, is based on a 6 m diameter passively cooled telescope enabling observations in the near and medium-IR, from 1 μm to 30 μm. The science instruments are an NIR camera and two Europe-led instruments, ESA's NIR low-resolution spectrograph (NIRSpec) and a MIR camera-spectrograph combination (MIRI). The first two instruments operate at 30 K (passive cooling), while MIRI makes use of a Joule-Thomson cooler, pre-cooled by a 3-stage pulse tube to achieve a base temperature of about 6 K. Other space missions with cryogenic equipment are being studied. On the ESA side we should mention *Darwin*, a near-infrared space observatory dedicated to the search of planets. In Japan the mission *SPICA* (a medium- and far-IR observatory scheduled for launch in 2017) is under study, assuming a cryogenic chain entirely based on mechanical coolers, with a 3.5 m diameter mirror operating below 5 K (Sugita et al 2006). Recently, ESA's *XEUS* (Rando et al 2006) and NASA's *Constellation-X* (Petre et al 2006) have been merged into *IXO*, an ESA-JAXA-NASA project. An overview of the main scientific cryogenic space programmes is given in Table 37.1.

Earth-observation satellites and telecom applications

Earth-observation missions require cryogenics because of the utilisation of medium infrared detectors (typically operating around or just below 100 K). Although in some cases cryogens have been used, for instance on NASA's *UARS*, launched in 1991 (Reber 1990), a growing number of missions are making use of mechanical coolers. ESA's *ERS-1 & -2* (launched in 1995 and 1999) and *Envisat* (launched in 2000) make use of Stirling coolers (Aminou et al 1998). The next-generation missions are likely to take advantage of the progress made in the field of pulse-tube refrigerators (Barnes et al 1998).

The recent progress in the area of high-temperature superconductors may open new perspectives for radio-frequency superconducting devices (filters, delay lines, resonators), of interest also for space applications. For instance, YBCO (yttrium-barium-copper-oxide) at a temperature of 77 K and at a frequency of 10 GHz has a surface resistance which is 30 times lower than that of copper, implying the possibility to improve the energy efficiency of telecommunications systems or to significantly reduce their size and weight (Chaloupka et al 1993).

Table 37.1: Summary of main scientific cryogenic space programmes.

Mission	Science domain	Type/class	Launch year	Cryogenic system	In-flight T/K	Lifetime	Orbit	Status
IRAS (NASA, NIVR, SERC)	IR	satellite (surveyor)	1983	^4He (λ) cryostat	3	290 d	near polar	post-ops.
COBE (NASA)	IR	satellite (surveyor)	1989	^4He (λ) cryostat	1.4 – 1.6	305 d	near-Earth	post-ops.
ISO (ESA)	IR	satellite (observat.)	1995	^4He (λ) cryostat	1.8	840 d	HEO	post-ops.
HST (NASA)	NIR	Nicmos, instrument	1997	sN_2 cryostat	60	700 d	LEO	post-ops.
WIRE (NASA)	IR	satellite (surveyor)	1999	dual, sH_2 cryostat	< 7.5	120 d	LEO	post-ops/lost
INTEGRAL (ESA)	γ-ray	instrument (observat.)	2002	Stirling cooler	85	2–5 a	HEO	operat.
RHESSI (NASA)	solar phys.	satellite (observat.)	2002	Stirling cooler	85	> 5 a	LEO	operat.
Spitzer (NASA)	IR	satellite (observat.)	2006	^4He (λ) cryostat	1.4	2.5 a	Earth trailing	operat.
Rosetta (ESA)	planet. sci.	instrument (probe)	2003	Stirling cooler	80	10 a	heliocentr.	operat.
Suzaku (ISAS, NASA)	X-ray	satellite (observat.)	2005	sNe+^4He cryost.+ADR	0.065	730 d	LEO	lost
Akari (ISAS)	IR	satellite (observat.)	2006	^4He (λ) cryost.+ cooler	1.8	550 d	LEO	operat.
Herschel (ESA)	FIR	satellite (observat.)	2009	^4He (λ) cryost.+^3He SC	0.3 & 1.7	4.5 a	Sun-Earth L2	developm.
Planck (ESA)	sub-mm	satellite (surveyor)	2009	H_2&^4He JT + DR	0.1 & 20	460 d	Sun-Earth L2	developm.
JWST (NASA)	NIR	satellite (observat.)	2013	passive rad.+ cooler	4 – 40	5–10 a	Sun-Earth L2	developm.
SPICA	MIR-FIR	satellite (observat.)	2017	passive rad. + cooler.	4.5	5 a	Sun-Earth L2	study
IXO (ESA, JAXA, NASA)	X-ray	instrument (observat.)	2020	Stirling cool.+ADR	0.05–0.3	> 5 a	Sun-Earth L2	study

Technology-demonstration missions

A number of space missions have been dedicated to the validation of specific cryogenic technological issues. In particular NASA has flown a number of cryogenic payloads on several space shuttle flights (*STS-77, 79* and *87*), including studies of superfluid helium (Chui et al 1994), test facilities (Bowman et al 1997) and coolers (Lipa et al 1996).

Cryo-electronics and large-scale applications

Electronics systems operating at cryogenic temperature have found application on board spacecraft. For example, in the case of ISOPHOT (on board *ISO*), a specific MOS IC (metal oxide semiconductor integrated circuit) was developed for the instrument front-end which demonstrated low-noise, low-dissipation, multiplexed operations, operating at cryogenic temperature (Dierickx et al 1996).

Another category of cryogenic device is the SQUID (superconducting quantum interference device). Such a device is the most sensitive magnetometer known to date, reaching sensitivities of the order of a few femtotesla per square root of hertz at frequencies of a few hertz. SQUIDs are used in very sensitive gravity gradiometers, fast digital electronics and detector read-out circuitry (Braginski 1999). Large-scale applications involving cryogenics in space include energy storage and gas storage for life support systems and propulsion purposes. Energy can be stored in the form of an intense magnetic field generated by a superconducting magnet to cope with sudden demands for large amounts of energy, which cannot be supplied by the combination of solar cells and batteries used on board spacecraft. The storage of gases in liquefied form is well known and offers obvious advantages for storage of large amounts in a limited volume tank. In addition this technique is relevant to life support systems for inter-planetary missions (e.g., future missions to Mars) or for the *ISS* (Kohout 1989). Required operating temperatures are less demanding in comparison with other cryogenics applications, depending on the different gases (H_2, N_2 and O_2), ranging between 20 K and 90 K.

Cryogenics and spacecraft engineering

Architecture of cryogenic spacecraft

A spacecraft is usually composed of a satellite bus (or service module) and a payload module. The payload module carries one or several instruments used to process signals coming from Earth or space, with either individual or common optics (main telescope) (Wertz and Larson 1999). Cryogenic installations have a strong impact on the architecture of both spacecraft and instrument. Key design factors are:

- A cooling system must be used with performances (heat lift, base temperature and hold time) compatible with both instrument requirements and spacecraft resources.

- The low-temperature equipment must be properly supported, insulated from the room temperature satellite bus and protected from solar, Earth or planet radiation.
- The cold parts have to be accessible (e.g., optical access to the focal plane) and wiring needs to be routed between cold payload and satellite bus.
- Cryogenic ancillary equipment is required to operate the cryogenic payload (e.g., heat links, heat switches, filters, thermometry).
- Activities related to assembly, integration and test (AIT) of cryogenic equipment must be taken into account already during the design phase.
- The complete system must survive the vibrations induced by the launcher.
- The cooler has to operate in zero gravity for a period of the order of a few years.
- The lifetime of the equipment should exceed the mission duration.

Space coolers

Coolers provide a cold heat sink, by removing the heat in the cold area and dissipating it into the warm area. Either the energy is directly radiated to space (via **radiators**), or work has to be performed to pump the energy between two temperature levels, from a cold to a warm level to then be more easily radiated away. Such an operation can be done using an open-cycle configuration or in a closed cycle.

The open cycle corresponds to the use of **stored cryogens**, where the work is performed before the mission, on the ground, by a liquefier. The cold heat sink is provided by evaporation of liquid or solid cryogens. The closed cycle corresponds to the use of **mechanical coolers**, where the work is done continuously during operations. Existing space coolers can provide about 1 W of cooling power in the temperature range 50 K to 100 K (Stirling coolers, pulse tubes), about 100 mW in the range 15 K to 20 K (double-stage Stirling), or a few milliwatt at 4 K (Joule Thomson). Very low-temperature coolers (e.g., ^{3}He cryo-sorption refrigerators, dilution or ADR) rely on the pre-cooling systems mentioned above to reach temperatures between 50 mK and 1 K. For base temperatures above 50 K a single stage can be sufficient. At lower temperature, multiple-stage systems with various types of coolers have to be used.

Types of coolers

Radiators are the most efficient, simplest and reliable space coolers. They are based on the fact that all objects emit infra-red radiation proportionally to their area S, emissivity ε, and to the fourth power of their temperature T, and on the fact that the ambient temperature (deep space) is very cold (black-body at $T_0 = 2.73$ K). The net cooling power is thus

$$Q_{\mathrm{rad}} = \sigma\, S\, F\, \varepsilon\, (T^4 - T_0^4) \cong \sigma\, S\, F\, \varepsilon\, T^4 \quad , \tag{37.1}$$

where σ is the Stefan constant and $F \cong 1$ is the shape factor (Bard 1984). Radiators are efficient above 100 K, but have limited performance at low temperatures and limitations related to their size (a few square metres on a spacecraft) and orientation: they need to be shaded from the solar radiation (1.4 $\mathrm{kW\,m^{-2}}$), and from the Earth or planet infra-red and albedo radiation (about 300 $\mathrm{W\,m^{-2}}$ for the Earth) and looking at dark space in order to efficiently radiate. For low Earth orbits, the temperature limit is about 100 K, with a cooling power lower than 1 $\mathrm{W\,m^{-2}}$. For far-away orbits (e.g., Lagrangian points), the radiator architecture becomes simpler, with lower temperature and better performance. In the case of *Planck*, it is expected to have a cooling power of about 2 W at about 50 K.

A **stored-cryogen cooler** is composed of a cryogen tank, a vacuum vessel (isolating the cryogen tank before and during launch), filling and venting lines, heat shields / multilayer insulations (MLI), and some interface or volume for instrument accommodation. In the absence of gravity, the fluid needs to be maintained inside the tank by a phase separator. Space dewars, to withstand the launch loads, have specific supports and a separate venting line to efficiently use the gas enthalpy to cool the shields, and to release the gas without applying momentum to the spacecraft. In space it is also possible to cool the whole vacuum vessel by radiation to space and by the venting line (the *Herschel* vessel is expected to be at 77 K). The bath equilibrium pressure is not 1 bar as on ground, but it is vented to the space vacuum. This allows to pump on the cryogen bath and to use solid cryogens, which usually have a sublimation heat much larger than their latent heat. The proper design of the exhaust nozzle allows tuning the base temperature (vapour pressure) of the cryogen bath by adjusting the pressure drop. The volume of cryogen to be carried depends on the mission duration and on the heat input. The choice of the cryogen to be used depends on the base temperature required. The most widely used are superfluid or supercritical helium, solid H_2 and solid Ne. Bi-cryogen systems, such as N_2/He, or H_2/He, optimise the cooler mass, but their design is more complex, as all lines and valves have to be doubled. Coolers based on cryogens are cooled on ground and topped up just prior to the launch, while the low vapour-pressure is maintained by pumping on the bath through the vent line.

In a **mechanical cooler** mechanical work is transformed into refrigerating power. Active coolers can be categorised into *regenerative cycles* (Stirling, pulse tube, Gifford coolers) and *recuperative cycles* (Joule-Thomson or Brayton coolers). *Regenerative coolers* are based on a pressure wave generated by a compressor (usually mechanical), and a cold finger, using a mobile (Stirling, Gifford) or a fixed (pulse tube) regenerator. The heat is extracted at the cold end when the gas expands, and rejected at the warm end when the gas is compressed. The recuperative cycles use the enthalpy difference between high- and low-pressure gas. The Brayton-cycle coolers use a cold turbine to expand the gas, whereas the Joule-Thomson coolers use the expansion through an orifice, and the properties of real gas. The JT cooler (normally coupled to Stirling units) is less efficient than the Brayton, but it is simpler. A lifetime of five years is a typical requirement for most space applications (Ross 1990). A typical power allocation for a space cryo-cooler is between 50 W and 200 W. Most mechanical coolers (typically based on the Stirling cycle) have an efficiency of the order of 2 % to 5 % of the ideal Carnot cycle, implying a cooling power of a few milliwatt at 4.2 K with an input power of about

Figure 37.3: 4 K sorption-cooler prototype under development (Univ. Twente).

100 W. Mass also is a critical parameter in the evaluation of space coolers, since the typical allocated values are of the order of 100 kg to 150 kg. Coolers should not export vibrations degrading the performance of the instruments. To date most space mechanical coolers are based on the Stirling cycle or on the Joule-Thomson expansion, but more recently pulse-tube refrigerators have become an interesting alternative. Finally, we should mention the closed-cycle, hydrogen or helium-filled sorption coolers, an interesting alternative to mechanical coolers, such as the one developed by JPL (US) for the *Planck* mission (base temperature of 20 K, no moving parts) (Jones et al 1990) and a similar unit being developed by ESA (Figure 37.3) (Burger et al 2002).

Very low-temperature coolers ($T < 1$ K)

In many scientific satellite applications it is necessary to achieve temperatures well below 1 K. Such a temperature (cf., Figure 37.1) can be achieved by using closed-cycle ^{3}He sorption coolers (down to 250 mK), by dilution refrigerators (50 mK to 100 mK) and by adiabatic demagnetisation refrigerators (50 mK to 300 mK).

^{3}He sorption coolers offer simplicity of operations, lack of moving parts and closed-cycle operations with an efficient duty cycle (^{3}He condensation phase vs. hold time at base temperature). Typical cooling power is of the order of 10 µW at 300 mK. Sorption coolers have flown aboard balloons (Boomerang, Maxima, Archeops), sounding rockets and on the satellite *SFU* (IRTS, Infrared Telescope in Space instrument) (Freund et al 1998). Sorption coolers are also used on board *Herschel* instruments SPIRE and PACS.

Dilution refrigerators, based on the quantum-mechanical properties of ^{3}He–^{4}He mixtures, are routinely used on ground to achieve temperatures below 100 mK, with cooling power exceeding 100 μW. This technique has been adapted for space applications for *Planck* (cooling power ≈ 0.1 μW, Benoit and Pujol 1991), avoiding circulation pumps by working in open loop, thus requiring a very large amount of gas mixture with a lifetime limited by the gas reservoirs. A closed-loop approach is under study.

Adiabatic demagnetisation refrigerators have been already used on board sounding rockets and scientific satellites (*ASTRO-E*) (Hagmann and Richards 1995). They produce base temperatures of the order of 50 mK to 100 mK by reducing the entropy associated with the electronic spins of the atoms of paramagnetic salt, aligned by a magnetic field of the order of a few tesla. Cooling powers of about 1 μW to 10 μW are achieved (Pobell 1996). ADRs offer very low base temperatures with simple operations and good duty cycle efficiency. The main challenges are the need for large magnetic fields and for high performance and high-reliability thermal switches. The use of an ADR system is proposed for a number of future science missions, including *IXO* and *SPICA*.

Solid-state coolers, equivalent to Peltier elements but operating below 1 K, are being investigated. Based on normal metal-insulator-superconductor (NIS) junctions, they provide cooling of the lattice by relying on phonon-electron coupling and removing the hottest electrons present in the normal metal electrode of the device (Nahum et al 1994). Cooling of membranes from 0.3 K to 0.1 K has already been achieved. Such coolers are developed with the aim of building self-cooling detectors (e.g., bolometers), capable of operating with simpler pre-coolers (e.g., ^{3}He sorption coolers instead of ADR).

Insulation technology and ancillary cryogenic equipment

Thermal insulation is used to limit the heat loads (due to conductive and radiative coupling) on the cold stage to a level compatible with the heat lift of the cooler.

Low-conductive supports are key elements of the structure of any cryogenic payload, in view of the launch loads. Steel and titanium alloys are still frequently utilised, but Kevlar, glass and carbon fibre based materials find ever more applications. Struts, tension straps, are used to support dewars and cold stages. Kevlar strings are preferred to ensure a high degree of stiffness combined with small contact surfaces. A review of cryogenic structural supports and materials can be found in Reed and Golda (1990).

Multilayer insulation is commonly used on ground as well as in space. MLI consists of a stack of polyester (Mylar) or polyamide (Kapton) foils which are embossed, crinkled or separated by a spacing material, such as a thin net. The foils are aluminised on one or both sides to reduce the radiative transfer, thus protecting the spacecraft from intense solar radiation and insulating the low-temperature stages (Bapat 1990).

V-groove shields are based on a few (< 5) angled and highly-reflective solid plates, open to space, rejecting radiation after a number of reflections between the angled shields (rather than trapping it between layers as with MLI). V-grooves play a critical role on *Planck*. Additional details can be found in Bard (1987).

Cryogenic ancillary equipment plays a critical role during ground testing as well as in flight. We list here the key items required on board spacecraft.

High thermal conductivity links ($K > 1$ W/K) are required to link the focal plane to the cold stage of the cooler. Such links need to be flexible and electrically insulating as well as damping any cooler vibrations. Cryogenic heat pipes are under development (Gilman et al 1995).

Heat switches have a key role in several units, including ^{3}He sorption coolers and ADR systems. To date gas-gap switches (cryo-sorption based) and electro-mechanical switches are used, but their performance and reliability need improvement (Torre and Chanin 1984).

Pressure, level and flow meters are required to monitor the performance of cryogenic equipment on board spacecraft. There are almost no such devices that are space-qualified.

Space-qualified temperature sensors are a critical item, especially at $T <$ 1 K. An overview of cryogenic thermometry is provided in Rubin (1997).

Cryogenic cables with low thermal conductivity, low electrical resistance and low capacitance are crucial for space instrumentation, especially at $T < 1$ K. Ribbon cables can provide 1000 lines (metal or superconductor), with a resistance $< 20\,\Omega$ and a total load at the lowest temperature stage of less than 20 μW at 0.3 K (Cunningham et al 1995).

Key technologies and programmatic aspects

Based on typical applications, a number of critical technologies can be identified. A short description of the main development needs is provided below.

Passive radiators (40 K $< T <$ 100 K) play a major role, reducing the requirements imposed on active cooling systems. Further improvements are needed to ensure high emissivity at low temperature and better thermal isolation solutions.

Active cooling systems — the few existing space-qualified coolers are very expensive (on the order of a million euros) and heavy; they remain a major source of vibrations and their efficiency needs to be improved.
20 K coolers are essential pre-cooling stages. Stirling coolers are difficult to accommodate, while Joule-Thomson coolers are less efficient. Modularity and cooling power scalability need to be improved. Cryo-sorption systems (e.g., H_2 based) are a valid option and need further development.
2 K to 4 K coolers should provide greater cooling power ($>$ 50 mW) to support lower temperature stages. Minimisation of vibrations is required for very low-temperature systems, sensitive detectors and high-accuracy spacecraft pointing and/or positioning (e.g., interferometry applications).

Very low-temperature coolers ($T <$ 1 K) are essential to future space missions. A large effort is required to develop closed-loop, space-qualified coolers (such as ADR, DR, sorption coolers), offering reliable performance and long lifetime (>5 years).

Finally the development of ancillary equipment and devices (e.g., heat pipes, heat switches, thermometry and cryogenic mechanisms) should not be neglected. The use of cryogenics in space projects also has a number of important programmatic implications. The complexity, duration and cost of integration and end-to-end test activities are key examples, with a large impact on any space project. In fact, typical turn-around times of complex cryogenic systems and test facilities are of the order of several weeks. In order to simplify the verification phase, timely development of the cryo-chain units is required (by the end of phase B). On this basis, a thorough scrutinity of the requirements and a vigorous technology demonstration plan are essential activities during the assessment study phase (phase 0 / phase A). Adequate resources must be made available in order to raise the technology readiness of key units in the early phases of the programme.

Conclusions

The use of cryogenic detectors on board spacecraft has allowed unprecedented results, especially in the field of astrophysics. Over the last 25 years, several missions have demonstrated the advantages of cryogenic instruments, although at the expense of additional system complexity, increased development risk and increased cost.

Cryogenics will continue to play a key role on board science missions. In fact, future astrophysics applications require large-format detectors operating at ≈ 50 mK to 100 mK, with cryogenic front-end electronics, and, in the case of medium and far infrared, cold-entrance optics. The development of such complex cryogenic payloads calls for a truly system approach, involving the complete spacecraft design from the assessment phase. Clear examples are set by several space-science observatories, which are built around their cryogen tanks, or by the crucial role played by mission operations control in the case of passively cooled instruments. Significant development effort is required to further improve the performance of cryogenic systems, to enhance cooling efficiency and reliability, and to reduce resource demands as well as cost. Such development must be undertaken in the early project phases, based on actual flight requirements, so as to minimise the risks during phase C/D when the impact of design changes is largest. The effort produced in this field by the leading space organisations shows without any doubt that cryogenics is going to play a strategic role on board future space missions.

Bibliography

Aminou DM, Jacquet B, Pasternak F (1997) Characteristics of the *Meteosat* Second Generation (MSG) radiometer/imager: SEVIRI. Proc SPIE 3221:19–31

Bapat SL (1990) Performance prediction of multilayer insulation. Cryogenics 30:700–710

Bard S (1984) Advanced passive radiator for spaceborne cyrogenic cooling. J Spacecraft Rockets 21:150–155

Bard S (1987) Development of a high-performance cryogenic radiator with V-groove radiation shields. J Spacecraft Rockets 24:193–197

Barnes WL, Pagano TS, Salomonson VV (1998) Prelaunch characteristics of the Moderate Resolution Imaging Spectroradiometer (MODIS) on *EOS*-AM1. IEEE Trans Geosci Remote Sensing 36:1088–1100

Bennett CL, Boggess NW, Cheng ES (plus four authors) (1992) Recent results from *COBE*. In: HA Thronson, JM Shull (eds) Proc Third Teton Summer School: The Evolution of Galaxies and their Environment

Benoit A, Pujol S (1991) A dilution refrigerator insensitive to gravity. Physica B 169:457–458

Bowman RC Jr., Karlmann PB, Bard S (1997) Brilliant eyes ten-kelvin sorption cryocooler Experiment (BETSCE), Final Report, JPL Publication 97-14

Braginski AI (1999) Superconducting electronics coming to market. IEEE Trans Appl Supercon 9:2825–2836

Burger JF, ter Brake HJM, Rogalla H, Linder M (2002) Vibration-free 5 K sorption cooler for ESA's *Darwin* mission. Cryogenics 42:97–108

Chaloupka H, Hein M, Müller G, Piel H (1993) Research and development of analog HTS microwave components for possible space applications. Proc ESA/ESTEC Space applications of High Tc superconductors, ESA WPP-0052, 9–25

Chui TCP, Lipa JA, Nissen JA, Swanson DR (1994) Lambda point experiment in microgravity. Cryogenics 34:341–347

Cihlar J, Belward A, Govaerts Y (1999) *Meteosat* second generation opportunities for land surface research and applications. EUMETSAT ISSN 1561-140X, ISBN 92-9110-031-5

Crone C, Elfving A, Passvogel T (plus two authors) (2006) Unveiling the Universe – Two missions to unlock the secrets of the cold Cosmos. ESA Bull 128:10–17

Cunningham CR, Hastings PR, Strachan JMD (1995) Woven ribbon cable for cryogenic instruments Cryogenics 35:399–402

Dierickx B, Seijnaeve J, Scheffer D (1996) Cryogenic readout amplifiers for *FIRST*'s stressed Ge:Ga array. ESA SP-388:61–65

Elliott D, Hacking P, Schember H (1994) Engineering design of the Wide-Field Infrared Explorer (*WIRE*). Proc SPIE 2268:188–195

Freund MM, Dubond L, Lange AE (plus four authors) (1998) Design and flight performance of a space borne ^{3}He refrigerator for the infrared telescope in space. Cryogenics 38:435–443

Gardner JP, Mather JC, Clampin M (plus 20 authors) (2006) The James Webb Space Telescope. Space Sci Rev 123:485–606

Gilman DC, Russo SC, Sugimura RS (1995) Lessons learned during the integration phase of the NASA CSE. Proc 8th Internat Cryocooler Conf, Plenum Press, pp.869–882

Hagmann C, Richards PL (1995) Adiabatic demagnetization refrigerators for small laboratory experiments and space astronomy. Cryogenics 35:303–309

Hannah I, Hurford G, Hudson H, Lin R (2007) A new method of observing weak extended X-ray sources with the Reuven Ramaty High-Energy Solar Spectroscopic Imager. Rev Sci Instrum 78:024501-1–6

Jones JA, Bard S, Schember HR, Rodriguez J (1990) Sorption cooler technology development at JPL. Cryogenics 30:239–245

Kessler MF, Steinz JA, Anderegg ME (plus eight authors) (1996) The Infrared Space Observatory (*ISO*) mission. Astron Astrophys 315:L27-L31

Kohout LL (1989) Cryogenic reactant storage for lunar base regenerative fuel cells. Space power ISSN 0883-6272, 1989, Vol 8, no 4, 443–457

Langford DL, Simmons JJ, Ozawa T (plus two authors) (1983) Infrared and visible detector electronics for the Infrared Astronomical Satellite (*IRAS*). Proc SPIE 445:244–253

Lipa JA, Swanson DR, Nissen JA (plus two authors) (1996) Heat capacity and thermal relaxation of bulk helium very near the lambda point. Phys Rev Lett 76:944–947

Nahum M, Eiles TM, Martinis JM (1994) Electronic microrefrigerator based on a normal-insulator-superconductor tunnel junction. Appl Phys Lett 65:3123–3125

Ogawara Y (1998) The *Astro-E* Mission. In: The Hot Universe, Koyama K, Kitamoto S, Itoh M (eds) Proc IAU 188, Kluwer Academic Publ, pp. 75–78

Petre R, Lehan J, O'Dell S (plus six authors) (2006) The *Constellation-X* Spectroscopy X-ray Telescope: recent technology development. Proc SPIE 6266:62661Q-1–9

Pilbratt G (2008) *Herschel* mission overview and key programmes. Proc SPIE 7010:1–14

Pobell F (1996) Matter and methods at low temperatures, Springer

Rando N, Lyngvi A, Gondoin A (plus six authors) (2006) ESA study of *XEUS*, a potential follow-on to *XMM-Newton*. Proc 6th Int Conf on Space Optics, ESA SP-621, session 8, ISBN 92-9092-932-4

Reber CA (1990) The Upper Atmosphere Research Satellite (*UARS*). EOS, Trans AGU 71:1867

Reed RP, Golda M (1997) Cryogenic composite supports: a review of strap and strut properties. Cryogenics 37:233–250

Ross RG Jr (1990) Requirements for long-life mechanical cryocoolers for space applications. Cryogenics 30:233–245

Rubin LG (1997) Cryogenic thermometry: a review of progress since 1982. Cryogenics 37:341–356

Shibai H (2007) *AKARI (ASTRO-F)*: Flight performance and preliminary results. Adv Space Res 40:595–599

Sugita H, Nakagawa T, Murakami H (plus five authors) (2006) Cryogenic infrared mission JAXA/*SPICA* with advanced cryocoolers. Cryogenics 46:149–157

Thompson RI, Rieke M, Schneider G (plus two authors) (1998) Initial on-orbit performance of *NICMOS*. Astrophys J 492:L95–L97

Torre JP, Chanin G (1984) Heat switch for liquid-helium temperatures. Rev Sci Instrum 55:213–215

Werner MW (2005) The *Spitzer* space telescope mission. Adv Space Res 36:1048–1049

Wertz JR, Larson WJ (1999) Space mission analysis and design. 3rd edition, Kluwer Academic Press, ISBN 1-881883-10-8

Wild W (2010) Coherent far-infrared / submillimetre detectors. ISSI SR-009:503–523

Winkler C (2004) *INTEGRAL*—status and outlook New Astron Rev 48:183–187

— 38 —

Laser-aligned structures in space

David H. Lumb[I]

Abstract

Aligning and maintaining separate spacecraft at a range of distances via laser metrology and micropropulsion creates the potential for exciting new applications in space. A selection of such formation-flying missions is described, and some of the requirements for technology developments that might be implemented are summarised.

Constellation and formation

A number of applications call for the operation of multiple spacecraft in a coordinated manner. Historically the first of these examples was rendezvous and docking in manned spaceflight, but we concentrate here on scientific applications of automatically operated spacecraft. A distinction is made between a *constellation* of satellites and a set of spacecraft flying in a *formation*. In a constellation the relative or absolute distances between satellites are not controlled, but still measured. In a formation, in contrast, the relative locations of the satellites are actively controlled.

An example of the first kind, a constellation, is *LISA* (ESA 2000, Bika et al 2007). Its scientific goal is to observe gravitational waves from astrophysical objects such as massive black holes in remote galaxies and closely orbiting binary stars in our own Galaxy. *LISA* consists of three spacecraft located at the corners of a near-equilateral triangle in an Earth-trailing heliocentric orbit with side lengths of 5 Gm. It measures the temporal variation of the strains in space-time along the three sides of the triangle by interferometrically monitoring changes in these baselines. The mission concept relies (i) on "undisturbed" masses — i.e., masses under the influence of gravitation only — that act as reliable endpoints of the baselines and (ii) on a measurement system that is able to monitor the changes in length of these baselines. The orbits of the *LISA* spacecraft are such that the near-equilateral triangle is inclined by 60° to the ecliptic, and this also leads to a rotation of the triangle by 360° in the course of one orbit, i.e., one year. The changing orientation of the side lengths enables a scan over the sky, and thus provides a capability for spatial resolution. These orbits do not require station-keeping to maintain their near-equilateral triangular formation. Active control of the relative locations of the

[I]Science and Robotic Exploration Projects Department, ESA, ESTEC, Noordwijk, The Netherlands

three spacecraft is not required. Nevertheless, there is a need to control disturbances, and this is a theme in common with formation applications. The distance measuring system of *LISA* is a continuous interferometric laser ranging scheme: lasers at each end of each arm operate in a "transponder" mode. A beam is sent out from one spacecraft to its distant partner at the other end of the baseline. The laser in the distant spacecraft is phase-locked to the incoming beam and returns a high-power phase replica. When that beam reaches the original spacecraft, it is brought to interference against the local laser. As the orbit of the constellation is, like that of the Earth, slightly elliptical, the arm lengths change continually — the system "breathes" — and, concomitantly, fringes of high frequency are observed. The presence of one or several gravitational waves, which have a markedly lower frequency, then generates one or several side-bands of the fringe frequency.

Besides this measurement, laser frequency noise and clock noise are determined. Laser metrology of this kind is also employed in a number of formation applications as we shall see in the section on optical laser metrologies.

Formation flying has been proposed for applications like geodesy, Earth observation with synthetic aperture radar (SAR) and security applications. However, we shall concentrate here on the two generic classes of astronomy payloads that need formation flying:

Interferometers with several telescopes, where the separation and alignment of the telescopes needs to be controlled and where, in addition, a change in the separation distances permits the sampling of different spatial frequencies, i.e., of the so-called (uv) plane (cf., Quirrenbach 2010).

Telescopes with two spacecraft, where the very long focal length requires telescope and detector to be located in separate spacecraft: here the spacecraft containing the detector needs to be kept at the focus of the telescope, i.e., at a proper distance and on axis, or in a controlled position relative to the axis.

Terminology

The introduction of two or more spacecraft leads to new definitions of pointing terms and budgets. For the purpose of formation manoeuvres, one of a pair of spacecraft can be designated as the target and the other as the chaser. The following reference frames can be used to define the pointing. (We use the nomenclature of the standard description of spacecraft pointing, ESA 1992.)

- **Optical frame:** This is a frame which is fixed with respect to the spacecraft body, with origin and axes chosen as is appropriate for the payload.

- **Rotating reference frame:** This defines the desired position and orientation of the optical frame.

- **Measured optical frame:** This represents the measured — or sometimes estimated — position and orientation of the optical frame.

A standard terminology for attitude exists only for attitude errors of a single spacecraft. For the current case of position and attitude errors for a two-spacecraft formation, some modifications are necessary. We use here the following definitions:

- An **error** is the difference between the actual value of a quantity and its nominal, i.e., the demanded value.
- A **measurement error** is the difference between the measured value of a quantity and its actual value.
- The **stability** of a quantity refers to its deviation from the average over a specified time interval.
- An **absolute** quantity (error, measurement error or stability) refers to the value for a single spacecraft, and hence can be defined for either spacecraft.
- A **relative** quantity refers to the value with respect to the target spacecraft, and hence can be defined only for the chaser spacecraft.

For a single spacecraft we have: absolute attitude error (AAE) — the instantaneous error in attitude with respect to the demanded value, absolute attitude stability (AAS) — the variation of attitude error (with respect to an average) over a given timescale and absolute attitude measurement error (AAME) — the instantaneous error in measured attitude with respect to the desired value.

Now in addition, the following should be considered for satellite control involving two or more spacecraft: relative attitude error (RAE), relative attitude stability (RAS), relative attitude measurement error (RAME), relative displacement error (RDE), relative displacement stability (RDS) and relative displacement measurement error (RDME). The coupling of these parameters depends upon the application as well as on the type and arrangement of the sensors, and on how they are aligned and maintained, in operation, with the reference axes. In the following section we discuss some of the critical performance requirements for these parameters in different missions.

Missions

In the ensuing descriptions of proposed missions requiring formation flying, we restrict ourselves to the specific requirements that need to be placed on the guidance and navigation aspects.

In the near future there are plans only for test missions that demonstrate some features of formation flying in orbit. These are proposed as low-cost missions that should "retire the risk" of the new techniques. Based on past experiences, however, the wisdom of relying on low-cost approaches is being questioned. In an experiment attempting formation flying by use of GPS in a rendezvous experiment between *SNAP* and *Tsinghua*, the *SNAP* satellite was supposed to chase *Tsinghua* and make images of it (Fortescue et al 2003). Differential launch injection prevented the close rendezvous of the satellites. Another example of a failed test of a low-cost mission is the *DART* mission (Rumford 2003): the NASA/OSC platform was lost after a combination of software errors and incorrect sensor readings led to a collision of the *DART* spacecraft with its target satellite. The exhaust of its limited propellant then resulted in the loss of a $ 110 million "test" satellite.

PRISMA

PRISMA (Lundahl and von Scheele 2004) is a technology mission to be launched in 2010. Its main aim is the test of different sensor technologies and guidance and navigation strategies. The project consists of two spacecraft: one, called *MAIN*, is advanced and highly manoeuvrable; the other, called *TARGET*, is a simpler, passive craft. *MAIN* is equipped with several sensor systems such as GPS, a camera sensitive to visible light (VBS) and a radio-frequency based navigation instrument (FFRF). Together with tests of advanced guidance, navigation and control algorithms, several high-level demonstrations are foreseen. The demonstrations are planned for a ten-month duration at an orbital altitude of about 700 km.

Proba-3

Proba-3 (Vivès et al 2006) is the third in ESA's series of missions for validating developments in space systems. It will comprise two independent, three-axis-stabilised spacecraft utilising either cold-gas or electrical thrusters for agile manoeuvring, and both radio-frequency and laser-based optical metrology techniques for accurate position measurement and control. The scientific payload is a solar coronagraph, where a remote external occulter needs to be aligned with the actual coronagraph with a positioning accuracy of the order of 100 µm for a separation range of 100 m to 250 m.

Simbol-X and *MAX*

Several national space agencies in Europe also feel a need to develop formation-flying capability. The most advanced studies are those by CNES. One example is *MAX* (Borde et al 2005), a concept for γ-ray astronomy, designed to observe in two spectral bands around important γ-ray energies (450 keV to 530 keV and 800 keV to 900 keV). The formation flight is supposed to maintain alignment of two spacecraft over a focal length of 80 m to 90 m. This is necessary to build a reasonably sized γ-ray telescope based on a Laue-crystal lens.

Simbol-X (Ferrando et al 2006) aims at increasing the sensitivity for hard X-ray astronomy by approximately three orders of magnitude over the present non-focusing optics, such as coded masks and collimators. This is accomplished by use of a grazing-incidence X-ray mirror with very long focal length (20 m to 25 m). This again requires two spacecraft, one for the optics and the other for the focal-plane instrumentation. CNES and ASI entered into an agreement to develop the two spacecraft separately, with a view to preparing many technology aspects for the ambitious *XEUS* mission that has been proposed for ESA's long-term science programme.

XEUS

XEUS (Parmar et al 2006) was proposed to the Cosmic Vision 2015–2025 long-term plan for ESA's Science Programme as the potential successor to ESA's *XMM-Newton* X-ray observatory. Light-weight optics with an angular resolution of 2″ and with an effective area of 5 m^2 at 1 keV and 2 m^2 at 7 keV on one spacecraft will be combined with advanced detectors on a second spacecraft, to achieve much improved imaging, spectroscopic and timing performances. The two spacecraft will be launched to the second Earth-Sun Lagrangian point (L2), where they will fly in formation, separated by a focal length of 35 m to 50 m. The focal-plane spacecraft will contain narrow- and wide-field imaging spectrometers and other specialized instruments.

GRI

The *Gamma-Ray Imager* (*GRI*, Knödlseder et al 2007) is a mission concept that will provide a sensitivity leap in the soft γ-ray domain by use of a focusing lens built of Laue diffracting crystals. The lens will transmit and focus in the energy band from 200 keV to 1.3 MeV with an effective area reaching 600 cm^2. It will be complemented by a single-reflection multilayer-coated mirror, which will extend the energy band into the hard X-ray range, down to about 10 keV. The photons will be detected by a position-sensitive CZT (CdZnTe) stack detector. *GRI* will have an angular resolution of better than 30″ within a field of view of roughly 5′ — an unprecedented achievement in the γ-ray domain. Owing to the large focal length of 100 m of the lens and the mirror, the optics and detector will again require two separate spacecraft flying in formation. A highly elliptical orbit is foreseen.

Darwin/TPF

The *Darwin* (Leger and Herbst 2007) and *TPF-I* (Lawson et al 2006) missions of ESA and NASA, respectively, are conceived as free-flying infrared interferometers based on nulling interferometry (Quirrenbach 2010). Their main objective is to detect and characterize other Earth-like planets, and analyze the composition of their atmospheres and thus their capability to sustain life. A number of mission architectures with three and four free-flying telescopes of 3.5 m diameter are being considered. Nulling interferometry of exo-solar planets requires a minimum of two telescopes, of which one is phase shifted by 180°, such that the radiation of an on-axis stellar object is interfering destructively, while the light from the off-axis planet interferes constructively. Interferometer performance improves with chopping that uses rapid signal modulation to detect a planet in the high background noise. This technique, however, requires at least three telescopes and a further separate spacecraft dedicated to beam recombination. Moreover, the formation must be rotated to identify the planet's azimuthal location in the celestial plane, and be resized to maximize sensitivity as a function of radius from the parent star.

Table 38.1 summarises the pertinent characteristics of the above formation-flying missions.

Table 38.1: Characteristics of the described formation-flying missions.

Mission	*Prisma*	*Proba-3*	*Simbol-X*	*MAX*	*XEUS*	*GRI*	*Darwin TPF-I*
Type	demonstration	demonstration/ solar coronagraph	hard X-ray astronomy	γ-ray astronomy	X-ray astronomy	γ-ray astronomy	extra-solar planet finder, life-marker
Launch	2008	2011	2013	?	2018 +	?	2020 +
Orbit	LEO	HEO	HEO	HEO	L2	L2	L2
No. of spacecraft	2	2	2	2	2	2	4
Intersat. distance	25 m to 500 m	150 m	25 m	80 m	35 m	100 m	25 m to 1000 m
Attitude accuracy	N/A	$8''$	$10''$	$15''$	$1''$	$15''$	$0.1''$
Relative separation accuracy (Lat/Long)	centimetres	6 mm; 10 mm	1 cm; 10 cm	1 cm; 10 cm	1 mm; 10 mm	2 cm; 16 cm	5 cm

Dynamics

The type of orbit chosen strongly determines the dynamical range of disturbances that must be handled. In low Earth orbit (LEO) the gravity gradient between the spacecraft must be overcome continually to maintain a fixed inertial position. If, for example, one spacecraft flies in an orbit without propulsion correction, the other must follow a "non-Keplerian" orbit. This requires continuous metrology and propulsion. The finite propellant mass available on board limits the total "delta-V" (i.e., the integral of differences, ΔV, in orbital speed). Accordingly, formation flying should occur only beyond some distance from Earth.

At perigee in a geostationary transfer orbit (GTO), the gravity gradient can be as low as ≈ 1 mN. Even lower values occur in a highly eccentric orbit (HEO). As an example: for two spacecraft of 200 kg at a distance of 500 m in an HEO with a perigee of 75 000 km, the relative force remains < 1 mN for ≈ 75 % of the orbit. But lower perigee heights lead to a dramatic increase in the force: to maintain the same formation for 75 % in an HEO with a perigee of 40 000 km, the upper limit of the relative force rises to 8 mN.

Operation at the L2 point minimises gravity gradients. The gravity gradient torque $T_{\mathrm{GG}} \leq 3\mu\Delta I/R^3$, where μ is the gravity constant (398 000 $\mathrm{km^3\,s^{-2}}$ for Earth), R is the distance from L2 to Earth (1.5×10^6 km) and ΔI is the maximum difference between the satellite principal moments of inertia. For a difference of say 300 $\mathrm{kg\,m^2}$ between principal moments of inertia, the maximum gravity gradient torque acting on one satellite at L2 is about 1 pN m. The gravity gradient disturbing force F_{GG} is $\leq 2\,\mu\,d/R^3$, where d is the inter-satellite distance and R is the distance to L2 and typically F_{GG} is $\approx 10^{-8}$ N.

At L2 a stronger disturbance is generated by solar radiation pressure, for which the acceleration is given by $a_{sp} = 4.5 \times 10^{-6}(1+r)\,A/M$, where A is the illuminated cross section area, M the satellite mass, r the reflectance (= 1 for specular normal reflection in the normal). For a satellite of mass 200 kg and with an area of 5 m^2, this is 2×10^{-7} $m\,s^{-2}$. For formation flying it is the *difference* in reflectance and area between satellites that forces a relative position drift and torque. For two 200 kg satellites with a factor of two difference in reflectance, the differential force is $\approx$ 10 µN. If this is not compensated for, the two satellites would drift apart or towards each other by about 1 cm in 10 min.

From these values one can appreciate the order of magnitude of the problem of measuring and correcting the relative location of two or more satellites in formation. In the following section we discuss the sensor and actuator technologies currently being developed to enable formation-flying activities.

Technology

Radio-frequency metrology

Omni-directional coverage permits the long-distance ranging that is required to bring the satellites into an initial formation configuration, and into the coarse sensor range. Therefore in each satellite of the proposed formations a radio-frequency terminal is installed, which is connected to sets of antennae. The terminals can receive and transmit or receive only (Feng 2002). A terminal may operate in TDMA (time division multiple access) mode, i.e., it transmits radio-frequency ranging signals toward and receives radio-frequency signals from other satellites. Ranging and angular measurements are extracted from received signals and are used for computing relative position, relative velocity and line of sight. The antenna sets (typically triplets) point in different directions, so that they provide quasi full-space coverage.

Optical laser metrologies

There are optical metrologies for coarse and fine measurements. Coarse measurements are made by two kinds of techniques: time-of-flight, and amplitude modulation.

In the time-of-flight method (Buller et al 2005) a light pulse is sent out to a reflective object and the time it takes for the pulse to travel back to the source is recorded. The benefits of this method are its simplicity and the ability to operate at very large distances. The accuracy is limited by the relative stability of the clocks used to measure the time of flight. The resolution is usually limited to around a millimetre, because of the finite resolution in measuring time intervals, but measuring the returning signal is possible even with as little as a few photons per pulse.

Distance measurements with a higher resolution can be made by modulating the intensity of a laser beam with a sinusoidal function of time (Payne et al 2002). The increase in resolution compared to the time-of-flight method arises from the advantage in the analogue signal processing, where a time measurement is substituted by a phase measurement. The reflected light is recorded with a fast detector, band-pass filtered and compared to the modulation signal in an electrical phase meter. This technique could thus be seen as a sort of electrical interferometer, with an optical delay line in one of the arms.

The latest implementation proposed uses a femtosecond fibre-laser, which emits a train of very sharply peaked pulses. The Fourier transform of this pulse train contains some very high harmonics. Isolating a single high harmonic of the received signal by proper bandpass filtering then yields a signal with a frequency that is a very large multiple of the pulse repetition frequency, thus allowing a measurement with a small synthetic wavelength. A resolution of several micrometres is achieveable at distances of a few hundred metres. Because the phase is measured modulo 2π the measured phase will repeat itself after a path-length change of one synthetic wavelength. The amplitude-modulation technique is therefore not a truly absolute distance method, since it requires a priori knowledge of the path length to better than a synthetic wavelength. The so called unambiguity range can be increased by lowering the modulation frequency, but with a constant error in the phase measurement which will cause a lower resolution in length.

Fine resolution metrology in the optical can be achieved by single- or double-wavelength interferometry. The most common way to accurately measure a distance is by interferometry with a single stabilized laser as a source (Baldwin et al 1971). Again the main problem is the very small unambiguity range. It is possible to measure a distance first to low resolution with one of the other techniques described and then improve the measurement with a single wavelength measurement. Since the low-resolution measurement should already have an accuracy of better than one optical wavelength, this is in most cases only possible at short distances. Single-wavelength interferometry can then be used for measuring changes in distance: fringe counting permits the measurement of large changes in distance. A requirement is that the wavelength remains stable during the measurement.

A technique to increase the unambiguity range is to measure the path length in the interferometer with two different wavelengths simultaneously, yielding two phases (Mahal and Arie 1996). This method is called double-wavelength interferometry (DWI). As with amplitude modulation, there is a synthetic wavelength defining the unambiguity range resulting from the frequency differences, i.e., the beat frequency.

Electric propulsion and cold gas thrusters

Cold gas propulsion is the simplest form of micropropulsion that might be used for formation flying (Köhler et al 2002). Typically, a piezo-actuated valve controls a gas flow in the thruster. Thrust levels reach up to 1 mN. The most common working fluid is nitrogen, which may be heated to a given temperature at the outlet, to ensure satisfactory specific impulse and to eliminate the effects of ambient temperature changes on the accuracy of the thrust.

The basic elements of electrostatic propulsion include a source of charged particles, an acceleration region and a method to neutralise the exhaust flow so that the spacecraft does not accumulate a net charge.

The range of FEEP (Field Emission Electric Propulsion) forces extends from 1 µN to 1 mN. Here low system mass and low propellant consumption (Marcuccio et al 1997) rather than power-to-thrust is the key issue. The thruster can be operated with cesium propellant or other liquid metals or alloys. The electrical efficiency of FEEP is very high, due to the absence of thermal losses: ionization occurs in the liquid phase by field effect, and favourable ion extraction geometry minimizes ion impingement losses. FEEP has no moving parts, no pressurized vessels, and no propellant feed lines.

In an ion thruster a heavy inert gas is ionised by electron bombardment or by radio-frequency excitation and then accelerated with an electrostatic potential. Ion engines producing a force of millinewtons have been used frequently on interplanetary probes and for station-keeping on telecommunications satellites. Significant development is, however, still required to achieve micronewton thrust levels (Leiterl et al 2007).

Outlook

The capability of aligning and maintaining separate spacecraft at a given distance by use of laser methods and micropropulsion will enable exciting new space missions. There is still the need for a large number of technology developments before such missions can be flown. Furthermore, the resulting systems must be tested thoroughly, also to minimize the possibility of collisions. The need to robustly verify the performance before embarking on a major observatory is a great challenge.

Bibliography

Baldwin RR, Gordon RR, Rude AF (1971) Remote laser interferometry. Hewlett-Packard J 23:14–20

Bika JJCM, Vissera PNAM, Jennrich O (2007) *LISA* satellite formation control. Adv Space Res 40:25–34

Borde J, vonBallmoos P, Soumagne R (2005) Small-sat platforms and formation flying: An opportunity for the gamma ray telescope *MAX*. Experimental Astron 20:465–482

Buller GS, Harkins RD McCarthy A (plus 11 authors) (2005) Multiple wavelength time-of-flight sensor based on time-correlated single-photon counting. Rev Sci Inst 76:083112-1–7

ESA (1992), Pointing Error Handbook ESA-NCR-502. ESA-ESTEC, Noordwijk, The Netherlands

ESA (2000) *LISA*: System and Technology Study Report. ESA-SCI(2000)11, ESA-ESTEC, Noordwijk, The Netherlands

Feng Y (2002) Applications of GPS in space programs. GPS Solutions 6:126–127

Ferrando P, Arnaud M, Briel U (plus 19 authors) (2006) *Simbol-X*: Mission overview. Proc SPIE 6266:62660F-1–12

Fortescue PW, Stark J, Swinerd G (2003) Spacecraft systems engineering. John Wiley and Sons, ISBN 0-470-85102-3

Knödlseder J, von Ballmoos P, Frontera F (plus four authors) (2007) *GRI*: Focusing on the evolving violent Universe. Proc SPIE 6688:668806-1–11

Köhler J, Bejhed J, Kratz H (plus four authors) (2002) A hybrid cold gas microthruster system for spacecraft. Sensors and Actuators A: Physical 97-98:587–598

Lawson PR, Ahmed A, Gappinger RO (plus seven authors) (2006) Terrestrial Planet Finder interferometer technology status and plans. Proc SPIE 6268:626828-1–8

Léger A, Herbst T (2007) *DARWIN* mission proposal to ESA. eprint arXiv:0707.3385L

Leiterl H, Killinger R, Boss M (plus seven authors) (2007) RIT-μX — High precision micro ion propulsion system based on RF-technology. Proc 43rd AIAA Joint Propulsion Conference AIAA 2007 pp 5250

Lundahl K, von Schéele F (2004) Swedish small satellites. ESA SP-571:1–6

Mahal V, Arie A (1996) Distance measurements using two frequency-stabilized Nd:YAG lasers. Appl Opt 35:3010–3015

Marcuccio S, Giannelli S, Andrenucci M (1997) Attitude and orbit control of small satellites and constellations with FEEP thrusters. IEPC-97-188:1152–1159

Parmar AN, Arnaud M, Barcons X (plus 14 authors) (2006) *XEUS*: The X-ray Evolving Universe Spectroscopy mission. Proc SPIE 6266:62661R-1–66

Payne JM, Parker D, Bradley RF (1992) Rangefinder with fast multiple range capability. Rev Sci Inst 63:3311–3316

Quirrenbach A (2010) Interferometric imaging from space. ISSI SR-009:293–311

Rumford TE (2003) Demonstration of autonomous rendezvous technology (DART) project summary. Proc SPIE 5088:10–19

Vivès S, Lamy P, Levacher P (plus two authors) (2006) Formation flying applied to solar coronal diagnostics: the ASPIICS coronagraph. Proc SPIE 6265:626524-1–9

— 39 —

Earth and planet observations

ANUSCHKA PAULUHN[I]

Abstract

Photon measurements are vital in astronomy but can also help to improve our understanding of our closer neighbourhood. The use of photon observations from space for Earth and planetary studies is outlined. In particular, some of the most important radar instruments for Earth observation and climate studies, as well as the corresponding satellite orbits, are described. The importance of monitoring our environment from space is emphasised.

Why observe the Earth — and why from space?

As mentioned in the introduction (Chapter 1, Huber et al 2010) the kind of instrumentation that has been developed for space astronomy can, in general, also be used for observing the Earth and planets. In addition, active measurements, i.e., radar, are employed for Earth and planetary observation from space as well. Both of these aspects have not been covered in detail in this book before. In this chapter, we discuss observing strategies that differ from space astronomy and stress the many applications to today's complex problems of the Earth's changing environment.

Nearly all space-faring nations maintain or are involved with Earth-observation missions.[1] A historic survey as well as an overview of the most important Earth-observation missions can be found in the article by Harries (2001). As may be concluded from Figure 1.1 of Chapter 1, the wavelengths suitable for Earth observation mainly belong to the visible, infrared and microwave parts of the electromagnetic spectrum. For investigations of certain atmospheric trace gases, UV wavelengths are employed as well. Sensors can be active radar instruments, e.g., synthetic aperture radars (SAR), altimeters, scatterometers, or passive instruments such as radiometers.

The applications are manifold, and nowadays many Earth-observation programmes have reached the status of "operational workhorses", like, e.g., EUMETSAT's *Meteosat* and *MSG* (Meteosat Second Generation) series. There are also larger networks, as for example the *Metop* series of polar orbiting satellites together with the *NOAA-N,N'* satellites, which form the Initial Joint Polar-Orbiting

[I]PSI—Paul Scherrer Institut, Villigen, Switzerland

[1]e.g., http://www.esa.int/esaEO/index.html, http://www.eorc.jaxa.jp, http://earth.esa.int/, http://neo.sci.gsfc.nasa.gov, http://www.dlr.de, http://www.eumetsat.int/

Operational Satellite System (IJPS) constellation, a collaboration project between EUMETSAT, the US National Oceanic and Atmospheric Administration (NOAA), ESA and the French space agency Centre National d'Etudes Spatiales (CNES).

In general, satellite-based measurements which offer access to the entire globe provide valuable opportunities for weather and climate research. In particular for oceanography, where large areas of ocean are not easily accessible, observations from space are crucial for the study of global ocean circulation, long-term, mid-term and short-term variability, El Niño and La Niña prediction, currents and waves (see, e.g., Johannessen et al 2001). Additionally, data from satellite measurements are used for assimilation into models (i.e., the statistical combination of data and model see, e.g., Daley 1991).

Imaging and non-imaging radars have proven extremely useful also for geophysical studies, environmental studies and pollution detection, or crop detection and (de-)forestation monitoring.

For climate research, dedicated atmospheric and oceanographic experiments need to be established and the appropriate satellites, orbits and instruments designed. Two particular series of satellites have provided excellent data for atmosphere and ocean physics, geophysics, and climate research, namely ESA's *ERS-1, ERS-2, Envisat* series, and the joint NASA/CNRS "Topographic Experiment", consisting of *TOPEX/Poseidon, Jason-1* and *Jason-2.* [2]

The choice of orbit depends on the observational demands, i.e., the area, the recurrence cycle, and the spatial resolution. The highest latitude of observed area is determined by the orbit inclination, and the recurrence cycle by the orbit altitude. Spatial resolution decreases as the orbit altitude increases. Preferred orbits for Earth observation are geostationary orbits, in which the satellite revolves at the same speed as the Earth and can always be seen at the same place from the ground; polar orbits with inclination 90°, from which the entire Earth surface can be observed; and Sun-synchronous orbits, for which the angle between the orbital plane and the Sun is constant, i.e., the satellite appears at the same local time every day at any location on the ground.

Radars and radiometers

Synthetic aperture radars (SAR) are two-dimensional imaging systems. The cross-track (or 'range') measurement and resolution are achieved in SAR in the same manner as in most other radars: Range is determined by precisely measuring the time from transmission of a pulse to receiving the echo from a target, and, in the simplest SAR, range resolution is determined by the transmitted pulse width, i.e., narrow pulses yield fine range resolution. To obtain fine along-track ('azimuth' direction) resolution, a physically large antenna is needed to focus the transmitted and received energy into a sharp beam. By taking the movement of the antenna into account via the Doppler-shifted echos of its targets, a larger ("synthetic") aperture is generated (cf., Franceschetti and Lanari 1999). An example SAR image of an oil spill, taken with *Envisat's* Advanced Synthetic Aperture Radar (ASAR), is shown in Figure 39.1.

[2] http://envisat.esa.int/, http://topex-www.jpl.nasa.gov/science/el-nino.html

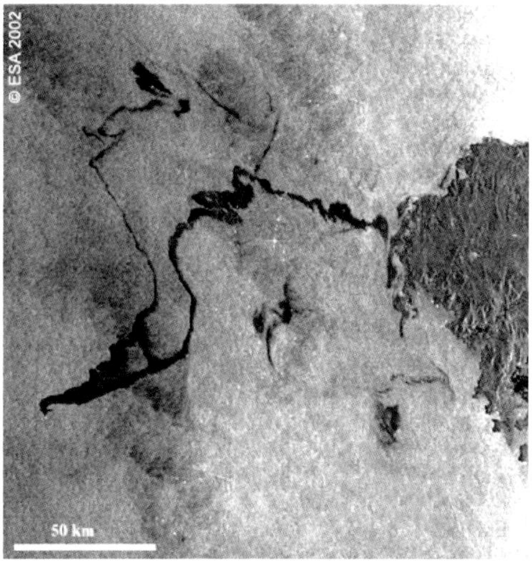

Figure 39.1: *Envisat's* Advanced Synthetic Aperture Radar (ASAR) image acquired 17 November 2002 shows an oil spill originating from a damaged tanker 100 km off the Spanish coast. (Credits: ESA.)

The *Cassini* mission to Saturn is currently using SAR to map the surface of the planet's major moon Titan, whose surface is partly hidden from direct optical inspection by atmospheric haze.

If two SAR images from slightly different viewing angles are considered, their phase difference (interferometric fringes) can be used to generate digital elevation maps, to monitor terrain changes and to improve the range resolution (interferometric SAR, InSAR).

A radar scatterometer is designed to determine the normalized radar cross section, σ_0, of the ocean surface. The primary application of spaceborne scatterometry has been the measurement of near-surface winds over the ocean (wind scatterometers). By combining σ_0 measurements from different azimuth angles, the near-surface wind vector over the ocean surface can be determined.

Altimeters measure the distance via the travel time of a radio signal. They are mainly used to monitor the sea level/ocean topography, which is crucial for the study of the large- and mesoscale ocean circulation, such as planetary waves (Kelvin waves, Rossby waves, eddies, see, e.g., Apel 1987) and associated climatic events such as El Niño. For a comprehensive approach see for example the volume edited by Fu and Cazenave (2001). Information derived from altimetry data can also be used for geophysical studies, especially geodesy, i.e., to study the Earth's shape and size, gravity anomalies, seafloor relief (bathymetry), tectonic plate motion and rifts, cf., Barlier and Lefebre (2001). To determine the sea surface variability, the difference of the measurements on the repeat tracks can be used. To determine the circulation, which defines the ocean topography, however, a reference surface is needed — the geoid. The latter is the equipotential level of the Earth's gravitational field which coincides with the ocean surface at rest. On the other hand, knowledge of the ocean currents can be used to improve geoid models (see also Wunsch and Stammer 2003). Today, the height of the geoid,[3] smoothed over horizontal distances greater than roughly 400 km, is known with an uncertainty of approximately 1 mm from data collected by the Gravity Recovery and Climate Experiment *GRACE* satellite mission.[4] ESA's recently-launched *GOCE* mission is intended to measure the global variations of the Earth's gravity field and to determine the geoid with an accuracy of (1 to 2) cm with a spatial resolution of 100 km.[5]

A precise knowledge of the satellite's orbital position is necessary in order to obtain measurements accurate to within a few centimetres over a range of several hundred kilometres. To achieve this, three locating systems (Doppler Orbitography and Radio-positioning Integrated by Satellite, DORIS; Global Positioning System Payload, GPSP; Laser Retroreflector Array, LRA), for example, are carried on board *Jason-2*, the long-term ocean-topography experiment continuing the uninterrupted time-series that originated with *TOPEX/Poseidon*. In order to ensure the continuity and optimal inter-calibration of observations over the long term, *Jason-2* is flying in the same orbit as *Jason-1* and *TOPEX/Poseidon* (1336 km, circular, 66° inclination, 10 d repeat of ground track with ± 1 km accuracy, orbital velocity 7.2 $\mathrm{km\,s^{-1}}$, ground scanning velocity 5.8 $\mathrm{km\,s^{-1}}$, ground track separation at Equator 315 km).[6] Water vapour and electrons in the atmosphere, sea state, and a range of other parameters can affect the signal round-trip time, thus distorting range measurements. Data from the Advanced Microwave Radiometer (AMR) are used to correct for these interference effects on the altimeter signal.

Since the 1980s satellites have been increasingly utilized to measure the sea surface temperature (SST)[7] and have provided an enormous leap in our ability to view its spatial and temporal variation. Satellite measurements of SST are far more consistent and, in some cases, accurate than the in-situ temperature measurements. The satellite measurement is made by sensing the ocean radiance at two or more wavelengths in the infrared or microwave part of the electromagnetic spectrum; this can then be empirically related to SST. Radiometers have been flown, e.g., on the Geostationary Orbiting Earth Satellites (*GOES*), on the *Terra* and *Aqua* satellites, on *ERS-1*, *ERS-2*, and *Envisat*.[8]

[3]The geoid varies in height by as much as 100 m over distances of several thousand kilometres due to uneven mass distribution within the planet's crust, mantle and core.

[4]http://www.csr.utexas.edu/grace/gravity/oceanographic_sciences.html

[5]http://www.esa.int/esaLP/ESAYEK1VMOC_LPgoce_0.html

[6]http://topex-www.jpl.nasa.gov/science/jason1-quick-look/

[7]SST is the water temperature close to the surface. However, the exact meaning of "surface" varies according to the measurement method used. A satellite infrared radiometer indirectly measures the temperature of a very thin layer of about 10 µm thickness (referred to as the "skin") of the ocean which leads to the phrase skin temperature (because infrared radiation is emitted from this layer). A microwave instrument measures subskin temperature at about 1 mm depth. A thermometer attached to a moored or drifting buoy in the ocean would measure the temperature at a specific depth, e.g., at 1 m below the sea surface.

[8]http://www.osdpd.noaa.gov/ml/index.html, http://modis.gsfc.nasa.gov/about/

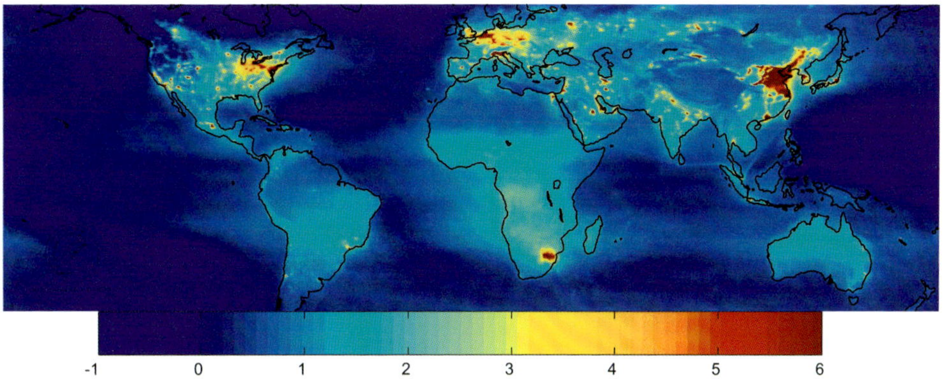

Figure 39.2: The global mean tropospheric NO_2 vertical molecule column density for 2003 to 2006, as measured by the SCIAMACHY instrument on ESA's *Envisat*. The colour scale is in 10^{15} cm^{-2}. (Image courtesy of Steffen Beirle, MPI für Chemie Mainz.)

Spectrometers for atmospheric monitoring

ESA's ten-instrument *Envisat*, the world's largest satellite for environmental monitoring, was launched in February 2002, and it has been decided to extend the *Envisat* mission to 2013. Its onboard Scanning Imaging Absorption Spectrometer for Atmospheric Chartography (SCIAMACHY) instrument performs global measurements of trace gases in the terrestrial troposphere and stratosphere. Figure 39.2 shows a global atmospheric map of nitrogen dioxide (NO_2) pollution. NO_2 is a mainly man-made gas (generated, e.g., in combustion engines or thermal power stations), excess exposure to which causes lung damage and respiratory problems. It also plays an important role in atmospheric chemistry, because it leads to the production of ozone in the troposphere.

Recently, SCIAMACHY was able to acquire spectra of the Venus atmosphere, too, showing detailed absorption features of carbon dioxide.[9]

Ozone measurements via satellite have been made by backscatter UV instruments since the launch of the *Nimbus-4* satellite in April 1970. The Solar Backscatter Ultraviolet (SBUV) instrument currently onboard the *NOAA-17* polar orbiting satellite measures the ultraviolet sunlight scattered by the Earth's atmosphere at several wavelengths ranging from 252 nm to 340 nm.[10] The ratios of the backscattered signals can be used to calculate the total columnar ozone.

Summary

In this short chapter, we have tried to describe some techniques and observing strategies that are slightly different from space astronomy, but nevertheless using photon observations from space. Monitoring our own habitat in the Solar System

[9] http://www.esa.int/esaEO/SEMGFC7CTWF_index_0.html

[10] http://www.ozonelayer.noaa.gov/action/sbuv2.htm

is of fundamental importance to helping ensure reasonable living conditions on our planet.

Bibliography

Apel JR (1987) Principles of Ocean Physics. International Geophysics Series Vol 38, Academic Press, 634 pp

Barlier F, Lefebvre M (2001) A new look at planet Earth: Satellite geodesy and geosciences. In The Century of Space Science (eds JAM Bleeker, J Geiss, MCE Huber) Vol 2, pp 1623–1652

Daley R (1991) Atmospheric data analysis. Cambridge University Press, 457 pp

Franceschetti G, Lanari R (1999) Synthetic Aperture Radar Processing. CRC Press, Boca Raton, London

Fu LL, Cazenave A (eds) (2001) Satellite Altimetry and Earth Sciences. International Geophysics Series Vol 69, San Diego: Academic Press

Harries JE (2001) Chemistry and physics of the atmosphere. In The Century of Space Science (eds JAM Bleeker, J Geiss, MCE Huber) Vol 2, pp 1653–1670

Huber MCE, Pauluhn A, Timothy JG (2010) Observing photons in space. ISSI SR-009:3–21

Johannessen JA, Sandven S, Durand D (2001) Earth science—oceanography. In The Century of Space Science (eds JAM Bleeker, J Geiss, MCE Huber) Vol 2, pp 1585–1622

Wunsch C, Stammer D (2003) Global ocean data assimilation and geoid measurements. ISSI Space Sci Ser 17:147–162

VII. Space Environment

— 40 —

Implications of the space environment

Barry J. Kent[I]

Abstract

This chapter addresses those issues that make detecting photons in space different from detecting photons on the Earth. Above the absorbing atmosphere of Earth, the wavelength range available to an orbiting spacecraft is limited only by the details of the photon-collecting systems. At the radio end of the electromagnetic spectrum the limit is the size of the available antenna and for γ-rays the limit is the detector stopping ability. Both high and low energies are thus constrained by the mass and volume of detector that can be put into space. In previous chapters the need to go into space to achieve particular scientific aims for astronomical purposes has been explained. This chapter is concerned with the environments encountered, both in space and where a space instrument is prepared and launched, and with the way in which these environments impact the photon-detection systems.

Introduction

The approach in this chapter is pragmatic, based on established space heritage rather than deriving practice by formal reduction from physical principles, and provides a basic appreciation of the issues involved. For detailed technical information there are many excellent textbooks (Fortescue et al 2003; Tribble 2003) but a recommended starting point is the web page and publications of the European Cooperation for Space Standardisation (ECSS, see http://www.ecss.nl/).

There are four environments that the space-instrument designer needs to consider:

- the preparation environment,
- the launch environment,
- the environment provided locally by the spacecraft,
- the environment of space itself.

[I]Independent consultant, previously with the Space Science and Technology Department, Rutherford Appleton Laboratory, Didcot, UK

The preparation environment

The preparation of instruments for space flight, which includes the processes of assembly, integration and verification (AIV), is carried out in an area appropriate to precision instrumentation. Assembly of the flight instrument from its component parts takes place after a long programme (often many years) of design, mathematical modelling and testing of various engineering models. The particular set of models used is known as the model philosophy, the different variations in philosophy being based on the level of testing needed or the reliability of the modelling software. For example, typical philosophies use non-flight engineering models such as structural thermal models (STM) and electrical models (EM) or proto flight (PF) in which the model evolves into the flight instrument, and more recently hardware-in-the-loop systems in which hardware elements are introduced to replace software in an overall software model. In all of these processes the aim is to ensure the instrument to be launched (flight model, FM) will have been tested for every eventuality that it will encounter during its scientific programme (or mission).

This thoroughness does not prevent the occasional unforeseen event or particularly the consequences of human activity. Stories of multimillion-dollar spacecraft coming to a catastrophic end due to human influence are part of space instrumentation folklore. Spacecraft have been dropped from test fixtures because the information to reattach the fixing bolts was not passed on during a technician shift change (http://www.nasa.gov/home/hqnews/2003/sep/HQ_03287_satellite_accident.html). Spacecraft have missed or slammed into target planets because of misunderstandings about the system of units being used (http://mars.jpl.nasa.gov/msp98/news/mco990930.html). Such stories reach the pages of the general newspapers. There is a much larger set of stories in which spacecraft have been subject to damaging events that result in compromised performance and are known only to the space community. There is however a vast set of near-miss stories known only to those directly involved in which potentially damaging events have been avoided by dumb luck. To ensure that the lessons from such experiences are learned and that future, limited, scientific funds are not squandered due to avoidable accidents, a documentation and process-monitoring system known as product assurance (PA) is enforced. One major remit of PA is contamination control (ECSS-Q-70-01A).

Contamination control

Contamination control is important in the building of space instrumentation to ensure that it functions as intended and is not troubled by trivial problems, as the remote operation means that there are no second chances (or at least very few). However, there is also a more direct relevance to the detection of photons in space than simple good practice. As noted, mass and volume are at a premium and thus every square millimetre of active surface in an optical telescope is essential for ideal performance. An area that is not actively contributing to the detection of photons due to particle-debris covering is essentially dead mass, as are surfaces that are not of the intended material but are covered with a thin coating of outgassed and re-condensed molecules. So for reasons of good workmanship but also for essential physics, the control of contamination during AIV of space instruments is very important.

Space-instrument design and assembly takes place according to a contamination control plan: a document that defines the cleanliness levels that must be met in order for an instrument to carry out its scientific investigation. It also apportions, by means of a contamination budget, the fractional levels of contamination that may be accumulated during the various phases assembly, test, launch and operation, for example, 60 % during ground operations and 40 % during and after launch.

Contamination-control requirements differ for different wavelength ranges and technologies: for wavelengths comparable in size to dust particles (a few micrometres) the sources of scattering dust will obviously need to be controlled; and for cryogenic detectors, which act as cryopumps trapping any free molecules, outgas control is important. Extreme-ultraviolet instruments using cryogenic detectors will have particularly challenging contamination-control procedures; whereas instruments operating in the radio spectrum with room-temperature detectors may not need any specialized contamination control other than high-quality workmanship.

Contamination is classified into two types, particulate and molecular, although in reality it is a continuous spectrum of unwanted debris which ranges in size from molecules to visible pieces of human skin and machining swarf. Volatile molecules that condense on instrument surfaces, from simple alcohols and water vapour to long-chain molecules of lubricants and plasticizers, are known as molecular contamination. Particles from the invisible 0.1 µm scale to up to 1 mm are known as particulate contamination. Particles larger than 1 mm are simply a consequence of poor housekeeping and are known as dirt.

Particulate contamination

Particle contamination is controlled by use of cleanrooms classified according to the quantities of particles of particular size that are allowed through their high-efficiency particle air (HEPA) filters, as listed in Table 40.1. The air flow from the filters is laminar and may be either vertically directed from the ceiling to the floor or horizontal, a configuration which is useful for clean assembly on an optical bench. Small laminar-flow benches may be used for particularly sensitive items, which require careful work protected from the main cleanroom volume.

Particulate contamination is monitored by two classes of systems: instantaneous air-sample systems that count the number of particles of a given size in an air sample drawn into the counter, and integrating particle fallout (PFO) systems that measure the obscuration of witness plates that have been exposed to the cleanroom environment for a given time. To monitor the instrument itself the level of particulate contamination present is measured by a tape-lift method in which the surface is sampled with a low-residue tape and the particles attached to the tape are counted under a microscope. This is an activity best carried out by a specialist.

Table 40.1: Cleanroom classification by particle number.

Cleanroom air class			Number of particles per cubic metre by size					
ISO 14644-1	FED STD 209E							
ISO class	English	Metric	0.1 µm	0.2 µm	0.3 µm	0.5 µm	1 µm	5 µm
1			10	2				
2			100	24	10	4		
3	1	M1.5	1000	237	102	35	8	
4	10	M2.5	10 000	2370	1020	352	83	
5	100	M3.5	100 000	23 700	10 200	3520	832	29
6	1000	M4.5	1 000 000	237 000	102 000	35 200	8320	293
7	10 000	M5.5				352 000	83 200	2930
8	100 000	M6.5				3 520 000	832 000	29 300
9						35 200 000	8 320 000	293 000

Molecular contamination

Molecular levels of contamination are quantified by means of the surface density of non-volatile residue (NVR) and are designated by the letters A to J in the standard MIL STD 1246C as listed in Table 40.2.

Molecular contamination in ambient cleanroom air is generally monitored by witness techniques in which the infrared or ultraviolet reflectivity of a witness mirror is compared before and after exposure to the environment for a few hours or days. Alternatively, large witness plates can be exposed to the environment and the resulting NVR is washed off using a solvent and its mass measured. Again this is best left to the specialist. A direct instantaneous measurement of ambient molecular contamination can be made using a quartz crystal micro balance (QCM). In a QCM a crystal exposed to the environment is used in an electrical oscillator circuit and as its mass changes due to the accumulation of surface contamination so the measured oscillator frequency changes. In vacuum test chambers the volatile species are unconstrained and are able to travel large distances from, for example, electronics to optics. This process can be monitored using either residual gas analyzers (RGA) or cooled (temperature controlled) quartz crystal micro balances (TQCM). The RGA is a mass spectrometer and measures the partial pressure of gaseous molecules, such as volatiles, whereas the TQCM measures condensed mass on a cooled crystal.

The practicalities of using cleanrooms

The book by Tribble (2000) covers cleanroom practicalities and theory in much more detail than can be covered here, where we discuss some practicalities. In approximate terms ISO class 8 (Table 40.1) might be thought of as the environment of a clean precision laboratory and achievable by normal working practices carried out by competent technicians. Particulate control to ISO class 6 and molecular control to level E may be achieved by trained staff with moderately careful use of a suitable cleanroom. ISO class 5 (FED STD 209, class 100), although a very common cleanroom designation, needs near heroic measures to achieve and maintain. Staff must be very well trained, be wearing full “bunny suits” including mask and be

Table 40.2: Molecular contamination nominal thickness compared to MIL STD 1246C cleanliness level.

Level	NVR σ/ 10 mg m^{-2}	Equivalent thickness of contaminating layer d/nm (assuming a density of 1 g cm^{-3})
A/100	0.01	0.1
A/50	0.02	0.2
A/20	0.05	0.5
A/10	0.1	1
A/5	0.2	2
A/2	0.5	5
A	1.0	10
B	2.0	20
C	3.0	30
D	4.0	40
E	5.0	50
F	7.0	70
G	10.0	100
H	15.0	150
J	25.0	250

constantly vigilant for breaches of protocol. Class 5, A/2 is the sort of environment used for the assembly of vacuum-ultraviolet (VUV) and cryogenic instrumentation.

Although cleanrooms are becoming increasingly common and part of the normal working routine, they need to be treated as specialised laboratory equipment and must be operated and maintained in perfect condition in order to provide the required performance. Working in a cleanroom requires a quite different mind set than in a standard laboratory, requiring advance planning, to clean and prepare tools, items of test equipment, storage boxes and note-taking devices. The advanced preparation and meticulous work mean that cleanroom operations incur labour and process costs, and at least double the time required in a standard laboratory.

Working in cleanroom clothing

Humans are normally the dirtiest things that we ever take into a cleanroom. They shed hair, dried skin and moisture droplets in copious quantities — a typical human will shed $\approx$ 1000 dried-skin particles per second. Smokers introduce an additional particulate source from the smoke residues in their breath and as a minimum two hours must elapse between smoking and entering a cleanroom. It is the function of cleanroom clothing to keep all of this human detritus either contained, as is the case for full bunny suits, or directed towards the floor, as in simple coat systems. Two things to be noted about this: firstly, that clothing to contain the large quantities of debris must be very efficient at trapping micron-scale particles and must thus be treated with care, and secondly, the cleanroom floor is certainly not clean. Ideally all work in a cleanroom should be conducted at about waist height. The floor debris must be removed by daily specialist cleaning and movement across the floor should be such as to minimise disturbing the layer of dust etc., i.e., walk carefully and deliberately, do not run.

Cleanroom clothes serve a very specific function and are designed for particular levels of particle control. In order to keep the outer surface of the clothing clean the order of dressing is important. If instead of invisible dust particles we imagined that the person dressing in the cleanroom suit was covered with a layer of wet paint, it would then be obvious that the action of dressing contributes to contaminating the exterior (cleanroom facing) surface of the suit. For the highest levels of cleanliness, class 100 (ISO 5) and better, an initial set of clothing is used. This consists of a set of disposable overshoes, hair covering and mask, and a pair of cleanroom gloves. The clean suiting is then added on top, consisting of in order: a hood with an incorporated face mask, a full bunny suit and overboots and finally an outer set of gloves. With such a suit in a suitably filtered room class 100 conditions can be achieved but clearly with a penalty in operator comfort. The presence of people in cleanrooms should thus be minimised and they should only be present for specific well-planned operations. For normal class 100 operations it is necessary to restrict the number of people in a cleanroom and a useful estimate is no more than one person per 10 m^3 of cleanroom volume or one per 5 m^2 of floor area.

Clean tools

The clean gloved hand holding a dirty tool or a tool constructed from materials exuding a plasticiser is lethal to space instrumentation. Just as the cleanroom and the cleanroom garment are specialised pieces of equipment, so the everyday laboratory tools of instrument construction must be processed for that special environment. Tools used in a cleanroom should undergo the same cleaning protocol as component parts of the instrument, and once clean should be maintained clean. The selection of suitable tools also involves as rigorous an appreciation of the materials used as is applied to the instrument. Thus PVC cables and cheap plastics are barred from use in cleanrooms as they enable plasticisers to be transferred from the surface of handles to the sensitive optical surface of instruments via gloved hands and outgassing. A set of cleaned stainless-steel tools which permanently reside in clean containers in the cleanroom solves most clean tool problems.

Asphyxiation

The cleaning advantage obtained by repeatedly passing the cleanroom air through a filter many times means that cleanroom air is used several times with a small amount of fresh top-up air added. Cleanrooms thus concentrate components unaffected by filtering, such as gases. When gases other than air (e.g., dry nitrogen used to refill vacuum chambers or offgassing from cryogenic liquids) are in use in cleanrooms great care needs to be taken to ensure that sufficient oxygen is present. For all such cases, including when a nitrogen purge is used to keep an instrument interior clean and dry, an oxygen monitor (low-oxygen alarm) must be included in the cleanroom.

Temporary storage

During instrument assembly there will be a need to store parts and components between build procedures or in anticipation of a future task. For short periods of time up to a few hours items can be stored in clean open racks in the cleanroom. However, because of the accumulation of fallout particles, for longer periods items need to be stored in clean enclosed containers. These boxes and bags are stored on open racks in the cleanroom. This also applies to the instrument under construction which, when not actively being worked on, should be protected from fallout by a clean cover.

Inconvenient materials

There are several materials and processes which must be used in an instrument build but which in general are incompatible with cleanroom practice, because they involve a component which at some time in its life cycle outgasses, such as lubricants and solvents. In such cases, if a non-outgassing alternative cannot be found, specialist procedures apply. These procedures are written with an understanding of the contamination budget so that the quantity (mass and area) of outgassing material is accounted for. Other "difficult" materials include composites CFRP, GFRP (carbon fibre and glass fibre reinforced plastic), because of the intrinsically unclean nature of their assembly and machining processes (dust and epoxies). When used in conjunction with honeycomb panels there is the additional problem of the long time required for pump down and removal of outgassing and particulates trapped in the numerous honeycomb cells. Long pumping paths and particle trapping is also a problem associated with multilayer insulation (MLI) used as thermal control blankets.

Incoming inspections

Inspection and monitoring are key to successfully operating a cleanroom. This includes controlling people: specifying access, specifying the working practices, and training to overcome ingrained culture based on previous cleanroom experience for different applications.

It is important that everything entering a cleanroom is known to be clean. Suppliers of clean components will need to supply a certificate of cleanliness stating the clean level achieved. In addition, an inspection for particulates must be carried out prior to cleanroom entry. This is most conveniently carried out using hand-held white and "black lamps" in a cleanroom antechamber. The location in which packaging is removed is obviously not a suitable location. A simple visual inspection is sufficient to establish if the supplier has made a suitable effort to meet cleanliness requirements. If all is well visually then inspection using hand-held, high-luminosity lamps will confirm cleanliness at higher levels. A white light is used to inspect for inorganic and metal particles such as machining swarf. A similar UV lamp (sometimes called a black lamp) is useful for inspecting for organic debris such as skin and hair fragments. In both cases the return reflection from a small particle is very much easier to see than the particle under normal lighting conditions.

Cleaning

Cleaning components for use in ISO class 5 / A or better is a multistage process, involving a coarse clean in a preparation area, precision cleaning in a class 6 room, and finally a vacuum bake to drive off volatiles, using an RGA to identify any failures in earlier cleaning. A set of cleaning procedures will be required to cover the wide range of materials that need to be cleaned. For reasons of economy and avoiding mistakes it is best if these different procedures do not use widely different materials. Cleaning based on deionised water, common solvents (isopropyl alcohol, IPA) and non-ionic detergents has been found to be very adaptable. For VUV instruments anhydrous ethyl alcohol is preferred over IPA (Timothy 2008). It is important that the initial coarse-cleaning part of the process be carried out thoroughly to ensure good results from the precision stage.

Vacuum bake

Vacuum bake makes use of the property that diffusion and outgassing rates greatly increase with temperature. A useful rule of thumb is that for each 10 % relative increase in absolute temperature the outgas rate increases by a factor 10. Although the highest tolerable temperature will achieve the fastest results, baking at moderate temperatures ($\approx 50\ ^\circ\mathrm{C}$) is beneficial and a temperature can be found at which even sensitive items can be baked. The end point of the bake is reached when a pre-defined outgas threshold has been reached, which can be based on either TQCM frequency measurement or an RGA partial pressure reading. A typical RGA-based end point is that when the partial pressure of mass 43 (as a proxy for all hydrocarbons) is reduced to below 5×10^{-9} Torr, then the system is molecularly clean. Using a TQCM, molecularly clean is achieved when the rate of change of frequency is not significantly different (say within 10 %) relative to the empty chamber rate at the bake temperature.

Vacuum chambers must be verified as clean over the full temperature cycle of intended use prior to use for flight instruments. Care must be taken in recovery from vacuum to atmospheric pressure to avoid introducing contaminants and to avoid turbulent gas flow, which might cause particle disturbance. Recovery after thermal vacuum test cycles must always take place after a hot cycle so that the instrument is warmer than the ambient chamber and hence cannot act as a contaminant sink.

For optical instruments the use of hydrocarbon-free (dry) vacuum systems is strongly recommended, as the deposition of depolymerized carbon and hydrocarbon fragments on optical surfaces has severe effects on optical performance. Also such hydrocarbons and depolymerised free carbon destroy the gain of open structure multipliers.

Humidity

During assembly, test and on completion, instruments are stored and handled in conditions of controlled humidity. In general this environment is provided by the cleanroom climate control to give relative humidity of 40 % to 50 %. If too high there are all of the problems associated with wet atmosphere: material degradation, surface films etc., and if too low then generation of electrostatic charge and the potential for damaging high-voltage sparks can become significant. Electrostatic discharge (ESD) is a major issue for CMOS electronics and is covered, e.g., on http://www.esda.org/basics/part6.cfm). Some open-structure detectors make use of moisture-sensitive photocathodes such as cesium iodide (CsI) and in these cases the detector needs to be sealed in a dry atmosphere and maintained dry throughout the build, test and launch phases until the instrument has completed outgassing in orbit (Barstow and Sansom 1990). Relative humidity can also affect the dimensional properties of structures made from some composite materials and so a constant level is required and thus a sealed environment should be provided to maintain that level during excursions out of the cleanroom, for example to test facilities or to the launch site.

The launch environment

Stack assembly / white room

A hard lesson for a space scientist to learn occurs when the carefully assembled and aligned instrument is installed with its host spacecraft onto the launch vehicle. This takes place in a semi-industrial environment in which scientific spacecraft are handled according to a detailed mechanical handling plan written by the instrument project teams and carried out by technicians usually from the launch-vehicle supplier. Various levels of cleanliness can be catered for but none will be as clean as the assembly laboratory — class 100 000 (ISO class 8) ventilated with dry air (molecular class E) is typical. Thus, contamination planning and budget must take account of this and equipment for more special precautions, such as dry purge gas, must be supplied by the project team.

In European and US launches, the assembly of the spacecraft onto the launch vehicle takes place in a specialized room, the white room, at the top of the launch tower. Here the spacecraft transportation and contamination control covers will be removed and the spacecraft attached to the launcher. For a few hours the instrument is exposed to an environment (ISO 8 / E) to which it may only be tolerant for a relatively short period of time. A launch delay for longer than this limited period could have damaging effects and in some very rare circumstances the environmental protection umbilical may be re-established if the delay is prolonged. However, this is costly, and there will be a trade off between cost and some loss of instrument performance.

Mechanical handling

Although a tedious document to write, the detailed handling plan allows the non-specialist vehicle assembly team to understand the important elements of the particular instrument and to quickly understand the instrument's vital elements and handling precautions.

Removal of support umbilicals

Among the numerous preparations to be considered for launch are the provision of power and other necessary supplies. These will be supplied by an umbilical, which will be removed prior to launch. The length of time the instrument is allowed to survive without these "essential" supplies is a subject of negotiation between the launch supplier and the spacecraft team, but will certainly be a compromise between what is desirable and what is essential. Typically a scientific instrument may be expected to last for a few hours to days without a supply of clean dry purge gas. Power can usually be supplied to within a few hours of launch. The details are dependent on the launch vehicle and supplier.

Rapid pressure change

During launch the pressure outside and inside large volumes of the spacecraft drops from atmospheric to 10^{-6} Torr in about two minutes, resulting in both vacuum and structural implications. The volume of air inside the instrument must be vented sufficiently fast that it closely follows the fall in pressure outside the spacecraft so that a large differential pressure is not allowed to build up. This calls for careful design and positioning of the vents — also taking into account that these vents are potential entry points for contamination during instrument assembly. Each case needs to be analyzed for its particular configuration but a useful rule of thumb to get approximate vent dimensions assumes 0.01 mm^2 of vent area for each cubic centimetre of volume to be vented. In orbit such vents allow external charged plasma access to the internal instrument volume, and so for instruments which employ open-structure detectors, some method of trapping or deflecting charged particles must be considered. It is often convenient to make the vent in the form of a labyrinth such that particles have to make many collisions with the labyrinth walls before entering the instrument, thus increasing the chance of trapping. Where a labyrinth device is not suitable, magnetic and electrostatic trapping or deflection can be used (Sumner and Lieu 1990). Labyrinth vents may also be used for instruments sensitive to out-of-band scattered light so that the photon losses at each wall reflection enable a total scattered light attenuation up to a factor of 10^{14}.

In the general instrument environment a good high vacuum of 10^{-6} Torr is not reached until several days after launch because of the outgassing of the spacecraft and instrument components. Time must be allowed for instruments to outgas before high voltages etc. are switched on. Depending on the instrument configuration this time can be from a few days to months.

Transition through corona pressure

A necessary consequence of the reducing atmospheric pressure is that at some time during the launch the spacecraft will transition the pressure regime in which, following the Paschen law, gas discharge at comparatively low voltages can take place. At around 1 mbar air breaks down and forms electrically conducting plasma at electric fields as low as $\approx 100\ \mathrm{V\,cm^{-1}}$. This has implications for the decision as to whether an instrument is launched with power on or off.

Effects of acoustic and mechanical vibration

The most extreme effect of the launch environment is the intense mechanical vibration. This arises from the mechanical components in the launcher, such as fuel pumps, combustion phenomena, and sloshing or "pogoing" of liquid fuel and oxidizer. These inputs will be directional with maximum effect along the thrust vector. However, the major input is isotropic arising from the aerodynamic interaction of the rocket with the surrounding atmosphere, particularly when transitioning through regions of maximum aerodynamic pressure such as moving into supersonic flight at relatively high atmospheric pressure. The instrument will be tested to validate the design during the AIV programme to ensure that it is capable of surviving, with some margin, the loads to which it will be subject during launch. The design is achieved through a combination of analysis and physical test. The loads presented to the space instrument depend on the specific properties of the launcher and the particular chain of mechanical linkage from the source of the vibration to the instrument attachment points. For a particular mission the launcher team will issue, through the project manager, a test specification and a low-frequency limit. The spacecraft must then be designed such that its lowest natural frequency is above the low-frequency limit and thus avoid resonant coupling to the launcher vibrations. A low-level sine sweep of the instrument on a shaker table, using an acceleration of about 0.25 g ($1\ g = 9.81\ \mathrm{m\,s^{-2}}$, the gravitational acceleration on the Earth's surface) over the frequency range from 20 Hz to 2000 Hz, will then confirm that this is the case in each of the three principal axes. To demonstrate survival of vibration loads at a high level the same shaker table is employed but this time using either a well crafted spectrum of random noise or high-level sine wave vibrations, and to represent the launcher input, 25 g accelerations are not atypical. The vibration environment increases in severity as the flight progresses and then falls away as the atmospheric pressure and hence the transfer medium decreases. The total duration is thus similar to the pressure-reduction profile, lasting between 1 min and 2 min.

Staging shocks

In addition to the semi-continuous mechanical acoustic noise of the launcher there are also specific times, in which the spacecraft will experience shocks; obviously the initial launch but also as the early launcher stages are jettisoned by means of explosive bolts and the rocket motors of subsequent stages ignite. These short-term shocks have a characteristic frequency spectrum and will also be modelled during design and tested during the AIV programme.

For optical instruments the danger from the launch mechanical acoustic environment is the possibility of misalignment or actual structural damage to the instrument. These effects are particularly hazardous to mechanisms, which must be free to move and scan during the instruments' scientific investigations. Such freedom must be constrained during launch by means of launch locks, which hold the mechanism stationary until unlocked on command when in the quiet environment of weightlessness. For most instruments some form of locking is also applied to fasteners. This can be in the form of special threaded inserts in tapped holes and nuts (helicoils), or application of a small blob of low-outgas epoxy to bolt heads to prevent vibration-induced rotation.

Particulate redistribution

A concern for an instrument in which contamination control is important is the redistribution, during launch, of debris that has eluded the contamination control processes. During the 1 g environment of assembly and test, these particles will have been confined to locations where their presence has caused no harm. Although rather high acceleration is required to move micron-sized particles by vibration alone, when coupled with gas flow during launch venting such particles can migrate and cause loss of scientific data by attaching to sensitive detector parts. Much more serious damage can occur if macroscopic objects such as solder blobs, small fasteners, or detached pieces of locking epoxy become free to be thrown around at high velocity. A rigorous contamination-control programme must ensure that such particles are not available for redistribution.

Also of concern are particles from the launcher shroud and its associated acoustic damping layers. These items will have inevitably been built to a lower cleanliness standard than the precision optics of the instrument and thus the instrument must be constructed in such a manner as to tolerate some level of external contamination, such as by enclosing the optics in a particle-proof enclosure and building in some margin to the area of thermal control surfaces.

The environment provided by the spacecraft

Provision of suitable operating environment

Space instruments are built in terrestrial laboratories and often they function best in the thermal conditions of such an environment which must be provided by the host spacecraft. The host spacecraft is the source of all of the essential energy necessary for the instrument to function and is also both a source and sink for its thermal control. In general, externally-mounted instruments are expected to maintain a semi-independent relationship with the host spacecraft. That is to say that the instrument is expected not to dissipate any excess heat generated by its electronics or received from solar irradiation to the spacecraft. In return, the spacecraft will not load or extract heat from the instrument. Such an arrangement makes for ease of design as both instrument and spacecraft design teams can progress through design and build in an independent manner, interacting only at formal design and exchange meetings. The externally-mounted instrument thus dissipates its excess heat via radiator panels into deep space, and conserves heat by means of thermal control hardware such as multilayer blankets. MLI blankets are constructed from many layers of aluminized plastic (often Mylar) spaced from each other by an intermediate layer of low thermal conductivity such as net or rigid spacers. This ensures that the thermal path between blankets is dominated by radiation and thus with a few layers a high degree of thermal isolation is achieved. But MLI also provides a vast area for the accumulation of both molecular and particulate contamination. Care must be taken to ensure that the multiple metal layers of MLI are all electrically connected together and also to an electrically conducting outer layer which must be connected to "spacecraft ground" to ensure that interactions with the surrounding plasma do not charge the exterior and allow build up of damaging flash-over voltages.

For instruments mounted internally within a spacecraft there is a greater need for a mutually-agreed thermal control protocol between instruments and between the instruments and spacecraft, and a much greater need for an interactive design process. Such a process invariably carries cost implications.

Contamination control and material screening, photo-polymerisation

The ideal construction materials for structural components of spacecraft and space instruments combine strength and rigidity with low mass per unit area. These properties can be met using composite materials such as carbon or glass fibre reinforced plastic (CFRP and GFRP) either as solid sheets or as faceplates on light-weight metal honeycomb. These materials involve the use of jointing compounds such as epoxies which outgas to a greater or lesser extent and which, if allowed, could recondense onto other components of the instrument. The design must ensure, through testing and analysis, that recondensing does not occur on optical elements. A particular hazard occurs if the outgassing recondenses onto a component with a direct view of the Sun. There is then a danger that the photochemically active component of the solar spectrum (UV and VUV) can cause the previously volatile components to polymerise to form a much less volatile layer which will permanently damage the component.

The mechanical environment – on-orbit microvibration

Astronomical spacecraft typically carry a number of instruments supplied by different institutions and designed by different teams. Each team thus has a responsibility not to disturb the environment in which the other instruments operate. The disturbance can be electromagnetic, outgassing, or vibrational. The relevance of microvibrations for the detection of photons is the equivalent of camera shake. The vibration arises because of the operation of mechanisms particularly if driven by stepper motors. The effects can be mitigated by judicious selection of stepper motor step rates to move the spectrum of the disturbance into a range in which other instruments are insensitive. The ultimate mitigation is to avoid mechanical cross talk and to schedule the most disturbing instruments to operate when others are not observing.

Mass and size limitations

For every kilogram of object placed in low Earth orbit 30 kg to 50 kg of launcher are required; for geostationary orbits this becomes 70 kg to 100 kg. Such mass amplification factors mean that instrument mass is a major constraint. The efforts of launcher designers have let spacecraft masses steadily increase, so that an observatory-class spacecraft, such as *SOHO*, can have an all up mass of $\approx$ 2000 kg.

When this is distributed over a dozen instruments each investigator is given a comfortable 160 kg. Although every investigator tries to achieve the maximum science return within his mass budget and will push for every available gram, major mass struggle is really only found for interplanetary landing probes in which a high degree of complexity, often with subsystems requiring mechanical movement (analysis carousels, trench diggers, rock grinders), are factored into the deep-space requirements of interplanetary travel. For example the *Philea* 100 kg lander on the 3000 kg *Rosetta* spacecraft was launched by an *Ariane 5* with an all up wet mass (including liquid fuel and oxygen) of 746 t (http://www.esa.int/SPECIALS/Launchers_Access_to_Space/SEM9UD67ESD_0.html).

Power

The power source for most space instrumentation is a battery pack, which is charged by means of photovoltaic solar cells on solar panels. Both these elements have significant mass and volume implications. A few spacecraft have made use of radioisotope thermoelectric generators (RTG). In these devices a voltage is generated by the thermoelectric effect, which occurs at the junction of a pair of differing materials held at a temperature higher than a similar reference junction. In RTGs the temperature difference is provided by the heat generated by radioactive decay of a nuclear isotope — typically plutonium. Environmental concerns over the potential hazard if this plutonium were to be involved in an explosive launch failure have essentially vetoed the use of this system until suitable methods to prevent this hazard have been found. All power sources and power harnesses produce heat, which must be accounted in the thermal control of the instrument and ultimately dissipated.

The external space environment

Orbit selection

The orbit to be used is selected for a number of interrelated reasons such as to maximise the observation time, optimise ground-contact/data-download time, to optimise solar-panel power availability, to ensure the visibility of navigation stars, to ensure optimum visibility of the observation target, Sun, Earth, or to avoid unwanted effects from view directions which include those objects. Thermal stability may require that the selected orbit avoids the spacecraft being eclipsed by the Earth. Orbit altitude influences interaction with potentially harmful radiation from the radiation belts and orbit inclination affects the susceptibility to high-energy cosmic charged particles. Data transfer rates are also orbit dependent. Each of these considerations has implications for the ability of the instrument used to observe photons.

Temperature regime

The temperature of a spacecraft surface is determined by the balance between heat input via radiant and conducted heat energy and heat losses by radiation to deep space. The input of radiant energy comes from external sources such as the Sun and Earth and also from local sources such as other parts of the spacecraft. Conducted heat comes from other parts of the spacecraft. The radiant heat loss will depend on the surface construction and its physical properties. The ability of a surface to radiate heat is defined by its emissivity, denoted by the symbol ε. The ability to absorb radiant energy is defined by its absorptivity, denoted by the symbol α. Both of these are surface properties of a particular material and may be changed by application of surface coatings and may be compromised by surface contamination.

The balance of heat input from external sources and loss of internally generated heat is a dynamic process for an orbiting spacecraft if it moves into and out of eclipse. Thus, good attitude control is important for good thermal control to ensure that the correct surfaces are viewing the appropriate external thermal sources and sinks. The temperature oscillations caused by eclipses can be smoothed by ensuring some thermal inertia. The problem of changing heat input is exacerbated for interplanetary probes due to the inverse square decrease in the solar heat input with distance from the Sun. Thus it becomes difficult for a spacecraft designer to use passive systems to accommodate the relentlessly changing heat input in order to maintain the scientific instrument at a particular optimum temperature.

Temperature stability is important for the detection of photons for a variety of reasons. In optical instruments the effect of temperature changes on structural dimensions is important as stability is required in order to maintain optical properties such as focal length, wavelength range, freedom from distortion etc. For cooled photon detectors there is an additional difficulty of maintaining the detector at its optimum working temperature in an environment with variable heat input. This is important when using bolometric detectors as in infrared telescopes or when using low-noise silicon detectors for the VUV wavelength region.

Vacuum regime

The atmospheric pressure decreases exponentially as the height above the surface of the Earth increases, and with decreasing pressure also comes change in composition. At ground level the atmosphere is dominated by molecular nitrogen, oxygen and water vapour, but at spacecraft altitudes between 300 km and 1000 km the dominant species is atomic oxygen. The reduced pressure has implications for the operation of electrical devices. High vacuum is a good electrical insulator as is air at atmospheric pressure, but between these two extremes in the intermediate pressure regime at a few millibars, gasses are very easily ionised and break down to produce conducting plasma. Thus if launched with power on, any instrument cables and connectors carrying a voltage above 50 V must be constructed to prevent arcing and leakage at vacuum and intermediate pressures. This inevitably involves a mass penalty.

An undesirable feature of the vacuum environment is that it allows the free passage of outgassed molecules from other parts of the spacecraft to the sensitive interiors of optical instruments. All materials have a vapour pressure of some sort; in most metals it is negligibly low, but an important exception is cadmium. However, most plastics do outgas some of their components and in a low-pressure environment an outgassed component may bounce around the various surfaces of a space instrument. This is relevant to the detection of photons in two specific cases. For VUV and soft X-ray wavelengths the normal-incidence reflectivity of most materials is very poor (much less than 1 %), thus optics are constructed from the few materials that have a measurable reflectivity, such as gold, osmium, aluminium and silicon carbide (SiC). It is thus important that the reflecting surface is actually composed of these materials and is not coated by organic compounds. The situation becomes drastic if any of the optical or detecting surfaces are cooled, such as CCDs and APS detectors (see Chapter 25, Schühle 2010). The freely bouncing molecules of contaminant can become trapped on the cold surface and at best degrade the detector performance or even stop its functioning completely.

Trapped volumes also contribute to the gas load (virtual leaks) that has to be pumped away into space before an instrument can be switched on and operated. Such trapped volumes can arise from numerous sources such as a closed volume between the base of a threaded blind hole and the inserted bolt. These trapped volumes must be provided with vent paths, such as using a through hole or slotting the bolt threads to allow the rapid removal of air from this volume.

Particle and radiation environment

The Earth sits in the extended outer atmosphere of the Sun, the solar wind, and is thus bathed in a flux of solar plasma. These particles and solar photons result in ionisation of the Earth's outer atmosphere producing a plasma of ions, protons and electrons. Much of this flux is trapped by the Earth's magnetic field in the Van Allen belts at high altitude above the Earth. Within the Van Allen belts proton energies range from 10 keV to 400 MeV with fluxes 10^8 $cm^{-2}s^{-1}$ to 600 $cm^{-2}\,s^{-1}$, respectively, and occur in a D-shaped torus, wrapped approximately around the equator, with the flat side of the D facing the Earth. The maximum proton flux occurs at altitudes between 1.5 and 2 Earth radii. The electron flux reaches a maximum at around 3 Earth radii (Fortescue et al 2003; Tribble 2003). Because of the tilt and offset of the geomagnetic axis relative to the Earth's rotation axis there is a region of enhanced radiation at lower altitudes known as the South Atlantic Anomaly (SAA) in which the radiation belts are brought closer to the Earth's surface and become a significant radiation hazard for spacecraft in low Earth orbit. The name SAA reflects the position of this structure above the South Atlantic centred at roughly latitude 45° S and longitude 30° W (ECSS-E-10-04A).

Spacecraft on interplanetary trajectories or in geosynchronous orbits, whilst beyond the range of the high-flux regions of the Van Allen belts, will encounter radiation from primary cosmic rays and from galactic cosmic rays, which stem from particle and γ-ray emission from galactic sources, such as novae and supernovae.

Micrometeorite and dust environment

In addition to the gaseous components of low Earth orbit there also exists a population of solid objects ranging in size from a few micrometres to tens of centimetres and a few larger. Clearly the impact of a body of centimetre dimensions will have a profound effect on a spacecraft's astronomy observing programme. Fortunately, objects larger than a few centimetres are tracked and their positions are known to the flight dynamics operators. However, because of the high relative velocities involved the energetics of a collision with even millimetre sized particles can cause significant damage. The individual orbits of these small objects are unknown although the statistical mapping of their orbital density is known.

Atomic oxygen

As mentioned the dominant species at the altitude of low Earth orbits is atomic oxygen. An orbiting spacecraft at 500 km impacts the essentially stationary atmospheric species at a relative velocity of $\approx 8\ \mathrm{km\,s^{-1}}$. This is equivalent to energy per impacting atom of ≈ 5 eV (Tribble 2003). This is enough to cause chemical effects. In the case of carbon-containing materials the chemical reaction with atomic oxygen is the production of volatile oxides. A plastic in an atomic oxygen beam will be eroded as its surface material oxidizes and becomes volatile and leaves the surface. In the case of metals the oxide may not itself be volatile but certainly has significantly different material properties than the parent material. In some metals (e.g., aluminium) the resulting oxide is resistant to further attack and thus creates a protective coating, however in others (such as silver) the oxide layer is porous and oxygen erosion to great depths is possible.

In terms of the effect of photon detection this is relevant to the structural integrity of space instruments, to the effectiveness of exposed electrical connectors, and to harmful degradation of thermal control surfaces and all of the consequent ramifications of lack of thermal stability. In the particular case of soft X-ray wavelengths, in which thin-film band-pass filters are used to define wavelength intervals, the erosion of these thin-film filters can have catastrophic effects for the science, either by complete removal of the filter or more perniciously by puncturing a small hole through it and thus swamping the detector system with radiation outside of the required detection band. Mitigation techniques include providing thin coatings of a protective material. In band-pass filters it is difficult to find materials that do not then compromise the required band pass, although in the case of the Lexan–carbon filter for the *ROSAT* WFC a coating of boron carbide was very effective (Kent et al 1991). A straightforward approach to minimizing atomic oxygen damage is to ensure that sensitive materials are not exposed to the most severe erosion effects by keeping such material inside the instrument and not exposed to the ram direction, i.e., not on the forward-facing velocity vector part of the spacecraft at any time as it moves in its orbit.

Effects of microgravity

For both terrestrial and ground-based instruments in which photons travel more than a few centimetres along an optical path, alignment of optical components is a critical topic. In the terrestrial environment one uses the presence of a very uniform gravitational force both as component lock and as precision alignment tool. Optical components built for stability are often massive and use gravity as an anchor mechanism. In the weightless environment of an orbiting spacecraft this component of gravity is eliminated. Thus the obvious requirement for space instruments is that items on optical benches must be actively secured, and this must be carried out in a manner that does not distort optical alignment. Equally important is that systems that are assembled in the 1 g of a terrestrial environment must be validated for use under 0 g. In many cases this can be achieved by an alignment check in $+1$ g and -1 g across each of the three principal axes, i.e., the instrument is checked both "correct way up" and inverted in each of the three principal axes. These confirmations must be carried out at final assembly and after vibration testing.

For very large structures with dimensions in excess of a few metres the effects of tidal gravity need to be considered and this also applies for optical systems implemented as separate free-flying components, e.g., separate telescope and detector spacecraft as proposed for the X-ray observatory *XEUS* (http://www.esa.int/science/xeus). Tidal gravity results from the difference in radial distance of instrument components from the idealised gravitational point source of the body being orbited. Both the radial and across-radius variation in gravitational force are important. Orbital variations in these distances cause tidally-induced variations in the forces on the components and thus must be accounted for and accommodated in alignment, locking devices and co-flying schemes.

Concluding remarks

Space-science instrumentation has provided us with a remarkable view of the Universe through its capability to observe photons in wavelength ranges not limited by the Earth's atmosphere — or even its gravitational effects in the case of gravitational wave detectors. It is important to remember that the ultimate aim of this specialist engineering is the detection of photons for scientific purposes. We would thus expect the same rigorous standards to apply to scientific space instruments as to terrestrial scientific equipment. Firstly, that it has been thoroughly tested as a complete device in an environment as close as possible to the use environment, including the data subsystem. This is perhaps best captured by the phrase *"test it like you fly it"*. And secondly, like all good scientific instruments, it has been effectively calibrated, and comes with a calibration certificate, so that measurements of photon spatial, spectral and radiance distribution can be unambiguously linked to the physical systems of units and ultimately the standards of mass, length and time (Pauluhn et al 2002).

Acknowledgements

It is a pleasure to acknowledge the numerous discussions with the talented specialist space engineers in many laboratories and organizations over many years, particularly at the Rutherford Appleton Laboratory Space Science and Technology Department. I also acknowledge the RAL Space Engineering and Technology Division for travel support for this project.

Bibliography

Barstow MA, Sansom AE (1990) The *ROSAT* WFC imaging detectors. Proc SPIE 1344:244–255

ECSS-E-10-04A Space engineering, space environment standard, see http://www.ecss.nl/

ECSS-Q-70-01A Cleanliness and contamination control standard, see http://www.ecss.nl/

Fortescue P, Stark J, Swinerd G (2003) Spacecraft Systems Engineering. John Wiley and Sons/Praxis, Chichester

Kent BJ, Swinyard BM, Maier HJ, Frischke D (1991) Boron carbide as atomic oxygen protection for the Lexan-carbon filters on the *ROSAT* Wide Field Camera. Proc SPIE 1546:312–321

Pauluhn A, Huber MCE, von Steiger R (eds) (2002) The Radiometric Calibration of *SOHO*. ISSI SR-002, ESA Publications Division, Noordwijk, The Netherlands

Schühle U (2010) Intensified solid state sensor cameras: ICCD and IAPS. ISSI SR-009:419–428

Sumner TJ, Lieu R (1990) Soft electron background in X-ray telescopes using grazing-incidence optics in near-Earth orbits. Opt Eng 29:1291–1295

Timothy JG (2008) Personal communication

Tribble AC (2000) Fundamentals of contamination control. The Society of Photo-Optical Instrumentation Engineers SPIE Press, Bellingham WA

Tribble AC (2003) The Space Environment. Princeton University Press, Princeton

VIII. Epilogue

— 41 —

Postscript

MARTIN C.E. HUBER[I], ANUSCHKA PAULUHN[I], J. LEN CULHANE[II], J. GETHYN TIMOTHY[III], KLAUS WILHELM[IV] AND ALEX ZEHNDER[I]

Abstract

We summarise the aim and content of the book 'Observing Photons in Space' (ISSI SR-009), comment on emerging national space activities around the globe, and point out the trends toward global collaborations in space astronomy.

Aim and content of the book

In the early days of space astronomy scientists who were intent on observing photons in space used to be able to build their own instruments and to interpret the observations (cf., e.g., Fredga 1969). This has become rare today: the number of astronomers concentrating exclusively on interpreting data has grown enormously, while the number of scientists capable of designing, building, testing and calibrating instrumentation for observations in space has dwindled. Therefore, preserving the knowledge and experience that has been gained in experimental space science over more than half a century, particularly with photon observations, was judged to be timely. Accordingly, this book is written with the practitioner in mind.

As editors of this book we are grateful that many of those who prepared the observing instruments for several earlier space-astronomy missions, as well as leading practitioners who were (and are) building observing facilities for the infrared and sub-millimetre domains, agreed to contribute chapters to this book. We also benefited from the cooperation of outstanding experts in the more future-oriented methods and techniques, like polarimetry and interferometry, as well as in technologies, such as cryogenics and laser-aligned structures in space. The editors themselves took pleasure in writing several chapters.

The early pursuit of astronomical observations, first from high altitude, later from outside the terrestrial atmosphere, and some of the difficulties to be overcome to enable the early space observations, are the topics of an introductory chapter. This is followed by a tutorial on the physics that governs generation, transmission

[I]PSI—Paul Scherrer Institut, Villigen, Switzerland
[II]UCL/MSSL—University College London/Mullard Space Science Laboratory, UK
[III]Nightsen, Inc., Tiverton RI, USA
[IV]MPS—Max-Planck-Institut für Sonnensystemforschung, Katlenburg-Lindau, Germany

and imaging — as well as spectroscopic and polarimetric analysis — of photons. The entire spectral range from high-energy radiation to microwaves is then covered by six articles, which describe the physical processes that can be investigated by observing in the different spectral domains and mention the main missions with their key results. Another 11 chapters address the principles employed in imaging and spectroscopic analysis as well as the instruments required to cover the wavelength range accessible in space. A brief overview of detection systems suitable for use in space then precedes a dozen chapters with in-depth descriptions of currently used detectors (as well as of some that are still under development). Chapters on polarimetry and on special techniques, such as calibration, cryogenics, and laser-aligned structures, along with a short summary on Earth and planet observations, then lead up to an essay on the practical precautions that one needs to take when developing space instrumentation — and to the present summary and outlook.

International collaborations and new national programmes

International collaborations have always played an important role in space research and more recently they have become global. Although space research was initially an element in the Cold War, some early collaboration across the iron curtain nevertheless took place between scientists in European countries and in the USSR. Later, joint projects between the US and the USSR, such as the *Apollo-Soyuz* mission, were undertaken as well; and after the division between 'East' and 'West' was overcome, the International Space Station emerged as a global project.

The Committee on Space Research (COSPAR) and the International Astronautical Federation (IAF) provided, from the start of the space age, a forum for encounters of scientists and engineers from East and West. These organisations continue to facilitate and promote progress in all space activities — also in space astronomy and Earth observations.

Today, we witness efforts toward establishing major space infrastructures by countries around the globe, such as India and China. Similarly, nations starting space programmes include, among others, Brazil and Israel. Private organizations are also starting space operations.

Plans for planetary and particularly lunar exploration seem to be the focus of many current national efforts. In space astronomy, on the other hand, giant international projects such as the *JWST* or the planned *IXO* (combining the ESA *XEUS* and the NASA *Constellation X* projects) — predominantly for observing the far-away Universe — now prevail. We may expect that a continuation of the search for, and later the detailed study of, exo-planets will dominate the space-astronomy scene in future.

Re-use of early techniques and a look beyond photons

The long lead-time of satellite missions and rocket flights turned out to be a major obstacle to rapid developments in space instrumentation. Some advanced projects, such as Boomerang (de Bernardis et al 2000), or more recently Sunrise (Gandorfer et al 2007), have taken advantage of balloon flights. Not only the relatively low cost of such flights has led to a renaissance of this early technique: observing time at altitudes of 35 km to 40 km, i.e., above the troposphere, can extend to several days, seeing is absent and atmospheric absorption is reduced.

The introductory chapter contains a recollection of the initially disapproving attitude of ground-based astronomers vis-à-vis space technology. Fortunately this later gave way to today's mutually beneficial co-existence of ground-based and space astronomy, given their complementarity. Although this book deals with the methods and techniques of photon observations in space, one may hope that 'photon observers' will remember this lesson in the coming decade, when the implementation of gravitational-wave observatories in space will offer the prospect of opening a fundamentally new window to the Universe and thus provide new, deeper insights into the Cosmos.

The transformation from an initially adversarial relationship between ground- and space-based astronomy into today's complementary and mutually supporting endeavour is a prospective paradigm: future opportunities to use gravitational-wave observations, particularly in space, are likely to extend the tools for investigating the Cosmos — beyond observing photons in space.

We look forward to further advances and exploits in space astronomy, and hope that this book helps in sustaining a community of instrumentalists who are able to advance the field by providing the required observational means. Ultimately, our understanding of the Universe will be nurtured not only by observing photons — both in space and from the ground — but also by a view of the World through the observation of gravitational waves.

Bibliography

de Bernardis P, Ade PAR, Bock JJ (plus 33 authors) (2000) A flat Universe from high-resolution maps of the cosmic microwave background radiation. Nature 404:955–959

Fredga K (1969) Spectroheliograms in the Mg II line at 2795.5 Å. Sol Phys 9:358–371

Gandorfer AM, Solanki SK, Barthol P (plus four main authors and the SUNRISE team) (2007) SUNRISE: High resolution UV/VIS observations of the Sun from the stratosphere. In: Modern Solar Facilities – Advanced Solar Science, eds Kneer F, Puschmann KG, Wittmann AD. Göttingen: Universitätsverlag Göttingen, pp. 69–76, ISBN 978-3-938616-84-0, also accessible through ⟨http://webdoc.sub.gwdg.de/univerlag/2007/solar_science_book.pdf⟩

IX. Appendices

SI units

Klaus Wilhelm[I], Martin C. E. Huber[II], J. Len Culhane[III], Anuschka Pauluhn[II], J. Gethyn Timothy[IV] and Alex Zehnder[II]

Introduction

The editors agreed that the Système International d'Unités (SI), the International System of Units, should be followed as closely as possible in order to arrive at a consistent presentation of the photon measurements. To achieve this goal the following guidelines have been provided, and their use has been strongly advocated. Nevertheless, some non-SI units that have been used in this book — although not recommended for use with SI — are defined in terms of SI units in this appendix.

Writing of physical quantities and equations

SI is based on the international metre convention, and is the legal system in most countries. The rules and style conventions for SI have been issued by the Bureau International des Poids et Mesures (BIPM 2006).[1] In this guide, the SI rules and conventions that are most relevant for the present book are categorized by *** : required; ** : recommended; * : avoid (needs explanation); — : not permitted.

Physical quantities:
*** The value of a physical quantity is the product of a number and a unit symbol, for instance, $P = 1$ W for a power of one watt.
*** 20 cm $\pm$ 2 cm or (20 $\pm$ 2) cm,
— not 20 $\pm$ 2 cm.

Units:
*** Use, for instance, ... 1 m to 10 m ... or ... (1 to 10) m,
— not 1 – 10 m.
** Symbols for the quantities are recommendations only,
*** but the correct form of the unit symbols is mandatory, i.e., they must not be modified by any indices etc. However, SI prefixes (see Table 41.1) can and should

[I]MPS—Max-Planck-Institut für Sonnensystemforschung, Katlenburg-Lindau, Germany
[II]PSI—Paul Scherrer Institut, Villigen, Switzerland
[III]UCL, MSSL—University College London/Mullard Space Science Laboratory, UK
[IV]Nightsen, Inc., Tiverton RI, USA

[1]Available at www.bipm.fr/utils/common/pdf/si_brochure_8_fr.pdf in the official French version and at ... _en.pdf in English. For the English style conventions see also the Special Publication 811 of the National Institute of Standards and Technology (NIST) at physics.nist.gov/Pubs/SP811/sp811.html.

Table 41.1: SI prefixes

Factor	Prefix name	Symbol[a]	Factor	Prefix name	Symbol[a]
10^1	deca	da	10^{-1}	deci	d
10^2	hecto	h	10^{-2}	centi	c
10^3	kilo	k	10^{-3}	milli	m
10^6	mega	M	10^{-6}	micro	µ
10^9	giga	G	10^{-9}	nano	n
10^{12}	tera	T	10^{-12}	pico	p
10^{15}	peta	P	10^{-15}	femto	f
10^{18}	exa	E	10^{-18}	atto	a
10^{21}	zetta	Z	10^{-21}	zepto	z
10^{24}	yotta	Y	10^{-24}	yocto	y

[a] A unit symbol must be added without space, e.g., 1 GW is one gigawatt.

Table 41.2: SI base units

Base quantity	Symbol	Unit name	Unit symbol	Definition
Length	l	metre[a]	m	1983[b]
Mass	m	kilogram[c]	kg	1889[b]
Time, duration	t	second[d]	s	1967[b]
Electric current	i	ampere[e]	A	1946[f]
Thermodynamic temperature	T	kelvin[g]	K	1954[b]
Amount of substance	n	mole[h]	mol	1971[b]
Luminous intensity	I_v	candela[i]	cd	1979[b]

[a] Length of the path travelled in vacuum during 1/299 792 458 s.

[b] CGPM—Conférence générale des poids et mesures.

[c] Mass of the international prototype of the kilogram.
The 3rd CGPM (1901) explicitly stated: "The kilogram is the unit of mass".

[d] The definition follows from $\nu = 9\,192\,631\,770$ Hz (exact) for the hyperfine splitting of the ground state of a ^{133}Cs atom at rest and at 0 K. The temps atomique international (TAI) is based on this second. Civil (legal) times are related to the coordinated Universal Time (UTC) derived from TAI by leap seconds.

[e] The definition follows from $\mu_0 = 4\,\pi \times 10^{-7}$ H/m (exact) for the permeability of free space.

[f] CIPM—Comité international des poids et mesures.

[g] The triple point of water is exactly at 273.16 K.

[h] The molar mass of ^{12}C is exactly 12 g/mol.

[i] Radiant intensity of 1/683 W sr^{-1} emitted in a certain direction at a frequency of $\nu = 540 \times 10^{12}$ Hz.

be added (as required) to the unit symbols. In the case of the mass unit, prefixes are applied to "g" and not to the base unit "kg" (cf., Table 41.2) with definitions of the SI base units; in addition, many quantities have special unit symbols).

** Unit symbols can be treated as mathematical entities, for which multiplication and division rules apply, e.g., $t/°\text{C} = T/\text{K} - 273.15$ defines the relationship between the thermodynamic and Celsius temperature scales (where 273.15 K is the

ice point of water). This is also useful for labelling coordinate axes of diagrams or columns of tables, e.g., "Temperature, T/K". It produces the numbers that are actually plotted or listed.
*** This concept is required for logarithmic expressions, such as $\lg(T/\mathrm{K}) = 5$, as a logarithm cannot be defined for physical quantities.
— Unit symbols cannot stand alone.
** Thus it should read, for instance: "a temperature of a few millikelvins" — and not "a few mK".
*** The names of units are written in lower case, even if they are derived from proper names, but then the corresponding unit symbols are capitalized (cf., Table 41.2).
*** Unit symbols and prefixes are written in roman style.
*** There must be a space between the number and the unit symbol of a physical quantity (e.g., $P = 1$ W), except for the plane angle units degree, minute and second in expression like $1° = 60' = 3600''$ or $\alpha = 5°40'30''$. There must also be a space between successive unit symbols. For instance, $1\ \mathrm{m\,s^{-1}}$ is totally different from $1\ \mathrm{ms^{-1}}$. It looks, in general, nicer if the second space and later ones are written as narrow spaces.
** The non-decimal angular units, degrees, minutes and seconds, are outside SI, but are accepted for use with SI. The SI unit name for plane angles is "radian": 1 rad = 1 m/m = 1, and the name for solid angles is "steradian": 1 sr = 1 $\mathrm{m^2/m^2}$ = 1.
*** The symbol "%" represents the number 0.01. Thus it should read, for instance, the "relative increase" of the power, P, is 10 % or the "fractional abundance" of the element is 10 %. It must always be clearly stated which dimensionless quantity is meant.
** The following non-SI unit symbols (and some others) are also accepted:
1 d = 24 h = 1440 min (= 86 400 s); 1 eV (= $1.602\,176\,487(40) \times 10^{-19}$ J);
1 ua, the astronomical unit, l'unité astronomique (= $1.495\,978\,706\,91(6) \times 10^{11}$ m).
Also accepted are 1 mas = $0.001''$ and 1 µas = $0.000\,001''$.
** Equations should preferably be written between quantities and not between numerical values. In any case, the equations must be correct in their dimensions.
— More than one solidus ("/") is not permitted in a single expression, unless parentheses are used.
The non-SI unit 1 a = 100 $\mathrm{m^2}$ defined in 1879 is still in use, but in this book the convention, in general, is that 1 a will be the length of time of a (tropical) year.
* Although 1 Å, 1 mÅ, etc. are acceptable, we tried to avoid them as much as possible.
— In any case, SI unit symbols of a certain quantity should not be mixed with non-SI symbols in one chapter.
* If, for special reasons, non-SI units have to be used, an explicit definition in terms of SI units is required in principle, e.g., $W = 1 \times 10^7$ erg (= 1 J), when employing the centimetre-gram-second (CGS) system. In order to avoid too many repetitions of such definitions, the most common deviations from SI are compiled in Table 41.3.

Table 41.3: Definition of non-SI units[a] used in this book

Quantity	Name	Unit symbol	SI conversion
Spectral irradiance	jansky	Jy	$1\ \mathrm{Jy} = 1 \times 10^{-26}\ \mathrm{W\,m^{-2}\,Hz^{-1}}$
Energy	erg	erg	$1\ \mathrm{erg} = 1 \times 10^{-7}\ \mathrm{J}$
Length	parsec	pc	$1\ \mathrm{pc} = 3.085\ldots \times 10^{16}\ \mathrm{m}$
Pressure	torr	Torr	$1\ \mathrm{Torr} = (101325/760)\ \mathrm{Pa}$
Pressure	millibar	mbar	$1\ \mathrm{mbar} = 1\ \mathrm{hPa}$
Magnetic flux density	gauss	G	$1\ \mathrm{G} = 1 \times 10^{-4}\ \mathrm{T}$

[a] X-ray irradiances are often given in 1 mCrab $\approx 2.4 \times 10^{-14}\ \mathrm{W\,m^{-2}}$ in the energy band from 2 keV to 10 keV (assuming a power-law spectrum).

Equations and relations:

** Approximately: $\approx$; proportional: $\sim$ or better $\propto$.

*** The differential symbol, "d" is written in roman style, i.e.,

$$\frac{\mathrm{d}\exp(x)}{\mathrm{d}x} = \mathrm{e}^x \quad \text{and} \quad \int \cos x\,\mathrm{d}x = \sin x \quad .$$

General:

** Differentiate between "radiance" for measurements with spatial resolution and "intensity" without spatial resolution (cf., Table 41.4).

** In the context of radiometry the quantity to describe a detector response is called "responsivity" and should be preferred over the much more general expression "sensitivity".

— A prefix cannot stand alone, nor can it be combined with another prefix (cf., Table 41.1).

*** The decimal marker is a point on the line in publications in English.

— No comma is inserted between groups of three digits,

** but a small space can be introduced (as in 11 222.333 444 55).

*** In the context of calibrated measurements an uncertainty definition is required. The distinction between "accuracy" and "precision" should be kept in mind. Precision is related to an inherent consistency of a data set, whereas accuracy refers back to a standard. The standard uncertainty (i.e., the estimated standard deviation) has a coverage factor of $k = 1$ ($1\ \sigma$ in old, but still acceptable, notation) and is denoted by u. Uncertainties can also be given with $k = 2$ ($2\ \sigma$) or $k = 3$ ($3\ \sigma$), as required, but this has to be indicated (conveniently by $U = k\ u$; $k = 2, 3, \ldots$). As an example, we consider the Newtonian constant of gravitation, G: after CODATA (Committee on Data for Science and Technology) 2006 it is
$G = 6.674\,28(67) \times 10^{-11}\ \mathrm{m^3\,kg^{-1}\,s^{-2}}$, which is equivalent to
$G = (6.674\,28 \pm 0.00067) \times 10^{-11}\ \mathrm{m^3\,kg^{-1}\,s^{-2}}$ or $u(G) = 6.7 \times 10^{-15}\ \mathrm{m^3\,kg^{-1}\,s^{-2}}$.
The same information can be conveyed in the form of a relative uncertainty
$u_\mathrm{r}(G) = u(G)/G = 0.00010 = 0.010\ \%$ or $U_\mathrm{r}(G) = 0.02\ \%$; $k = 2$.
Contrary to many statements in the literature, an accuracy and not a precision is defined by the relative standard uncertainty.

** In order to assign lines to the spectrum of a given element and ionisation stage, one uses the convention that the spectrum of an atom (e.g., carbon) is called C I,

Table 41.4: Examples of physical quantities without special unit symbols

Quantity and symbol	Unit name	Unit symbol
Heat flux density[a], irradiance, E	watt per square metre	$\mathrm{W\,m^{-2}}$
Energy density[b], w	joule per cubic metre	$\mathrm{J/m^3}$ or $\mathrm{J\,m^{-3}}$
Radiance, L	watt per square metre and steradian	$\mathrm{W\,m^{-2}\,sr^{-1}}$
Spectral radiance, L_λ	watt per square metre, steradian, and metre	$\mathrm{W\,m^{-2}\,sr^{-1}\,m^{-1}}$
Radiant intensity, I	watt per steradian	$\mathrm{W\,sr^{-1}}$

[a] In the official French document: flux thermique surfacique; [b] energie volumique.

the first spectrum of the element. The singly-charged carbon ion, C^+, emits the C II spectrum, etc. A spectral line is identified by the spectrum and its wavelength. Astronomers use brackets to indicate that a line belongs to a forbidden transition (usually involving a quadrupole rather than a dipole transition); thus one may speak of a [S IV] line at 10.5 µm.

Final remark

The nature of this publication, with many figures and diagrams taken from the literature, unfortunately but necessarily led to many deviations from SI rules since it is impractical to redraw existing figures. Nevertheless, the editors want to stress that SI defines the rules of the future, and must be respected as far as possible under given circumstances. Common usage within a discipline in the past is certainly not a sufficient justification for ignoring SI rules.

Bibliography

Bureau International des Poids et Mesures (BIPM) (2006) Le Système International d'Unités (SI) (8e édition). Sèvres

Mohr PJ, Taylor BN, Newell DB (2008) CODATA recommended values of the fundamental physical constants: 2006. J Phys Chem Ref Data 37:1187–1284

List of missions and acronyms

Missions .....

ACE .... Advanced Composition Explorer
AGILE .... Astro-rivelatore Gamma a Immagini Leggero, Italian high-energy astrophysics mission
AKARI .... Japanese space mission for infrared astronomy, formerly *Astro-F*
ASCE .... Advanced Spectroscopic and Coronagraphic Explorer
Aura .... NASA mission for atmospheric physics
AXAF .... Advanced X-ray Astrophysics Facility, now *Chandra*
BeppoSAX ...... Italian-Dutch satellite for X-ray astronomy
CGRO .... Compton Gamma-Ray Observatory
Chandra .... X-ray observatory, formerly *AXAF*
CHASE .... Coronal Helium Abundance Spacelab Experiment
COBE .... Cosmic Background Explorer
CORONAS .... Complex Orbital Near-Earth Observations of the Solar Activity
COROT .... Convection, Rotation and planetary Transits
DART .... Demonstration for Autonomous Rendezvous Technology
ERS-1&2 .... Earth Remote Sensing satellites
EURECA .... European Retrievable Carrier
EUVE .... Extreme Ultraviolet Explorer
EXOSAT .... European X-ray Observatory Satellite
Fermi .... Gamma-ray Space Telescope, formerly *GLAST*
FIRI .... Far Infrared Interferometer
FUSE .... Far-Ultraviolet Spectroscopic Explorer
GALEX .... Galaxy Evolution Explorer
GLAST .... Gamma-ray Space Telescope, now *Fermi*
GOCE .... Gravity field and steady-state Ocean Circulation Explorer
GOES .... Geostationary Orbiting Earth Satellite
GRACE .... Gravity Recovery and Climate Experiment
GRANAT .... Russian X-ray satellite
GRI .... Gamma-Ray Imager
HALCA .... Highly Advanced Laboratory for Communications and Astronomy
HEAO .... High Energy Astronomy Observatories
HETE .... High Energy Transient Explorer
Hinode .... Solar observatory, formerly *Solar-B*
Hipparcos .... High Precision Parallax Collecting Satellite
HSM .... *HST* Servicing Mission
HSM .... *Hubble* Servicing Mission
HST .... Hubble Space Telescope
IMAGE .... Imager for Magnetopause-to-Aurora Global Exploration

IMP .... Interplanetary Monitoring Platform
INTEGRAL ...... International Gamma-Ray Astrophysics Laboratory
IRAS .... Infrared Astronomy Satellite
IRTS .... Infrared Telescope in Space
ISEE .... International Sun Earth Explorer
ISO .... Infrared Space Observatory
ISS .... International Space Station
IUE .... International Ultraviolet Explorer
IXO .... International X-ray Observatory
JWST .... James Webb Space Telescope
LISA .... Laser Interferometer Space Antenna
MIDEX .... Medium-class Explorer Mission
MSG .... *Meteosat* Second Generation
MSX .... Midcourse Space Experiment
NuSTAR .... Nuclear Spectroscopic Telescope Array
OAO .... Orbiting Astronomical Observatory
OSO .... Orbiting Solar Observatory
Proba .... Project for On-Board Autonomy
RHESSI .... Reuven Ramaty High Energy Solar Spectroscopic Imager
ROSAT .... Röntgensatellit
RXTE .... Rossi X-ray Timing Explorer
SAS .... Small Astronomy Satellite
SDO .... Solar Dynamics Observatory
SIM .... Space Interferometry Mission, *SIM PlanetQuest*
SIRTF .... Space Infrared Telescope Facility, now *Spitzer*
SMART .... Small Missions for Advanced Research in Technology
SMEX .... Small Explorer Mission
SMM .... Solar Maximum Mission
SNAP .... Supernova Acceleration Probe
SOHO .... Solar and Heliospheric Observatory
SORCE .... Solar Radiation and Climate Experiment
Spacelab .... Laboratory for use on *Space Shuttle* flights
SPARTAN .... *Shuttle*-launched satellites for solar studies
SPECS .... Submillimeter Probe of the Evolution of Cosmic Structure
SPICA .... Space Infrared telescope for Cosmology and Astrophysics
SPIRIT .... Space Infrared Interferometric Telescope
Spitzer .... Space Infrared Telescope Facility, formerly *SIRTF*
SRG .... Spectrum-Röntgen-Gamma
STEREO .... Solar Terrestrial Relations Observatory
STS .... Space Transportation System
Suzaku .... Japanese X-ray astronomy mission, formerly *Astro-E2*
SWAS .... Submillimeter Wave Astronomy Satellite
Swift .... NASA Gamma-Ray Burst Mission
TDRSS .... Tracking and Data Relay Satellite System
TIMED .... Thermosphere, Ionosphere and Mesosphere Energetics and Dynamics mission
TOPEX .... Topology Ocean Experiment

TRACE .... Transition Region and Coronal Explorer
TSRSS .... Tracking and Data Relay Satellite System
UARS .... Upper Atmosphere Research Satellite
WIRE .... Wide Field Infrared Explorer
WMAP .... Wilkinson Microwave Anisotropy Probe
XEUS .... X-Ray Evolving Universe Spectroscopy Mission
XMM-Newton .... X-ray Multi-Mirror Mission
Yohkoh .... Solar X-ray observatory

General .....

AAE .... Absolute attitude error
AAME .... Absolute attitude measurement error
AAS .... Absolute attitude stability
AC .... Autocorrelator and alternating current
ACBAR .... Arcminute Cosmology Bolometer Array Receiver
ACRIM .... Active Cavity Radiometer Irradiance Monitor
ACS .... Advanced Camera for Surveys
ADC .... Analogue-to-digital converter
ADR .... Adiabatic demagnetization refrigerator
ADS .... Astrophysics Data System
AFB .... Airforce base
AGC .... Automatic gain control
AGN .... Active galactic nucleus
AIA .... Atmospheric Imaging Assembly
AIV .... Assembly, integration and verification
AKR .... (Terrestrial) auroral kilometric radiation
ALMA .... Atacama Large Millimeter/submillimeter Array
AMR .... Advanced Microwave Radiometer
AO .... Adaptive optics
AOS .... Acousto-optical spectrometer
APEX .... Atacama Pathfinder Experiment
APS .... Active pixel sensor
AR .... Active region and anti-reflective
ASAR .... Advanced Synthetic Aperture Radar
ASI .... Agenzia Spaziale Italiana
ASIC .... Application specific integrated circuit
ATM .... Apollo Telescope Mount
BAT .... Burst Alert Telescope
BATSE .... Burst And Transient Source Experiment
BCS .... Bragg Crystal Spectrometer and bent crystal spectrometer
BGO .... Bismuth germanate, $Bi_4 Ge_3 O_{12}$
BIB .... Blocked impurity band
BIPM .... Bureau International des Poids et Mesures
BLAST .... Balloon-borne Large-Aperture Submillimetre Telescope
BMDO .... Ballistic Missile Defense Organization
BOLD .... Blind to optical light detector
BRAIN .... B-mode Radiation Interferometer

BUSS .... Balloon-borne Ultraviolet Stellar Spectrograph
cBN .... Cubic BN, cubic boron nitride
CCD .... Charge-coupled device
CDS .... Coronal Diagnostic Spectrometer
CEM .... Channel electron multiplier
CERN .... Conseil Européen pour la Recherche Nucléaire
CFC .... Chlorofluorocarbon
CFRP .... Carbon fibre reinforced plastic
CG .... Centre of gravity
CHASE .... Coronal Helium Abundance Spacelab Experiment
CHIANTI .... An atomic database for spectroscopic diagnostics of astrophysical plasmas
CID .... Charge induction device
CIRS .... Composite Infrared Spectrometer
CLOVER .... Ground-based telescope to detect the imprint of inflationary gravity waves on the polarisation of the CMB
CMA .... Channel multiplier array
CMB .... Cosmic Microwave Background
CME .... Coronal mass ejection
CMOS .... Complementary metal oxide superconductor
CNES .... Centre National d'Etudes Spatiales
CNM .... Cold neutral medium
CNRS .... Centre National de la Recherche Scientifique
CODACON .... Coded anode converter
COMPTEL .... Imaging Compton Telescope
COS .... Cosmic Origins Spectrograph
COSPAR .... Committee on Space Research
COSTAR .... Corrective Optics Space Telescope Axial Replacement
CSIRO .... Commonwealth Scientific and Industrial Research Organisation, Australia
CTE .... Charge transfer efficiency and thermal expansion coefficient
CTI .... Charge transfer inefficiency
CTIA .... Capacitive transimpedance amplifier
CVD .... Chemical vapour deposition
CZT .... Cadmium zinc telluride ($Cd_{1-x}Zn_xTe$)
DARPA .... Defense Advanced Research Project Agency
DASI .... Degree Angular Scale Interferometer
DC .... Direkt current
DDL .... Double-delay line
DECT .... Digital enhanced cordless telecommunications
DEM .... Differential emission measure
DF .... Dielectric filter
DIRBE .... Diffuse Infrared Background Experiment
DMR .... Differential Microwave Radiometer
DORIS .... Doppler Orbitography and Radio-positioning Integrated by Satellite
DPU .... Data processing unit

DQE .... Detective quantum efficiency
DROID .... Distributed read-out imaging device
DRT .... Decoupled ring technique
EBIT .... Electron Beam Ion Trap (facility, LLNL)
EBS .... Electron-bombarded silicon
ECR .... Electrically calibrated radiometer
ECS .... EBIT calorimeter spectrometer
EE .... Encircled energy
EGRET .... Energetic Gamma Ray Experiment Telescope
EGS .... Extreme-ultraviolet Grating Spectrograph
EIS .... EUV Imaging Spectrometer
EISM .... Equal index solid Michelson (interferometer)
EIT .... Extreme-ultraviolet Imaging Telescope
EM .... Emission measure
EOF .... Experimenters' operation facility
EPIC .... European Photon Imaging Camera
ePSF .... Effective PSF
ESA .... European Space Agency
ESRO .... European Space Research Organisation, now part of ESA
ESTEC .... European Space Research and Technology Centre, ESA
ETH .... Eidgenössische Technische Hochschule, Zürich
EUNIS .... (Solar) EUV Normal-Incidence Spectrometer
EUV .... Extreme ultraviolet
EUVI .... EUV Imager
FEEP .... Field emission electric propulsion
FFT .... Fast Fourier transform
FIFI .... Far Infrared Imaging Fabry–Perot Interferometer
FILM .... Far-Infrared Line Mapper
FIP .... First-ionization potential
FIR .... Far infrared
FIRAS .... Far Infrared Absolute Spectrophotometer
FIRP .... Far IR Photometer
FIS .... Far Infrared Surveyor
FOC .... Faint Object Camera
FOT .... Flight operations team
FOV .... Field of view
FP .... Fabry–Perot
FPA .... Focal plane array
FPGA .... Field programmable gate array
FPI .... Fabry–Perot interferometer
FTS .... Fourier transform spectrometer
FUSP .... Far-Ultraviolet Spectropolarimeter
FUV .... Far ultraviolet
FWHM .... Full width at half maximum
GEANT .... Geometry and Tracking
GEM .... Gas electron multiplier
GEO .... Geosynchronous orbit

GEOC .... Galileo Electro-Optics Corporation
GFRP .... Glass fibre reinforced plastic
GHRS .... Goddard High Resolution Spectrograph
GI .... Grazing incidence
GP .... Goniopolarimetry
GPS .... Global Positioning System
GPSP .... Global Positioning System Payload
GRAPE .... Gamma Ray Polarimeter Experiment
GRB .... Gamma ray burst
GSFC .... Goddard Space Flight Center
GSO .... Gadolinium orthosilicate, $Gd_2\,Si\,O_5$
GTO .... Geosynchronous transfer orbit
HAWC .... High-resolution Airborne Wideband Camera
HCO .... Harvard College Observatory
HEB .... Hot electron bolometer
HEFT .... High Energy Focusing Telescope
HEMT .... High Electron Mobility Transistor
HEO .... Highly eccentric orbit
HEPA .... High-efficiency particle air (filter)
HESS .... High Energy Stereoscopic System
HEXTE .... High-Energy X-ray Timing Experiment
HFI .... High Frequency Instrument
HIFI .... Heterodyne Instrument for the Far-Infrared
HMI .... Helioseismic and Magnetic Imager for *SDO*
HRC .... High Resolution Camera
HRDI .... High Resolution Doppler Imager
HRI .... High Resolution Imager
HRTS .... High Resolution Telescope and Spectrograph
HV .... High voltage
HXD .... Hard X-ray Detector
IAF .... International Astronautical Federation
IAPS .... Intensified active pixel sensor
IBIS .... Imager on Board the *INTEGRAL* Spacecraft
ICCD .... Intensified charge coupled device
IF .... Intermediate frequency
IFU .... Integral field unit
IMO .... Inverted-mode operation
InSAR .... Interferometric synthetic aperture radar
IP .... Ionization potential
IPS .... Instrument pointing system
IR .... Infrared
IRAM .... Institut de Radioastronomie Millimétrique
IRMB .... Institut Royal Météorologique de Belgique
ISAS .... Institute of Space and Astronautical Science, Japan
ISM .... Interstellar Medium
ISOCAM .... Camera on *ISO*
ISOPHOT .... Imaging photo-polarimeter on *ISO*

ISRO .... Indian Space Research Organisation
ISSI .... International Space Science Institute
J-PEX .... Joint astrophysical Plasmadynamic Experiment
JAXA .... Japan Aerospace Exploration Agency
JFET .... Junction field effect transistors
JPL .... Jet Propulsion Laboratory
KAO .... Kuiper Airborne Observatory
KOG .... Kilometric optical gyro
LASCO .... Large Angle Spectroscopic Coronagraph
LASP .... Laboratory for Atmospheric and Space Physics
LAT .... Large Area Telescope
LEO .... Low Earth orbit
LEP .... Laboratoires d'Electronique et de Physique Appliquée
LESIA .... Laboratoire d'études spatiales et d'instrumentation en astrophysique
LFI .... Low Frequency Instrument
LLNL .... Lawrence Livermore National Laboratory
LMSAL .... Lockheed Martin Solar and Astrophysics Laboratory
LNA .... Low-noise amplifier
LO .... Local oscillator
LOS .... Line of sight
LRA .... Laser Retroreflector Array
LSF .... Line spread function
LWIR .... Long-wave IR
LWS .... Long Wavelength Spectrometer
LYRA .... Lyman Alpha Radiometer
MAGIC .... Major Atmospheric Gamma Imaging Cherenkov telescope
MAMA .... Multi-Anode Microchannel Array
MBE .... Molecular beam epitaxy
MBI .... Millimeter-Wave Bolometric Interferometer
MCP .... Microchannel plate
MCS .... Multi-channel spectrometer
MDI .... Michelson Doppler Imager
MEGA .... Medium Energy Gamma-Ray Astronomy experiment
MEM .... Magnetic electron multiplier
MEMS .... Micro-electro-mechanical systems
MI .... Michelson interferometer
MIPS .... Multiband Imaging Photometer for *Spitzer*
MIR .... Mid infrared
MIRI .... Mid-Infrared Instrument
MKID .... Microwave kinetic inductance detector
MLI .... Multilayer insulation
MLS .... Microwave Limb Sounder
MMC .... Metallic magnetic calorimeter
MOSFET .... Metal oxide semiconductor field-effect transistor
MPG .... Max-Planck-Gesellschaft
MPP .... Multi-pinned phase

MPS	Max-Planck-Institut für Sonnensystemforschung, formerly Max-Planck-Institut für Aeronomie (MPAE)
MSDP	Multi channel subtraction double pass
MSM	Metal-semiconductor-metal
MSSL	Mullard Space Science Laboratory
MSSTA	Multi-Spectral Solar Telescope Array
MUV	Medium ultraviolet
MWIR	Mid-wave MW
NASA	National Aeronautics and Space Administration (US)
NBS	National Bureau of Standards, now NIST
NEP	Noise equivalent power
NETD	Noise equivalent temperature difference
NGC	New General Catalogue (of nebulae and clusters of stars)
NI	Normal incidence
NICMOS	Near Infrared Camera and Multi-Object Spectrometer
NIR	Near infrared
NIRSpec	Near infrared multiobject dispersive spectrograph to be flown on the *JWST*
NIST	National Institute of Standards and Technology (US)
NIXT	Normal Incidence X-ray Telescope
NOAA	National Oceanic and Atmospheric Administration (US)
NOAO	National Optical Astronomy Observatory
NRL	Naval Research Laboratory (US)
NRL	Naval Research Laboratory
NSF	National Science Foundation
NSSDC	National Space Science Data Center (US)
NTD	Neutron-transmutation-doped
NUV	Near ultraviolet
OIR	Optical infrared
OMT	Orthomode-transducer
OPD	Optical path difference
OSC	Orbital Science Corporation
OSSE	Oriented Scintillation Spectrometer Experiment
PA	Power amplifier
PACS	Photodetector Array Camera and Spectrometer for *Herschel*
PAH	Polycyclic aromatic hydrocarbon
PDF	Probability density function
PDR	Photodissociation region
PEM	Piezoelastic modulator
PFO	Particle fall out
PHOTICON	Resistive anode encoder (with special photocathodes)
PI	Proportional-integral
PIN	p-type/intrinsic/n-type semiconductor (diode)
PIXIE	Pixel Imaging Experiment
PMOD/WRC	Physikalisch-Meteorologisches Observatorium Davos / World Radiation Center
PMT	Photo-multiplier tube

POGO .... Polarized Gamma-ray Observer
POLAR .... Space-borne hard X-ray polarimeter dedicated to the polarisation measurement of GRBs
POLRAD .... Experiment to measure power spectra and polarisation of the auroral kilometric radiation (AKR)
PSB .... Polarization-Sensitive Bolometer
PSD .... Phase-sensitive detector
PSF .... Point spread function
PSI .... Paul Scherrer Institut
QCL .... Quantum-cascade laser
QCM .... Quartz crystal micro balance
QE .... Quantum efficiency
QP .... Quasi-particle
RAE .... Resistive anode encoder and relative attitude error
RAISE .... Rapid Acquisition Imaging Spectrograph
RAL .... Rutherford Appleton Laboratory
RAME .... Relative attitude measurement error
RANICON .... Resistive anode encoder
RAS .... Relative attitude stability
RDE .... Relative displacement error
RDME .... Relative displacement measurement error
RDS .... Relative displacement stability
RESIK .... Rentgenovsky Spektrometr s Izognutymi Kristalami
RGA .... Residual gas analyzer
RGS .... Reflection Grating Spectrometer
RMC .... Rotation Modulation Collimator
RMS .... Root mean square
ROIC .... Readout integrated circuit
RTG .... Radioisotope thermoelectric generators
RTS .... Random telegraph signal
SAA .... South Atlantic Anomaly
SAFARI .... *SPICA* Far-infrared Instrument
SAFIRE .... Submillimeter and Far-Infrared Experiment
SAO .... Smithsonian Astrophysical Observatory
SAR .... Synthetic aperture radar
SBI .... Solar Bolometric Imager
SCD .... Swept charge device
SCIAMACHY .... Scanning Imaging Absorption Spectrometer for Atmospheric Chartography
SDD .... Silicon drift detector
SECCHI .... Sun Earth Connection Coronal and Heliospheric Investigation
SEE .... Solar EUV Experiment
SEL .... Single-event latch-up
SEM .... Scanning electron microscope and Solar Extreme-ultraviolet Monitor
SERTS .... Solar Extreme-ultraviolet Research Telescope and Spectrograph

SFU .... Space flyer unit
SGR .... Soft gamma repeater
SI .... Système International d'Unités, International System of Units
SICDH .... Science instruments command and data handling system
SIGMA .... French telescope aboard *GRANAT*
SIS .... Superconductor-insulator-superconductor
SLAC .... Stanford Linear Accelerator Center
SN .... Supernova
SNR .... Signal-to-noise ratio
SOC .... Science operations coordinator
SOFIA .... Stratospheric Observatory For Infrared Astronomy
SOLSTICE .... Solar-Stellar Irradiance Comparison Experiment
SPAN .... Spiral anode
SPENVIS .... Space Environment Information System
SPI .... Spectrometer on *INTEGRAL*
SPIFI .... South Pole Imaging Fabry–Perot Interferometer
SPIRE .... Spectral and Photometric Imaging Receiver
SPOrt .... Sky Polarization Observatory
SQUID .... Superconducting quantum interference device
SRC .... Science Research Council (UK)
SRON .... Netherlands Institute for Space Research
SSI .... Solar spectral irradiance
SSPP .... Solar/Stellar Pointing Platform
SST .... Sea surface temperature and Swedish Solar Telescope
STFC .... Science and Technology Facilities Council, UK
STIS .... Space Telescope Imaging Spectrograph
STJ .... Superconducting tunnel junction
STR .... Special theory of relativity
SUMER .... Solar Ultraviolet Measurements of Emitted Radiation
SUMI .... Solar Ultraviolet Magnetograph Investigation
SURF .... Synchrotron Ultraviolet Radiation Facility
SUSIM .... Solar Ultraviolet Spectral Irradiance Monitor
SVD .... Singular value decomposition
SWB .... Spider-Web Bolometer
SWFM .... Solid wide field Michelson (interferometer)
SWIR .... Short-wave IR
SWS .... Short Wavelength Spectrometer
SXT .... Soft X-ray Telescope
TAI .... Temps atomique international, international atomic time
TDI .... Time, delay and integrate (mode)
TDM .... Time division multiplexer
TDMA .... Time division multiple access
TEMS .... Transition-Edge Microcalorimeter Spectrometer
TES .... Transition-edge sensor and Tropospheric Emission Spectrometer
TGRS .... Transient Gamma-Ray Spectrometer

THEMIS .... Telescopio Heliografico para el Estudio del Magnetismo y de las Inestabilidades Solares
TID .... Total ionising dose
TIM .... Total Irradiance Monitor
TOF .... Time of flight
TPF .... Terrestrial Planet Finder / *Darwin*
TQCM .... Temperature-controlled quartz crystal micro balance
TRC .... Transition Region Camera
TSI .... Total solar irradiance
TXI .... Tunable X-ray Imager
UC/UVS .... University of Colorado/Ultraviolet Spectrometer
UCS .... Ultraviolet Coronal Spectrometer
ULIRG .... Ultraluminous infrared galaxy
URA .... Uniformly Redundant Array
UTC .... Coordinated universal time
UV .... Ultraviolet
UVCS .... Ultraviolet Coronagraph Spectrometer
UVISI .... Ultraviolet and Visible Imagers and Spectrographic Imagers on *MSX*
UVOT .... Ultraviolet and Spectroscopic Telescope
UVSP .... Ultraviolet Spectrometer and Polarimeter
VAULT .... Very high Angular-resolution Ultraviolet Telescope
VDF .... Velocity distribution function
VIRGO .... Variability of Solar Irradiance and Gravity Oscillations
VLA .... Very Large Array
VLBI .... Very Long Baseline Interferometry
VLT .... Very Large Telescope
VLWIR .... Very long-wave IR
VPG .... Volume phase grating
VUV .... Vacuum ultraviolet
WBG(M) .... Wide band gap (material)
WBS .... Wide Band Spectrometer
WFC .... Wide Field Camera
WFM .... Wide field Michelson (interferometer)
WFPC .... Wide Field and Planetary Camera
WFS .... Wavefront sensor
WIM .... Warm ionized medium
WRR .... World radiometric reference
WSA .... Wedge-and-strip array
XDL .... Cross-delay line
XQC .... X-ray Quantum Calorimeter
XS .... Crossed strip
XUV .... Extreme ultraviolet
YBCO .... Yttrium barium copper oxide, YBa_2Cu_3O7
ZIMPOL .... Zurich Imaging Polarimeter

Authors' addresses

Peter A.R. Ade
Cardiff University
School of Physics and Astronomy
Cardiff CF24 3AA
Wales, UK
peter.ade@astro.cf.ac.uk

Bernd Aschenbach
Mozartstr. 8
85591 Vaterstetten, Germany
via gok@mpe.mpg.de

Steven Beckwith
University of California
1111 Franklin St.
Oakland CA 94607, USA
steven.beckwith@ucop.edu

Ettore Carretti
CSIRO
Australia Telescope National Facility
Parkes Observatory
PO Box 276
Parkes, NSW 2870, Australia
ettore.carretti@csiro.au

Baptiste Cecconi
LESIA
Observatoire de Paris
5, place Jules Janssen
F-92195 Meudon Cedex, France
baptiste.cecconi@obspm.fr

John Leonard (Len) Culhane
Mullard Space Science Laboratory
University College London
Holmbury St. Mary
Dorking, Surrey, RH5 6NT, UK
jlc@mssl.ucl.ac.uk

Harry A.C. Eaton
Johns Hopkins University
Applied Physics Laboratory
1110 Johns Hopkins Road
Laurel MD 20723, USA
harry.eaton@jhuapl.edu

Claus Fröhlich
Physikalisch-Meteorologisches
Observatorium Davos /
World Radiation Center (PMOD/WRC)
7260 Davos Dorf, Switzerland
claus.froehlich@pmodwrc.ch

Matt J. Griffin
Cardiff University
School of Physics and Astronomy
Cardiff CF24 3AA
Wales, UK
matt.griffin@astro.cf.ac.uk

Wojtek Hajdas
Paul Scherrer Institut
5232 Villigen PSI, Switzerland
wojtek.hajdas@psi.ch

Jean-François Hochedez
Royal Observatory of Belgium
Solar Physics Department
Circular Avenue 3
B-1180 Brussels, Belgium
hochedez@oma.be

Andrew Holland
Planetary and Space Sciences Research Institute
Open University, Walton Hall
Milton Keynes MK7 6AA, UK
a.d.holland@open.ac.uk

Martin C.E. Huber
Paul Scherrer Institut
OVGA/404
CH-5232 Villigen PSI, Switzerland
mceh@bluewin.ch

Gordon J. Hurford
Space Sciences Laboratory
University of California
7 Gauss Way
Berkeley CA 94720, USA
ghurford@ssl.berkeley.edu

Kevin C. Hurley
Space Sciences Laboratory
University of California
7 Gauss Way
Berkeley CA 94720, USA
khurley@ssl.berkeley.edu

Gottfried Kanbach
Max-Planck-Institut für
extraterrestrische Physik
Giessenbachstraße
85748 Garching, Germany
gok@mpe.mpg.de

Barry J. Kent
Independent Consultant
previously with
Rutherford Appleton Laboratory
Didcot, Oxfordshire OX11 0QX, UK
barry_kent1@btinternet.com

Jean-Michel Lamarre
LERMA
Observatoire de Paris and CNRS
61 Av. de l'Observatoire
75014 Paris, France
jean-michel.lamarre@obspm.fr

Philippe Lemaire
Institut d'Astrophysique Spatiale
Université de Paris XI
91405 Orsay Cedex, France
philippe.lemaire@ias.u-psud.fr

Lennart Lindegren
Lund Observatory
Box 43
22100 Lund, Sweden
lennart@astro.lu.se

David H. Lumb
ESA/ESTEC
P.O. Box 299
2200 AG Noordwijk, The Netherlands
david.lumb@esa.int

Didier D.E. Martin
ESA/ESTEC
P.O. Box 299
2200 AG Noordwijk, The Netherlands
didier.martin@esa.int

Anuschka Pauluhn
Paul Scherrer Institut
CH-5232 Villigen PSI, Switzerland
anuschka.pauluhn@psi.ch

F. Scott Porter
NASA
Goddard Space Flight Center
Code 662
Greenbelt MD 20771, USA
frederick.s.porter@nasa.gov

Andreas Quirrenbach
ZAH, Landessternwarte
Königstuhl
D-69117 Heidelberg, Germany
a.quirrenbach@lsw.uni-heidelberg.de

Walfried Raab
Max-Planck-Institut für
extraterrestrische Physik
Giessenbachstraße
85748 Garching, Germany
raab@mpe.mpg.de

Nicola Rando
ESA/ESTEC
P.O. Box 299
2200 AG Noordwijk, The Netherlands
nicola.rando@esa.int

Cyrille Rosset
Laboratoire Astroparticules et
Cosmologie
Université Paris 7 Denis Diderot
10, rue Alice Domon et Léonie Duquet
75205 Paris Cedex 13, France
rosset@apc.univ-paris-diderot.fr

Volker Schönfelder
Max-Planck-Institut für
extraterrestrische Physik
Giessenbachstraße
85748 Garching, Germany
vos@mpe.mpg.de

Udo Schühle
Max-Planck-Institut für
Sonnensystemforschung
Max-Planck-Straße 2
37191 Katlenburg-Lindau, Germany
schuehle@mps.mpg.de

John F. Seely
Space Science Division
Naval Research Laboratory
Washington D.C. 20375-5320, USA
john.seely@nrl.navy.mil

David M. Smith
Physics Department and Santa Cruz Institute for Particle Physics
University of California, Santa Cruz
Santa Cruz CA 95064, USA
dsmith@scipp.ucsc.edu

Jan O. Stenflo
Institute of Astronomy
ETH Zürich
8093 Zürich, Switzerland
stenflo@astro.phys.ethz.ch

Estela Suarez-Garcia
DPNC & ISDC
University of Geneva
24, Quai Ernest-Ansermet
1211 Geneva 4, Switzerland
estela.suarez@psi.ch

Bruce M. Swinyard
Space Science and Technology Department
Rutherford Appleton Laboratory
Harwell Science and Innovation Campus
Didcot, Oxfordshire OX11 0QX, UK
bruce.swinyard@stfc.ac.uk

Alexander G.G.M. Tielens
University Leiden Observatory
P.O. Box 9513
NL-2300 RA Leiden, The Netherlands
tielens@strw.leidenuniv.nl

J. Gethyn Timothy
Nightsen, Inc.
P.O. Box 296
Tiverton RI 02878, USA
nightsen@earthlink.net

Alan M. Title
Lockheed Martin Space Systems
Advanced Technology Center (ATC)
Palo Alto CA 94304, USA
title@lmsal.com

Peter Verhoeve
ESA/ESTEC
P.O. Box 299
2200 AG Noordwijk, The Netherlands
peter.verhoeve@esa.int

Nick Waltham
Space Science and Technology Department
Rutherford Appleton Laboratory
Harwell Science and Innovation Campus
Didcot, Oxfordshire OX11 0QX, UK
nick.waltham@stfc.ac.uk

Wolfgang Wild
European Southern Observatory
Karl-Schwarzschild-Str. 2
D-85748 Garching, Germany
wwild@eso.org

Klaus Wilhelm
Max-Planck-Institut für Sonnensystemforschung
Max-Planck-Str. 2
37191 Katlenburg-Lindau, Germany
wilhelm@mps.mpg.de

Lidong Xia
School of Space Science and Physics
Shandong University at Weihai
Weihai, Shandong 264209, China
xld@sdu.edu.cn

Alex Zehnder
Paul Scherrer Institut
OVGA/404
5232 Villigen PSI, Switzerland
azehnder@sy-silmaril.ch

Index